AF784103

Progress in Systems and Control Theory
Volume 5

Series Editor

Christopher I. Byrnes, Washington University

Associate Editors

S.-I. Amari, University of Tokyo, Japan
B.D.O. Anderson, Australian National University, Canberra, Australia
Karl J. Äström, Lund Institute of Technology, Lund, Sweden
Jean-Pierre Aubin, CEREMADE, Paris, France
H.T. Banks, University of Southern California, Los Angeles, California
John S. Baras, University of Maryland, College Park, Maryland
A. Bensoussan, INRIA, Paris, France
John Burns, Virginia Polytechnic Institute, Blacksburg, Virginia
Han-Fu Chen, Beijing University, People's Republic of China
M.H.A. Davis, Imperial College of Science and Technology, London, England
Wendell Fleming, Brown University, Providence, Rhode Island
Michel Fliess, CNRS-ESE, Gif-sur-Yvette, France
Keith Glover, University of Cambridge, Cambridge, England
D. Hinrichsen, University of Bremen, Federal Republic of Germany
Alberto Isidori, University of Rome, Italy
B. Jakubczyk, Academy of Sciences, Warsaw, Poland
Hidenori Kimura, Osaka University, Japan
Arthur Krener, University of California, Davis
H. Kunita, Kyushu University, Japan
Alexandre Kurzhansky, IIASA, Laxenburg, Austria
Harold M. Kushner, Brown University, Providence, Rhode Island
Anders Lindquist, Royal Institute of Technology, Stockholm, Sweden
Andrzej Manitius, George Mason University, Fairfax, Virginia
Clyde F. Martin, Texas Tech University, Lubbock, Texas
Sanjoy Mitter, Massachusetts Institute of Technology, Cambridge, Massachusetts
Giorgio Picci, LADSEB-CNR, Padova, Italy
Boris Pshenichnyj, Glushkov Institute of Cybernetics, Kiev
H.J. Sussmann, Rutgers University, New Brunswick, New Jersey
T.J. Tarn, Washington University, St. Louis, Missouri
V.M. Tikhomirov, Institute for Problems in Mechanics, Moscow
Pravin P. Varaiya, University of California, Berkeley
Jan C. Willems, University of Gröningen, The Netherlands
W.M. Wonham, University of Toronto

M.A. Kaashoek
J.H. van Schuppen
A.C.M. Ran
Editors

Signal Processing, Scattering and Operator Theory, and Numerical Methods

Proceedings of the International Symposium MTNS-89, Volume III

1990

Birkhäuser
Boston · Basel · Berlin

M.A. Kaashoek
Faculteit Wiskunde en Informatica
Vrije Universiteit
De Boelelaan 1081
1081 HV Amsterdam, The Netherlands

J.H. van Schuppen
Centre for Mathematics & Computer Science
P.O. Box 4079
1009 AB Amsterdam, The Netherlands

A.C.M. Ran
Faculteit Wiskunde en Informatica
Vrije Universiteit
De Boelelaan 1081
1081 HV Amsterdam, The Netherlands

Library of Congress Cataloging-in-Publication Data
Proceedings of the International Symposium MTNS-89 / M.A. Kaashoek,
J.H. van Schuppen, A.C.N. Ran, editors.
p. cm. — (Perspectives in control theory ; v. 3–5)
"The 1989 International Symposium on the Mathematical Theory of Networks and Systems (MTNS) . . . held in Amsterdam, The Netherlands, June 19–23, 1989"—Pref.
Contents : v. 1. Realization and modelling in system theory — v. 2. Robust control of linear systems and nonlinear control — v. 3. Signal processing, scattering and operator theory, and numerical methods.
1. System analysis—Congresses. 2. Control theory—Congresses.
I. Kaashoek, M. A. II. Schuppen, J. H. van. III. Ran, A. C. M.
IV. International Symposium on the Mathematical Theory of Networks and Systems (9th : 1989 : Amsterdam, Netherlands) V. Series.
QA402.P766 1990
003—dc20 90-1003
CIP

Printed on acid-free paper.

© Birkhäuser Boston, 1990
All rights reserved. No part of this publication may be reproduced, stored in a retrieval system, or transmitted, in any form or by any means, electronic, mechanical, photocopying, recording or otherwise, without prior permission of the copyright owner.
Permission to photocopy for internal or personal use, or the internal or personal use of specific clients, is granted by Birkhäuser Boston, Inc., for libraries and other users registered with the Copyright Clearance Center (CCC), provided that the base fee of $0.00 per copy, plus $0.20 per page is paid directly to CCC, 21 Congress Street, Salem, MA 01970, U.S.A. Special requests should be addressed directly to Birkhäuser Boston, Inc., 675 Massachusetts Avenue, Cambridge, MA 02139, U.S.A.

ISBN 0-8176-3471-1
ISBN 3-7643-3471-1

For three-volume set:
ISBN 0-8176-3468-1
ISBN 3-7643-3468-1

Camera-ready text provided by the individual authors.
Printed and bound by Edwards Brothers, Inc., Ann Arbor, Michigan.
Printed in the U.S.A.

9 8 7 6 5 4 3 2 1

Preface

This volume is the third of the three volume publication containing the proceedings of the *1989 International Symposium on the Mathematical Theory of Networks and Systems (MTNS-89),* which was held in Amsterdam, The Netherlands, June 19-23, 1989.

The International Symposia MTNS focus attention on problems from system and control theory, circuit theory and signal processing, which, in general, require application of sophisticated mathematical tools, such as from function and operator theory, linear algebra and matrix theory, differential and algebraic geometry. The interaction between advanced mathematical methods and practical engineering problems of circuits, systems and control, which is typical for MTNS, turns out to be most effective and is, as these proceedings show, a continuing source of exciting advances.

The third volume contains invited papers and a large selection of other symposium presentations on signal processing, scattering and operator theory, and numerical methods for systems and control. A large spectrum of powerful mathematical methods from operator theory, including interpolation techniques, are discussed in connection with problems appearing in scattering theory. Also new advances in signal processing, circuit theory, and neural networks appear here. The numerical part concerns large scale and parallel matrix computations, and recent results in numerical linear algebra.

The titles of the two other volumes are: *Realization and Modelling in System Theory* (volume 1) and *Robust Control of Linear Systems and Nonlinear Control* (volume 2).

The Editors are most grateful to the about 300 reviewers for their help in the refereeing process. The Editors thank Ms. G. Bijleveld and Ms. L.M. Schultze for their professional secretarial assistance, and Mr. K. van 't Hoff for his programming support.

M.A. Kaashoek
J.H. van Schuppen Amsterdam
A.C.M. Ran February 1990

Contents

INVITED PAPERS

FACTORIZATION METHODS

INFINITE-DIMENSIONAL SYSTEMS

Contributors to volume 3

D. Alpay
Fakulteit Elektrotechniek, TUD - Technische Universiteit Delft, Postbus 5031, 2600 GA Delft, The Netherlands

G. Ammar
Department of Mathematics, Northern Illinois University, DeKalb, IL 60115, U.S.A.

W.N. Anderson
Fairleigh Dickinson University, Madison NJ-07940, U.S.A.

T. Ando
Research Institute of Appl. Electricity Division of Applied Mathematics, Hokkaido University, Kita-ku, Sapporo Hokkaido 060, Japan

J. Arazy
Dept. of Math. & Computer Science, University of Haifa, Haifa 31999, Israel

J.A. Ball
Virginia Polytechnic Institute Department of Mathematics, Virginia State University, Blacksburg VA-24861, U.S.A.

H. Bercovici
Indiana University, Bloomington, IN 47405, U.S.A.

J.F. Boehme
Fakultaet fuer Elekrotechnik, Ruhr - Universitaet Bochum, Universitaetsstrasse 150, D-4630 Bochum 1, B.R.D.

R. Boite
Departement d'Electricité, Faculte Polytechnique de Mons, Rue de Houdain, 9, B-7000 Mons, Belgium

D. Boley
Dept. of Computer Science, University of Minnesota, Minneapolis MN 55455, U.S.A.

P. Bruinsma
Mathematisch Instituut, RUG - Rijksuniversiteit Groningen, Postbus 800, 9700 AV Groningen, The Netherlands

A. Bunse-Gerstner
Department of Mathematics, University of Bielefeld, Bielefeld, B.R.D.

R. Byers
Department of Mathematics, University of Kansas, Lawrence, Kansas 66045, U.S.A.

D. Callaerts
Departement Elektrotechniek, Katholieke Universiteit Leuven, Kardinaal Mercierlaan 94, B-3030 Heverlee, Belgium

F.M. Callier
Departement de Mathematiques, Universitaires de Namur, Rempart de La Vierge 8, B-5000 Namur, Belgium

L.E. Carpenter
Department of Mathematics, Virginia Technic, Blackburg, VA 24061, U.S.A.

Chin Chang
Department of Electrical and Computer Engineering, Iowa State University, Ames, IA 50011, U.S.A.

C. Corduneanu
Department of Mathematics, University of Texas, Arlington, TX 76019, U.S.A.

G. Cybenko
Center for Supercomputing Research and Development, University of Illinois, 104 South Wright Street, Urbana, IL 61801-2932, U.S.A.

E. de Doncker
c/o Prof. P. Messina, California Institute of Technology, Mail code 158-79, Pasadena CA 91125, U.S.A.

E.F. Deprettere
Faculteit Elektrotechniek, TUD - Technische Universiteit Delft, 2600 GA Delft, The Netherlands

P. Dewilde
Faculteit Elektrotechniek, TUD - Technische Universiteit Delft, 2600 GA Delft, The Netherlands

A. Dijksma
Mathematisch Instituut, RUG - Rijksuniversiteit Groningen, Postbus 800, 9700 AV Groningen, The Netherlands

J. Ding
Department of Mathematics, Michigan State University, East Lansing, MI 48824, U.S.A.

M. Dominguez
Apartada Postal 47.159, Caracas 1041-A, Venezuela

P. van Dooren
Philips Research Laboratory, Avenue van Becelaere 2, B-1170 Brussel, Belgium

H. Dym
Department of Theoretical Mathematics, Weizmann Institute of Science, Rehovot 76100, Israel

P. Eberlein
Computer Science Department, State University of New York, Buffalo, NY, U.S.A.

A. Fettweis
Lehrstuhl fuer Nachrichtentechnik, Ruhr - Universitaet Bochum, Postfach 2148, D-4630 Bochum, B.R.D.

C. Foias
Department of Mathematics, Indiana University, Bloomington, IN 47405, U.S.A.

P. Gader
Algorithms and Robot-Vision Dept., Environmental Research Institute of Michigan, P.O. Box 8618, Ann Arbor, MI 48107, U.S.A.

M. Gassó
Dept. Sistemas Informatica, Universidad Politecnica de Valencia, Apdo 22012, E-46071 Valencia, Spain

T.T. Georgiou
Department of Electrical Engineering, University of Minnesota, 200 Union Street S.E., Minneapolis MN 55455, U.S.A.

D.S. Gilliam
Department of Mathematics, Texas Technical University, Lubbock, TX 79409, U.S.A.

I. Gohberg
Department of Mathematics, University of Tel Aviv, Tel Aviv, Israel

W.L. Green
Department of Mathematics, Georgia Institute of Technology, Atlanta, GA 30332, U.S.A.

G. Heinig
Sektion Mathematik Hauptgebaude, Karl Marx Universitaet, Karl-Marx-Platz, DDR-7010 Leipzig, D.D.R.

V. Hernández
Dept. Sistemas Informatica, Universidad Politecnica de Valencia, Apdo 22012, E-46071 Valencia, Spain

S. van Huffel
Faculteit Electrotechniek, Katholieke Universiteit Leuven, Kardinaal Mercierlaan 94, B-3030 Heverlee, Belgium

H. Ichikawa
Dept. of Mechanical Engineering, Osaka University, 2-1 Yamada-oka Suita Osaka, Japan

M. Iwatsuki
Dept. of Information Engineering, Tohoku University, Sendai 980 Aoba, Aramaki, Japan

B. Kågström
Institute of Information Processing, University of Umeå, S-90187 Umeå, Sweden

J. Kapenga
Computer Science Dept., Western Michigan University, Kalamazoo, MI 49008, U.S.A.

M.G. Karpovsky
College of Engineering, Boston University, Boston, MA 02215, U.S.A.

T. Katayama
Dept. of Applied Mathematics and Physics Faculty of Engineering, Kyoto University, Kyoto 606, Japan

M. Kawamata
Dept. of Information Engineering, Tohoku University, Aoba, Aramaki Sendai 980, Japan

H. Kimura
Department of Mechanical Engineering, Osaka University, Suita 2-1 Yamada-oka, Osaka 565, Japan

I. Koltracht
Dept. of Mathematics, University of Connecticut, Storrs, CT, U.S.A.

F. Kubo
Dept. of Mathematics, Toyama University, Toyama 930, Japan

A. Kummert
Lehrstuhl fuer Nachrichtentechnik, Ruhr - Universitaet Bochum, P.O. Box 102148, D-4630 Bochum 1, B.R.D.

P. Kunkel
Fachbereich Mathematik, Universitaet Oldenburg, Postfach 2503, D-2900 Oldenburg, B.R.D.

Ph. Loubaton
Ecole Nationale Supérieure des Telecommunications, 46, Rue Barrault, F-75634, Paris Cedex 13, France

R. Maier
Computer Science Department, University of Minnesota, Minneapolis, MN 55455, U.S.A.

B. Mair
Department of Mathematics, Texas Technical University, Lubbock, TX 79409, U.S.A.

C. Martin
Department of Mathematics, Texas Technical University, Lubbock, TX 79409, U.S.A.

G. Maskarinec
Dept. of Elect. & Comp. Engineering, Drexel University, Philadelphia, PA 19104, U.S.A.

J.G. McWhirter
Royal Signals and Radar Establishment, St. Andrews Road, Great Malvern WR14 3PS, United Kingdom

V. Mehrmann
Department of Mathematics, University of Bielefeld, Bielefeld, B.R.D.

M. Moonen
ESAT, Katholieke Universiteit Leuven, Kardinaal Mercierlaan 94, B-3030 Heverlee, Belgium

T.D. Morley
Department of Mathematics, Georgia Institute of Technology, Atlanta, GA 30332, U.S.A.

N. Nagai
Research Institute of App. Electricity, Hokkaido University, Sapporo 060, Japan

M.S. Nakhla
Dept. of Electronics, Carleton University, K1S 5B6 Ottawa, Canada

H. Nelis
Department of Electrical Engineering, TUD - Technische Universiteit Delft, Mekelweg 4, 2628 CD Delft, The Netherlands

N.K. Nichols
Department of Reading, University of Reading, P.O. Box 220, Reading, Berks. RG6 2AX, United Kingdom

L. Pandolfi
Dept. di Mathematica, Politecnico di Torino, Corso Duca Degli Abruzzi 24, I-10129 Torino, Italy

N.S. Papageorgiou
1015 Department of Mathematics, University of California, Davis CA 95616, U.S.A.

H. Park
Computer Science Department, University of Minnesota, Minneapolis, MN 55455, U.S.A.

B.N. Parlett
Dept. of Mathematics, University of California, Berkeley, CA 94720, U.S.A.

R.J. Plemmons
Center for Research in Scientific Computation, North Carolina State University, Raleigh NC-27695-8205, U.S.A.

C.V.K. Prabhakara Rao
Dept. of Electrical & Computer Eng., Drexel University, Philadelphia, PA 19104, U.S.A.

I. Proudler
Royal Signals and Radar Establishment, St. Andrews Road, Great Malvern WR14 3PS, United Kingdom

P.A. Regalia
Departement Electronique et Communications, Institut National de Telecommunications, 9 rue Charles Fourier, F-91011 Evry, France

T. Roziner
College of Engineering, Boston University, MA 02215 Boston, U.S.A.

Y. Saad
RIACS mail stop 230-5, NASA Ames Research Center, Moffett Field, CA 94035, U.S.A.

A. Sameh
Ctr. for Supercomputing, Res. & Devel., University of Illinois, Urbana, IL 61801, U.S.A.

Tan Shao Hua
Mathematics Department, Beijing University, Beijing, People's Republic of China

T.J. Shepherd
Royal Signal and Radar Establishment, St. Andrews Road, Great Malvern WR14 3PS, United Kingdom

H.S.V. de Snoo
Mathematisch Instituut, RUG - Rijksuniversiteit Groningen, Postbus 800, 9700 AV Groningen, The Netherlands

P. Sommen
Faculteit Elektrotechniek, TUE - Technische Universiteit Eindhoven, Postbus 513, 5600 MB Eindhoven, The Netherlands

R. Su
Dept. of Elect. & Comp. Engineering, University of Colorado, Boulder, CO 80309, U.S.A.

A. Tannenbaum
Department of Electrical Engineering, Technion, Haifa 32000, Israel

L.A. Trachtenberg
Dept. of Electrical and Computer Engineering, Drexel University, Philadelphia, PA 19104, U.S.A.

G.E. Trapp
Department of Mathematics, West Virginia University, Morgantown, WV 26506, U.S.A.

R. Trenary
Computer Science Department, Western Michigan University, Kalamazoo, MI 49008, U.S.A.

L. Vandenberghe
Faculteit Electrotechniek, Katholieke Universiteit Leuven ESAT-Laboratorium, Kardinaal Mercierlaan 94, B-3030 Heverlee, Belgium

J. Vandewalle
ESAT, Electrotechniek, Katholieke Universiteit Leuven, Kard. Mercierlaan 94, B-3030 Heverlee, Leuven, Belgium

Xu Wenli
Dept. of Elect. & Computer Engineering, University of Colorado, Boulder, CO 80309, U.S.A.

H.A.G. Wijshoff
Centre for Supercomputing, University of Illinois, Urbana, IL 61801, U.S.A.

J. Winkin
Department of Mathematics, Universitaires de Namur, Rempart de La Vierge 8, B-5000 Namur, Belgium

B. Yang
Fakultaet fur Elektrotechnik, Ruhr - Universitaet Bochum, Universitaetsstrasse 150, D-4630 Bochum 1, B.R.D.

Yang Xin-Qiang
Faculte Polytechnique de Mons, 31 Boulevard Dolez, B-7000 Mons, Belgium

H.J. Zwart
Faculteit der Toegepaste Wiskunde, UT - Universiteit Twente, Postbus 217, 7500 AE Enschede, The Netherlands

TIME–VARYING SIGNAL APPROXIMATION AND ESTIMATION*

D.Alpay and P.Dewilde

Time–varying signals may be approximated and estimated much as stationary signals when use is made of an algebra in which diagonal operators act as constants. Much of the classical approximation theory, the Schur algorithm and Nevanlinna–Pick interpolation carry through. We give a survey of the main properties and theorems.

1. INTRODUCTION

We shall consider time–discrete, non–stationary linear systems as bounded operators from $l^2(\mathbf{Z})$ to $l^2(\mathbf{Z})$. If T is such an operator, and ϵ_n is the natural basis for $l^2(\mathbf{Z})$, i.e. $\epsilon_n(k) = 0$ for $n \neq k$ and $\epsilon_n(n) = 1$, then T can be represented by a doubly infinite matrix (by convention we let an operator apply to the left of an element)

$$T_{ij} = \langle \epsilon_i T, \epsilon_j \rangle .$$

Elements in $l^2(\mathbf{Z})$ can be represented as (row–)vectors on which such operators act (vector–matrix and matrix–matrix multiplications must be handled with care). Let P_α denote the orthogonal projection of $l^2(\mathbf{Z})$ on the closed linear span of the $\epsilon_n, n \geq \alpha$. A bounded operator T from $l^2(\mathbf{Z})$ into itself is *upper triangular* if for any $\alpha \in \mathbf{Z}$, $P_\alpha T P_\alpha = P_\alpha T$. We shall denote by $\mathcal{U}$ the set of such operators. The space $\mathcal{L}$ of lower triangular operators is defined as the set of adjoints of members of $\mathcal{U}$ and the space of diagonal operators is $\mathcal{D} = \mathcal{L} \cap \mathcal{U}$. An element $W \in \mathcal{D}$ may be written as:

$$W = diag(\ldots, W_{-1}, \boxed{W_0}, W_1, W_2 \ldots).$$

where the $\boxed{}$ distinguishes the central element. Likewise, an operator $T \in \mathcal{U}$ can be represented by:

$$\begin{bmatrix} & & \cdots & & \\ \cdots & T_{-1,-1} & T_{-1,0} & T_{-1,1} & \cdots \\ \cdots & 0 & \boxed{T_{00}} & T_{0,1} & \cdots \\ \cdots & 0 & 0 & T_{1,1} & \cdots \\ & & \cdots & & \end{bmatrix}$$

For $W \in \mathcal{D}$, $|W|$ will denote the element in $\mathcal{D}$ whose entries are the absolute values of the elements of W:

$$|W| = diag(\ldots, |W_{-1}|, \boxed{|W_0|}, |W_1|, \ldots).$$

We now introduce Σ and Θ matrices characteristic for what, in analogy with the classical theory, we may consider to be a *lossless system*. Such a system is characterized e.g. by an operator which is upper triangular and unitary, as an operator from $l^2(\mathbf{Z})$ to $l^2(\mathbf{Z})$. Such an operator will be called *inner*. A *two-port* linear, lossless system will be a map from $\mathcal{U} \times \mathcal{U}$ into $\mathcal{U} \times \mathcal{U}$, characterized by a scattering operator Σ:

$$\Sigma = \begin{bmatrix} S_i & T_i \\ T_0 & S_0 \end{bmatrix}$$

with $\Sigma_{ij} \in \mathcal{U}$. The input-output relation defined by Σ is expressed by the block-diagram of fig.1.

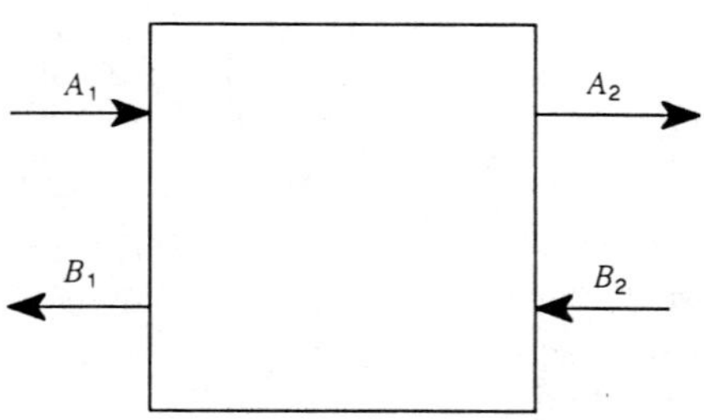

$$[B_1, A_2] = [A_1, B_2]\ \Sigma$$

Fig.1. Block diagram expressing the input-output relation for Σ.

Related to Σ is the *Chain Scattering Matrix* Θ defined by the functional relation:

$$[A_1\ ,\ B_1]\ \Theta\ =\ [A_2\ ,\ B_2] \tag{1}$$

Direct calculation shows that Θ will exist if and only if T_0^{-1} exists, in which case (we denote by 1 the identity in $\ell^2(\mathbf{Z})$)

$$\Theta = \begin{bmatrix} 1 & -S_i \\ 0 & 1 \end{bmatrix} \begin{bmatrix} T_i & 0 \\ 0 & T_0^{-1} \end{bmatrix} \begin{bmatrix} 1 & 0 \\ S_0 & 1 \end{bmatrix} \tag{2}$$

In this paper we shall restrict ourselves to the case where T_0^{-1} is bounded and upper–triangular. We shall say that Θ is *J–inner* if Σ is inner. With $J = \begin{pmatrix} 1 & 0 \\ 0 & -1 \end{pmatrix}$ the necessary and sufficient conditions for that are:

(1) $$\Theta^* J \Theta = J$$

(2) the corresponding S_i , S_0 , T_i , T_0 are upper–triangular.

An alternative, "immitance" form for Σ and Θ can be given when $1 - S_i$ is invertible on a dense domain, and is as follows. Let

$$C_i = (1 + S_i)(1 - S_i)^{-1}$$

$$F_i = T_0(1 - S_i)^{-1}$$

$$R_i = (1 - S_i)^{-1} T_i$$

then

$$\Theta = \begin{bmatrix} T_i^{-*} & -S_i T_0^{-1} \\ -S_i^* T_i^{-*} & T_0^{-1} \end{bmatrix} = \begin{bmatrix} \frac{1 + C_i^*}{2} R_i^{-*} & \frac{1 - C_i}{2} F_i^{-1} \\ \frac{1 - C_i^*}{2} R_i^{-*} & \frac{1 + C_i}{2} F_i^{-1} \end{bmatrix} \tag{3a}$$

$$G_i = \frac{1}{2}(C_i + C_i^*) = F_i^* F_i = R_i R_i^* . \tag{3b}$$

We shall only consider cases where Θ is "causal" and where all the entries of Θ are bounded and upper, F_i^{-1} is then itself an upper operator so that both F_i and F_i^{-1} are upper. Interestingly, it follows then that R_i and R_i^{-*} are upper! (In engineering terminology: F_i is minimal phase while R_i is maximal phase).

In the sequel we shall need an extension of the notion of *incoming* and *reflected* waves. Let $[\,A,\ B\,] \in \mathcal{U}\times\mathcal{U}$. We shall say that $[A,\ B]$ is strictly *admissible* if:

1) A is boundedly invertible from $\ell^2(\mathbf{Z})$ into $\ell^2(\mathbf{Z})$, (but A^{-1} need not be in $\mathcal{U}$),

and

2) the operator $A^{-1}B$ is in $\mathcal{U}$ and of norm strictly less than 1.

The Lossless Inverse Scattering (LIS) problem for an admissible pair of waves $[A,\ B]$ consists in finding a chain scattering matrix Θ of a lossless system such that $[A,\ B]\,\Theta$ is still admissible. To avoid trivial solutions a more stringent requirement is that $[A,\ B]\,\Theta = U\,[A_1,\ B_1]$ where U is upper unitary and $[A_1\ ,\ B_1]$ admissible.

Let

$$Z = \begin{bmatrix} & \vdots & \vdots & \vdots & \vdots & \\ \cdots & 1 & 0 & 0 & 0 & \cdots \\ \cdots & 0 & 1 & 0 & 0 & \cdots \\ \cdots & 0 & \boxed{0} & 1 & 0 & \cdots \\ \cdots & 0 & 0 & 0 & 1 & \cdots \\ & \vdots & \vdots & \vdots & \vdots & \end{bmatrix}$$

be the "right shift" in $\ell^2(\mathbf{Z})$. Z is obviously unitary.

The notion of admissibility was first considered in [1] in the context of time–invariant estimation problems. In [2] – [5] the theory is further developed, approximation and optimization properties are proven and reproducing kernels are introduced. In [6] – [8] the general solution of the LIS problem is derived for the stationary case and the theory is placed in the context of de Branges spaces. The first papers hinting at LIS for time–varying systems, can be found in [11] – [14] . Recent papers have been devoted to various extensions of classical theories related to our topic, e.g. Beurling–Lax theory for matrix spaces [15], generalizations of the Wiener–Hopf theory [16] – [18] or of the Schur algorithm [18]. The present paper extends the classical LIS mathematics to an even more general case: the space of bounded operators on $\ell^2(\mathbf{Z})$. It is amazing that many properties go through, although results are not obtained without pain. One can go further and consider $\ell^2(\mathcal{H})$, with $\mathcal{H}$ a separable Hilbert space. This is done in [19], a paper which was in preparation at the time of submission of the present and which contains complete proofs of the properties presented here.

Solutions of the LIS problem for the case $U = Z^n$ were described in [18] for the case of finite operators. We shall review these here in our context in section 2. New solutions "of finite degree" will be presented in section 3. In section 4 we give an important application of the theory to approximation theory. Finally, section 5 gives a sketch of a broader theory based on Reproducing Kernel Hilbert Spaces which form a natural habitat for the elementary theory developed in the earlier sections. Finally, we place the Nevanlinna-Pick problem in the present context.

2. ELEMENTARY SOLUTIONS OF THE LIS PROBLEM

The solutions to the LIS problem to be reviewed briefly in this section were presented in [18]. We define the Halmos extension $H(E)$ of a contraction E from $l^2(\mathbf{Z})$ to $l^2(\mathbf{Z})$ by

$$H(E) = \begin{bmatrix} (1-EE^*)^{-1/2} & E(1-E^*E)^{-1/2} \\ E^*(1-EE^*)^{-1/2} & (1-E^*E)^{-1/2} \end{bmatrix} \tag{4}$$

In the remainder of the paper we shall always choose $E \in \mathfrak{D}$, so that the entries in $H(E)$ are diagonal. If $A \in \mathfrak{U}$, then $A^+ = Z^{-1}AZ$ is also in $\mathfrak{U}$. Moreover, A^+ is in $\mathfrak{D}$ if and only if $A \in \mathfrak{D}$.

THEOREM 2.1

Let $[A, B]$ be an admissible pair of waves and let $S = A^{-1}B$. Let S_0 be the element in $\mathfrak{D}$ defined by the first diagonal of S,

$$S_0 = diag(\ldots, s_{-1,-1}, \boxed{s_{00}}, s_{11}, \ldots)$$

Then $\| S_0 \| < 1$ and the operator Θ defined by

$$\Theta = H(-S_0) \begin{bmatrix} Z & 0 \\ 0 & I \end{bmatrix} \tag{5}$$

is a solution to the lossless scattering problem associated to $[A, B]$. More precisely

$$[A, B]\Theta = Z[A', B']$$

where $[A', B']$ *is still an admissible pair of waves.*

The procedure leading to theorem 2.1 may be iterated on $[A', B']$. In this way, we can associate an infinite sequence of diagonal operators to any admissible pair of waves. These generalize the classical Schur coefficients.

3. SOLUTIONS OF RATIONAL TYPE.

To every element $S \in \mathcal{U}$ we associate a function $\mathcal{D} \to \mathcal{D} : W \mapsto S(W)$, its W-transform, defined for $W \in \mathcal{D}$ of norm strictly less than one and with values in $\mathcal{D}$. We start out by writing a given element $S \in \mathcal{U}$ as a formal series:

$$S = \sum_{n=0}^{\infty} Z^n S_n$$

where

$$S_n = diag(\ldots s_{-n-1,-1}, \boxed{s_{-n,0}}, s_{-n+1,1}, \ldots).$$

For any bounded operator S let $r(s)$ denote its spectral radius:

$$r(S) = \lim_{n \to \infty} \| S^n \|^{1/n}$$

Then it is well-known that $r(S) \leq \| S \|$ and that $(\lambda - S)^{-1}$ is a bounded operator when $|\lambda| > r(S)$. Furthermore we define $S^{[n]}$ recursively by:

$$S^{[0]} = I, \quad S^{[n+1]} = S(S^{[n]})^{+}.$$

Let $l_S = r(ZS)$ be the spectral radius of the operator ZS, then we have that

$$l_S = r(ZS) = \lim_{n \to \infty} \| S^{[n]} \|^{1/n}$$

DEFINITION

Let $S \in \mathcal{U}$ *with formal expansion* $S = \sum_{n=0}^{\infty} Z^n S_n$ *and* $W \in \mathcal{D}$ *with* $l_W < 1$.
Then, the W*-transform of* S *is defined by:*

$$S(W) = \sum_{n=0}^{\infty} W^{[n]} S_n \tag{6}$$

where the sum converges in the operator norm.

The sum in (6) will converge a fortiori when $\| W \| < 1$ in which case

$$\| S(W) \| \leq \| S \| / (1 - \| W \|).$$

When $W = wI$, $w \in \mathbf{C}$, then $S(W) = \sum_{n=0}^{\infty} w^n S_n$ and the W–transform formally looks like the z–transform. If, moreover, S is Toeplitz and represented by the symbol $s(z)$, then the two transforms virtually coincide – $S(wI)$ is a diagonal matrix with identical entries equal to $s(w)$. Also $\| W \| = |w|$.

However, the properties of the W–transform are quite different. The W–transform lacks a number of obvious properties of the z–transform because of the lack of commutativity of Z with elements in $\mathcal{D}$. For instance $(S_1 S_2)(W)$ and $S_1(W) S_2(W)$ are in general different. Moreover, there may be convergence difficulties. $\| S \| < 1$ and $\| W \| < 1$ do not guarantee that $\| S(W) \| < 1$, as can be shown by counterexamples.

THEOREM 3.1

Let $S = [\, s_{ij} \,]$ be in $\mathcal{U}$. Then the following two conditions are equivalent:

(1) $$\sup_{W \in \mathcal{D}, \| W \| < 1} \| S(W) \| \leq 1$$

(2) *for every k,* $$\sum_{k=0}^{\infty} |s_{k-j,j}| \leq 1.$$

Let $S \in \mathcal{U}$. Then $SZ \in \mathcal{U}$ and $ZS \in \mathcal{U}$ and their W–transforms are related by

$$(SZ)(W) = WS(W)^{+}$$

$$(ZS)(W) = WS(W^{+})$$

THEOREM 3.2

Let $W \in \mathcal{D}$. Then for every positive integer n there exits a polynomial P_n in Z with coefficients in $\mathcal{D}$, such that

$$Z^n - W^{[n]} = (Z - W) P_n(Z). \tag{7}$$

Moreover,

$$\| P_n(Z) \| \leq 1 + \| W \| + \ldots + \| W \|^n$$

THEOREM 3.3

Let $W \in \mathcal{D}$ with $l_W < 1$, $S \in \mathcal{U}$ and $D \in \mathcal{D}$. Then the operator $(Z-W)^{-1}(S-D)$ is in $\mathcal{U}$ if and only if $D = S(W)$.

We now define the analogues of Blaschke factors $\left(\frac{z-w}{1-zw^*}\right)$ for elements in $\mathcal{U}$. We shall need these to define other solutions of the LIS problem of "rational type".

THEOREM 3.4

Let $W \in \mathcal{D}$ of norm strictly less than one and let

$$K = \sqrt{\frac{1-|W^{+}|^2}{1-|W|^2}}$$

Then $(K-W^*Z)^{-1}$ and $(Z-KW)^{-1}$ exist as bounded operators and

$$U = (Z-KW)(K-W^*Z)^{-1}$$

is inner, i.e. it is in $\mathcal{U}$ and unitary.

PROOF

The proof follows by observing first that

$$r(KWZ^*) = r(K^{-1}W^*Z) = r(W^*) = r(W)$$

which in turns follows from:

$$K^{[n]} = \sqrt{\frac{1-|W^{+n}|^2}{1-|W|^2}}$$

Hence $(K-W^*Z)^{-1}$ and $(Z-KW)^{-1}$ exist as bounded operators. That $U^*U = I$ follows from direct computations, using theorem 3.1.

THEOREM 3.5

Let $V \in \mathcal{D}$ with $l_V < 1$. Then there exists $W \in \mathcal{D}$ of norm strictly less than one such that

$$W\sqrt{\frac{1-|W^+|^2}{1-|W|^2}} = V. \tag{8}$$

Moreover,

$$U_V = (Z-V)(I-L_V V^* Z)^{-1} L_V^{1/2} \tag{9}$$

with

$$L_V = \frac{1-|W|^2}{1-|W^+|^2}$$

is inner.

PROOF

The proof starts by constructing a solution to equation (8). Then one shows that the solution W satisfies the hypotheses of theorem 3.4 from which (9) will follow by identifying $L_V = K^{-2}$. The solution to (8) proceeds in two steps. First square (8) to obtain an equation of the type

$$X\,\frac{1-X^+}{1-X} = B \tag{10}$$

where $B = |V|^2$ is given and $X = |W|^2$ is to be found. Define

$$N = \sum_{k=1}^{\infty} B^{[k]}$$

then N is well defined as a bounded operator due to the condition $l_V < 1$, since

$$\lim_{k\to\infty} \| B^{[k]} \|^{1/k} = \lim_{k\to\infty} \| (|V|^2)^{[k]} \|^{1/k} = l_V^2.$$

N is clearly uniformily positive as well. We claim that

$$X = \frac{N}{I+N}$$

solves (10). This follows by direct calculation, the observation that $BN^+ = N-B$ and the fact that $(I+N)^{-1}$ exists.

THEOREM 3.6

Let $[A, B]$ *be an admissible pair of elements in* $\mathfrak{U}$ *and let* $V \in \mathfrak{D}$ *be such that*

(1) $l_V < 1$

(2) $A(V)$ *is invertible*

(3) $S_V = A(V)^{-1}B(V)$ *is strictly contractive, (i.e.* $\| S_V \| < 1$ *) and let*

$$W = V.\frac{A^{+}(V)}{A(V)}.\sqrt{\frac{1-|S_V^{+}|^2}{1-|S_V|^2}} \tag{11}$$

Then

$$l_W = l_V$$

and the lossless chain operator

$$\Theta = H(-S_V).\begin{bmatrix} U_W & 0 \\ 0 & 1 \end{bmatrix} \tag{12}$$

solves the LIS problem for the pair $[A, B]$

$$[A, B]\,\Theta = U_V[A', B'] \tag{13}$$

where $[A', B']$ *is an admissible pair, and* U_V *is defined in* (9).

PROOF

The proof is by direct calculation in which theorem 3.3 plays an essential role.

In contrast to the Toeplitz case, the condition $l_V < 1$ will not guarantee the existence of the LIS solution, unless $S = A^{-1}B$ satisfies extra boundedness conditions as given in theorem 3.1. The strict contractiveness of S_V seems essential. In the Toeplitz case it is automatic. However, theorem 3.5 shows that a large number of solutions "of rational type" may indeed be found.

4. APPLICATION TO APPROXIMATION THEORY.

The procedure of the two previous sections may be repeated recursively to yield

$$[A,B]\Theta = U[A',B'] \tag{14}$$

where $\Theta = \Theta_1 \Theta_2 \ldots \Theta_n$ and $U = U_1 U_2 \ldots U_n$. The operator Θ will be J-inner as a product of J-inner operator; $U \in \mathcal{U}$ and is unitary as a product of unitary operators. In section 1 we have indicated general expressions for the Θ matrix, see (2) and (3). Let G be a positive definite operator on $l^2(\mathbf{Z})$ (typically a covariance matrix), and assume that $G = F^*F$ for some $F \in \mathcal{U}$ which has the additional property that $F^{-1} \in \mathcal{U}$ as well (in the classical theory one merely assumes that F is outer. Our theory may be extendable to that case, but this avenue has not been explored yet). Our goal will be to approximate F with approximations of rational type which are optimal with respect to a norm compatible with the algebra of diagonals which we have utilized so far. We will assume further that we can write

$$G = \frac{1}{2}(C+C^*)$$

with $C \in \mathcal{U}$. We apply the previous LIS theory to

$$[A,B] = [C+1, \; C-1]$$

which is clearly admissible, to obtain

$$[C+I, \; C-I]\Theta = U[A', \; B'] \tag{15}$$

where Θ is represented by the quantities in (3).

THEOREM 4.1

Let the Θ matrix of the LIS solution (15) be represented by (3), then

$$C - C_i = UR$$

for some $R \in \mathcal{U}$.

PROOF

The proof is by direct calculation.

Let X, Y be bounded operators on $l^2(\mathbf{Z})$. We define the diagonal inner product or "D-inner-product" of X and Y with respect to G and values in $\mathcal{D}$ by

$$[X, Y]_G = P_0 Y^* G X \tag{16}$$

where P_0 be the projection on the main diagonal. We define also the space

$$\mathcal{H}_U = \{X \in \mathcal{U} : U^* X \in Z^* \mathcal{L}\} \tag{17}$$

Suppose that $A = A_1 + A_2$ with $A_1 \in \mathcal{H}_U$ and $A_2 \in U\mathcal{U}$ (so that $A_1 \perp A_2$ in the D-sense), then $[A - A_1,\ A - A_1]_I$ is *entrywise* smaller than any $[A - X,\ A - X]_I$ with $X \in \mathcal{H}_U$.
This follows from the fact that $[X,\ Y]_I = 0$ for $X \in \mathcal{H}_U$ and $Y \in U\mathcal{U}$ since

$$[X,\ Y]_I = P_0 Y^* X = P_0((U^* Y)^* (U^* X)) = 0$$

and from the equality, for $X \in \mathcal{H}_U$,
$[A - X,\ A - X]_I = [A - A_1,\ A - A_1]_I + [A_1 - X,\ A_1 - X]_I$.
A_1 can be viewed as an optimal approximation to A in $\mathcal{H}_U$. We shall see that the LIS theory provides us with an approximation.

DEFINITION 4.2

Let Θ be a J-inner chain scattering operator and U an inner operator. We shall say that Θ is minimal for U if $U^* R_i^{-*} \in \mathcal{L}$ and $U^* F_i^{-1} \in \mathcal{L}$.

DEFINITION 4.3

Let Θ be a J-inner chain scattering generator. We shall say that Θ is normalized if $P_0(\theta_{11} + \theta_{21}) = 0$.

THEOREM 4.4

Let (15) be a solution to the LIS problem, and suppose that $A'^{-1} \in \mathcal{U}$, then Θ is minimal.

PROOF

The proof is somewhat elaborate and technical. It uses the representation (3) for Θ and its properties.

THEOREM 4.5

Let (15) be a solution to the LIS problem. Θ will be normalized if $P_0 B' = 0$, in which case $U^* F_i^{-1} \in Z^{-1} \mathcal{L}$.

PROOF

Here also the proof is technical and uses the fact that $V_i = F_i R_{\bar{i}}^*$ is inner.

Our main approximation result is the following:

THEOREM 4.6

Let Θ be a minimal, normalized solution of the LIS problem with inner function U. Then the solution of the optimization problem:

$$\text{(pointwise)} \inf_{X \in \mathcal{H}_U} [I - FX, \ I - FX]_I$$

exists and is given by

$$X = F_{\bar{i}}^{-1} P_0 (F_{\bar{i}}^* F^*) \tag{18}$$

and the minimum is given by

$$I - \frac{|P_0 F|^2}{|P_0 F_i|^2} . \tag{19}$$

PROOF

A proof goes via the following steps:
(1) If Θ solves the LIS problem with U, then

$$[X, Y]_G = [X, Y]_{G_i}$$

This is a consequence of theorem 4.1.
(2) If Θ is normalized and minimal w.r. U, then for any $X \in \mathcal{H}_U$ it holds that

$$[X, F_{\bar{i}}^{-1}(P_0 F_{\bar{i}}^*)]_G = P_0 X.$$

This is a reproducing property which will be made more explicit in the following section.
(3) Let X_1 be defined by (18), then it can be seen that $F^{-1} - X_1 \in U\mathcal{U}$. For all $Y \in \mathcal{H}_U$ we have

$$[F^{-1} - X_1, \ Y]_G = [F^{-1}, \ Y]_G - [X_1, \ Y]_{G_i} = (P_0 Y^*)(P_0 F^*) - (P_0 Y^*)(P_0 F^*) = 0$$

It follows that

$$[I-FX,\ I-FX]_I \ = \ [F^{-1}-X,\ F^{-1}-X]_G \ = \ [F^{-1}-X_1,\ F^{-1}-X_1]_G \ + \ [X_1-X,\ X_1-X]_G$$

which will be minimal pointwise when $X = X_1$ since the last term will always be positive pointwise.

The "pointwise" approximation shows that an infinite collection of approximation problems are solved in one stretch.

5. OPERATOR THEORY INTERPRETATION

The classical Schur algorithm and associated LIS problem both have an interpretation in term of reproducing kernel spaces: this was done in [3] – [8] where also the theory of de Branges spaces was used. In this section we give a brief overview of a theory which is analogous of the theory of de Branges spaces in the present framework and we give an interpretation of the lossless inverse scattering problem in terms of operator ranges. The notion of complementation is a generalization of orthogonality: see [9] and [10]. Let $(H, <,>)$ and $(H_1, <, >_1)$ be two Hilbert spaces with inner products as shown and related norms. The Hilbert space H_1 is contractively included in H if $H_1 \subset H$ as sets and if the inclusion map is a contraction from H_1 into H. There is then a uniquely defined Hilbert space $(H_2, <,>_2)$ associated to H_1 with the following properties: every element x in H can be written as $x = x_1 + x_2,\ x_i \in H_i,\ i = 1, 2,$ and for every such decomption

$$\| x \|^2 \ \le \ \| x_1 \|_1^2 \ + \ \| x_2 \|_2^2$$

and there exists a unique decomposition where equality holds. The space H_2 is called the complementary space of H_1 in H. It coincides with the orthogonal complement of H_1 if and only if H_1 is isometrically included in H.

THEOREM 5.1

Let H be a Hilbert space with norm $\| \ \|$ and let Γ be a bounded self adjoint positive operator in $L(H)$. Let P denote the orthogonal projection onto Ker Γ. Then,
(1) The range of $\Gamma^{1/2}$, Ran $\Gamma^{1/2}$, endowed with the norm $\| \Gamma^{1/2} u \|_\Gamma = \| (I-P)u \|$ is a Hilbert space, and Ran Γ is dense in Ran $\Gamma^{1/2}$ in the norm $\| \ \|_\Gamma$.
Moreover, $\| \Gamma u \|_\Gamma = \| \Gamma^{1/2} u \|$ for u in H.
(2) If $\Gamma = \Gamma_1 + \Gamma_2$ is a decomposition of Γ into a sum of two positive self adjoint operators, then Ran $\Gamma^{1/2}$ is the sum, in the sense of complementation, of Ran $\Gamma_1^{1/2}$ and Ran $\Gamma_2^{1/2}$.

Let $\mathcal{H}_2$ define the space of bounded operators X which have finite Hilbert–Schmidt norm, i.e. for which trace $X^*X < \infty$, endowed with the Hilbert–Schmidt inner product

$$<X, Y>_{HS} = trace\,(Y^*X)$$

Let X be in $\mathcal{U}$ and let M_X denote the multiplication operator $F \to XF$. It is a bounded operator from $\mathcal{H}^2$ into itself. When $\|X\| < 1$, M_X is easily seen to be a contraction. If $[A, B]$ is an admissible pair of elements of $\mathcal{U}$, then the operator $\Gamma = M_A M_A^* - M_B M_B^*$ is a positive self adjoint bounded operator from $\mathcal{H}^2$ into itself. We will denote by $\mathcal{B}(A, B)$ the range of $\Gamma^{1/2}$ endowed with the norm $\|\ \|_\Gamma$. In order to define a Hilbert space associated to a chain scattering operator Θ we first define $\mathcal{H}_2^2$ to be the Hilbert space of pairs $\binom{f}{g}$ of elements in $\mathcal{H}^2$ with norm

$$\|\binom{f}{g}\|^2_{\mathcal{H}_2^2} = \|f\|_2^2 + \|g\|_2^2$$

LEMMA 5.2

Let Σ be the scattering operator of a lossless medium, such that there exists an associated chain scattering operator Θ. Suppose moreover that Θ can be written as a 2x2 matrix with block $\theta_{ij} \in \mathcal{U}$, $i, j \in 1, 2$. Then

$$J - M_\Theta J M_\Theta^* \geq 0 \tag{20}$$

where M_Θ denotes the muliplication operator by Θ from $\mathcal{H}_2^2$ into itself.

We will denote by $\mathcal{H}(\Theta)$ the space $Ran\ \Gamma^{1/2}$ with norm $\|\ \|_\Gamma$ where $\Gamma = J - M_\Theta J M_\Theta^*$. Let $\mathcal{H}_J^2$ be the space $\mathcal{H}_2^2$ endowed with the indefinite inner product

$$< F, G >_J = < F, JG >_{\mathcal{H}_2^2} \tag{21}$$

THEOREM 5.3

Let Θ be a chain scattering operator as in theorem 4.2. Then

$$\mathcal{H}(\Theta) = \mathcal{H}_J^2 \ominus \Theta \mathcal{H}_J^2 = Ran(J - M_\Theta J M_\Theta^*) \tag{22}$$

where the $\ominus$ refers to the J–inner product, and

$$\| F \|_{\mathcal{H}(\Theta)} = \| F \|_J \tag{23}$$

The main result of this section generalizes to the present non stationary framework a result of [8] and may be formulated as follows.

THEOREM 5.4

Let $[A, B]$ be an admissible pair of waves and let Θ be the solution of a LIS problem described in the previous sections. Then

$$\mathfrak{B}(A,\ B) = [A,\ B]\,\mathcal{H}(\Theta) \oplus \mathfrak{B}([A, B]\Theta). \tag{24}$$

Let $\varrho_V = I - ZV^*$. We note that $\varrho_V^{-1}D$ belongs to $\mathcal{H}_2$ for every $V \in \mathfrak{D}$, $\| V \| < 1$ and $D \in \mathfrak{D}_2 = \mathfrak{D} \cap \mathcal{H}_2$. Moreover, for F in $\mathcal{H}_2$,

$$< F,\ \varrho_V^{-1}D >_{\mathcal{H}_2} = trace(D^*F(V)) \tag{25}$$

which is the analogue of Cauchy's formula in the present framework.

For U and V in $\mathfrak{D}$ of norm strictly less than one the function

$$K(U, V) = \varrho_V^{-1}(U) = \sum_0^\infty U^{[n]} V^{[n]*} \tag{26}$$

is positive in the sense that

$$trace\ (\sum_{i,j=1}^n D_i^* K(V_i, V_j) D_j) \geq 0$$

for any choice of $D_1, \ldots, D_n$ *in* $\mathfrak{D}_2$.

The span of functions $\{ U \to K(U, V)D,\ V \in \mathfrak{D}, \| V \| < 1,\ D \in \mathfrak{D}_2\}$ endowed with the inner product

$$< K(.,V_2)\ D_2, K(.,V_1)\ D_1 > = trace(D_1^*\ K(V_1, V_2)\ D_2) \tag{27}$$

is a prehilbert space. Its (unique) completion, called $\hat{\mathcal{H}}_2$, is the set of functions of the form

$$\hat{F}(V) = \sum_0^\infty V^{[n]} D_n$$

where the D_n are in $\mathcal{D}$ and such that

$$trace\ (\sum_0^\infty D_n D_n^*) < \infty$$

and thus coincide with the set of W-transform of elements of $\mathcal{H}_2$.

It is the reproducing kernel Hilbert space with reproducing kernel (26) i.e., it has the properties that

(1) for every $V \in \mathcal{D}$, $\|V\| < 1$ and D in $\mathcal{D}_2$, $U \to K(U, V)D \in \hat{\mathcal{H}}_2$,

(2) for every $\hat{F}$ in $\hat{\mathcal{H}}_2$,

$$< \hat{F}, K(., V)D >_{\hat{\mathcal{H}}_2} = TrD^* \hat{F}(V) \tag{28}$$

Note that $\hat{F}\ (V)$ can be interpreted as the W-transform of the element

$$F = \sum_{n=0}^\infty Z^n D_n \in \mathcal{H}_2 .$$

THE NEVANLINNA-PICK PROBLEM

The Nevanlinna-Pick problem translates to the present situation. Suppose that we are given strictly contractive "points" $\{V_1, \ldots, V_n\}$ in $\mathcal{D}$ and n "values" $\{s_1, \ldots, s_n\}$ also in $\mathcal{D}$. One may wonder whether there exists a contractive operator $S \in \mathcal{U}$ such that

$$S(V_k) = s_k$$

where $S(V_k)$ is the value of the W-transform of S at the point V_k. In the remainder of this paper we shall show a technique which solves this generalized Nevanlinna-Pick problem. Further properties can be found in a report and forthcoming paper [19].

Recall that $\mathcal{D}_2 = \mathcal{H}_2 \cap \mathcal{D}$ indicates diagonals with finite Hilbert-Schmidt norm and $\varrho_V = (1 - ZV^*)$. Suppose that we succeed in finding a Θ with the property that

$$\mathcal{H}(\Theta) \;=\; span\left\{\begin{bmatrix} 1 \\ -s_j^* \end{bmatrix} \frac{1}{\varrho V_j} D\right\} \tag{29}$$

for $j = 1 \ldots n$ and $D \in \mathcal{D}_2$, then the corresponding S_i function - see (2) - solves the extended Nevanlinna–Pick problem defined above. This follows from theorem 5.3, since for all $f, g \in \mathcal{H}_2$ and $D \in \mathcal{D}_2$ it will be true that

$$\begin{aligned} 0 \;&=\; < \begin{bmatrix} 1 \\ -s_j^* \end{bmatrix} \frac{1}{\varrho V_j} D, \quad J\Theta \begin{bmatrix} f \\ g \end{bmatrix} > \\ &=\; < \frac{1}{\varrho V_j} D, \; [T_i \quad (s_j - S_i)\, T_0^{-1}] \begin{bmatrix} f \\ g \end{bmatrix} >, \end{aligned}$$

specializing to $f=0$ and $g=T_0\, D_1$ for any $D_1 \in \mathcal{D}_2$ produces:

$$0 \;=\; < \frac{1}{\varrho V_j} D, \; (s_j - S_i) D_1 > \tag{30}$$

and hence

$$S_i(V_j) \;=\; s_j.$$

It follows that S_i "interpolates" the s_j at the "points" V_j . Since S_i is contractive, it solves the interpolation problem, which is reduced to the construction of a Θ with (29) as its $\mathcal{H}(\Theta)$.

This now can be done using a recursive procedure akin to the solution of the LIS problem presented in section 3. We shall put ourselves in the somewhat restricted setting where we assume that a "Pick operator" associated with the basis (29) for $\mathcal{H}(\Theta)$ is strictly positive.

Let

$$f_j \;=\; \begin{bmatrix} 1 \\ -s_j^* \end{bmatrix} \frac{1}{\varrho V_j}$$

and define (see definition (16) with $G=I$)

$$\begin{aligned} P_{ij} \;&=\; [f_j,\, Jf_i]_I \\ &=\; P_0\, \frac{1}{\varrho V_i^*}\, (1 - s_i s_j^*)\, \frac{1}{\varrho V_j} \end{aligned} \tag{31}$$

We shall assume that the operator matrix $P = [P_{ij}]$ aching on $\mathcal{H}_2^n$ is strictly positive, i.e. that there exists a scalar ϵ such that

$$P - \epsilon I \geq 0 \tag{32}$$

In that case all s_i will be strictly bounded away from 1, in fact

$$|s_i| < 1 - \epsilon \qquad (pointwise)$$

The construction proceeds recursively. We start out with

$$f_1 = \begin{bmatrix} 1 \\ -s_1^* \end{bmatrix} \frac{1}{\varrho_{V_1}}$$

and search for an elementary Θ_1 of type (5) such that

$$J\Theta_1^* J f_1 \in Z^{-1} \begin{bmatrix} \mathcal{L} \\ \mathcal{L} \end{bmatrix}. \tag{33}$$

This is found to be

$$\Theta_1 = H(-s_1) \begin{bmatrix} U_{W_1} & 0 \\ 0 & I \end{bmatrix} \tag{34}$$

where:

$$W_1 = V_1 \sqrt{\frac{1 - |s_1^+|^2}{1 - |s_1|^2}} \tag{35}$$

Next we apply $M_{J\Theta_1^* J}$ to the remaining $f_1 \ldots f_n$ to obtain

$$f_k^{(1)} = M_{J\Theta_1^* J} f_k \,, \qquad k = 2 \ldots n \tag{36}$$

It turns out that these $f_k^{(1)}$ have a form similar to the orginal, i.e.

$$f_k^{(1)} = \begin{bmatrix} 1 \\ -(s_k^{(1)})^* \end{bmatrix} \frac{1}{\varrho_{V_k^{(1)}}} \tag{37}$$

for some new values of $s_k^{(1)}$ and $V_k^{(1)}$.

Also, the new Pick matrix $P^{(1)}$ (of one dimension less) is again strictly positive, due to an extension of Schur's theorem for matrices. The construction can be repeated on $f_2^{(2)}$ to yield Θ_2 etc..., i.e. etc....$\Theta = \Theta_1 \Theta_2 .. \Theta_n$ is the result sought. The construction is in effect a Gramm–Schmidt orthogonalization of the original set of vectors. Just as in the classical case, all solutions of the interpolation problem can be expressed as

$$S = T_\Theta(S_L) = -(AS_L - B)(CS_L - D)^{-1}$$

where

$$\Theta = \begin{bmatrix} A & B \\ C & D \end{bmatrix}.$$

Acknowledgement
The authors gratefully acknowledge contributions of Prof. H. Dym of the Weizmann Institute concerning sharpening of conditions of several theorems in this paper, in particular theorem 3.4 and the usage of the spectral radius in it, as well as a careful reading of the original version.

*This research has received partial support by the Commission of the EEC under the ESPRIT NANA project and by FOM under DEL 77.1260

Bibliography

[1] P.Dewilde, A.Viera and T.Kailath, On a generalized Szego–Levinson realization algorithm for optimal linear prediction based on a network synthesis approach, IEEE *Transactions on Circuits and Systems, 25,* 1988.

[2] P.Dewilde and H.Dym, "Schur Recursions, Error Formulas and Convergence of Rational Estimators for Stationary Stochastic Sequences", *IEEE Trans. Info. Theory* IT–27(4) pp.446–461 (Jul 1981).

[3] P.Dewilde and H.Dym, "Lossless Chain Scattering Matrices and Optimum Linear Prediction: The Vector Case", *Intl. J. Circuit Theory and Appln.* 9, pp. 135–175 (1981).

[4] P.Dewilde and H.Dym. "Lossless Inverse Scattering, Estimation Theory and Digital Filters", *IEEE TRans. Info. Theory* IT–30(4) pp. 644–662 (July 1984).

[5] D.Alpay, P.Dewilde and H.Dym, "On the existence and construction of solutions to the partial lossless inverse scattering problem with applications to estimation theory", *IEEE Trans. Info. Theory,* 1989 (to appear).

[6] D.Alpay, Reproducing kernel Krein spaces of analytic functions and inverse scattering, Ph.D. Thesis, The Weizmann Institute of Science, Rehovot, Israel, (1985).

[7] D.Alpay and H.Dym, Hilbert spaces of analytic functions, inverse scattering and operator models I, *Integral Equations and Operator Theory, 8,* 589–641 (1984).

[8] D.Alpay and H.Dym, On applications of reproducing kernel spaces to the Schur algorithm and rational J–unitary factorization: Operator Theory: Advances and Applications OT18, (1986).

[9] L.de Branges, Square summable power series, Preprint.

[10] L.de Branges, Complementation in Krein spaces, *Transactions of the American Mathematical Society,* 305, 277–291, (1988).

[11] H.Dym and I.Gohberg, "Extensions of Band Matrices with Band Inverses", *Linear Algebra and its Applications,* pp. 1–24 Elsevier, (1981).

[12] E.Deprettere, "Mixed Form Time–Variant Lattice Recursions", *Outils et Modeles Mathématiques pour l'Automatique, l'Analyse de systemes et le traitement du signal,* pp. 545–562 CNRS, Paris (1981).

[13] H. Lev–Ari and T.Kailath, "Schur and Levinson Algorithms for Nonstationary Processes", *Proceedings ICASSP,* (1981).

[14] Ph.Delsarte, Y.Genin and Y.Kamp, "A Method of Matrix Inverse Triangular Decomposition, Based on Contiguous Principal submatrices", *J. Linear Algebra and its Applications* 31, pp. 199–212 (June 1980).

[15] J.Ball and I. Gohberg, "Shift Invariant Subspaces, Factorization and Interpolation of Matrices. I. The Canonical Case", *Lin. Algebra and its Applications* 74, pp. 87–150 (1986).

[16] H.Dym and I.Gohberg, A new class of contractive interpolants and maximum entropy principles, *Operator Theory: Advances and Applications,* 29, 1988.

[17] H.J.Woerdeman, *Strictly contractive and positive completions for block matrices,* Technical Report WS 337, Vrije Universiteit Amsterdam, November 1987.

[18] H.Nelis, P.Dewilde and E.Deprettere, "Inversion of Partially Specified Positive Definite Matrices by Inverse Scattering", *Operator theory: Advances and Applications,* 40 pp. 325–357 (1989)

[19] D.Alpay, P.Dewilde and H.Dym, *Lossless Inverse Scattering Theory for Upper Operators,* Internal Report, Delft University of Technology, 1989 (to be published).

SPECTRAL VARIANTS OF THE NEVANLINNA-PICK INTERPOLATION PROBLEM

H. Bercovici, C. Foias and A. Tannenbaum

Abstract

In this paper we give two spectral versions of the commutant lifting theorem in which one bounds the spectral radius of the interpolant and not the norm. We relate this to spectral analogues of classical matricial Nevanlinna-Pick interpolation.

It was shown in [9] that for SISO systems, the gain margin optimization can be reduced to the classical scalar Nevanlinna-Pick interpolation problem. However, the natural similar reduction of the gain optimization problem for MIMO systems is not the classical matrix Nevanlinna-Pick interpolation problem, but, as shown in [10] it is a new spectral variant of the matrix Nevanlinna-Pick interpolation problem. Indeed in [10] it is shown that the construction of a controller $C(s)$ which internally stabilizes the closed loop

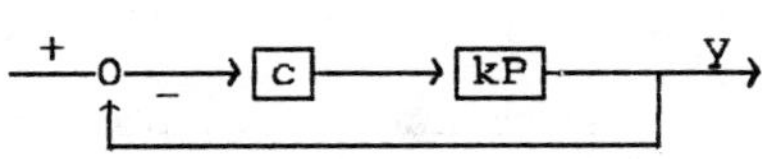

for all $k = (1+\lambda)^{-1}$, $|\lambda| \leq r$ reduces (under some suitable technical conditions on the $N \times K$-matrix valued plant $P(s)$) to the finding of the $N \times N$-matrix valued sensitivity function $S = (I + P(s)C(s))^{-1}$ satisfying the interpolation conditions

$$S(s_j)^* \nu_j = \omega_j \quad (j=1,2,\ldots,n)$$

(where the s_j's are the transmission zeros of P assumed simple and $\nu_j, \omega_j \in C^N$ $(j=1,2,\ldots,n)$) and also (since $(1+kP(s)C(s))^{-1}$ must be invertible for all s in the closed right half plane $\bar{H}$)

$$\det(1+\rho S(s)) \neq 0 \quad \text{for all } |\rho| \leq r,\ s \in \bar{H}\ .$$

Through an adequate fractional transformation on the space of the $N \times N$-matrices and a conformal map from H to the unit disk D the problem of finding this sensitivity function becomes precisely the spectral tangential Nevanlinna-Pick interpolation problem (see [10], [3] and Section 4 below).

The present article is a brief presentation of the mathematical theory of the spectral Nevanlinna-Pick interpolation problem as developed by the authors in [2] and [3]. Our approach is based on operator theoretical techniques, but in the present article we omitted most of the operator theoretical proofs. We present however the operator theoretic framework of two spectral Nevanlinna-Pick interpolation problems for matrix valued functions, as well as the propositions and computations which may lead to an algebraic algorithm for solving the tangential problem.

1. Preliminary Facts

Throughout this paper, by "Hilbert space" we will mean "complex separable Hilbert Space." By "operator" we shall always mean "bounded linear operator," unless explicitly stated otherwise. For $\mathcal{H}$ a Hilbert space, let $B(\mathcal{H})$

denote the set of operators on $\mathcal{H}$. For $\mathcal{H}_1$ and $\mathcal{H}_2$ Hilbert spaces, we set

$$B(\mathcal{H}_1,\mathcal{H}_2) := \{A : \mathcal{H}_1 \to \mathcal{H}_2 : A \text{ an operator}\}.$$

Let $\mathcal{E}$ be a Hilbert space, and denote by

$$H^2(\mathcal{E}) := H^2 \otimes \mathcal{E}$$

the Hilbert space of square-summable $\mathcal{E}$-valued power series (see [1],[5]). Given a bounded analytic function $F : D \to B(\mathcal{E})$, we can consider the multiplication operator $M_F : H^2(\mathcal{E}) \to H^2(\mathcal{E})$ defined by

$$(M_F f)(z) := F(z)f(z), \; f \in H^2(\mathcal{E}), \; z \in D .$$

The operator M_F commutes with the unilateral shift S given by

$$(Sf)(z) := zf(z), \; f \in H^2(\mathcal{E}), \; z \in D,$$

and it satisfies the norm equality

$$\|M_F\| = \|F\|_\infty = \sup\{\|F(z)\| : z \in D\} .$$

By slight abuse of notation we shall sometimes identify F and M_F in what follows when no confusion will be possible.

Given an operator A, $\|A\|_{sp}$ will denote its spectral radius. We will now state without proof some results from [2] which we will be using implicitly throughout the paper:

Proposition 1. *Let* $F : \bar{D} \to B(\mathcal{E})$ *be a continuous function, analytic in* D . *Then*

$$\|M_F\|_{sp} = \sup_{z\in D}\|F(z)\|_{sp} = \max_{z\in \bar{D}}\|F(z)\|_{sp} .$$

Proposition 2. *If* $\mathcal{E}$ *is finite dimensional, and* $F : D \to B(\mathcal{E})$ *is a bounded analytic function, then*

$$\|M_F\|_{sp} = \sup\{\|F(z)\|_{sp} : z \in D\} .$$

For $L : \mathcal{H} \to \mathcal{H}$ a contraction ($\mathcal{H}$ is a Hilbert space), let $D_L := (I - L^*L)^{1/2}$, and $\mathcal{D}_L := \overline{D_L H}$. Then we will be using the following proposition in Section 5 whose proof may be easily derived from the results of [6]:

Proposition 3. *Let* $A : \mathcal{H}_1 \oplus \mathcal{H}_2 \to \mathcal{H}_1 \oplus \mathcal{H}_2$ *be an operator (where* $\mathcal{H}_1, \mathcal{H}_2$ *are Hilbert spaces). Suppose that* $A\mathcal{H}_1 \subset \mathcal{H}_1$, *so that we can express*

$$A = \begin{bmatrix} B & X \\ O & Y \end{bmatrix},$$

for $B := A|\mathcal{H}_1$, $Y := (A^*|\mathcal{H}_2)^*$, *and* $X : \mathcal{H}_2 \to \mathcal{H}_1$. *Then we have that:*

(i) A *is a contraction if and only if* $\|B\| \leq 1$, $\|Y\| \leq 1$, *and* $X = D_{B^*} C D_Y$ *for some operator* $C : \mathcal{D}_Y \to \mathcal{D}_{B^*}$, $\|C\| \leq 1$.

(ii) If $\|B\| < 1$, $\|Y\| < 1$, $X = D_{B^*} C D_Y$ *with* $C : \mathcal{D}_Y \to \mathcal{D}_{B^*}$ *and* $\|C\| < 1$, *then* $\|A\| < 1$.

Remark 1. In [2] we have shown that it is not possible to significantly weaken the hypotheses Propositions 1 and 2. Indeed, one can show that if the analytic function $f : D \to B(\mathcal{E})$ fails to be continuous on $\bar{D}$, and if $\mathcal{E}$ is infinite dimensional, then we may have

$$\sup_{z \in D} \|F(z)\|_{sp} < \|M_F\|_{sp} .$$

2. Generalized Spectral Commutant Lifting Theorem

In this section, we will define the first object of study of this paper, and prove one of our main results. We will use the notation and terminology of Section 1 here.

In order to motivate our results, let us briefly first review the set-up from [2]. Accordingly, let H be a Hilbert space, $T \in B(H)$, and let $A \in \{T\}'$, where $\{T\}'$ denotes the commutant of T. In [2], we defined the T-spectral radius $\rho_T(A)$ of A as

$$\rho_T(A) := \inf\{\|X^{-1}AX\| : X \text{ invertible}, \ X \in \{T\}'\} .$$

We are now ready to state (again without proof) the main results of [2] and [3].

Theorem 1 (Spectral Commutant Lifting Theorem). *Let* $T \in B(\mathcal{H})$ *be a contraction with minimal isometric dilation* $U \in B(\mathcal{K})$. *Let* $A \in \{T\}'$ *be fixed, and set*

$$\mathrm{Dil}(A) := \{B \in \{U\}' : P_{\mathcal{H}}B = AP_{\mathcal{H}}\}$$

where $P_{\mathcal{H}} : \mathcal{K} \to \mathcal{H}$ *denotes orthogonal projection.*

(i) If $\mathcal{K} \ominus \mathcal{H}$ *is hyperinvariant for* U, *then*

$$\rho_T(A) \leq \inf\{\rho_U(B) : B \in \mathrm{Dil}(A)\} .$$

(ii) If $\mathcal{H}$ *is finite dimensional, then*

$$\rho_T(A) \leq \inf\{\rho_U(B) : B \in \mathrm{Dil}(A)\} .$$

(iii) If $\mathcal{H}$ *is finite dimensional, and* $\mathcal{K} \ominus \mathcal{H}$ *is hyperinvariant for* U, *then*

$$\rho_T(A) \leq \inf\{\|B\|_{sp} : B \in \mathrm{Dil}(A)\} .$$

For the study of the tangential Nevanlinna-Pick problem, we will need a generalization of $\rho_T(A)$. For $T \in B(\mathcal{H})$, let $\mathcal{M} \subset \mathcal{H}$ be a subspace which is T^*-invariant. We denote by $P_{\mathcal{M}}$ the orthogonal projection of $\mathcal{H}$ onto $\mathcal{M}$, and by $T_{\mathcal{M}}$ the compression of T to $\mathcal{M}$, i.e.,

$$T_{\mathcal{M}} := P_{\mathcal{M}}T|\mathcal{M} = (T^*|\mathcal{M})^*.$$

Let now $A \in L(\mathcal{H})$ be an operator such that $A\mathcal{H} \subset \mathcal{M}$

and $T_{\mathcal{M}}A = AT$ or equivalently

(1) $P_{\mathcal{M}}TA = AT$.

We introduce a quantity analogous to $\rho_T(A)$ namely

$\rho_{T,\mathcal{M}}(A) := \inf\{\|P_{X^*\mathcal{M}}X^{-1}AX\| : X$ is invertible and $X \in \{T\}'\}$.

Note that $X^*\mathcal{M} = \mathcal{H} \ominus X^{-1}(\mathcal{H} \ominus \mathcal{M})$ and that for $\mathcal{H} = \mathcal{M}$, we have

$$\rho_{T,\mathcal{H}}(A) = \rho_T(A) .$$

Given an operator $A : \mathcal{H} \to \mathcal{M}$ such that

$$P_{\mathcal{M}}TA = AT ,$$

the commutant lifting theorem [6], [7] shows that there exist operators $B \in \{U\}'$ with

$$P_{\mathcal{M}}B = AP_{\mathcal{H}} .$$

For such A, we set

$$\mathrm{Dil}(A) := \{B \in \{U\}' : P_{\mathcal{M}}B = AP_{\mathcal{H}}\} .$$

Now let

$$\tau(A) := \inf\{\|B\|_{sp} : B \in \mathrm{Dil}(A)\}$$

where $\|B\|_{sp}$ denotes as above the spectral radius of B.

The key to our solution of the tangential spectral Nevanlinna-Pick problem is the following result for the proof of which we refer to [3].

Theorem 2 (Spectral Tangential Commutant Lifting Theorem). *Let* $U \in B(\mathcal{K})$ *be an isometry,* $\mathcal{H} \subset \mathcal{K}$ *a finite dimensional hyperinvariant subspace for* U^*, $T = U_{\mathcal{H}}$, *and* $\mathcal{M} \subset \mathcal{H}$ *an invariant subspace for* T^* *. Then for every operator* $A \in B(\mathcal{H})$ *such that* $A\mathcal{H} \subset \mathcal{M}$ *and* $P_{\mathcal{M}}TA = AT$, *we have*

$$\tau(A) = \rho_{T,\mathcal{M}}(A) .$$

We would like to make some comments now on the applicability of Theorem 2. Notice that for T completely non-unitary (i.e., T has no non-zero reducing subspaces on which it is unitary; see [1],[6]), U must be a shift of finite multiplicity. In this case, up to unitary equivalence, we may assume that U is the canonical shift on $H^2(\mathcal{E}) = H^2 \otimes \mathcal{E}$ where $\mathcal{E}$ is a finite dimensional (complex) Hilbert space. The hyperinvariant subspaces of $H^2(\mathcal{E})$ are of the form $H^2(\mathcal{E}) \ominus mH^2(\mathcal{E})$ with $m \in H^\infty$ inner (see [1] and [6]); moreover the space $H^2(\mathcal{E}) \ominus mH^2(\mathcal{E})$ is finite dimensional if and only if m is a finite Blaschke product. Thus the operators T to which Theorem 2 applies have the form $S(m) \otimes I_{\mathcal{E}}$, with $\mathcal{E}$ a finite dimensional Hilbert space and m is a finite Blaschke product.

3. Spectral Nevanlinna-Pick Theory

In this section, we apply Theorem 1 to a spectral version of the Nevanlinna-Pick interpolation problem. In order to do this, we first put the Nevanlinna-Pick interpolation theory into the commutant lifting framework [5] (see also [2] and [3]). First let us recall the classical Nevanlinna-Pick problem in the matrix case.

Let $\mathcal{E}$ be a finite dimensional vector space, let $z_1, \ldots, z_n \in D$ be mutually distinct, and let $F_1, \ldots, F_n \in B(\mathcal{E})$. Then we want necessary and sufficient conditions for the existence of an analytic function $F : D \to B(\mathcal{E})$ with $\|F\|_\infty \leq 1$ such that

$$F(z_j) = F_j$$

for $j = 1,\ldots,n$.

Define

$$m(z) := \prod_{j=1}^{n} \frac{z - z_j}{1 - \bar{z}_j z}$$

and

$$f_j := \left[\prod_{k \neq j} \frac{z - z_j}{1 - \bar{z}_k z}\right] \frac{1}{1 - \bar{z}_j z} .$$

We set $T := S(m) \otimes I_{\mathcal{E}}$, and note that

$$\mathcal{H} = f_1 \otimes \mathcal{E} + f_2 \otimes \mathcal{E} + \cdots + f_n \otimes \mathcal{E} .$$

This sum is direct but not orthogonal. For the given interpolation data above, we define $A : \mathcal{H} \to \mathcal{H}$ by linearity and by

$$A(f_j \otimes \xi) = f_j \otimes F_j \xi$$

for all $\xi \in \mathcal{E}$, $j = 1,2,\ldots,n$. Note that

$$T(f_j \otimes \xi) = z_j f_j \otimes \xi$$

for all $\xi \in \mathcal{E}$, $j = 1,2,\ldots,n$. Thus $A \in \{T\}'$. It is easy to check that F satisfies the interpolation conditions

(2) $F(z_j) = F_j, \quad \forall j = 1,\ldots,n$

if and only if

$$P_{\mathcal{H}} M_F = A P_{\mathcal{H}} .$$

Thus, the commutant lifting theorem implies that there exists an F satisfying (2) and $\|F\|_\infty < 1$, if and only if $\|A\| < 1$. In the spectral Nevanlinna-Pick problem one replaces the condition

$$\|F\|_\infty = \sup\{\|F(z)\| : z \in \mathbb{D}\} < 1$$

with the condition

$$\|M_F\|_{sp} = \sup\{\|F(z)\|_{sp} : z \in \mathbb{D}\} < 1 .$$

It is now obvious that the spectral Nevanlinna-Pick problem can be given the following solution, based on the spectral commutant lifting theorem 1.

Theorem 3. *The following properties are equivalent:*

(i) There exists a bounded analytic function $F : D \to B(\mathcal{E})$ *satisfying*

(3) $F(z_j) = F_j \quad (1 \leq j \leq n)$

and such that $\|M_F\|_{sp} < 1$.

(ii) There exists a bounded rational function $F : \overline{D} \to B(\mathcal{E})$, *satisfying the interpolation conditions* (3) *and such that*

$$\|F(z)\|_{sp} < 1$$

for all $z \in \overline{D}$.

(iii) $\rho_T(A) < 1$.

4. Spectral Tangential Nevanlinna-Pick Theory

In this section, we apply Theorem 2 to a spectral version of the tangential Nevanlinna-Pick interpolation problem. In order to do this, we first put the tangential Nevanlinna-Pick problem [4] into the commutant lifting framework, along a similar line to that in the previous section; see [3]. Accordingly, we start by recalling the general problem of tangential Nevanlinna-Pick interpolation.

Let $z_1, \ldots, z_n \in D$ be distinct, let $u_1, u_2, \ldots, u_n \in C^N$ be non-zero vectors, and let $v_1, v_2, \ldots, v_n \in C^N$ be arbitrary vectors. We are interested in those bounded analytic functions $F : D \to B(C^N)$ with which satisfy the interpolation conditions

(4) $F(z_j)u_j = v_j$

for $j = 1,\ldots,n$. The classical tangential Nevanlinna-Pick problem (first considered in [4]) asks for such functions F with

$$\|F\|_\infty := \sup\{\|F(z)\| : z \in D\} < 1,$$

while the spectral problem asks for such a function F satisfying

$$\|M_F\|_{sp} = \sup\{\|F(z)\|_{sp} : z \in D\} < 1 .$$

We will construct an isometry $U \in L(\mathcal{K})$, a finite dimensional hyperinvariant subspace $\mathcal{H}$ for U^*, an invariant subspace $\mathcal{M}$ for $T^* := U^*|\mathcal{H}$, and an operator $A : \mathcal{H} \to \mathcal{M}$ such that $T_{\mathcal{M}}A = AT$ and with the following property: there exists an isometric bijection between $\mathrm{Dil}(A)$ and the interpolating functions satisfying (4).

The first observation is that replacing F by $G(z) := F(\bar{z})^*$ condition (4) becomes

(5) $G(\bar{z}_j)^*u_j = v_j$.

Let

$$\mathcal{K} := H^2 \otimes C^N,$$

$$m(z) := \prod_{j=1}^{n} \frac{z - \bar{z}_j}{1 - \bar{z}_j z}$$

and set $\mathcal{H} := \mathcal{H}(m) \otimes C^N$ where $\mathcal{H}(m) = H^2 \ominus mH^2$. Further, let U denote the canonical shift on $\mathcal{K}$, and T the compression of U to $\mathcal{H}$. To define the space $\mathcal{M}$, we need to consider the vectors $y_j \in \mathcal{H}(m)$ given by

$$y_j(z) := \frac{1}{1 - \bar{z}_j z}, \quad z \in D .$$

We have

$$S(m)^* y_j = z_j y_j, \quad j = 1, \ldots, n .$$

where as above S is the canonical shift on H^2, and $S(m)$ is the compression of S to $\mathcal{H}(m)$. Note that $U = S \otimes I_{C^N}$, and $T = S(m) \otimes I_{C^N}$.

Define now

$$\mathcal{M} := Cy_1 \otimes u_1 + \cdots + Cy_n \otimes u_n .$$

Finally, we define an operator

$$A(z_1, \ldots, z_n; u_1, v_1; \ldots; u_n, v_n) : \mathcal{H} \to \mathcal{M}$$

by setting

$$(6) \quad A(z_1, \ldots, z_n; u_1, v_1; \ldots; u_n, v_n)^* (y_j \otimes u_j) := y_j \otimes v_j, \quad 1 \le j \le n.$$

We will sometimes set $A(y_1, v_1; \ldots; u_n, v_n) := A(z_1, \ldots, z_n; u_1, v_1; \ldots; u_n, v_n)$ when the dependence on the z_j is clear, or even $A := A(z_1, \ldots, z_n; u_1, v_1; \ldots; u_n, v_n)$ when the dependence on the z_j, u_j, v_j is understood. Since A^* takes eigenvectors of $T^*|\mathcal{M}$ to eigenvectors corresponding to the same eigenvalue of T^*, the relation

$$T^* A^* = A^* T^* |\mathcal{M}$$

is immediate. Thus we have that

$$P_{\mathcal{M}} T A = A T .$$

It is well-known that the commutant $\{U\}'$ consists of analytic multiplication (Toeplitz) operators of the form

M_G, where $G : D \to B(C^N)$ is bounded and analytic. We need the following fact, the proof of which is similar to that given in [5] for the classical Nevanlinna-Pick interpolation problem.

Lemma 1. *With the above notation, let* $G : D \to B(C^N)$ *be analytic and bounded. Then the interpolation conditions* (5) *are satisfied if and only if* $M_G \in Dil(A)$.

Next note that if $X \in \{T\}'$, then necessarily

$$X(y_j \otimes \eta) = y_j \otimes X_j \eta$$

for all $\eta \in C^N$ and where $X_j \in B(C^N)$ for $j = 1, \ldots n$. The operator X is invertible if and only if each X_j $(1 \le j \le n)$ is invertible. Moreover, we have that

$$XA(u_1,v_1;\ldots;u_n,v_n)^*X^{-1}|X\mathcal{M} = A(X_1u_1,X_1v_1;\ldots;X_nu_n,X_nv_n)^*.$$

Clearly

$$\rho_{T,\mathcal{M}}(A) = \inf\{\|A(X_1u_1,X_1v_1;\ldots;X_nu_n,X_nv_n)\| : X_j \in B(C^N), X_j \text{ invertible}\} \tag{7}$$

Now define the *tangential Nevanlinna-Pick matrix* as follows:

$$N(z_1,\ldots,z_n;u_1,v_1;\ldots;u_n,v_n;\rho) :=$$

$$= \left[\frac{\langle \rho u_j, \rho u_k\rangle - \langle v_j, v_k\rangle}{1 - \bar{z}_j z_k}\right]_{1 \le j,k \le n}$$

By using Theorem 2 and Lemma 1 we can easily prove:

Theorem 4 (Spectral Tangential Nevanlinna-Pick Theorem). *There exists* $F \in H^{\infty}(C^N)$ *with* $\|M_F\|_{sp} < \rho$ *satisfying the interpolation conditions* (4) *if and only if there exist* $X_j \in B(C^N)$ $(1 \le j \le n)$ *invertible such that*

$$N(z_1,\ldots,z_n;X_1u_1,X_1v_1;\ldots;X_nu_n,X_nv_n;\rho) > 0 .$$

Indeed this condition expresses the fact that

$$\sigma(u_1,v_1;u_2,v_2;\ldots;u_n,v_n) := \rho_{T,M}(A) < \rho .$$

Using this theorem, it is easy to check that in the spectral tangential Nevanlinna-Pick interpolation, we can ignore points z_j such that u_j and v_j are linearly independent vectors (since in this case $\{[X_ju_j,\ X_jv_j] \in C^{2N}: X_j\in B(C^N),$ X_j invertible$\}$ is dense in C^{2N}).

Therefore from now on without loss of generality we will assume that

$$v_j = \lambda_j u_j,\ \lambda_j \neq 0,\ j = 1,\ldots,n.$$

We now have the following result:

Proposition 4. *Suppose that* $n \le N$. *Then*

$$\sigma_n := \sigma(u_1,\lambda_1u_1;,,,;u_n,\lambda_nu_n) = \max\{|\lambda_1|,\ldots,|\lambda_n|\} .$$

Proof. Clearly

$$\sigma_n \ge \max\{|\lambda_1|,\ldots,|_n|\} .$$

We can get the required equality by choosing the X_j such that the X_ju_j are orthogonal. □

Remark 2. (i) Thus we see that as long as the number of dependent pairs (u_j,v_j) is $\le N$, we have a rather easy way of computing σ_n. In the next section, we consider the case in which the number of dependent vector pairs exceeds N.

(ii) We would like to give now a simple example which shows that the infimum might not be a minimum in the definition of $\rho_{T,\mathcal{M}}$ when some of the pairs u_j, v_j are linearly independent. Specifically, let $z_1 = 0$, $z_2 = 1/2$, let u_1 be any non-zero vector, $v_1 = 0$, and let u_2 and v_2 be linearly independent. Clearly, in this case $\rho_{T,\mathcal{M}} = 0$. On the other hand, the zero interpolating function is obviously not a solution. We will see in Section 5 that when all the pairs are linearly dependent, then in fact it is possible to replace the "inf" with "min" in (7).

5. Optimal Solutions

In this section we consider the spectral tangential Nevanlinna-Pick interpolation problem in which u_j and v_j are linearly dependent for each j. As we saw before, the general case actually reduces to this one. Fix therefore $z_1, \ldots, z_n \in D$ and $\lambda_1, \ldots, \lambda_n \in C$ such that $v_j = \lambda_j u_j$, $\lambda_j \neq 0$ for $1 \leq j \leq n$. We set

$$A_{u_1,\ldots,u_n} := A(z_1,\ldots,z_n; u_1,v_1;\ldots;u_n,v_n)$$

in this case. As in Section 4, we have that

$$\rho_{T,\mathcal{M}} := \rho_{T,\mathcal{M}}(A_{u_1,\ldots,u_n}) =$$

$$= \inf\{\|A_{X_1u_1,\ldots,X_nu_n}\| : X_j \in B(C^N),\ X_j \text{ invertible}\}.$$

In the latter infimum only the vectors $w_j = X_j u_j$ count, so that

$$\rho_{T,\mathcal{M}}:=(A_{u_1,\dots,u_n}) = \inf\{\|A_{w_1,\dots,w_n}\| : w_1,\dots,w_n \in C^N\setminus\{0\}\}$$
$$= \inf\{\|A_{w_1,\dots,w_n}\| : w_j \in C^N, \|w_j\|=1, 1\le j\le n\}.$$

Since the unit sphere in C^N is compact, the latter infimum is actually attained. The operator $A_{u_1,\dots,u_n}$ will be said to be *optimal* if

$$\rho_{T,\mathcal{M}}(A_{u_1,\dots,u_n}) = \|A_{u_1,\dots,u_n}\| .$$

Our discussion shows that for the case at hand (in which all the pairs u_j, v_j are linearly dependent) optimal operators do indeed exist. See also Remark 2(ii)

Fix for the moment $u_1,\dots,u_n$, and as above let $\mathcal{M}$ be the space generated by

$$\{y_i \otimes u_i : 1 \le i \le n\} .$$

We have that

$$A^*_{u_1,\dots,u_n} = \lambda_i y_i \otimes u_i$$

so that $A^*_{u_1,\dots,u_n}$ can be considered for all practical purposes to be an operator acting on $\mathcal{M}$.

Our idea now is to fix j, $1 \le j \le n$ and vary the vector u_j while we keep $\{u_i : i \ne j\}$ fixed. In order to do this, we let $\mathcal{M}_j$ denote the space generated by $\{y_i \otimes u_i : i \ne j\}$, and note that the restriction

$$B_j := A^*_{u_1,\dots,u_{j-1},u,u_{j+1},\dots,u_n}|\mathcal{M}_j$$

does not depend on u for each $1 \le j \le n$. Therefore, if we write

$$\mathcal{M}_u = \mathcal{M}_j + Cy_j \otimes u = \mathcal{M}_j \oplus \mathcal{N}_u$$

for each $u \in C^N$, $u \ne 0$, we see that

$A^*_{u_1,\dots,u_{j-1},u,u_{j+1},\dots,u_n}$ has decomposition of the form

$$A^*_{u_1,\ldots,u_{j-1},u,u_{j+1},\ldots,u_n} = \begin{bmatrix} B_j & D_u \\ 0 & C_u \end{bmatrix}$$

Let $P_{\mathcal{M}_j} : H^2 \otimes C^N \to \mathcal{M}_j$ denote orthogonal projection. Then to determine C_u and D_u, we note that $\mathcal{N}_u$ is generated by the single vector $(I-P_{\mathcal{M}_j})(y_j \otimes u)$, and

$$\begin{aligned} A^*_{u_1,\ldots,u_{j-1},u,u_{j+1},\ldots,u_n}(I-P_{\mathcal{M}_j})(y_j\otimes u) &= \\ &= \lambda_j(y_j\otimes u) - B_jP_{\mathcal{M}_j}(y_j\otimes u) \\ &= \lambda_j(I-P_{\mathcal{M}_j})(y_j\otimes u) + (\lambda_j - B_j)P_{\mathcal{M}_j}(y_j\otimes u). \end{aligned}$$

We conclude that C_u is simply multiplication by λ_j, while D_u sends the unit vector

$$(I-P_{\mathcal{M}_j})(y_j\otimes u)/\|(I-P_{\mathcal{M}_j})(y_j\otimes u)\|$$

to

$$(\lambda_j - B_j)P_{\mathcal{M}_j}(y_j\otimes u)/\|(I-P_{\mathcal{M}_j})(y_j\otimes u)\|.$$

Identifying $\mathcal{M}_j \oplus \mathcal{N}_u$ with $\mathcal{M}_j \oplus C$, we may write

$$A^*_{u_1,\ldots,u_{j-1},u,u_{j+1},\ldots,u_n} = \begin{bmatrix} B_j & \dfrac{(\lambda_j - B_j)P_{\mathcal{M}_j}(y_j\otimes u)}{\|(I-P_{\mathcal{M}_j})(y_j\otimes u)\|} \\ 0 & \lambda_j \end{bmatrix}.$$

Using Proposition 2 we easily obtain the following results.

Lemma 2. *Let* $\rho > 0$. *We have* $\|A^*_{u_1,\ldots,u_{j-1},u,u_{j+1},\ldots,u_n}\| \le \rho$ *if and only if the following conditions are satisfied:*

(i) $\|B_j\| \le \rho$;

(ii) $|\lambda_j| \le \rho$;

(iii) there exists a vector $f_{\rho,u} \in \mathcal{M}_j$ *such that* $\|f_{\rho,u}\| \le 1$ *and*

$$(\lambda_j - B_j)P_{\mathcal{M}_j}(y_j \otimes u)/\rho\|(I-P_{\mathcal{M}_j})(y_j \otimes u)\| =$$
$$= (1-|\lambda_j|^2/\rho^2)^{1/2}(1-B_jB_j^*/\rho^2)^{1/2} f_{\rho,u}.$$

Note that in case $\rho > \max\{\|B_j\|, |\lambda_j|\}$, then

$$f_{\rho,u} = \frac{\rho(\rho^2 - B_jB_j^*)^{-1/2}(\lambda_j - B_j)P_{\mathcal{M}_j}(y_j \otimes u)}{(\rho^2 - |\lambda_j|^2)^{1/2}\|(I-P_{\mathcal{M}_j})(y_j \otimes u)\|} .$$

Since $y_j \otimes u \notin \mathcal{M}_j$, we have that $(I-P_{\mathcal{M}_j})(y_j \otimes u) \neq 0$ if $u \neq 0$.

Lemma 3. *Let* $u_j \in C^N \setminus \{0\}$ *be such that*

$$\|A_{u_1,u_2,\dots,u_n}\| = \inf\{\|A_{u_1,\dots,u_{j-1},u,u_{j+1},\dots,u_n}\| : u \in C^N \setminus \{0\}\} =: \rho_j.$$

Suppose that

$$\rho_j > \max\{\|B_j\|, |\lambda_j|\} .$$

Then

$$\rho_j^2 P_{y_j \otimes C^N}(\lambda_j - B_j)^*(\rho_j^2 - B_jB_j^*)^{-1}(\lambda_j - B_j)(y_j \otimes u_j) =$$
$$= (\rho_j^2 - |\lambda_j|^2)P_{y_j \otimes C^N}(I-P_{M_j})(y_j \otimes u_j).$$

Lemma 4. *Under the hypotheses of Lemma 3, we have*

$$(8) \quad y_j \otimes u_j - \frac{(\rho_j^2 - \bar{\lambda}_j B_j)(\rho_j^2 - B_j^*B_j)^{-1}(\rho_j^2 - \lambda_j B_j^*)}{\rho_j^2 - |\lambda_j|^2} P_{\mathcal{M}_j}(y_j \otimes u_j) \perp y_j \otimes C^N.$$

Now we can write out a rather precise expression for

(8) using linear algebra. To do this, let us denote by Γ_j the self-adjoint (Grammian) matrix given by

$$(\Gamma_j)_{ik} = \langle y_k \otimes u_k, y_i \otimes u_i \rangle, \ 1 \le i, k \le n, \ i,k \ne j .$$

Let Λ_j be the $(n-1) \times (n-1)$ diagonal matrix defined by

$$(\Lambda_j)_{ik} = \lambda_i \delta_{ik}, \ 1 \le i, k \le n, \ i,k \ne j$$

where δ_{ik} denotes the Kronecker delta. Note that Λ_j is precisely the matrix of B_j in the basis $\{y_i \otimes u_i : 1 \le i \le n, \ i \ne j\}$. Moreover, it is a standard fact in linear algebra that the matrix of B_j^* in this basis is $\Gamma_j^{-1}\Lambda_j^*\Gamma_j$.

Write

$$P_{\mathcal{M}_j}(y_j \otimes u_j) = \sum_{i \ne j} \alpha_i^{(j)} y_i \otimes u_i .$$

To calculate $\alpha_i^{(j)}$, we note that

$$\langle y_k \otimes u_k - P_{\mathcal{M}_j}(y_j \otimes u_j), y_k \otimes u_k \rangle = 0$$

for $k \ne j$, so that

$$\mu_k^{(j)} := \frac{\langle u_j, u_k \rangle}{1 - \bar{z}_k z_j} = \sum_{i \ne j} \alpha_i^{(j)} (\Gamma_j)_{ki}$$

and hence we have that

$$\alpha^{(j)} = \Gamma_j^{-1} \mu^{(j)}$$

where $\alpha^{(j)}$ denotes the column vector with components $\alpha_i^{(j)}$, and similarly for $\mu^{(j)}$.

Next set

$$(\rho_j^2 - \bar{\lambda}_j B_j)(\rho_j^2 - B_j^* B_j)(\rho_j^2 - \lambda_j B_j) P_{\mathcal{M}_j}(y_j \otimes u_j) = \sum_{i \ne j} \eta_i^{(j)} y_i \otimes u_i .$$

Using the preceding computations, the column vector $\eta^{(j)}$ with components $\eta_i^{(j)}$ can be explicitly calculated as follows:

$$\eta^{(j)}=(\rho_j^2-\overline{\lambda}_j\Lambda_j)(\rho_j^2-\Gamma_j^{-1}\Lambda_j^*\Gamma_j\Lambda_j)^{-1}(\rho_j^2-\lambda_j\Gamma_j^{-1}\Lambda_j^*\Gamma_j)\Gamma_j^{-1}\mu^{(j)}$$

$$(9)\qquad =(\rho_j^2-\overline{\lambda}_j\Lambda_j)(\rho_j^2-\Gamma_j^{-1}\Lambda_j^*\Gamma_j\Lambda_j)^{-1}\Gamma_j^{-1}(\rho_j^2-\lambda_j\Lambda_j^*)\mu^{(j)}$$

$$=(\rho_j^2-\overline{\lambda}_j\Lambda_j)(\rho_j^2\Gamma_j-\Lambda_j^*\Gamma_j\Lambda_j)^{-1}(\rho_j^2-\lambda_j\Lambda_j^*)\mu^{(j)} .$$

We can now write out the conclusion of Theorem 3 in a rather simple form. Indeed we have shown that

$$(\rho_j^2 - |\lambda_j|^2)y_j \otimes u_j + \sum_{i\neq j}\eta_i^{(j)}y_i \otimes u_i \perp y_j \otimes C^N$$

where the $\eta_i^{(j)}$ are computed from equation (9). But from this it is easy to compute that in fact

$$(10)\quad \sum_{i\neq j}\frac{\eta_i^{(j)}}{1-\overline{z}_jz_i} + (\rho_j^2 - |\lambda_j|^2)u_j = 0 .$$

The above argument then proves that if $\rho_j > \max\{\|B_j\|, |\lambda_j|\}$, then ρ_j and u_j satisfy the system (10). This is precisely Theorem 3 stated in matrix form.

It is possible to re-write the equations (10) in a slightly modified form and to thereby extend their validity to the case in which

$$\rho_j = \|B_j\| > |\lambda_j| .$$

Set

$$D_j := \rho_j^2 - \lambda_j\Gamma_j^* ,$$

$$N_j := \rho_j^2\Gamma_j - \Lambda_j^*\Gamma_j\Lambda_j .$$

Then if we multiply both sides of equation (10) by $\det N_j$, we get that

$$(11)\quad (\det N_j)\sum_{i\neq j}\frac{\eta_i^{(j)}}{1-\bar{z}_j z_i} + (\det N_j)(\rho_j^2 - |\lambda_j|^2)u_j = 0;$$

whence, after some elementary computations,

$$(12)\quad \sum_{i,k\neq j}\frac{(\rho_j^2-\bar{\lambda}_j\lambda_i)}{1-\bar{z}_j z_i}(N_j^{alg})_{ik}(\rho_j^2-\lambda_j\bar{\lambda}_k)\frac{\langle u_j,u_k\rangle}{1-\bar{z}_j z_k}u_i + (\rho_j^2 - |\lambda_j|^2)u_j = 0.$$

Define now

$$Q_j := Q_j(\rho_j; z_1,\dots,z_n;\lambda_1,\dots,\lambda_n;u_1,\dots,u_n) :=$$

$$\begin{bmatrix} \cdots & \frac{\rho_j^2-\bar{\lambda}_{j-1}\lambda_1}{1-\bar{z}_{j-1}z_1}\langle u_1,u_{j-1}\rangle & \frac{\rho_j^2-\bar{\lambda}_j\lambda_1}{1-\bar{z}_j z_1}u_1 & \frac{\rho_j^2-\bar{\lambda}_{j+1}\lambda_1}{1-\bar{z}_{j+1}z_1}\langle u_1,u_{j+1}\rangle & \cdots \\ \vdots & \vdots & \vdots & \vdots & \vdots \\ \cdots & \frac{\rho_j^2-\bar{\lambda}_{j-1}\lambda_1}{1-\bar{z}_{j-1}z_1}\langle u_n,u_{j-1}\rangle & \frac{\rho_j^2-\bar{\lambda}_j\lambda_n}{1-\bar{z}_n z_1}u_n & \frac{\rho_j^2-\bar{\lambda}_{j+1}\lambda_n}{1-\bar{z}_{j+1}z_n}\langle u_n,u_{j+1}\rangle & \cdots \end{bmatrix}.$$

Note that Q_j is regarded as an operator from C^N into $C^{n\times n}$ given by

$$Q_j v :=$$

$$\begin{bmatrix} \cdots & \frac{\rho_j^2-\bar{\lambda}_{j-1}\lambda_1}{1-\bar{z}_{j-1}z_1}\langle u_1,u_{j-1}\rangle & \frac{\rho_j^2-\bar{\lambda}_j\lambda_1}{1-\bar{z}_j z_1}\langle u_1,v\rangle & \frac{\rho_j^2-\bar{\lambda}_{j+1}\lambda_1}{1-\bar{z}_{j+1}z_1}\langle u_1,u_{j+1}\rangle & \cdots \\ \vdots & \vdots & \vdots & \vdots & \vdots \\ \cdots & \frac{\rho_j^2-\bar{\lambda}_{j-1}\lambda_1}{1-\bar{z}_{j-1}z_1}\langle u_n,u_{j-1}\rangle & \frac{\rho_j^2-\bar{\lambda}_j\lambda_n}{1-\bar{z}_n z_1}\langle u_n,v\rangle & \frac{\rho_j^2-\bar{\lambda}_{j+1}\lambda_n}{1-\bar{z}_{j+1}z_n}\langle u_n,u_{j+1}\rangle & \cdots \end{bmatrix}$$

for $v \in C^N$. Then it is easy to show that the system (12) is equivalent to the system

$$(13)\quad \det Q_j(\rho_j; z_1,\ldots,z_n; \lambda_1,\ldots,\lambda_n; u_1,\ldots,u_n) := 0.$$

Notice that equation (13) makes sense when $\det N_j = 0$, which occurs in case $\rho_j = \|B_j\| > |\lambda_j|$.

From the above discussion, we can infer:

Theorem 5. *If*

$$(14)\quad \rho_{T,\mathcal{M}} > \max_{1\le j\le n} \{|\lambda_j|\},$$

then $\rho_{T,\mathcal{M}}$ *and the corresponding optimal* $u_1,\ldots,u_n$ *satisfy the system*

$$(15)\quad \det Q_j(\rho_{T,\mathcal{M}}; z_1,\ldots,z_n; u_1,\ldots,u_n) = 0 \quad \textit{for} \quad 1 \le j \le n.$$

Proof. We have only explicitly worked out the details of the proof in case

$$\rho_{T,M} > \max_{1\le j\le n} \{\|B_j\|, |\lambda_j|\}.$$

A careful perusal of the above computations reveals however, that they do indeed extend to the situation when $\rho_{T,\mathcal{M}}$ satisfies the inequality (14). The details are left to the interested reader. □

Remark 3. Let

$$\overline{\rho} := \inf\{\|f\|_\infty : f \in H^\infty,\ f(z_j) = \lambda_j \quad \text{for} \quad 1 \le j \le n\}.$$

Clearly, we have that

$$\bar{\rho} \geq \rho_{T,\mathcal{M}} \geq \max_{1 \leq j \leq n} \{|\lambda_j|\} .$$

Thus in an algorithm for the computation of $\rho_{T,\mathcal{M}}$ one need only look for ρ's in a finite interval. In fact, we plan to organize the above results into such an algorithm in a future engineering oriented paper.

Acknowledgments

This work was supported in part by grants from the Research Fund of Indiana University, Department of Energy DE-FG02-86ER25020, National Science Foundation DMS-8858149, DMS-8521683, DMS-8802596, ECS-8704047, DMS-8811084, and by the Air Force Office of Scientific Research AFOSR-88-0020.

References

[1] H. Bercovici, "*Operator Theory and Arithmetic in H^∞*," Mathematical Monographs and Surveys, No.26, Amer. Math. Soc., Rhode Island, 1988.

[2] H. Bercovici, C. Foias, and A. Tannenbaum, "A spectral commutant lifting theorem," to appear in *Transactions Amer. Math. Soc.*

[3] H. Bercovici, C. Foias, and A. Tannenbaum, "On spectral tangential Nevanlinna-Pick interpolation," to appear in *Journ. of Math. Anal. and Applic.*

[4] I.P. Fedcina, "A description of the solution of the Nevanlinna-Pick tangent problem," *Akad. Nauk. Armjan. SSR Dokl.*, 60(1975), 37-42.

[5] D. Sarason, "Generalized interpolation in H^∞," *Transactions Amer. Math. Soc.*, 127(1967), 179-203.

[6] B. Sz.-Nagy and C. Foias, "Dilatation des commutants d'operateurs," *C. R. Acad. Sci. Paris, Serie A*, 266(1968), 493-495.

[7] B. Sz.-Nagy and C. Foias, *Harmonic Analysis of Operators on Hilbert Space*, North Holland, Amsterdam, 1970.

[8] B. Sz.-Nagy and C. Foias, "Forme triangulaire d'une contraction et factorisation de la fonction caracteristique," *Acta Sci. Math. Szeged*, 28(1967), 201-212.

[9] A. Tannenbaum, "Feedback stabilization of linear dynamical plants with uncertainty in the gain factor," *Intern. Journ. Control*, 32(1980), 1-16.

[10] A. Tannenbaum, "Spectral Nevanlinna-Pick interpolation theory and robust stabilization," *Proceedings of the 26th Conference on Decision and Control, Los Angeles, CA (Dec.87)*, 1987 IEEE, 1635-1636.

Mathematical Problems in Neural Computing

G. Cybenko *

Abstract

Neural computing techniques are receiving significant attention from the engineering research and development community. Focusing on signal processing applications in pattern recognition and classification, we present some analytic and experimental results that indicate promising directions for future research. At the same time, these results show that in a well defined sense, a certain class of neural computing techniques are universally applicable if feasibility constraints are not an issue. In particular, we should not be surprised that neural network techniques work reasonably well in many applications - instead, we should examine closely the design methodology, if any, that was used to select the network architecture, select the size of the training set and terminate the training process. Explicit examples are given where failure to adequately address these design issues can lead to very poor performance.

1 Introduction

The last five years has witnessed a significant revival in neural computing research with a growing interest in the application of such techniques to engineering problems. So-called *connectionism* advocates the use of numerous simple processing elements in parallel to solve a variety of computational problems in a manner that is suggestive of biological neurocomputing. In spite of this great interest and potential, neural computing remains controversial. Critics from the biological and behavioral sciences often argue that connectionist models are not faithful to some fundamental and critical details of how biological brains work. Critics from the engineering and mathematical sciences often point out that there are serious gaps in our understanding of how such systems work or that the performance of neural computing is not altogether very compelling when compared with conventional approaches.

In order for a computational technique to become useful in science or engineering, there must be some confidence in how that technique performs and what kinds of resources are required for its use. For instance, if a certain phenomenon can be modelled by a known partial differential equation, then theoretical results would likely indicate what kinds of initial conditions should be given, what degrees of discretization are appropriate and what kind of computing resource is necessary to achieve solutions with a desired degree of accuracy.

Not only does no such corresponding body of theory exist for the application of neural computing techniques, but to complicate matters even more, many of the problems for which neural computing seem promising are patently devoid of succinct mathematical models. To give the reader some idea of scope of the applications under study we briefly describe four problems in pattern recognition and classification. With the exception of one problem (sonar signal classification), it seems highly unlikely that formal mathematical models exist for describing the classification process.

In this paper we describe some analytic and experimental results with the goal of identifying fruitful directions for future research in this area. To some extent, the analytic results we present will hopefully strip away the mystic surrounding the use of neural networks in classification and other signal processing problems. By the end of this paper, we hope that the reader with be less surprised at hearing how neural networks are being used in some novel application but more sensitive to the kinds of questions that have to be answered if neural computing is being used reliably from an engineering point of view.

Section 2 describes the class of neural networks that we focus on in this paper, namely multilayered feedforward networks. Section 3 reviews 4 applications of such networks to problems in pattern recognition and classification. These four case studies were taken from the published literature and are often cited as successful applications of these techniques. Section 4 presents analytic results and some interpretation of the applications from the preceding section. Section 5 presents results from a series of simulations and experiments carried out by the author over the past four months.

Section 6 summarizes the issues and questions that challenge the field.

2 Basic Notions

In this section, we briefly review the basic ingredients of multilayered, feed-forward artificial neural networks. The reader is encouraged to consult other sources with more complete descriptions and historical accounts [5,13,7,1].

2.1 Perceptrons

The notion of a *perceptron* is usually attributed to F. Rosenblatt [17]. A perceptron is a simple computational element that accepts as input some number of real quantities, computes an affine linear transformation of those inputs and then applies a sigmoidal nonlinearity. More precisely, let $x_1, x_2, ..., x_n$ be real valued inputs so that

$$\sum_{j=1}^{n} y_j x_j + \theta$$

is an affine function of these inputs. Here y_j and θ are real constants whose values are the free parameters of the system. Let σ be a real valued function with the property that

$$\sigma(t) \rightarrow \begin{cases} 1 & \text{as} \quad t \rightarrow +\infty \\ 0 & \text{as} \quad t \rightarrow -\infty \end{cases} \tag{1}$$

Such a σ is called a *sigmoidal* function. We are not specifying whether σ is continuous or monotonic at this point, however the practice has been to use either of the following

$$\sigma(t) = \begin{cases} 1 & \text{if} \quad t \geq 0 \\ 0 & \text{if} \quad t < 0 \end{cases} \quad \text{or} \quad \sigma(t) = \frac{1}{1+e^{-t}} \tag{2}$$

which are discontinuous and monotonic, continuous respectively.

Combining the above constructions, we see that the functional output of a perceptron with inputs $x = (x_1, ..., x_n)$ is merely

$$P(x) = \sigma(\sum_{j=1}^{n} y_j x_j + \theta) = \sigma(y^T x + \theta)$$

A useful observation about the action of a single perceptron is that it roughly partitions the input space into two half spaces - the half space where $P(x)$ is approximately 0 and the half space where $P(x)$ is approximately 1. In the case of the discontinuous, hard limiting sigmoidal function, this partitioning is quite precise but when using the continuous sigmoidal activation function, there is a transitional region centered about the hyperplane $y^T x + \theta = 0$. Clearly, a scaling of y and θ makes this transition region arbitrarily small without changing the orientation of the half spaces. Under this scaling, in the limit, the two types of activation functions are essentially identical.

The capabilities and limitations of such perceptrons were explored in the seminal works [15,17]. Perceptrons can be viewed as an implementation of a binary decision rule based on linear discrimination. The goal is to classify inputs as belonging to one of two classes, 0 or 1. In practice, empirical data is used to determine the values of the various coefficients that constitute the affine transformation. This process of estimating the appropriate parameters is typically called *training* or *learning* in the connectionist literature. One of the fundamental results for classical discontinuous perceptrons is the so-called *Perceptron Convergence Theorem* which states that if a linear discriminant exists for a problem as specified by data, then a certain training procedure will converge to that answer. The situation for continuous perceptrons is not quite as simple since the continuous transition from 0 to 1 requires dealing with errors that arise from the fact that 0 and 1 are not attained for finite input values. Those anomalies have been studied in [19].

2.2 Multilayered Networks

A multilayered neural network is merely a hierarchy of perceptrons, arranged in layers so that the output of one layer is taken as the input for the next. Typically, the outputs of the last layer are weighted and summed together to produce the output of the network. A network can have multiple outputs by allowing different weightings and summations at the last layer. We describe the network architecture according to the distribution of Perceptron processing elements at each level together with the number of outputs. Thus, when we say a *a-b-c-d* network, we mean a network with a inputs, b and c perceptron nodes at the first and second internal, hidden layers respectively and finally d outputs. Much of the resurgence of interest in neural network research can be attributed to the discovery of tractable training algorithms for multilayered networks [8].

While the next section discusses some specific examples of the application of artificial neural networks to pattern classification and recognition problems, we outline the general principle of their use here.

A pattern classification or recognition problem is informally specified as follows. A collection of data $\{(x^i, f^i), 1 \leq i \leq M\}$ is given. The goal is to produce a function that replicates, as well as possible, the given data - that is we seek a function $f(x)$ so that $f(x^i) \approx f^i$. Described as such, this is quite similar to classical interpolation and estimation problems - given data, we seek a function from a class of functions that interpolates the data as well as possible. One of the key features of promising neural network applications is that they are patently hard to model using conventional mathematical modelling techniques such as differential equations or other formal devices. Instead, the empirical data obtained from the application serves to define the problem *without* a formal model.

A multilayered artificial neural network implements some function of n variables and by an appropriate choice of network parameters, that function might interpolate the given data to some desired degree. Let $N(x, \Pi)$ be the function implemented by a fixed network for input x and parameter values Π. "Good" values of Π are values that tend to give small errors as determined by the mean square error expression

$$\sum_{j=1}^{M}(N(x^j, \Pi) - f^j)^2 \tag{3}$$

This quantity is highly nonlinear in Π so that the best that can be hoped for typically is a good local optimum.

There are a variety of techniques being used to find a local minimum of (3) but we will only mention two, namely back-propagation and a more traditional nonlinear least squares solver.

Back-propagation is essentially a stochastic optimization method that uses gradient descent as the basis for improving the approximate solution at each iteration [8]. The particular form of the network function $N(x, \Pi)$ allows for simple application of what is called "automatic differentiation" (see [10]) to compute explicit gradients efficiently. This differentiation method requires evaluation of the function $N(x, \Pi)$, storing intermediately computed quantities and then a reverse accumulation using the intermediate quantities to obtain the gradient - this reverse accumulation suggested the name "back-propagation". Hence, the use of back-propagation in an application requires the specification of an ordering of the data samples (x^j, f^j) and some termination criterion. The sample applications that we discuss below used this technique with random ordering of the sample data but no explicitly specified termination criteria were given in the papers. Some of our numerical experiments suggest that this termination criterion is of fundamental importance since it determines the quality of the extrapolation to other data.

The minimization procedure that we used in our experiments was based on the Levenberg-Marquardt nonlinear least squares solver as implemented by Jorge More, called *lmstr*, and available from the *netlib* facility. This approach operates directly on the objective function given by (3) and so computes a local minimum of that explicit function.

In summary, most if not all applications of multilayered feedforward neural networks to pattern recognition and classification problems can be viewed as parametric modelling with a specialized parametric model suggestive of biological neural networks at a very abstract level. Most of the application work described in the literature has assumed the following form:

1. An application problem is specified by a collection of empirical training data $\{(x^i, f^i), 1 \leq i \leq M\}$.

2. A network architecture is specified according to how many nodes are used at various levels and what the connectivity of those nodes is.

3. Using a random starting point as a guess for the network parameters, a minimization procedure (typically back-propagation) is used to obtain some locally optimal parameter values.

An important measure of neural networks' suitability in a specific application is the network's ability to "generalize" - that is to perform well on data that was not in the training set. Generalization is thus seen as a synonym for extrapolation in classical approximation theory.

3 Applications

In this section, we briefly describe four representative applications of neural networks that have appeared in the technical literature over the past two years. These applications were chosen because of their interesting problem domains and the relative detail provided in the original papers.

3.1 NETtalk

Sejnowski and Rosenberg [18] describe a neural network implementation of written English text to phoneme mapping. The goal of such a system is to take English text and produce a sequence of phoneme representations that can be fed to a speech synthesizer resulting in an overall system for "reading" aloud written text. NETtalk addresses the mapping from text to phonemes part of the problem. We briefly describe the main features of that implementation.

The input data for NETtalk was derived from an existing archive of transcriptions of 1000 words spoken by a child. The transcriptions gave a letter by letter mapping of letters in those 1000 words to phonemes. The training data for the study consisted of these transcriptions together with the contextual information provided by the preceding and following 3 letters. Finally, the phonemes were coded by 26 articulatory features using the fact that all phonemes could be uniquely represented by some combination of those features.

The network architecture used in the study had $7 \times 29 = 203$ inputs (7 letters were presented at one time and 29 different input variables were used for each letter to account for the 26 alphabetic letters and 3 spacing and punctuation marks). There were 80 nodes at the intermediate, hidden layer and there were 26 output units. In our terminology, this is a 203-80-26 network. All tolled there were over 16,000 parameters (degrees of freedom) in the network used.

The training method used was back-propagation with random ordering of the 1000 words. Upon convergence (the convergence criterion was not explicitly stated), the network achieved over 95% accuracy on the training data with less than 77% accuracy on a larger dictionary of about 20,000 words.

3.2 Target Classification from Sonar Data

Gorman and Sejnowski [9] describe a signal classification problem wherein a sonar echo signal is the input data and the output determines whether that signal came from either a rock or a cylindrical object. Preprocessing of the sonar signal consisted of computing a spectral envelope first and then normalizing (to the interval [0,1]) 60 samples of the resulting continuous envelope. There are two outputs of the network with the interpretation that a (0,1) output denotes a rock and a (1,0) output denotes a cylinder.

The network architectures used therefore had 60 inputs and two outputs with a varying number of hidden units all in one layer. Networks with 0,2,3,6,12 and 24 hidden units were tested. We describe the performance of the network with 24 hidden units, namely a 60-24-2 network. This network has 1,620 parameters (degrees of freedom). A total of 192 sonar samples were used to train the networks with another set of 16 signals being used to check generalization. Back propagation was used to train the network - the stopping criterion was not described in the paper.

The 24 hidden node network was found to be 99.8% accurate on the training set with 84.5% accuracy on the 16 extra signal samples. (These figures were obtained by averaging over numerous runs).

3.3 Handwritten Character Recognition

Burr [3] describes a problem based on recognizing handwritten upper case alphabetic characters. A handwritten character is scaled and centered in a template that extracts projections of the written character on 13 line segments arranged in the template. This results in 13 real valued inputs. There were 26 outputs, one for each alphabetic character, with the convention that the output corresponding to a letter should be 1 if the character is detected and 0 otherwise.

The networks tested had 12, 16, 20 and 24 hidden nodes all on one layer. These networks have between 480 and 960 parameters (degrees of freedom) to be determined.

The training data used consisted of 104 samples of handwritten characters, 4 of each upper case letter. Another 104 samples were used to test generalization. Again, back propagation was used for training. No figures are given for the level of accuracy attained upon convergence (criterion not given) but the trained networks performed at levels of between 90% and 95% accuracy on an additional set of 104 written samples used specifically to test generalization.

3.4 Spoken Numeral Recognition

In the same paper as above [3], Burr describes another application to recognition of spoken digit in isolated speech. The problem is to classify an isolated speech signal as one of the ten digits. The input data is obtained by taking 12 cepstral coefficients from the appropriately transformed signal. There are 10 outputs, one for each digit, using the same conventions as in the previous three examples.

A total of 50 speech signals (obtained from a single speaker with 5 samples per digit) were used for training networks with between 5 and 64 hidden nodes at one internal layer. Those networks have between 115 and 1,420 parameters.

Using back propagation for training with these 50 samples, the resulting trained networks achieved over 95% accuracy on an additional 50 speech samples used solely for testing generalization.

4 Analysis

The neural network applications described above demonstrate a remarkable breadth and level of performance. Are the results purely serendipitous or is there some deeper principle at work? In this section, we will attempt to put some perspective on this question.

Is it indeed remarkable that neural computing techniques can admirably address such disparate problems as handwritten digit recognition and sonar signal classification? Abstracting these applications, the underlying problem becomes one of data interpolation using a parametric family of functions, so we should ask whether this parametric family of functions has the power to interpolate an arbitrary data set.

The following result from [6] appears definitive.

Let I_n denote the n-dimensional unit cube, $[0,1]^n$. The space of continuous functions on I_n is denoted by $C(I_n)$ and we use $\|f\|$ to denote the supremum (or uniform) norm of an $f \in C(I_n)$.

The interpolation question stated above can be formulated as determining conditions under which sums of the form

$$G(x) = \sum_{j=1}^{N} \alpha_j \sigma(y_j^T x + \theta_j) \tag{4}$$

are dense in $C(I_n)$ with respect to the supremum norm. It should be noted that the form of the function G in (4) is precisely the form of the output function of an n-N-1 network.

Theorem 1 (Cybenko [6]) *Let σ be any continuous (but not necessarily monotonic) sigmoidal function. Then finite sums of the form*

$$G(x) = \sum_{j=1}^{N} \alpha_j \sigma(y_j^T x + \theta_j)$$

are dense in $C(I_n)$. In other words, given any $f \in C(I_n)$ and $\epsilon > 0$, there is a sum, $G(x)$, of the above form, for which

$$|G(x) - f(x)| < \epsilon \quad \textit{for all} \quad x \in I_n.$$

Additional related results for different norms and functions can be found in [6,11,4,12]. Although such a result asserts the universality of neural networks with only a single hidden layer it makes no claims about the number of nodes needed. Nonetheless, the results described in the above application studies appear less profound in light of this theorem. Specifically, given an arbitrary collection of input-output data, the above theorem implies that by using a network with sufficiently many hidden nodes, all at one layer, the network parameters can be chosen to achieve any desired modelling error criterion.

Thus stated, we do not believe that reports describing the ability of a neural network to "learn" a given input-output relationship taken from some application to be interesting in their own right. In fact, a close look at the above four applications examples shows that in each case, the parameter estimation problem is severely *underdetermined* - that is, far fewer data points are given than there are degrees of freedom in the parametric model! This fact coupled with the above theorem leads one to anticipate such results with some confidence.

The remainder of this paper is devoted to a discussion of the questions that need to be addressed in order to make neural computing techniques useful from an engineering point of view. We list these now:

1. What is the distribution of local minima and how do we guarantee or encourage the computation of provably good solutions?
2. What is the relationship between network architectures and application problems?
3. What are appropriate termination criteria and what are their relationships to generalization (the quality of interpolation)?

5 Numerical Experiments and Mathematical Problems

We are in the process of performing extensive numerical experiments using a simple two variable classification problem that lends itself to easy visualization. Our motivation rests with getting a better understanding of questions about how local minima are distributed, how generalization is affected by convergence criteria and how network architectures might be designed in a way that reflects the complexity of the underlying recognition problem.

As mentioned before, these experiments used a general nonlinear least squares solver as opposed to back-propagation for computing parameter estimates. This method was used to ensure the computation of local minima for the objective function (3) but it raises the question of whether these results really bear any relationship with results obtained by back-propagation. We are not aware of any work addressing this matter.

5.1 The Model Problem

The model problem used in this experiment was a synthetic two input variable problem that could be easily visualized. The inputs were real numbers between 0 and 1 and the output was 1 if the inputs were inside circles with centers at the four points (0.25,0.25), (0.25,0.75), (0.75,0.25) and (0.75,0.75) with radii 0.15. The output was 0 if the input was outside all of the circles. A density plot of the function is given below in Figure 1 where white indicates 1 and black indicates 0. There are therefore two classes, with one class being described by a disconnected set.

Figure 2 shows the density plot of the function resulting from training a 2-6-3-1 network on 200 data points drawn uniformly randomly. Figure 2 is a particularly good example of training.

5.2 Distribution of Local Minima

As already noted, the four case study applications used networks with single internal hidden layers and the analytic result of Theorem 1 suggests that such networks are sufficiently general for virtually all interpolation problems. On the other hand, there are numerous convincing heuristic arguments that suggest networks with more than one internal layer may be more parsimonious and easier to train. Our experiments below involve networks with two internal layers.

Problem 1 *What tradeoffs exit between networks with depth (more than one internal layer) and networks with breadth (only one internal layer but that layer has more nodes) with respect to both representability and trainability?*

A key difficulty with nonunimodal optimization problems is that they do not have unique local minima. Since the training procedures used for

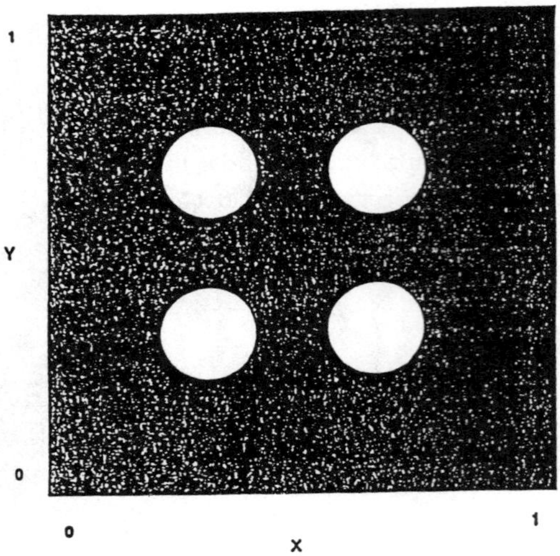

Figure 1: Test Pattern

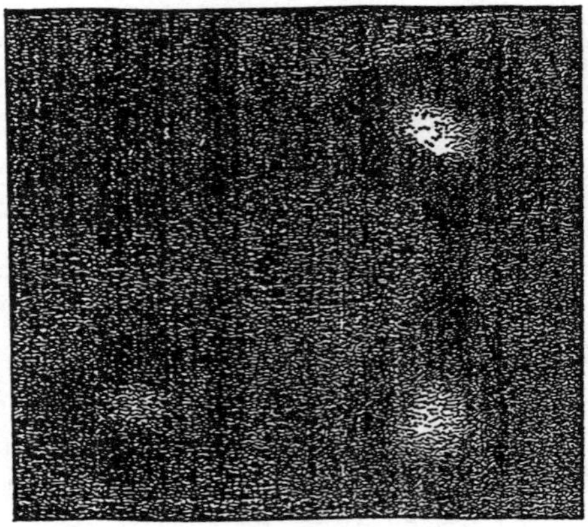

Figure 2: Trained 2-6-3-1 Network Output

neural networks typically find local minima starting from randomly selected starting guesses for the parameters, it should be expected that local minima of varying quality will be found. We were interested in examining the distribution of local minima when using random starting guesses. Figures 3 through 5 show four histograms for 2-4-3-1, 2-6-3-1 and 2-8-4-1 networks using 200 data points. These histograms describe the distribution of the errors at local minima when the minimization procedure's starting points are randomly generated.

The nonlinear least squares solver was terminated when the objective function decreased by less than 10^{-4} in relative value and the parameters ceased to decrease by more than 10^{-7}. The procedure also terminated when the Jacobian became sufficiently rank deficient, indicating coincidence of parameters for nodes at the same. Geometrically, this happens when the discriminating hyperplanes become almost parallel and in the same position in the input space. Computations were done in double precision Fortran on a SUN 3/60 workstation. To get an idea of the problem scale, each solution for a 2-8-4-1 network took about an hour of CPU time so that the histogram for the 2-8-4-1 example required over a week of CPU time to compute alone.

These histograms indicate a tendency towards a bimodal distribution with the bimodality decreasing as the network complexity increased. Visual inspection of the *bad* solutions indicated that they corresponded to strange

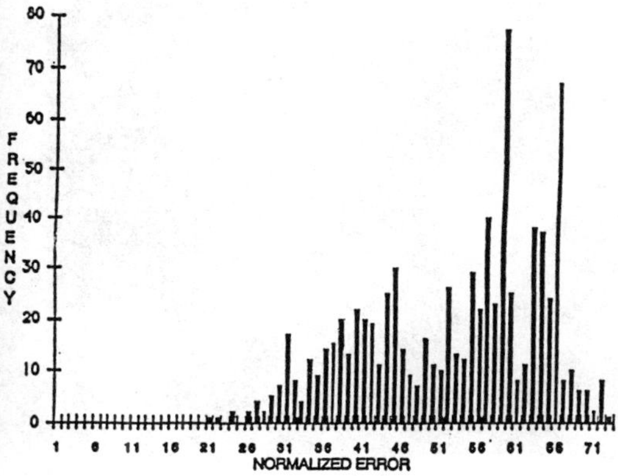

Figure 3: Local Minima in 2-4-2-1 Nets

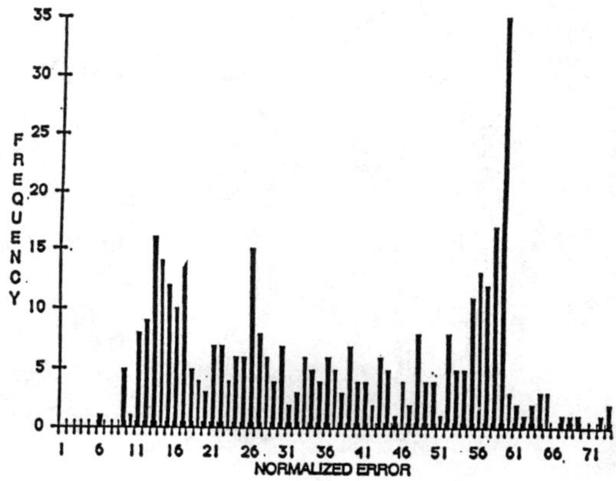

Figure 4: Local Minima in 2-6-3-1 Nets

local minima where the discriminating hyperplanes lined up around some anomaly of the data set. Figure 6 shows the density plot of a bad 2-4-2-1 network solution for example.

These experiments suggest that the more complex the architecture, the more likely it is for a randomly generated starting point to result in a reasonably good solution. Taken to extremes, this observation can be formalized as follows.

Theorem 2 *Suppose that we are given a collection of data* $\{(x^j, f^j), 1 \leq j \leq M\}$ *and* $\epsilon > 0$. *Then the probability of obtaining, by linear methods, a solution within* ϵ *of an optimal solution (which has residual 0 for sufficiently large networks by Theorem 1) of the minimization problem (3) approaches 1 as the number of nodes in the first layer of a network increases unboundedly. The initial guesses for the network parameters are independent and identically distributed with the distribution having infinite tails.*

A sketch of the proof goes as follows. Start by interpolating the data by any continuous function. By Theorem 1, this function can be unformly approximated arbitrarily well by some finite network. Under the probabilistic assumptions of the theorem, for sufficiently large networks, the probability that the starting guesses for some of the nodes at the first layer will be close to the values of the network given by Theorem 1, will be 1. Then by solving

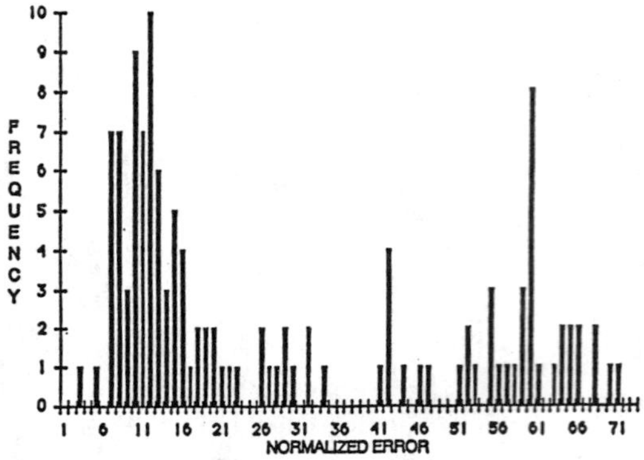

Figure 5: Local Minima in 2-8-4-1 Nets

Figure 6: Typical Bad Minimum

the strictly linear optimization problem in the coefficients α_j in equation (4), we obtain a sufficiently close solution.

This result is certainly quite different in detail from the situation we are dealing with because we are not in the asymptotic regime for sure and the optimization is not purely linear. Nonetheless, the spirit of the observation made above seems to be captured by Theorem 2.

Problem 2 *What is the relationship between the distribution of local minima, the geometry of the classification problem and the network architecture?*

Problem 3 *Can some simple preprocessing be done on the data to obtain starting guesses that ensure better distributions of local minima?*

Some interesting results in this direction that relate partitioning properties of hyperplanes to the number of nodes required in a network can be found in [14].

5.3 Generalization and Termination Criteria

A well known fact in approximation theory is that underdetermined approximation runs a high risk of providing a terrible extrapolation to new data. However, the four sample applications given above are all underdetermined and yet do give reasonable extrapolations. Is this some useful and unique

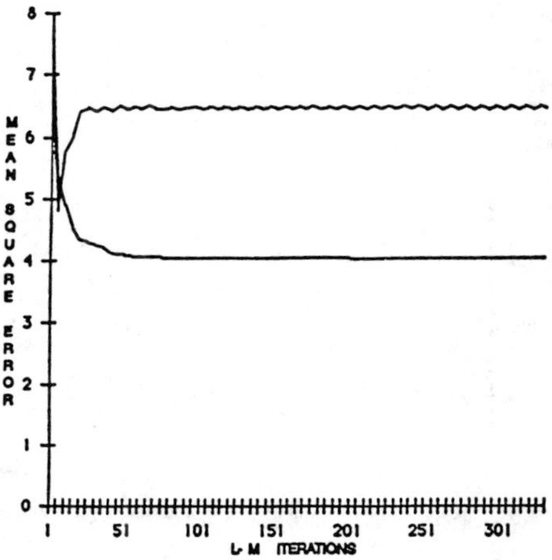

Figure 7: 2-2-2-1 Net with 30 Samples

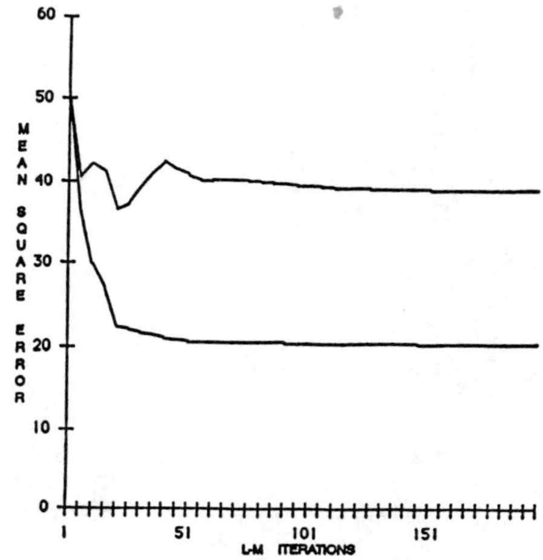

Figure 8: 2-2-2-1 Net with 200 Samples

property of neural networks that they do not share with say polynomial or trigonometric interpolation? Or is it the combined nature of the applications problems structure and neural network approximation that accounts for the unexpectedly good extrapolation?

We have performed experiments that monitor the decrease in the objective function together with the mean square error of the network applied to a complimentary data set of the same size as the training set. Those results are depicted in Figures 7 through 10 for the test problem with differing network architectures and training sets of various sizes. In each case, the graph is a plot of the mean square error vs the iteration count in the minimization routine. The two lines in each graph represent the mean square error on the training set, which is monotonically decreasing since the minimization routine guarantees this, and the mean square error on an additional data set of the same size as the training set. The mean square error on the training set exhibits erratic behavior with occasional large deviations to very large errors.

Notice that in Figures 9 and 10, there are exceedingly large, but transient, deviations from a reasonable extrapolation error. This raises the question of how to define a termination criterion for the training process, whatever it may be? Letting the process run too long could lead to modelling of error and fine detail that might be idiosyncratic for a particular data set.

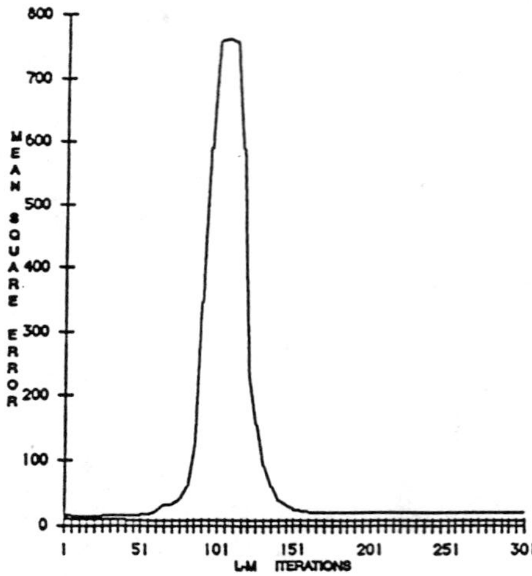

Figure 9: 2-4-4-1 Net with 60 Samples

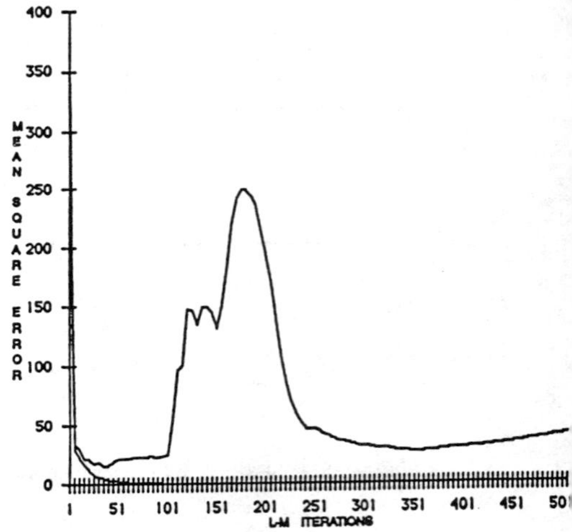

Figure 10: 2-8-8-1 Net with 200 Samples

Problem 4 *Is there some explanation for the transient lack of generalization capability as seen in Figures 9 and 10? Can preprocessing of the data and suitable initial conditions avoid such dangerous transients?*

Problem 5 *Is there some form of cross validation similar to statistical cross validation that is suitable as a termination criterion for the training process?*

6 Summary

We have presented a collection of case studies taken from the published literature, some analytic and experimental results. The analytic results indicate that neural network techniques can in principle have wide applicability and that the research community should not necessarily be impressed with demonstrations of that applicability. At the same time, our experimental results point out some serious potential pitfalls with the practical use of neural networks. We have presented and discussed a few relatively well posed mathematical problems that arise from these investigations.

The perspective of this paper has been from the approximation theoretic point of view. Discussions of neural computing and training from the perspectives of computation theory (see [16]) and statistics (see [2]) also exist and provide alternate views of the subject.

References

[1] J. Anderson and E Rosenfeld. *Neurocomputing.* M. I. T. Press, Cambridge, 1988.

[2] A.R. Barron and R.L. Barron. Statistical learning networks: a unifying approach. In *Proceedings of 1988 Symposium on the Interface: Statistics and Computing Science, Reston VA*, April 1988.

[3] D.J. Burr. Experiments on neural net recognition of spoken and written text. *IEEE Trans. Acoust. Speech and Signal Process.*, 36:1162–1168, 1988.

[4] S.M. Carroll and B.W. Dickinson. Construction of neural nets using the Radon transform. preprint, 1989.

[5] J.D. Cowan and D.H. Sharp. Neural nets and artificial intelligence. *Daedalus*, 117:85–122, Winter 1988.

[6] G. Cybenko. Approximation by superpositions of a single function. *Mathematics of Control, Signals and Systems*, to appear in August 1989.

[7] J. McClelland D. Rumelhart. *Parallel Distributed Processing.* M. I. T. Press, Cambridge, 1987.

[8] G.E. Hinton D.E. Rumelhart and J.L. McClelland. *A General Framework for Parallel Distributed Processing.* MIT Press, 1986.

[9] R.P. Gorman and T.J. Sejnowski. Learned classification of sonar targets using a massively parallel network. *IEEE Trans. Acoust. Speech and Signal Process.*, 36:1135–1140, 1988.

[10] A. Griewank. On automatic differentiation. to appear in Mathematical Programming 88, Kluwer Academic Publishers, 1989.

[11] K. Hornik, M. Stinchcombe, and H. White. Multi-layer feedforward networks are universal approximators. preprint, 1988.

[12] L. K. Jones. Constructive approximations for neural networks by sigmoidal functions. preprint, 1988.

[13] S. Lippmann. An introduction to computing with neural nets. *Computer Architecture News*, 16:7–25, 1988.

[14] J. Makhoul, R. Schwartz, and A. El-Jaroudi. Classification capbilities of two-layer neural nets. In *Proceedings of IEEE International Conference on Acoustics, Speech and Signal Processing, Glasgow, Scotland*, 1989.

[15] M. Minsky and S. Papert. *Perceptrons.* M. I. T. Press, Cambridge, 1988.

[16] Proceedings of the 1st workshop on computational learning theory. August 1988.

[17] F. Rosenblatt. *Principles of Neurodynamics.* Spartan Books, Washington D.C., 1961.

[18] T.J. Sejnowski and C.M. Rosenberg. Parallel networks that learn to pronounce english text. *Complex Systems*, 1:145–168, 1987.

[19] E.D. Sontag and H.J. Sussmann. Backpropagation can give rise to spurious local minima even for networks without hidden layers. *Complex Systems*, 3, to appear 1989.

Supported in part by NSF grant DCR-8619103, ONR contract N000-86-G-0202 and DOE grant DE-FG02-85ER25001.

RKHS, LIS AND INTERPOLATION

Harry Dym

Abstract

A quick introduction to reproducing kernel Hilbert spaces and the application of certain classes of them to the lossless inverse scattering problem of network theory and some associated problems of interpolation is given.

1. Introduction

The purpose of this paper is to give a quick introduction to the application of reproducing kernel Hilbert space methods to solving the lossless inverse scattering (LIS) problem of network theory and some related problems of interpolation. Although this paper can be read independently, it is really meant to be read in conjunction with a companion paper by P. Dewilde which treats a number of related topics from a different point of view. Because of limitations of space many topics are only touched upon lightly, or are only illustrated by special cases. Almost everything which appears here can be found in more expanded form in the monograph [7] or in the paper [8]. The story, however, begins earlier: The LIS problem was extensively studied in [12], [13] and [14], by variants of the Schur algorithm and some global techniques. Reproducing kernel space methods closely related to the ones discussed here were introduced and applied to the LIS problem in [1], though in a somewhat different setting. Some generalizations of the methods discussed here to the "PLIS" problem will appear in [3].

The paper is divided into a number of short sections, the headings of which give a pretty good idea of the contents.

The notation is fairly standard: We shall use $\mathbb{C}$ for the complex numbers, $\mathbb{D}$ for the open unit disc and $\mathbb{T} = \{\lambda \in \mathbb{C} : |\lambda| = 1\}$ for its boundary. A^* stands for the conjugate transpose of a matrix A (or simply the complex conjugate if A is a number) and $|A|$ for its maximum s-number. $L^2_{p\times q}$ [resp. $H^2_{p\times q}$] stands for the space of $p \times q$ matrix valued functions with entries which belong to the ordinary L^2 [resp. H^2] space on $\mathbb{T}$, where H^2, as usual, denotes the Hardy space of $f \in L^2$ with one sided Fourier series: $f = \sum_{n=0}^{\infty} a_n e^{in\theta}$. L_k^2 [resp. H_k^2] is short for $L_{k\times 1}^2$ [resp. $H_{k\times 1}^2$] equipped with the standard inner product:

$$< f, g >= \frac{1}{2\pi} \int_0^{2\pi} g(e^{i\theta})^* f(e^{i\theta}) d\theta \ . \tag{1.1}$$

The symbol $\underline{\underline{p}}$ designates the orthogonal projection of L_k^2 onto H_k^2; the value of k will always be clear by context and is not indicated explicitly in the last two notations.

$\mathcal{S}_{p\times q}$ stands for the set of $p \times q$ matrix valued functions which are analytic and contractive in $\mathbb{D}$.

2. Reproducing kernel Hilbert spaces

A Hilbert space $\mathcal{H}$ of $m \times 1$ vector valued functions which are defined in a set Ω, with inner product $< \ , \ >_{\mathcal{H}}$, is said to be a reproducing kernel Hilbert space if there exists an $m \times m$ matrix valued function $K_\omega(\lambda)$ such that for every $f \in \mathcal{H}$, $v \in \mathbb{C}^m$ and $w \in \Omega$

(1) $K_\omega v \in \mathcal{H}$ (as a function of λ), and

(2) $< f, K_\omega v >_{\mathcal{H}} = v^* f(\omega)$.

We shall refer to $K_\omega(\lambda)$ as the reproducing kernel for $\mathcal{H}$. The phrase "the reproducing kernel" is justified because there is only one such: If also $L_\omega(\lambda)$ meets (1) and (2), then

$$v^* L_\alpha(\beta) u =< L_\alpha u, K_\beta v >_{\mathcal{H}} =< K_\beta v, L_\alpha u >_{\mathcal{H}}^* = v^* K_\beta(\alpha)^* u$$

for every choice of α, β in Ω and u, v in $\mathbb{C}^m$. This proves that

$$L_\alpha(\beta) = K_\beta(\alpha)^*$$

and hence yields both the asserted uniqueness and the supplementary fact that

$$K_\alpha(\beta) = K_\beta(\alpha)^* \ .$$

EXAMPLE 1. The Hardy space H_m^2 of $m \times 1$ vector valued functions with components in the Hardy space H^2 over the disc $\mathbb{D}$. Here $\Omega = \mathbb{D}$ and

$$K_\omega(\lambda) = \frac{I_m}{\rho_\omega(\lambda)} \ , \qquad \text{where} \qquad \rho_\omega(\lambda) = 1 - \lambda\omega^*$$

and we work with the standard inner product. The verification rests essentially on Cauchy's formula for H^2 functions.

EXAMPLE 2. Let $f_1, \ldots, f_n$ be any given set of linearly independent $m \times 1$ vector valued functions which are analytic in some open nonempty set Ω and let $\mathbb{P}$ be any $n \times n$ positive definite matrix. Then the space $\mathcal{H} = \text{span}\{f_1, \ldots, f_n\}$ endowed with the inner product

$$< f_j, f_i >_{\mathcal{H}} = \mathbb{P}_{ij} \ , \qquad i, j = 1, \ldots, n \ ,$$

is a reproducing kernel Hilbert space with reproducing kernel

$$K_\omega(\lambda) = \sum_{i,j=1}^{n} f_i(\lambda)(\mathbb{P}^{-1})_{ij} f_j(\omega)^* \ . \tag{2.1}$$

The verification is a straightforward calculation.

EXAMPLE 3. Let $S \in \mathcal{S}_{p\times q}(\mathbb{D})$ and, for $f \in H_p^2$, let

$$\kappa(f) = \sup\{\|f + Sg\|^2 - \|g\|^2 : \ g \in H_q^2\} \ .$$

Then the space

$$\mathcal{H}(S) = \{f \in H_p^2 : \ \kappa(f) < \infty\}$$

with inner product based on

$$\|f\|^2_{\mathcal{H}(S)} = \kappa(f)$$

is a reproducing kernel Hilbert space with reproducing kernel

$$\Lambda_\omega(\lambda) = \frac{I_p - S(\lambda)S(\omega)^*}{\rho_\omega(\lambda)} \ . \tag{2.2}$$

This example is due to de Branges and Rovnyak [10, 11].

It is readily verified that:

(1) $\{(I - S\underline{\underline{p}}S^*)f : \ f \in H_p^2\} \subset \mathcal{H}(S)$. (2.3)

(2) $\|(I - S\underline{\underline{p}}S^*)f\|^2_{\mathcal{H}(S)} =< (I - S\underline{\underline{p}}S^*)f, f >$ for every $f \in H_p^2$. (2.4)

(3) If S is also isometric on $\mathbb{T}$, i.e., if $S(e^{i\theta})^* S(e^{i\theta}) = I_q$ for a.e. point $e^{i\theta} \in \mathbb{T}$, then

$$\mathcal{H}(S) = H_p^2 \ominus SH_q^2 \quad \text{and} \quad \|f\|_{\mathcal{H}(S)} = \|f\|$$

for every $f \in \mathcal{H}(S)$.

For present applications we shall be particularly interested in reproducing kernel Hilbert spaces of $m \times 1$ vector valued functions with reproducing kernels of the special form

$$K_\omega(\lambda) = \frac{J - \Theta(\lambda)J\Theta(\omega)^*}{\rho_\omega(\lambda)} \ , \tag{2.5}$$

where J is an $m \times m$ signature matrix (i.e., J is constant and both selfadjoint and unitary with respect to the standard inner product in $\mathbb{C}^m$) and Θ is a rational $m \times m$ matrix valued function which is J inner. This means that

(1) $\Theta(\omega)^* J \Theta(\omega) \leq J$, for $\omega \in \mathbb{D}$, and

(2) $\Theta(\omega)^* J \Theta(\omega) = J$, for $\omega \in \mathbb{T}$,

where in the foregoing ω is taken in the domain of analyticity of Θ so that the statements make sense. Such spaces will be termed $\mathcal{H}(\Theta)$ spaces.

It is readily checked that every $m \times m$ signature matrix J is unitarily equivalent to

$$J_{pq} = \begin{bmatrix} I_p & 0 \\ 0 & -I_q \end{bmatrix}, \qquad \text{where } p + q = m .$$

For ease of exposition we shall fix $J = J_{pq}$ from now on, since this will suffice for present needs. There are applications where other choices are more convenient. The simplest examples of J $(= J_{pq})$ inner functions (assuming that $p \geq 1$ and $q \geq 1$) are

$$\Theta_1(\lambda) = \begin{bmatrix} b_\alpha(\lambda) & 0 \\ 0 & I_{m-1} \end{bmatrix} \quad \text{and} \quad \Theta_2(\lambda) = \begin{bmatrix} I_{m-1} & 0 \\ 0 & 1/b_\alpha(\lambda) \end{bmatrix} ,$$

where

$$b_\alpha(\lambda) = \frac{\lambda - \alpha}{1 - \lambda\alpha^*}$$

is the more or less standard Blaschke factor with $\alpha \in \mathbb{D}$. The latter example illustrates the fact that Θ may have poles in $\mathbb{D}$.

We leave it to the reader to check that

$$\mathcal{H}(\Theta_1) = \text{span}\left\{\frac{v_1}{\rho_\alpha}\right\} \quad \text{and} \quad \mathcal{H}(\Theta_2) = \text{span}\left\{\frac{v_2}{\rho_{\alpha'}}\right\} \qquad \text{for } \alpha \neq 0 ,$$

where $\alpha' = 1/\alpha^*$, $v_1 = \begin{bmatrix} 1 \\ 0 \\ \vdots \\ 0 \end{bmatrix}$ and $v_2 = \begin{bmatrix} 0 \\ \vdots \\ 0 \\ 1 \end{bmatrix}$.

This illustrates the general fact that the McMillan degree of Θ is equal to the dimension of $\mathcal{H}(\Theta)$; see Theorem 5.6 of [2] for a proof.

3. Characterization of $\mathcal{H}(\Theta)$ spaces

A basic problem is to characterize the class of reproducing kernel Hilbert spaces which admit a reproducing kernel of the form (2.5), based on a J inner Θ. This problem was solved by de Branges [9] for Θ's which are J inner with respect to the open upper half plane and with $\rho_\omega(\lambda) = -2\pi i(\lambda - \omega^*)$, up to a technical point which was resolved by Rovnyak. These results were adapted to the disc by Ball [6]. We next state a finite dimensional version of this theorem. The notation

$$(R_\alpha f)(\lambda) = \frac{f(\lambda) - f(\alpha)}{\lambda - \alpha}$$

for the generalized backwards shift will be useful.

THEOREM 3.1. *Let $\mathcal{M}$ be a finite dimensional reproducing kernel Hilbert space of $m \times 1$ vector valued functions which are analytic in some nonempty open subset Ω of $\mathbb{D}$. Then $\mathcal{M}$ is an $\mathcal{H}(\Theta)$ space if and only if*

(1) $\mathcal{M}$ is R_α invariant for some (and hence every) point $\alpha \in \Omega$, and

(2) the identity

$$< f, g >_{\mathcal{M}} + \alpha < R_\alpha f, g >_{\mathcal{M}} + \beta^* < f, R_\beta g >_{\mathcal{M}} - (1 - \alpha\beta^*) < R_\alpha f, R_\beta g >_{\mathcal{M}}$$
$$= g(\beta)^* J f(\alpha) \tag{3.1}$$

holds for every choice of f, g in $\mathcal{M}$ and α, β in Ω.

To better understand the significance of the conditions in the theorem it is instructive to consider a one dimensional example. If $f \not\equiv 0$ and $R_\alpha f$ both belong to a one dimensional space, then they must be proportional:

$$(R_\alpha f)(\lambda) = \frac{f(\lambda) - f(\alpha)}{\lambda - \alpha} = \mu f(\lambda)$$

for some constant μ. But this in turn implies that

$$f(\lambda) = \frac{f(\alpha)}{1 - \mu(\lambda - \alpha)}$$

and hence that either

$$f(\lambda) = \frac{v}{\rho_\omega(\lambda)} \quad \text{or} \quad f(\lambda) = \frac{v}{\lambda}$$

for some choice of $v \in \mathbb{C}^m$ and $\omega \in \mathbb{C}$. Thus we see that condition (1) serves to specify the form of the elements in the space. Furthermore, in the first case

$$(R_\alpha f)(\lambda) = \frac{\omega^*}{\rho_\omega(\alpha)} f(\lambda) \qquad (\text{providing } \alpha\omega^* \neq 1) \ ,$$

whereas in the second

$$(R_\alpha f)(\lambda) = -\frac{1}{\alpha} f(\lambda) \qquad (\text{providing } \alpha \neq 0) \ .$$

Substituting the first case into the identity (3.1) with $f = g$ yields

$$(1 - |\omega|^2) < f, f >_{\mathcal{M}} = v^* J v \ ,$$

which in turn implies that

$$< f, f >_{\mathcal{M}} = \frac{v^* J v}{\rho_\omega(\omega)} \quad \text{if } |\omega| \neq 1 \tag{3.2}$$

(and $v^* J v = 0$ if $|\omega| = 1$).

Thus we see that if the underlying space $\mathcal{M}$ is one dimensional, then the role of condition (1) in Theorem 3.1 is to specify the form of the elements in the space, whereas the role of condition (2) is to specify the inner product (at least if $|\omega| \neq 1$). The same holds true in higher dimensional spaces: Condition (1) insures that $\mathcal{M}$ has a basis which is made up of chains of vector valued functions which are either of the form

$$\frac{v_1}{\rho_\omega(\lambda)}, \frac{\lambda v_1}{\rho_\omega(\lambda)^2} + \frac{v_2}{\rho_\omega(\lambda)}, \ldots, \frac{\lambda^{k-1}}{\rho_\omega(\lambda)^k} v_1 + \ldots + \frac{v_k}{\rho_\omega(\lambda)}$$

or of the form

$$\frac{v_1}{\lambda}, \frac{v_1}{\lambda^2} + \frac{v_2}{\lambda}, \ldots, \frac{v_1}{\lambda^k} + \ldots + \frac{v_k}{\lambda},$$

where $v_1, \ldots, v_k$ belong to $\mathbb{C}^m$. In particular, if $\mathcal{M}$ has a basis consisting of a single chain of length n of the first type:

$$f_1(\lambda) = \frac{v_1}{\rho_\omega(\lambda)}, \qquad f_j(\lambda) = \frac{\lambda}{\rho_\omega(\lambda)} f_{j-1}(\lambda) + \frac{v_j}{\rho_\omega(\lambda)}, \qquad j = 2, \ldots, n,$$

then the identity (3.1) holds if and only if the $n \times n$ matrix $\mathbb{P}$ with entries

$$\mathbb{P}_{ij} = < f_j, f_i >_{\mathcal{M}}$$

is a solution of the matrix equation

$$\mathbb{P} - (Z_{\omega^*})^* \mathbb{P} Z_{\omega^*} = V^* J V, \tag{3.3}$$

where V is the $m \times n$ matrix with columns $v_1, \ldots, v_n$,

$$Z_\omega = \omega I_n + Z \quad \text{and} \quad Z = \begin{bmatrix} 0 & 1 & 0 & \ldots & 0 & 0 \\ 0 & 0 & 1 & \ldots & 0 & 0 \\ \vdots & & & & & \vdots \\ 0 & 0 & & \ldots & 0 & 1 \\ 0 & 0 & & \ldots & 0 & 0 \end{bmatrix}$$

is the backwards shift on $\mathbb{C}^n$.

More information on such chains may be found in [8]; see Theorem 3.1 therein for a proof of the last assertion and Theorem 5.1 for the corresponding statement when $\mathcal{M}$ has a basis consisting of many such chains.

It is well known that (3.3) admits a unique solution if $\omega \notin \mathbb{T}$. In particular, if $\omega \in \mathbb{D}$, then the solution of (3.3) is given by the matrix with ij entry

$$\mathbb{P}_{ij} = < Jf_j, f_i > , \qquad i,j = 1,\ldots,n ,$$

based on the standard inner product (1.1). In fact in general, if $\mathcal{M} \subset H_m^2$, then it turns out that the identity is satisfied if and only if the inner product for $\mathcal{M}$ is equal to the J inner product $< J\cdot, \cdot >$, and hence, in this special setting, the identity need not be mentioned at all, and the space

$$\mathcal{M} = H_m^2 \boxminus \Theta H_m^2 , \tag{3.4}$$

where the symbol $\boxminus$ indicates that the orthogonal complement of ΘH_m^2 is taken with respect to the J inner product. The matrix Θ is unique, up to a right constant J unitary factor and may be expressed in terms of a basis for $\mathcal{M}$ via (2.1); see e.g., (4.1) of [7].

4. Elementary factors

If $\mathcal{M}$ is the one dimensional $\mathcal{H}(\Theta)$ space corresponding to multiples of the function $v/\rho_\omega(\lambda)$ with $|\omega| \neq 1$, then it follows from Theorem 3.1, formulas (2.1) and (3.2) and the uniqueness of the reproducing kernel that

$$K_\beta(\lambda) = \frac{v}{\rho_\omega(\lambda)} \frac{\rho_\omega(\omega)}{v^*Jv} \{\frac{v}{\rho_\omega(\beta)}\}^* = \frac{J - \Theta(\lambda)J\Theta(\beta)^*}{\rho_\beta(\lambda)} .$$

This in turn leads to the conclusion that

$$\Theta(\lambda) = I_m + \{b_\omega(\lambda) - 1\}v(v^*Jv)^{-1}v^*J , \tag{4.1}$$

up to a J unitary constant factor on the right, and is indicative of the general fact that one dimensional $\mathcal{H}(\Theta)$ spaces correspond to elementary J inner factors in the sense of Potapov [15]. It is instructive to check that both of the examples considered in Section 2 can be written in this form. For a more complete analysis, see Theorems 1.4 and 2.10 of [7]; formula (1.19) therein is particularly useful.

5. Linear fractional transformations

Let

$$\Theta = \begin{bmatrix} A & B \\ C & D \end{bmatrix} \tag{5.1}$$

be a J inner function with blocks A of size $p \times p$, B of size $p \times q$, C of size $q \times p$ and D of size $q \times q$ and let Ω denote the set of points in $\mathbb{D}$ at which Θ is analytic. Then it follows readily from the 22 blocks of the inequalities

$$\Theta^*J\Theta \leq J \quad \text{and} \quad \Theta J\Theta^* \leq J \tag{5.2}$$

that

$$B^*B + I_q \le D^*D \quad \text{and} \quad CC^* + I_q \le DD^* \tag{5.3}$$

at every point in Ω. In particular this implies that D is invertible in Ω and that

$$\sigma_1 = BD^{-1} \in \mathcal{S}_{p\times q} \quad \text{and} \quad \sigma_2 = D^{-1}C \in \mathcal{S}_{q\times p} \ .$$

In fact the preceding inequalities imply that σ_1 and σ_2 are both strictly contractive in Ω and hence that the linear fractional transformation

$$T_\Theta[S_o] = (AS_o + B)(CS_o + D)^{-1} \tag{5.4}$$

is well defined in Ω for every choice of $S_o \in \mathcal{S}_{p\times q}$ since $CS_o + D$ is invertible.

THEOREM 5.1. *If* Θ *is* J $(= J_{pq})$ *inner and* $S_o \in \mathcal{S}_{p\times q}$*, then*

(1) $T_\Theta[S_o] \in \mathcal{S}_{p\times q}$ *and*

(2) $T_\Theta[S_o] - T_\Theta[0] = (A - BD^{-1}C)S_o(CS_o + D)^{-1}$. (5.5)

(3) If also Ψ *is* J *inner, then*

$$T_{\Theta\Psi}[S_o] = T_\Theta[T_\Psi[S_o]] \ . \tag{5.6}$$

The proof is by direct calculation. For help, if need be, and more information see Chapter 3 of [7].

6. The LIS problem

For a given $S \in \mathcal{S}_{p\times q}$, the LIS problem is to find a J inner Θ which is analytic in $\mathbb{D}$ such that

$$S = T_\Theta[S_o] = (AS_o + B)(CS_o + D)^{-1}$$

for some choice of $S_o \in \mathcal{S}_{p\times q}$.

With the help of a theorem of de Branges and Rovnyak [11] (see Theorems 2.3 and 2.4 of [2]) the LIS problem can be restated in terms of reproducing kernel Hilbert spaces:

THEOREM 6.1. *If* $S \in \mathcal{S}_{p\times q}$ *and* Θ *is* J *inner, then* Θ *is a solution of the LIS problem if and only if*

(1) $[I_p \quad -S]f \in \mathcal{H}(S)$ *and*

(2) $\|[I_p \quad -S]f\|_{\mathcal{H}(S)} \le \|f\|_{\mathcal{H}(\Theta)}$

for every choice of $f \in \mathcal{H}(\Theta)$.

We shall consider here only the simplest case of this problem which seeks to identify "elementary factors" Θ which are analytic in $\bar{\mathbb{D}}$. By the preceding analysis all such Θ correspond to one dimensional spaces $\mathcal{H}(\Theta)$ which are based on multiples of the vector valued function $f(\lambda) = v/\rho_\omega(\lambda)$, with $\omega \in \mathbb{D}$. Upon writing $v^* = [\xi^* \quad \eta^*]$ with $\xi \in \mathbb{C}^p$ and $\eta \in \mathbb{C}^q$, it turns out by a reasonably straightforward calculation that f will meet the conditions of Theorem 6.1 if $\eta = S(\omega)^*\xi$, or equivalently if

$$\frac{\eta}{\rho_\omega} = \underline{\underline{p}}S^*\frac{\xi}{\rho_\omega} ,$$

in which $\underline{\underline{p}}$ denotes the orthogonal projection of L_p^2 onto H_p^2. This in fact captures the flavor of the general result which identifies rational solutions Θ of the LIS problem which are analytic in $\bar{\mathbb{D}}$ with finite dimensional R_α invariant subspaces of H_p^2. For every such subspace $\mathcal{M}_+$, with basis say $g_1, \ldots, g_n$ it turns out that

$$\mathcal{M} = \operatorname{span}\left\{\begin{bmatrix} g_1 \\ \underline{\underline{p}}S^*g_1 \end{bmatrix}, \ldots, \begin{bmatrix} g_n \\ \underline{\underline{p}}S^*g_n \end{bmatrix}\right\} ,$$

endowed with the J inner product, is an $\mathcal{H}(\Theta)$ space when the Gram matrix "$\mathbb{P}$" (with respect to the J inner product) is positive definite and that in this case the corresponding Θ is a solution of the LIS problem. Moreover, every rational solution of the LIS problem which is analytic in $\bar{\mathbb{D}}$ is generated this way. Details of this and more are provided in Chapter 8 of [7]. For other approaches see [13] and [1]. A description of the rational solutions of the more general PLIS problem (which are analytic in $\bar{\mathbb{D}}$) is furnished in [3]. Solutions of the LIS problem with poles on the boundary were first treated in [14] for scalar S. For more on this case see [1] and [7].

7. Interpolation

The simplest matrix interpolation problem is the one-sided Nevanlinna-Pick problem. Therein one is given a set of points $\omega_1, \ldots, \omega_n$ in $\mathbb{D}$, a set of vectors $\xi_1, \ldots, \xi_n$ in $\mathbb{C}^p$ and a set of vectors $\eta_1, \ldots, \eta_n$ in $\mathbb{C}^q$, where $p + q = m$ and it is assumed that the functions $\xi_1/\rho_{\omega_1}, \ldots, \xi_n/\rho_{\omega_n}$ are linearly independent. The objective is to (1) find necessary and sufficient conditions on the data to insure the existence of an $S \in \mathcal{S}_{p\times q}$ such that $\xi_j^* S(\omega_j) = \eta_j^*$ for $j = 1, \ldots, n$, and (2) to obtain a description of the set of all solutions when these conditions are met.

It is convenient to let

$$v_j = \begin{bmatrix} \xi_j \\ \eta_j \end{bmatrix} \quad \text{and} \quad f_j = \frac{v_j}{\rho_{\omega_j}} , \qquad j = 1, \ldots, n .$$

THEOREM 7.1. *The Nevanlinna-Pick problem is solvable if and only if the $n \times n$ matrix $\mathbb{P}$ with ij entry*

$$\mathbb{P}_{ij} = < Jf_j, f_i > = \frac{v_i^* J v_j}{\rho_{\omega_j}(\omega_i)}$$

is positive semidefinite.

PROOF OF NECESSITY. If S is a solution, then for any choice of constants $c_1, \ldots, c_n$,

$$\sum_{i,j=1}^{n} \frac{c_i^* v_i^* J v_j c_j}{\rho_{\omega_j}(\omega_i)} = \sum_{i,j=1}^{n} c_i^* \xi_i^* \left\{ \frac{I_p - S(\omega_i)S(\omega_j)^*}{\rho_{\omega_j}(\omega_i)} \right\} \xi_j c_j$$

$$= \sum_{i,j=1}^{n} c_i^* \xi_i^* \Lambda_{\omega_j}(\omega_i) \xi_j c_j$$

$$= \left\| \sum_{j=1}^{n} c_j \Lambda_{\omega_j} \xi_j \right\|_{\mathcal{H}(S)}^2 \geq 0 \, .$$

∎

The proof of sufficiency is obtained by showing first that the condition $\mathbb{P} > 0$ guarantees the existence of a solution and then invoking a limiting argument. Full details are given in [7]; the sufficiency when $\mathbb{P} > 0$ is also sketched out below.

THEOREM 7.2. *If* $\mathbb{P} > 0$, *then the space*

$$\mathcal{M} = \operatorname{span}\{f_1, \ldots, f_n\}$$

endowed with the J *inner product is an* $\mathcal{H}(\Theta)$ *space and*

$$\{T_\Theta[S_o] : \ S_o \in \mathcal{S}_{p \times q}\}$$

is a complete list of the solutions to the Nevanlinna-Pick problem.

PROOF. The proof is divided into steps.

STEP 1. $\mathcal{M}$, *endowed with the* J *inner product, is an* $\mathcal{H}(\Theta)$ *space.*

PROOF OF STEP 1. This is a consequence of Theorem 3.1.

STEP 2. $v_j^* J \Theta(\omega_j) = 0$ *for* $j = 1, \ldots, n$.

PROOF OF STEP 2. The reproducing formulas

$$x^* f_j(\alpha) = < J f_j, \frac{J - \Theta J \Theta(\alpha)^*}{\rho_\alpha} x >$$

$$= < J f_j, \frac{Jx}{\rho_\alpha} > - < J f_j, \frac{\Theta J \Theta(\alpha)^*}{\rho_\alpha} x >$$

$$= x^* f_j(\alpha) - < \frac{J \Theta J \Theta(\alpha)^*}{\rho_\alpha} x, f_j >^*$$

$$= x^* f_j(\alpha) - \left\{ v_j^* \frac{J \Theta(\omega_j) J \Theta(\alpha)^*}{\rho_\alpha(\omega_j)} x \right\}^*$$

lead readily to the asserted conclusion since x and α are arbitrary elements of $\mathbb{C}^m$ and $\mathbb{D}$ respectively.

STEP 3. $\sigma_1 = BD^{-1}$ *is a solution of the NP problem.*

PROOF OF STEP 3. By Step 2,

$$[\xi_j^* \quad \eta_j^*]J \begin{bmatrix} A(\omega_j) & B(\omega_j) \\ C(\omega_j) & D(\omega_j) \end{bmatrix} = [0 \quad 0] \; .$$

But this in turn implies that

$$\xi_j^* A(\omega_j) - \eta_j^* C(\omega_j) = 0 \tag{7.1}$$

and

$$\xi_j^* B(\omega_j) - \eta_j^* D(\omega_j) = 0 \; . \tag{7.2}$$

The latter yields the asserted implication, since D is invertible in $\mathbb{D}$ and $\sigma_1 = BD^{-1} \in \mathcal{S}_{p\times q}$, as was shown in Section 5.

STEP 4. $T_\Theta[S_o]$ *is a solution of the NP problem for every* $S_o \in \mathcal{S}_{p\times q}$.

PROOF OF STEP 4. By (5.5), (7.1) and (7.2)

$$\xi_j^* \{T_\Theta[S_o] - T_\Theta[0]\}(\omega_j) = 0 \; ,$$

which yields the asserted result by Step 3.

STEP 5. *Every solution* S *of the NP problem is of the form* $S = T_\Theta[S_o]$ *for some choice of* $S_o \in \mathcal{S}_{p\times q}$.

PROOF OF STEP 5. If S is a solution, then

$$f_j = \begin{bmatrix} g_j \\ \underline{\underline{p}} S^* g_j \end{bmatrix} \; , \quad \text{with } \; g_j = \frac{\xi_j}{\rho_{\omega_j}} \; , \quad \text{for } \; j = 1, \ldots, n \; .$$

Therefore,

$$[I \quad -S]f_j = (I - S\underline{\underline{p}}S^*)g_j \in \mathcal{H}(S)$$

by (2.3) and thus, for any $f = \sum c_j f_j$ in $\mathcal{H}(\Theta)$,

$$\begin{aligned} \|[I \quad -S]f\|^2_{\mathcal{H}(S)} &= < (I - S\underline{\underline{p}}S^*)g, g > \\ &= < g, g > - < \underline{\underline{p}}S^* g, \underline{\underline{p}}S^* g > \\ &= < Jf, f > \; , \end{aligned}$$

where $g = \sum c_j g_j$. The desired conclusion is now immediate from Theorem 6.1. ∎

Many other interpolation problems can be handled in much the same way; see e.g., [7] and [8]. There are also other approaches of which perhaps the most far reaching is that of Ball and Helton; a detailed list of references to their work is provided in [7].

8. Approximation

If, for a given $S \in \mathcal{S}_{p\times q}$, Θ is a rational solution of the LIS problem which is analytic in $\bar{\mathbb{D}}$, then, as noted earlier,

$$\mathcal{M}_+ = \{[I_p \quad 0]f : \ f \in \mathcal{H}(\Theta)\}$$

is an R_α invariant subspace of H_p^2 for every choice of $\alpha \in \mathbb{D}$. Therefore, by the Beurling Lax Theorem (or by another application of an appropriate specialization of Theorem 3.1), it is equal to $H_p^2 \ominus WH_p^2$ for some $p \times p$ finite Blaschke-Potapov product W. Thus, every $f \in \mathcal{H}(\Theta)$ can be expressed in the form

$$f = \underline{\underline{p}}JX^*g \ ,$$

where

$$X = [I_p \quad -S] \ \text{ and } \ g \in H_p^2 \ominus WH_p^2 \ .$$

Therefore, by the analysis in the proof of Step 2 of Theorem 7.2,

$$\begin{aligned} 0 &= < J\underline{\underline{p}}JX^*g, \frac{\Theta J\Theta(\alpha)^*}{\rho_\alpha}x > \\ &= < g, \frac{X\Theta J\Theta(\alpha)^*}{\rho_\alpha}x > \ , \qquad \text{for } \ \alpha \in \mathbb{D} \ , \end{aligned}$$

which in turn implies that

$$\frac{X\Theta J\Theta(\alpha)^*}{\rho_\alpha}x \in WH_p^2$$

for every choice of $x \in \mathbb{C}^m$ and $\alpha \in \mathbb{D}$. Thus

$$X\Theta = WY$$

for some choice of $Y \in H^2_{p\times m}$. In fact $Y \in H^\infty_{p\times m}$. Upon writing $Y = [Y_1 \quad Y_2]$ with components $Y_1 \in H^\infty_{p\times p}$ and $Y_2 \in H^\infty_{p\times(m-p)}$ it follows that

$$A - SC = WY_1 \ , \qquad B - SD = WY_2$$

and hence that

$$A - BD^{-1}C = A - SC - (B - SD)D^{-1}C = W(Y_1 - Y_2D^{-1}C) \ .$$

But, with the help of (5.5), this in turn implies that the difference between the true $S = T_\Theta[S_o]$ and the approximant $S = T_\Theta[0]$:

$$T_\Theta[S_o] - T_\Theta[0] = WF \ , \tag{8.1}$$

for some $F \in H^{\infty}_{p\times q}$. The formula

$$F = W^{-1}\{T_\Theta[S_o] - T_\Theta[0]\}$$

supplies the auxiliary bound $|F(\lambda)| \leq 2$ at every point $\lambda \in \mathbb{D}$ via the maximum modulus principle for matrix valued functions. Therefore, formula (8.1) serves to exhibit $T_\Theta[0]$ as an approximant to $S = T_\Theta[S_o]$ (as well as an interpolant) in those "directions" at which $W(\lambda) \to 0$ as the number of factors therein (or equivalently the dimension of $\mathcal{M}_+$) increases. For a more detailed discussion see [13].

9. Maximum entropy

THEOREM 9.1. *Let Θ be a rational J inner function which is analytic in $\bar{\mathbb{D}}$ and let*

$$S = T_\Theta[S_o] = (AS_o + B)(CS_o + D)^{-1}\ , \qquad S_o \in \mathcal{S}_{p\times q}\ .$$

Then

$$\frac{1}{2\pi}\int_0^{2\pi} \log\ \det\{I_p - S(e^{i\theta})S(e^{i\theta})^*\}\frac{1-|\omega|^2}{|e^{i\theta}-\omega|^2}d\theta$$

$$\leq \log\ \det\{D(\omega)^*D(\omega) - C(\omega)^*C(\omega)\}^{-1}$$

with equality if and only if

$$S_o(\lambda) = -\{D(\omega)^{-1}C(\omega)\}^*$$

for every point $\lambda \in \mathbb{D}$.

Perhaps the first proof of this fact, though in a somewhat different formulation, was given in [13]. Burg's famous result corresponds to a special choice of Θ and $\omega = 0$. For a proof in the style of Arov and Krein [4], [5], and a geometric interpretation, see Chapter 11 of [7].

We remark that by introducing an appropriate constant J unitary factor V, the maximum entropy in Theorem 9.1 will be achieved by $T_{\Theta V}[0]$. (In fact you have only to invoke (5.6) together with the formula $K = T_{H(K)}[0]$ which is valid for any strictly contractive $p \times q$ matrix K with $H(K)$ as in (1.11) of [7].)

References

[1] D. Alpay and H. Dym, *Hilbert spaces of analytic functions, inverse scattering, and operator models I*, Integral Equations and Operator Theory **7** (1984), 589–641.

[2] —, *On applications of reproducing kernel spaces to the Schur algorithm and rational J unitary factorization*, in: I. Schur Methods in Operator Theory and Signal Processing (I. Gohberg, ed.), Operator Theory:

Advances and Applications, **OT18**, Birkhäuser Verlag, Basel, 1986, pp. 89-159.

[3] D. Alpay, P. Dewilde and H. Dym, *On the existence and convergence of solutions to the partial lossless inverse scattering problem with applications to estimation theory*, IEEE Trans. Inf. Theory, in press.

[4] D.Z. Arov and M.G. Krein, *Problems of search of the minimum of entropy in indeterminate extension problems*, Funkts. Anal. Prilozhen., **15** (1981), 61-64 (Russian). English translation: Funct. Anal. Appl. **15** (1981), 123-126.

[5] —, *On computations of entropy functionals and their minimums in indeterminate continuation problems*, Acta Sci. Math. **45** (1983), 33-50 (Russian).

[6] J.A. Ball, *Models for non contractions*, J. Math. Anal. Appl. **52** (1975), 235-254.

[7] H. Dym, *J Contractive Matrix Functions, Reproducing Kernel Hilbert Spaces and Interpolation*, CBMS Regional Conference Series in Mathematics, Amer. Math. Soc., Providence, 1989.

[8] —, *On reproducing kernel spaces, J unitary matrix functions, interpolation and displacement rank*, in: *The Gohberg Anniversary Collection* (H. Dym, S. Goldberg, M.A. Kaashoek and P. Lancaster, eds.), Operator Theory: Advances and Applications, **OT41**, Birkhäuser Verlag, Basel, 1989, pp. 173-239.

[9] L. de Branges, *Some Hilbert spaces of analytic functions I*, Trans. Amer. Math. Soc. **106** (1963), 445–468.

[10] L. de Branges and J. Rovnyak, *Square summable power series*, Holt, Rinehart and Winston, New York, 1966.

[11] —, *Canonical models in quantum scattering theory*, in: Perturbation Theory and its Applications in Quantum Mechanics (C. Wilcox, ed.), Wiley, New York, 1966, pp. 295-392.

[12] P. Dewilde and H. Dym, *Schur recursions, error formulas and convergence of rational estimators for stationary stochastic processes*, IEEE Trans. Inf. Theory, **27** (1981), 446-461.

[13] — *Lossless chain scattering matrices and optimum linear prediction: The vector case*, Circuit Theory and Applications **9** (1981), 135-175.

[14] —, *Lossless inverse scattering for digital filters*, IEEE Trans. Inf. Theory **30** (1984), 644-662.

[15] V.P. Potapov, *The multiplicative structure of J-contractive matrix functions*, Trudy Moskov. Mat. Obšč. **4** (1955), 125-236; English transl. in Amer. Math. Soc. Transl. (2) **15** (1960), 131–243.

Department of Theoretical Mathematics
The Weizmann Institute of Science
Rehovot 76100, ISRAEL

FORMULAS AND ALGORITHMS FOR BLOCK HANKEL MATRIX INVERSION AND PARTIAL REALIZATION

Georg Heinig

Abstract

A survey of an algebraic theory of block Hankel matrices is presented. It includes kernel description, inversion formulas, formulas for partial realization, and fast algorithms. Special attention is given to parallel and superfast algorithms.

INTRODUCTION

Block Hankel $[S_{i+j}]$ and block Toeplitz $[T_{i-j}]$ matrices play a considerable role in many fields of mathematics and engineering application. But comparing with the large literature for the scalar case (see [8,23] and references therein) there is, as far as we know no comprehensive treatment for the block case.

The present paper is aimed to give a survey of an algebraic theory of block Hankel matrices generalizing the corresponding theory for the scalar case in the version of [23]. Our main concerns are fast inversion algorithms for block Hankel matrices and algorithms for multivariate partial realization or, what is amost the same, for matrix Pade approximation. The proofs and the details will be published elsewhere. For the scalar case the algorithms are treated in [22]. Part of the results are joint with P.Jankowski.

Algorithms for block Toeplitz and Hankel inversion have been presented in [1,9,14,17] and other papers. All these algorithms work only under some restrictions as the strong regularity of the matrix. The algoritms presented below do not need such an additional assumption. We present two kinds of algorithms. The first one is of the Levinson/Trench type and the second one is of the Schur type. The Schur-type algorithm is suited for parallel processing and admits a further complexity reduction via a divide and conquer strategy.

The inversion algorithms are based on the following facts. Firstly, the inverse matrix can be represented with the help of so-called fundamental matrices and, secondly, the fundamental matrices can be computed recursively. The fundamental matrices are closely related with the solution of certain homogeneous Hankel equations. For this reason the study of the kernel structure will be the initial point of our consideration.

The concept of fundamental matrix is also useful for solving the partial realization problem. With its help all solutions can be described and computed recursively. In this way a purely linear algebra approach of this problem is obtained (for other approaches see [2,15,29]).

NOTATIONS

$\mathbb{F}$ given field; $\mathbb{F}^q = \mathbb{F} \times \ldots \times \mathbb{F}$ = column vectors

$\mathbb{F}^{p\times q}$ $p\times q$ matrices; I_q $q\times q$ identity matrix

$x(\lambda) = \sum_i x_i\lambda^i$ if $x = (x_i)$, $x_i \in \mathbb{F}^q$

$A(\lambda,\mu) = \sum_{i,j} a_{ij}\lambda^i\mu^j$ if $A = [a_{ij}]$

$\mathbb{F}^q_n(\lambda) = \{x(\lambda): x=(x_i)_0^{n-1},\ x_i \in \mathbb{F}^q\}$

(The indeterminate λ will be often omitted).

1. KERNEL DESCRIPTION

Throughout the paper let S denote a sequence $(S_0, \ldots, S_{N-1})$ of $p\times q$ matrices with entries from a given field $\mathbb{F}$. We associate S with the family $\{H_k\}$ $(k=1,\ldots,N)$ of block Hankel matrices

$$H_k = H_k(S) = [S_{i+j}]_0^{l-1}\ {}_0^{k-1} \qquad (k+l=N+1).$$

The initial point of our consideration is a description of the kernels of the matrices H_k and their transposed. It is convenient to do it in polynomial language. We define

$$\mathcal{K}_k = \{u(\lambda): H_k u=0\},\quad \hat{\mathcal{K}}_k = \{\hat{u}(\lambda): H_k^T\hat{u} = 0\}$$

and, for convenience

$$\mathcal{K}_{N+1} = \mathbb{F}^q_{N+1}(\lambda) \quad , \qquad \hat{\mathcal{K}}_{N+1} = \mathbb{F}^p_{N+1}(\lambda).$$

The basic result is the following one.

THEOREM 1.1. <u>There exist a uniquely defined (p+q)-tuple of integers</u> $\deg S := (d_1,\ldots,d_{p+q})$, $0 \le d_1 < \ldots \le d_{p+q} < N+1$, <u>and polynomial vectors</u> $u_i(\lambda)$ $(i=1,\ldots,p+q-t)$ <u>and</u> $\hat{u}_j(\lambda)$ $(j=\hat{t},\ldots,p+q)$, <u>where</u> t <u>and</u> $\hat{t}$ <u>denote the defects of the matrices</u> H_N <u>and</u> H_1, <u>respectively, such that</u>:

(1) $\deg u_i \le d_i$, $\deg \hat{u}_j \le \hat{d}_j := N+1-d_j$

(2) <u>For any</u> $k=1,\ldots,N+1$, <u>the polynomials</u> $u_i(\lambda),\ldots,$

$\lambda^{k-d_i-1} u_i(\lambda)$, <u>where</u> i <u>runs over all indices with</u> $k>d_i$, <u>forms a basis of</u> $\mathcal{A}_k$.

(3) <u>The polynomials</u> $\hat{u}_j(\lambda),\ldots,\lambda^{k-\hat{d}_j-1}\hat{u}_j(\lambda)$ $(k>\hat{d}_j)$ <u>form a basis of</u> $\hat{\mathcal{A}}_k$.

This theorem is a consequence of the abstract results in [21]. In vector language Theorem 2.1 states, in particular, that the kernel of block Hankel matrices are spanned by not more than p+q shift chains.

DEFINITION. The integers d_i will be called right and $\hat{d}_j= N+1-d_j$ left <u>characteristic degrees</u> of the sequence S. Let $\{u_i\}$ and $\{\hat{u}_j\}$ be systems satisfying (2) and (3) and let U and $\hat{U}$ be polynomial matrices with the columns u_i or $\hat{u}_j$, respectively. Then U will be called right and $\hat{U}$ left <u>fundamental matrix (f.m.)</u> of S. Since we mainly have to do with the "right" case we shall often omit this attribute.

We consider now a fixed member H_n of the family $\{H_k\}$. The integers $k_i=n-d_i$ $(i=1,\ldots p+q)$ will be called <u>partial indices</u> of H_n . They have analogous properties as the classical partial indices of the theory of Toeplitz and Wiener-Hopf operators (cf.[16]). For example, Theorem 1.1 implies the following.

COROLLARY 1.1. <u>The following relations hold true</u>:

$$\dim \ker H_n = \sum_{k_i>0} k_i \quad , \qquad \dim \ker H_n = -\sum_{k_i<0} k_i$$

<u>In particular</u>, H_n <u>is regular iff all</u> k_i <u>vanish</u>.

Also semi-stability of the partial indices with respect to small pertubations can be proved. The system of partial indices is stable iff $|k_i- k_j|\leq 1$ for all i,j.

Clearly, a sequence S possess many f.m. If U_0 is a f.m. of S, then the general form of a f.m. is given by

$$U = U_0 C, \tag{1.1}$$

where $C = [c_{ij}]$ is an unimodular polynomial matrix with $\deg c_{ij}\leq d_j-d_i$ if $d_j\geq d_i$ and $c_{ij}=0$ if $d_j<d_i$. An important feature of our approach will be to switch from one f.m. to another one. In order to characterize the different f.m. we introduce constant matrices E(U) and F(U), which will be called indicator matrices and defined now.

Let u_i denote the columns of U and $(u_i)_k$ the coefficient of λ^k in u_i. Then E(U) is, by definition, the matrix with the columns $(u_i)_{d_i}$. The columns f_i of F(U) are defined by

$$f_i = S_{N-1}(u_i)_{d_i-1} + \ldots + S_{N-d_i}(u_i)_\emptyset . \tag{1.2}$$

2. INVERSION FORMULAS

We assume in this section that H_n is a regular mxn block Hankel matrix, mp=nq. In this case any basis of $\mathcal{A}_{n+1}$ provides a right and any basis of $\hat{\mathcal{A}}_{m+1}$ a left f.m.

THEOREM 2.1. _Let_ U _and_ $\hat{U}$ _be a right and a left f.m., respectively. Then the constant_ (p+q)x(p+q) _matrix_

$$M := \begin{bmatrix} F(\hat{U}) \\ E(\hat{U}) \end{bmatrix}^T \begin{bmatrix} \emptyset & I_q \\ -I_p & \emptyset \end{bmatrix} \begin{bmatrix} F(U) \\ E(U) \end{bmatrix} \tag{2.1}$$

is regular and the inverse of H_n _is given by_

$$H_n^{-1}(\lambda,\mu) = U(\lambda)M^{-1}\hat{U}(\mu)^T/(\lambda-\mu) . \tag{2.2}$$

Relation (2.2) can be written in the form of a Gohberg/Semencul formula

$$H_n^{-1} = \begin{bmatrix} U_o & & \emptyset \\ \vdots & \ddots & \\ U_{n-1} & \cdots & U_o \end{bmatrix} M^{-1} \begin{bmatrix} \hat{U}_1 & \cdots & \hat{U}_m \\ \vdots & \cdot^{\cdot^{\cdot}} & \\ \hat{U}_m & & \emptyset \end{bmatrix}^T \tag{2.3}$$

The importance of formula (2.3) consists in the fact that for its application FFT can be utilized. This leads to an $O(n \log^2 n)$ complexity algorithm to solve block Hankel (or Toeplitz) systems if a right and left f.m. are known.

REMARKS. 1) It can be shown that (in the regular case) there are f.m. U and $\hat{U}$ such that

$$\begin{bmatrix} F(U) \\ E(U) \end{bmatrix} = \begin{bmatrix} F(\hat{U}) \\ E(\hat{U}) \end{bmatrix} = I_{p+q} .$$

In this case we represent $U = [U^1 \; U^2]$ and $\hat{U} = [\hat{U}^1 \; \hat{U}^2]$. Now (2.2) goes over into

$$H_n^{-1}(\lambda,\mu) = (U^2(\lambda)\hat{U}^1(\mu) - U^1(\lambda)\hat{U}^2(\mu))/(\lambda-\mu). \tag{2.4}$$

The right-hand side of (2.4) represents a Bezoutian in the sense of Anderson/Jury [2].

2) Fundamental matrices can readily be constructed with the help of some special "fundamental" equations or with the help of columns and rows of the inverse matrix. Using construction of this kind one can deduce all inversion formulas presented in the papers [4,17,28].

3) The inverse of a matrix of the form (2.2) is block Hankel (see [24]).

3. EXTENSION AND REDUCTION

In this section we deal with the following extension and reduction problems: Given a f.m. U for $S=(S_\emptyset,\dots,S_{N-1})$. How to construct a f.m. for

(a) $S' := (S_\emptyset,\dots,S_{N-2})$

(b) $(S,S_N) = (S_\emptyset,\dots,S_{N-1},S_N)$, where S_N is given ?

The solution of (b) provides a fast recursive algorithm for constructing a f.m.

To solve these problems we shall need the indicator matrix E(U) defined in Section 1 and a generalization of F(U). For given U and S_N we denote by $F(S_N,U)$ the matrix with the columns

$$f_i = S_N(u_i)_{d_i}+\dots+ S_{N-d_i}(u_i)_\emptyset .$$

DEFINITION. 1) Let C be a constant matrix with the columns c_i. C will be called <u>column regular</u> iff there is a set of indices J such that $c_i = \emptyset$ if $i \notin J$ and the c_i with $i \in J$ form a basis of the column space. J will be referred to as the set of proper indices of C.

2) A f.m. U is said to be <u>reduction matched</u> iff E(U) is column regular and S_N-<u>extension matched</u> iff $F(S_N,U)$ is column regular.

It is obvious that, for any matrix C, an upper triangular matrix T can be constructed by Gaussian elimination so that CT is column regular.

For simplicity we assume now that S is such that the matrix H_N has full rank, i.e. $t=\emptyset$. Later we shall explain how to proceed in the case $t>\emptyset$.

THEOREM 3.1. (1) <u>If</u> U <u>is reduction matched, then</u> U <u>is also a f.m. of</u> S'. <u>The characteristic degrees</u> d_i' <u>are given by</u>

$$d_i' = \begin{cases} d_i - 1 & : i \notin J \\ d_i & : i \in J , \end{cases}$$

<u>where</u> J <u>denotes the set of proper indices of</u> E(U).

(2) <u>If</u> U <u>is</u> S_N-<u>extension matched, then</u> U <u>is also a f.m. of</u> (S,S_N). <u>The characteristic degrees</u> D_i <u>of</u> (S,S_N) <u>are given by</u>

$$D_i = \begin{cases} d_i + 1 & : i \in J \\ d_i & : i \notin J , \end{cases}$$

<u>where</u> J <u>denotes the set of proper indices of</u> $F(S_N,U)$.

We show now how to construct reduction and extension matched f.m.

THEOREM 3.2. <u>Let</u> T <u>be a regular upper triangular matrix such that</u> $E(U)T$ <u>or</u> $F(S_N,U)T$ <u>is column regular</u>. <u>Define</u>

$$D := \operatorname{diag}(\lambda^{d_1},\dots,\lambda^{d_{p+q}}) \text{ and } X(\lambda)=D(\lambda)^{-1}TD(\lambda). \qquad (3.1)$$

<u>Then</u> UX <u>is reduction or</u> S_N<u>-extension matched, respectively</u>.

Now we discuss shortly how to proceed in case that the matrix H_N has a defect $t>0$. Let U be a f.m. with the columns u_i $(i=1,\dots,p+q-t)$ in reduction matched form and let J denote the set of proper indices of $E(U)$. Furthermore, let $T=[t_{ij}]$ be a regular upper triangular matrix such that $F:=F(S_N,U)T$ is S_N-extension matched and let J' denote the set of proper indices of F and t' the defect of $H_{N+1}(S,S_N)$. Introduce now the index set

$$K = \{i \in J:\ t_{ij}= 0 \text{ for all } j \in J'\}.$$

Then the number of elements in K coincides with $t-t'$. Moreover, if W denotes the matrix with the columns λu_i, $i \in K$, then a f.m. of (S,S_N) is given by $[\ UX\ \ W\ ]$, where X is defined by (3.1).

4. SCHUR-TYPE ALGORITHM

To start with we introduce the notations $S(n) = (S_0, \dots, S_{n-1})$ $(n=1,\dots,N)$, $H_k^n = H_k^n(S(n))$, $\mathcal{A}_k^n=\{u(\lambda):\ H_k^n u=0\}$. If $u \in \mathbb{F}_k^q(\lambda)$, then we denote $u^{\#}(\lambda) := u(\lambda^{-1})\lambda^k$. For $u \in \mathcal{A}_k^n$ the polynomial vector

$$r(\lambda) = S(\lambda)u^{\#}(\lambda) \bmod \lambda^{N+1} \quad (S(\lambda)=S_0+\dots+S_{N-1}\lambda^{N-1})$$

will be called (following A.Bultheel) residual of u. The vector $r \in \mathbb{F}^{(N+1-n)q}$ corresponding to $r(\lambda)$ consists of 3 parts, $r^T = [v^T\ 0\ \ w^T]^T$, where $v \in \mathbb{F}^{(k-1)q}$, $w \in \mathbb{F}^{(N+1-n)q}$ and 0 is a zero vector of length $(n-k+1)q$.

Now let U_n be a f.m. for $S(n)$ with the columns u_i. The residual matrix R_n of U_n is, by definition, the matrix the columns of which are the residuals of u_i. If R_n is the residual matrix of some f.m. then we shall say that R_n is a residual matrix of S of the level n. The crucial point for constructing a parallel algorithm to obtain a f.m. is the following.

OBSERVATION. <u>The coeffient</u> R_{nn} <u>of</u> λ^n <u>in</u> R_n <u>equals the indicator matrix</u> $F(S_n,U_n)$.

THEOREM 4.1. *Let R_n be a n th level residual matrix of* S *such that R_{nn} is column regular. Then a residual matrix of level* n+1 *is given by*

$$R_{n+1} = R_n Y_n ,$$

where

$$Y_n = \mathrm{diag}(\lambda^{h(1)},\dots,\lambda^{h(p+q)}),\ h(i) = \begin{cases} 0 & : i \notin J \\ 1 & : i \in J \end{cases}$$

and J *denotes the set of proper indices of R_{nn}. For the characteristic degrees of* S(n) *and* S(n+1) *the relation*

$$d_i(n+1) = d_i(n) + h(i)$$

holds.

For a general given R_n we determine a regular upper triangular T such that $R_{nn}T$ is column regular. Then R_nT fulfils the assumption of the theorem.

Theorem 4.1 provides an O(N) complexity algorithm to find a f.m. on a parallel computer. Let us remark that it is not necessary to compute in each step the f.m. but only at the very end. For this the following fact is important.

Let Y_{nm} be a transformation matrix transforming an n th level residual matrix R_n into an m th level matrix, m>n: $R_m = R_n Y_{nm}$. Introduce X_{nm} by

$$X_{nm} = D_n^{-1} Y_{nm}(\lambda^{-1}) D_m ,$$

where

$$D_k = \mathrm{diag}(\lambda^{d_1(k)},\dots,\lambda^{d_{p+q}(k)}) \qquad (k=n,m).$$

and $d_i(k)$ denote the characteristic degrees of S(k).
If now U_n corresponds with R_n then $U_m = U_n X_{nm}$ is a f.m. corresponding to R_m.

5. SUPERFAST ALGORITHM

Another advantage of the Schur type algorithm is that a divide and conquer strategy can be applied. This leads to an algorithm with O(N log²N) complexity for computing a f.m. and together with formula (2.3) to an algorithm with this complexity to solve Hankel or Toeplitz systems of equations.

We introduce projections P_{1m} by

$$P_{1m} \sum_i x_i\lambda^i = \sum_{i=1}^{m-1} x_i\lambda^i$$

If R_n is a n th level residual matrix of S, then we denote

$R_n(l,m) = P_{lm} R_n$

The superfast algorithm rests on the following facts:

1) $R_l(l,m) = P_{lm}R_n(n,m)$ $(n<l<m)$ (5.1)

2) Suppose that $R_n(n,m)$ $(m>n)$ and deg S(n) are given. Then Y_{nm} and deg S(m) can be evaluated with the help of Theorem 4.1.

Now we may apply the following.

PROCEDURE. Input $R_n(n,m)$, deg S(n).
Output Y_{nm}, deg S(m)

(I) If m=n+1 then find Y_{nm} and deg S(m) by Theorem 4.1.

(II) If m>n+1 then carry out the following steps:

(1) Choose l, n<l<m.

(2) Call the procedure for $R_n(n,l)$, deg S(n) (the result is Y_{nl}, deg S(l)).

(3) Compute $R_l(l,m)$ by (5.1)

(4) Call the procedure for $R_l(l,m)$ and deg S(l) (the result is Y_{lm}, deg S(m)).

(5) Multiply $Y_{nm} = Y_{nl} Y_{lm}$.

For scalar Toeplitz systems other divide and conquer algorithms were presented in [5,7,10].

6. PARTIAL REALIZATION

The partial realization problem can be posed as follows. Given a sequence S; find a rational pxq matrix T such that in a neighbourhood of infinity

$$T(\lambda) = S_0\lambda^{-1} + \dots + S_{N-1}\lambda^{-N} + o(\lambda^{-N}) \qquad (6.1)$$

If n is the McMillan degree of T then (6.1) is called partial realization (p.r) of order n (n equals the state space dimension).

For the scalar (i.e.SISO) case this problem has a long history, since it coincides with the Padé approximation problem and is related to continuous fractions, orthogonal polynomials and others. We note only the survey [20] for references. The importance of p.r. in system theory was first recognized by Kalman. During the last decade also the multivariate (i.e. matrix or MIMO) case found considerable interest, for example in the papers [6,12,14,18]. Especially we point out Antoulas' paper [3].

We show now how the p.r. problem is related to our approach of fundamental matrices. Represent T of (6.1) in the form of a matrix fraction $T = PA^{-1}$, where A is column reduced (see [25]).It can easily shown that the columns a_i

of A belong to $\mathcal{A}_n$, where n_i denote the column degrees of A. Furthermore, the columns of P are just the residuals of the a_i (except for the last component) and E(A) is a regular matrix. From these facts it becomes clear that the results and algorithms described above can be applied to partial realization.

For the sake of simplicity we restrict ourselves to S of the form such that H_N has full rank.

In order to present all p.r. we introduce the concept of a V-matrix. This concept was first introduced in different form and with a system interpretation in [3]. Let U be a f.m. of S with the columns u_i and characteristic degrees d_i. We introduce polynomial vectors w_i by

$$w_i^\# = S(\lambda)u_i^\# \quad \mod \lambda^N$$

where $w_i^\#(\lambda):=w_i(\lambda^{-1})\lambda^{d_i-1}$ (i.e. $w_i^\#$ is the residual of u_i, except for the last component) and the matrix W with the columns w_i. The (p+q)x(p+q) matrix $V = \begin{bmatrix} W \\ U \end{bmatrix}$ will be called (right) V-matrix of S. Analogously left V-matrices are defined.

THEOREM 6.1. <u>Let</u> V <u>and</u> $\hat{V}$ <u>be a right and a left V-matrix of</u> S, <u>respectively, the columns of which are ordered so that the i th columns of</u> V <u>and</u> $\hat{V}$ <u>correspond to the same characteristic degree</u>. <u>Then</u>

$$\hat{V} \begin{bmatrix} \emptyset & I_q \\ -I_p & \emptyset \end{bmatrix} V = M \quad ,$$

<u>where</u> M <u>is the matrix defined by (2.1) appearing in the inversion formula</u>. <u>In particular, V-matrices are unimodular</u>.

It is easily checked that V-matrices transform in the same way as f.m. after extension and reduction, i.e. if $U_m = U_n X_{nm}$, then $V_m = V_n X_{nm}$, where V_k denotes the V-matrix corresponding to U_k (k=n,m).

Now we consider the p.r. problem. From the remarks at the beginning of this section and the results of Sect.1 it follows that any p.r. of S has the form $WZ(UZ)^{-1}$, where Z is a (p+q)xq polynomial matrix. More precisely the following holds true.

THEOREM 6.2. <u>Let</u> $V = \begin{bmatrix} W \\ U \end{bmatrix}$ <u>be a V-matrix of</u> S, <u>where</u> U <u>is in reduction matched form, and let</u> J <u>denote the set of proper indices of</u> E(U). <u>Further let</u> d_i <u>be the</u>

characteristic degrees of S and n_j given integers. The general form of a p.r. $T=PA^{-1}$ with column degrees n_j of A is given by

$$P = WZ \qquad A = UZ ,$$

where $Z=[z_{ij}]$ fulfils the following conditions:

(a) $\deg z_{ij} \le n_j - d_i$ (if $n_j \ge d_i$) ; $z_{ij}=0$ (if $n_j<d_i$),

(b) The qxq matrix consisting of the coefficients of $\lambda^{n_j-d_i}$ in z_{ij} (i J; j=1,...,q) is regular.

The order of the p.r. equals the sum of all n_j . Minimal order is achieved iff the n_j coincide with the d_i, $i \in J$ or, equivalently, iff the z_{ij} with $i \in J$ are constants.

COROLLARY 6.1. The minimal order of a p.r. equals the sum of all d_i with $i \in J$.

From this corollary one can also deduce the formula (0.2) in [18].

7. CONCLUDING REMARKS

The algorithms presented above can also be applied for block Toeplitz matrices. However, there are modifications which are more fitted to the Toeplitz structure.

Our approach can be transmitted and generalized to other types of structured matrices, e.g. to close-to-Toeplitz and Toeplitz-plus-Hankel (T+H-) matrices. For T+H-matrices a theorem analogous to Theorem 1.1 holds true, where the powers λ^k have to replaced by the powers $(\lambda^2+1)^k$.

Some of the results of the theory of scalar Hankel matrices (see [23]) are not yet generalized to the block case. Let us formulate some open problems in this field.

1) Can the f.m. be used for constructing generalized inverses of block Hankel matrices (as in the scalar case)? (cf.[23])

2) How to generalize Frobenius' signature rule to the block case? How to compute the signature effectively? What about generalized LU-factorization ?

3) Is there a matrix generalization of the Routh and Schur-Cohn algorithms for root location ? (For some results see [29])

REFERENCES

[1] Akaike,H., Block Toeplitz matrix inversion. SIAM J.Appl.Math.24,2 (1973),140-156.

[2] Anderson,B. and Jury,E., Generalized Bezoutian and

Sylvester matrices in multivariable linear control. IEEE Trans. AC-21,4 (1976),551-556.

[3] Antoulas,A.C., On recursiveness and related topics in linear systems.IEEE Trans. AC-31,12 (1986),1121-1135.

[4] Ben-Artzi,A.and Shalom,T.,On inversion of block Toeplitz matrices. Int.Equ.and Oper.Th.8,6 (1985),173-192.

[5] Bitmead,R.R. and Anderson,B.D.O., Asymptotically fast solution of Toeplitz and related systems of linear equations. Linear Alg.Appl. 34 (1980),259-295.

[6] Bosgra,O.H.,On parametrization for the minimal partial realization problem. Syst.Contr.Lett.3 (1983), 181-187.

[7] Brent,R.P., Gustavson,F.G.,Yun,D.Y.Y., Fast solution of Toeplitz systems of equations and computation of the Pade approximation. J.Algor.1 (1980),259-295.

[8] Bultheel,A.,„Laurent series and their Padé approximations. Birkhäuser Basel,Boston 1987.

[9] Bultheel,A., Recursive relations for block Hankel and Toeplitz systems. J.of Comp.and Appl.Math.10 (1984), 301-354.

[10] de Hoog,F., A new algorithm for solving Toeplitz systems of equations. Lin.Alg.Appl. 88/89 (1987),123-138.

[11] Delsarte,P.,Genin,Y.,Kamp,Y., On the Toeplitz embedding of an arbitrary matrix. Lin.Alg.Appl.51 (1983), 97-119.

[12] Dickinson,B.W.,Morf,M.,Kailath,T., A minimal relization algorithm for matrix sequences. IEEE Trans. AC-19 (1974),31-38.

[13] Fuhrmann,P., Algebraic system theory: an analyst's point of view. J.Franklin Inst.301 (1976),521-540.

[14] Fuhrmann,P.,Block Hankel matrix inversion: a polynomial approach. In: Oper.Theory and Syst.,Birkhauser Basel,Boston,Stuttgart 1986, 207-230.

[15] Fuhrmann,P., A matrix Euclidean algorithm and matrix continued fraction expansions. Syst.Contr.Lett.3 (1983),263-271.

[16] Gohberg,I. and Feldman,I., Convolution equations and projection methods for their solution. Transl.Math. Mon.,vol.41,1974.

[17] Gohberg,I. and Heinig,G., Inversion of finite Toeplitz composed of elements of a noncommutative algebra (in Russian). Rev.Roumaine Math.Pures et Appl.19,5 (1974),623-663.

[18] Gohberg,I.,Kaashoek,M.A.,Lerer,L., On minimality in the partial realization problem. Syst.Contr.Lett.9 (1987),97-104.

[19] Gohberg,I., Kailath,T., Koltracht,I., Lancaster,P., Linear complexity parallel algorithms for linear systems of with recursive structure. Lin.Alg.Appl. 88/89 (1987),271-316.

[20] Gragg,W. and Linquist A., On the partial realization problem. Lin.Alg.Appl.50 (1983),277-320.

[21] Heinig,G., Partial indices for Toeplitz-like operators. Int.Equ.and Oper.Theory 8 (1985),805-824.

[22] Heinig,G. and Jankowski,P. 1) Fast inversion inver-

sion algorithms for Hankel matrices. 2) Parallel and superfast inversion algorithms for Hankel matrices. Preprints 108 and 110, TU Karl-Marx-Stadt, 1989.

[23] Heinig,G. and Rost,K., Algebraic methods for Toeplitz-like matrices and operators. Akademie-Verlag Berlin and Birkhäuser Basel,Boston,Stuttgart 1984.

[24] Heinig,G. and Tewodros,A. On the inverses of Hankel and Toeplitz mosaic matrices. Sem.Anal.1987/88, Karl-Weierstrass-Inst.,Berlin 1988,53-65.

[25] Kailath,T. Linear systems. Prentice-Hall 1980.

[26] Kailath,T. A theorem of I.Schur and its impact on modern signal processing. In: I.Schur Methods in Oper.Theory and Sign.Proc., Birkhäuser Basel, Boston, Stuttgart 1986, 31-50.

[27] Kalman,R.E., On minimal partial realizations of a linear input/output map.In: Aspects of Networks and Syst. Theory. Holt,Rinehart and Winston, New-York 1971, 385-407.

[28] Lerer,L. and Tismenetsky,M., Generalized Bezoutian and the inversion problem for block Toeplitz matrices. Int.Equ.and Oper.Theory 9,6 (1986),790-819.

[29] Lerer,L. and Tismenetsky,M., The Bezoutian and the eigenvalue separation problem for matrix polynomials. Int.Equ.and Oper.Theory 5,3(1982),386-445.

[30] Shamir,T. An algebraic approach to the partial realization problem. Lin.Alg.Appl.73 (1986),171-196.

Elementary symmetric functions and positive functions in several variables

Alfred Fettweis

Abstract

It is shown that the so-called elementary symmetric functions in k variables p_1 to p_k are reactance Hurwitz polynomials, that the ratio of two consecutive elementay symmetric functions in the same variables is a reactance function in irreducible form and that the sum of two such elementary symmetric functions is a scattering Hurwitz polynomial. Various other such polynomials and functions as well as related polynomials and more general positive functions, in particular for any values of the partial degrees in the p_1 to p_k, can be generated by appropriate simple transformations. All positive functions obtained can be synthesized by means of circuits in series-parallel connection.

1. Introduction

For certain approaches to multidimensional signal processing there exists a pronounced interest in reactance functions and, more generally, positive functions $F(\mathbf{p})$, and in Hurwitz polynomials of various types, in particular in so-called scattering Hurwitz polynomials $g(\mathbf{p}), \mathbf{p} = (p_1, \cdots, p_k)$ being a k-tuple of complex frequencies p_1 to p_k [1-3]. However, no general method for generating such functions is known. Furthermore, it is known that a general method for synthesizing a positive function $F(\mathbf{p})$ in terms of an electric circuit with passive elements cannot exist for $k > 2$, not even in the case that $F(\mathbf{p})$ is a reactance function [4]. Hence, there is a definite interest in approaches for which subclasses of such functions can be generated and for which corresponding synthesis solutions can be offered.

Such a subclass can be obtained by using the well-known elementary (or fundamental) symmetric functions, which can indeed easily be generated. It is shown that elementary symmetric functions are reactance Hurwitz polynomials, that the ratio of two consecutive elementary symmetric functions in the same variables is a reactance function in these variables, that a corresponding circuit realization always exists, and that the sum of two consecutive elementary symmetric functions is a scattering Hurwitz polynomial. While the functions thus obtained are of first degree in each one of the variables, functions of higher degree can easily be generated by starting from elementary symmetric functions of a sufficiently larger number of variables and then putting equal or inversely equal to one another some of these variables. Further functions can be generated by means of elementary transformations. Also, positive functions more general than reactance functions can be generated by starting again from elementary symmetric functions of a sufficiently larger number of variables and then setting some of these variables equal to positive constants or to complex constants having a positive real part.

2. Basic concepts and notation

We designate by $\mathbf{p}, -\mathbf{p}$, and $\mathbf{p}^*$ the k-tuples $\mathbf{p} = (p_1, \cdots, p_k)$, $-\mathbf{p} = (-p_1, ..., -p_k)$, and $\mathbf{p}^* = (p_1^*, \cdots, p_k^*)$, the asterisk designating complex conjugation. The paraconjugate, g_*, of a function, g, is defined by $g_*(\mathbf{p}) = g^*(-\mathbf{p}^*)$; thus, if g is a real function, then $g_*(\mathbf{p}) = g(-\mathbf{p})$. If $g_* = Cg, C = const.$ (necessarily $|C| = 1$), g is said to be self-paraconjugate. If $C = 1$, g is *paraeven*; if $C = -1$, g is *paraodd*.

The following definitions are used ([1-3]):

1. A polynomial g is a *widest-sense Hurwitz polynomial* (WSHP) if $g(\mathbf{p}) \neq 0$ for $\operatorname{Re}\mathbf{p} > \mathbf{0}$.
2. A polynomial is a *scattering Hurwitz polynomial* (SHP) if g is a WSHP and if g and g_* are relatively prime.
3. A *self-paraconjugate Hurwitz polynomial* (SPHP) is a WSHP that is self-paraconjugate.
4. A *reactance Hurwitz polynomial* (RHP) is the paraeven or paraodd part of an SHP times an arbitrary constant.
5. A rational function, $F(\mathbf{p})$, is said to be *positive* if, for $\operatorname{Re}\mathbf{p} > \mathbf{0}, F(\mathbf{p})$ is holomorphic and such that $\operatorname{Re} F(\mathbf{p}) \geq 0$.
6. A positive rational function, $F(\mathbf{p})$ is said to be a *reactance* function if it is paraodd, i.e., if $F_*(\mathbf{p}) = -F(\mathbf{p})$.
7. We designate by s_{km} the the *elementary* (or fundamental) *symmetric function* of order m in $k \geq 1$ variables, with $1 \leq m \leq k$. Thus

$$s_{k1}(\mathbf{p}) = p_1 + p_2 + \cdots + p_k,$$
$$s_{k2}(\mathbf{p}) = p_1 p_2 + \cdots + p_1 p_k + \cdots + p_{k-1} p_k,$$
$$\cdots$$
$$s_{kk}(\mathbf{p}) = p_1 p_2 \cdots p_k,$$

and in general

$$s_{km}(\mathbf{p}) = \sum_{1 \leq i_1 < i_2 < \cdots < i_m \leq k} p_{i_1} p_{i_2} \cdots p_{i_m}.$$

For convenience, we also use the definition

$$s_{k0}(\mathbf{p}) = 1.$$

8. We designate by F_{km} the rational function defined by

$$F_{km}(\mathbf{p}) = \frac{s_{km}(\mathbf{p})}{s_{k,m-1}(\mathbf{p})}, \quad 1 \leq m \leq k. \tag{1}$$

9. We also designate by I the index set

$$I = \{1, 2, \cdots, k\}. \tag{2}$$

10. For any $i \in I$, the notation $\mathbf{p}\backslash p_i$ is defined as follows:
    $\mathbf{p}\backslash p_i = (k{-}1)$ - tuple composed of all p_1 to p_k except p_i. (3)
11. We also use the following simplified notation

$$\mathbf{p}\prime = \mathbf{p}\backslash p_k. \tag{4}$$

3. Fundamental properties

Fundamental Theorem (Theorem 1). The function F_{km} defined by (1) has the following properties:

1. It is a real reactance function in $\mathbf{p}$.
2. Circuits can be given by means of which $F_{km}(\mathbf{p})$ can be synthesized as an impedance or an admittance.
3. The circuits just mentioned have series-parallel structures.
4. The expression for $F_{km}(\mathbf{p})$ obtained according to (1) is in irreducible form, i.e., $s_{km}(\mathbf{p})$ and $s_{k,m-1}(\mathbf{p})$ are relatively prime.

Before approaching the proof in detail, we give a few comments, followed by several helpful lemmas:

1. Since $F_{km}(\mathbf{p})$ is real and odd in $\mathbf{p}$, the proof of the first statement of the Fundamental Theorem amounts to proving that $F_{km}(\mathbf{p})$ is a positive function.
2. The proof of the latter aspect can be done by direct mathematical induction. Alternatively, it follows implicitly from the construction of a synthesis procedure as required by statements 2 and 3.
3. Statement 3 implies that the structures mentioned can be translated by simple means into wave-digital-filter (WDF) structures [5].

Lemma 1. For $k \geq 2, 1 \leq m \leq k$, and any $i \in I$, we have (*proof* easy):

$$s_{km}(\mathbf{p}) = s_{k-1,m}(\mathbf{p}\backslash p_i) + p_i s_{k-1,m-1}(\mathbf{p}). \tag{5}$$

Lemma 2. We have (*proof* elementary):

$$s_{km}(1, 1, \cdots, 1) = \binom{k}{m} = \frac{k!}{m!(k-m)!} \tag{6}$$

Lemma 3. For $k \geq 2$ and $1 \leq m \leq k$, we have

$$m s_{km}(\mathbf{p}) = \sum_{i=1}^{k} p_i s_{k-1,m-1}(\mathbf{p}\backslash p_i) \tag{7}$$

Proof. $s_{km}(\mathbf{p})$ and the sum in the right-hand side of (7) are both composed of *all* terms of the form $p_{i_1}p_{i_2}\cdots p_{i_k}$, with $1 \le i_1 < \cdots < i_m \le k$. Due to the symmetry in all p_1 to p_k, we must thus have

$$Ks_{km}(\mathbf{p}) = \sum_{i=1}^{k} p_i s_{k-1,m-1}(\mathbf{p}\backslash p_i)$$

where K is a constant. Applying Lemma 2 to this expression, one finds indeed $K = m$.

Lemma 4. The function F_{k1} is a positive function.

Proof. We have indeed $F_{k1} = s_{k1}/s_{k0}$, thus

$$F_{k1}(\mathbf{p}) = p_1 + p_2 + \cdots + p_k. \tag{8}$$

Proof of the Fundamental Theorem.

1. Statement 1 is true for $k \ge m = 1$ (Lemma 4). From Lemmas 1 and 3, one finds, for $k \ge m \ge 2$,

$$\begin{aligned} F_{km}(\mathbf{p}) &= \frac{1}{m}\sum_{i=1}^{k} \frac{p_i s_{k-1,m-1}(\mathbf{p}\backslash p_i)}{s_{k,m-1}(\mathbf{p})} \\ &= \frac{1}{m}\sum_{i=1}^{k} \frac{p_i s_{k-1,m-1}(\mathbf{p}\backslash p_i)}{s_{k-1,m-1}(\mathbf{p}\backslash p_i) + p_i s_{k-1,m-2}(\mathbf{p}\backslash p_i)} \\ &= \sum_{i=1}^{k} \frac{p_i F_{k-1,m-1}(\mathbf{p}\backslash p_i)}{p_i + F_{k-1,m-1}(\mathbf{p}\backslash p_i)} \end{aligned} \tag{9}$$

   The proof of statement 1 can then be completed by using Lemma 4, with k replaced by $k-m+1$, thus with F_{k1} replaced by $F_{k-m+1,1}$, and then applying induction by proceeding from $F_{k-m+1,1}$ to $F_{k-m+2,2}$ etc. until F_{km}.
2. Interpreting $F_{km}(\mathbf{p})$ as an impedance, one finds, from (9), for $F_{km}(\mathbf{p})$ the realization shown in Fig. 1. In this structure, we can then in turn replace each one of the k impedances $F_{k-1,m-1}(\mathbf{p}\backslash p_i)$, $i = 1$ to k, by a realization of the same type as that of Fig. 1, but with $F_{k-1,m-1}$ replaced by $F_{k-2,m-2}$ etc. until one arrives at impedances $F_{k-m+1,1}$. There are altogether $k!/(k-m+1)!$ impedances of this latter type, but they all depend on different combinations of $k-m+1$ of the variables p_1 to p_k. In any case, according to Lemma 4, they can all be synthesized very easily in form of a series connection of $k-m+1$ positive inductances. Clearly, if we interpret $F_{km}(\mathbf{p})$ as an admittance, a second realizing structure can be obtained; it is the dual of that obtained by means of Fig. 1.

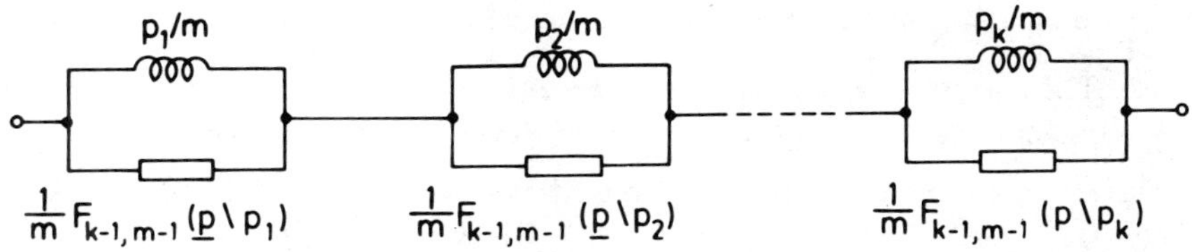

FIGURE 1. Basic series-parallel structure for realizing $F_{km}(\mathbf{p})$ as an impedance.

3. The structure of Fig. 1, clearly, consists only of series and parallel connections, and this remains true for the complete structure obtained by the process just described.
4. Clearly, $s_{k1}(\mathbf{p})$ and $s_{k0}(\mathbf{p})$ are relatively prime. Thus, if $s_{km}(\mathbf{p})$ and $s_{k,m-1}(\mathbf{p})$, for $m \geq 2$, had a common factor, $f(\mathbf{p})$, it would, by definition, be a polynomial involving at least one of the $p_i, i \in I$. In view fo the symmetry in all $p_i, i = 1$ to k, we may assume this to be p_k. Using the notation (4) and Lemma 1, we can write

$$s_{km}(\mathbf{p}) = p_k s_{k-1,m-1}(\mathbf{p}') + s_{k-1,m}(\mathbf{p}'),$$

$$s_{k,m-1}(\mathbf{p}) = p_k s_{k-1,m-2}(\mathbf{p}') + s_{k-1,m-1}(\mathbf{p}'). \tag{10}$$

Existence of a common factor would thus imply that [6, 7] the resultant of (10) with respect to p_k is zero, i.e., that

$$s_{k-1,m}(\mathbf{p}') \cdot s_{k-1,m-2}(\mathbf{p}') = s_{k-1,m-1}^2(\mathbf{p}').$$

This, however, is impossible as can be verified by applying Lemma 2 with $\mathbf{p}$ replaced by $\mathbf{p}'$.

Some further results are the following:

Corollary 1. For $1 \leq m \leq k$, the $s_{km}(\mathbf{p})$ are reactance Hurwitz polynomials.

The *proof* of this follows from the 2nd statement of Theorem 29 in [3], recalling that $s_{km}(\mathbf{p})/s_{k,m-1}(\mathbf{p})$ is a reactance function in irreducible form.

Corollary 2. For $1 \leq m \leq k$, the polynomial

$$g(\mathbf{p}) = s_{km}(\mathbf{p}) + s_{k,m-1}(\mathbf{p}) \tag{11}$$

is scattering Hurwitz.

The *proof* of this follows by invoking Theorem 7 in [1].

Theorem 2. For $1 \leq m \leq k$, $F_{km}(\mathbf{p})$ is holomorphic and has Re $F_{km}(\mathbf{p}) > 0$ for all $\mathbf{p}$ with Re $\mathbf{p} \geq \mathbf{0}$ provided that Re $\mathbf{p} \neq \mathbf{0}$. (*Proof* omitted).

4. Generation of further desirable functions

The polynomials/functions obtained so far are of very simple type, but more general ones can be obtained by means of simple transformations:

Theorem 3. Let us choose positive constants a_i, $i = 0$ to k, and replace p_i by $a_i p_i$, $i = 1$ to k. The following holds for $1 \leq m \leq k$:

a) The function f defined by

$$f(\mathbf{p}) = a_o F_{km}(a_1 p_1, \cdots, a_k p_k)$$

is a reactance function in irreducible form.

b) The polynomial a defined by

$$a(\mathbf{p}) = s_{km}(a_1 p_1, \cdots, a_k p_k)$$

is reactance Hurwitz.

c) The polynomial g defined by

$$g(\mathbf{p}) = s_{km}(a_1 p_1, \cdots, a_k p_k) + a_0 s_{k,m-1}(a_1 p_1, \cdots, a_k p_k)$$

is scattering Hurwitz.

Proof.

a) Like $F_{km}(\mathbf{p})$, $f(\mathbf{p})$ is a paraodd rational positive function. Furthermore, $s_{km}(a_1 p_1, \cdots, a_k p_k)$ and $s_{k,m-1}(a_1 p_1, \cdots, a_k p_k)$ are relatively prime at the same time as $s_{km}(\mathbf{p})$ and $s_{k,m-1}(\mathbf{p})$.

b) The proof of statement b) follows from that of statement a) and Theorem 29 in [3].

c) The proof of statement c) follows from that of statement a) and Theorem 7 in [1].

Theorem 4. Freeze one or several of the p_i, $i \in I$, at (real or complex) values p_{i_0} with Re $p_{i_0} > 0$, then, for $1 \leq m, k$, assuming the variables frozen to be the ones numbered from $k' + 1$ to k, with $1 < k' < k$, and designating by a_i, $i = 0$ to k', arbitrary positive constants:

a) The polynomial a defined by

$$a(p_1, \cdots, p_{k'}) = s_{km}(a_1 p_1, \cdots, a_{k'} p_{k'}, p_{k'+1,0}, \cdots, p_{k0})$$

is widest-sense Hurwitz.

b) The function f defined by

$$f(p_1, \cdots, p_{k'}) = a_0 F_{km}(a_1 p_1, \cdots, a_{k'} p_{k'}, p_{k'+1,0}, \cdots, p_{k0})$$

is a positive function.

c) The polynomial g defined by

$$g(p_1,\cdots,p_{k'}) = s_{km}(a_1p_1,\cdots,a_{k'}p_{k'},p_{k'+1,0},\cdots,p_{k0})+$$

$$+a_0s_{k,m-1}(a_1p_1,\cdots,a_{k'}p_{k'},p_{k'+1,0},\cdots,p_{k0})$$

is scattering Hurwitz.

The *proofs* of statements a) and c) follow from Theorem 12 and Lemma 5 in [3]; statement b) is obvious. Note that the functions generated by the method of Theorem 4 are usually not real.

Theorem 5: Let us replace one or several of the p_i by $1/p_i$, say for $i = k' + 1$ to k, with $1 \leq k' < k$. We have:

a) The polynomials a and b defined by

$$a(\mathbf{p}) = p_{k'+1}\cdots p_k s_{km}(p_1,\cdots,p_{k'},\frac{1}{p_{k'+1}},\cdots,\frac{1}{p_k}) \tag{12}$$

$$b(\mathbf{p}) = p_{k'+1}\cdots p_k s_{k,m-1}(p_1,\cdots,p_{k'},\frac{1}{p_{k'+1}},\cdots,\frac{1}{p_k}) \tag{13}$$

are relatively prime and reactance Hurwitz.

b) The function f defined by

$$f(\mathbf{p}) = a(\mathbf{p})/b(\mathbf{p}),$$

where a and b are defined by (12) and (13), is a reactance function in irreducible form.

c) The polynomial g defined by

$$g(\mathbf{p}) = a(\mathbf{p}) + a_0b(\mathbf{p}),$$

where a_0 is a positive constant, is scattering Hurwitz.

The *proof* of these statements is quite straightforward.

We like to stress that the transformation implied by the present theorem can also be combined with the transformations of Theorems 3 and 4.

So far, all polynomials/functions considered are of 1st degree in each one of the p_i, $i \in I$. However, since the generation of an elementary symmetric function is easy, we can start out from a sufficiently large number, k, of variables p_i, after which we may apply any of the transformations discussed so far. We are then still free to make equal to one another any of the variables p_i, thus reducing the original polynomials/functions to polynomials/functions in fewer variables but of correspondingly larger degree:

Theorem 6. After applying the procedure just described, we have the following:

a) Polynomials resulting from widest-sense Hurwitz polynomials are still widest-sense Hurwitz.

b) Polynomials resulting from reactance Hurwitz polynomials are still reactance Hurwitz.

c) Rational functions resulting from positive functions are still positive functions.

d) Rational functions resulting from reactance functions are still reactance functions.

Proof. Preservation of the widest-sense-Hurwitz property and of the positivity property is obvious. The rest follows by observing that self-paraconjugacy and paraoddness are also preserved.

If g is a widest-sense Hurwitz polynomial obtained this way, one can determine the greatest common divider, d, of g and g_* by standard procedures [6,7] (although these may be cumbersome to apply in practice). Then g/d is scattering Hurwitz.

Important remarks. The positive functions obtained by any of the procedures described in this section can always be synthesized by a series-parallel structure obtained from that discussed in relation with Fig. 1. A further possibility of generating positive (or reactance) functions that are of higher degree in the individual p_1 to p_k consists in replacing any of the p_i by positive (or reactance) functions generated earlier.

Some remaining open questions are:

1. Which are the additional requirements that must be imposed in order to ensure that the polynomial g obtained as discussed subsequently to Theorem 6 is still scattering Hurwitz?
2. How wide are the classes of Hurwitz polynomials (of various types), positive functions, and reactance functions obtained by the procedures described in this paper?

Bibliography

[1] A. Fettweis, "Some properties of scattering Hurwitz polynomials", Arch. Elektr. Übertr., vol. 38, No. 3, pp. 171–176, 1984.

[2] A. Fettweis, "A new approach to Hurwitz polynomials in several variables", Circuits, Systems, and Signal Processing, vol. 5, No. 4, pp. 405–417, 1986.

[3] A. Fettweis and S. Basu, "New results on multidimensional polynomials — Part I: Continuous case", IEEE Trans. Circuits and Systems, vol. CAS-34, pp. 1221–1232, Oct. 1987.

[4] A. Kummert, "Beiträge zur Synthese mehrdimensionaler Reaktanzmehrtore", Dissertation, Ruhr-Universität Bochum, W-Germany, February 1988.

[5] A. Fettweis, "Wave digital filters: theory and practice", Proc. IEEE, vol. 74, pp. 270–327, Feb 1986.

[6] M. Bôcher, "Introduction to Higher Algebra", Macmillan, 1907, and Dover Publications, New York, 1964.

[7] O. Perron, Algebra, Part 1, W. de Gruyter, Berlin, 1932.

Synthesis of Two-Dimensional Passive One-Ports with Lumped Elements

Anton Kummert

Abstract

Two-dimensional passive networks are of interest e.g. for use as reference filters for two-dimensional wave digital filters [1, 2]. Necessary properties of the immittance of such one-ports have been established, but not yet been shown to be also sufficient for a given two-variable rational function to be the impedance or admittance of such a system. In the design of 2-D passive networks it will be however of great interest whether this mentioned features can be used as a basis for a general synthesis procedure.

In this paper, it is shown that this is the case. The presented method for the synthesis of 2-D one-ports is based mainly on a paraunitary bordering of the given reflectance of the desired network to obtain the scattering matrix of a lossless 2-D multiport, which can be realized with a known procedure [3, 12, 13]. The representation of a real nonnegative definite two-variable polynomial as a sum of squares of real polynomials plays the crucial role in the presented design approach. No restrictions are made concerning the coefficients of the given rational immittance; they may be either real or complex, so as to include even complex networks which are of special interest for multidimensional systems [4].

I. Introduction

The theory of multidimensional passive networks has been drawing attention for several years. Due to applications to multidimensional wave digital filters (WDF) realizations, this field of research has attained new importance.

Once a suitable reference structure is known, the design of the corresponding WDF is rather straightforward. The major problem is that of synthesizing the reference filter. Until now, a general synthesis procedure only exists for a restricted, nevertheless important, class of passive 2-D networks, namely lossless multiports [3, 12, 13], which are also of special interest in the context of this paper.

Necessary and sufficient properties of a mathematical description of a system are usually the basis for a synthesis procedure. But yet, only necessary conditions concerning the immittance and reflectance, respectively, of 2-D passive one-ports have been established, since attempts to find a general synthesis method for this class of networks have failed, or turned out to be incorrect (cf. [5] together with [6, chap. 5.3.2]). But now, in Section III it is shown that this mentioned properties are also sufficient for a given two-variable rational function to be the immittance of such a system, and a general design procedure is presented in order to generate a corresponding network.

Basic definitions, properties and notations are outlined in Section II, and an important Theorem, known from literature, is summarized in the Appendix to have it available.

II. Definitions and Notations

By a rational function or matrix we mean expressions in $k \geq 1$ complex variables, p_i, $i = 1$ to k. For convenience of notation we summarize them in a vector $\underline{p} = (p_1, \ldots, p_k)$.

A notation such as Re $\underline{p} > \underline{0}$ designating Re $p_i > 0$, $i = 1$ to k.

The variables considered will be assumed to belong either to the field of complex numbers or to the field of real numbers. The letter ω will be used exclusively to designate real numbers.

The paraconjugate, $\underline{S}_*(\underline{p})$, of a given rational matrix $\underline{S}$ is defined by

$$\underline{S}_*(\underline{p}) = \underline{S}^*(-\underline{p}^*),$$

where the superscript asterisk designates complex conjugation of all entries and transposition of the matrix. For $\underline{p} = j\underline{\omega}$ we have

$$\underline{S}_*(j\underline{\omega}) = \underline{S}^*(j\underline{\omega}).$$

Definition 1:

Ω_p is the set of all matrices, $\underline{S}(\underline{p})$, having the properties

i, $\underline{S}(\underline{p})$ is rational in $\underline{p}$,

ii, for Re $\underline{p} > \underline{0}$, the entries of $\underline{S}$ are holomorphic,

iii, $\underline{S}$ is paraunitary, i.e.,

$$\underline{S}\,\underline{S}_* = \underline{S}_*\,\underline{S} = \underline{1},$$

where $\underline{1}$ is the unit matrix of appropriate order.

Next, we recall the definitions of two different types of Hurwitz polynomials [10].

Definition 2:

A polynomial, $g = g(\underline{p})$, is widest sense Hurwitz if $g(\underline{p}) \neq 0$ for $\text{Re } \underline{p} > \underline{0}$.

Definition 3:

A polynomial, $g = g(\underline{p})$, is scattering Hurwitz if the following conditions simultaneously hold true:

i, g is widest sense Hurwitz,

ii, g and g_* are relatively prime polynomials.

Definition 4:

A rational function, $F(\underline{p})$, is called positive if $\text{Re } F(\underline{p}) \geq 0$ in $\text{Re } \underline{p} > \underline{0}$.

Definition 5:

By a k-dimensional passive multiport with lumped elements we understand in the following a network in the classical sense, obeying Kirchhoff's laws for currents and voltages. It can consist of

— ideal connecting wires

— nonenergic, constant building elements like ideal transformers, gyrators, imaginary resistances, etc.

— real constant resistances

— two types of frequency-dependent elements, namely inductances and capacitances, each of them assigned to a certain complex frequency variable p_i, $i = 1, \ldots, k$.

III. Synthesis Procedure

In the literature it has been shown that the impedance and admittance, respectively, of a k-D passive one-port with lumped elements is always a rational positive function. Conversely, every one-variable rational positive function can be realized as the impedance of a passive one-port [8, 9].

Now, we want to investigate whether the last conclusion can be extended to two-variable rational positive functions. But before the actual synthesis method can be developed, we have to consider some preliminary results.

Let us assume the rational function, $F = \frac{a}{b}$, to be positive, where a and b are relatively prime two-variable polynomials of p_1, p_2. Then, due to [10, Theorem 25],

$$g(\underline{p}) = a(\underline{p}) + b(\underline{p}) \tag{1}$$

is a scattering Hurwitz polynomial. Furthermore, let us define

$$d(\underline{p}) = a(\underline{p}) - b(\underline{p}), \tag{2}$$

and consider the expression

$$g\, g_* - d\, d_* = 2(a\, b_* + a_*\, b) = 2b\, b_*(F + F_*), \tag{3}$$

i.e.,

$$|g(j\underline{\omega})|^2 - |d(j\underline{\omega})|^2 = 2|b(j\underline{\omega})|^2\, 2\,\mathrm{Re}\; F(j\underline{\omega}) \geq 0, \tag{4}$$

since $F(\underline{p})$ is positive. Thus, the term

$$g(j\underline{\omega})g_*(j\underline{\omega}) - d(j\underline{\omega})d_*(j\underline{\omega}) = |g(j\underline{\omega})|^2 - |d(j\underline{\omega})|^2$$

is obviously a nonnegative definite real polynomial in the variables ω_1, ω_2. Hence, according to Theorem A1 (cf. the Appendix), this polynomial can be written as

$$|g(j\underline{\omega})|^2 - |d(j\underline{\omega})|^2 = \frac{\sum_{i=1}^4 Q_i^2(\omega_1, \omega_2)}{h^2(\omega_1)}, \tag{5}$$

where the quantities Q_i are real polynomials of ω_1, ω_2, and $h(\omega_1)$ is a real one-variable polynomial. Define

$$R_i(p_1, p_2) = Q_i\left(\frac{p_1}{j}, \frac{p_2}{j}\right), \quad i = 1, \ldots, 4, \tag{6}$$

$$c(p_1) = h\left(\frac{p_1}{j}\right), \tag{7}$$

i.e.,

$$R_{i*} = R_i, \quad c_* = c. \tag{8}$$

With (5), (6), (7), and (8) we obtain

$$g(\underline{p})g_*(\underline{p}) - d(\underline{p})d_*(\underline{p}) = \frac{\sum_{i=1}^4 R_{i*}(\underline{p})R_i(\underline{p})}{c_*(p_1)c(p_1)} \tag{10}$$

for $\underline{p} = j\underline{\omega}$, and by analytic continuation for all $\underline{p} \in C^2$. The zeroes of the polynomial $c_*\, c$ can be ordered always in such a way that

$$e_*\, e = c_*\, c, \tag{11}$$

where $e(p_1)$ is a polynomial of p_1 with

$$e(p_1) \neq 0, \quad \text{for} \quad \mathrm{Re}\; p_1 > 0. \tag{12}$$

Additionally summarize the polynomials R_i to the polynomial vector

$$\underline{R} = (R_1, R_2, R_3, R_4)^T, \tag{13}$$

and we obtain in view of (10), (11), and (13)

$$g\, g_*\, e\, e_* - d\, d_*\, e\, e_* = \underline{R}_*\, \underline{R}. \tag{14}$$

Now, we are able to develop the mentioned synthesis procedure for 2-D passive one-ports.

Let $\underline{S}(\underline{p})$ be the scattering matrix of a lossless m-port, N', with lumped elements, then, according to [7] and Definition 1, $\underline{S}(\underline{p})$ is an element of Ω_p. We partition $\underline{S}$ as

$$\underline{S} = \begin{pmatrix} S_{11} & \underline{S}_{12} \\ \underline{S}_{21} & \underline{S}_{22} \end{pmatrix} \begin{matrix} 1 \\ m-1 \end{matrix} \qquad \begin{matrix} 1 & m-1 \end{matrix} \tag{15}$$

and terminate the $m-1$ last ports of N' on the corresponding port-resistances. We know that the reflectance of the resulting passive one-port, N, is the entry S_{11} of (15) (cf. FIGURE 1).

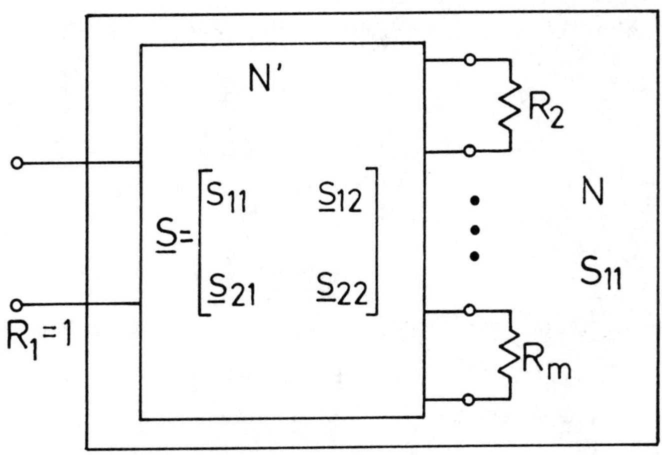

FIGURE 1

Conversely, if it is necessary to synthesize a passive one-port of prescribed impedance F or reflectance

$$S_{11} = \frac{F-1}{F+1} = \frac{a-b}{a+b} = \frac{d}{g}, \tag{16}$$

respectively, one may try to border S_{11} by the additional submatrices in (15) to form a matrix $\underline{S}$ which is an element of Ω_p. The synthesis is then achieved by realizing the lossless m-port N' of scattering matrix $\underline{S}$ with a known method [3, 12, 13] and terminating its $(m-1)$ last ports on the corresponding port-resistances. The synthesis of general passive one-ports is thus reduced to the synthesis of lossless networks. This simplified problem requires only the matrix $\underline{S}$ to be an element of Ω_p (cf. [3, 12, 13]).

Theorem 1:

Given the already defined rational positive function F and the polynomials a, b, g, d, respectively, the above-mentioned bordering of

$$S_{11} = \frac{F-1}{F+1} = \frac{d}{g}$$

can be achieved by calculating the polynomial vector $\underline{R}$, the one-variable polynomial e (cf. (5 — 14)), and using the expressions

$$S_{11} = \frac{d}{g}, \tag{17a}$$

$$\underline{S}_{21} = \frac{\underline{R}}{g\,e}, \tag{17b}$$

$$\underline{S}_{12} = \frac{\underline{R}_*}{g\,e}, \tag{17c}$$

$$\underline{S}_{22} = \frac{\underline{R}\,\underline{R}_*}{g\,e^2 2a} - \frac{e_*\,a_*}{e\,a}\,\underline{1}. \tag{17d}$$

With other words, the matrix

$$\underline{S} = \begin{pmatrix} S_{11} & \underline{S}_{12} \\ \underline{S}_{21} & \underline{S}_{22} \end{pmatrix}$$

is an element of Ω_p.

Proof:

i, $\underline{S}$ is obviously rational in $\underline{p}$, since $\underline{R}, g, e, d, a$ are polynomial expressions.

ii, Since $F = \frac{a}{b}$ is a rational positive function in irreducible form and due to [10, Theorem 25], the polynomial $a(\underline{p})$ is widest sense Hurwitz, i.e.,

$$a(\underline{p}) \neq 0 \quad \text{for} \quad \text{Re}\ \underline{p} > \underline{0}, \tag{18}$$

and the polynomial $g = a + b$ is scattering Hurwitz, i.e.,

$$g(\underline{p}) \neq 0 \quad \text{for} \quad \text{Re}\ \underline{p} > \underline{0}. \tag{19}$$

Thus, in view of (12), (18), (19), and (17), the matrix $\underline{S}$ is holomorphic in Re $\underline{p} > \underline{0}$.

iii, The paraunitarity of $\underline{S}$ can be shown by inserting (17) in (20) and using the relations (14), (1), and (2),

$$\underline{S}\,\underline{S}_* = \begin{pmatrix} S_{11}\,S_{11*} + \underline{S}_{12}\,\underline{S}_{12*} & S_{11}\,\underline{S}_{21*} + \underline{S}_{12}\,\underline{S}_{22*} \\ \underline{S}_{21}\,S_{11*} + \underline{S}_{22}\,\underline{S}_{12*} & \underline{S}_{21}\,\underline{S}_{21*} + \underline{S}_{22}\,\underline{S}_{22*} \end{pmatrix} = \underline{1}. \qquad (20)$$

The proof is completed by comparing i, — iii, with Definition 1.

q.e.d.

Theorem 1 shows that the bordering of S_{11} can be done always in such a way that the resulting matrix $\underline{S}$ is an element of Ω_p and therefore can be realized as the scattering matrix of a lossless 2-D multiport. The vector $\underline{R}$ has at most four nonzero entries which implies the need of at most four resistances for terminating this lossless network N'. Nothing can be said about the number of required reactive elements, which depends on the partial degrees with respect to p_1 and p_2 of the determinant of $\underline{S}$.

IV. Conclusion

It has been shown that the general problem of synthesizing 2-D passive one-ports can be always reduced to the synthesis of lossless multiports, which has been solved already. An extension of this result to passive multiports is nontrivial and in preparation.

Appendix

Theorem A1:

Let $f(\omega_1, \omega_2)$ be a nonnegative definite real polynomial in the variables ω_1, ω_2, then it can be written always as a sum of at most four squares

$$f = \sum_{i=1}^{4} w_i^2,$$

where the quantities w_i are polynomials in ω_2 with real rational coefficients in ω_1.

Proof: See [11, Part II].

References

[1] A. Fettweis, "Principles of Multidimensional Wave Digital Filtering", in Digital Signal Processing, Point Lobos Press, Hollywood, CA, pp. 262 - 282, 1979.

[2] A. Fettweis, "Multidimensional Wave Digital Filters", Proc. of the 1976 European Conference on Circuit Theory and Design, Genova, vol. II, pp. 409 - 416, 1976.

[3] A. Kummert, "Minimal Synthesis of Two-Dimensional Reactance Multiports", Proc. 29th Midwest Symposium on Circuits and Systems, Lincoln, August 1986, pp. 674 - 677.

[4] A. Fettweis, "Multidimensional Digital Filters with Closed Loss Behaviour Designed by Complex Network Theory Approach", IEEE Trans. on Circuits and Systems, vol. CAS-34, No. 4, April 1987, pp. 338 - 344.

[5] T. Koga, "Synthesis of finite passive n-ports with prescribed positive real matrices of several variables", IEEE Trans. Circuit Theory, CT-15, March 1968, pp. 2 - 23.

[6] N.K. Bose, "Applied Multidimensional Systems Theory", Van Nostrand Reinhold Company, New York, 1982.

[7] A. Fettweis, "On the Scattering Matrix and the Scattering Transfer Matrix of Multidimensional Lossless Two-ports", AEÜ, vol. 36, September 1982, pp. 374 - 381.

[8] V. Belevitch, "Classical Network Theory", Holden-Day, 1968.

[9] R.W. Newcomb, "Linear Multiport Synthesis", McGraw-Hill, 1966.

[10] A. Fettweis, S. Basu, "New results on Stable Multidimensional Polynomials - Part I: Continuous Case", IEEE Trans. on Circuits and Systems, vol. CAS-34, No. 10, October 1987, pp. 1221 - 1232.

[11] E. Landau, "Über die Darstellung definiter Funktionen durch Quadrate", Mathem. Annalen, vol. 62, 1906, pp. 272 - 285.

[12] A. Kummert, "Beiträge zur Synthese mehrdimensionaler Reaktanzmehrtore", Doctoral Dissertation, Ruhr-Universität Bochum, Bochum, 1988.

[13] A. Kummert, "Synthesis of Two-Dimensional Lossless m-Ports with Prescribed Scattering Matrix", accepted for publication in Circuits Systems Signal Processing, 1989.

ANALYSIS OF NONLINEAR NETWORKS WITH FREQUENCY DEPENDENT COMPONENTS

MICHEL NAKHLA

ABSTRACT

A new method is presented for the transient analysis of nonlinear networks which contain frequency-dependent components such as multiconductor lossy transmission line systems. The problem in simulating such networks arises from the fact that these components can be analyzed only in the frequency domain, whereas the analysis of the nonlinear network is possible only in the time domain. The proposed method is based on piecewise decomposition techniques where the network is decomposed into linear and nonlinear subnetworks. Signals at the interconnection of the subnetworks are represented by piecewise linear functions. Newton's method is used to obtain parameters of the piecewise functions which satisfy the resulting topological and constitutive constraints

I - INTRODUCTION

An important problem in the computer-aided design of nonlinear systems is the time domain analysis of nonlinear networks which contain frequency-dependent components. Examples of such networks are in the area of microwave circuits [3] and multiconductor lossy transmission line systems [1] &[9]. It is common for these circuits to have a number of distributed components. The analysis of these components is possible only in the frequency domain. Whereas, the analysis of the nonlinear networks, in the general case, is possible only in the time domain. In [5], a piecewise harmonic balance technique has been presented for the analysis of nonlinear networks. The method is applicable for networks with frequency-dependent components [2] & [8]. However, it is limited to the steady-state case with periodic inputs. In this paper , a new method for the transient analysis is presented. It is based on a piecewise technique where the nonlinear network is decomposed into linear and nonlinear subnetworks. The linear subnetworks contain all the frequency-dependent components. The terminals of the subnetworks are excited by piecewise time-dependent sources. Through solution of the linear subnetworks in the frequency domain, the nonlinear subnetworks in the time domain, and by applying Newton's method , the parameters of the piecewise waveforms are adjusted so that the topological and constitutive relations of each subnetwork as well as the topological equations resulting from the interconnection of these subnetworks are satisfied.

II- BASIS OF THE METHOD

The proposed method will be demonstrated assuming a single nonlinear two-terminal subnetwork. The analysis of the general case is described in [6-7] . Consider a network "T" which consists of linear lumped components, lossy transmission lines and nonlinear components. Decompose "T" into linear and nonlinear subnetworks such that all the distributed components are contained within the linear subnetwork (Fig.1a). Augment the subnetworks with independent voltage source e(t) (Fig.1b). Let Q(t) and W(t) be the resulting currents at the terminals A-B and A' -B' , respectively. Define an error function $\partial(t) = \| Q(t)-W(t)\|$. The problem is reduced to finding the waveform e(t) which reduces the error function to zero for all t >0 . In the following e(t) is approximated by a piecewise linear function (Fig. 2) in the form

(1) $$e(t) = \sum_{k=1}^{N}(b_k + a_k t_{k-1})\,[u(t - t_{k-1}) - u(t - t_k)]$$; where

u(.) is the unit step function

N is the number of linear segments

a_k and b_k are parameters defining the segment k in the interval $t_{k-1} \le t \le t_k$

$$a_k = \frac{e_k - e_{k-1}}{t_k - t_{k-1}}$$

$$b_k = \frac{e_{k-1} t_k - e_k t_{k-1}}{t_k - t_{k-1}}$$

The parameters e_k ; $k = 1,....,N$ are obtained by solving set of nonlinear equations at each time point. These equations are obtained by discretizing the equation Q(t) - W(t) =0 at N time points.

III- FORMULATION OF THE NONLINEAR EQUATION

Eq.(1) can be rewritten in the form

(2) $$e(t) = \sum_{k=1}^{k=N}(b_k + a_k t_{k-1})[u(t - t_{k-1}) - u(t - t_k)] + \sum_{k=1}^{k=N} a_k (t - t_{k-1})[u(t - t_{k-1}) - u(t - t_k)]$$

Since the subnetwork "L" is linear time-invariant, the current Q(t) at terminals A-B can be written in the form

(3) $$Q(t) = \sum_{k=1}^{k=N} (b_k + a_k t_{k-1}) h(t - t_{k-1}) + \sum_{k=1}^{k=N} a_k g(t - t_{k-1}) + \eta(t)$$

where

h(t) is the current at terminals A-B corresponding to input voltage

$$\tilde{e}(t) = [u(t) - u(t - \Delta)]; \text{where } \Delta = t_k - t_{k-1}$$

g(t) is the current at terminals A-B corresponding to input voltage

$$\hat{e}(t) = t[u(t) - u(t - \Delta)]$$

$\eta(t)$ is current at terminals A – B corresponding to independent input sources within "L"

h(t) and g(t) can be obtained by applying the inverse fast Fourier Transform or the inverse Laplace Transform to the frequency response of the linear subnetwork "L".

In the next step , the nonlinear subnetwork "N" is excited by the augmenting voltage source e(t) described in (1). Assume that the subnetwork "N" is described by the following set of differential-algebraic equations

(4a) $$\frac{\partial \mathbf{x}}{\partial t} = \mathbf{f}(\mathbf{x}, t, \mathbf{e})$$

(4b) $$W(t) = g(\mathbf{x}, t)$$

where

x is a vector representing the state variables

$\mathbf{e} = [e_0\ e_1 \ldots\ldots\ldots\ldots e_N]^{\dagger}$;where † denotes transpose

W(t) is the current at terminals A`-B`

For consistent solution

$$Q(t) - W(t) = 0 ; \qquad 0 \le t \le t_N$$

or

(5) $$Q(t_r) - W(t_r) = 0 ; \qquad r = 0, 1, \ldots\ldots, N$$

Eq.(5) represents a set of (N+1) nonlinear equations in the (N+1) unknowns e_k ; $k = 0, 1, \ldots\ldots, N$ and can be solved using Newton's as follows

(6) $$\mathbf{e}^{i+1} = \mathbf{e}^{i} - \mathbf{J}^{-1} \mathbf{F}^{i}$$

where

the superscript i denotes the i<u>th</u> iteration,

$\mathbf{F} = [F_0\ F_1 \ldots\ldots\ldots\ldots F_N]$; and where

$F_r = Q(t_r) - W(t_r)$

$$\mathbf{J} = \frac{\partial \mathbf{F}}{\partial \mathbf{e}}$$

In the following section the evaluation of the Jacobian matrix **J** is described. It is shown that **J** is a lower triangular matrix and solution of (6) can be obtained using forward substitution. This reduces the computational complexity of Newton's iterations. It should also be noted that it is required to evaluate the functions h(t) and g(t) ONLY once (independent on the number of iterations).

IV- JACOBIAN EVALUATION

Eq. (3) can be rewritten in the form

(7) $$Q(t) = \sum_{k=1}^{k=N} [\lambda_k(t) + h(t - t_{k-1})]e_k - \sum_{k=1}^{k=N} \lambda_{k-1}(t)e_{k-1} + \eta(t)$$

where

$$\lambda_k(t) = \frac{g(t - t_{k-1})}{\Delta} - h(t - t_{k-1})$$

From (5) , the Jacobian matrix **J** is given by

$$\mathbf{J} = [j_{r,k}]; \qquad r = 0, 1, \ldots\ldots, N$$
$$k = 0, 1, \ldots\ldots, N$$

where

(8) $$j_{r,k} = \frac{\partial Q(t_r)}{\partial e_k} - \frac{\partial W(t_r)}{\partial e_k}$$

Using (7)

(9) $$\frac{\partial Q(t_r)}{\partial e_k} = \lambda_k(t_r) + h(t_r - t_{k-1}) - \lambda_{k+1}(t_r) \qquad ; r = 0, 1, \ldots\ldots, N$$
$$k = 1, 2, \ldots\ldots, N-1$$

(9 a) $$\frac{\partial Q(t_r)}{\partial e_0} = -\lambda_1(t_r)$$

(9 b) $$\frac{\partial Q(t_r)}{\partial e_N} = \frac{g(t_r - t_{n-1})}{\Delta}$$

From the definition of h(t) and g(t)

(1 0) $$\frac{\partial Q(t_r)}{\partial e_k} = 0 \qquad \text{for } k \geq (r+1)$$

To find the second term in (8) , differentiate (4) with respect to **e** to obtain

(1 1a) $$\frac{\partial \mathbf{x}'}{\partial e_k} = \frac{\partial \mathbf{f}}{\partial \mathbf{x}} \frac{\partial \mathbf{x}}{\partial e_k} + \frac{\partial \mathbf{f}}{\partial \mathbf{e}} \frac{\partial \mathbf{e}}{\partial e_k}$$

(1 1b) $$\frac{\partial W(t)}{\partial e_k} = \frac{\partial g}{\partial \mathbf{x}} \frac{\partial \mathbf{x}}{\partial e_k}$$

Eq.(11) represents the following set of variational equations

(1 2a) $$\mathbf{z}_k' = \left.\frac{\partial \mathbf{f}}{\partial \mathbf{x}}\right|_{\Gamma} \mathbf{z}_k + \mu_k \quad ;k = 0, 1, \ldots\ldots, N$$

(1 2b) $$y = \frac{\partial W(t)}{\partial e_k} = \left.\frac{\partial g}{\partial \mathbf{x}}\right|_{\Gamma} \mathbf{z}_k$$

where

$$\mathbf{z}_k = \frac{\partial \mathbf{x}}{\partial e_k}$$

Γ is the trajectory $\mathbf{x}(t;\mathbf{e})$

μ_{k-1} is the kth column of the matrix $\frac{\partial \mathbf{f}}{\partial \mathbf{e}}$

Since (12) has the same Jacobian as (4) , the (N+2) sets of differential equations can be solved simultaneously. In this case , one L/U factorization is needed at each time point along the trajectory.

From (4)

(1 3) $$\frac{\partial W(t_r)}{\partial e_k} = 0 \qquad \text{for} \quad k \geq (r+1)$$

from (10) &(13) it follows that the Jacobian **J** is a lower triangular matrix.

In the special case where the nonlinear subnetwork is resistive

$W(t) = F(e(t))$; and

(1 4) $$\frac{\partial W(t_r)}{\partial e_k} = \frac{\partial F}{\partial e} \frac{\partial e(t_r)}{\partial e_k}$$

$$= 0 \quad \text{if} \quad r \neq k$$

$$= \frac{\partial F}{\partial e} \quad \text{if} \quad r = k$$

V- EXAMPLE

Figure 3 shows a nonlinear network which contains a lossy transmission line. The line length is assumed to be 0.5 m and the inductance , capacitance, resistance ,and conductance are 160nH/m, 100 pF/m,75 Ω/m , 1 mS/m ,respectively. The output response is shown in Fig.4. To verify the method , the output response for the lossless case was compared with SPICE [4] results . Agreement within 0.5% was obtained.

VI-CONCLUSION

A new method has been presented for finding the response of nonlinear networks which contain frequency -dependent components. The method is based on a piecewise decomposition technique. The linear subnetworks , which contain the frequency-dependent components , are solved in the frequency domain while the nonlinear subnetworks are solved in the time domain. Newton's algorithm is used to obtain the parameters of the piecewise linear augmenting sources which satisfy the topological and constitutive relations of each subnetwork as well as the topological equations resulting from the interconnection of these subnetworks. The Jacobian matrix of the resulting nonlinear equations proved to be lower triangular matrix. This significantly reduces the computational complexity of the method. The proposed technique has been tested on several examples. An example of lossy transmission line with nonlinear termination has been presented in this paper. The method can be easily extended to the general case of multi-nonlinear subnetworks.

ACKNOWLEDGMENT

This work was supported through the Computer-Aided Engineering Chair funded by Bell-Northern Research and the Natural Sciences and Engineering Research Council of Canada. The author is the current holder of the Chair.

REFERENCES

[1] A.R.Djordjevic,T.K.Sarkar,and R.F.Harrington, "Time-domain response of multiconductor transmission lines, " *Proceedings IEEE* ,vol. 75, pp.743-764 , June 1987.

[2] K.S. Kundert and A. Sangiovanni-Vincentelli, "Simulation of nonlinear circuits in the frequency domain," *IEEE Trans. Computer-aided Des.*, vol CAD-5, pp.521-535, Oct.1986.

[3] A.Matezrka and T.Kocprzak, "Computer calculation of large-signal GaAs FET amplifier characteristics," *IEEE Trans. Microwave Theory Tech.*, vol MTT-33, pp.129-135,Feb.1985.

[4] L. Nagel, " SPICE2 : a computer program to simulate semiconductor circuits," Ph.D. dissertation ,Univ. California at Berkeley, May 1975.

[5] M. Nakhla and J. Vlach ," A piecewise harmonic balance technique for determination of periodic response of nonlinear systems," *IEEE Trans. Circuits Syst.*, vol.CAS-23, pp.85-91, Feb. 1976.

[6] M. Nakhla, "Analysis of pulse propagation on high-speed VLSI chips",*IEEE Journal of Solid-State Circuits* (to be published) , Apr.1990.

[7] M. Nakhla, " A Piecewise decomposition technique for analysis of nonlinear networks with distributed components," *Proceedings IEEE International Symp. Circuits &systs.* , pp.606-609,May 1989

[8] V.Rizzoli and A.Neri ," State of the art and present trends in nonlinear microwave CAD techniques ," *IEEE Trans. Microwave Theory Tech.*, vol. 36,pp.343-365, Feb. 1988.

[9] S. Seki and H.Hasegawa," Analysis of crosstalk in very high-speed LSI/VLSI's using a coupled multiconductor MIS microstrip line model," *IEEE Trans. Electron Dev.* , vol. ED-31pp. 1948-1953 ,Dec. 1984.

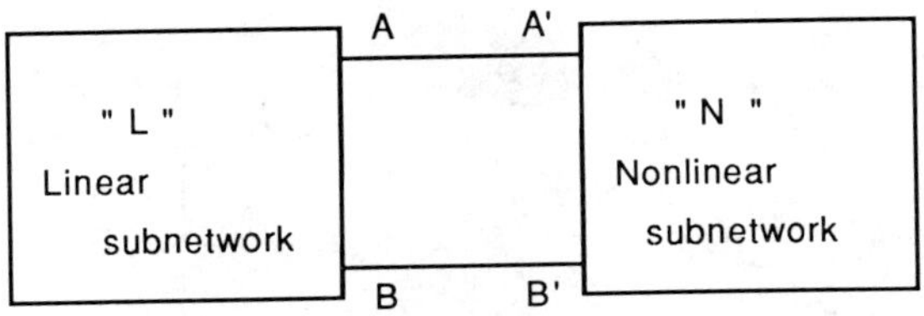

Figure 1a **Arbitrary network "T"**

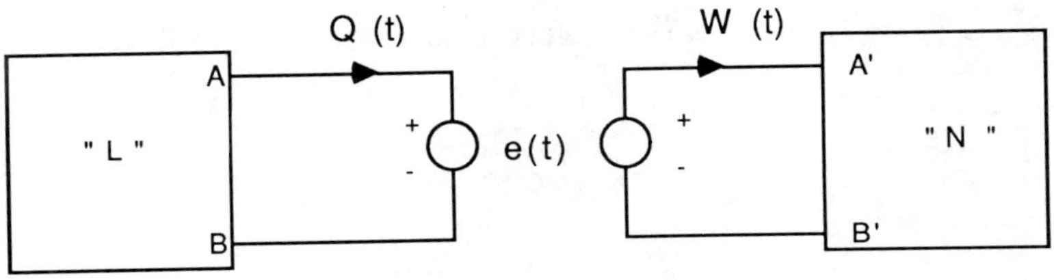

Figure 1b **Same network separated into linear and nonlinear subnetworks , each augmented by a voltage source**

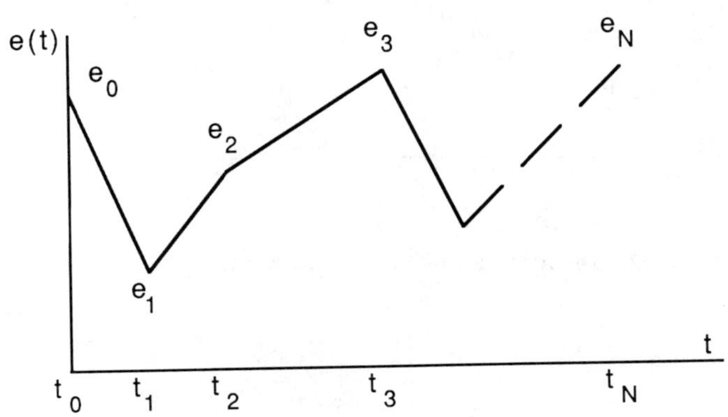

Figure 2 **Piecewise linear augmenting source e(t)**

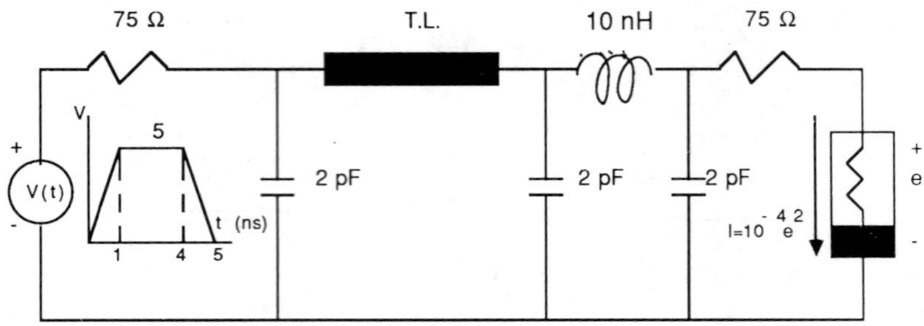

Figure 3 : A nonlinear network with a lossy transmission line

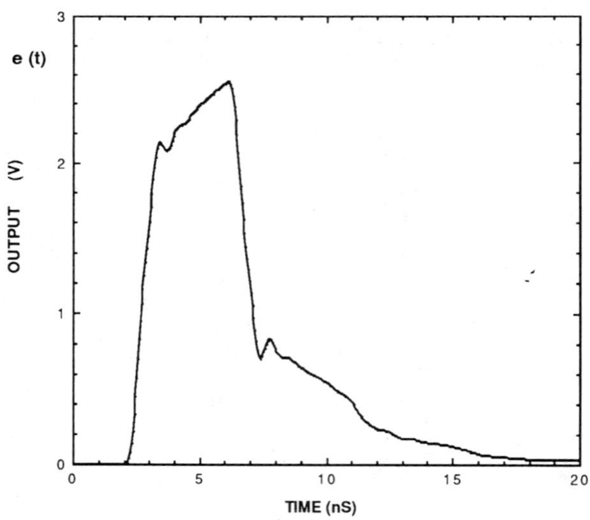

Figure 4 : Output response of the nonlinear network of Fig.3

A HILBERT SPACE ASSOCIATED WITH A NEVANLINNA FUNCTION

Daniel Alpay, Piet Bruinsma, Aad Dijksma, Henk de Snoo

To any matrix valued Nevanlinna function there is associated in a natural way a reproducing kernel Hilbert space. Such a Hilbert space provides a model for a simple symmetric, not necessarily densely defined operator. Via its so called Q function we connect the results of de Branges [dB1,2] with those of Kreĭn and Langer [KL1,2] and of Langer and Textorius [LT].

We define the Nevanlinna class $\mathbf{N}^n$ as the class of all $n\times n$ matrix functions $N(\ell)$, which are holomorphic in $\mathbb{C}\backslash\mathbb{R}$, satisfy $N(\ell)^*=N(\bar{\ell})$, $\ell\in\mathbb{C}\backslash\mathbb{R}$, and for which the kernel

$$\mathsf{K}_N(\ell,\lambda)=\frac{N(\ell)-N(\lambda)^*}{\ell-\bar{\lambda}},\qquad \ell,\lambda\in\mathbb{C}\backslash\mathbb{R},\qquad \ell\neq\bar{\lambda},$$

is nonnegative. The reproducing kernel Hilbert space associated with the kernel $\mathsf{K}_N(\ell,\lambda)$ (see [D], [dB2]) is denoted by $\mathfrak{L}(N)$. For general information concerning reproducing kernel Hilbert spaces we refer to Aronszajn's paper [A]. It is well known that for each $N(\ell)\in\mathbf{N}^n$ there exist $n\times n$ matrices $\mathcal{A}$ and $\mathcal{B}$ with $\mathcal{A}=\mathcal{A}^*$ and $\mathcal{B}=\mathcal{B}^*\geq 0$, and a nondecreasing $n\times n$ matrix function Σ on $\mathbb{R}$ with $\int_{\mathbb{R}}(t^2+1)^{-1}d\Sigma(t)<\infty$ such that

$$(1)\qquad N(\ell)=\mathcal{A}+\mathcal{B}\ell+\int_{\mathbb{R}}\Big(\frac{1}{t-\ell}-\frac{t}{t^2+1}\Big)\,d\Sigma(t),\qquad \ell\in\mathbb{C}\backslash\mathbb{R}.$$

With this Riesz–Herglotz–Nevanlinna representation the kernel $\mathsf{K}_N(\ell,\lambda)$ takes the form

$$(2)\qquad \mathsf{K}_N(\ell,\lambda)=\mathcal{B}+\int_{\mathbb{R}}\frac{d\Sigma(t)}{(t-\ell)(t-\bar{\lambda})},\qquad \ell,\lambda\in\mathbb{C}\backslash\mathbb{R}.$$

We define the space $\mathbb{C}^n_{\mathcal{B}}$ to be the range $\mathfrak{R}(\mathcal{B})$ of $\mathcal{B}$ in $\mathbb{C}^n$ endowed with the inner product $[\mathcal{B}c,\mathcal{B}d]=d^*\mathcal{B}c$. We note that $\mathbb{C}^n_{\mathcal{B}}$ is equal to the space $\mathfrak{L}(\mathcal{B}\ell)$. As usual $L^2(d\Sigma)$ is the Hilbert space of all $n\times 1$ vector functions f defined on $\mathbb{R}$, such that $\|f\|^2_{\Sigma}=\int_{\mathbb{R}}f(t)^*d\Sigma(t)f(t)<\infty$. Recall that by the Stieltjes–Livšic inversion formula the functions $(t-\ell)^{-1}c$, $c\in\mathbb{C}^n$ and $\ell\in\mathbb{C}\backslash\mathbb{R}$, are dense in

$L^2(d\Sigma)$. The scalar version of the next result can be found in [dB2, Theorem 5].

PROPOSITION 1. *Let $N(\ell)\in\mathbf{N}^n$ have the integral representation* (1). *The Hilbert space $\mathfrak{L}(N)$ is isomorphic to the space of all $n\times 1$ vector functions $F(\ell)$ of the form*

$$(3)\qquad F(\ell)=\mathcal{B}c+\int_{\mathbb{R}}\frac{d\Sigma(t)f(t)}{t-\ell},\qquad \ell\in\mathbb{C}\backslash\mathbb{R},$$

where $c\in\mathbb{C}^n$, $f\in L^2(d\Sigma)$ are uniquely determined by $F(\ell)$, with norm given by

$$(4)\qquad \|F\|^2=c^*\mathcal{B}c+\|f\|^2_{\Sigma}.$$

Proof. Let $F(\ell)$ be an element of the form (3). We first check that $\mathcal{B}c$ and f are uniquely determined by $F(\ell)$. Indeed, if $F(\ell)$ admits two such representations with $f\in L^2(d\Sigma)$, $c\in\mathbb{C}^n$ and $\tilde{f}\in L^2(d\Sigma)$, $\tilde{c}\in\mathbb{C}^n$, we have

$$\mathcal{B}(\tilde{c}-c)=\int_{\mathbb{R}}\frac{d\Sigma(t)(f(t)-\tilde{f}(t))}{t-\ell}=[f(t)-\tilde{f}(t),\ (t-\bar{\ell})^{-1}]_{\Sigma}.$$

Letting $\ell\to\infty$ we obtain $\mathcal{B}c=\mathcal{B}\tilde{c}$ and $f=\tilde{f}$. The set of functions of the form (3) with norm (4) is easily seen to be a Hilbert space, which we denote by $\mathfrak{H}$. The representation (2) shows that the function $\ell\to\mathsf{K}_N(\ell,\lambda)d$ belongs to $\mathfrak{H}$ for any $d\in\mathbb{C}^n$ and $\lambda\in\mathbb{C}\backslash\mathbb{R}$. Moreover, for $F(\ell)$ of the form (3) we have the reproducing property of the kernel $\mathsf{K}_N(\ell,\lambda)$:

$$(5)\qquad [F,\mathsf{K}_N(.,\lambda)d]=d^*F(\lambda),\qquad d\in\mathbb{C}^n,$$

where $[.,.]$ denotes the inner product associated with the norm (4). The uniqueness of the reproducing kernel Hilbert space with reproducing kernel $\mathsf{K}_N(\ell,\lambda)$ implies that $\mathfrak{H}$ and $\mathfrak{L}(N)$ are isomorphic.

Note that we may write $\mathfrak{L}(N)=\mathfrak{L}(\mathcal{B}\ell)\oplus\mathfrak{L}(N-\mathcal{B}\ell)=\mathbb{C}^n_{\mathcal{B}}\oplus\mathfrak{L}(N-\mathcal{B}\ell)$. The mapping defined by

$$f\to\int_{\mathbb{R}}\frac{d\Sigma(t)f(t)}{t-\ell},\qquad f\in L^2(d\Sigma),$$

is an isometry from $L^2(d\Sigma)$ onto $\mathfrak{L}(N-\mathcal{B}\ell)$ and the mapping defined by (3) is an isometry from $\mathbb{C}^n_{\mathcal{B}}\oplus L^2(d\Sigma)$ onto $\mathfrak{L}(N)$. The elements of $\mathfrak{L}(N)$ are $n\times 1$ vector functions, which are defined and holomorphic on $\mathbb{C}\backslash\mathbb{R}$. This follows from the fact that $\mathfrak{L}(N)$ is a reproducing kernel Hilbert space, but also from (3). We shall identify $\mathfrak{L}(N)$ with the space $\mathbb{C}^n_{\mathcal{B}}\oplus L^2(d\Sigma)$. For any $n\times 1$ vector function $F(\ell)$, holomorphic on $\mathbb{C}\backslash\mathbb{R}$, and any $\lambda\in\mathbb{C}\backslash\mathbb{R}$, we define the operator R_λ by

$$(R_\lambda F)(\ell)=\frac{F(\ell)-F(\lambda)}{\ell-\lambda},\qquad \ell\in\mathbb{C}\backslash\mathbb{R},\qquad \ell\neq\lambda,\qquad (R_\lambda F)(\lambda)=F'(\lambda).$$

For Hilbert spaces $\mathfrak{H},\mathfrak{K}$ we denote by $\mathbf{L}(\mathfrak{H},\mathfrak{K})$ the space of all bounded linear operators from $\mathfrak{H}$ to $\mathfrak{K}$; we write $\mathbf{L}(\mathfrak{H})=\mathbf{L}(\mathfrak{H},\mathfrak{H})$.

PROPOSITION 2. *(i) For all* $\lambda\in\mathbb{C}\backslash\mathbb{R}$, $R_\lambda\in\mathbf{L}(\mathfrak{L}(N))$. *In fact, we have* $\|R_\lambda F\|\le|\mathrm{Im}\,\lambda|^{-1}\|F\|$, $F\in\mathfrak{L}(N)$. *(ii) For all* $\lambda\in\mathbb{C}\backslash\mathbb{R}$, $(R_\lambda)^*=R_{\bar\lambda}$. *(iii) The resolvent identity* $R_\lambda-R_\mu=(\lambda-\mu)R_\mu R_\lambda$ *holds for all* $\lambda,\mu\in\mathbb{C}\backslash\mathbb{R}$.

Proof. Let $F(\ell)\in\mathfrak{L}(N)$ have representation (3), then for $\lambda\in\mathbb{C}\backslash\mathbb{R}$

$$(6)\qquad (R_\lambda F)(\ell)=\int_{\mathbb{R}}\frac{d\Sigma(t)f(t)}{(t-\ell)(t-\lambda)},$$

which on account of Proposition 1 implies that $R_\lambda F\in\mathfrak{L}(N)$, $\lambda\in\mathbb{C}\backslash\mathbb{R}$. By (6)

$$\|R_\lambda F\|^2=\|f(t)/(t-\lambda)\|_\Sigma^2=\int_{\mathbb{R}}\Big(\frac{f(t)}{t-\lambda}\Big)^*d\Sigma(t)\Big(\frac{f(t)}{t-\lambda}\Big)\le$$

$$\le|\mathrm{Im}\,\lambda|^{-2}\int_{\mathbb{R}}f(t)^*d\Sigma(t)f(t)=|\mathrm{Im}\,\lambda|^{-2}\|f\|_\Sigma^2\le|\mathrm{Im}\,\lambda|^{-2}\|F\|^2,$$

for $\lambda\in\mathbb{C}\backslash\mathbb{R}$, which shows that R_λ is a bounded operator in $\mathfrak{L}(N)$. This proves *(i)*. Items *(ii)* and *(iii)* also follow from (6). This completes the proof.

Combining *(ii)* and *(iii)* of Proposition 2 we obtain the identity

$$(7)\qquad R_\lambda-R_\mu^*=(\lambda-\bar\mu)R_\mu^*R_\lambda,\qquad\qquad \lambda,\mu\in\mathbb{C}\backslash\mathbb{R}.$$

Conversely, if $R_\lambda\in\mathbf{L}(\mathfrak{L}(N))$ satisfies (7), then *(ii)*, *(iii)* and the inequality in *(i)* follow.

For a selfadjoint relation (i.e., multivalued linear operator) A in a Hilbert space $\mathfrak{H}$, $A(0)=\{\varphi\in\mathfrak{H}\,|\,\{0,\varphi\}\in A\}$ stands for the multivalued part of A and $A_s=A\cap(\mathfrak{H}\ominus A(0))^2$ is the operator part of A, which is a (densely defined) selfadjoint operator in $\mathfrak{H}\ominus A(0)$ with $\mathfrak{D}(A_s)=\mathfrak{D}(A)$. In the sequel we shall use without further explanation the various notions and results concerning linear relations developed in, e.g., [C] and [DS], where other references can be found. In the following theorem we assume that $N(\ell)\in\mathbf{N}^n$ and that it has the integral representation (1).

THEOREM 3. *The operator* R_λ, $\lambda\in\mathbb{C}\backslash\mathbb{R}$, *is the resolvent operator of the selfadjoint relation* A *in* $\mathfrak{L}(N)$ *given by*

$$(8)\qquad A=\{\{F,G\}\in\mathfrak{L}(N)^2\,|\,G(\ell)-\ell F(\ell)=c,\ \textit{all }\ell\in\mathbb{C}\backslash\mathbb{R},\ \textit{for some }c\in\mathbb{C}^n\}.$$

The multivalued part of A *is equal to* $A(0)=\mathfrak{L}(N)\cap\mathbb{C}^n=\mathfrak{R}(\mathcal{B})$. *Representing* $F(\ell)\in\mathfrak{L}(N)$ *by* (3), *we have*

$$(i)\qquad F\in\mathfrak{D}(A)\Leftrightarrow F(\ell)=\int_{\mathbb{R}}\frac{d\Sigma(t)f(t)}{t-\ell}\ \textit{for some } f\in L^2(d\Sigma)\textit{ with } tf(t)\in L^2(d\Sigma),$$

$$(ii)\qquad (A_sF)(\ell)=\ell F(\ell)+\int_{\mathbb{R}}d\Sigma(t)f(t)=\int_{\mathbb{R}}\frac{d\Sigma(t)\,t\,f(t)}{t-\ell},\qquad F\in\mathfrak{D}(A).$$

Proof. Proposition 2 implies that $B=\{\{R_\mu H,(I+\mu R_\mu)H\}\mid H\in\mathfrak{L}(N)\}$ is a selfadjoint relation in $\mathfrak{L}(N)$ which is independent of $\mu\in\mathbb{C}\backslash\mathbb{R}$. We have $R_\lambda=(B-\lambda)^{-1}$, i.e., R_λ is the resolvent of B. It is easy to see that B is contained in the righthand side of (8). To prove the inclusion $A\subset B$, take $\{F,G\}\in\mathfrak{L}(N)^2$ for which $G(\ell)-\ell F(\ell)=c$ for some $c\in\mathbb{C}^n$ and put $H(\ell)=G(\ell)-\mu F(\ell)$. Then $H(\ell)\in\mathfrak{L}(N)$, $(R_\mu H)(\ell)=F(\ell)$ and $((I+\mu R_\mu)H)(\ell)=G(\ell)$, and hence, $\{F,G\}\in B$. This proves $B=A$ and the first part of the theorem. The item about the multivalued part of A is clear. Let $F(\ell)\in\mathfrak{D}(A)$ have the representation (3). Then $\mathcal{B}c=0$, since $\mathfrak{D}(A)$ is orthogonal to $A(0)$. By (8) there exists a constant $d\in\mathbb{C}^n$ so that $\ell F(\ell)+d\in\mathfrak{L}(N)$, and hence for some $e\in\mathbb{C}^n$, $g\in L^2(d\Sigma)$

$$\int_{\mathbb{R}}\frac{d\Sigma(t)\ell f(t)}{t-\ell}+d=\mathcal{B}e+\int_{\mathbb{R}}\frac{d\Sigma(t)\,g(t)}{t-\ell}.$$

Applying R_α with $\alpha\in\mathbb{C}\backslash\mathbb{R}$ to both sides, we obtain

$$\int_{\mathbb{R}}\frac{d\Sigma(t)(g(t)-tf(t))}{(t-\ell)(t-\alpha)}=0,$$

which implies $g(t)=tf(t)$ in $L^2(d\Sigma)$, and hence $tf(t)\in L^2(d\Sigma)$. Therefore F has the representation in (*i*), the integral $\int_{\mathbb{R}}d\Sigma(t)f(t)$ exists and

$$\ell F(\ell)=\int_{\mathbb{R}}\frac{d\Sigma(t)\ell f(t)}{t-\ell}=-\int_{\mathbb{R}}d\Sigma(t)f(t)+\int_{\mathbb{R}}\frac{d\Sigma(t)\,t\,f(t)}{t-\ell}.$$

This shows that $\ell F(\ell)+\int_{\mathbb{R}}d\Sigma(t)f(t)\in\mathfrak{L}(N)$ and is orthogonal to $A(0)$, so that (*ii*) has been proved. As to the converse in (*i*), let $F(\ell)\in\mathfrak{L}(N)$ have the indicated representation for some $f\in L^2(d\Sigma)$ with $tf(t)\in L^2(d\Sigma)$. Then $\int_{\mathbb{R}}d\Sigma(t)f(t)$ exists and

$$\ell F(\ell)+\int_{\mathbb{R}}d\Sigma(t)f(t)=\int_{\mathbb{R}}\frac{d\Sigma(t)\,t\,f(t)}{t-\ell},$$

where the righthand side belongs to $\mathfrak{L}(N)$. Therefore, by (8), $F(\ell)\in\mathfrak{D}(A)$.

COROLLARY 4. *If* $F(\ell)\in\mathfrak{L}(N)$ *has representation* (3), *then* $\ell F(\ell)\in\mathfrak{L}(N)$ *if and only if* $\mathcal{B}c=0$, $tf(t)\in L^2(d\Sigma)$ *and* $\int_{\mathbb{R}}d\Sigma(t)f(t)\in\mathfrak{R}(\mathcal{B})$. *Consequently, if* S *stands for multiplication by* ℓ *in* $\mathfrak{L}(N)$, *i.e.*,

$$S=\{\{F,G\}\in\mathfrak{L}(N)^2\mid G(\ell)=\ell F(\ell),\ \textit{all}\ \ell\in\mathbb{C}\backslash\mathbb{R}\},$$

then S *is a closed symmetric operator in* $\mathfrak{L}(N)$ *and a restriction of* A.

Except for the closedness of S, Corollary 4 follows from Theorem 3 and its proof. The reproducing property (5) of the kernel $\mathsf{K}_N(\ell,\lambda)$ implies that S is closed.

It follows from Theorem 3 and Corollary 4, that in the Hilbert space $\mathbb{C}^n_{\mathcal{B}} \oplus L^2(d\Sigma)$ the symmetric operator S and the selfadjoint relation A take the form

$$S = \{\{f, d+g\} \in (\mathbb{C}^n_{\mathcal{B}} \oplus L^2(d\Sigma))^2 \mid f(t),\ tf(t) \in L^2(d\Sigma),\ \int_{\mathbb{R}} d\Sigma(t) f(t) \in \Re(\mathcal{B}),$$
$$d = -\int_{\mathbb{R}} d\Sigma(t) f(t),\ g(t) = tf(t)\},$$

$$A = \{\{f, d+g\} \in (\mathbb{C}^n_{\mathcal{B}} \oplus L^2(d\Sigma))^2 \mid f(t),\ tf(t) \in L^2(d\Sigma),\ d \in \Re(\mathcal{B}),\ g(t) = tf(t)\}.$$

Hence, if $\mathcal{B} = 0$, then A is multiplication by t in $L^2(d\Sigma)$, and S is the restriction of A, given by the domain restriction $\int_{\mathbb{R}} d\Sigma(t) f(t) = 0$.

We recall the notion of a Q function, see [KL1,2], [LT]. Let H_0 be a closed simple symmetric operator in a Hilbert space $\mathfrak{H}$ with finite and equal defect numbers (n,n) and let H be a canonical selfadjoint extension of H_0. Let Γ_ℓ be a function on $\rho(H)$ with values in $\mathbf{L}(\mathbb{C}^n, \mathfrak{H})$ which for $\ell, \lambda \in \rho(H)$ satisfies

(a) $\quad \Gamma_\ell$ maps $\mathbb{C}^n$ one to one and continuously onto the null space $\nu(H_0^* - \ell)$,

(b) $\quad \dfrac{\Gamma_\ell - \Gamma_\lambda}{\ell - \lambda} = (H-\ell)^{-1}\Gamma_\lambda = (H-\lambda)^{-1}\Gamma_\ell.$

As $I + (\ell - \lambda)(H-\ell)^{-1}$ is a boundedly invertible mapping from $\mathfrak{H}$ onto $\mathfrak{H}$ and in particular from $\nu(H_0^* - \lambda)$ onto $\nu(H_0^* - \ell)$, the existence of such a Γ_ℓ follows from the first equality in (b), after choosing one Γ_λ satisfying (a). The Q function associated with H_0 and H (and the choice of Γ_ℓ) is the $n{\times}n$ matrix function $Q(\ell)$ defined on $\rho(H)$ which is a solution of the equation

$$\frac{Q(\ell) - Q(\lambda)^*}{\ell - \bar{\lambda}} = \Gamma_\lambda^* \Gamma_\ell, \qquad \ell, \lambda \in \rho(H).$$

It follows that $Q(\ell)$ is defined up to a constant hermitian $n{\times}n$ matrix, $Q(\ell) \in \mathbf{N}^n$ and $\operatorname{Im} Q(\ell)$ is invertible for all $\ell \in \mathbb{C}\backslash\mathbb{R}$. (Note that if $N(\ell) \in \mathbf{N}^n$ and $\operatorname{Im} N(\ell)$ is invertible for one $\ell \in \mathbb{C}\backslash\mathbb{R}$, then $\operatorname{Im} N(\ell)$ is invertible for all $\ell \in \mathbb{C}\backslash\mathbb{R}$). For a fixed $\mu \in \mathbb{C}\backslash\mathbb{R}$ $Q(\ell)$ can be written as

$$(9) \qquad Q(\ell) = Q(\mu) + (\ell - \mu)\Gamma_\mu^*\left(I + (\ell - \bar{\mu})(H-\ell)^{-1}\right)\Gamma_\mu, \qquad \ell \in \mathbb{C}\backslash\mathbb{R}.$$

If a simple symmetric operator H_0 and a canonical selfadjoint extension H, and a simple symmetric operator H_0' and a canonical selfadjoint extension H', define the same Q function, then H and H', and H_0 and H_0', are unitarily equivalent, see [LT], in particular the argument preceding Corollary 2.3. Now we return to the situation of a function $N(\ell) \in \mathbf{N}^n$ with symmetric operator S and selfadjoint extension A in the Hilbert space $\mathfrak{L}(N)$, as described in Theorem 3 and Corollary 4.

THEOREM 5. *Let $N(\ell)\in\mathbf{N}^n$ with* $\operatorname{Im}N(\alpha)$ *invertible for some* $\alpha\in\mathbb{C}\backslash\mathbb{R}$. *Then S is simple with defect numbers (n,n) and $N(\ell)$ is the Q function of S and its canonical selfadjoint extension A.*

Proof. On account of (5) the spaces $\Re(S-\alpha)$ and $\mathsf{K}_N(.,\alpha)\mathbb{C}^n$, $\alpha\in\mathbb{C}\backslash\mathbb{R}$, are orthogonal. Their sum is all of $\mathfrak{L}(N)$, since each element $F(\ell)$ in $\mathfrak{L}(N)$ can be written as

$$(10)\qquad F(\ell)=(\ell-\alpha)G_\alpha(\ell)+\mathsf{K}_N(\ell,\alpha)\mathsf{K}_N(\alpha,\alpha)^{-1}F(\alpha),$$

for some $G_\alpha(\ell)\in\mathfrak{L}(N)$. To see this, define $G_\alpha(\ell)$ by

$$G_\alpha(\ell)=\frac{F(\ell)-\mathsf{K}_N(\ell,\alpha)\,\mathsf{K}_N(\alpha,\alpha)^{-1}F(\alpha)}{\ell-\alpha}$$
$$=\frac{F(\ell)-F(\alpha)}{\ell-\alpha}+\frac{\mathsf{K}_N(\alpha,\alpha)-\mathsf{K}_N(\ell,\alpha)}{\ell-\alpha}\mathsf{K}_N(\alpha,\alpha)^{-1}F(\alpha).$$

Then, since $\mathfrak{L}(N)$ is resolvent invariant, $G_\alpha(\ell)\in\mathfrak{L}(N)$ and by (10) $\ell G_\alpha(\ell)\in\mathfrak{L}(N)$. Hence, we obtain

$$(11)\qquad \mathfrak{L}(N)=\Re(S-\alpha)\oplus\mathsf{K}_N(.,\alpha)\mathbb{C}^n,\qquad \alpha\in\mathbb{C}\backslash\mathbb{R}.$$

It follows that the null space $\nu(S^*-\bar{\alpha})$, $\alpha\in\mathbb{C}\backslash\mathbb{R}$, is spanned by the column vectors of $\mathsf{K}_N(.,\alpha)$, and is therefore n dimensional. This implies that the defect numbers of S are (n,n). By (11), to prove the simplicity of S, it suffices to show that $\mathfrak{L}(N)$ is equal to the closed linear span of $\mathsf{K}_N(.,\alpha)\mathbb{C}^n$, where α runs through $\mathbb{C}\backslash\mathbb{R}$. But this is a direct consequence of the reproducing kernel property. In order to show that $N(\ell)$ is a Q function for S and A we define the mapping Γ_ℓ from $\mathbb{C}^n$ onto $\nu(S^*-\ell)$ by $c\to\mathsf{K}_N(.,\bar{\ell})c$. It is clear that (a) and (b) are satisfied and that

$$[\Gamma_\ell c,\Gamma_\lambda d]_{\mathfrak{L}(N)}=[\mathsf{K}_N(.,\bar{\ell})c,\mathsf{K}_N(.,\bar{\lambda})d]_{\mathfrak{L}(N)}=d^*\frac{N(\ell)-N(\lambda)^*}{\ell-\bar{\lambda}}c,\quad c,d\in\mathbb{C}^n,$$

which shows that $N(\ell)$ is a Q function for S and A. This concludes the proof.

The following corollary is related to the results of Kreĭn (see [K1], Theorems 1 and 2), which state that a closed simple symmetric operator with equal and finite defect numbers is isomorphic to multiplication by the complex variable in a resolvent invariant Hilbert space of locally meromorphic functions on $\mathbb{C}\backslash\mathbb{R}$. However, Kreĭn derived these results using module spaces and in his representation all the functions representing the elements of the space can in certain cases be entire, whereas in the representation of $\mathfrak{L}(N)$ here these functions must have singularities on $\mathbb{R}$.

COROLLARY 6. *Let H_0 be a simple closed symmetric operator in some Hilbert space with equal and finite defect numbers (n,n), and let H be a canonical selfadjoint extension. Then H_0 and H are isomorphic to S and A in* *the space $\mathfrak{L}(N)$, where $N(\ell)$ is the Q function of H_0 with respect to H (and hence belongs to the class $\mathbf{N}^n$).*

The fact that a function $N(\ell)\in\mathbf{N}^n$, $\operatorname{Im}N(\ell)$ invertible, is a Q function of a symmetric operator H_0 and a selfadjoint extension H is proved in [KL1,2] under the extra hypothesis $N(iy)/y\to 0$ as $y\to\infty$ (which ensures that H is densely defined) and for the general case in [LT].

We want to compare our model in $\mathfrak{L}(N)$, or equivalently in $\mathbb{C}^n_B\oplus L^2(d\Sigma)$ with the model developed in [KL1,2] and [LT], where a procedure is used that seems to go back at least to [K2]. Since $\nu(\operatorname{Im}N(\ell))$ is independent of $\ell\in\mathbb{C}\backslash\mathbb{R}$, a suitable decomposition of $\mathbb{C}^n$ makes it possible to write $N(\ell)=A+N_1(\ell)\oplus O$, where $A=A^*$ is a constant $n{\times}n$ matrix, O is the $(n-k){\times}(n-k)$ zero matrix and $N_1(\ell)\in\mathbf{N}^k$ with $\operatorname{Im}N_1(\alpha)$ invertible for all $\alpha\in\mathbb{C}\backslash\mathbb{R}$. Thus without loss of generality we may assume that $\operatorname{Im}N(\alpha)$ is invertible for some (and hence for all) $\alpha\in\mathbb{C}\backslash\mathbb{R}$. We consider the linear space $\mathfrak{L}$ of finite sums of the form $\sum\varepsilon_\ell x_\ell$, where $\ell\in\mathbb{C}\backslash\mathbb{R}$, $x_\ell\in\mathbb{C}^n$ and ε_ℓ is a symbol associated with each ℓ, endowed with the possibly degenerate nonnegative inner product

$$[\sum\varepsilon_\ell x_\ell,\sum\varepsilon_\lambda y_\lambda]_{\mathfrak{L}}=\sum y_\lambda^*\frac{N(\ell)-N(\lambda)^*}{\ell-\bar{\lambda}}x_\ell.$$

In $\mathfrak{L}$ we define the linear relations

$$S_0=\{\{\sum\varepsilon_\ell x_\ell,\sum\ell\varepsilon_\ell x_\ell\}\mid\sum x_\ell=0,\ \sum N(\ell)x_\ell=0\},$$

$$A_0=\{\{\sum\varepsilon_\ell x_\ell,\sum\ell\varepsilon_\ell x_\ell\}\mid\sum x_\ell=0\}.$$

In the Hilbert space completion $\tilde{\mathfrak{L}}$ of the quotient space $\mathfrak{L}/(\mathfrak{L}\cap\mathfrak{L}^\perp)$ these relations give rise to a simple closed symmetric operator $\tilde{S}$ with defect numbers (n,n) and a canonical selfadjoint extension $\tilde{A}$ of $\tilde{S}$, respectively, cf. [LT] and [DLS], p.74. The mapping $\iota:\sum\varepsilon_\ell x_\ell\mapsto\sum\mathsf{K}_N(.,\bar{\ell})x_\ell$ can be extended in the usual way to an isomorphism from $\tilde{\mathfrak{L}}$ onto $\mathfrak{L}(N)$ and with this isomorphism $\tilde{S}$ and $\tilde{A}$ are isomorphic to S and A. This follows from relations like

$$[\iota(\varepsilon_\ell x),\iota(\varepsilon_\lambda y)]_{\mathfrak{L}(N)}=y^*\frac{N(\ell)-N(\lambda)^*}{\ell-\bar{\lambda}}x=[\varepsilon_\ell x,\varepsilon_\lambda y]_{\mathfrak{L}}$$

and

$$\iota(S_0(\sum\varepsilon_\lambda x_\lambda))(\ell)=\iota(\sum\lambda\varepsilon_\lambda x_\lambda)(\ell)=\sum\frac{N(\ell)-N(\bar{\lambda})^*}{\ell-\lambda}\lambda x_\lambda=\ell\,\iota(\sum\varepsilon_\lambda x_\lambda)(\ell),$$

where we have used $N(\lambda)^*=N(\bar{\lambda})$, $\sum N(\lambda)x_\lambda=0$ and $\sum x_\lambda=0$. It is clear from the above that $d\to\varepsilon_\ell d$, $d\in\mathbb{C}^n$ induces a mapping Γ_ℓ from $\mathbb{C}^n$ to $\nu(\tilde{S}^*-\ell)$, which satisfies (a) and (b) (with H_0 and H replaced by $\tilde{S}$ and $\tilde{A}$, respectively). Hence, fixing $\mu\in\mathbb{C}\backslash\mathbb{R}$, we obtain from (9) that

$$N(\ell)=N(\mu)+(\ell-\mu)\Gamma_\mu^*(I+(\ell-\bar{\mu})(\tilde{A}-\ell)^{-1})\Gamma_\mu.$$

Finally, if we denote by $E_s(t)$ the spectral family of the selfadjoint operator part $\widetilde{A}_s$ in the Hilbert space $\widetilde{\mathfrak{L}}\ominus\widetilde{A}(0)$ and by P_0 the orthogonal projection of $\widetilde{\mathfrak{L}}$ onto $\widetilde{A}(0)$ and if, for simplicity, we take $\mu=i$, we find that the above representation for $N(\ell)$ leads to the Riesz–Herglotz–Nevanlinna representation (1) with $\mathcal{A}=\mathrm{Re}N(i)$, $\mathcal{B}=\Gamma_i^*P_0\Gamma_i$ and $d\Sigma(t)=(t^2+1)d\Gamma_i^*E_s(t)\Gamma_i$.

In the reproducing kernel Hilbert space $\mathfrak{L}(N)$ the operator S of multiplication by ℓ is closed and symmetric. It is possible to describe the adjoint S^* of S, and hence all the selfadjoint extensions of S, thereby generalizing Theorem 3. This leads to a description of all possible Q functions by means of fractional linear transformations, cf. [KL1], [LT]. As a further generalization one may consider in an analogous fashion the operator of multiplication by ℓ in de Branges spaces. This will yield an operator theoretic interpretation of several results in [dB2]. We hope to treat these matters elsewhere.

References

[A] N. Aronszajn, "Theory of reproducing kernels", Trans. Amer. Math. Soc., 68 (1950), 337–404.

[dB1] L. de Branges, "Perturbations of self-adjoint transformations", Amer. J. Math., 84 (1962), 543–560.

[dB2] L. de Branges, *Hilbert spaces of entire functions*, Prentice Hall, Englewood Cliffs, N.J., 1968 (French translation: *Espaces Hilbertiens de fonctions entières*, Masson et Cie, Paris 1972).

[C] E.A. Coddington, *Extension theory of formally normal and symmetric subspaces*, Mem. Amer. Math. Soc., 134, 1973.

[D] H. Dym, *J contractive matrix functions, reproducing kernel Hilbert spaces and interpolation*, Regional conference series in mathematics, 71, Amer. Math. Soc., Providence, R.I., 1989.

[DLS] A. Dijksma, H. Langer, H.S.V. de Snoo, "Hamiltonian systems with eigenvalue depending boundary conditions", Operator Theory: Adv. Appl., 35 (1988), 37–83.

[DS] A. Dijksma, H.S.V. de Snoo, "Self-adjoint extensions of symmetric subspaces", Pacific J. Math., 54 (1974), 71–100.

[K1] M.G. Kreĭn, "Fundamental aspects of the representation theory of hermitian operators with deficiency index (m,m)", Ukrain. Math. Zh., 1, no. 2, (1949), 3–66 (Russian) (English translation: Amer. Math. Soc. Transl., (2) 97 (1970), 75–143).

[K2] M.G. Kreĭn, "Hermitian-positive kernels on homogeneous spaces I", Ukrain. Math. Zh., 1, no. 4, (1949), 64–98 (Russian) (English translation: Amer. Math. Soc. Transl., (2) 34 (1963), 69–108).

[KL1] M.G. Kreĭn, H. Langer, "Über die Q-Funktion eines π-hermiteschen Operators im Raume Π_κ", Acta Sci. Math. (Szeged), 34 (1973), 191–230.

[KL2] M.G. Kreĭn, H. Langer, "Über einige Fortsetzungsprobleme, die eng mit der Theorie hermitescher Operatoren im Raume Π_κ zusammenhängen. I: Einige Funktionenklassen und ihre Darstellungen", Math. Nachr., 77 (1977), 187–236.

[LT] H. Langer, B. Textorius, "On generalized resolvents and Q-functions of symmetric linear relations (subspaces) in Hilbert space", Pacific J. Math., 72 (1977), 135–165.

MULTIVARIATE PREDICTION THEORY AND WEIGHTED INEQUALITIES

Marisela Domínguez

ABSTRACT

We give a continuous multivariate version of the Helson-Szegö theorem (Theorem A-1) [11] and we obtain an extension to multivariate processes of Ibragimov-Hayashi's theorem for the spectral density of a strongly mixing stationary gaussian process (Theorem A-2) [9,13], which is the continuous version of the Helson-Sarason theorem [10]. We also study two weighted inequalities for the Hilbert Transform for matricial weights, considering functions whose Fourier Transform vanishes in an interval.

Let R = {real numbers}, $R_1 = \{t \in R / t \geq 0\}$, $R_2 = \{t \in R / t < 0\}$.

Let C_{qxq} = {hermitian complex qxq matrices}. Denote the adjoint of A by A^*. C_{qxq} is a Hilbert space with scalar product

$$\langle\langle A, B \rangle\rangle_q = tr\ (AB^*).$$

Let $M_{qxq} = \left\{ C_{qxq} \text{ - valued finite Borel measures } \mu \text{ in } R \right\}$,

If $\mu \in M_{qxq}$, $d\mu = d\mu_s + w(x)dx$, where μ_s is singular with respect to Lebesgue measure. w will always be $(w_{kj})_{k,j=1,\dots,q}$

We will use the following notations:

$(\mu_{\alpha\beta})_{\alpha,\beta=1,2}$ is an hermitian matrix $\begin{pmatrix} \mu_{11} & \mu_{12} \\ \mu_{21} & \mu_{22} \end{pmatrix}$ where $\mu_{\alpha\beta} \in M_{qxq}$,

H^1 is the space of Hardy functions in the upper half plane,

H^1_{qxq} = {matrices $(h_{ik})_{i,k=1,\dots,q}$ such that $h_{ik} \in H^1$} ,

$e_t : R \to C$, $e_t(x) = e^{itx}$,

$\hat{\phi}$ is the Fourier transform of ϕ.

$$E(C_{qxq}) = \left\{ \phi : R \to C_{qxq} \ / \ \phi(x) = \sum_t e_t(x) \hat{\phi}(t) \text{ and } \hat{\phi} : R \to C_{qxq} \text{ has finite support} \right\}$$

$E_1(C_{qxq}) = \{\phi \in E(C_{qxq});\ \hat{\phi}(t) = 0 \text{ for all } t \in R_2\}$,

$E_2(C_{qxq}) = \{\phi \in E(C_{qxq});\ \hat{\phi}(t) = 0 \text{ for all } t \in R_1\}$ and

$E_{(t)}(C_{qxq}) = \{\phi = e_t\phi_1 + \phi_2 \text{ for } \phi_1 \in E_1(C_{qxq});\ \phi_2 \in E_2(C_{qxq})\}$ for $t \in R_1$.

Let $\mu \in M_{qxq}$ be fixed and $tr\mu = \text{trace}(\mu)$, following [18] we define

$$L^2_q(R,\mu) = \left\{\phi : R \to C_{qxq} \text{ such that } \int_R \phi(t)d\mu(t)\phi^*(t) \text{ exists}\right\}$$

where:

$$\left(\int_R \phi\ d\mu\ \psi^*\right)_{11} = \int_R \left(\phi \frac{d\mu}{dtr\mu} \psi^*\right)_{11} dtr\mu,$$

$$\frac{d\mu}{dtr\mu} = \left(\frac{d\mu_{kj}}{dtr\mu}\right)_{k,j=1,\dots,q}$$

Around 1963, independently, Rosenberg and Rozanov [16] proved that $L^2_q(R,\mu)$ is complete under the inner product

$$\langle\langle\phi\ ,\ \psi\rangle\rangle_\mu = tr\int_R \phi\ d\mu\ \psi^*\ .$$

Let $\mu \in M_{qxq}$ then $E(C_{qxq}) \subset L^2_q(R,\mu)$.

Definition. Let $\mu \in M_{qxq}$ μ is matricially positive if for every Borel set $\Delta \subset R$ $\langle\langle A.\mu(\Delta), A\rangle\rangle \geq 0$.

Definition. $(\mu_{\alpha\beta})_{\alpha,\beta=1,2}$ is matricially positive if for all $\Delta \subset R$, for all $A_1, A_2 \in C_{qxq}$

$$\sum_{\alpha=1}^{2} \sum_{\beta=1}^{2} \langle\langle A_\alpha \cdot \mu_{\alpha\beta}(\Delta); A_\beta\rangle\rangle_q \geq 0.$$

For given $(\mu_{\alpha\beta})_{\alpha,\beta=1,2}$, for every $\phi_1, \phi_2 \in E(C_{qxq})$, let

$$Mq(\phi_1, \phi_2) = \sum_{\alpha=1}^{2} \sum_{\beta=1}^{2} \langle\langle\phi_\alpha, \phi_\beta\rangle\rangle_{\mu_{\alpha\beta}}$$

Proposition. $(\mu_{\alpha\beta})_{\alpha,\beta=1,2}$ is matricially positive if $Mq(\phi_1, \phi_2) \geq 0$ for all $\phi_1, \phi_2 \in E(C_{qxq})$.

Definition. $(\mu_{\alpha\beta})_{\alpha,\beta=1,2}$ is matricially weakly positive if $Mq/E_1(C_{qxq}) X E_2(C_{qxq}) \geq 0$.

Lemma 1. [8] $(\mu_{\alpha\beta})_{\alpha,\beta=1,2}$ is matricially positive if and only if for every $k,j\in\{1,\dots,q\}$

$$\begin{pmatrix} \mu_{11_{kk}} & \mu_{12_{kj}} \\ \mu_{21_{jk}} & \mu_{22_{jj}} \end{pmatrix} \geq 0 .$$

The property holds for matricially weakly positive matrices.

Theorem 1. [2,4,5] $(\mu_{\alpha\beta})_{\alpha,\beta=1,2}$ is matricially weakly positive if and only if there is $h\in H^1_{qxq}$ such that

$$\begin{pmatrix} \mu_{11} & \mu_{12} + h.dx \\ \mu_{21} + h^*.dx & \mu_{22} \end{pmatrix} \geq 0.$$

Let H be the Hilbert transform defined linearly in $E(C_{qxq})$ by $H(\phi) = -i\phi$ if $\phi\in E_1(C_{qxq})$ and $H(\phi) = i\phi$ if $\phi\in E_2(C_{qxq})$.

L^2 weighted problems for the Hilbert transform have been studied by several authors [1,6,11,12,14,15]. Arocena and Cotlar[3] established a relation between the deep work of P. Koosis and weak positivity.

Let $a \geq 1$, $\mu,\upsilon\in M_{qxq}$, $\mu \geq 0$, $\upsilon \geq 0$ we will say that $(\mu,\upsilon)\in R_{qxq}(t,a)$ when for all $\phi\in E_{(t)}(C_{qxq})$

$$tr\int_R \left(H\phi\right)d\mu\left(H\phi\right)^* \leq a.tr\int_R \phi\, d\upsilon\, \phi^*,$$

We will characterize the elements of $R_{qxq}(t,a)$.

This can be reformulated as a prediction theory problem. Let (Ω,A,P) be a probability space, $(X_t)_{t\in R}$ a q-variated weakly stationary process, $X_t=(X_{t(1)},\dots,X_{t(q)})$, $X_{t(j)}\in L^2(\Omega,A,P)$ for $j=1,\dots,q$ and the covariance depends only on t-s: $\Gamma_{t-s} = (E(X_{t(i)}\overline{X_{s(j)}}))_{i,j=1,\dots,q}$ The Bochner theorem [18] assures the existence of the spectral measure of the process $\mu\in M_{qxq}$, which is positive hermitian. The maximal correlation coefficient is

$$\rho_t(\mu) = \sup_{\substack{\phi_1\in E_1(C_{qxq});\ \phi_2\in E_2(C_{qxq}) \\ <<\phi_1;\phi_1>>_\mu = <<\phi_2;\phi_2>>_\mu = 1}} |<<e_t\phi_1;\phi_2>>_\mu|$$

$\rho_t(\mu)$ can be defined for any $\mu\in M_{qxq}$. The problem is to characterize the $\mu\in M_{qxq}$ such that $\rho_t(\mu) \leq r$ for some $r\in(0,1)$.

Definition. $(X_t)_{t\in R}$ is linearly completely regular if $\lim_{t\to\infty} \rho_t(\mu) = 0$.

Definition. g is an Ibragimov-Hayashi function if there exist real bounded functions u and v on R with $||v||_\infty < \pi/2$, and an entire function Γ of exponential type bounded on the real axis and zero free in the upper halfplane such that

$$w(x) = |q(x)|^2 = (x^2 + 1)|\Gamma(x)|^2 \exp(u(x) + \tilde{v}(x))$$

where $q\in H^2$ is outer and $\tilde{v}$ denotes the harmonic conjugate of v.

Let IH = {Ibragimov-Hayashi functions}, [9,13].

Lemma 2. [7] $w\in IH$ if and only if

$$w(x) = |h(x)|\ e^{u(x)}$$

where $h\in H^1$, u and v are real bounded functions on R,

$$v(x) = -\arg(e_{-t}(x)\ h(x))$$

for a $t > 0$ and $||v||_\infty < \pi/2$.

Let $V(r) = \arcsin r < \pi/2$ and

$U(r,v) = \operatorname{arccosh}\left(\frac{\cos v}{\sqrt{1-r^2}}\right)$ if $|v| \le V(r)$.

Theorem A. Let $\mu\in M_{qxq}$, $\mu \ge 0$, and $d\mu = d\mu_s + w(x)dx$. Let

$$\omega_{\mu kj}(x) = \frac{\sqrt{w_{kk}(x)w_{jj}(x)}}{|w_{kj}(x)|},\quad m_\mu = \operatorname*{ess\,inf}_{k;j;x}\{\omega_{\mu kj}(x)\},\quad M_\mu = \operatorname*{ess\,sup}_{k;j;x}\{\omega_{\mu kj}(x)\}.$$

1) Let $r\in(0,1)$, $t \in R_1$, and $r_{kj}(x) = r\ \omega_{\mu kj}(x)$. If $r_{kj}(x) < 1$ then the following conditions are equivalent:

a) $(\mu,\mu) \in R_{qxq}(t,a)$ where $a = \frac{1+r}{1-r}$.

b) $\rho_t(\mu) \le r$.

c) $\begin{pmatrix} r\mu & e_t\mu \\ e_{-t}\mu & r\mu \end{pmatrix}$ is weakly positive.

d) $d\mu = w(x)dx$ and for all $k,j\in\{1,\dots,q\}$ there exist functions

$$g_{kj}(x) = |h_{kj}(x)|\, e^{u_{kj}(x)}$$

where $h_{kj} \in H^1$ such that:

$$g_{kk} \in IH, \qquad w_{kk}(x) = g_{kk}(x),$$

$$|w_{kj}(x)| = \sqrt{(1-r^2)\ (g_{kj}(x))^2 + r^2\ g_{kk}(x)g_{jj}(x)},$$

$|u_{kj}(x)| \le U(r_{kj}(x), v_{kj}(x))$ and $|v_{kj}(x)| \le V(r_{kj}(x)) < \pi/2$.

where $\quad v_{kj}(x) = -\arg(e_{-t}(x)h_{kj}(x)e^{-i.\arg w_{kj}(x)})$.

2) Let $0 < m_\mu \le M_\mu < \infty$. We have that

$\lim_{t\to\infty} \rho_t(\mu) = 0$ if and only if

for all $\varepsilon > 0$, there exist $t \in R_1$ and $r_\varepsilon < 1/M_\mu$ such that for all $r < r_\varepsilon$, for all $k,j\in\{1,\dots,q\}$ there exist functions

$$g_{\varepsilon kj} = |h_{\varepsilon kj}|\, e^{u_{\varepsilon kj}}$$

where $h_{\varepsilon kj} \in H^1$ that satisfy

$$g_{\varepsilon kk} \in IH, \qquad w_{kk}(x) = g_{\varepsilon kk}(x),$$

$$|w_{kj}(x)| = \sqrt{(1 - r^2)(g_{\varepsilon kj}(x))^2 + r^2\, g_{\varepsilon kk}(x)g_{\varepsilon jj}(x)}$$

and for $v_{\varepsilon kj}(x) = -\arg(e_{-t}(x)h_{\varepsilon kj}(x)e^{-i.\arg w_{kj}(x)})$

$$||u_{\varepsilon kj}||_\infty + ||v_{\varepsilon kj}||_\infty < \varepsilon.$$

3) Let $0 < m_\mu \le M_\mu < \infty$, $\{r_t\}_{t\in R_1} \subset (0,1)$ with $\lim_{t\to\infty} r_t = 0$.

$\rho_t(\mu) = O(r_t)$, $\quad t \to \infty$, $\quad$ if and only if

there exist s and $t_o \in R_1$ such that for every $t \ge t_o$, for $k,j\in \{1,\dots,q\}$ there exist functions

$$g_{tkj} = |h_{tkj}(x)|\; e^{u_{tkj}(x)}$$

where $h_{tkj} \in H^1$ such that:

$$g_{tkk} \in IH, \qquad w_{kk}(x) = g_{tkk}(x),$$

$$|w_{kj}(x)| = \sqrt{(1 - (sr_t)^2)(g_{tkj}(x))^2 + (sr_t)^2\, g_{tkk}(x)g_{tjj}(x)},$$

$$||v_{tkj}||_\infty = O(r_t) \quad\text{and}\quad \frac{\cosh|u_{tkj}|}{\cos v_{tkj}} - 1 = O(r_t^2), \qquad t \to \infty.$$

where $v_{tkj}(x) = -\arg(e_{-t}(x)h_{tkj}(x)e^{-i.\arg w_{kj}(x)})$.

4) If $w_{kj}(x) = |w_{kj}(x)|e^{ixT_{kj}}$ for a real number T_{kj}, then $g_{kj} \in IH$, $g_{\varepsilon kj} \in IH$ and $g_{tkj} \in IH$.

Proof. We will only prove part 1).The equivalence between a), b) and c) is proved as in the onedimensional case [4]. Suppose c) holds, by Lemma 1 and Theorem 1 there exist $h = (h_{kj}) \in H^1_{qxq}$ such that for every $\Delta \subset R$

$$\left| e_t\mu_{kj}(\Delta) + \sqrt{1 - r^2}\int_\Delta h_{kj}(x)dx \right|^2 \le r^2\mu_{kk}(\Delta)\mu_{jj}(\Delta)$$

If Δ is a set of Lebesgue measure zero then $\mu_{kk}(\Delta) \le r\ \mu_{kk}(\Delta)$. Thus $\mu_{kk}(\Delta) = 0$, and then $\mu_{kj}(\Delta) = 0$.

$$\left| e_t(x)w_{kj}(x) + \sqrt{1 - r^2}\ h_{kj}(x) \right|^2 \le r^2 w_{kk}(x)w_{jj}(x) = (r_{kj}(x)|w_{kj}(x)|)^2$$

$$0 \ge |w_{kj}|^2 - 2\mathrm{Re}(\sqrt{1 - r^2}\ h_{kj}e_{-t}\overline{w_{kj}}) + (1 - r^2)|h_{kj}|^2 - r_{kj}^2|w_{kj}|^2 =$$

$$= (1 - r_{kj}^2)|w_{kj}|^2 - 2|w_{kj}|\sqrt{1 - r^2}\ (\mathrm{Re}\phi_{kj}) + (1 - r^2)|h_{kj}|^2$$

where $\phi_{kj}(x) = e_{-t}(x)h_{kj}(x)e^{-i.\arg w_{kj}(x)}$.

Let $v_{kj} = -\arg\phi_{kj}$, $u_{kj}(x) = \log\left(\dfrac{\sqrt{1 - r_{kj}^2}\ |w_{kj}|}{\sqrt{1 - r^2}\ |h_{kj}|}\right)$ and

$$\zeta_{\alpha kj} = \frac{\sqrt{1 - r^2}}{\sqrt{1 - r_{kj}^2}}\ |h_{kj}|\ \exp\left((-1)^\alpha \operatorname{arcosh}\left(\frac{\cos v_{kj}}{\sqrt{1 - r_{kj}^2}}\right)\right) \quad \text{for } \alpha = 1,2$$

then $\zeta_{1kj} \le |w_{kj}| \le \zeta_{2kj}$

that is

$$|u_{kj}(x)| \le \operatorname{arcosh}\left(\frac{\cos v_{kj}(x)}{\sqrt{1 - r_{kj}^2}}\right) = U\big(r_{kj}(x); v_{kj}(x)\big).$$

Let $g_{kj}(x) = |h_{kj}(x)|e^{u_{kj}(x)}$ then

$$\sqrt{1 - r^2}\ g_{kj} = \sqrt{1 - r_{kj}^2}\ |w_{kj}|.$$

If $k = j$ then $r_{kj} = r$, and so $g_{kk} = w_{kk}$.

Also

$$\sqrt{1 - r^2}\ g_{kj} = \sqrt{|w_{kj}|^2 - r_{kj}^2|w_{kj}|^2} = \sqrt{|w_{kj}|^2 - r^2 w_{kk}w_{jj}}$$

thus

$$|w_{kj}(x)| = \sqrt{(1 - r^2)\ (g_{kj}(x))^2 + r^2\ g_{kk}(x)g_{jj}(x)}\ .$$

And $\left| e_t|w_{kj}|e^{i\arg w_k} - \sqrt{1 - r^2}\ h_{kj} \right| \le r_{kj}|w_{kj}|$,

thus

$$\left|\frac{1}{r_{kj}} - \sqrt{1 - r^2}\ \phi_{kj}/r_{kj}|w_{kj}|\right| \le 1.$$

We have that

$$v_{kj} = -\arg(\sqrt{1 - r^2}\ \phi_{kj}/r_{kj}|w_{kj}|).$$

Then $|v_{kj}| < \pi/2$ and $\sin|v_{kj}| \le r_{kj}$. So $|v_{kj}(x)| \le V(r_{kj}(x)) < \pi/2$.

And the proof that c) implies d) is finished. That d) implies c) follows in a similar way.

If there exists $T_{kj} \in R$ such that $w_{kj}(x) = |w_{kj}(x)| e^{ixT_{kj}}$, then $v_{kj}(x) = - \arg(e_{-T_{kj}-t}(x) h_{kj}(x))$. Applying lemma 2 the required representation is obtained.

Theorem B. Let $\mu, \nu \in M_{qxq}$, $\mu \geq 0$, $\nu \geq 0$, $d\mu = w(x)dx$ and $d\nu = y(x)dx$. Let $a > 1$, $t \in R_1$ and

$$r_{a;kj}(x) = \frac{\sqrt{(ay_{kk}(x) - w_{kk}(x))(ay_{jj}(x) - w_{kjj}(x))}}{|ay_{kj}(x) + w_{kj}(x)|} < 1.$$

$(\mu, \nu) \in R_{qxq}(t, a)$ iff for all $k, j \in \{1, \ldots, q\}$ there exist $g_{kj} = |h_{kj}| e^{u_{kj}}$ where $h_{kj} \in H^1$ such that

$$g_{kk} \in IH, \qquad \sqrt{y_{kk} w_{kk}} = g_{kk},$$

$$|ay_{kj} + w_{kj}| = \sqrt{4ag_{kj}^2 + (ay_{kk} - w_{kk})(ay_{jj} - w_{jj})},$$

$$|u_{kj}(x)| \leq U(r_{a;kj}(x), v_{kj}(x)) \quad \text{and} \quad |v_{kj}(x)| \leq V(r_{a;kj}(x)) < \pi/2$$

where $v_{kj}(x) = - \arg(e_{-t}(x) h(x) e^{-ix.\arg(ay_{kj}(x) + w_{kj}(x))})$.

If $\arg(ay_{kj}(x) + w_{kj}(x)) = xT_{kj}$ for $T_{kj} \in R$, then $g_{kj} \in IH$.

Onedimensional versions of theorems A and B are given in [6,7]. In [6], using the classical lifting theorem of weakly positive matrices [2,4,5], we gave a generalization of the theorem of Helson-Sarason [10] in a form similar to the extension given by Pourahmadi [17] and solve this problems for tempered Borel measures in R. In the latter case we do not obtain Ibragimov-Hayashi functions.

All of the preceding analysis which was carried out in R can also be done in Z and $[0, 2\pi]$. In [8] it was used to give extensions of the Helson-Szegö theorem and the Helson-Sarason theorem for multivariate processes [6,10,11].

This research was supported by the Universidad Central de Venezuela.

References

[1] ADAMS, E.; On weighted norm inequalities for the Hilbert Transform of functions with moments zero; Transactions of the American Mathematical Society, Vol. 272, no.2,(1982), 487 -500.

[2] AROCENA, R., and M. COTLAR; Continuous generalized Toeplitz kernels in R; Portugaliae Mathematica; Vol.39, Fasc. 1-4, (1980), 419-434.

[3] AROCENA, R., and M. COTLAR; A generalized Herglotz-Bochner theorem and L^2 - weighted inequalities with finite measures; in Conference Harmonic Analysis in honor of Prof. A. Zygmund, Chicago 1981; Eds. A.P.; Intl. Math. Ser. 1982; 258 - 269.

[4] AROCENA, R., M. COTLAR and C. SADOSKY; Weighted inequalities in L^2 and lifting properties; Adv. Math, Sup.Studies,V.7A,(1981), 95-128.

[5] COTLAR, M. and C. SADOSKY; The Helson-Szegö theorem and a related class of modified Toeplitz kernels, Proc.Symp. Pure Math.AMS. 35: I (1979), 383-407

[6] DOMINGUEZ, M.; Weighted inequalities for the Hilbert Transform and the adjoint operator in the continuous case, Studia Mathematica, 1990.

[7] DOMINGUEZ, M.; Rate of convergence of the maximal correlation coefficient in the continuous case, Revista Brasileira de Probabilidade e Estatística; Vol. 3, Nº 2, (1989).

[8] DOMINGUEZ, M.; A matricial extension of the Helson-Sarason theorem and a characterization of some multivariate linearly completely regular processes; Journal of Multivariate Analysis, Vol.31, Nº2, (1989), 289-310.

[9] HAYASHI, E.; The spectral density of a strongly mixing stationary gaussian process; Pac. Jour. of Math.; Vol.96, Nº2, (1981), 343-359.

[10] HELSON, H., D. SARASON; Past and Future;Math.Scand.,21,(1967), 5 - 16.

[11] HELSON, H., and G. SZEGÖ; A problem in prediction theory; Am. Math. Pura Appl.51, (1960), 107-138.

[12] HUNT, R., B. MUCKENHOUPT and R. L.WHEEDEN; Weighted norm inequalities for the conjugate function and Hilbert Transform. Trans. Amer. Math. Soc. 176, 227 - 251.

[13] IBRAGIMOV, I.A.; On conditions of strong mixture for stationary Gaussian processes. Dokl. AN SSSR, 161, 1, (1965), 33-36 (Russian).

[14] KOOSIS, P.; Weighted quadratic means of Hilbert transforms. Duke Math. J. 38, (1971), 609 - 634.

[15] KOOSIS, P.; Moyennes quadratiques de transformées de Hilbert et fonctions de type exponentiel. Comp. Rend. Acad. Sc. Paris 276 (1973), A-1201- 1204.

[16] MASANI, P.; Recent trends in multivariate prediction theory. Multivariate Analysis (P. R. Krishnaiah, Ed.), Ac. Press, New York; (1966); 351 - 382.

[17] POURAHMADI, M.; A matricial extension of the Helson-Szegö theorem and its application in multivariate prediction; Journal of Multivariate Analysis 16; (1985); 265-275.

[18] ROSENBERG, M.; The square integrability of matrix valued functions with respect to a non negative hermitian measure; Duke Math. J. 31, (1964), 291-298.

SOME PROPERTIES OF THE REGULAR MINIMAL UNITARY DILATION OF A PAIR OF COMMUTING CONTRACTIVE OPERATORS. APPLICATION TO THE QUARTER PLANE MARKOVIAN REPRESENTATION PROBLEM OF STATIONARY PROCESSES ON $\mathbb{Z}^2$.

PHILIPPE LOUBATON

ABSTRACT.

Let T_1 and T_2 be two commuting contractions on the Hilbert space X and let (U_1,U_2,K) denote the regular minimal unitary dilation of (T_1,T_2,X). Then, we show that the sequence of spaces $(U_1^m U_2^n X)_{(m,n)\in Z^2}$ possesses Markovian properties analogous to those introduced in [8]. Starting from this observation, we establish some new structural properties of (U_1,U_2,K). Finally, we treat from the geometric point of view the quarter-plane Markovian representation problem of a two-parameter stationary process.

INTRODUCTION.

Several works have been devoted to the study of the minimal unitary dilations of a pair of commuting contractive operators (T_1,T_2). The existence of such a dilation was established in [1]. Unlike with a single contraction, there is no unicity of the minimal unitary dilation and some recent works have dealt with the construction of the set of these dilations ([4]). The notion of regular minimal unitary dilation was introduced by Brehmer ([3]) who showed that any pair of commuting contractions has a regular minimal unitary dilation, unique up to an isomorphism. Its structure was studied by Halperin ([7]).

In the first Part of this paper, we establish some new structural properties of the regular minimal unitary dilation (U_1,U_2) of (T_1,T_2). We denote by X and K the Hilbert spaces on which (T_1,T_2) and (U_1,U_2) are defined respectively. Our work is based on the fact that X, as a subspace of K, possesses splitting properties analogous to those of two-parameter Markov processes defined by the following state equation

$$(0\text{-}1)\quad X_{m+1,n+1} = F_1 X_{m,n+1} + F_2 X_{m+1,n} - F_1 F_2 X_{m,n} + L \nu_{m,n}$$

where F_1 and F_2 are two commuting asymptotically stable matrices and ν is a white noise. These processes was used by Attasi in [2] and studied in a more general frame in [8]. They are characterized by the fact that the variables $X_{m,0}$ for $m\geq 0$ (resp. $X_{0,n}$ for $n\geq 0$) are conditionally orthogonal given the space generated by the variables $X_{k,l}$ for $k\leq 0$ and $l\in\mathbb{Z}$ (resp. for $k\in\mathbb{Z}$ and $l\leq 0$). Starting from the above mentionned splitting properties, we show that some of the results by Slocinski ([14]) concerning the case of doubly commuting T_1 and T_2 remain valid. In Part II, we consider the case where T_1 and T_2 belong to the class $C_{0,.}$ and we derive some interesting properties of the associated functional models. Part III is devoted to the representation problem of a multivariate stationary process $y = (y_{m,n})_{(m,n)\in Z^2}$ on $\mathbb{Z}^2$ by means of a process satisfying a state equation of type (0-1). The following weak stochastic realization problem was solved by Attasi ([2]): given the autocorrelation function of y, find minimal dimension matrices H, F_1, F_2, L such that

F_1 and F_2 are two commuting asymptotically stable matrices,

there exists a process $(X_{m,n})_{(m,n)\in Z^2}$ and a white noise $(\nu_{m,n})_{(m,n)\in Z^2}$ with

$$X_{m+1,n+1} = F_1 X_{m,n+1} + F_2 X_{m+1,n} - F_1 F_2 X_{m,n} + L \nu_{m,n}$$

(0-2)

$$y_{m,n} = H X_{m,n}$$

Following Ruckebusch ([13]) and Lindquist-Picci ([9]), we present a geometrical formulation of the problem based on the fact that any representation (0-2) of y can be characterized by the space $\{X_{0,0}\}$ generated by the components of $X_{0,0}$. Let U_1 and U_2 be commuting unitary operators defined on a Hilbert space K and prolonging the horizontal and vertical shift operators associated to y and let X be a subspace of K. Then, there exists a state process $(X_{m,n})_{(m,n)\in Z^2}$ representing y such that $X = \{X_{0,0}\}$ iff X contains $\{y_{0,0}\}$ and if $(U_1 \mid H(X), U_2 \mid H(X), H(X))$ where $H(X) = \bigvee_{(m,n)\in Z^2} U_1^m U_2^n X$ is the regular minimal unitary dilation of a pair of two commuting contractions defined on X. Using the results of Part I, we give a geometric characterization of spaces X satisfying these conditions. Finally, we study the functional models associated to the regular and co-regular representations by means of the results of Part II.

NOTATIONS AND DEFINITIONS.

If K is a Hilbert space, we shortly say that S is a subspace of K if S is a closed vector space included in K. For a family $(S_i, i\in I)$ of subspaces of K, $\bigvee_{i\in I} S_i$ denotes the smallest subspace of K containing every S_i. If S is a subspace of K, then E^S denotes the orthogonal projection operator onto S and $\overline{E}^S$ the closure of its range; two subspaces S_1 and S_2 are said to intersect orthogonally if $\overline{E}^{S_1}(S_2) = S_1 \cap S_2$ or equivalently if $E^{S_1}E^{S_2} = E^{S_2}E^{S_1}$. Given three subspaces A,B,C of K, A and B are said to be conditionally orthogonal given C (written $A \perp B \mid C$) if $A \vee C \ominus C \perp B \vee C \ominus C$. If M is a continuous linear operator defined on K, then M^* denotes its adjoint. Let U be a unitary operator on K; a subspace W is said to be wandering for U if $U^k W \perp W$ $\forall k\neq 0$; if (U_1, U_2) is a pair of commuting unitary operators defined on K, then W is said to be wandering for (U_1,U_2) if $U_1^k U_2^l W \perp W$ $\forall (k,l)\neq(0,0)$.

Let X be a Hilbert space and let (T_1, T_2) be a pair of commuting contractions defined on X. Let (U_1,U_2) be a pair of commuting unitary operators defined on a Hilbert space K containing X; then, (U_1,U_2,K) is said to be a regular unitary dilation of (T_1,T_2,X) if $E^X U_1^m U_2^n \mid X = T_1^m T_2^n$ and $E^X U_1^m U_2^{-n} \mid X = T_2^{*n} T_1^m$ $\forall (m,n)\in \mathbb{N}^2$; the dilation is said to be minimal if $K = \bigvee_{(m,n)\in Z^2} U_1^m U_2^n X$. It is well known [3] that for every pair of commuting contractions there exists a unique regular minimal unitary dilation.

For a fixed pair (U_1,U_2) of commuting unitary operators defined on the Hilbert space K and for every subspace X of K, we denote by X^i, X^i_-, X^i_+ $i = 1,2$ the subspaces $\bigvee_{m\in Z} U_i^m X$, $\bigvee_{m\leq 0} U_i^m X$, $\bigvee_{m\geq 0} U_i^m X$ $i = 1,2$ respectively and by $H(X)$, $H^i_-(X)$, $H^i_+(X)$ $i = 1,2$ the subspaces $\bigvee_{(m,n)\in Z^2} U_1^m U_2^n X$, $\bigvee_{m\leq 0}\bigvee_{n\in Z} U_i^m U_j^n X$, $\bigvee_{m\geq 0}\bigvee_{n\in Z} U_i^m U_j^n X$ $j\neq i$ $i=1,2$ respectively; finally, we put $H_{++}(X) = \bigvee_{m\geq 0}\bigvee_{n\geq 0} U_1^m U_2^n X$; $H_{+-}(X)$, $H_{-+}(X)$ and $H_{--}(X)$ are defined in a similar way.

We denote by $\mathbb{C}^{p\times q}$ the set of all (p,q) matrices with complex coefficients; if A is a matrix, then A^T (resp. A^*) is the transpose (resp. the conjugate transpose) of A . I_p denotes the identity matrix of $\mathbb{C}^{p\times p}$. For a complex number a, a^* is the conjugate.

We put I =[-1/2, 1/2] and we denote by du the Lebesgue measure on I. Let $\Phi(z)$ be a analytic function in the open unit disc of $\mathbb{C}$; if Φ admits a radial limit (i.e if $\lim_{r\to 1} \Phi(re^{-2i\pi u})$ exists), we denote by $\Phi(u)$ (and not by $\Phi(e^{-2i\pi u})$) the limit in order to shorten the notations. For $(p,q)\in \mathbb{N}^2$, we denote by H^2_{pxq} and $\overline{H}^2_{pxq}$ the classical Hardy spaces of square integrables functions defined on I = [-1/2, 1/2] with values in $\mathbb{C}^{p\times q}$:

$$H^2_{pxq} = \left\{ \Phi(u) = \sum_{k=0}^{\infty} \Phi_k\, e^{-2ik\pi u} \right\}, \quad \overline{H}^2_{pxq} = \left\{ \Phi(u) = \sum_{k=0}^{\infty} \Phi_k\, e^{2ik\pi u} , \right\}.$$

The spaces H^2_{1xq} and $\overline{H}^2_{1xq}$ are denoted H^2_q and $\overline{H}^2_q$. A rigid function is an element Q(u) of H^2_{pxq} such that $QQ^*(u) = I_p$; Q is said to be inner if q=p.

$\mathbb{L}^2_{pxq}(dudv)$ is the set of all square integrable (with respect to the Lebesgue measure) functions defined on I^2 with values in $\mathbb{C}^{pxq}$. $H^{2,1}_{pxq}$ (resp. $H^{2,2}_{pxq}$) is the following subspace $\mathbb{L}^2_{pxq}(dudv)$

$$H^{2,1}_{pxq} = \left\{ \Phi(u,v) = \sum_{k=0}^{\infty} \Phi_k(v)\, e^{-2ik\pi u} \right\} \quad (\text{resp. } H^{2,2}_{pxq} = \left\{ \Phi(u,v) = \sum_{l=0}^{\infty} \Phi_l(u)\, e^{-2il\pi v} \right\})$$

and $\overline{H}^{2,1}_{pxq}$ (resp. $\overline{H}^{2,2}_{pxq}$) is defined as

$$\overline{H}^{2,1}_{pxq} = \left\{ \Phi(u,v) = \sum_{k=0}^{\infty} \Phi_k(v)\, e^{2ik\pi u} \right\} \quad (\text{resp. } \overline{H}^{2,2}_{pxq} = \left\{ \Phi(u,v) = \sum_{l=0}^{\infty} \Phi_l(u)\, e^{2il\pi v} \right\})$$

Moreover, we put $H^{2,--}_{pxq} = H^{2,1}_{pxq} \cap H^{2,2}_{pxq}$. $H^{2,-+}_{pxq}$, $H^{2,+-}_{pxq}$ and $H^{2,++}_{pxq}$ are defined similarly. We shall denote by $H^2_{pxq}(du)$ (resp. $H^2_{pxq}(dv)$) the subset of all functions of $H^{2,1}_{pxq}$ (resp. $H^{2,2}_{pxq}$) which do not depend on v (resp. u); $\overline{H}^2_{pxq}(du)$ (resp. $\overline{H}^2_{pxq}(dv)$) are defined in a similar way. Finally, the spaces $H^{2,1}_{1xq}$, $H^{2,2}_{1xq}$,...are denoted $H^{2,1}_q$, $H^{2,2}_q$ ...

I. Some properties of the regular minimal unitary dilation of a pair of commuting contractions (T_1, T_2).

We begin this section by a result implicitely contained in previous works concerning the construction of the pairs of Ando dilation of (T_1, T_2) ([4], [11], [12]).

Proposition I-1.

Let (U_1, U_2) be the regular minimal unitary dilation of (T_1, T_2, X) defined on the space H(X). Then, the following property holds:

(I-1) $H^1_-(X) \perp X^1_+ \mid X$

(I-1') $H^2_-(X) \perp X^2_+ \mid X$

Conversely, if (U_1, U_2) is a pair of commuting unitary operators defined on H(X) for which (I-1) and (I-1') hold, then $T_1 = E^X U_1|_X$ and $T_2 = E^X U_2|_X$ are commuting contractions and $(U_1, U_2, H(X))$ is the regular minimal unitary dilation of (T_1, T_2, X).

Proof.

Let (U_1, U_2) be a pair of commuting unitary operators defined on H(X) for which (I-1) and (I-1') hold. Then, $X^1_- \perp X^1_+ \mid X$ and $X^2_- \perp X^2_+ \mid X$ so that ([15]), $(U_1|_{X^1}, X^1)$ and $(U_2|_{X^2}, X^2)$ are the minimal unitary dilation of (T_1, X) and (T_2, X) respectively. Moreover, $\forall x \in X$, $\forall m \geq 0$, $E^{H^i_-(X)} U_1^m x \in X$ and consequently, $E^{H^i_-(X)} U_1^m x = T_1^m x$ i = 1,2. $\forall (m,n) \in \mathbb{N}^2$, $E^X U_1^m U_2^n x = E^X E^{H^1_-(X)} U_1^m U_2^n x$; as $U_2 H^1_-(X) = H^1_-(X)$, $E^{H^1_-(X)} U_1^m U_2^n x = U_2^n E^{H^1_-(X)} U_1^m x = U_2^n T_1^m x$; consequently, $E^X U_1^m U_2^n x = T_2^n T_1^m x$; similarly, $E^X U_1^m U_2^n x = T_1^n T_2^m x$. Moreover, $E^X U_1^m U_2^{-n} x = E^X E^{H^1_-(X)} U_1^m U_2^{-n} x = E^X U_2^{-n} E^{H^1_-(X)} U_1^m x = T_2^{*n} T_1^m x$. Hence, (T_1, T_2) is a pair of commuting contractions and (U_1, U_2) is its regular minimal unitary dilation.

Let us show that the regular minimal unitary dilation (U_1, U_2) of (T_1, T_2) satisfies (I-1) and (I-1').

Following [4], (U_1, U_2) can be constructed as follows. Let V_1 be the minimal unitary dilation of T_1 defined on the space $K = \bigvee_{m\in\mathbb{Z}} V_1^m X$; let T_V be the extension of T_2 defined on K by

(I-2) $T_V(V_1^m x) = V_1^m T_2 x \quad \forall m\in\mathbb{Z}, \forall x\in X,$

and let us consider U_2 the minimal unitary dilation of T_V, defined on $\bigvee_{n\in\mathbb{Z}} U_2^n K$ and denote by U_1 the unique unitary extension of V_1 commuting with U_2. Then, (U_1, U_2) is the regular minimal unitary dilation of (T_1, T_2) and $T_V^n = E^{X^1} U_2^n |_{X^1} = E^{H_-^2(X)} U_2^n |_{X^1}$ $\forall n\in\mathbb{Z}$ because $(U_2, H(X))$ is the minimal unitary dilation of (T_V, X^1) (we recall that $X^1 = \bigvee_{m\in\mathbb{Z}} U_1^n X$). But, by (I-2), $T_V^n |_X = T_2^n$ and $E^{H_-^2(X)} U_2^n x = E^X U_2^n x$ $\forall n\geq 0, \forall x\in X$; this implies (I-1'). (U_1, U_2) can also be constructed similarly by exchanging T_1 and T_2 ; consequently, (I-1) holds. □

-

In the following proposition, we give some known results concerning the structure of the regular minimal unitary dilation of (T_1, T_2) ([7]).

Proposition I-2.

Let Z_1 and Z_2 be the subspaces defined by

(I-3) $Z_1 = \overline{(U_1 - T_1)X}$, $Z_2 = \overline{(U_2 - T_2)X}$.

Then, Z_1 (resp. Z_2) is wandering for U_1 (resp. for U_2) and is given by

(I-4) $Z_1 = U_1 X_-^1 \ominus X_-^1$ (resp. $Z_2 = U_2 X_-^2 \ominus X_-^2$)

and

(I-5) $(Z_i)_+^i \perp H_-^i(X)$ $i = 1,2$.

The space Z_1 is invariant under the operator $T_V = E^{X^1} U_2 |_{X^1}$ previously defined and the space Z_2 is invariant under the operator $T_h = E^{X^2} U_1 |_{X^2}$. Moreover, $(Z_1)^2 = \overline{(U_1 - T_h)X^2} = U_1 H_-^1(X) \ominus H_-^1(X)$, $(Z_2)^2 = \overline{(U_2 - T_V)X^1} = U_2 H_-^2(X) \ominus H_-^2(X)$, $(Z_1)_-^2 \perp (Z_1)_+^2 | Z_1$ i.e $(U_2 |_{(Z_1)^2}, (Z_1)^2)$ is the minimal unitary dilation of $(T_V |_{Z_1}, Z_1)$, $(Z_2)_-^1 \perp (Z_2)_+^1 | Z_2$ i.e $(U_1 |_{(Z_2)^1}, (Z_2)^1)$ is the minimal unitary dilation of $(T_h |_{Z_2}, Z_2)$. The space

(I-6) $\Delta = \overline{(U_1U_2 - U_1T_2 - U_2T_1 + T_1T_2)X}$

is wandering for (U_1, U_2) and satisfies

$$
\begin{aligned}
\text{(I-6')} \quad \Delta &= \overline{(U_1 - T_h)Z_2} = U_1(Z_2)_-^1 \ominus (Z_2)_-^1 = \overline{(U_2 - T_V)Z_1} = U_2(Z_1)_-^2 \ominus (Z_1)_-^2 \\
&= \left(U_1 H_-^1(X) \ominus H_-^1(X)\right) \cap (U_2 H_-^2(X) \ominus H_-^2(X))
\end{aligned}
$$

(I-7) $H_{++}(X) = X \oplus (\overline{(U_1 - T_1)X})_-^1 \oplus (\overline{(U_2 - T_2)X})_-^2 \oplus H_{++}(\Delta).$ □

The following properties will be useful in the sequel.

Proposition I-3.

(U_1, U_2) satisfies the following properties:

(I-8) $E^{H^1_-(X)} E^{H^2_-(X)} = E^{H^2_-(X)} E^{H^1_-(X)}$

(I-9) $H^1_-(X) \cap H^2_-(X) = H_{--}(X)$

(I-10) $X = H_{--}(X) \cap H_{++}(X) = \overline{E}^{H_{--}(X)}(H_{++}(X))$. □

We omit the proof which is based on elementary arguments.

In general, $T_h^{*m} x = E^{H^1_+(X)} U_1^{-m} x$ and $T_v^{*n} x = E^{H^1_+(X)} U_2^{-n} x$ do not belong to X, that is (I-1) and (I-1') do not hold if U_1 and U_2 are exchanged in U_1^{-1} and U_2^{-1} respectively. In fact, we have the following result.

Proposition I-4.

The following equalities hold:

(I-11) $E^{H^1_+(X)}(X^1_-) = E^{X^2}(X^1_-)$,
(I-12) $E^{H^2_+(X)}(X^2_-) = E^{X^1}(X^2_-)$.

Proof.

We just show (I-11). (I-11) holds iff $\forall m \geq 0$, $\forall x \in X$, $T_h^{*m} x = E^{H^1_+(X)} U_1^{-m} x$ is orthogonal to the space $X^2 \ominus X^2_-$; but $X^2 \ominus X^2_- = (Z_2)^2_+$ because $(U_2|_{X^2}, X^2)$ is the minimal unitary dilation of (T_2, X); consequently, we have to show that

(I-13) $< T_h^{*m} x , U_2^n (U_2 y - T_2 y) > = 0 \quad \forall n \geq 0, \forall y \in X.$

$< T_h^{*m} x , U_2^n (U_2 y - T_2 y) > = < x , T_h^m U_2^n (U_2 y - T_2 y) > = < x , U_2^{n+1} T_1^m y > - < x , U_2^n T_1^m T_2 y > =$
$< x , T_2^{n+1} T_1^m y > - < x , T_2^n T_1^m T_2 y > = 0$ because $T_1 T_2 = T_2 T_1$. □

We deduce from this result that $H^1_+(X)$ and $H^2_-(X)$ meet orthogonally and that $H^2_+(X)$ and $H^1_-(X)$ meet orthogonally; this and (I-8) will be of particular interest in the next section.

Proposition I-5.

We have the following commutation properties:

(I-14) $E^{H^1_+(X)} E^{H^2_-(X)} = E^{H^2_-(X)} E^{H^1_+(X)}$
(I-14') $E^{H^2_+(X)} E^{H^1_-(X)} = E^{H^1_-(X)} E^{H^2_+(X)}$.

Moreover,

(I-15) $H^1_+(X) \cap H^2_-(X) = H_{+-}(X)$ and $H^2_+(X) \cap H^1_-(X) = H_{-+}(X)$.

Proof.

Let us show (I-14). (I-14) is obvious on $H_{+-}(X)$. For $(k,l) \in \mathbb{N}^2$ and $x \in X$, $E^{H^1_+(X)} E^{H^2_-(X)} U_1^k U_2^l x = U_1^k T_2^l x$ and $E^{H^2_-(X)} E^{H^1_+(X)} U_1^k U_2^l x = E^{H^2_-(X)} U_1^k U_2^l x = U_1^k T_2^l x$ so that (I-14) holds on $H_{++}(X)$. $E^{H^1_+(X)} E^{H^2_-(X)} U_1^{-k} U_2^{-1} x = E^{H^1_+(X)} U_1^{-k} U_2^{-1} x = U_2^{-1} E^{H^1_+(X)} U_1^{-k} x = U_2^{-1} T_h^{*k} x$;on the other hand, $E^{H^2_-(X)} E^{H^1_+(X)} U_1^{-k} U_2^{-1} x = E^{H^2_-(X)} U_2^{-1} E^{H^1_+(X)} U_1^{-k} x = E^{H^2_-(X)} U_2^{-1} T_h^{*k} x = U_2^{-1} T_h^{*k} x$ because T_h^*

belongs to X_-^2 .

Finally, $E^{H_+^1(X)} E^{H_-^2(X)} U_1^{-k} U_2^1 x = T_h^{*k} T_2^1 x$;$E^{H_-^2(X)} E^{H_+^1(X)} U_1^{-k} U_2^1 x = E^{H_-^2(X)} U_2^1 E^{H_+^1(X)} U_1^{-k} x = E^{H_-^2(X)} U_2^1 T_h^{*k} x$; $U_2^1 T_h^{*k} x \in U_2^1 X_-^2$ because $T_h^{*k} x \in X_-^2$; then, $U_2^1 T_h^{*k} x = E^{X_-^2} U_2^1 T_h^{*k} x + E^{U_2^1 X_-^2 \ominus X_-^2} U_2^1 T_h^{*k} x$; but $U_2^1 X_-^2 \ominus X_-^2 = \bigoplus_{n=0}^{l-1} U_2^n Z_2$ and because $(Z_2)_+^2 \perp H_-^2(X)$, $E^{H_-^2(X)} U_2^1 T_h^{*k} x = E^{X_-^2} U_2^1 T_h^{*k} x$. $U_2^1 x = T_2^1 x + E^{[H_-^2(X)]^\perp} U_2^1 x$ and $E^{X_-^2} U_2^1 T_h^{*k} x = T_h^{*k} T_2^1 x + E^{X_-^2} T_h^{*k} E^{[H_-^2(X)]^\perp} U_2^1 x$. Let us show that $T_h^{*k} E^{[H_-^2(X)]^\perp} U_2^1 x$ is orthogonal to X_-^2 ; $\forall\, z \in X_-^2$, $< T_h^{*k} E^{[H_-^2(X)]^\perp} U_2^1 x\, ,\, z > = < U_1^{-k} E^{[H_-^2(X)]^\perp} U_2^1 x\, ,\, z > = < E^{[H_-^2(X)]^\perp} U_2^1 x\, ,\, U_1^k z > = 0$ because $U_1^k z \in H_-^2(X)$. The proof of (I-14) is now complete. (I-14') is shown similarly and (I-15) is easy. □

Remark I-6.

In the proof of the Proposition (I-5), we have shown that $T_h^{*k} T_2^1 x = E^{X_-^2} U_2^1 T_h^{*k} x$ $\forall x \in X$ and $\forall (k,l) \in \mathbb{N}^2$. Then, $\forall x \in X$, $T_h^* (U_2 x - T_2 x) = U_2 T_h^* x - E^{X_-^2} U_2 T_h^{*k} x$; because $T_h^{*k} x \in X_-^2$, $T_h^* (U_2 x - T_2 x) \in U_2 X_-^2 \ominus X_-^2$; moreover, by (I-4), $U_2 X_-^2 \ominus X_-^2 = (Z_2)$; this and Proposition I-2 implies that Z_2 reduces T_h ; finally, it is easy to show that the space

$$\text{(I-16)} \quad \Delta^{+-} = \overline{U_1^{-1} Z_2 - T_h^* Z_2}$$

is wandering for (U_1,U_2) and is also given by $\Delta^{+-} = \left(U_1^{-1} H_+^1(X) \ominus H_+^1(X)\right) \cap (U_2 H_-^2(X) \ominus H_-^2(X))$. Let us remark that Δ^{+-} do not coincide with $\overline{U_1^{-1} Z_2 - T_1^* Z_2}$ as in the case where T_1 and T_2 doubly commute (i.e $T_1 T_2 = T_2 T_1$ and $T_1 T_2^* = T_2^* T_1$)([14]). Similarly, Z_1 reduces T_v and

$$\text{(I-17)} \quad \Delta^{-+} = (\overline{U_2^{-1} - T_v^*})Z_1 = \left(U_2^{-1} H_+^2(X) \ominus H_+^2(X)\right) \cap (U_1 H_-^1(X) \ominus H_-^1(X))$$

is wandering for (U_1,U_2). □

<u>II. Functional models.</u>

Now, we are going to apply the previous results in order to derive functional models. For sake of simplicity, we suppose that T_1 and T_2 belong to the class $C_{0,.}$ (i.e T_i^k converges strongly to 0 if $k \to \infty$ $i = 1,2$) and that the space Δ given by (I-6) is of finite dimension q. Then, it is well known ([6]) that $H_-^i(X) = H_-^i(\Delta)$ $i = 1,2$; consequently, T_h, T_v, $T_h|_{Z_2}$ and $T_v|_{Z_1}$ belong also to the class $C_{0,.}$; then, we have the following decompositions:

$$\text{(II-1)} \quad (Z_2)_-^1 = U_1^{-1}(\Delta)_-^1 = U_1^{-1}(\Delta^{+-})_-^1 \oplus Z_2 \quad , \quad (Z_1)_-^2 = U_2^{-1}(\Delta)_-^2 = U_2^{-1}(\Delta^{-+})_-^2 \oplus Z_1$$

Let us consider a fixed orthonormal basis $(v_1, v_2, ..., v_q)$ of Δ and put $v = (v_1, v_2, ..., v_q)^T$. Let us denote by $E_1(du)$ and $E_2(dv)$ the spectral families of U_1 and U_2 defined on I; then, $\hat{v}(du,dv) = E_1(du)E_2(dv)v$ is a q-variate orthogonally scattered measure such that $<\hat{v}(du,dv), \hat{v}(du,dv)> = I_q$. Put $q_{+-} = \dim \Delta^{+-}$ and let us consider a q_{+-}-variate orthogonally scattered measure $\hat{v}^{+-}(du,dv)$ defined as previously by means of an orthonormal basis of Δ^{+-} ; then, by (II-1), there exists a rigid function $Q_1(u)$ of $H^2_{q_{+-}xq}$ such that

$$\text{(II-2)} \quad \hat{v}^{+-}(du,dv) = Q_1(u)\, \hat{v}(du,dv) .$$

If we denote by Σ_v the unitary operator defined by

$$\Sigma_v \begin{cases} H(X) = H(\Delta) \longrightarrow \mathbb{L}_q^2(dudv) \\ \int_{I^2} \phi(u,v)\, \hat{v}(du,dv) \longrightarrow \phi \end{cases}$$

then, by (II-1)

(II-3) $\Sigma_v(Z_2) = e^{-2i\pi u}\,(H^2_q(du) \ominus H^2_{q_{+-}}(du)Q_1(u))\ ,\ \Sigma_v(X^2) = e^{-2i\pi u}\,(H^{2,1}_q \ominus H^{2,1}_{q_{+-}}Q_1(u)).$

Similarly, if we put $q_{-+} = \dim \Delta^{-+}$, there exists a rigid function $Q_2(v)$ of $H^2_{q_{-+}xq}$ such that

(II-4) $\Sigma_v(Z_1) = e^{-2i\pi v}\,(H^2_q(dv) \ominus H^2_{q_{-+}}(dv)Q_2(v))\ ,\ \Sigma_v(X^1) = e^{-2i\pi v}\,(H^{2,2}_q \ominus H^{2,2}_{q_{-+}}Q_2(v))$

Moreover, T_h and T_v are unitarily equivalent to the operators $E^{\Sigma_v(X^2)}\, e^{2i\pi u} \mid \Sigma_v(X^2)$ and $E^{\Sigma_v(X^1)}\, e^{2i\pi v} \mid \Sigma_v(X^1)$ respectively.
Unfortunately, it is not possible to obtain simple functional models for (T_1,T_2,X) in the general case.

III. Application to the markovian representation problem of stationary two-parameter processes.

Now, we are going to study the realization problem of a p-variate two-parameter wide sense stationary process by means of the recursive models introduced by Attasi ([2]). In a geometric framework, the problem can be formulated as follows.

Let K be a fixed Hilbert space and let U_1,U_2 be two commuting unitary operators defined on K; Y is a p-dimensional subspace of K ($p<\infty$). Then, a subspace X of K is said to be a forward-forward quarter plane Markovian representation (FFQMR in short) of Y if

(III-1) $X \supset Y\ ,\ \ H^i_-(X) \perp X^i_+ \mid X \quad i = 1,2.$

Backward-backward, backward-forward, forward-backward quarter-plane Markovian representations could be defined in a similar way by exchanging U_1 with U_1^{-1} and U_2 with U_2^{-1}, U_1 with U_1^{-1}, U_2 with U_2^{-1} respectively (with corresponding replacements of F by B in FFQMR).

By Proposition (I-1), X is a FFQMR of Y iff X contains Y and if $(U_1\,|_{H(X)}, U_2\,|_{H(X)}, H(X))$ is the regular minimal unitary dilation of two commuting contractions T_1 and T_2 defined on X. When T_1 and T_2 doubly commute, it can be shown ([10]) that X is both a FFQMR and a BBQMR; the properties of such state-spaces are studied in ([10]); due to the lack of space, we do not consider these kind of Markovian representations. In the sequel, we shall use all the previous notations and definitions related to (T_1,T_2,X) and to its regular minimal unitary dilation.

The geometric characterization of the set of all FFQMR is the following.

Théorème III-1.

A subspace X is a FFQMR of Y iff

(III-2) $X = \bar{E}^{S_1 \cap S_2}(\bar{S})$

where S_1, S_2, $\bar{S}$ are subspaces such that

(III-3) $\bar{S} \supset U_1\bar{S}\ ,\ \bar{S} \supset U_2\bar{S}\ ,\ \bar{S} \supset Y,$
(III-4) $S_i \supset U_i^{-1}S_i\ ,\ U_jS_i = S_i \quad j \neq i \quad i = 1,2,\ S_i \supset Y,$
(III-5) S_1 and S_2 meet orthogonally i.e $E^{S_1}E^{S_2} = E^{S_2}E^{S_1}$. □

The proof can be found in [10].

By Theorem (III-1) and Proposition (I-3), if X is a FFQMR of Y, so is $\mathring{X} = E^{H^1_-(X) \cap H^2_-(X)}(H_{++}(Y)) = E^X(H_{++}(Y))$; moreover, $X \supset \mathring{X}$; this remark allows us to reduce the state space and motivates the following definition.

Definition III-2.

Let X be a FFQMR of Y; then, X is said to be observable if $X = E^X(H_{++}(Y))$ or equivalently if $X \cap [H_{++}(Y)]^{\perp} = \{0\}$. □

In the one-parameter Markovian representation theory, the introduction of the concepts of observability and constructibility makes possible to characterize the set of all minimal Markovian representations (in the sense that they do not contain another representation) ([9], [13]). Unfortunately, we have not succeeded to obtain such characterization for FFQMR. However, we have obtained some results related to this question by using functional models ([10]). Now, we are going to present these models; the results concerning the minimal realizations characterization will be presented in a forthcoming paper.

From now on, we assume that there exists a finite dimensional subspace L, wandering for (U_1,U_2), such that $K = \bigoplus_{(k,l)\in \mathbb{Z}^2} U_1^k\, U_2^l\, L$. Let us consider a fixed basis $(y_1,y_2,.......,y_p)$ of Y; put $y_{0,0} = (y_1,y_2,.......,y_p)^T$ and $y_{m,n} = U_1^m\, U_2^n\, y_{0,0}$; then, $y = (y_{m,n})_{(m,n)\in \mathbb{Z}^2}$ is a p-variate two-parameter stationary sequence; let us denote by $\hat{y}(du,dv)$ the orthogonally scattered measure defined on I^2 by $\hat{y}(du,dv) = E_1(du)E_2(dv)y_{0,0}$ where E_1 and E_2 are the spectral families of U_1 and U_2 respectively; then, the spectral measure of $(y_{m,n})_{(m,n)\in \mathbb{Z}^2}$ is the positive matrix valued measure defined by the distribution function $<\hat{y}(du,dv),\hat{y}(du,dv)>$; as $K = \bigoplus_{k,l\in \mathbb{Z}^2} U_1^k\, U_2^l\, L$, the spectral measure of y is absolutely continuous with respect to the Lebesgue measure; the associated density is called the spectral density of y and is denoted by f(u,v).

In the following, we are just concerned with the so-called regular and co-regular representations. A FFQMR X of Y is said to be regular if $\bigcap_{m\in \mathbf{Z}} U_i^m\, H^i_-(X) = \{0\}$ i = 1,2 or equivalently if T_1 and T_2 (or T_h and T_v) belong to the class $C_{0,.}$. X is said to be co-regular if $\bigcap_{m\in \mathbf{Z}} U_i^m\, H^i_+(X) = \{0\}$ i = 1,2 or equivalently if T_h and T_v belong to the class $C_{.,0}$.

Let X be a FF-observable FFQMR; then, it is clear that X is regular iff $X = \bar{E}^{U_1^{-1}U_2^{-1}H_{--}(\Delta)}(H_{++}(Y))$ where Δ is defined by (I-6). Put q = dimΔ and let us consider $v = (v_1, v_2, ...,v_q)^T$ a fixed orthonormal basis of Δ; then, as $U_1^{-1}U_2^{-1}\, H_{--}(\Delta) \supset X \supset Y$, there exists a sequence $(\Phi_{k,l})$ such that

$$y = \sum_{k=1}^{\infty} \sum_{l=1}^{\infty} \Phi_{k,l}\, v_{-k,-l} \quad \text{where } v_{m,n} = U_1^m\, U_2^n\, v \text{ and where } \sum_{k=1}^{\infty} \sum_{l=1}^{\infty} \operatorname{tr}(\Phi_{k,l}\, \Phi_{k,l}^*) < \infty \,;$$

then , $\hat{y}(du,dv) = \Phi(u,v)\, \hat{v}(du,dv)$ where $\Phi(u,v) = \sum_{k=1}^{\infty} \sum_{l=1}^{\infty} \Phi_{k,l}\, e^{-2i\pi(ku+lv)}$ belongs to $e^{-2i\pi(u+v)}\, H^{2,--}_{pxq}$, and $f(u,v) = (\Phi\Phi^*)(u,v)$

Moreover, if Σ_v is the unitary operator defined in section II, we have

$$\text{(III-6)} \quad \Sigma_v(X) = \bar{E}^{e^{-2i\pi(u+v)}H^{2,--}_q}(\, H^{2,++}_p \Phi).$$

Finally, we have the following result.

Theorem III-3.

There exist regular FFQMR iff there exist $q \in \mathbb{N}$ and $\Phi \in e^{-2i\pi(u+v)} H^{2,--}_{pxq}$ such that $f(u,v) = (\Phi\Phi^*)(u,v)$. Moreover, a subspace X is a regular FF-observable FFQMR iff

$$\text{(III-7)} \quad X = \Sigma_v^* \left[\overline{E}^{e^{-2i\pi(u+v)} H_q^{2,--}} (H_p^{2,++} \Phi) \right]$$

where Φ and ν are such that $\hat{y}(du,dv) = \Phi(u,v)\, \hat{\nu}(du,dv)$. □

In the sequel, we shall say that the function Φ associated to a regular FF-observable FFQMR is the forward-forward (FF) factorization associated to X.

Now, we give the conditions under which a regular FF-observable FFQMR is co-regular. We begin by the following Lemma.

Lemma III-4.

Let X be a regular FFQMR and let Δ, Δ^{+-}, Δ^{-+} be the spaces defined by (I-6), (I-15) and (I-16) respectively. Then, X is co-regular iff $\dim \Delta = \dim \Delta^{+-} = \dim \Delta^{-+}$.

Proof.

X is co-regular iff $T_h \mid \overline{(U_2 - T_2)X}$ and $T_v \mid \overline{(U_1 - T_1)X}$ belong to the class $C_{0,0}$; but, it is well known that it is the case iff $\dim \Delta^{+-} = \dim \Delta$ and $\dim \Delta^{-+} = \dim \Delta$. □

Finally, we have the following result which is the functional version of Lemma III-4.

Theorem III-5.

Let X be a regular FF-observable FFQMR and $\hat{\nu}(du,dv)$, $\Phi(u,v)$ the q-variate orthogonally scattered measure and the factorization associated to X as previously. Then, X is co-regular iff there exist two inner functions $Q_1(u)$ and $Q_2(v)$ of H^2_{qxq} such that

$$\text{(III-8)} \quad \Phi(u,v) = \Phi_{+-}(u,v) Q_1(u) = \Phi_{-+}(u,v) Q_2(v)$$

where Φ_{+-} (resp. Φ_{-+}) belongs to $e^{-2i\pi v} H^{2,+-}_{pxq}$ (resp. $e^{-2i\pi u} H^{2,-+}_{pxq}$).

Proof.

Suppose that X is co-regular. Then, $\dim \Delta^{+-} = \dim \Delta = q$. By (II-7), there exists an inner function $Q_1(u)$ such that $\Sigma_v(X^2) = e^{-2i\pi u} (H_q^{2,1} \ominus H_q^{2,1} Q_1(u))$. As $X^2 \supset X \supset Y$, each row Φ^j of Φ belongs to $e^{-2i\pi u} (H_q^{2,1} \ominus H_q^{2,1} Q_1(u))$; but, this space can also be written as $e^{-2i\pi u} H_q^{2,1} \cap \overline{H}_q^{2,1} Q_1(u))$; this implies that there exists $\Phi_{+-} \in e^{-2i\pi v} H^{2,+-}_{pxq}$ such that $\Phi(u,v) = \Phi_{+-}(u,v) Q_1(u)$. The existence of Φ_{-+} is shown similarly.

Conversely, suppose that (III-8) holds. Then, $\overline{E}^{e^{-2i\pi u} H_q^{2,1}} (\overline{H}_q^{2,1} Q_1(u)) \supset \Sigma_v(X^2)$; but, $\overline{E}^{e^{-2i\pi u} H_q^{2,1}} (\overline{H}_q^{2,1} Q_1(u)) = e^{-2i\pi u} (H_q^{2,1} \ominus H_q^{2,1} Q_1(u))$; consequently, $\overline{H}_q^{2,1} Q_1(u) \supset \Sigma_v(X^2)$ and $\overline{H}_q^{2,1} Q_1(u) \supset \Sigma_v(H^1_+(X))$; this implies that $\bigcap_{m \in Z} U_1^m H^1_+(X) = \{0\}$; $\bigcap_{n \in Z} U_2^n H^2_+(X) = \{0\}$ is shown similarly. □

The set of all functions for which (III-8) holds can be seen as an extension of the set of strictly non cyclic functions ([5]) and has been studied in [10].

REFERENCES.

[1] T. ANDO, *On a pair of commutative contractions*, Acta. Sci. Math., 24 (1963), 88-90.

[2] S. ATTASI , *Modelling and recursive estimation for double indexed sequence* , In System Identification : Advances and Studies , ed. R.K. Mehra and D.G. Lainiotis , Mathematics in Science and Engineering , vol 126 , 1976 , 289-348.

[3] S. BREHMER , *Über vertauschbare kontraktionen des Hilbertschen raumes* , Acta. Sci. Math., 22 (1961), 106-111.

[4] Z. CEAUCESCU, I. SUCIU , *Isometric dilation of commuting contractions. IV* , In Advances in Invariant Subspaces and Other Results of Operator Theory , Birkhäuser Verlag (vol OT 17) , 1986 , 55-77.

[5] P.A. FUHRMANN , *Linear Systems and Operators in Hilbert Space* , Mac-Graw-Hill (1981).

[6] I. HALPERIN , *Unitary dilations which are orthogonal bilateral shift operators* , Duke Math. J. , 29 (1962), 573-580

[7] I. HALPERIN , *Intrinsic description of the Sz-Nagy–Brehmer dilation* , Studia Math. , 22 (1963), 211-219.

[8] H. KOREZLIOGLU, P. LEFORT, G. MAZZIOTTO , *Une propriété markovienne et diffusions associées* , Lect. Notes in Math. 863, Springer-Verlag, 1981, 245-274.

[9] A. LINDQUIST, G. PICCI , *Realization theory for multivariate stationary Gaussian processes* , Siam J. Contr. Optim. , vol 23 , n°6 , November 1985.

[10] P. LOUBATON , *Prédiction et Representation Markovienne des Processus Stationnaires Vectoriels sur* $\mathbb{Z}^2$ *. Utilisation de Techniques d'Estimation Spectrale 2-D en Traitement d'Antenne* , Phd Thesis , Ecole Nationale Supérieure des Télécommunications , 1988.

[11] W. MLAK , *Note on unitary dilation of two commuting contractions* , Bull. Acad. Polon. Sci., 21 (1973) , 941-943.

[12] S. PARROT , *Unitary dilations for commuting contractions* , Pac. J. Math. , 34 (1970) , 481-490.

[13] G. RUCKEBUSCH , *Théorie géométrique de la représentation markovienne* , Ann. Inst. Henri Poincaré , 16 (1980) , 225-297.

[14] M. SLOCINSKI , *Models for doubly commuting contractions* , Ann. Polon. Math. , 45 (1985) , 23-42.

[15] B. Sz-NAGY, C. FOIAS , *Harmonic Analysis of Operators on Hilbert Space* , North-Holland (1970).

A CONSIDERATION ON COMPLEX REFLECTION COEFFICIENTS

Nobuo Nagai

Abstract

The scattering matrix which is a basic tool in the classical circuit theory, is introduced from the incident and reflected waves travelling on a transmission line. The incident and reflected waves used in the scattering matrix are in general normalized waves, not voltage waves or current waves. The normalized waves, however, are derived from voltages and currents [10]. When the internal impedance of a voltage generator is given by a complex impedance, the voltage and current reflection coefficients for a scattering matrix involve the complex conjugate number of the internal impedance. On the other hand, the voltage and current reflection coefficients for two transmission lines never involve the complex conjugate number of the characteristic impedance [8].

In this paper we explain clearly the discrepancy for complex reflection coefficients. By using these considerations, we propose a pseudo transmission line whose complex reflection coefficient involve the conjugate complex impedance. And we also show that the circuit constituted with the transmission lines satisfies the condition of "positive function" [1],[3].

1. Voltage Reflection Coefficients for Transmission-Line Circuits

The cascade matrix for a uniform lossy line, whose propagation constant, characteristic impedance and length are γ, Z_0 and l respectively, is represented as follows:

$$\begin{bmatrix} V_1 \\ I_1 \end{bmatrix} = \begin{bmatrix} \cosh\gamma l & Z_0 \sinh\gamma l \\ Z_0^{-1}\sinh\gamma l & \cosh\gamma l \end{bmatrix} \begin{bmatrix} V_2 \\ -I_2 \end{bmatrix} \tag{1}$$

Let us consider the circuit constituted with the line shown in Fig.1. The terminal condition for the circuit is given by

$$V_1 + Z_1 I_1 = E, \quad V_2 = Z_2(-I_2) \tag{2}$$

Therefore, the voltage V_1 is obtained as:

$$V_1 = E\frac{Z_0}{Z_0 + Z_1}\left\{\frac{1 + m_2 e^{-2\gamma l}}{1 - m_1 m_2 e^{-2\gamma l}}\right\} \tag{3}$$

Where m_1 and m_2 represent voltage reflection coefficients at port 1 and 2 looked from the line, and

$$m_1 = (Z_1 - Z_0)/(Z_1 + Z_0), \quad m_2 = (Z_2 - Z_0)/(Z_2 + Z_0) \tag{4}$$

Among microwave circuits, there is a commensurate transmission line (CTL) circuit [11] which is composed of lengths of lossless transmission line that has commensurate one-way delay and only propagates the transverse electromagnetic (TEM) mode. The CTL section is lossless and uniform. The most basic element for CTL circuits is a UE whose cascade matrix is represented as:

$$\begin{bmatrix} V_1 \\ I_1 \end{bmatrix} = \frac{z^{1/2}}{2R_0} \begin{bmatrix} R_0(1+z^{-1}) & R_0^2(1-z^{-1}) \\ 1-z^{-1} & R_0(1+z^{-1}) \end{bmatrix} \begin{bmatrix} V_2 \\ -I_2 \end{bmatrix} \tag{5}$$

where $z^{-1}=e^{-sT}$ *or* $z^{-1}=e^{-j\omega T}$, s is the complex variable and $T/2$ is the one-way delay.

The cascade matrix for the circuit shown in Fig 2, which constituted with UE and reactances jX_0 and $-jX_0$, is represented as follows:

$$\begin{bmatrix} V_1 \\ I_1 \end{bmatrix} = \frac{z^{1/2}}{2R_0} \begin{bmatrix} Z_0+Z_0^* z^{-1} & Z_0 Z_0^*(1-z^{-1}) \\ 1-z^{-1} & Z_0^*+Z_0 z^{-1} \end{bmatrix} \begin{bmatrix} V_2 \\ -I_2 \end{bmatrix} \tag{6}$$

Substituting Eq.(2) into (6), we obtain the voltage V_1 as follows:

$$V_1=E\frac{Z_0}{Z_0+Z_1}\left\{\frac{1+\Gamma_2 z^{-1}}{1-\Gamma_1\Gamma_2 z^{-1}}\right\} \tag{7}$$

where

$$\Gamma_1=\frac{Z_0}{Z_0^*}\frac{(Z_1-Z_0^*)}{(Z_1+Z_0)},\quad \Gamma_2=\frac{Z_0^*}{Z_0}\frac{(Z_2-Z_0)}{(Z_2+Z_0^*)} \tag{8}$$

Γ_1 and Γ_2 represent voltage reflection coefficients, but they are not equal to m_1 and m_2 given by Eq.(4). Reactances are circuit elements in steady state. The fact shows that we should reconsider about reflection coefficients in steady state for CTL circuit.

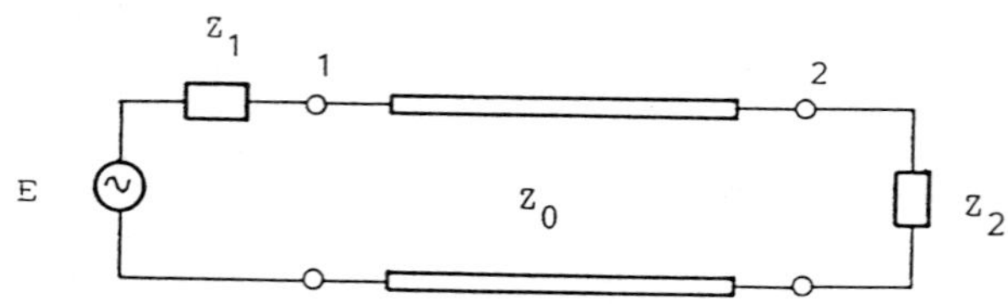

FIGURE 1 A lossy transmission − line circuit.

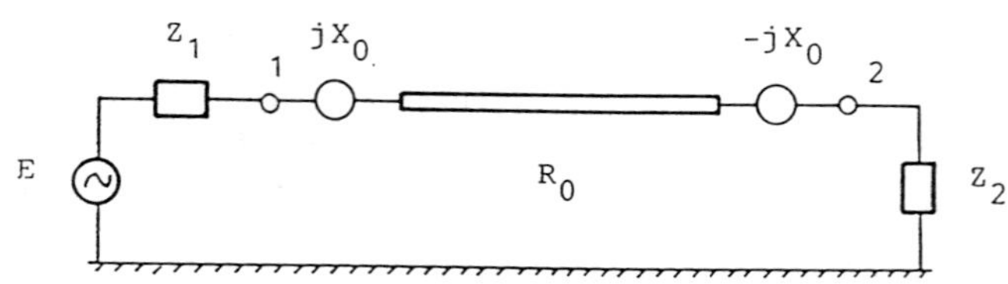

FIGURE 2 A lossless transmission line circuit with two reactances.

Belevitch [1] shows that a complex circuit can be obtained by adding imaginary resistances. An imaginary resistance absorbs no real instantaneous power and is abstractly lossless. Therefore we can assume the reactance jX_0 and $-jX_0$ as imaginary resistances for complex CTL circuit. As a result, we should reconsider the reflection coefficient for complex CTL circuit.

2. Definition of Voltage and Current Waves, and Assumptions for Matching

We consider a circuit of Fig.3 to be loaded by a voltage source E_G and a complex impedance $Z_G=R_G+jX_G$ ($R_G>0$). And the load is also a complex impedance $Z_L=R_L+jX_L$ ($R_L\geq 0$).

In order to obtain the voltage and current reflection coefficients and voltage and current transfer coefficients, each wave is defined as presented in Fig.3. Where V_i and I_i are incident voltage and current waves, V_t and I_t are transfer voltage and current waves, and V_r and I_r are reflected voltage and current waves.

DEFINITION 1. (Definition for a voltage source).

$$V_t = V_i + V_r, \quad I_t = I_i - I_r,$$
$$V_t/I_t = Z_L, \quad V_r/I_r = Z_G,$$
$$E_G = (Z_G + Z_L)I_t.$$

In the above definition, we should notice that the impedance for the incident wave V_i/I_i is not involved.

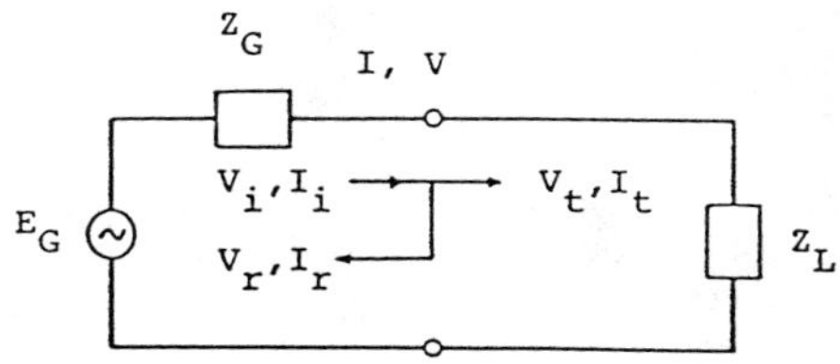

FIGURE 3 *Representation of incident*, *reflected* and *transfer voltage* and *current waves*.

At the arbitrary point on the uniform transmission line, no reflected wave is produced. We assume this fact to be a matching condition.

ASSUMPTION 2. (Equal matching condition)

If $Z_G = Z_L$,
then $V_r = I_r = 0$.

Under Assumption 2, Voltage Reflection Coefficient (VRC) V_r/V_i and Current Reflection Coefficient (CRC) I_r/I_i are obtained as follows:

$$\frac{V_r}{V_i} = \frac{I_r}{I_i} = \frac{Z_L - Z_G}{Z_L + Z_G} = m \tag{9}$$

And Voltage Transfer Coefficient (VTC) V_t/V_i and Current Transfer Coefficient (CTC) I_t/I_i are

$$V_t/V_i = 2Z_L/(Z_L + Z_G) = 1 + m \tag{10a}$$
$$I_t/I_i = 2Z_G/(Z_L + Z_G) = 1 - m \tag{10b}$$

By using Def.1 and Eqs.(9),(10), we obtain the following relationship among each voltage and current.

$$V_r I_r + V_t I_t = V_i I_i \tag{11}$$

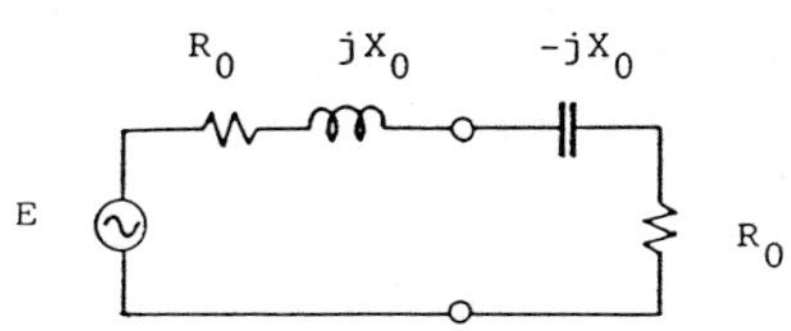

FIGURE 4 *A circuit* for *Example* 1.

[Example 1] Let us obtain the reflection coefficient for the circuit shown in Fig.4.

We obtain V_t and I_t from the circuit shown in Fig.4 as follows:

$$V_t=\frac{E(R_0-jX_0)}{2R_0}, \quad I_t=\frac{E}{2R_0}$$

We obtain V_i and I_i from the assumption as follows:

$$V_i=\frac{E}{2}, \quad I_i=\frac{E}{2(R_0+jX_0)}$$

Then we obtain V_r and I_r as follows:

$$V_r=V_t-V_i=\frac{-jX_0E}{2R_0}, \quad I_r=\frac{-jX_0E}{2R_0(R_0+jX_0)}$$

Therefore, m (VRC and CRC) is obtained as:

$$m=-jX_0/R_0$$

That is, if $|X_0|>R_0$, then the absolute value for the reflection coefficient is larger than unity.

It is well known that the maximum power is supplied, when a conjugate match of impedances is given. So let us assume that no reflection wave is produced when a conjugate match is achieved in the circuit shown in Fig.3.

ASSUMPTION 3. (Conjugate matching condition)

If $Z_L=Z_G^*$,

then $V_r^{(II)}=I_r^{(II)}=0$.

Substituting Assumption 3 into Def.1, we obtain

$$V_i^{(II)}/I_i^{(II)}=Z_G^*, \quad I_i^{(II)}=E_G/2R_G. \tag{12}$$

Eq.(12) means that the impedance for the incident wave is given by Z_G^*, although such an impedance does not exist in the circuit shown in Fig.3.

VRC and CRC yield under Assumption 3:

$$\frac{V_r^{(II)}}{V_i^{(II)}}=\frac{Z_G}{Z_G^*}\frac{Z_L-Z_G^*}{Z_L+Z_G}, \quad \frac{I_r^{(II)}}{I_i^{(II)}}=\frac{Z_L-Z_G^*}{Z_L+Z_G} \tag{13}$$

We can get the average dissipated power for the incident wave as follows:

$$\mathrm{Re}(V_i^{(II)}I_i^{(II)*})=|E_G|^2/4R_G \tag{14}$$

This power represents the maximum available power.

Let us obtain VTC and CTC.

$$\frac{V_t^{(II)}}{V_i^{(II)}}=\frac{2R_GZ_L}{Z_G^*(Z_L+Z_G)}, \quad \frac{I_t^{(II)}}{I_i^{(II)}}=\frac{2R_G}{Z_L+Z_G} \tag{15}$$

By using Eqs.(13)-(15), we obtain the following relations for average dissipated power of waves.

$$\mathrm{Re}(V_r^{(II)}I_r^{(II)*})+\mathrm{Re}(V_t^{(II)}I_t^{(II)*})=\mathrm{Re}(V_i^{(II)}I_i^{(II)*}) \tag{16}$$

It should be noted that a reflection coefficient with an amplitude larger than unity is never obtained from the relation of Eq.(16).

The complex reflection coefficient given by Eq.(13) is important, because it relates to average dissipated power, so we call it as RC.

3. Application of Thévenin Theory

We consider power transfer for a generalized lossless 2-port. Fig.5(a) shows a lossless 2-port inserted between resistive terminations. We assume that the cascade matrix of the lossless 2-port is presented as:

$$\begin{bmatrix} V_1 \\ I_1 \end{bmatrix}=\begin{bmatrix} A & B \\ C & D \end{bmatrix}\begin{bmatrix} V_2 \\ -I_2 \end{bmatrix} \tag{17}$$

To make the 2-port lossless, it is necessary and sufficient for the cascade matrix to satisfy the following condition[5].

[Lossless condition]

$$\begin{bmatrix} A^* & C^* \\ B^* & D^* \end{bmatrix} \begin{bmatrix} 0 & 1 \\ 1 & 0 \end{bmatrix} \begin{bmatrix} A & B \\ C & D \end{bmatrix} = \begin{bmatrix} 0 & 1 \\ 1 & 0 \end{bmatrix} \tag{18}$$

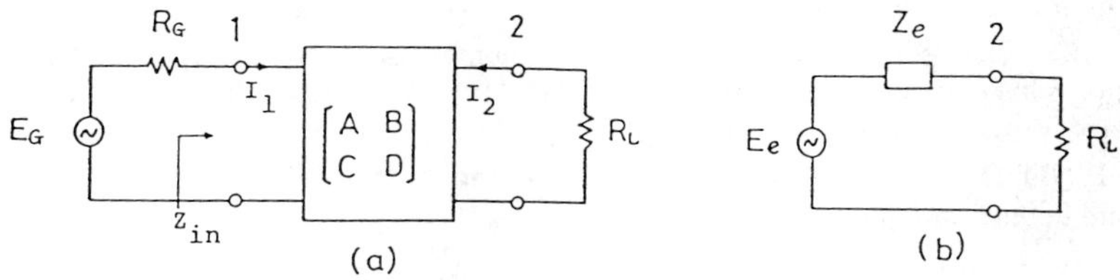

FIGURE 5 (*a*) *A circuit with a lossless transmission* 2−*port*.
(*b*) *The equivalent circuit at the port* 2.

In order to obtain the current I_1 and voltage V_1 at port-1 for the circuit shown in Fig.5(a), we should get the impedance Z_{in} looking toward the load from port-1.

$$Z_{in} = (R_L A + B)/(R_L C + D) \tag{19}$$

Next, in order to obtain the current I_2 and voltage V_2 at port-2 shown in Fig.5(a), we use the Thévenin theorem. That is, the equivalent circuit at port-2 is shown in Fig.5(b), and the voltage E_e and the internal impedance Z_e for the equivalent generator are

$$E_e = E_G/(A + R_G C) \tag{20a}$$

$$Z_e = (B + R_G D)/(A + R_G C) \tag{20b}$$

I_2 and V_2 are obtained by using Eq.(20) and R_L.

The maximum available power for the generator shown in Fig.5(a) is $|E_G|^2/4R_G$, and the maximum available power for the equivalent generator shown in Fig.5(b) is obtained from Eq.(20) as:

$$\frac{|E_e|^2}{4Re(Z_e)} = \frac{|E_G|^2}{4Re(BA^* + R_G^2 DC^* + R_G(DA^* + BC^*))} = |E_G|^2/4R_G \tag{21}$$

Thus, the maximum available power for Thévenin equivalent generator is equal to the one for the original generator of Fig.5(a), provided the 2-port is lossless.

Now we consider RC for the Thévenin equivalent circuit. RC at ports 1 and 2 (we present with Γ_1 and Γ_2) for the circuit shown in Fig.5 is obtained as:

$$\Gamma_1 = \frac{Z_{in} - R_G}{Z_{in} + R_G}, \quad \Gamma_2 = \frac{R_L - Z_e^*}{R_L + Z_e} \tag{22}$$

and we can prove

$$|\Gamma_2| = |\Gamma_1| \tag{23}$$

Thus, we have proved that the maximum available power for Thévenin equivalent generator is equal to the one for the given generator, and that the absolute value of RC for the Thévenin equivalent circuit is equal to the one at the input port, provided the 2-port is lossless. If the circuit is at a steady state after multiple reflections, these properties can be applied to lossless transmission line circuits, because a cascade matrix is a representation of the circuit at a steady state.

We can also employ the Thévenin theorem to lossy 2-port circuits. For that case, the absolute values of RC at port-1 and port-2 are not always equal. Therefore, it is necessary that conjugate matching is usually achieved at port-2, but not at port-1, in order that the load dissipates the maximum available power it can dissipate. The conjugate matching at port-2 is achieved when the load impedance is equal to conjugate complex of the internal impedance for the Thévenin equivalent generator.

4. Complex Voltage Scattering Matrix

Fettweis has introduced WDF [6], [7]. The forward and backward voltage waves for WDF are defined by using voltage, current and a port resistance.

If a given port impedance is a complex number as shown in Fig.3, it is possible to define voltage waves in two different ways [9], [10]. If we apply the definition of Ref.[9], voltage waves are defined as follows:

DEFINITION 4. Incident voltage waves α_1 and α_2, and reflected voltage waves β_1 and β_2 are defined as follows:

$$\alpha_1=(V+Z_G I)/2,\quad \beta_1=(V-Z_G I)/2,$$
$$\alpha_2=(V+Z_L(-I))/2,\quad \beta_2=(V-Z_L(-I))/2.$$

The circuit shown in Fig.3 satisfies $\alpha_2=0$, therefore we obtain the following waves based on Assumption 2.

$$\alpha_1=Z_G I_i=V_i,\quad \beta_1=Z_G I_r=V_r,\quad \beta_2=Z_L I_t=V_t. \tag{24}$$

That is, α_1, β_1 and β_2 are coincident with the incident, reflected and transfer voltage waves, V_i, V_r and V_t obtained under Assumption 2, respectively.

5. Complex Scattering Matrix

Let's consider the other complex voltage wave flow [10]. We can define voltage waves as:

DEFINITION 5a. Incident voltage waves α_1 and α_2, and reflected voltage waves β_1 and β_2 are defined as:

$$\alpha_1=\alpha_G=(V+Z_G I)/2,\quad \beta_1=\beta_G=(V-Z_G^* I)/2,$$
$$\alpha_2=\alpha_L=(V+Z_L(-I))/2,\quad \beta_2=\beta_L=(V-Z_L^*(-I))/2.$$

Substituting the equations based on Assumption 3 into Def.5a, we obtain

$$\alpha_G=R_G I_i^{(II)},\quad \beta_G=R_G I_r^{(II)},\quad \beta_L=R_L I_t^{(II)}. \tag{25}$$

Although the incident and reflected waves obtained by the substitution are represented by voltages, they do not agree with the incident and reflected voltage waves $V_i^{(II)}$ and $V_r^{(II)}$ defined by Assumption 3, but each wave is presented by a current wave multiplied by the real part of impedance. This yields that the waves defined by Def.5a are concerned with the average dissipated power in complex power.

Now, we define the following waves as:

$$a_G=\alpha_G/\sqrt{R_G}=\sqrt{R_G}I_i^{(II)}, \tag{26a}$$

$$b_G=\beta_G/\sqrt{R_G}=\sqrt{R_G}I_r^{(II)}, \tag{26b}$$

$$b_L=\beta_L/\sqrt{R_L}=\sqrt{R_L}I_t^{(II)}. \tag{26c}$$

Then we get the following relations between the waves and the average dissipated power.

$$a_G a_G^*=Re[V_i^{(II)}I_i^{(II)*}], \tag{27a}$$

$$b_G b_G^*=Re[V_r^{(II)}I_r^{(II)*}], \tag{27b}$$

$$b_L b_L^*=Re[V_t^{(II)}I_t^{(II)*}]. \tag{27c}$$

And we get the following relationship among the waves:

$$|b_L|^2 = |a_G|^2 - |b_G|^2 \tag{28}$$

In this case, the waves are concerned with the average dissipated power, so we change Def.5a to the following definition in order to fit a complex scattering matrix described in Ref.[10].

DEFINITION 5b. Incident waves a_1 and a_2, and reflected waves b_1 and b_2 for the circuit shown in Fig.3 are defined as:

$$a_1=(V+Z_G I)/2\sqrt{R_G},\quad b_1=(V-Z_G^* I)/2\sqrt{R_G},$$

$$a_2=(V+Z_L(-I))/2\sqrt{R_L},\quad b_2=(V-Z_L^*(-I))/2\sqrt{R_L}.$$

From Def.5b, we get the following complex scattering matrix:

$$\begin{bmatrix} b_1 \\ b_2 \end{bmatrix} = \begin{bmatrix} \dfrac{Z_L-Z_G^*}{Z_L+Z_G} & \dfrac{2\sqrt{R_G R_L}}{Z_L+Z_G} \\ \dfrac{2\sqrt{R_G R_L}}{Z_L+Z_G} & \dfrac{Z_G-Z_L^*}{Z_G+Z_L} \end{bmatrix} \begin{bmatrix} a_1 \\ a_2 \end{bmatrix} \tag{29}$$

The (1,1) element of this scattering matrix coincides with RC which is given by Eq.(13).

We use a chain scattering matrix for the synthesis of digital filters [3],[4]. The form is defined by making use of incident and reflected waves as:

$$\begin{bmatrix} b_2 \\ a_2 \end{bmatrix} = \begin{bmatrix} \Theta \end{bmatrix} \begin{bmatrix} a_1 \\ b_1 \end{bmatrix} \tag{30}$$

The complex scattering matrix given by Eq.(29) can be transformed to the chain scattering matrix defined in Eq.(30) as follows:

$$\begin{bmatrix} \Theta \end{bmatrix} = diag\left\{d_1, d_2\right\} \frac{1}{\sqrt{1-|\rho|^2}} \begin{bmatrix} 1 & \rho \\ \rho^* & 1 \end{bmatrix} \tag{31}$$

where

$$\rho=(Z_G-Z_L^*)/(Z_G^*+Z_L^*),$$

$$d_1=(Z_G^*+Z_L^*)/|Z_G^*+Z_L^*|,$$

$$d_2=(Z_G+Z_L)/|Z_G+Z_L|.$$

This equation satisfies the condition of "bounded function" or "positive function".[1],[3],[4]

6. Complex Unit Element

A complex filter can be derived through transform on a usual real filter [2]. The complex filter is given by replacing z^{-1} with $z^{-1}e^{j\phi}$ in the complex variable of a real filter. Thus we consider an element whose cascade matrix is obtained by replacing z^{-1} with $z^{-1}e^{j\phi}$ in Eq.(6).

$$\frac{z^{1/2}e^{-j\phi/2}}{2R_0} \begin{bmatrix} Z_0+Z_0^* e^{j\phi}z^{-1} & Z_0 Z_0^*(1-e^{j\phi}z^{-1}) \\ 1-e^{j\phi}z^{-1} & Z_0^*+Z_0 e^{j\phi}z^{-1} \end{bmatrix} \tag{32}$$

Since this element is one of expansions of the usual UE, we call it complex unit element (CUE). Ref[12] shows an analog circuit representation for CUE. The CUE has the

following properties:
(i) It is lossless,
(ii) It is reciprocal,
(iii) It has transmission zeros at $z^{-1}=0$ and $z^{-1}=\infty$,
(iv) The impedance looking right from port 1 is Z_0, if we connect an impedance Z_0 at the load (port 2). This means that the impedance for a forward wave is given by Z_0. The impedance looking left from port 2 is Z_0^*, if we connect an impedance Z_0^* at the input port (port 1). This means that the impedance for a backward wave is given by Z_0^*.

Thus, the impedances of the forward and the backward waves for CUE are conjugate complex at each other. Therefore, we can use the conjugate matching assumed in Assumption 3 for CUE.

7. Conclusion

We have given some new considerations for complex impedances, in order to explain clearly the discrepancy for complex reflection coefficients. That is, this paper examines properties of reflection coefficients for equal matching and conjugate matching. As a result, we show that reflection coefficients for conjugate matching relate to maximal available power and average dissipated power. Therefore, we should use them for the circuits containing reactances in steady state and for complex CTL circuits containing imaginary resistances.

References

[1] V.Belevitch, "Classical network theory", Holden-Day, San Francisco (1968)

[2] T.H.Crystal and L.Ehrman, "The Design and Applications of Digital Filters with Complex Coefficients", IEEE Trans. Audio & Electroacoustics, AU-16, 3, pp.315-320 (Sept. 1968)

[3] E.Deprettere, P.Dewilde and C.V.K.Prabhakara Rao, "Cascade orthogonal filter design", Delft Univ. of Tech. (1984)

[4] P.Dewilde and H.Dym, "Lossless Inverse Scattering Digital Filters, and Estimation Theory", IEEE Trans. Information Theory, IT-30, 4, pp.644-662 (July 1984)

[5] A.V.Efimov and V.P.Potapov, "J-expanding matrix functions and their role in the analytical theory of electrical circuits", Uspekhi Mat. Nauk, pp.65-130, 1973.

[6] A.Fettweis, "Digital filter networks related to classical filter networks", Arch. Elektr. Ubertrag., 25, pp.5-10 (1971)

[7] A.Fettweis, "Wave Digital Filter: Theory and Practice", Proc. IEEE, 74, 2, pp.270-327 (Feb. 1986)

[8] F.E.Gardiol, "Lossy transmission lines", Artech House, Inc.,Norwood, MA (1987)

[9] W.Klein, "Grundlagen der Theorie Elektrischer Schaltungen", Akademie-Verlag, Berlin (1961)

[10] E.S.Kuh and R.A.Rohrer, "Theory of linear active networks", Holden-Day, Inc., San Francisco (1967)

[11] A.Matsumoto (ed), "Microwave Filters and Circuits", Academic Press, New York, 1970.

[12] N.Nagai and M.Suzuki, "Complex Wave Digital Filter for the Analysis of Complex Signals", Proc. ICASSP 86, pp.2603-2606 (April 1986)

[13] N.Nagai and M.Suzuki, "Complex Wave Digital Filter for the Analysis of Complex Signals --Part II", Proc. EUSIPCO 86, pp.155-158 (Sept. 1986)

Optimal Sparse Approximations of Inverse Matrices

H. Nelis, P. Dewilde, and E. Deprettere

In this paper we consider the problem of approximating the inverse of a positive definite matrix by a sparse matrix with a prescribed sparsity pattern, and which meets the condition of monotone transitivity.

1 Introduction

Suppose that A is a positive definite matrix, and let L be the upper triangular matrix with positive diagonal entries that satisfies $A = LL^*$. In this paper we address the problem of determining the upper triangular matrix F that is such that the Frobenius norm $\|I - FL\|$ is minimal and F^*F is zero on the complement of a support set $\mathcal{S}$. We generalize the results in [1], and consider the case where $\mathcal{S}$ is monotone transitive. We show that the minimum occurs for $F = (D_{ME}^{-1}D)L_{ME}^{-1}$, where L_{ME} is defined in a way similar to L, A_{ME} is the maximum entropy extension of the part of A with support on $\mathcal{S}$, and D and D_{ME} are the projections of L and L_{ME} on the diagonal.

2 Preliminaries

Suppose that $A = [a_{ij}]$, $i, j = 1, \ldots, n$, is specified only on a subset $\mathcal{S}$ of the set of index pairs $\{(i,j) \mid i,j = 1,\ldots,n\}$, and let $(i,i) \in \mathcal{S}$ for all i — the diagonal entries in A are specified. We shall always assume that the diagonal entries are equal to one. A matrix that coincides with A on $\mathcal{S}$ is called an extension of A. The following result is due to Grone et al.

Theorem 1 ([2]) *Suppose that A is a positive definite matrix that is specified on a set $\mathcal{S}$ that contains the diagonal pairs, but is arbitrary* *otherwise, and let $\mathcal{E}$ be the set of positive definite extensions of A. Then, there is a unique matrix B in $\mathcal{E}$ such that*

$$\det B = \max\{ \det E \mid E \in \mathcal{E} \}.$$

Moreover, B is the unique positive definite extension whose inverse satisfies

$$(B^{-1})_{ij} = 0 \quad \forall (i,j) \notin \mathcal{S}.$$

Due to the analogy with the maximum entropy inequality in spectral estimation theory, B is called the maximum entropy extension. We denote it by A_{ME}.

L and M refer to the (unique) upper triangular matrices with positive diagonal entries that satisfy $A = LL^* = M^*M$. L_{ME} and M_{ME} are defined in a similar way.

$\underline{\underline{\mathrm{P}}}_0$ is the operator that projects a matrix on the diagonal:

$$(\underline{\underline{\mathrm{P}}}_0 A)_{ij} = \begin{cases} a_{ij} & \text{if } i = j; \\ 0 & \text{otherwise.} \end{cases}$$

The symbols $\det A$, $\mathrm{tr} A$, and $\|A\|$ denote the determinant, the trace, and the Frobenius norm of A. We denote the zero matrix and identity matrix by 0 and I.

A subset $\mathcal{S}$ of the set of index pairs $\{ (i,j) \mid i,j = 1,\ldots,n \}$ is called a

- *staircase band* if for all $i \le k \le l \le j$ $[i \ge k \ge l \ge j]$ $(i,j) \in \mathcal{S}$ implies that $(k,l) \in \mathcal{S}$ — see Figure 1(a);
- *monotone transitive set* if for all $i \le j \le k$ $[i \ge j \ge k]$ $(i,j) \in \mathcal{S}$ and $(i,k) \in \mathcal{S}$ implies that $(j,k) \in \mathcal{S}$ — see Figure 1(b).

3 Results

The notion of monotone transitive sets was introduced by Rose [3], who examined the problem of fill-ins in the triangular factors of sparse, positive definite matrices, and showed that if a positive definite matrix C is zero on the complement of a set $\mathcal{S}$, then its triangular factor M vanishes on the upper triangular part of the complement of the set, irrespective of the values of the entries in C, if, and only if, $\mathcal{S}$ is monotone transitive. Hence, if F is upper triangular, and F^*F vanishes on the complement of a monotone transitive set, then F must be zero on the upper triangular part of the complement of the set. Clearly, staircase bands are monotone transitive.

(a)

$$\begin{bmatrix}
\times & \times & \times & & & & & & & & & \\
\times & \times & \times & & & & & & & & & \\
\times & \times & \times & \times & & & & & & & & \\
 & & \times & \times & \times & \times & \times & \times & & & & \\
 & & & \times & \times & \times & \times & \times & & & & \\
 & & & \times & \times & \times & \times & \times & & & & \\
 & & & \times & \times & \times & \times & \times & \times & & & \\
 & & & \times & \times & \times & \times & \times & \times & & & \\
 & & & & & \times & \times & \times & \times & \times & \times & \times \\
 & & & & & & & & \times & \times & \times & \times \\
 & & & & & & & & \times & \times & \times & \times \\
 & & & & & & & & \times & \times & \times & \times
\end{bmatrix}$$

(b)

$$\begin{bmatrix}
\times & \times & \times & \times & & & & & & & & \times \\
\times & \times & \times & \times & & & & & & & & \times \\
\times & \times & \times & \times & & & & & & & & \times \\
\times & \times & \times & \times & \times & & & & & \times & & \times \\
 & & & \times & \times & & & & & \times & & \times \\
 & & & & & \times & \times & \times & \times & \times & & \\
 & & & & & \times & \times & \times & \times & \times & & \\
 & & & & & \times & \times & \times & \times & \times & & \\
 & & & & & \times & \times & \times & \times & \times & & \times \\
 & & & \times & \times & \times & \times & \times & \times & \times & \times & \times \\
 & & & & & & & & & \times & \times & \times \\
\times & \times & \times & \times & \times & & & & \times & \times & \times & \times
\end{bmatrix}$$

Figure 1: A matrix that is specified on (a) a staircase band and (b) a monotone transitive set. Specified entries are marked with an '×'; unspecified entries are blank.

With $\mathcal{S}$ Rose associated a graph $\mathcal{G}$ that has a vertex for every index i and an undirected edge between vertex i and j if $(i,j) \in \mathcal{S}$. He proved that if, and only if, every cycle in $\mathcal{G}$ that consists of four edges or more contains a chord, i.e., an edge joining two nonconsecutive vertices in the cycle, then there is a permutation matrix P such that the support of PCP^* is monotone transitive. Graphs with this property are said to be *chordal* or *triangulated*. For example, if C is as shown in Figure 2, then the cycle 2–7–10–5–2 has four edges yet no chord — we cannot permute C to a matrix with monotone transitive support. If, however, we partition C into blocks of size 4×4, then the graphs of the principal submatrices $\mathbf{C}(1,2)$ and $\mathbf{C}(2,3)$ are chordal — see Figure 3. Arranging the rows and columns of $\mathbf{C}(1,2)$ according to the order $1-5-2-6-3-7-4-8$,

$$
C \;=\; \begin{array}{c} \\ 1\\2\\3\\4\\5\\6\\7\\8\\9\\10\\11\\12 \end{array}
\begin{array}{c}
\begin{array}{cccc|cccc|cccc} 1&2&3&4&5&6&7&8&9&10&11&12 \end{array}\\
\left[\begin{array}{cccc|cccc|cccc}
\times&\times&&&\times&\times&&&&&&\\
\times&\times&\times&&\times&\times&\times&&&&&\\
&\times&\times&\times&&\times&\times&\times&&&&\\
&&\times&\times&&&\times&\times&&&&\\
\hline
\times&\times&&&\times&\times&&&\times&\times&&\\
\times&\times&\times&&\times&\times&\times&&\times&\times&\times&\\
&\times&\times&\times&&\times&\times&\times&&\times&\times&\times\\
&&\times&\times&&&\times&\times&&&\times&\times\\
\hline
&&&&\times&\times&&&\times&\times&&\\
&&&&\times&\times&\times&&\times&\times&\times&\\
&&&&&\times&\times&\times&&\times&\times&\times\\
&&&&&&\times&\times&&&\times&\times
\end{array}\right]
\end{array}
$$

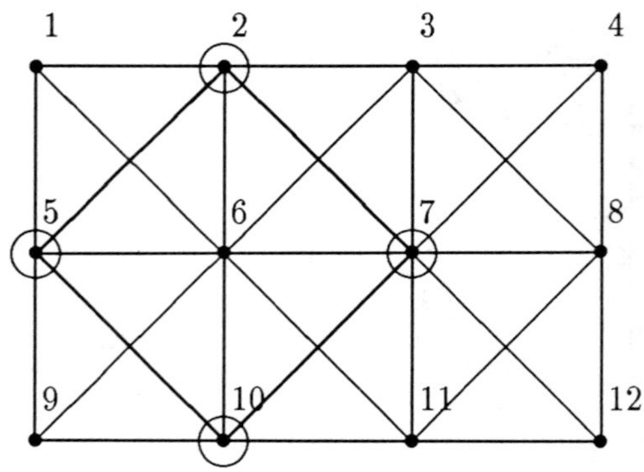

Figure 2: A matrix with support on a multiple band and its graph. Nonzero entries are marked with an '×'; vanishing entries are blank.

$$
\begin{array}{c|cccc|cccc|}
 & 1 & 2 & 3 & 4 & 5 & 6 & 7 & 8 \\
1 & \times & \times & & & \times & \times & & \\
2 & \times & \times & \times & & \times & \times & \times & \\
3 & & \times & \times & \times & & \times & \times & \times \\
4 & & & \times & \times & & & \times & \times \\
\hline
5 & \times & \times & & & \times & \times & & \\
6 & \times & \times & \times & & \times & \times & \times & \\
7 & & \times & \times & \times & & \times & \times & \times \\
8 & & & \times & \times & & & \times & \times
\end{array}
\qquad
\begin{array}{c|cccccccc|}
 & 1 & 5 & 2 & 6 & 3 & 7 & 4 & 8 \\
1 & \times & \times & \times & \times & & & & \\
5 & \times & \times & \times & \times & & & & \\
2 & \times & \times & \times & \times & \times & \times & & \\
6 & \times & \times & \times & \times & \times & \times & & \\
3 & & & \times & \times & \times & \times & \times & \times \\
7 & & & \times & \times & \times & \times & \times & \times \\
4 & & & & & \times & \times & \times & \times \\
8 & & & & & \times & \times & \times & \times
\end{array}
$$

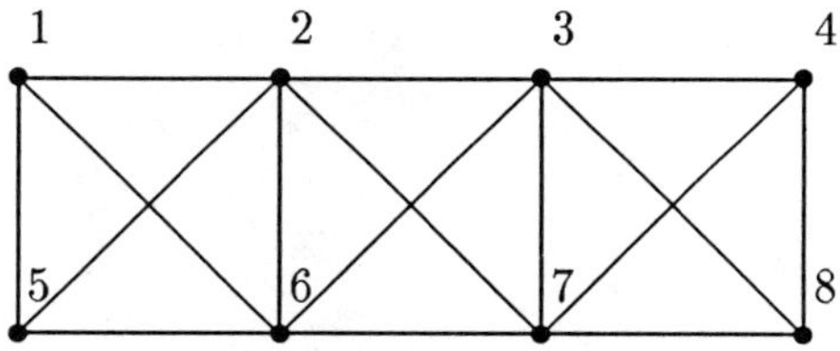

Figure 3: The principal submatrix $\mathbf{C}(1,2)$, the permuted matrix, and their graph.

we see that the support of the permuted matrix is staircase and hence monotone transitive.

Chordal graphs also have import on the existence of positive definite extensions of partially specified, not necessarily positive definite Hermitian matrices. In [2] Grone et al. showed that if $\mathcal{G}$ is chordal, and all completely specified principal submatrices with support on $\mathcal{S}$ are positive definite, then positive definite extensions necessarily exist, and if $\mathcal{G}$ is not chordal, then they do not in general exist.

The following results are generalizations of the approximation theorems in [1] to the case of monotone transitive sets. $\mathcal{F}$ denotes the set of upper triangular matrices that vanish on the upper triangular part of the complement of $\mathcal{S}$. D and D_{ME} are defined as $D = \underline{\underline{\mathrm{P}}}_0 L$ and $D_{ME} = \underline{\underline{\mathrm{P}}}_0 L_{ME}$.

Lemma 1 *Suppose that* $A = [a_{ij}]$, $i, j = 1, \ldots, n$, *is a positive definite matrix that is specified on a monotone transitive set* $\mathcal{S}$, *and let* $F \in \mathcal{F}$ *and* $H \in \mathcal{F}$. *Then,*

$$\underline{\underline{\mathrm{P}}}_0(FAH^*) = \underline{\underline{\mathrm{P}}}_0(FA_{ME}H^*).$$

Proof Suppose that $B = A - A_{ME}$. Because for all $i \leq j \leq k$ $(i,j) \in \mathcal{S}$ and $(i,k) \in \mathcal{S}$ implies that $(j,k) \in \mathcal{S}$ and hence $(k,j) \in \mathcal{S}$ — $\mathcal{S}$ is monotone transitive — and B vanishes on $\mathcal{S}$, we have that for all (i,j) in the upper triangular part of $\mathcal{S}$

$$(FB)_{ij} = \sum_{k=i}^{n} f_{ik} b_{kj} = 0.$$

Because H is upper triangular and zero on the upper triangular part of the complement of $\mathcal{S}$, it follows that

$$\underline{\underline{\mathrm{P}}}_0(FBH^*) = 0$$

and hence

$$\underline{\underline{\mathrm{P}}}_0(FAH^*) = \underline{\underline{\mathrm{P}}}_0(FA_{ME}H^*).$$

■

Lemma 2 *Let A and F be as defined in Lemma 1. Then,*

1. $\underline{\underline{\mathrm{P}}}_0(L_{ME}^{-1} A L_{ME}^{-*}) = I$;

2. $\underline{\underline{\mathrm{P}}}_0\left(FA(D_{ME}^{-1} L_{ME}^{-1})^*\right) = \underline{\underline{\mathrm{P}}}_0 F$.

Proof Because A_{ME}^{-1} vanishes on the complement of $\mathcal{S}$, and $\mathcal{S}$ is monotone transitive, we have that $L_{ME}^{-1} \in \mathcal{F}$, and the first equality in the lemma follows directly from Lemma 1. Furthermore, by the same lemma,

$$\begin{aligned} \underline{\underline{\mathrm{P}}}_0\left(FA(D_{ME}^{-1} L_{ME}^{-1})^*\right) &= \underline{\underline{\mathrm{P}}}_0\left(FA_{ME}(D_{ME}^{-1} L_{ME}^{-1})^*\right) \\ &= \underline{\underline{\mathrm{P}}}_0(FL_{ME} D_{ME}^{-*}), \end{aligned}$$

which evaluates to $\underline{\underline{\mathrm{P}}}_0 F$, as F and L_{ME} are upper triangular and D_{ME}^{-*} is diagonal. ■

The second equality in Lemma 2 is a reproducing property.

Let Π denote the operator that projects a matrix on $\mathcal{F}$ with respect to the inner product $\langle B, C\rangle_A = \mathrm{tr}(BAC^*)$, and suppose that $\|B\|_A = \sqrt{\langle B, B\rangle_A}$.

Lemma 3 *Let A be as defined in Lemma 1, F upper triangular, and $H \in \mathcal{F}$. Then,*

$$\underline{\underline{\mathrm{P}}}_0(FAH^*) = \underline{\underline{\mathrm{P}}}_0(\Pi(F)AH^*).$$

Proof Suppose that $K = F - \Pi F$. It follows from the definition of Π that

$$\langle K, H\rangle_A = \mathrm{tr}(KAH^*) = 0 \quad \forall\, H \in \mathcal{F}.$$

Specializing H to E_{ij}, for all (i,j) in the upper triangular part of $\mathcal{S}$, where

$$\left(E_{ij}\right)_{kl} = \begin{cases} 1 & \text{if } (k,l) = (i,j); \\ 0 & \text{otherwise,} \end{cases}$$

we obtain that KA is zero on the upper triangular part of $\mathcal{S}$. Hence, we have $KA = N + O$, for some strictly lower triangular N and an upper triangular O that vanishes on the upper triangular part of $\mathcal{S}$.

Because NH^* is strictly lower triangular, and OH^* vanishes on the diagonal — O is upper triangular and zero on the upper triangular part of $\mathcal{S}$, and H vanishes on the upper triangular part of the complement of $\mathcal{S}$ — it follows that

$$\underline{\underline{\mathrm{P}}}_0(KAH^*) = \underline{\underline{\mathrm{P}}}_0(NH^*) + \underline{\underline{\mathrm{P}}}_0(OH^*) = 0$$

and hence

$$\underline{\underline{\mathrm{P}}}_0(FAH^*) = \underline{\underline{\mathrm{P}}}_0(\Pi(F)AH^*).$$

■

Theorem 2 *Let A be as defined in Lemma 1. Then,*

$$\min\{\, \|I - FL\| \mid F \in \mathcal{F}\,\}$$

occurs for $F = (D_{ME}^{-1}D)L_{ME}^{-1}$. Moreover,

$$\|I - ((D_{ME}^{-1}D)L_{ME}^{-1})\, L\| = \sqrt{\mathrm{tr}\left(I - (D_{ME}^{-1}D)^2\right)}.$$

Proof Suppose that $F \in \mathcal{F}$. Because by Lemma 3

$$\underline{\underline{\mathrm{P}}}_0\left(\Pi(L^{-1})AF^*\right) = \underline{\underline{\mathrm{P}}}_0(D^*F^*),$$

and by a proof similar to the proof of Lemma 2(2)

$$\underline{\underline{\mathrm{P}}}_0\left(((D_{ME}^{-1}D)L_{ME}^{-1})\, AF^*\right) = \underline{\underline{\mathrm{P}}}_0(D^*F^*),$$

we have

$$\mathrm{tr}\left(\Pi(L^{-1})AF^*\right) = \mathrm{tr}\left(((D_{ME}^{-1}D)L_{ME}^{-1})\, AF^*\right) \quad \forall\, F \in \mathcal{F}.$$

Since A is invertible, it readily follows that

$$\Pi L^{-1} = (D_{ME}^{-1}D)L_{ME}^{-1}.$$

Hence,

$$\min\{ \|L^{-1} - F\|_A \mid F \in \mathcal{F} \}$$

and, consequently,

$$\min\{ \|I - FL\| \mid F \in \mathcal{F} \}$$

— $\|L^{-1} - F\|_A = \|I - FL\|$ — occur for $F = (D_{ME}^{-1} D) L_{ME}^{-1}$.

Because by Lemma 2(1)

$$\underline{\underline{P}}_0 (L_{ME}^{-1} A L_{ME}^{-*}) = I,$$

we have

$$\operatorname{tr}\left(\left((D_{ME}^{-1} D) L_{ME}^{-1} \right) A \left((D_{ME}^{-1} D) L_{ME}^{-1} \right)^* \right) = \operatorname{tr}\left((D_{ME}^{-1} D)^2 \right),$$

so that

$$\begin{aligned} \|I - \left((D_{ME}^{-1} D) L_{ME}^{-1} \right) L\| &= \sqrt{\operatorname{tr}\left(I - 2(D_{ME}^{-1} D)^2 + (D_{ME}^{-1} D)^2 \right)} \\ &= \sqrt{\operatorname{tr}\left(I - (D_{ME}^{-1} D)^2 \right)}. \end{aligned}$$

■

The case where $\mathcal{S}$ is arbitrary is still an open problem.

Algorithms for computing $F = (D_{ME}^{-1} D) L_{ME}^{-1}$, or approximations to it, can be found in [4].

References

[1] P. Dewilde and E. Deprettere, The generalized Schur algorithm: approximation and hierarchy, *Operator Theory: Advances and Applications*, **29**:97–116, 1988.

[2] R. Grone, C. Johnson, E. Sá, and H. Wolkowicz, Positive definite completions of partial Hermitian matrices, *Linear Algebra and its Applications*, **58**:109–124, 1984.

[3] D. Rose, A graph-theoretic study of the numerical solution of sparse positive definite systems of linear equations, In *Graph Theory and Computing*, pages 183–217, Academic Press, 1972.

[4] H. Nelis, *Sparse Approximations of Inverse Matrices*, PhD thesis, Dept. of Electrical Engineering, Delft University of Technology, Delft, The Netherlands, 1989.

ON THE FREE OVERFLOW OSCILLATIONS IN DIGITAL FILTERS

R. BOITE
YANG XIN-QIANG

Abstract
The overflow oscillation in digital filter with fixed-point 2's complement realization is a well-known problem. The influence of the initial state on the free response of a second degree direct section is investigated; some properties of the oscillations with and without overflow are described.

1 Introduction and description of the system

Consider the second order direct structure of a digital filter with zero input (fig.1), realized in fixed-point 2's complement. It is linearly stable for parameters located in the triangle of stability (fig.2). However, due to the adder characteristic (fig.3), overflow oscillations can occur for some initial conditions; these oscillations may or may not be maintained. The non linear model of the stucture of fig.1 satisfy the equation:

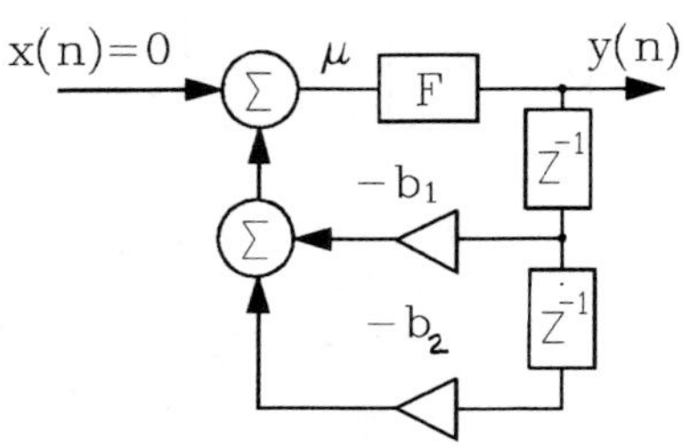

FIGURE 1 Second order direct structure.

$$y(n) = F[-b_1 . y(n-1) - b_2 . y(n-2)] = F[u] \quad (1)$$

where: $F[u]=u, \quad |u|<1$ (no overflow) (2a)

$=u-2, \quad 1<u$ (2b)

$=u+2, \quad u<-1$ (2c)

The linear model is characterized by $F(u)=u$, all u. For b_1 and b_2 in the triangle of stability, one has:

$$|u| = |-b_1.y(n-1)-b_2.y(n-2)| < 3 \qquad (3)$$

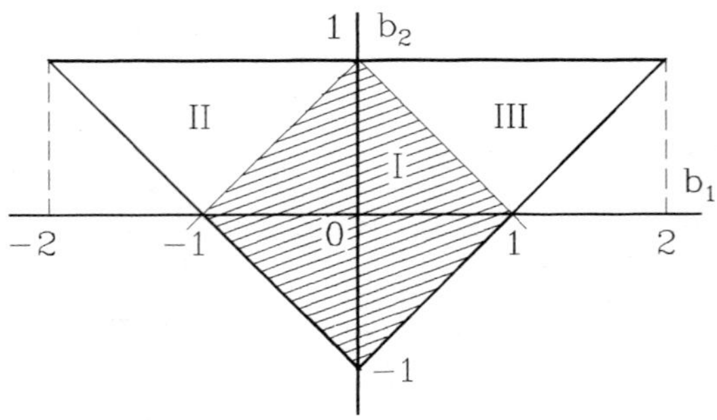

Figure 2 Triangle of stability.

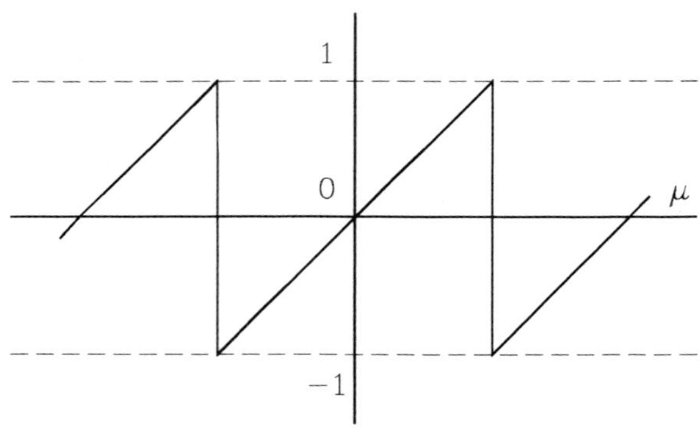

FIGURE 3 Adder characteristic (2's complement).

Moreover, in the hatched region (region I) of fig.2, one has necessarily $|u|<1$ and overflow never happen [2]. It is thus sufficient to consider b_1 and b_2 in the regions II and III of the triangle and even only in region II for symmetry reasons. The state variables are:

$$w_1(n)=y(n-1), \quad w_2(n)=y(n-2) \qquad (4)$$

2 The overflow free domain of the state plane

Overflow do not occur iff $|u(n)|<1$; the necessary and sufficient conditions for the absence of overflow for any n are thus:

$$|w_1(0)|<1 \quad , \quad |w_2(0)|<1 \quad ; \quad |u(n)|<1 \quad , \quad n=0,1,..,\infty \tag{5}$$

For complex conjugate poles:

$$z_{1,2} = r.\exp(j\theta) \quad ; \quad b_2=r^2 \quad , \quad b_1=-2r.\cos\theta \tag{6}$$

one easily finds:

$$u(n) = r^{n+1}/\sin\theta.[\sin(n+2)\theta.w_1(0)-r.\sin(n+1)\theta.w_2(0)] \tag{7}$$

Thus we must have:

$$|\alpha_1(n).w_1(0) + \alpha_2(n)w_2(0)| < 1 \quad , \quad n=0,1,..,\infty \tag{8}$$

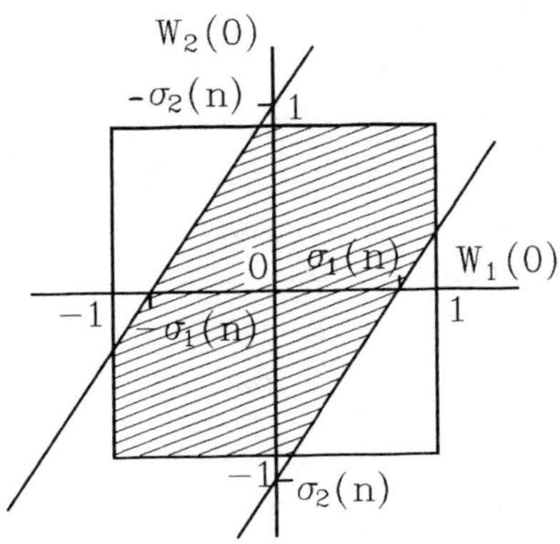

Figure 4 Domain of initial conditions ensuring no overflow at time n.

The hatched region of fig.4 corresponds to the initial conditions such that overflow does not occur at instant n and the overflow free domain corresponds to the intersection of such regions for $n=0,1,..,\infty$.

Numerical examples are illustrated at fig. 5 For $b_2=0.9$; the overflow free domain is delimited by straight lines but for $b_2=0.9999$, it looks like a perfect ellipse considered by CHUA and LIN [1] for $b_2=1$.

a) $b_1=-1.82$ $b_2=-0.9$ b) $b_1=-1.82$ $b_2=0.9999$

Figure 5 Overflow free regions.

3 Characterization of the overflow oscillations

For initial conditions located outside the overflow free domain, overflows can occur: there exists regions in the state plane corresponding to steady-state oscillations with period N (N=2,3,..). For initial conditions located outside these regions and outside the overflow free domain, the free response exhibits overflows, but after some time, the trajectories may enter into the overflow free domain. The domain of such initial conditions will be called **non-linear stable domain.**

The free response of the filter is the solution of (1), for prescribed initial conditions $[w_1(0),w_2(0)]$. Given (1,2), we can write:

$$y(n) + b_1.y(n-1) + b_2.y(n-2) = a(n) \qquad (9)$$

with:
$$a(n)=0 \quad \text{if} \quad |u|<1 \qquad (10a)$$
$$=2 \quad \text{if} \quad u<-1 \qquad (10b)$$
$$=-2 \quad \text{if} \quad 1<u \qquad (10c)$$

and consider {a(n)} as an input sequence, which of course is a priori unknown. Now, given {a(n)} we can solve (9), compute u and verify that the chosen sequence satisfy

(10a,b,c). This is due to EBERT et al.[2] who showed that it is sufficient to verify that the solution satisfy $|y(n)|<1$ for all n; an accepted sequence $\{a(n)\}$ is said to be **admissible**. The solution of (9) is the convolution:

$$y(n) = h(n)*a(n) \tag{11}$$

where h(n) is the impulse response:

$$h(n) = (z_1-z_2)^{-1}.(z_1^{n+1} - z_2^{n+1}) \ , \ n\geq 0 \tag{12}$$

4 The periodic oscillations

Consider a periodic oscillation of period N. From (11):

$$y_N(n) = h(n)* \sum_{k=-\infty}^{0} a(n+kN) = \sum_{i=0}^{N-1} a(i).h_N(n-i), \ n=0,1,.,N-1 \tag{13}$$

$$\text{with: } h_N(n)= \sum_{k=-\infty}^{0} h(n+kN)= \frac{1}{(z_1-z_2)}.\left[\frac{z_1^{n+1}}{1-z_1^N} - \frac{z_2^{n+1}}{1-z_2^N}\right] \tag{14}$$

We must verify that $|y_N(n)|<1$, (n=0,1,..,N-1).The expression (13) can be written as:

$$\begin{vmatrix} y_N(0) \\ y_N(1) \\ \\ \\ y_N(N-1) \end{vmatrix} = \begin{vmatrix} h_N(0) & h_N(N-1) & \ldots\ldots & h_N(1) \\ h_N(1) & h_N(0) & \ldots\ldots & h_N(2) \\ & & & \\ & & & \\ h_N(N-1) & & & h_N(0) \end{vmatrix} . \begin{vmatrix} a(0) \\ a(1) \\ \\ \\ a(N-1) \end{vmatrix} \tag{15}$$

or:

$$y_N = H_N.a_N \tag{16}$$

We have used the fact that $h_N(n)=h_N(n+N)$ for all n. H_N is a non symmetric TOEPLITZ matrix: to any cyclic permutation of the components of the vector a_N correspond the same cyclic permutation for y_N.

The search for periodic oscillations is thus straightforward: given a period N, all vectors a_N with $a_N(n) \in \{0,2,-2\}$ are tested; a vector is admissible if $|y_N(n)|<1$ for n=0,1,...,N-1; the periodic oscillation is the trajectory of the state vector $w_N(n)$:

$$w_{N,1}(n) = y_N(n-1) = y_N(n-1+N)$$
$$w_{N,2}(n) = y_N(n-2) = y_N(n-2+N) \;, \quad n=0,1,\ldots,N-1 \qquad (17)$$

In principle, for a period N, N^3 vectors a_N should be tested but a great number of tests can be avoided. Such an investigation has been done for b_1=-1.82, b_2=.99, and for N=2 to 7. Observe that if a_N is admissible, $-a_N$ is also admissible. Observe also that some vectors a_N include sub-vectors $a_{N/k}$ and do not generate new periodic solutions. The corresponding trajectories are given at figure 6.

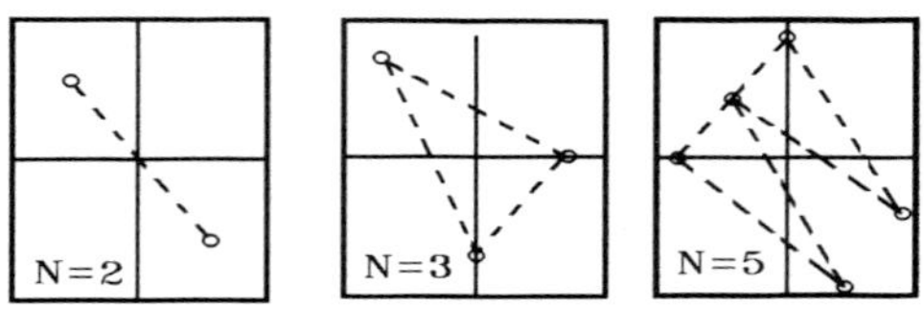

Figure 6 Steady-state oscillations

5 Initial states leading to periodic oscillations

Consider the periodic oscillation corresponding to an admissible vector a_N; it obeys equation (16) for n=0,1,..,N-1 if the initial conditions are $w_{N,1}(0)=y_N(-1)$, $w_{N,2}(0)=y_N(-2)$, with $w_N(n)$ given by (17). Let us look for the region of accessibility to that limit cycle. If the initial conditions are perturbed, the corresponding perturbation of the response is given by an expression similar to (7) but with $\delta w(0)$ instead of w(0).
It is expected to converge asymptotically to 0, if $b_2<1$. The admissible perturbations $\delta w(0)$ is such that the pertubed response satisfy:

$$|y_N(n)+\delta y(n)|<1 \quad , \quad n=0,1,\ldots,\infty \tag{18}$$

The domain of initial conditions such that the trajectories converge asymptotiquely to the periodic response $y_N(n)$ is thus determined by the following set of conditions, similar to (7):

$$|y_N(n) + \alpha_1(n).\delta w_1(0) + \alpha_2(n)\delta w_2(0)| < 1 \ , \ n=0,1,..,\infty \tag{19}$$

$$|w_{N,i}(0) + \delta w_i(0)|<1 \ , \quad i=1,2 \tag{20}$$

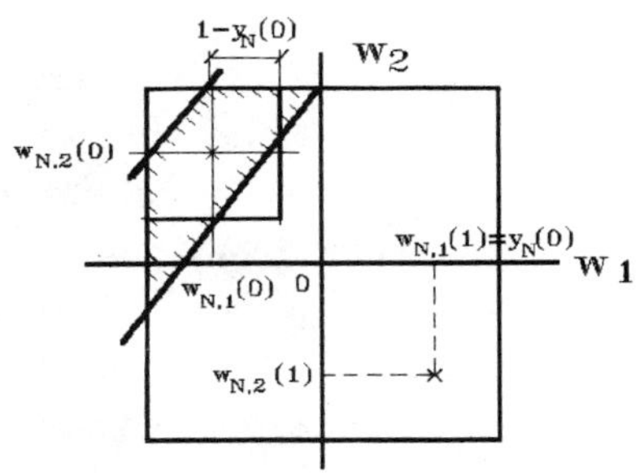

Figure 7 Determination of an oscillation region (N=2).

It is thus delimited by a set of straight lines in the same way as was delimited the overflow free domain, as shown at figure 8a,b; however, due to the presence in (19) of $y_N(n)$, the shapes of the two domains can differ, exept for special cases. This is illustrated at figures 8.

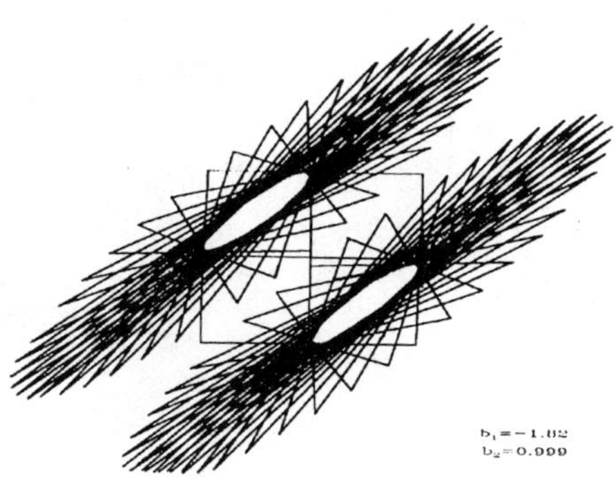

Figure 8 Determination of region of period 2.

8 Conclusions and open problems

Overflow oscillations of the free response of a digital filter realized in fixed point, 2's complement, have been investigated; sets of initial conditions assuring the absence of overflows or leading to periodic overflow oscillations have been located in the state plane. The analysis method is more intuitive than that developped in [1]; moreover we mainly consider the case $b_2<1$, and we show how our results converge to those of [1] when b_2 tends to unity.

First of all, the border of the overflow free domain converges to an ellipse whose equations is derived in [1]; this is not the case for $b_2<1$ and the border is no longer the enveloppe of a trajectory. Regions of initial states leading asymptotically to periodic oscillations are delimited in a way similar but not identical to that used for finding the border of the overflow free domain. It remains also to examine whether there exists an upper bound for the allowed periods N and more generally, whether beside the regions of periodic oscillations, the overflow free domain and the non linear stable domain there is room for chaos.

7 References

[1] L.O.CHUA, TAO LIN,"Chaos in digital filters", IEEE Trans. on Circ.and Syst., CAS-22, June 1988, pp 648-658.

[2] P.M. EBERT et al, "Overflow oscillations in digital filters", Bell Syst.Techn.J.,pp 2999-3020, Nov. 1969.

FILTERING WITH NEURAL MODELS

Elise de Doncker, John Kapenga and Robert Trenary

Abstract

As an implementation and simulation results of a back propagation network for noise filtering will be outlined, this paper illustrates the intimate relationship between neurocomputing and filtering. Simulations will be described in detail, with a focus on the adaptation of the network to time-varying signals and on the smoothing of high noise by a network trained with noisy signals. Aspects of parallel implementations of the network will be given.

1. Introduction

One of the purposes of this presentation is to draw the attention of researchers in signal processing, to the promises of neural networks in the area of noise filtering.

The foundations of neural learning models are described extensively by Rumelhart and McClelland in [13]. Stubbs [14] discusses the relation of the computational structure of neural networks to the brain. McClelland and Rumelhart [9] and NeuralWare, Inc. [1], provide canned software for experimentation and demonstration (the latter provides object code), together with an overview of background material and discussions of the applications. A back propagation model is used by Guiver and Klimasauskas [6] for noise filtering and is shown able to learn from noisy data, for an example of repetitive EKG data.

We shall give simulation results of a back propagation network, in order to study its learning behaviour in the presence of large noise on the training signals. Furthermore we demonstrate its ability to adapt to a time-varying signal. We give an implementation of the learning procedure and describe the influence of variations in the parameters. Aspects of parallel implementations are briefly discussed.

Widrow and Hoff [17] proposed a network for analog filtering (without bearing to neural nets), in the form of an adaptive switching network, their *Linear Adaptive Combiner* (ALC). Their *Least Mean Square* (LMS) learning rule is a special case of the now commonly used *delta rule* in neural models. The LMS rule is also known as the *Widrow-Hoff learning rule*. In the ALC, error correction resulted in alterations of the resistances in the circuit, based on the observed discrepancy between a target output and the circuit's.

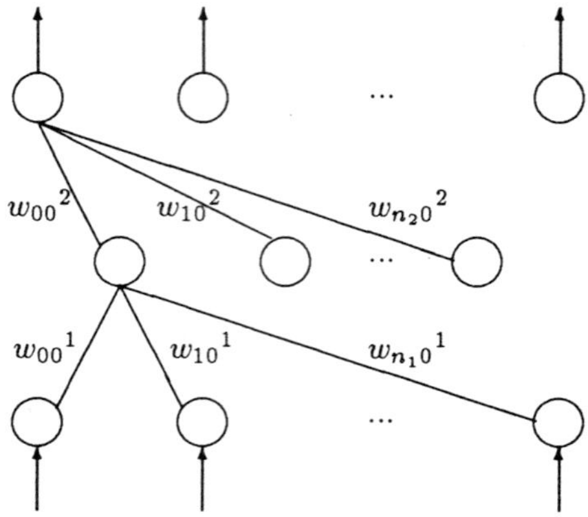

FIGURE 1 Standard back propagation network with three layers

The standard back propagation network and generalized delta rule [11, 12, 13] are described in section 2. The error correction scheme known as back propagation is bases on the work of Werbos [16]. Our version and implementation of it are presented in section 3. Section 4 contains simulation results. Modifications for parallel implementation and further avenues are suggested in section 5.

We aim to provide carefully done outlines, in order that the reader with some mathematical background and understanding of algorithms would gain insight in the procedures used and be able to easily interpret and extract information from the results.

2. The back propagation network

Picture a classical back propagation network, containing an input and an output layer and one or more *hidden* or *associative* layers. We shall label the layers from 0 (input) to L (output). Figure 1 should help clarifying the notations. Each layer contains a number of neurons also referred to as *processing elements* or PEs [14]. For the filtering application, the input and output layers have the same number of PEs [6], so that training of the network can be aimed at reproducing of the input signal at the output layer.

Let us call $a_j{}^k$ the *output* or *activation* of neuron j in layer k. We use a superscript to denote the layer as in [6]. Furthermore let $w_{ij}{}^k$ denote the *weight* on the connection between neuron i in layer $k-1$ and neuron j in k; it represents the connection strength between the two processing elements [14]. Neuron j in layer $k = 1, 2, ..., L$ receives as input a linear combination $s_j{}^k$, of the activations a_i^{k-1} from the PEs it is connected with

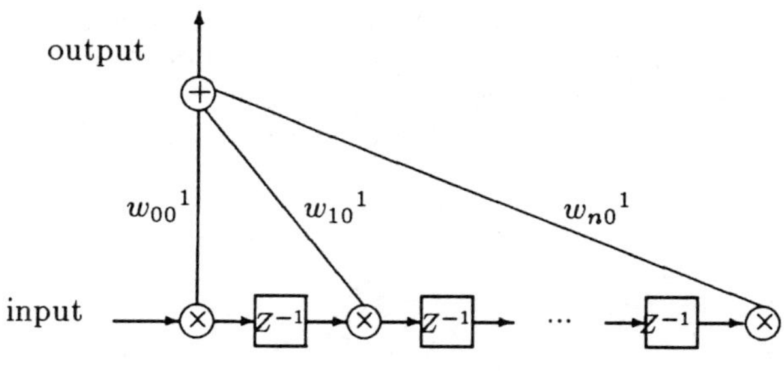

FIGURE 2 Standard FIR

on the previous layer, each weighted with the corresponding connection strength, $w_{ij}{}^k$, i.e.

$$(1) \quad s_j{}^k = \sum_i w_{ij}{}^k a_i{}^{k-1}, \quad k = 1, 2, ..., L.$$

For the neurons in layer 0, we define $s_j{}^0$ as the input i_j to neuron j,

$$(2) \quad s_j{}^0 = i_j.$$

We compute the output of the PEs on layers $k = 1, 2, ..., L - 1$ as the function value $a_j{}^k = \varphi(s_j{}^k)$ with $\varphi(x)$ given by

$$(3) \quad \varphi(x) = \frac{1}{1 + e^{-x}},$$

the *sigmoid* or *logistic* function. Note that this can be considered as an adaptation of a threshold function which would set the output to 1 or 0 depending on whether or not the input reaches a given threshold value. This corresponds to either firing or not firing of the neuron, as is typical for the *perceptron* [10]. At the input and the output layer, set

$$(4) \quad a_j{}^k = s_j{}^k, \quad k = 0 \ or \ k = L;$$

i.e. here the activation function φ is the identity function.

As long as φ is non-decreasing and differentiable it is termed a *semi-linear* activation function [12, 13]. In the original network by Widrow and Hoff [17], with only an input and an output layer, φ is a linear function. Consequently it is a special case of the back propagation network of Figure 1, where the weights from the hidden layer to output unit j, say, are all 0 except $w_{ij}^L = 1$. We also note that it resembles (structurally and functionally for forward propagation) the classical FIR network represented in Figure 2.

The learning phase, which precedes the actual use of the network, is characterized by the back propagation. Learning is achieved by applying weight changes on the connections between the PEs, while moving through the layers in reverse order ($k = L, L - 1, ..., 1$). The learning phase thus consists of successive cycles, each with a forward and backward propagation through the network. From the input signal, the forward propagation results in the computation of output values ${a_j}^L$ which, upon comparison with the desired or target signal $t_j = i_j$, yield the errors at the output units,

$$(5) \quad {\delta_j}^L = t_j - {a_j}^L.$$

According to the Widrow-Hoff learning rule [17] or its generalizations for 'semilinear' functions [11, 12, 13], the objective is to minimize the total LMS error at the output layer, with respect to the weights. For semilinear activation functions, the back propagation procedure corresponds to a gradient descent method.

A measure of the LMS error is

$$(6) \quad E = \frac{1}{2}\sum_j ({\delta_j}^L)^2$$

for a given signal. This quantity is summed over the signals presented; i.e. denoting the LMS error for signal σ by E_σ, the total LMS error at the output units is

$$(7) \quad E = \sum_\sigma E_\sigma$$

and the weight update corresponding to ${w_{ij}}^k$ will be given by

$$(8) \quad \Delta {w_{ij}}^k = -\lambda \sum_\sigma \frac{\partial E_\sigma}{\partial {w_{ij}}^k}, \quad k = 1, 2, ..., L,$$

where the constant λ is termed a *learning factor*. As a consequence, the weights can be updated accumulatively during the learning phase. Dropping the subscript σ again, an individual update is given by [13, 6] as

$$(9) \quad \Delta {w_{ij}}^k = \lambda {\delta_j}^k \frac{\partial {s_j}^k}{\partial {w_{ij}}^k},$$

where

$$(10) \quad \frac{\partial {s_j}^k}{\partial {w_{ij}}^k} = {a_i}^{k-1}$$

and

$$(11) \quad {\delta_j}^k = -\frac{\partial E}{\partial {s_j}^k}$$

corresponds to (5) above for $k = L$, since φ is the identity function at the output layer. Then for $k = L-1, L-2, ..., 1$, ${\delta_j}^k$ is computed by a backward recursion based on

$$(12)\quad \delta_j{}^k = -\sum_i \frac{\partial E}{\partial s_i{}^{k+1}} \frac{\partial s_i{}^{k+1}}{\partial a_j{}^k} \frac{\partial a_j{}^k}{\partial s_j{}^k} \quad = \sum_i \delta_i{}^{k+1} w_{ji}{}^{k+1} \; \varphi'(s_j{}^k).$$

According to [6], we will call $\delta_j{}^k$ a *local error*, in view of the interpretation for $k = L$.

Furthermore we note that the actual use of the network, with one forward propagation, requires Ln^2 multiplications, $Ln(n-1)$ additions and $(L-1)n$ evaluations of the activation function, if n neurons are used on each of the L layers.

3. Implementation

During the learning phase, the weights on the network connections are adjusted in order to minimize the total error at the output layer, using the method described in section 2.

Figure 4 provides an algorithmic description of the learning phase in our implementation of the back propagation algorithm for the noise filtering application. It is given in the form of a C program section. The parameters used are explained in Figure 3.

In the implementation presented here, the numbers of neurons on the hidden layers equal the number on the input and output layer. Because of this, and also in view of the relatively small numbers of processing elements needed for this application, a sequential data structure is used (a two-dimensional array of structures in C) to store and manipulate the network information, as opposed to a linked data structure and dynamic memory allocation and de-allocation. The latter would be appropriate in applications which need a much larger neural network, as is probably the case for two-dimensional (image) noise filtering.

The structure elements describing a neuron are its input s, output a, the local error d (δ) used in the computation of its weight updates, the array of weights w on its connections with the previous layer and corresponding arrays of weight updates dw (Δw) and of cumulative weight updates cw.

The algorithm section listed in Figure 4 is to be located after the input of the network (and other) parameters and the initializations needed. We initialized the weights with random values between 0 and 1.

Note that each iteration of the outer loop comprises a number of cycles through the network (the inner loop) and is ended with a cumulative update of the weights.

The network is trained for a class of signals by performing every cycle of the learning phase with a new signal picked at random from the given class, sampled over a window of constant length (*window*), at *lw* equidistant points including the endpoints and contaminated with noise. In Figure 4, the sampling of a class of signals is simulated by using a one-parameter class of functions (specified in *isignal*(), for parameter b). The (noisy) values obtained are presented as inputs to the network and also constitute

```
#define L 4
#define LW 65
#define T31 2147483647.0

  struct node {          /* neuron structure */
    float a;             /* neuron output (activation) */
    float d;             /* local error (delta) */
    float s;             /* weighted sum */
    float w[LW];         /* weights from neurons on previous layer
                            (if current layer > 0);
                            net[k][j].w[i] = weight connecting the i-th
                               neuron in layer k-1 with the j-th neuron in
                               layer k     = {w_{ij}}^{[k]},   k > 0    */
    float dw[LW];        /* corresponding weight updates */
    float cw[LW];        /* corresponding cumulative weight updates */
  } net[L][LW];

  int j,               /* neuron index in a layer */
      k,               /* layer index */
      l,               /* specified number of layers in the net */
      l1,              /* l-1 */
      l2,              /* l-2 */
      lw,              /* specified number of neurons in a layer */
      i,               /* neuron index in a layer */
      cum,             /* loop index */
      iter,            /* loop index */
      it,              /* number of iterations (= number of accum. weight
                                                 updates) */
      accum,           /* number of signals for one accum. weight update */
      rand();          /* random number generator */
  float sum,           /* sum accumulator */
        lambda,        /* learning factor */
        b,             /* signal parameter */
        shift,         /* max. shift of the signal over (0,window) */
        window,        /* size of window for all signals */
        beta,          /* noise level during training */
        noise,         /* actual noise put on signal */
        t[LW],         /* target signal */
        isignal(),     /* input signal */
        phi();         /* activation function */
```

FIGURE 3 Parameters used in the algorithm description

```
/*
* Learning phase
*/
  for(iter=0;iter<it;iter++) {  /* iteration loop */
    for(cum=0;cum<accum;cum++) {/* loop over the signals in between
                                   two cumulative weight updates */
      b = (float)(rand())*shift/T31;
      for(j=0;j<lw;j++) {       /* set target signal */
        if(j%2==1)              /* this signal will have noise on
                                   every other sampling point */
          t[j] = isignal(j,lw,b,window);
        else {
          noise = beta*(2.0*(float)(rand())/T31-1.0);
          t[j] = isignal(j,lw,b,window)+noise;
        }
        net[0][j].a = t[j];     /* apply input signal */
      }
      for(k=1;k<l;k++)          /* forward propagation */
        for(j=0;j<lw;j++) {
          net[k][j].s = 0;      /* calculate weighted sum for each neuron */
          for(i=0;i<lw;i++)
            net[k][j].s += net[k-1][i].a*net[k][j].w[i];
          if(k!=l1)
            net[k][j].a = phi(net[k][j].s); /* calculate output */
          else
            net[k][j].a = net[k][j].s;
        }
      for(j=0;j<lw;j++) {
        net[l1][j].d = t[j]-net[l1][j].a; /* local errors at output layer */
        for(i=0;i<lw;i++)       /* weight updates at output layer */
          net[l1][j].dw[i] = lambda*net[l1][j].d*net[l2][i].a;
      }
      for(k=l2;k>0;k--)         /* back propagation */
        for(j=0;j<lw;j++) {
          sum = 0;              /* local error at neuron j on layer k */
          for(i=0;i<lw;i++)
            sum += net[k+1][i].d*net[k+1][i].w[j];
          net[k][j].d = net[k][j].a*(1.0-net[k][j].a)*sum;
          for(i=0;i<lw;i++)     /* weight updates for neuron j on
                                   layer k from the preceding layer */
            net[k][j].dw[i] = lambda*net[k][j].d*net[k-1][i].a;
        }
      for(k=1;k<l;k++)          /* accumulate the weight updates */
        for(j=0;j<lw;j++)
          for(i=0;i<lw;i++)
            net[k][j].cw[i] += net[k][j].dw[i];
    } /* end of  for(cum  */
    for(k=1;k<l;k++)            /* update all weights */
      for(j=0;j<lw;j++)
        for(i=0;i<lw;i++) {
          net[k][j].w[i] += net[k][j].cw[i];
          net[k][j].cw[i] = 0.0;
        }
  } /* end of  for(iter  */
```

FIGURE 4 Back propagation network - implementation

the target signal for the current cycle.

4. Simulation results

We shall illustrate the learning aspect for the class of signals $\sin(x+\beta)$ (shifted with respect to one another), where the class parameter β varies between 0 and a given maximum, *shift*. In this case, the sampling process corresponds of course to sampling of the sine function over different portions of a larger interval. The actual use of the network for removing noise from a signal consists of one forward propagation of the noisy signal through the network.

After training we selected a signal from the target class, generated with a new random value of the parameter, for propagation through the network. The values resulting at the output layer were compared to the noise-free signal.

In the examples of Figures 5.1, 5.2, 5.3 and 5.4, the filtered signals are shown in -◇-, the noisy input signals in -+-. The noise-free function is $\sin(x)$, shown in dashed line. The following parameters were used: number of layers (including input and output layer) $l = 3$, learning factor $\lambda = 0.005$, number of iterations $it = 500$, number of cycles per weight update $accum = 16$. For Figures 5.1, 5.2 and 5.3, the maximum shift was set at $shift = 2\pi$, i.e. $0 \leq \beta \leq 2\pi$, the window length $window = 2\pi$ and the number of neurons per layer $lw = 33$. For the training, noise between -0.5 and +0.5, was put on every other sample value. After the training, with the parameter β drawn at $\beta = 5.035534$, Figure 5.1 displays the output obtained with noise put on every other sample of the input signal, Figure 5.2 with noise on all sample values. For Figure 5.3, noise between -1 and +1 was put on every other sample value after the training. Figure 5.4 was obtained with noise between -1 and 1 on 10% of the sampling points (every other 10) during the training, and ranging between -3 and 3 after the training. In this example, the number of neurons on a layer was set to 65 and $shift = window = 16\pi$. The latter three figures indicate that good results can be obtained with noise on the signal for which the network was not trained.

In our experience, the larger the windows and noise, the more neurons are required on a layer and the more iterations are needed. Use of a network with two hidden layers instead of one gave similar results (and is less efficient).

Another type of testing we want to discuss involves the network's ability of adapting to a slowly time-varying signal. First, the network is trained for a set of signals $\sin(\alpha x+\beta)$, with $\alpha = 1$ (as above), with a small amount of noise on the samples. After the training, we simulate sliding a window over the signal $\sin(\alpha x)$; i.e. β is changed linearly in $\sin(\alpha x + \beta)$. In the mean time, the value of α is incremented slowly; e.g. $\alpha = \alpha + 0.0001$ every cumulative weight update. This means that only one cumulative weight update is used for retraining when α is incremented. Evidently, the network

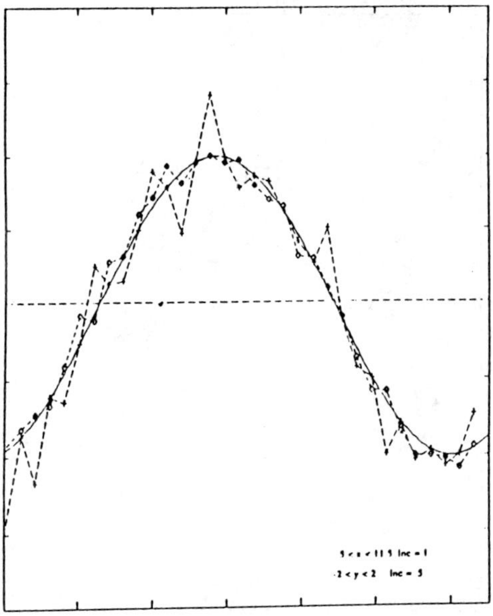

FIGURE 5.1

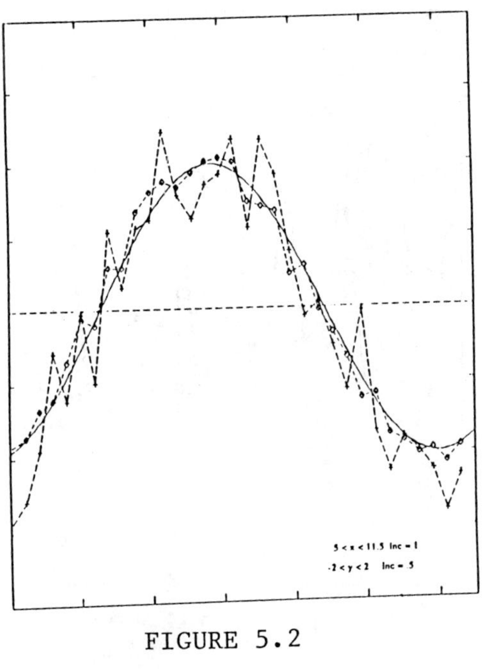

FIGURE 5.2

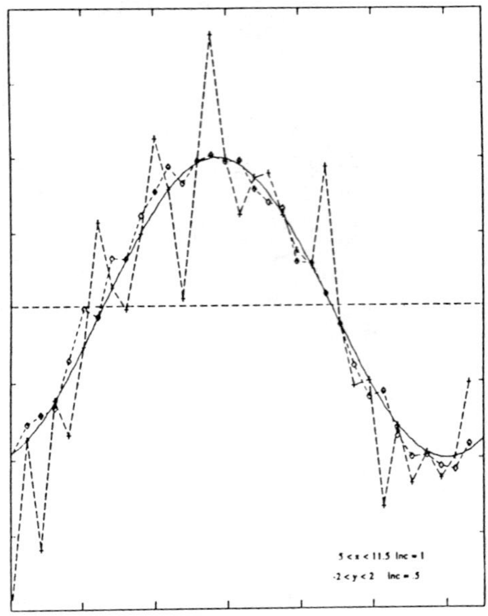

FIGURE 5.3

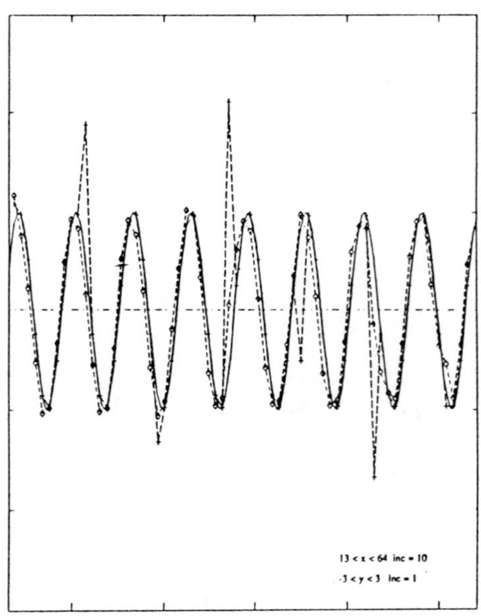

FIGURE 5.4

♯ weight upd.	α	E_c	A_c
10	1.0010	1.776667	0.543328
20	1.0020	2.406686	0.654893
30	1.0030	1.809769	0.582372
40	1.0040	1.926032	0.425525
50	1.0050	2.247852	0.428488
60	1.0060	2.384508	0.502885
70	1.0070	2.361742	0.532923
80	1.0080	2.353770	0.634051
90	1.0090	2.214882	0.615868
100	1.0100	2.056097	0.555917
110	1.0110	2.344906	0.685429
120	1.0120	2.305990	0.548513
130	1.0130	2.246117	0.557813
140	1.0140	2.200148	0.432416
150	1.0150	2.220258	0.511397
160	1.0160	1.975616	0.548478
170	1.0170	2.277874	0.528028
180	1.0180	1.833099	0.363498
190	1.0190	2.127633	0.513209
200	1.0200	1.744851	0.440522

TABLE 1

adapts easily to the varying frequency of the signal.

Table 1 shows the sum of squares (over one cumulative weight update) of the local errors calculated by the network at the output layer, with respect to the noisy target signal, as well as the corresponding sum of the actual errors, calculated by the test program with respect to the target sine function, as obtained every 10th increment of the α parameter. The first column lists the number of the weight update, the second shows the value of α, the third corresponds to the quantity

$$(13)\quad E_c = 2\sum_{\sigma} E_{\sigma},$$

where the summation runs over the cycles performed for one cumulative weight update. Let us represent the values in the fourth column by

$$(14)\quad A_c = 2\sum_{\sigma} A_{\sigma},$$

with the summation over the cycles for one cumulative weight update and

$$(15)\quad A_{\sigma} = \frac{1}{2}\sum_{j} (\varsigma_j - a_j{}^L)^2,$$

♯ weight upd.	α	E_c	A_c
10	1.002	1.882626	0.516800
20	1.004	1.937717	0.551943
30	1.006	2.045270	0.649179
40	1.008	2.276646	0.689945
50	1.010	2.550281	0.565884
60	1.012	2.182388	0.511017
70	1.014	2.292424	0.501673
80	1.016	2.082074	0.484799
90	1.018	2.065418	0.577531
100	1.020	2.252659	0.591228
110	1.022	2.053045	0.582565
120	1.024	2.238371	0.577627
130	1.026	2.579532	0.710853
140	1.028	2.435675	0.839676
150	1.030	3.043192	1.107599
160	1.032	2.649813	1.029902
170	1.034	2.287527	0.625830
180	1.036	2.186112	0.524063
190	1.038	2.452674	0.956077
200	1.040	2.032807	0.685024

TABLE 2

where the summation runs over all output elements and ς_j is the value of the noise-free signal corresponding to the output element indexed j.

The numbers were obtained for a 3-layer network, including the input and output layer, with 33 neurons each. The window was 16π long, as well as the maximum shift parameter used for the initial training. Cumulative weight updates were done every 16 forward-and-back propagation cycles, during and after the initial training. During and after the training, noise between -0.3 and 0.3 was put on 10% of the sample values in a window. In this example with a very small time variation, it appeared that the number of cycles per weight update could be reduced to 1 for each retraining of the network, producing similar results. In a run with one cycle per weight update and 4 updates per retraining, a plot of the output signal for $\alpha = 0.02$ showed that the difference with the noise-free signal was within 0.08 at all output units.

Table 2 gives the corresponding table for bigger increments of α, i.e. $\alpha = \alpha + 0.001$. In this case, we used 5 cumulative weight updates (of 16 cycles) for retraining the network after each increment of α. As expected, larger variations are handled easier with a longer learning phase in between the variations. For example, with $\alpha = \alpha + 0.01$ for the time variation, retraining can be done with 10 to 15 cumulative weight updates.

Observation of the estimated local errors calculated by the network at

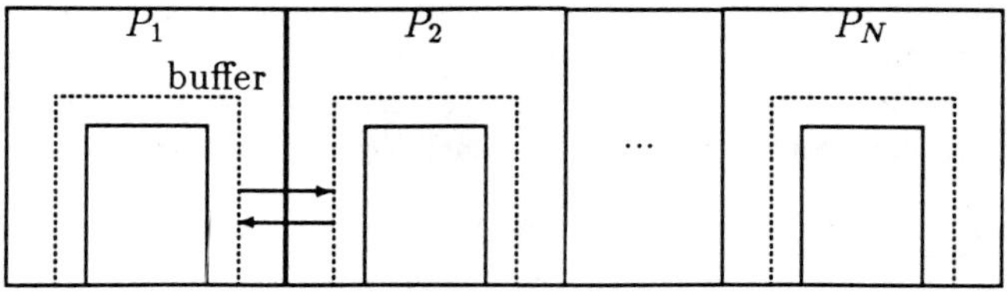

FIGURE 6 Back propagation network on a set of processors

the output layer (E_c) indicates that they follow the behaviour of the actual errors (A_c) closely. This suggests that an adaptive procedure can be used for terminating each retraining process, by comparing the current value of E_c with the value obtained at the end of the initial learning phase.

5. Parallel implementation and further avenues

We studied the problem of mapping a neural net (with fixed weights) onto an MIMD machine or a network of workstations. In multidimensional filtering, where the number of inputs is large, this should prove useful. It is a multiple resource allocation problem, where neurons are assigned to processors in such a way that the traffic of information is in some sense minimized [5]. In its general form, this leads to a nonlinear integer programming problem in combinatorial optimization. Branch and bound methods, techniques of dynamic programming and approximation methods like simulated annealing have been investigated for its solution. Alternatively, the problem can be simplified in various ways and for specific applications, allowing more efficient solution methods.

In particular, for a practical implementation of a back propagation network, we propose an approach which can be used when the weigths are not necessarily fixed (i.e. during as well as after training of the network). An illustration is given in Figure 6. In this implementation, a neuron stores the weights on the connections with the neurons on the next layer, as opposed to those of the previous layer in the serial approach presented earlier. The processors, presented by P_1, P_2, ..., each contain part of the $L+1$ layers. The contributions from layer k in processor i to the weighted sums needed on layer $k+1$ in the other processors, can then be calculated and buffered in P_i. During the training, the buffered sums are communicated only occasionally. In Figure 6, the buffers are represented by dashed lines. If a neural network with $L+1$ layers and n neurons on each layer would be put on N processors and each layer would be split up about equally among the processors, this means that the number of weighted sum contributions buffered in a processor would be about $(n - n/N)L$.

Note that the communication and data transport between Sun workstations on a network, as well as the execution of remote processes, is greatly facilitated by the use of systems such as ISIS [4] and SR [3, 2]. We are also considering using MACH for this purpose. Although current ethernet speeds are slow, advances may make such computations practical.

Another modification of the serial implementation of section 3 involves a change of the learning model, to work with the ℓ_1 norm of the error measured at the output layer. In this case, (6) is replaced by

$$(16) \quad E = \sum_j |\delta_j{}^L|$$

and subsequent formulas are changed accordingly. In particular, the quantity (11) reduces to ± 1 when $k = L$; i.e. only the sign of the discrepancy with respect to the target signal at the output layer is taken into account for the training. This bears similarity to the models known as 'reinforcement learning' [15, 7, 8]. However, the results obtained for the present applications are inferior to those of the learning model with the ℓ_2 error measure.

Other possible avenues, which we did not or not yet explore, include

- the derivation of theorems on the evolution of 'good' local minima in a slowly time-varying setting,
- an investigation of questions on the existence of 'equivalent' forms of the neural net, for different sets of weights, and of transfomations into 'standard forms' (that is, by re-assignment of the weights), and
- establishing more connections between learning rules and fast filter adaptation methods.

References

[1] *Neuralworks Explorer, revision 2.00.* Neuralware, Inc., Pittsburgh Philadelphia, 1988.

[2] G.R. Andrews and R.A Olsson. *Revised report on the SR programming language.* Technical Report, Dept. of Computer Science, The Univ. of Arizona, 1987. TR 87-27.

[3] G.R. Andrews, R.A. Olsson, M. Coffin, I. Elshoff, K. Nilsen, T. Purdin, and G. Townsend. An overview of the SR language and implementation. *ACM Trans. on Prog. Lang. and Systems*, 10(1), 1988.

[4] K. Birman, T. Joseph, K. Kane, and F. Schmuck. *The ISIS system manual.* Technical Report, The ISIS Project, Department of Computer Science, Cornell University, 1988.

[5] E.H. de Doncker and J.A. Kapenga. The mapping of neural networks into workstation networks. In *Second International Conference on Graph Theory, Combinatorics, Algorithms and Applications, San Francisco*, Abstracts, July 1989.

[6] J.P. Guiver and C.C. Klimasauskas. Networks II. In *Neuralworks Explorer, revision 2.00*, pages 435–501, Neuralware, Inc., Pittsburgh Philadelphia, 1988.

[7] A.H. Klopf. A drive-reinforcement model of single neuron function: an alternative to the Hebbian neuronal model. In J.S. Denker, editor, *AIP Conference Proceedings 151: Neural Networks for Computing*, pages 265–270, American Institute of Physics, 1986.

[8] A.H. Klopf. A neuronal model of classical conditioning. *Psychobiology*, 16(2):85–125, 1988.

[9] J.L. McClelland and D.E. Rumelhart. *Explorations in Parallel Distributed Processing.* The MIT Press, Cambridge Massachusetts, London England, 1988.

[10] M.L. Minsky and S. Papert. *Perceptrons.* The MIT Press, Cambridge Massachusetts, London England, 1969.

[11] D.B. Parker. *Learning-logic.* Technical Report, MIT, Center for Computational Research in Economics and Management Science, 1985. Report TR-47.

[12] D.E. Rumelhart, G.E. Hinton, and R.J. Williams. Learning representations by back-propagating errors. *Nature*, 323:533–536, 1986.

[13] D.E. Rumelhart and J.L. editors McClelland. *Parallel Distributed Processing: Explorations in the Microstructure of Cognition. Volume I, Foundations.* The MIT Press, Cambridge Massachusetts, London England, 1986.

[14] D.F. Stubbs. Neurocomputers. *MD Computing*, 5(3):1–12, 1988.

[15] R.S. Sutton and A.G. Barto. Toward a modern theory of adaptive networks: expectation and prediction. *Psychological Review*, 88:135–170, 1981.

[16] P.J. Werbos. *Beyond Regression: New Tools for Prediction and Analysis in the Behavioural Sciences, Doctoral dissertation.* PhD thesis, Harvard University, 1974.

[17] B. Widrow and M.E. Hoff. Adaptive switching circuits. *IRE WESCON Convention Record*, 4:96–104, 1960.

COMPUTATION OF DISCRETE FOURIER TRANSFORMS OVER FINITE ABELIAN GROUPS USING PIPELINED AND SYSTOLIC ARRAY ARCHITECTURES*

M. G. KARPOVSKY, E. A. TRACHTENBERG, T. ROZINER

ABSTRACT

The paper discusses two methods for computation of Generalized Discrete Fourier Transforms (GDFT's) over finite groups in a multiprocessor environment. We note that GDFT is simply the multidimensional DFT if the underlying group is Abelian. We use a precise algorithm for GDFT computation by a multiprocessor SIMD system with local non-shared memory. Tradeoffs between hardware complexity and computation time are established for unibus and complete communication networks. Bounds on the number of operations, data transfers and communication links are presented. The GDFT algorithm is then implemented using a pipelined architecture. Bounds are given for numbers of operations, data transfers, delay of the whole pipeline and clock period for unibus and complete communication networks. In order to use standard systolic arrays for GDFT computation,it is necessary to consider its direct computation instead of the sequential FFT-like algorithm used for the pipelined architecture. In case of Abelian groups we discuss such efficient systolic array implementations of GDFT which can be made adoptable to the structure of the underlying group, and we estimate their overall speed and hardware complexity.

1. INTRODUCTION

Recently, there has been an increasing interest in implementing different types of algorithms on systolic array structures, because these architectures are extremely suitable for VSLI technology [4-6]. Discrete Fourier Transform (DFT) is widely used in signal and image processing in the analysis, synthesis and optimization of linear time invariant systems [1-3]. However, the corresponding fast butterfly-like algorithm is inherently sequential. That restricts full utilization of parallel computations and systolic array architectures for real time applications. In [1] a method was presented for an optimal implementation of Generalized Discrete Fourier Tranform (GDFT) algorithms over finite groups in a multiprocessor environment. A SIMD multiprocesor system with local non-shared memory was used and tradeofs were established between computation time and hardware complexity for unibus and complete communication networks. In the next section we present that algorithm together with bounds on the number of operations, data transfers and communication links. It is shown that using non-Abelian (e.g. quaternion) groups results in the optimal (i.e. fastest) GDFT performance. In section 3 the GDFT algorithm is implemented using a pipelined architecture. Bounds are given for numbers of operations, data transfers, the delay of the whole pipeline and the clock period for unibus and complete communication networks. In section 4 we discuss implementation of two-dimensional GDFT (when the underlying group is a direct product of two cyclic groups) by a systolic array. Such an array consists of identical processing elements (PEs), where only the neighboring PEs are allowed to communicate with each other. The PEs are identical in the number of input/output lines as well as in their internal structure. In order to use standard systolic arrays, it is necessary to consider direct GDFT computation instead of the sequential FFT-like algorithm used for the pipelined architecture. Also in that section we consider an approach to efficient systolic array implementations of GDFT which can be made adoptable to the strucutre of the underlying group $G=G_1xG_2$, where G_1,G_2 are arbitrary Abelian groups.

2. COMPUTATION OF GENERALIZED DISCRETE FOURIER TRANSFORM (GDFT).

Given a set $\{0,1,...,N-1\}$ of N integers, let G be an arbitrary finite group of order N imposed on it. Let R_ω be the ωth irreducible unitary representation of G of dimensional d_ω. The set $\{R_\omega\} = \hat{G}$ of all these representations is called the dual object of G and its elements constitute an orthogonal basis in the space $\{f: G \rightarrow \mathbb{C}\}$, where $\mathbb{C}$ is the field of complex numbers. The direct and inverse Fourier transform over G are defined as follows

$$\hat{f}(\omega) = N^{-1} d_\omega \sum_{i \in G} f(i) R_\omega(i^{-1}),\ R_\omega \in \hat{G};\quad f(i) = \sum_{R_\omega \in \hat{G}} \text{Trace}\,(\hat{f}(\omega)\ R_\omega(i)),\ i \in G; \quad (1)$$

where $\hat{f}(\omega)$ is a $(d_\omega \times d_\omega)$ - matrix over $\mathbb{C}$.

The transform (1) has the usual properties of linearity, group transalation, convolution, Plancherel and Poisson theorems, etc., [7]. Without loss of generality, consider the computation of the direct transform only. We note that the algorithm described in [7] computes (1) in at most $N\sum_{j=0}^{m-1} n_j$ computer operations using N memory locations for intermediate data storage in case when G is a direct product of groups G_j (of order n_j), $G = \prod_j G_j$, $0\le j\le m-1$, $N=n_0\, n_1 \dots n_{m-1}$. We are going to consider a precise fast algorithm [1] for computing (1). Naturally, (see Table 2) it is always much faster than the upper bound discussed above. Let $x = (x_0,x_1,\dots,x_{m-1}) \in G$, $x_j\in G_j$, $0\le j\le 1$. For $R_\omega \in \hat{G}$ we have $R_\omega(x) = \bigotimes_{j=1}^{m=1} R_{\omega_j}(x_j)$, where $R_{\omega_j} \in \hat{G}_j$, $\otimes$ stands for Kronecker product of matrices and $\omega = (\omega_0, \omega_1,\dots,\omega_{m-1})$. In that case we have by (1):

$$\hat{f}(\omega) = \hat{f}(\omega_0,\omega_1,\dots,\omega_{m-1}) = \frac{d_\omega}{N}\sum_{x_0}\sum_{x_1}\dots\sum_{x_{m-1}} f(x_0,x_1,\dots,x_{m-1}) \bigotimes_{j=0}^{m-1} R_{\omega_j}(x_j^{-1})$$

$$= \frac{d_\omega}{N}\sum_{x_{m-1}}\left(\dots\left(\sum_{x_1}\left(\sum_{x_0} f(x_0,x_1,\dots,x_{m-1})\otimes R_0(x_0^{-1})\otimes\dots\right)\otimes R_{m-1}(x_{m-1}^{-1})\right)\right). \quad (2)$$

The calculation of spectrum $\hat{f}(\omega)$ is performed in m steps (Step j over G_j, $0\le j\le m-1$). For Step j: $f_{j+1}(\omega_0,\omega_1,\dots\omega_j,x_{j+1}\dots x_{m-1}) = \sum_{x_j} f_j(\omega_0,\omega_1,\dots,\omega_{j-1},x_j,\dots,x_{m-1})\otimes R_{\omega_j}(x_j^{-1})$.

The algorithm, described in [1] is based on that m-step procedure. In step j, N/n_j subsets of G which contain n_j elements of input data f(0), f(1),...,f(N-1) are used for the butterfly algorithm of the constituent group G_j. Denote the number of operations needed to perform the algorithm of G_j on a single input data set (one out of N/n_j) as $L(G_j)$. (In general, the algorithm is characterized by a pair of numbers - the number of additions/subtractions and the number of multiplications. We shall assume $L(G_j)$ is the equivalent number of additions). A variety of examples of butterfly algorithms is given in [1]. The numbers L(G) for some groups are given in Table 1, where C_N stands for the cyclic group of order N with respect to addition mod N, S_3 denotes the symmetric of order 6, and Q_2 denotes the quaternion group of order 8.

Groups	C_2	C_3	C_4	C_8	S_3	Q_3
Additions	2	7	8	24	14	20
multiplications		4		10	4	

Table 1: Computation of Fourier transform, Single Processor

To perform the spectrum calculation over $G=G_0xG_1x\dots xG_{m-1}$ by a single processor, L(G) operations are needed:

$$L(G) = \sum_{j=0}^{m-1} L(G_j)\prod_{\substack{i=0\\ i\ne j}}^{m-1} n_i = N\sum_{j=0}^{m-1}\frac{L(G_j)}{n_j} \quad (3)$$

In Table 2 below we summarize and compare computation of the Fourier Transform (1), (2) using the direct method, the upper bound algorithm [7] and the precise fast algorithm [1].

Group / Algoritm		C_8	C_3^2	S_3	Q_2	$S_3 \times Q_2$	$Q_2 \times Q_2$
N^2	adds.	64	64	36	64	2304	4096
	multpl.	64	0	36	0	2304	0
$N \sum_j n_j$	adds.	48	48	n.a.	n.a.	672	1024
	multpl.	48	0	n.a.	n.a.	48	0
L(G) by (3)	adds.	24	24	14	20	336	384
	multpl.	10	0	4	0	32	0

Table 2: Comparison of Fourier Transform Algorithms Single Processor

Consider n>1 processor working simultaneously on the ordered data f(0), f(1),...,f(N-1), $x \in G$, $x=(x_0,x_1,...,x_{m-1})$, $x_j \in G_j$; , $0 \le j \le m-1$. The order is established by the following rule: x_0 is the most significant position and x_{m-1} is the least significant position and the carry over in position j is modulo n_j, where n_j is the order of G_j. In that case, by (3)

$$L^{(n)}(G) = \frac{L(G)}{n} = \frac{N}{n} \sum_{j=0}^{m-1} \frac{L^{(n)}(G_j)}{n_j}. \qquad (4)$$

However, for the number of processors n>1, the data transfers among the processors are needed (at least in one of the algorithm steps). Therefore, the total time $T^{(n)}$ (G) needed for spectrum calculation over G with n processors is:

$$T^{(n)}(G) = L^{(n)}(G).\ t_a + M^{(n)}(G).\ t_c, \qquad (5)$$

where t_a is the time needed for one addition/subtraction, t_c is the time needed for one data transfer, and $M^{(n)}$ (G) is the total number of transfers needed for spectrum calculation. The value of $M^{(n)}$ (G) is not only a function of n and of the group structure for a fixed N; it depends also on the type of the processor communication network. We will consider two "extreme" communication networks: the taxi and unibus connections. In a taxi network, each processor is allowed to communicate to any other processor during the communication time t_c. The network requires a large number of links and $M^{(n)}$ (G) is, of course, minimized. In a unibus network all processors communicate through a single bus. Hence during the communication time t_c only one data transfer is allowed between two selected processors. We assume that:

(i) Each processor has local memory but there is no global shared memory.

(ii) There are no idle processors at any step $j \in \{0,1,...,m-1)$ of the algorithm which means that $n.n_j$ divides N for all j's.

(iii) The ordered input data f(0), f(1),...,f(N-1) are entered into local memories of n processors in blocks: the first $N/_n$ elements are given to the 1st processor, the next $N/_n$ elements to the 2nd, etc.

(iv) At any stage j of the algorithm, any set of n_j elements is delivered to a single processor, $j \in \{0,1,..., m-1\}$. Even though that requirement is not necessary for executing the algorithm, it maybe shown [1] that its violation results in drastic increase in both the number of transfers and the number of total computations.

Under these assumptions, (5) was evaluated analytically for an arbitrary group in [1] and the decrease in execution time was calculated due to use of n processors as compared to a single processor. The relative speed-up due to n processors depends on the ratio t_c/t_a which characterizes the speed of the bus. It was shown in [1] that the bus must be sufficiently fast ($t_c \ll t_a$) to have an affect on the increase in the number of processors. For example, for $G=Q_2 \times Q_2$ and for a unibus connection, for k=1.5 doubling the number of processors (from 4 to 8) results in only about 18% of overall execution time decrease. Similar results were obtained for other groups.

3. PIPELINED IMPLEMENTATION OF FOURIER TRANSFORM

Consider a pipelined implemention of the Fourier transform on $G = G_0 \times G_1 \times ... \times G_{m-1}$ by multiple processors. The algorithm (2) is implemented under the assumptions (i) - (iv). The pipeline is an array of n=m•N processors depicted below. The "width" of the array is N and N data points f(0),...f(N-1) are entered simultaneously from the left into the 1st stage of N processors that compute the Fourier Transform over G_0. The "length" of the array is m since there are m stages in the algorithm.

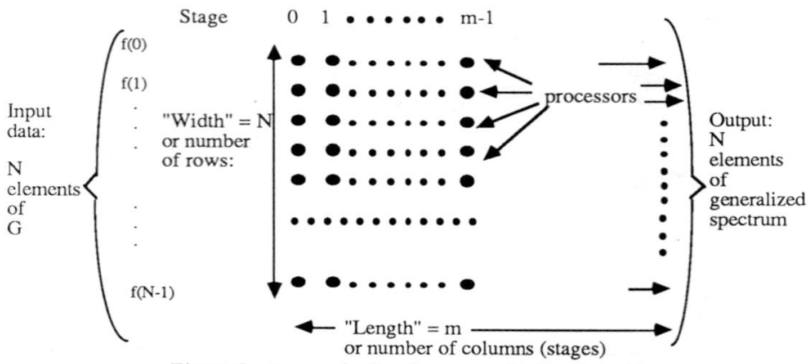

Figure 1: Array pipeline for group $G=G_0xG_1x...G_{m-1}$

At the stage j ($0 \le j \le m-1$), the intermediate spectrum on G_j is computed by the jth column of processors. Each processor in column j adds n_j numbers. It also may have to do some multiplications, depending on the G_j, see e.g. Table 1. Since $R_{\omega_0}(x) = 1$ for any $x \in G_j$, the number of multiplications is not larger than n_j-1 for each processor. It can therefore be shown that the total number of operations (both additions and multiplications) and of transfers satisfies

$$\sum_{j=0}^{m-1} (n_j-1) \le L^{(m\bullet N)}(G) \le 2\sum_{j=0}^{m-1} (n_j-1), \quad M^{(m\bullet N)}(G) = N\sum_{j=0}^{m-1} n_j \ . \qquad (6)$$

The delay of the whole array pipeline (i.e. time from the moment of input of N elements of data up to the moment of obtaining the N elements of the spectrum output) is given for the unibus connection as:

$$D^{(m\bullet N)}_{unibus}(G) = T^{(m\bullet N)}_{unibus}(G) \le \sum_{j=0}^{m-1} (n_j-1)\bullet(t_a + t_m) + N\sum_{j=1}^{m-1} n_j \bullet t_c, \qquad (7)$$

where t_a is one addition time, t_m is one multiplication time and t_c is the communication time. In case of a taxi connection, when all processors can communicate simultaneously, the number of transfers for stage j is n_j, $1 \le j \le m-1$ because each of the N processors at that stage transfers its output to n_j different processors of the next stage. No transfers are needed at stage 0. The delay is evaluated as follows:

$$D^{(mN)}_{taxi}(G) = T^{(mN)}_{taxi}(G) \le \sum_{j=0}^{m-1} (n_j-1)\bullet(t_a + t_m) + \left(\sum_{j=1}^{m-1} n_j\right)\bullet t_c \ . \qquad (8)$$

Note that the number of communication links among n processors in case of taxi connection is $N\sum_{j=1}^{m-1} n_j$. The evaluation of the number of operations and of the pipeline delay do not take into account the normalization of the spectrum which can be done after the last stage.

Example: $G = C_4xC_4$, m=2, N=16, $n_0=n_1=4$. The number of processors n = 32. It follows by (1) that for C_4 we have: $\hat{f}(0) = f(0)+f(1)+f(2)+f(3)$; $\hat{f}(1) = f(0)+if(1)-f(2)-if(3)$; $\hat{f}(2) = f(0)-f(1)+f(2)-f(3)$; $\hat{f}(3) = f(0)-if(1)-f(2)+if(3)$, $i = \sqrt{-1}$. At Stage 0, the data f(0) to f(3) is broadcast to processors 0 to 3, data f(4) to f(7) is broadcast to processors 4 to 7, ... , data f(12) to f(15) is broadcast to processors 12 to 15. Also, at Stage 0 three additions/subtractions are performed simultaneously in each of the 16 processors (we do not take into account multiplication by $\pm i$) as follows:

in Processor 0: $\hat{f}(0) = f(0)+f(1)+f(2)+f(3)$; in Processor 1: $\hat{f}(1) = f(0)+if(1)-f(2)-if(3)$;...;
in Processor 4: $\hat{f}(4) = f(4)+f(5)+f(6)+f(7)$;...; in Processor 15: $\hat{f}(15) = f(12)-if(13)-f(14)+if(15)$.
At stage 1, the input data for each processor are chosen "skipping over n_0=4" that is, each of the 16 processors from stage 0 performs four transfers.

Processors at Stage 1	16	...	19	20	...	23	...	28		31
Transfers from Processors at Stage 0	0,4,8,12	...	0,4,812	1,5,9,15	...	1,5,9,13	...	3,7,11,15	...	3,7,11,15

In each Processor of Stage 1, the number of additions/substractions is also $n_1-1=3$. The total number of operations is $\sum_{j=0}^{1}(n_j-1) = 3\bullet 2 = 6$ and the total number of communications is $16 \cdot 4 = 64 = N \cdot n$, for a unibus and $4 = n_1$ for a taxi connection. The pipeline delay is $6t_a+64t_c$ for the unibus and $6t_a+4t_c$ for the taxi connection. The number of required communication links for the taxi connection is $N \cdot n_1 = 64$.

In conclusion, we note that the clock period of the array pipeline is:

$$\tau_{unibus} \leq \max_j \{ (n_j - 1) \cdot (t_a + t_m) + N \cdot n_j \cdot t_c \}, \quad \tau_{taxi} \leq \max \{ (n_j - 1) \cdot (t_a + t_m) + n_j \cdot t_c \}. \quad (9)$$

To avoid having bottlenecks in the pipeline, it is reasonable to form G as a direct product of groups of the same type.

4. IMPLEMENTATION OF TWO-DIMENSIONAL DFT BY A SYSTOLIC ARRAY.

Let $G = C_N x C_N$. Then $R_{(i_1, i_0)}(l_1, l_0) = W_N^{i_1 l_1 + i_0 l_1}$, $W_N = \exp(2\pi\sqrt{-1}/N)$ and by (1):

$$\hat{f}(i_1, i_0) = \sum_{l_1=0}^{N-1} \sum_{l_0=0}^{N-1} f(l_1, l_0) W_N^{i_1 l_1 + i_0 l_0} = \sum_{l_1=0}^{N-1} P(l_1, i_0) W_N^{i_1 l_1} \quad (10)$$

$$\text{where } P(l_1, i_0) = \sum_{l_0=0}^{N-1} f(l_1, l_0) W_N^{i_0 l_0}. \quad (11)$$

It can be shown that algorithm (10), (11) can be written in the form of a FORTRAN - like program as follows.

Procedure 1.

DO 10 $\quad k_1 = 0, N-1$

DO 10 $\quad k_2 = 0, N-1$

$\quad P(k_1, k_2, -1) = 0$ (initialize)

I {

DO 20 $k_3 = 0, N-1$

$\quad f(k_1, -1, k_3) = f(k_1, k_3)$

$\quad f(k_1, k_2, k_3) = f(k_1, k_2-1, k_3)$

$\quad P(k_1, k_2, k_3) = P(k_1, k_2, k_3-1) + f(k_1, k_2, k_3)\, W_N^{k_2 k_3}$

}

20 CONTINUE

II {

DO 30 $k_3 = N, 2N-1$

$\quad \hat{f}(k_3-N, k_2, -1) = 0$ (initialize)

$\quad P(k_1, k_2, k_3) = P(k_1, k_2, k_3-1)$

$\quad \hat{f}(k_3-N, k_2, k_1) = \hat{f}(k_3-N, k_2, k_1 - 1) + P(k_1, k_2, k_3)\, W_N^{k_1 k_3}$

}

30 CONTINUE

10 CONTINUE

This program may be obtained using a permutation of the original index set in (10), (11). The algorithm of Procedure 1 is suitable for systolic/semi-systolic implementation with the use of the accepted method in [6], (see Figure 2). Note that in part I the index k_2 stands for the number of steps of accumulation of $f(\bullet, \bullet)$ in the right hand side of (10) whereas in II the index k_1 has the same role for the spectrum $\hat{f}(\bullet,\bullet)$ in the left hand side of (10).

The method of the algorithm transformation can be extended to the case of any number of dimensions. This method combined with the one in [6] allows to implement the multidimensional DFT with a systolic macropipeline (for an M-dimensional N^M DFT, the pipeline consists of M or M-1 hypercubes each containing N^M cells). For any number of dimensions, the throughput of the macropipeline is equal to N pipeline clocks - one spectrum set of N^M values every N pipeline clocks. The dependence matrix for the Procedure 1 is

$$D = \begin{bmatrix} 0 & 0 & 1 \\ 1 & 0 & 0 \\ 0 & 1 & 0 \end{bmatrix} \begin{matrix} k_1 \\ k_2 \\ k_3 \end{matrix}$$

$$\begin{matrix} f & P & \hat{f} \end{matrix}$$

The method developed in [4,6] is based on the idea of a nonsingular space-time mapping T of the index set (k_1, k_2, k_3) which maps a given problem to many possible systolic designs. That is, $\hat{D} = TD$; $(t\ s_0\ s_1)^T = T(k_1\ k_2\ k_3)^T$, $\hat{D}$ is the new dependence matrix and $(t\ s_0\ s_1)^T$ is the new index set, where t is the time coordinate with integer values ≥ 0; s_0, s_1 - the row and the column number, respectively, of the processing element $PE_{s_0 s_1}$, in the systolic array. One possible design for the (NxN) DFT requires N^2 $PE_{s_0 s_1}$; f moves through the array by s_0 ("north to south"),

P accumulates in the PEs during N array clocks and $\hat{f}$ moves by s_1 ("west to east") during N array clocks. And for this design we have

$$\tilde{D} = \begin{bmatrix} 1 & 1 & 1 \\ 1 & 0 & 0 \\ 0 & 0 & 1 \end{bmatrix} \begin{matrix} t \\ s_0 \\ s_1 \end{matrix} \text{ (columns } f,\ P,\ \hat{f}) ; \qquad T = \begin{bmatrix} 1 & 1 & 1 \\ 0 & 1 & 0 \\ 1 & 0 & 0 \end{bmatrix} ; \ t = k_1 + k_2 + k_3 ; \quad s_0 = k_2 ; \ s_1 = k_1 . \qquad (12)$$

Figure 2 illustrates the performance of the array for N = 2.

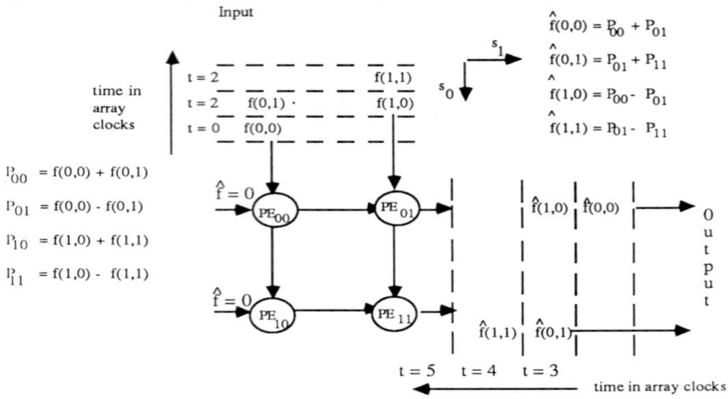

Figure 2: Systolic array of 4 PEs for 2x2 DFT

Each PE of the systolic array is a "double" multiply-add cell that calculates in parallel P and $\hat{f}$ values. The $PE_{s_0 s_1}$ is activated in the beginning of the array operation on array clock number $(s_0 + s_1)$ (the first clock is referred to as clock 0). During the first N active clocks, the PE performs only one multiply-add operation accumulating value of $P_{s_1 s_0}$. After that, the calculated value of $P_{s_1 s_0}$ is stored in the internal storage register during next N clocks; on each one of these clocks, the stored value of $P_{s_1 s_0}$ is used in multiply-add operations generating the values of the spectrum elements. In other words, after the first N active clocks, two operations occur in parallel in each PE: multiply-add operation with input data f and multiply-add operation with the stored value of $P_{s_1 s_0}$, to calculate in N steps the values of the spectrum. Figure 3 shows the general structure of the PE for the systolic array described above. The PE has the built-in counter, to calculate "the local time" modulo N, that is, to control the proper timing of the values $P_{s_1 s_0}$ to be accumulated, stored, or changed to a new value. Note that instead of the multipliers operating with the powers of W_N, the CORDIC rotation modules [8] can be used since the multiplications performed are rotations in the comlex plane. The CORDIC module uses only addition/subtraction and shift operations; having the same throughput and complexity as the complex multiplier, it occupies less space in a silicon chip. However since a standard CORDIC module performs rotations in the range of (-90°,90°) only, additional hardware would be needed for the rotation angles outside this range. The hardware complexity of one PE (in terms of the equivalent number of two-input gates) for the general case of N^2 DFT ($N = 2^n$ and the data lines have 2r bits) can be evaluated as:

$$L_{PE} = L_{PE}(n, r) = 6n^2 - 3n + 2^{n+1}(n-1) + 4r(2^n - 1) + 48r^2 + 100r + 294\left[\frac{r}{2}\right] + 158\left[\frac{n}{4}\right], \ n \geq 4 \quad (13)$$

The total complexity of the array of N^2 PE's is $L(N^2) = 2^{2n} L_{PE}$; the clock of the array is $C = (416 + 3n) \cdot t_g$, where t_g is the average delay of one gate; the total hardware x time complexity is $H = L(N^2) \times C \times N$. Note that the number of I/O pins of one PE is 4n + 12r for 2r - bit data lines, where $n = \log_2 N$. Table 3 shows these complexity parameters for r=8 (16-bit data lines).

n	N^2	L_{PE}	L	C	H	Number of pins per PE (data lines only)
4	256	5866	1.50×10^6	428	1.03×10^{10}	112
5	1024	6747	6.91×10^6	431	9.52×10^{10}	116
6	4096	8218	3.36×10^7	434	9.35×10^{11}	120

Table 4: Complexity of Systolic Arrays for NxN DFT, $n = \log_2 N$

Note that the throughput of such an array is one spectrum set (N^2 values) per N array clocks.

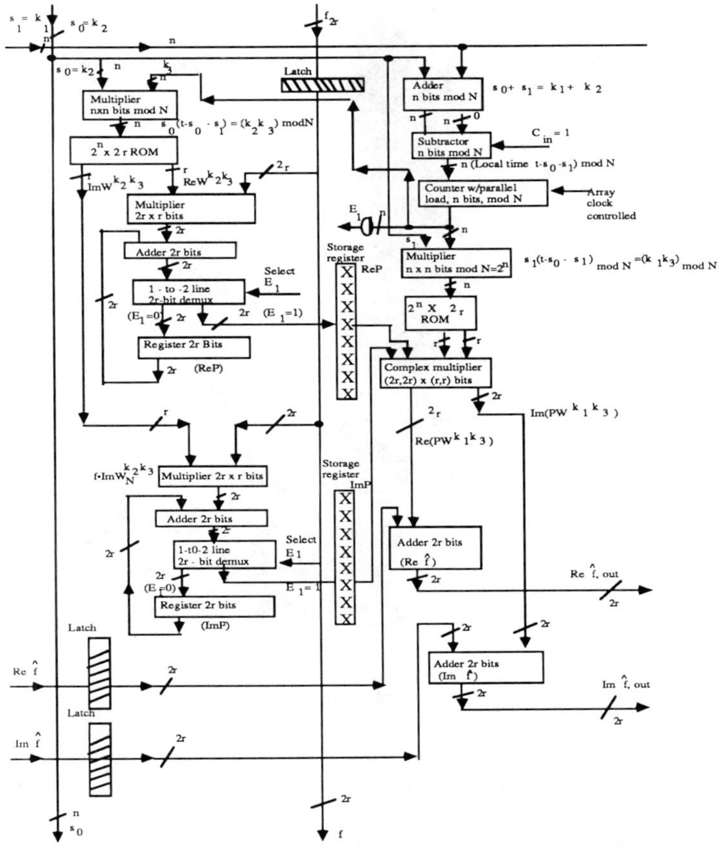

Figure 3: General structure of a PE for NxN systolic array (f moves by s_0, P accumulates, $\hat{f}$ moves by s_1)

Finally we outline our approach to the adaptation of the systolic array to any given group G. This can be achieved by adopting the PEs. We will illustrate our approach by a design of a systolic array which can be adapted to any group $G=G_1 \times G_2$, where $G_1, G_2 \in \{C_1, C_2, C_{2^2}, \ldots, C_{2^{k-1}}\}$ and $k=2^{\alpha}-1$. The design for a processing element is given in Figure 4.

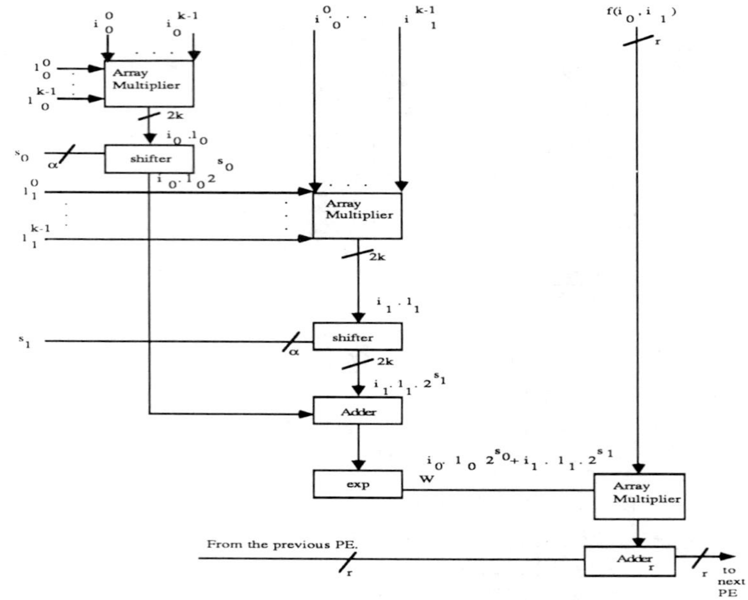

Figure 4: Adaptive Processing Element, $G=G_1 \times G_2$.

For this design $i = (i_0,i_1)$, $i_0 \in G_0$, $i_1 \in G_1$, $l = (l_0,l_1)$, $l_0 \in G_0$, $l_1 \in G_1$, $i_0 = (i_0^0,\dots,i_0^{k-1})$, $i_1 = (i_1^0,\dots,i_1^{k-1})$, $l_0 = (l_0^0,\dots,l_0^{k-1})$, $l_1 = (l_1^0,\dots,l_1^{k-1})$, $i_0^s, i_1^s, l_0^s, l_1^s \in \{0,1\}$, $W = \exp(2\pi \cdot 2^{-k+1}\sqrt{-1})$, and a shifter with the left input p computes a product by 2^p. Total number of input/output lines for a processing element is $4k+\log_2 k+1 + 3r$ $\alpha=\log_2 k+1$, r- is the number of bits in data). To adapt this processing element to $G = C_2a \times C_2b$, where $a,b \in \{0,1,\dots,k-1\}$, one should set $i_0^0=\dots=i_0^{k-1-a}=l_0^0=\dots=l_0^{k-a-1}=0$, $i_1^0=\dots=i_1^{k-b-1}=l_1^0=\dots=l_1^{k-b-1}=0$ and $s_0=k-a-1$, $s_1=k-b-1$.

A similar adaptation technique can be used in case when G is product of more than two groups.

*The work was supported by the Office of Naval Research Under Naval Research Contract N00014-87-K-0735 and by the National Science Foundation under Grant ECS-8512748.

5. References

1. T. Roziner, M. Karpovsky and E. A. Trachtenberg, "Fast Fourier Transforms over Finite Groups by Multiprocessor Systems," IEEE Trans. on ASSP, (To Appear, September, 1989).
2. E. A. Trachtenberg, "Fault Tolerant Computing and Reliable Communication: A Unified Approach," Information and Computation, 79, 3,pp. 257-279, December, 1988.
3. E. A. Trachtenberg and M. G. Karpovsky, "Optimal Varying Dyadic Structure Models of Time Invariant Systems," Proc. 1988 Int'l Symp. on Circuits and Systems, pp. 1111-1114, Espoo, Findland, June, 1988.
4. D.I. Moldovan, J. A. B. Fortes. "Partitioning and Mapping Algorithms into Fixed Size systolic arrays", IEEE Trans. Comp., vol C-35, 1, pp. 1-13, Jan. 1986.
5. W. L. Miranker and A. Winkler. "Space-time Representation of Computational Structures", Computing, vol. 32, pp.sssdd 93-114, 1984.
6. D. I. Moldovan. "On the Design of Algorithms for VLSI Systolic Arrays". proc. IEEE, vol. 71, 1, pp. 113-120, Jan. 1983.
7. E. A.Trachtenberg, M.G. Karpovsky, "Fourier Transforms Over Finite Groups for Error Detection and Error Correction in Computation Channels". Information and Control, 40, 3, pp. 335-358, March 1979.
8. A. M. Despain. "Fourier Transform Computers Using CORDIC Iterations". IEEE Trans. Comp., vol. C-23, 10, Oct. 1974, pp. 993-1001.
9. T. Roziner, "Multidimensional Fourier Transforms by Systolic Architectures", Proc. PARBASE Intern. conference Miami, Florida, March 6-9, 1990.

APPLICATION OF RELAXATION METHODS TO THE ADAPTIVE TRAINING OF NEURAL NETWORKS.

L. Vandenberghe, J. Vandewalle

Abstract

This text describes a systematic approach to derive adaptive training rules for several classes of neural networks. The basic observation is that a variety of design specifications can be translated into sets of linear inequalities with the coefficients of the network as unknowns. Applying classical relaxation methods for solving sets of inequalities then allows to go through the training set of the neural network in an adaptive way. Combined with some insight in the dynamic behaviour of the Hopfield model, this principle can be used for suppressing the occurrence of spurious or undesired invariant states. It is also applicable to the recently introduced class of *cellular neural networks*. A couple of examples to illustrate this are also included.

1 Introduction : relaxation methods for linear inequalities

The central problem in the use of neural networks for pattern recognition, image and signal processing is the development of training and learning algorithms. In this paper the problem of formulating *adaptive* learning algorithms will be addressed, i.e. algorithms with the ability to forget old, unused patterns and to learn newly imposed ones. The basic idea will be to translate the design specifications into linear inequalities in the network coefficients and to apply the *relaxation method* for solving these sets of linear inequalities. Let us therefore start with a brief account of this class of methods.

The relaxation method was first proposed by Agmon, Motzkin and Schoenberg (see [2]) as an iterative algorithm for solving sets of linear inequalities

$$a_i^t.x \geq b_i, \quad i = 1 \ldots m, \tag{1}$$

in n unknowns $x \in R^n$. It proceeds as follows. Starting from an arbitrarily chosen x^0, an approximate solution x^k will be obtained after the $(k-1)$-th iteration. Assume that x^k violates some constraint $a_j.x^k \geq b_j$ of the system (1). Then a correction will be made by taking

$$x^{k+1} = x^k - \lambda.\frac{a_j^t x^k - b_j}{a_j^t a_j}.a_j, \quad 0 < \lambda < 2. \tag{2}$$

For $\lambda = 1$, this step projects x^k orthogonally on the hyperplane $a_j^t x = b_j$. For $\lambda < 1$, one has underrelaxation, for $\lambda > 1$, overrelaxation. Several variants exist, according to which inequality is selected from the m inequalities (1). One possibility is to process the inequalities in a cyclical order.

Convergence of this method has been extensively studied [4, 6]. Interesting results from this analysis are :

Finite convergence. It was shown that, if $\lambda \in (0, 2)$, the relaxation method produces a sequence of iterates that is either finite or converges to a solution on the boundary of the solution polyhedron to (1).

Rate of convergence. Convergence is geometric with a rate depending on a condition number defined for the system (1).

Influence of overrelaxation. A condition number is introduced measuring the *obtuseness* of the solution polyhedron. If the polyhedron is full dimensional, convergence is finite for any $\lambda \in [1, 2]$ if the polyhedron is obtuse, and for any $\lambda \in (\lambda^*, 2]$ if it is not obtuse, where $\lambda^* > 1$ depends on the condition number.

In the subsequent sections, the cyclical version of this basic algorithm will be applied to the training of neural networks.

2 Single layer perceptrons

In a single-layer perceptron with n inputs, an n-dimensional vector x is applied to the input and the output of one neuron is computed as

$$y = \text{sign}(w^t.x - \theta), \tag{3}$$

where $w \in R^n$ is the coefficient vector and θ is the threshold. Training proceeds by applying a sequence of input vectors and specifying the corresponding output. Following the LMS rule, if w_k is the present value of the weight vector, x is the present input pattern vector, and y the desired response, the next value of the weight vector w_{k+1} is computed as

$$w_{k+1} = w_k + \lambda.\frac{\epsilon}{x^t x}.x, \tag{4}$$

where $\epsilon = y - w_k^t x + \theta$. Alternatively, this rule can be viewed as the relaxation step applied to the inequality $w^t.x - \theta \geq 1$ if $y = 1$ or $w^t.x - \theta \leq -1$ if $y = -1$.

3 Suppressing spurious invariant states in Hopfield neural networks

The mathematical structure of the network considered here is as follows. It consists of n interconnected neurons, with state vector $x \in R^n$. Its evolution in time is governed by an $n \times n$ synaptic weight matrix W according to the asynchronous update rule :

One component x_i^k is chosen among the n components of the state vector at time k. Then one takes $x_i^{k+1} = \text{sign}(w_i^t.x^k)$ and $x_j^{k+1} = x_j^k$ for $j \neq i$, where w_i^t is the ith row of W.

Application of these networks is mainly in pattern recognition : the coefficients w_{ij} are chosen in such a way that the network possesses a number of prespecified equilibrium states. If a perturbed pattern is applied as initial state, the network is expected to converge to the nearest equilibrium state and thus to "recognize" the stored pattern.

A problem common to all existing methods for the training of Hopfield neural networks (e.g. the outer product rule or the spectral algorithm [7]) is that these methods mainly rely on imposing the desired equilibria onto the network and do not succeed in avoiding the occurrence of undesired ("spurious") invariant states, and this can result in bad convergence. We will now review a number of interesting properties concerning the dynamic behaviour of this neural network, and based on these properties try to deduce a training rule from the general relaxation algorithm.

3.1 Dynamic properties

A neural net is called *globally stable* if whatever the initial state the network converges to an invariant state. The importance of symmetry is generally acknowledged and this goes back to the energy function defined by Hopfield [5]. By making use of the energy function $E = -x^t W x$, it can easily be proved that the condition that W be symmetric with positive diagonal elements is sufficient for global stability.

We now examine how the number of invariant states is related to the coefficients of the matrix W. A state vector x is an invariant point of the neural net

$$\text{sign}(w_i^t.x) = x_i, \quad i = 1 \dots n.$$

In other words the sign of $w_i^t.x$ must be equal to the sign of $e_i^t.x$ for $i = 1 \dots n$ (e_i is the ith unit vector). This condition can easily be interpreted geometrically. The two vectors w_i and e_i define two hyperplanes $\{x | w_i^t.x = 0\}$ and $\{x | e_i^t x = 0\}$ in n-space. These hyperplanes divide the state space in four regions, according to the 4 combinations of the sign of $w_i^t x$ and $e_i^t x$. The probability that for an arbitrary vector x the signs of $w_i^t x$ and $e_i^t x$ are equal, can be calculated from the angle θ_i between w_i and e_i. The cosine of this angle is given by $\cos\theta_i = w_{ii}/\|w_i\|$. From this observation it can easily be seen that if the diagonal elements of W are zero, and the other elements are randomly chosen, the expected number of invariant states is 1.

This is confirmed by the following experiment. We compute the average number of invariant states when the diagonal elements w_{ii} are fixed and equal for each i. The other elements of W are randomly chosen but in such a way that $\|w_i^t\| = 1, \quad i = 1 \dots n$ and, hence, that $\cos\theta_i = w_{ii}$. The results are summarized in the following table for $n = 5, 10$, and 15.

	Average number of invariant states		
$w_{ii},\ i = 1 \dots n$	$n = 5$	$n = 10$	$n = 15$
0.0	1	1	1
0.1	1	1	3
0.2	2	4	10
0.3	3	7	26
0.4	4	14	61
0.5	5	30	166
0.6	8	67	578
0.7	11	154	1906
0.8	18	368	7149
0.9	32	897	25564
1.0	32	1024	32768

3.2 Adaptive training

In the remainder of this section it will be understood that the matrix W is kept symmetric by updating w_{ij} and w_{ji} in the same way. First assume a state vector v is specified to be an equilibrium of the neural network. This can be achieved by making the coefficients W satisfy the inequalities

$$\begin{array}{ll} w_i^t.v \geq 1 & \text{if} \quad v_i = 1 \\ w_i^t.v \leq -1 & \text{if} \quad v_i = -1 \end{array} \tag{5}$$

Application of the relaxation method results in the updating rule

$$w_{ij}^{k+1} = w_{ij}^k - \lambda_1 \frac{\gamma_i}{v^t.v} v_j, \tag{6}$$

with $\gamma_i = \min(0, w_i^t v - 1)$ if $v_i = 1$ and $\gamma_i = \max(0, w_i^t v + 1)$ if $v_i = -1$.

In order to guarantee a sufficient domain of attraction around the equilibrium v, one can apply perturbed equilibrium patterns z and specify that convergence should be in the direction of v. This can be done by another set of inequalities

$$\begin{array}{ll} w_i^t.z \geq 1 & \text{if} \quad v_i = 1 \\ w_i^t.z \leq -1 & \text{if} \quad v_i = -1 \end{array} \tag{7}$$

This leads to the updating rule

$$w_{ij}^{k+1} = w_{ij}^k - \lambda_2 \frac{\delta_i}{z^t.z} z_j, \tag{8}$$

with $\delta_i = \min(0, w_i^t z - 1)$ if $v_i = 1$ and $\delta_i = \max(0, w_i^t z + 1)$ if $v_i = -1$. The relaxation parameters λ_1 and λ_2 allow to assign a different weight to the inequalities (5) and (7). The inequalities (5) are "hard" constraints and therefore λ_1 will be chosen larger than λ_2.

Finally, in order to reduce the diagonal elements one can alternate steps (6) and (8) with

$$w_{ii}^{k+1} = \max(0, \lambda_3 w_{ii}^k), \tag{9}$$

with $0 \leq \lambda_3 < 1$, depending on the desired number of equilibria.

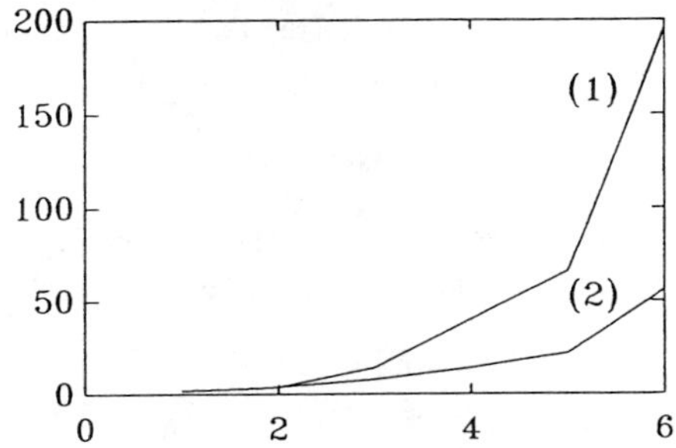

Figure 1: Total number of invariant states versus the specified number of invariant states with the outer-product rule (1) and the relaxation algorithm (2).

3.3 Example

As an example, fig. 2 compares the total number of equilibria in a 16-neuron network after training with the outer-product rule (curve 1) and with the relaxation algorithm (curve 2), with $\lambda_1 = 2$, $\lambda_2 = 0.5$ and $\lambda_3 = 0$. In formula (8), z consists of equilibrium patterns where 40% of the components are reversed.

4 Cellular neural networks

Cellular neural networks were introduced recently [3], but can be seen as a variant of the so-called *brain-state-in-a-box* model [1] defined as

$$dx/dt = f(x - Dx - c) - x, \tag{10}$$

in the continuous time case, and by

$$x[k+1] = f(x[k] - Dx[k] - c), \tag{11}$$

in the discrete time case. Here D is an $n \times n$ matrix, c is an input vector and f is a diagonal mapping defined by

$$\begin{aligned} f_i(x) &= 1 \quad \text{if} \quad x_i \geq 1, \\ f_i(x) &= x_i \quad \text{if} \quad -1 \leq x_i \leq 1, \\ f_i(x) &= -1 \quad \text{if} \quad x_i \leq -1. \end{aligned}$$

In a cellular neural network, neurons are arranged in a 2-dimensional array and interconnections are made only locally. In [3] this structure was used for performing several image processing tasks such as noise removal and edge extraction.

4.1 Example : pattern recognition

As an example we take the design of a neural network to recognize the 10 characters in fig.2. The size of the pictures is 15×20, and the vector c is taken to be zero. Each neuron is connected to other neurons in a symmetric 11×11 neighbourhood. Symmetry of the coefficients was not explicitly assumed.

One can distinguish between imposing desired equilibria and desired transitions. If an equilibrium v is to be specified, the requirement

$$v = f(v - Dv - c)$$

can be written as a set of linear inequalities in the coefficients of D. For each row d_i^t of D, $i = 1 \dots n$, we have one of the following inequalities :

$$\begin{aligned} d_i^t.v \leq 0 &\quad \text{if} \quad v_1 = 1 \\ d_i^t.v = 0 &\quad \text{if} \quad -1 < v_i < 1 \\ d_i^t.v \geq 0 &\quad \text{if} \quad v_i = -1 \end{aligned}$$

Applying the relaxation algorithm leads to the following update rule for the coefficients of D (only applicable for those (i, j) where an interconnection exists) :

$$d_{ij}^{k+1} = d_{ij}^k - \frac{\lambda_1}{v^t v} \gamma_i . v_j, \tag{12}$$

with γ an n-vector given by

$$\begin{aligned} \gamma_i = \max(0, (d_i^k)^t v) &\quad \text{if} \quad v_i = 1 \\ \gamma_i = (d_i^k)^t v &\quad \text{if} \quad -1 < v_i < 1 \\ \gamma_i = \min(0, (d_i^k)^t v) &\quad \text{if} \quad v_i = -1. \end{aligned}$$

In order to guarantee a sufficient domain of attraction around the equilibrium v, one can apply perturbed equilibrium patterns z and specify that convergence should be in the direction of v. This corresponds to the set of inequalities :

$$\begin{aligned} z_i - d_i^t.z \geq 1 &\quad \text{if} \quad v_1 = 1 \\ z_i - d_i^t.z = v_i &\quad \text{if} \quad -1 < v_i < 1 \\ z_i - d_i^t.z \leq -1 &\quad \text{if} \quad v_i = -1. \end{aligned}$$

This can be achieved by the updating rule

$$d_{ij}^{k+1} = d_{ij}^k - \frac{\lambda_2}{z^t z} \gamma_i . z_j, \tag{13}$$

$$\begin{aligned} \gamma_i = \max(0, (d_i^k)^t z - z_i + 1) &\quad \text{if} \quad v_i = 1 \\ \gamma_i = (d_i^k)^t z - z_i + v_i &\quad \text{if} \quad -1 < v_i < 1 \\ \gamma_i = \min(0, (d_i^k)^t z - z_i - 1) &\quad \text{if} \quad v_i = -1. \end{aligned}$$

In the experiment of fig. 1 these two steps are alternated for all stored patterns with $\lambda_1 = 1.1$ and $\lambda_2 = 0.3$. For the perturbed patterns 30% of the pixels were reversed. The figure shows the results of a simulation after convergence.

4.2 Example : edge extraction

In this application the input is a linear transformation $F.u$ of the input u to the network :

$$x[k+1] = f(x[k] - Dx[k] - F.u - \theta). \tag{14}$$

In [3] the coefficients of F and D were chosen from the analogy with digital filters. Here an attempt will be made to determine the weights in an adaptive way,

Figure 2: Recognition of characters with the network (11) and trained with the relaxation algorithm. In the noisy patterns, 30% of the pixels are reversed.

by applying the relaxation algorithm. The mode of operation is different from the previous example. Operation of the network will consist in applying an input pattern u to the network and keeping it fixed until the network converges. The offset θ will be chosen as $\theta_i = 0.5, \quad i = 1 \ldots n$, and also D will be fixed as $D = 0.5I_n$ with I_n the identity matrix. This choice of D ensures convergence to a unique equilibrium for all possible input vectors u, as shown in [8]. The network will be trained to display the edge of the picture presented at the input. Neurons are arranged in a 10×15 array, and are only connected to components of the input within a 3×3 neighbourhood. The matrix F can then be determined adaptively by applying the relaxation algorithm, every time an input pattern and the corresponding equilibrium are specified :

$$f_{ij}^{k+1} = f_{ij}^{k} - \frac{\lambda}{u^t u}\gamma_i . u_j, \tag{15}$$

for those pairs of indices (i, j) where an interconnection exists, and with

$$\begin{aligned} \gamma_i &= \max(0, d_i^t v + (f_i^k)^t u + \theta_i) \quad \text{if} \quad v_i = 1 \\ \gamma_i &= d_i^t v + (f_i^k)^t u + \theta_i \quad \text{if} \quad -1 < v_i < 1 \\ \gamma_i &= \min(0, d_i^t v + (f_i^k)^t u + \theta_i) \quad \text{if} \quad v_i = -1. \end{aligned}$$

where v is the equilibrium corresponding to input u. The results are shown in fig. 3.

5 Conclusion

The relaxation algorithm for linear inequalities provides a systematic way to include a variety of specifications in the adaptive training of different types of neural networks. Further work will focus on acceleration of the convergence and on the use of variants to the basic relaxation scheme. Robustness and convergence of these methods will undoubtedly benefit from exploiting the relationship with relaxation methods for systems of linear equations and with subgradient optimization.

Acknowledgement : L. Vandenberghe is supported by the Belgian National Fund for Scientific Research (N.F.W.O.).

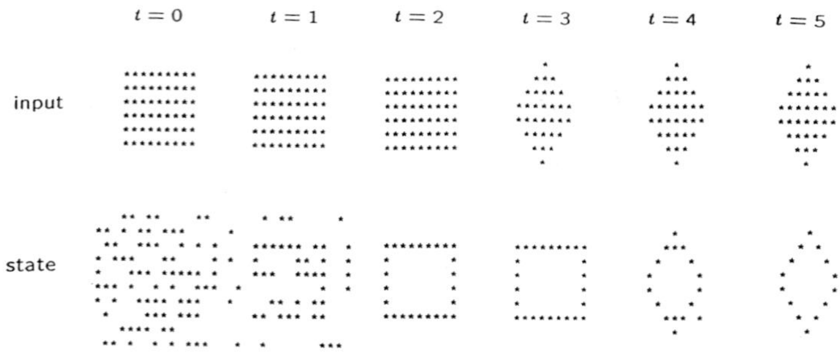

Figure 3: Edge extraction with network (14) : input and state transitions from an arbitrary initial state.

References

[1] J.A. Anderson, "Cognitive and psychological computation with neural models", *IEEE Transactions on Systems Man and Cybernetics*, SMC-13 (1983) 799-815.

[2] Y. Censor, "Row-action techniques for huge and sparse systems and their applications", *SIAM Review* 23, 1980, 444-466.

[3] L.O. Chua and L. Yang, "Cellular neural networks : Theory", and "Cellular neural networks : Applications", *IEEE Transactions on Circuits and Systems*, CAS-35 (1988) 1257-1272 and 1273-1290.

[4] J.L. Goffin, "The relaxation method for solving systems of linear inequalities", *Mathematics of Operations Research* 5, 1980, 388-414.

[5] J. Hopfield, "Neural networks and physical systems with emergent collective abilities", *Proc. Natl. Acad. Sci. USA* 79, 1982, 2554-2558.

[6] J. Mandel, "Convergence of the cyclical relaxation method for solving linear inequalities", *Mathematical Programming* 30, 1984, 218-228.

[7] S.S. Venkatesh, "Computation with neural networks", *Proc. Ninth Annual Conference of the IEEE Engineering in Medicine and Biology Society* 1987, 1364-1365.

[8] L. Vandenberghe, J. Vandewalle, "Brain-state-in-a-box neural networks with asymmetric coefficients", *International Joint Conference on Neural Networks*, June 18-22 1989, Washington.

[9] B. Widrow, R.G. Winter, R.A. Baxter, "Layered neural nets for pattern recognition ", *IEEE Trans. on Acoustics, Speech and Signal Processing* 36, 1988, 1109-1118.

SYSTOLIC ARRAYS FOR ADAPTIVE LEAST SQUARES ALGORITHMS

B. Yang and J.F. Böhme

Abstract

In this paper, we re-examine earlier works and present new results concerning parallel algorithms and systolic arrays implementing solutions of different types of adaptive least squares problems. In particular, we investigate the canonical least squares problem, the multichannel least squares lattice (MLSL) filter and the linearly constrained least squares (LCLS) problem. Aspects of computational complexity, numerical stability, efficiency etc. are discussed.

1. Introduction

Adaptive least squares problems arise in many disciplines of science and engineering. Particular applications are beamforming and direction finding, Kalman filtering, system identification and control, noise cancellation, channel equalization and speech processing. Two different techniques have been developed for solving these problems, the least mean square (LMS) algorithm based on the gradient-search method and the recursive least squares (RLS) algorithm [1]. While the former approximative method is quite computational simple, the latter offers better convergence behavior and parameter tracking capability due to its exact solution at each time step. In this paper, we restrict our investigation to the RLS approach.

Adaptive algorithms based on this technique require a large amount of computations. High data throughput implementations such as systolic arrays for real-time signal processing are desirable. This leads to the development of stable, highly parallel adaptive algorithms and appropriate array architectures.

In section 2, we first re-examine the canonical least squares problem that has been investigated by McWhirter [2]. Then, we discuss two generalizations of this problem, the MLSL filter and the LCLS problem, in sections 3 and 4, respectively. We improve known algorithms, develop new methods and present systolic arrays. Finally, we conclude this paper with some discussions.

2. The canonical least squares problem

Let $x(t) \in \mathbb{R}^N$ and $y(t) \in \mathbb{R}$ be the samples of N reference channels and a single primary channel at the discrete time t, respectively. We assume that they are available for $t \geq 1$. The canonical least squares problem is defined as to find the optimum weight vector $w(t) \in \mathbb{R}^N$ that minimizes the sum of exponentially weighted residual squares :

$$\text{minimize} \quad \sum_{i=1}^{t} \lambda^{t-i} e(t,i)^2 = \sum_{i=1}^{t} \lambda^{t-i} [y(i)+w(t)^T x(i)]^2 \qquad \text{over } w(t) . \tag{1}$$

The forgetting factor $0<\lambda<1$ has the function of progressively weighting against past residual values and providing adaptive properties.

Note that in many applications, we are more interested in the *a posteriori* residual

$$e(t,t) = y(t) + w(t)^T x(t) \tag{2}$$

rather than in the optimum weight vector. In array processing, for example, we need the beamformer output for approriate post-processing operations while w(t) is only an intermediate computing quantity. The main subject of this paper is to investigate how to efficiently compute the residual (2) without explicitly determining the weight vector. For applications such as system identification, where the weight vector itself is the desired information, a method for extracting the weight vector will be discussed at the end of this paper.

McWhirter [2] has proposed a parallel algorithm for computing (1) and (2). It is based on updating the QR decomposition of the data matrix $[x(1), x(2), ..., x(t)]^T$. We use a slightly different, but equivalent approach to re-examine his method. We define

$$R(t) = \sum_{i=1}^{t} \lambda^{t-i} x(i)x(i)^T \quad \text{and} \quad \Delta^y(t) = \sum_{i=1}^{t} \lambda^{t-i} x(i)y(i) \tag{3}$$

which are estimates of the covariance matrix of x(t) and the cross covariance vector between x(t) and y(t), respectively. The solution of (1) and (2) is given by

$$e(t,t) = y(t) - \Delta^y(t)^T R(t)^{-1} x(t) , \tag{4}$$

where R(t) and $\Delta^y(t)$ can be computed by the following rank-one updates

$$R(t) = \lambda R(t-1) + x(t)x(t)^T \quad \text{and} \quad \Delta^y(t) = \lambda \Delta^y(t-1) + x(t)y(t) . \tag{5}$$

We assume R(t) being positively definite. This is true if we use a positively definite initial value R(0) in the recursion (5).

A straightforward implementation of (4) and (5) by computing the matrix inverse $R(t)^{-1}$ leads to $O(N^3)$ operations. A more efficient method which is based on recursively updating the Cholesky decomposition $R(t)=L(t)L(t)^T$ involves only $O(N^2)$ operations. By using the lower triangular Cholesky factor L(t), we define the following variables,

$$\tilde{x}(t) = L(t)^{-1} x(t) , \quad \tilde{\Delta}^y(t) = L(t)^{-1} \Delta^y(t) . \tag{6}$$

Equation (4) can then be expressed by

$$e(t,t) = y(t) - \tilde{\Delta}^y(t)^T \tilde{x}(t) . \tag{7}$$

Now, an orthogonal transformation U(t) ($U(t)^T U(t)=I$) is used to both update quantities in (6) and to compute, except for a scaling factor $\alpha(t)$, the desired residual e(t,t) :

$$U(t) \cdot \begin{bmatrix} \sqrt{\lambda} L(t-1)^T & 0 & \sqrt{\lambda} \tilde{\Delta}^y(t-1) \\ x(t)^T & 1 & y(t) \end{bmatrix} = \begin{bmatrix} L(t)^T & \tilde{x}(t) & \tilde{\Delta}^y(t) \\ 0 & \alpha(t) & e(t,t)/\alpha(t) \end{bmatrix} . \tag{8}$$

Equation (8) can be verified by noting the identity $\tilde{A}^T \tilde{A} = A^T A$ for $\tilde{A}=U(t)A$. The orthogonal matrix U(t) is designed to eliminate the row vector $x(t)^T$ while preserving the upper triangular structure of $L(t-1)^T$. For parallel implementations, U(t) can be realized by N Givens rotations. In this case, the last diagonal element of U(t) is $\alpha(t)$ and can be computed as the product of the N cosine parameters of the associated Givens rotations [2].

Figure 1 shows the principle architecture of McWhirter's triangular systolic array for implementing the canonical least squares algorithm. The quantities $L(t-1)^T$ and $\tilde{\Delta}^y(t-1)$ are stored in the processing elements (PEs) of the array and are updated at each time step while input data x(t) and y(t) are put into the array and propagate from top to bottom. The sine and cosine parameters are calculated in the diagonal PEs and propagate from left to right. After computing the product of the cosine parameters along the diagonal and applying Givens rotations in the interior PEs, $\alpha(t)$ and $e(t,t)/\alpha(t)$ appear at the diagonal and the most right column outputs, and the residual e(t,t) is calculated by a single multiplication. For more details, we refer to McWhirter's paper [2].

3. The adaptive multichannel least squares lattice filter

The adaptive MLSL filter generalizes the above canonical least squares problem in the sense that we estimate y(t) by a linear combination of the last p samples x(t), x(t–1), ..., x(t–p+1) of the N reference channels. As characteristic for lattice filters, solving this joint process estimation problem (p>1) requires to solve two related linear prediction problems, where we have to estimate x(t+1) (forward prediction) and x(t–p) (backward prediction) by using linear combinations of the same data.

Let e(p,t+1), r(p,t) and $e^y(p,t)$ denote the forward prediction, backward prediction and joint estimation errors, respectively. By applying similar weighted least squares criteria as (1), Lewis [3] has derived the following set of time and order recursions for computing e(p,t+1), r(p,t) and $e^y(p,t)$:

<u>Forward and backward prediction</u>

For. pred. error : $$e(p,t) = e(p-1,t) - \Delta(p-1,t)R^r(p-1,t-1)^{-1}r(p-1,t-1)\ , \qquad (9a)$$

Back. pred. error : $$r(p,t) = r(p-1,t-1) - \Delta(p-1,t)^T R^e(p-1,t)^{-1}e(p-1,t)\ , \qquad (9b)$$

For. error covar. : $$R^e(p,t) = \lambda R^e(p,t-1) + e(p,t)e(p,t)^T/\gamma(p,t-1)\ , \qquad (9c)$$

Back. error covar. : $$R^r(p,t) = \lambda R^r(p,t-1) + r(p,t)r(p,t)^T/\gamma(p,t)\ , \qquad (9d)$$

Cross error covar. : $$\Delta(p,t) = \lambda\Delta(p,t-1) + e(p,t)r(p,t-1)^T/\gamma(p,t-1)\ , \qquad (9e)$$

Likelihood var. : $$\begin{aligned}\gamma(p,t) &= \gamma(p-1,t) - r(p-1,t)^T R^r(p-1,t)^{-1}r(p-1,t) \\ &= \gamma(p-1,t-1) - e(p-1,t)^T R^e(p-1,t)^{-1}e(p-1,t)\ , \qquad (9f)\end{aligned}$$

<u>Joint process estimation</u>

Joint estim. error : $$e^y(p,t) = e^y(p-1,t) - \Delta^y(p-1,t)^T R^r(p-1,t)^{-1}r(p-1,t)\ , \qquad (9g)$$

Cross error covar. : $$\Delta^y(p,t) = \lambda\Delta^y(p,t-1) + r(p,t)e^y(p,t)/\gamma(p,t)\ . \qquad (9h)$$

In equations (9c), (9d), (9e) and (9h), $R^e(t)$ and $R^r(t)$ are the estimated covariance matrices of e(p,t) and r(p,t) while $\Delta(p,t)$ and $\Delta^y(p,t)$ are the estimated cross covariance matrices between e(p,t), r(p,t–1) and r(p,t), $e^y(p,t)$, respectively.

Again, a straightforward implementation of (9) requires $O(pN^3)$ operations for each time-update of the p-th order multichannel lattice filter. This computational complexity can be reduced to $O(pN^2)$ in different ways. By using the matrix inversion lemma, equations (9c) and (9d) can be rewritten in terms of their inverses. This leads to a serial Kalman-type MLSL algorithm [3]. On the other side, Lewis has also shown a QR based MLSL algorithm for parallel implementations. In the following, we present our improved version of his algorithm leading to additional savings of computations and improved numerical properties.

We assume $R^e(p,t)$ and $R^r(p,t)$ being positively definite. Let

$$R^e(p-1,t) = L^e(p,t)L^e(p,t)^T \quad \text{and} \quad R^r(p-1,t-1) = L^r(p,t)L^r(p,t)^T \tag{10}$$

be the Cholesky decompositions with $L^e(p,t)$ and $L^r(p,t)$ being lower triangular matrices. By defining the following variables,

$$\beta^e(p,t) = L^e(p,t)^{-1}e(p-1,t)\ , \quad \beta^r(p,t) = L^r(p,t)^{-1}r(p-1,t-1)\ , \tag{11}$$

$$X^e(p,t)=L^e(p,t)^{-1}\Delta(p-1,t)\ , \ X^r(p,t)=L^r(p,t)^{-1}\Delta(p-1,t)^T\ , \ X^y(p,t)=L^r(p,t)^{-1}\Delta^y(p-1,t-1)\ , \tag{12}$$

equations (9) can be reformulated. Similar to (8), we apply two orthogonal transformations $U^e(p,t)$ and $U^r(p,t)$ in terms of Givens rotations to two suitably constructed matrices. After considerable algebraic manipulations, we can show :

$$U^e(p,t)\cdot\begin{bmatrix} \sqrt{\lambda}L^e(p,t-1)^T & 0 & \sqrt{\lambda}X^e(p,t-1) \\ e(p-1,t)^T/\sqrt{\gamma(p-1,t-1)} & \sqrt{\gamma(p-1,t-1)} & r(p-1,t-1)^T/\sqrt{\gamma(p-1,t-1)} \end{bmatrix}$$

$$= \begin{bmatrix} L^e(p,t)^T & \beta^e(p,t) & X^e(p,t) \\ 0 & \sqrt{\gamma(p,t)} & r(p,t)^T/\sqrt{\gamma(p,t)} \end{bmatrix}, \tag{13}$$

$$U^r(p,t)\cdot\begin{bmatrix} \sqrt{\lambda}L^r(p,t-1)^T & 0 & \sqrt{\lambda}X^r(p,t-1) & \sqrt{\lambda}X^y(p,t-1) \\ \dfrac{r(p-1,t-1)^T}{\sqrt{\gamma(p-1,t-1)}} & \sqrt{\gamma(p-1,t-1)} & \dfrac{e(p-1,t)^T}{\sqrt{\gamma(p-1,t-1)}} & \dfrac{e^y(p-1,t-1)}{\sqrt{\gamma(p-1,t-1)}} \end{bmatrix}$$

$$= \begin{bmatrix} L^r(p,t)^T & \beta^r(p,t) & X^r(p,t) & X^y(p,t) \\ 0 & \sqrt{\gamma(p,t-1)} & e(p,t)^T/\sqrt{\gamma(p,t-1)} & e^y(p,t-1)/\sqrt{\gamma(p,t-1)} \end{bmatrix}. \tag{14}$$

In equations (13) and (14), $U^e(p,t)$ and $U^r(p,t)$ are determined to eliminate the row vectors $e(p-1,t)^T/\sqrt{\gamma(p-1,t-1)}$ and $r(p-1,t-1)^T/\sqrt{\gamma(p-1,t-1)}$ of the left hand side matrices while preserving the upper triangular structures of $L^e(p,t-1)^T$ and $L^r(p,t-1)^T$, respectively. We see that, instead of computing a set of 8 matrix and vector recursions (9), we can use two orthogonal matrix transformations to directly compute scaled versions of the desired prediction and joint estimation errors. Because the scaling factors $\sqrt{\gamma(p,t)}$ and $\sqrt{\gamma(p,t-1)}$ are also available, $e(p,t)$, $r(p,t)$ and $e^y(p,t-1)$ can be extracted from the second rows of the right hand side matrices of (13) and (14) by simple multiplications. Moreover, all desired input data for the next time and order recursions can be recursively obtained from the output data of (13) and (14).

Two special cases of the MLSL filter are of particular interest. In the single tap case (p=1), the MLSL problem reduces to the canonical least squares one. The joint process estimation problem can now be solved independently from the prediction problems. By noting the initial conditions,

$$e(0,t) = r(0,t) = x(t)\ , \quad e^y(0,t) = y(t)\ , \quad \gamma(0,t) = 1\ , \tag{15}$$

equation (14) incorporates equation (8). On the other side, when the number of reference channels is reduced to one (N=1), all quantities in (9) are scalars. This is the conventional single channel least squares lattice (LSL) filter. In this case, equations (13) and (14) provide an interesting alternative to the standard implementation of the adaptive LSL algorithm [1] involving only stable rotations and multiplications.

Figure 2 shows a linear array with the well known topology for a lattice filter of order p. In the multichannel case, each of the p filter sections implements equations (13) and (14) and the last multiplications for extracting the prediction and joint estimation errors. These are matrix operations having a very regular structure so that they can be themselves implemented by systolic arrays. Depending on the performence requirement, there are many space-time tradeoffs available for the complexity distribution [3]. A structure allowing maximum parallelism is depicted in Figure 3, where equation (13) is implemented on a trapezoidal systolic array. A detailed discussion about this problem can be found in [4].

4. The adaptive linearly constrained least squares problem

Another important class of least squares problems is the adaptive LCLS problem. Typical applications are linearly constrained adaptive beamforming, also known as minimum variance distortionless response (MVDR) beamforming [5,6]. In these applications, all channel outputs are weighted prior to linear combining so that we have y(t)=0 in (1) and (2). Moreover, K linear constraints are imposed on the weight vector w(t)

$$C^T w(t) \; = \; g\ , \tag{16}$$

where C is a NxK (K<N) full rank constraint matrix and g is a Kx1 constraint vector. Notice that if equation (16) takes the special form $[0 \cdots 0\ 1]\cdot w(t)=1$, the LCLS problem simplifies to the canonical least squares one.

A well known method for solving (1) and (16) is the generalized sidelobe canceller (GSC) [5]. It reduces the constrained least squares problem to the canonical one which is then solved by methods described in section 2. This approach is based on decomposing the weight vector w(t) into two orthogonal components,

$$w(t) \; = \; w_q + W_s w_a(t)\ , \tag{17}$$

where the quiescent vector w_q lies in the range space, and the N-K column vectors of the blocking matrix W_s span the null space of the constraint matrix C. This means,

$$w_q = C(C^T C)^{-1} g \quad \text{and} \quad C^T W_s = 0\ . \tag{18}$$

Clearly, equation (16) is always satisfied. The residual elements in (1) are now given by

$$e(t,i) \; = \; w(t)^T x(i) \; = \; w_q^T x(i) + w_a(t)^T \cdot W_s^T x(i)\ . \tag{19}$$

By identifying $w_q^T x(i)$ and $W_s^T x(i)$ as the samples of a single primary channel and N–K reference channels, respectively, equation (1) and (19) describe a canonical least squares problem of reduced dimension. Notice that $w_a(t)$ is the adaptive weight vector while w_q and W_s are fixed.

At this point, it is interesting to note that, while w_q is uniquely determined by C and g, the Nx(N–K) blocking matrix W_s is arbitrary subject to the restriction in (18) and that it is of full rank. Therefore, there are different possibilities to implement the pre-processing operations $W_s^T \cdot x(t)$ depending on the choice of W_s. The trapzoidal pre-processor array given by McWhirter and Shepherd [6] is a special realization of these operations (Figure 4).

In the following, we present an alternative method without using the GSC structure. The solution of (1), (2) and (16) is given by

$$e(t,t) = g^T(C^T R(t)^{-1} C)^{-1}(C^T R(t)^{-1} x(t)) , \tag{20}$$

where R(t) is defined in (3) and can be updated according to (5). We assume R(t) being positively definite. Again, we avoid to compute $R(t)^{-1}$ and develop an algorithm based on updating the Cholesky decomposition $R(t)=L(t)L(t)^T$. By using the lower triangular Cholesky factor L(t), we define

$$\tilde{C}(t) = L(t)^{-1} C , \quad D(t) = \tilde{C}(t)^T \tilde{C}(t) , \quad \tilde{x}(t) = L(t)^{-1} x(t) . \tag{21}$$

Equation (20) can then be reformulated by

$$e(t,t) = g^T D(t)^{-1} \tilde{C}(t)^T \tilde{x}(t) . \tag{22}$$

Similar to sections 2 and 3, we apply an orthogonal transformation U(t) to a suitably constructed matrix for updating L(t), $\tilde{C}(t)$ and computing the matrix expression $\tilde{C}(t)^T \tilde{x}(t)$:

$$U(t) \cdot \begin{bmatrix} \sqrt{\lambda} L(t\text{-}1)^T & \tilde{C}(t-1)/\sqrt{\lambda} & 0 \\ x(t)^T & 0 & 1 \end{bmatrix} = \begin{bmatrix} L(t)^T & \tilde{C}(t) & \tilde{x}(t) \\ 0 & z(t)^T & \alpha(t) \end{bmatrix} . \tag{23}$$

Then, by exploiting the orthonormality of U(t), we can show $\tilde{C}(t)^T \tilde{x}(t) = -\alpha(t) z(t)$ so that equation (22) becomes

$$e(t,t) = -\alpha(t) g^T D(t)^{-1} z(t) . \tag{24}$$

Moreover, the unknown quantity D(t) in (24) can be computed either by using the definition (21) or more efficiently via the following rank-one downdating operation,

$$D(t) = D(t-1)/\lambda - z(t) z(t)^T . \tag{25}$$

Equations (24) and (25) involve only scalar operations for the single constraint case (K=1). When we have more than one linear constraints (K>1), however, $O(K^3)$ matrix operations are required. This computational complexity can be reduced to $O(K^2)$ if we use a similar method as above to recursively update the Cholesky decomposition $D(t)=F(t)F(t)^T$. By defining the variables

$$\tilde{g}(t) = F(t)^{-1} g \quad \text{and} \quad \tilde{z}(t) = F(t)^{-1} z(t) , \tag{26}$$

the residual expression (24) simplifies to

$$e(t,t) = -\alpha(t) \tilde{g}(t)^T \tilde{z}(t) . \tag{27}$$

Now, we apply a hyperbolic transformation H(t) in terms of K hyperbolic rotations, where

$$H(t)^T \begin{bmatrix} I & 0 \\ 0 & -1 \end{bmatrix} H(t) = \begin{bmatrix} I & 0 \\ 0 & -1 \end{bmatrix}, \tag{28}$$

to the following composed matrix for eliminating the row vector $z(t)^T$,

$$H(t)\cdot \begin{bmatrix} F(t-1)^T/\sqrt{\lambda} & 0 & \sqrt{\lambda}\tilde{g}(t-1) \\ z(t)^T & \alpha(t) & 0 \end{bmatrix} = \begin{bmatrix} F(t)^T & -\alpha(t)\tilde{z}(t) & \tilde{g}(t) \\ 0 & \beta(t) & \gamma(t) \end{bmatrix}. \tag{29}$$

We can show that the desired residual is now given by

$$e(t,t) = \beta(t)\gamma(t)\,. \tag{30}$$

Figure 5 shows the principle of a linear systolic array implementing the above adaptive algorithm. It mainly consists of N Givens rotors and K hyperbolic rotors. While the recursively used quantities $L(t-1)$, $\tilde{C}(t-1)$, $F(t-1)$ and $\tilde{g}(t-1)$ pass through the array from top to bottom, the input data $x(t)$ enters from left into the array. The residual $e(t,t)$ appears at the most right output. A detailed discussion of this linear array can be found by Yang and Böhme [7].

5. Discussions

Now, we briefly discuss aspects of numerical properties, choice of processing elements, how to extract the weight vector and space-time tradeoffs for the adaptive least squares algorithms and the systolic implementations.

The numerical stability of a recursive algorithm is determined by two factors, the rounding errors due to the finite arithmetic precision and the error propagation and accumulation due to the recursive nature of the algorithm. As shown in sections 2 and 3, the canonical least squares algorithm and the MLSL filter require only numerically well behaved orthogonal transformations (8), (13) and (14) and multiplications in which the multipliers $\alpha(t)$ and $\sqrt{\gamma(p,t)}$ are limited by one in magnitude. Moreover, all recursively used quantities and therefore also their accumulated errors are weighted by the factor $0<\sqrt{\lambda}<1$ at each time step preventing an unlimited growth of the errors. These two algorithms are thus numerically stable. In comparison, if we realize lattice filters in the usual way [1,3] by employing (9), divisions (9c)–(9e) and (9h) in conjunction with the subtraction (9f) may cause numerical problems because the divisor $\gamma(p,t)$ becomes smaller for every time-update.

In the LCLS applications, care must be taken. When using the GSC approach, realizing the pre-processing operations $w_0^T x(t)$ and $C_n^T x(t)$ plays an important role. In McWhirter and Shepherd's adaptive algorithm [6], where Gauss eliminations are applied to remove the linear constraints, relatively large rounding errors may occur if the pivoting elements are small in magnitude. Similar problems arise also with the alternative method. It is known [8] that the hyperbolic transformation (29) is stable in a certain "mixed" sense provided that it is carefully implemented and the downdating problem itself is not ill conditioned. The underlying recursion (25), however, is unstable because of the factor $1/\lambda>1$. One possible solution to this problem is to use the GSC approach with a more stable implementation of the pre-processing operations. On the other side, further investigations of the alternative method to improve its numerical properties are also recommended.

Another important aspect in systolic implementations is the choice of processing elements. We see that the basic operations in all presented algorithms are Givens rotations and, more seldom, hyperbolic rotations. Thus, we propose to use CORDIC (COordinate Rotation DIgital Computer) processors. They are dedicated processors for performing Givens and hyperbolic rotations requiring only a sequence of iterative shift-add operations. They have a regular structure for easy implementation, a high data throughput in a pipelined realization and can be easily programmed. A brief overview about this topic is given in [9].

In section 2, we have pointed out that the weight vector w(t) is sometimes needed explicitly. A method for extracting it without additional processors is given by McWhirter and Shepherd [6]. By re-examining equation (2), we see that the computed residual e(t,t) equals the elements of w(t) if we "freeze" adaptions and choose y(t)=0 and x(t) successively as the column vectors of the identity matrix. This procedure requires, however, a computation time as that for N residual elements.

Finally, we mention that all adaptive algorithms presented in this paper can be implemented by both two-dimensional and linear systolic arrays. The choice depends only on the costs and performence requirements. While linear arrays involve a lower hardware cost and are superior with respect to fault tolerance, synchronizing and wafer-scale integration, two-dimensional arrays offer a higher data throughput and a higher efficiency because of the exact mapping of the matrix structures onto the systolic array structures.

References

[1] B. Friedlander, "Lattice Filters for Adaptive Filtering", Proc. IEEE 70(8), pp.829–867, 1982

[2] J.G. McWhirter, "Recursive Least-Squares Minimization Using a Systolic Array", Proc. SPIE 431, Real-Time Signal Processing, pp.105–112, San Diego, 1983

[3] P.S. Lewis, "QR Algorithm and Array Architectures for Multichannel Adaptive Least Squares Lattice Filters", Proc. ICASSP, pp.2041–2044, New York, April 1988

[4] B. Yang and J.F. Böhme, "On a Parallel Implementation of the Adaptive Multichannel Least Squares Lattice Filter", to appear in Proc. Int. Symp. on Signals, Systems and Electronics, Erlangen, Sept. 1989

[5] L.J. Griffiths and C.W. Jim, "An Alternative Approach to Linearly Constrained Adaptive Beamforming", IEEE Trans. on Antennas and Propagation 30(1), pp.21982

[6] J.G. McWhirter and T.J. Shepherd, "A Systolic Array for Linearly Constrained Least-Squares Problems", Proc. SPIE 696, Advanced Algorithms and Architectures for Signal Processing, pp.80–87, San Diego, Aug. 1986

[7] B. Yang and J.F. Böhme "Linear Systolic Arrays for Constrained Least-Squares Problems", Proc. IMA Conf. on Mathematics in Signal Processing, Warwick, Dec. 1988

[8] A.W. Bojanczyk et al. "Downdating the Cholesky Factorization", Algorithms and Applications on Vector and Parallel Computers, Elsevier Science Publishers B.V. (North-Holland), ed. H.J.J. te Riele et al., pp.307–323, 1987

[9] J. Böhme and B. Yang, "Highly Parallel and Pipelined Computing Using CORDIC Processors and Applications to Computer Graphics, Signal Processing and Linear Algebra", to appear in Proc. German-Chinese Electronics Week, Shanghai, VDE Verlag Düsseldorf, Sept. 1989

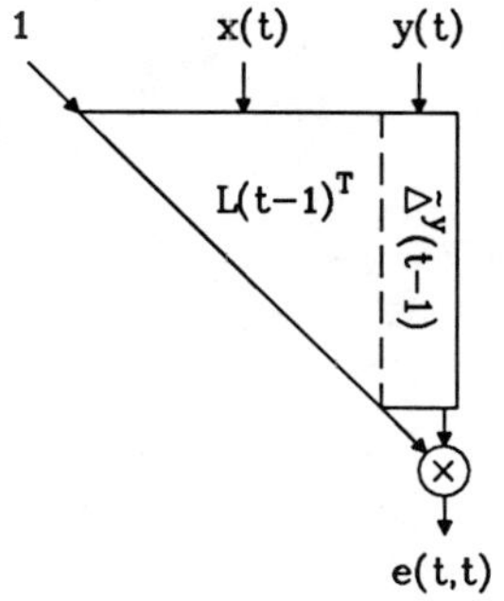

FIGURE 1: McWhirter's triangular systolic array implementing the canonical least squares algorithm

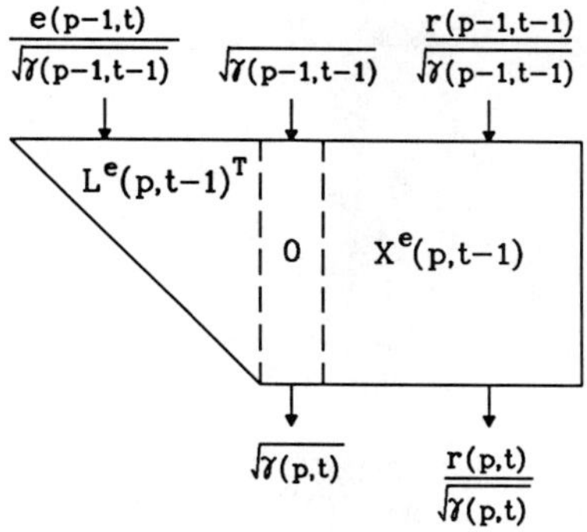

FIGURE 3: A trapezoidal systolic array implementing equation (13)

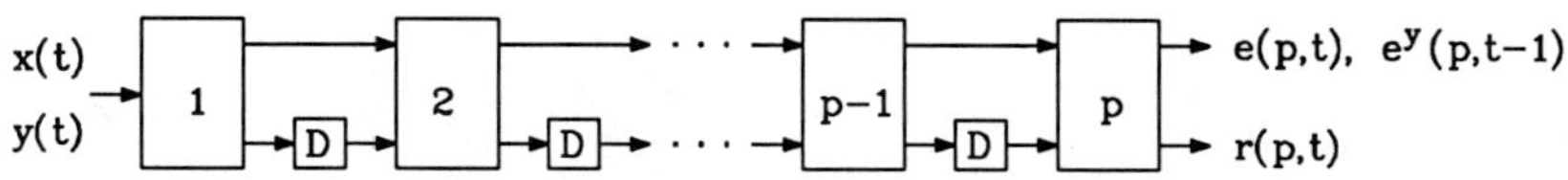

FIGURE 2: A linear array implementing a lattice filter of order p

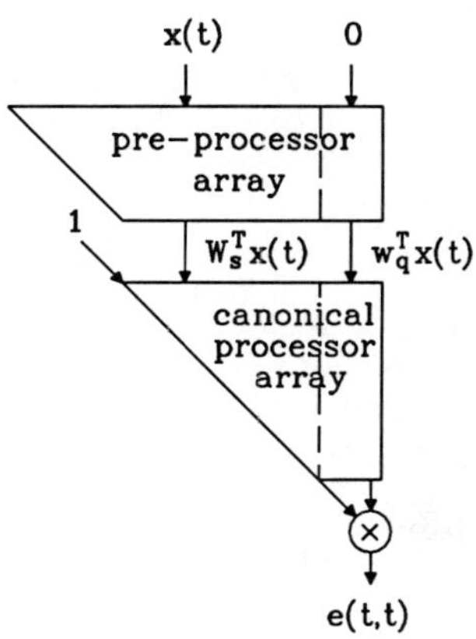

FIGURE 4: A triangular systolic array for solving the LCLS problem by using the GSC approach

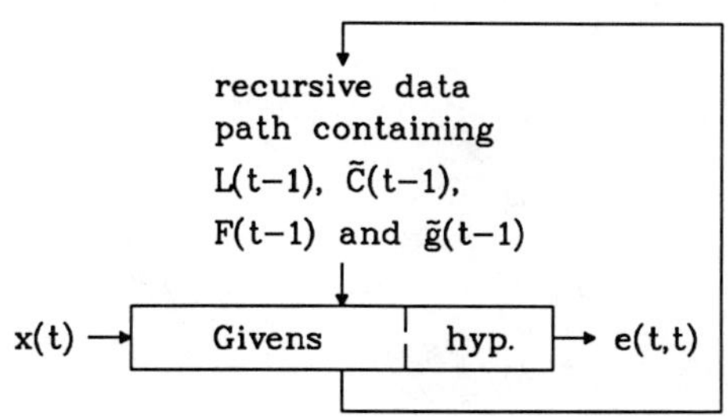

FIGURE 5: A linear systolic array for solving the LCLS problem

The Schur Algorithm: Connections with the Difference Riccati Equation, the Chandrasekhar Algorithm, and Square-root filtering

Chin Chang and Tryphon T. Georgiou

Abstract: This paper points out a connection between the Schur Algorithm and the Difference Riccati Equation arising in Linear Optimal Filtering. In particular a version of the Schur Algorithm leads to a Square-root Chandrasekhar-type Algorithm studied by Morf and Kailath.

Keywords: Schur algorithm, spectral factorization, difference Riccati equation, Chandrasekhar algorithm, square-root filtering, optimal filtering

0. NOTATION

$\mathbb{D}$ the open unit disc in $\mathbb{C}$,
X^* the complex conjugate transposed of X
X' the transposed of X
$X^{1/2}$ the Hermitian square root of a Hermitian matrix X
$f(\lambda)^{\sim}$ the paraconjugate $f^*(\lambda^{-1})$.
λ $(= z^{-1})$ complex variable
$[A, B, C, D] := D + \lambda C(I - \lambda A)^{-1}B$

1. INTRODUCTION

The purpose of this paper is to present a connection between the Schur algorithm of analytic interpolation theory and the difference Riccati equation which arises in linear optimal filtering. More specifically, we will show that a version of the Schur algorithm given in Georgiou [7], corresponds directly to a square-root version of a Chandrasekhar-type algorithm that was presented by Morf and Kailath [12].

The application of the Schur algorithm to the problem of Spectral factorization has been discussed in [6, 7, 8, 9, 10]. The steps of the algorithm allow for a certain degree of freedom in the choice of certain relevant parameters. This direction has been considered in the above publications and has a potential advantage in terms of speed of convergence as well as accuracy and conditioning of the computations. In [7], a special version of the Schur algorithm was considered and the iterations were formulated in state-space. It turns out that there is a close connection between this form of the Schur algorithm and the standard discrete-time Riccati equation that arises in linear optimal filtering and the quadratic regulator problem. For instance the Kalman filter gains are computed as by-products of the Schur recursions. In particular, we will show that the version of the Schur algorithm presented in [7], corresponds directly to the square-root version of the Chandrasekhar algorithm presented by Morf and Kailath [12]. But, the actual implementation of the steps in these two algorithms is different.

2. OPTIMAL FILTERING AND THE RICCATI EQUATION

Consider the standard finite-dimensional discrete-time Markovian model (see Anderson and Moore [1])

$$\begin{aligned} x_{k+1} &= Ax_k + w_k \\ z_k &= Cx_k + v_k \end{aligned}$$

where $k = 0, 1, \cdots,$ $\begin{pmatrix} w_k \\ v_k \end{pmatrix}$ is zero-mean Gaussian white noise with covariance

$$E\{\begin{pmatrix} w_k \\ v_k \end{pmatrix}(w'_k \quad v'_k)\} = \begin{pmatrix} Q & S \\ S' & R \end{pmatrix},$$

A is a stable matrix (i.e., all its eigenvalues have magnitude less than unity), and the state covariance $P = E\{x_0 x'_0\}$ is the unique nonnegative-definite solution of the Lyapunov equation

$$P = APA' + Q \tag{1}$$

The power spectrum of the stationary stochastic process z_k is given by

$$\begin{aligned} \Phi(\lambda,\lambda^{-1}) &= \begin{pmatrix} \lambda C(I-\lambda A)^{-1} & I \end{pmatrix} \begin{pmatrix} Q & S \\ S' & R \end{pmatrix} \begin{pmatrix} \lambda^{-1}(I-\lambda^{-1}A')^{-1}C' \\ I \end{pmatrix} \\ &= 2D + C\lambda(I-\lambda A)^{-1}B + B'\lambda^{-1}(I-\lambda^{-1}A')^{-1}C' \end{aligned}$$

where $B = APC' + S$ and $2D = CPC' + R$. In the sequel for simplicity of the presentation we restrict our attention to the case where R is nonsingular.

With A, B, C, and D as above, the function $F(\lambda) := [A, B, C, D]$ is in the *Caratheodory class* (i.e., it is analytic with positive real part in $\mathbb{D}$). The power spectrum $\Phi(\lambda,\lambda^{-1}) = F(\lambda) + F(\lambda)^{\sim}$ admits a factorization $\Phi(\lambda,\lambda^{-1}) = G(\lambda)G(\lambda)^{\sim}$ with $G(\lambda) = [A, L, C, W]$ being a square matrix function that is invertible in $\mathbb{D}$. The factor G is uniquely defined up to a right constant unitary factor and is the *(left) canonical spectral factor* of Φ (or, of F).

Consider the Kalman filter corresponding to the Markovian model given above. The state equation of the filter is

$$\hat{x}_{k+1/k} = A\hat{x}_{k/k-1} + K_k(z_k - C\hat{x}_{k/k-1})$$

where

$$K_k = (A\Sigma_k C' + S)(C\Sigma_k C' + R)^{-1} \tag{2}$$

$$\Sigma_{k+1} = A\Sigma_k A' - (A\Sigma_k C' + S)(C\Sigma_k C' + R)^{-1}(A\Sigma_k C' + S)' + Q \tag{3}$$

for $k = 0, 1, \cdots$, and $\Sigma_0 = P$. Σ_k is the error covariance matrix. It is a standard result that Σ_k tends to a limit as $k \to \infty$ (see for instance [1, Chapter 4]). This limit, denoted by Σ, is the unique "stabilizing" solution of the algebraic Riccati equation

$$\Sigma = A\Sigma A' - (A\Sigma C' + S)(C\Sigma C' + R)^{-1}(A\Sigma C' + S)' + Q \tag{4}$$

Then, with

$$W = (C\Sigma C' + R)^{1/2}$$

and

$$L = (A\Sigma C' + S)(C\Sigma C' + R)^{-1/2},$$

$[A, L, C, W]$ is a (left) canonical spectral factor of Φ (or, of F) (see [1, Chapter 4] and [5]).

3. THE SCHUR ALGORITHM

The Schur algorithm was originally introduced to solve certain problems in interpolation with functions analytic in $\mathbb{D}$ (see [3], [4], [11] and the references therein). It turns out that starting from say, a function $F(\lambda)$ of the Caratheodory class, it is possible to use the Schur algorithm in a "reverse" fashion and obtain the corresponding spectral factor $G(\lambda)$. This approach has been introduced and studied in [6, 7, 8, 9, and 10]. Here we present a version of the Schur algorithm as applied to functions of the Caratheodory class. It is formulated so as to process state-space data, and it follows directly from [7, Theorem 2.3].

THEOREM (3.1) Let $F(\lambda) = [A, B, C, D]$ be a function in the Caratheodory class (i.e., positive-real) and $D = D'$ be a Hermitian matrix. Set

$$\Gamma_0 = -BD^{-1/2}$$
$$M_0 = BD^{-1/2}$$
$$N_0 = 2D^{1/2}$$

and compute iteratively Γ_k, M_k, N_k for $k \geqslant 1$ using

$$\begin{aligned}
\beta_k &= N_k^{-1} C \Gamma_k \\
T_{k,right} &= (I - \beta_k \beta_k^*)^{-1/2} \\
T_{k,left} &= (I - \beta_k^* \beta_k)^{-1/2} \\
\Gamma_{k+1} &= (A \Gamma_k - M_k \beta_k) T_{k,left} \\
M_{k+1} &= (M_k - A \Gamma_k \beta_k^*) T_{k,right} \\
N_{k+1}^{-1} &= T_{k,right} N_k^{-1}
\end{aligned} \tag{5}$$

for $k = 0, 1, 2, \cdots$. Then as $k \to \infty$, the limits

$$\begin{aligned}
\Gamma_\infty &:= \lim_{k \to \infty} \Gamma_k \\
M_\infty &:= \lim_{k \to \infty} M_k \\
N_\infty &:= \lim_{k \to \infty} N_k
\end{aligned}$$

exist and it holds that $\Gamma_\infty = 0$ while $G(\lambda) = \frac{1}{\sqrt{2}}[A, M, C, N]$ is a (left) canonical spectral factor for $F(\lambda)$.

4. CONNECTION WITH THE RICCATI EQUATION

In this section we present a connection of the Schur algorithm given in (3.1) with the difference Riccati equation (3).

THEOREM (4.1). Let Σ_k, $k = 0, 1, \cdots$, be the solution of the difference Riccati equation (3) with Σ_0 the nonnegative definite solution of the Lyapunov equation (1), let

$$\begin{aligned}
W_k &= (C \Sigma_k C' + R)^{1/2} \\
L_k &= (A \Sigma_k C' + S)(C \Sigma_k C' + R)^{-1/2}
\end{aligned}$$

and the Kalman gain $K_k = L_k W_k^{-1}$. Let also Γ_k, M_k, and N_k, for $k = 0, 1, \cdots$, be obtained as in Theorem (3.1). Then

$$\begin{aligned}
\Sigma_{k+1} - \Sigma_k &= -\frac{1}{2} \Gamma_k \Gamma'_k \\
W_k &= \frac{1}{\sqrt{2}} N_k U_k \\
L_k &= \frac{1}{\sqrt{2}} M_k U_k \\
K_k &= M_k N_k^{-1}
\end{aligned}$$

for $k = 1, 2, \cdots$, with U_k being a constant unitary matrix.

The proof of this theorem follows by induction and is included in [2].

5. CONNECTION WITH THE CHANDRASEKHAR ALGORITHM AND SQUARE-ROOT FILTERING

It has been noted (see [1]) that it is often advantageous to derive formulae in terms of the difference $\Sigma_{k+1}-\Sigma_k$ to replace the Riccati equation (3). This often leads to a significant reduction in the computational effort. Algorithms of this type are usually termed after Chandrasekhar. From Theorem (4.1) it is now evident that the Schur algorithm is related to a Chandrasekhar-type algorithm.

Computational inaccuracies when iterating the difference Riccati equation may lead to a covariance matrix that fails to be nonnegative definite. A technique to cope with this problem introduced by Potter (see [1]), relies on propagating the covariance matrix in "square-root" form. Square-root type algorithm received considerable attention because they have an additional advantage in terms of the numerical conditioning of the computations. In certain cases however, this is traded off with an increase in computational effort. The Schur algorithm given in (3.1) involves computing square-roots of of certain positive definite matrices. Below we will present an identity satisfied by the quantities M_k, N_k, and Γ_k, produced in Theorem (3.1). This will reveal the connection between the Schur algorithm and a square-root version of the Chandrasekhar-type algorithm considered by Morf and Kailath [12].

THEOREM (5.1). Let Γ_k, M_k, N_k, β_k, $T_{k,left}$, and $T_{k,right}$, for $k = 1, 2, \cdots$, be computed as in Theorem (3.1). Then

$$\begin{pmatrix} N_{k-1} & \sqrt{-1}C\,\Gamma_{k-1} \\ M_{k-1} & \sqrt{-1}A\,\Gamma_{k-1} \end{pmatrix} J_k = \begin{pmatrix} N_k & 0 \\ M_k & \sqrt{-1}\Gamma_k \end{pmatrix}$$

where

$$J_k = \begin{pmatrix} I & -\sqrt{-1}\beta_{k-1} \\ \sqrt{-1}\beta^*{}_{k-1} & I \end{pmatrix} \begin{pmatrix} T_{k-1,right} & 0 \\ 0 & T_{k-1,left} \end{pmatrix}$$

is a unitary matrix.

In [12], Morf and Kailath presented the following.

THEOREM (5.2). Let $R^{\epsilon}{}_k$, $\hat{L}_k$, and $\hat{K}_k$, for $k = 1, 2, \cdots$, be computed recursively from $R^{\epsilon}{}_0 = C\,\Sigma_0 C' + R$, $\hat{K}_0 = (A\,\Sigma_0 C' + S)(C\,\Sigma_0 C' + R)^{-1/2}$, and $\hat{L}_0 = -\hat{K}_0$, and

$$\begin{pmatrix} R^{\epsilon}{}_{k-1}{}^{1/2} & \sqrt{-1}C\hat{L}_{k-1} \\ \hat{K}_{k-1} & \sqrt{-1}A\hat{L}_{k-1} \end{pmatrix} \hat{J}_k = \begin{pmatrix} R^{\epsilon}{}_k{}^{1/2} & 0 \\ \hat{K}_k & \sqrt{-1}\hat{L}_k \end{pmatrix} \qquad (6)$$

where $\hat{J}_k$ is a complex unitary matrix chosen so as to create a 0-block as the (1,2)-entry in the right hand side of the above equation. Then,

$$\Sigma_{k+1} - \Sigma_k = -\hat{L}_k \hat{L}'_k$$
$$R^{\epsilon}_k = C \Sigma_k C' + R$$

and

$$K_k = \hat{K}_k R^{\epsilon\,-1/2}_k$$

where Σ_k is the solution of (3) with Σ_0 the nonnegative definite solution of (1), and K_k is given by (2).

This result was presented in [12] under the simplifying assumption that $S = 0$. However, it holds in complete generality. The computation of the steps of the iteration relies on creating the required pattern with the 0-block, starting from the matrix on the left hand side in (6). This is done by a sequence of elementary transformations (typically Householder transformations). It should be noted that $\hat{J}_k$ is a complex matrix. However, as pointed out in [12], complex arithmetic is not required.

PROPOSITION (5.3) With Γ_k, M_k, N_k, $\hat{L}_k$, $\hat{K}_k$ and R^{ϵ}_k as in Theorems (5.1) and (5.2), the following hold:

$$\hat{L}_k = \frac{1}{\sqrt{2}} \Gamma_k U_k$$
$$\hat{K}_k = \frac{1}{\sqrt{2}} M_k U_k$$

and

$$R^{\epsilon\,1/2}_k = \frac{1}{\sqrt{2}} N_k U_k$$

where U_k is a constant unitary matrix.

The proof of Theorem (5.1) follows from equations (5) and is included in [2]. Proposition (5.3) follows from Theorems (4.1) and (5.2).

Thus the Schur-type algorithm of Theorem (3.1) corresponds directly with the square-root version of the Chandrasekhar algorithm of Theorem (5.2). However, the steps of the iteration in (3.1) are quite different from the way the algorithm in Theorem (5.2) proceeds. Also, Theorem (5.1) gives a form for the unitary matrix $\hat{J}_k$ required in Theorem (5.2).

This research was supported in part by NSF under Grant No. ECS-8705291 and MIP-8708811.

6. REFERENCES

[1] B. D. O. Anderson and J. B. Moore, **Optimal Filtering**, Prentice-Hall, Englewood Cliffs, N.J., 1979.

[2] C. Chang and T. T. Georgiou, *The Schur-algorithm and the Discrete-time Riccati equation*, preprint (in preparation).

[3] Ph. Delsarte, Y. Genin, and Y. Kamp, *The Nevanlinna-Pick problem for matrix-valued functions*, SIAM J. Appl. Math., 36(1): 47-61 (1979).

[4] P. Dewilde and H. Dym, *Lossless chain scattering matrices and optimum linear prediction: the vector case*, Circuit Theory and Appl., 9: 135-175 (1981).

[5] P. Faurre, M. Clerget, and F. Germain, **Operateurs Rationnels Positifs: Application a l' Hyperstabilite at aux Processus Aleatoires**, Dunod, Paris, 1978.

[6] T. T. Georgiou, *Computational aspects of spectral factorization and the tangential Schur algorithm*, IEEE Transactions on Circuits and Systems, CAS-36(1): 103-108 (1989).

[7] T. T. Georgiou, *On a Schur-algorithm based approach to spectral factorization: State-space formulae*, Systems & Control Letters, 10(2): 123-129 (1988).

[8] T. T. Georgiou and P. P. Khargonekar, *Linear fractional transformations and spectral factorization*, IEEE Transactions Automatic Control, AC-31: 345-348 (1986).

[9] T. T. Georgiou and P. P. Khargonekar, *Spectral factorization and Nevanlinna-Pick interpolation*, SIAM J. Control and Opt., 25(3): 754-766 (1987).

[10] T. T. Georgiou and P. P. Khargonekar, *Spectral factorization of matrix-valued functions using interpolation theory*, IEEE Transactions on Circuits and Systems, CAS-36(4): 568-574 (1989).

[11] T. Kailath, *A theorem of I. Schur and its impact on modern signal processing*, in **Operator Theory: Advances and Appl.**, 18: 9-30 (1986).

[12] M. Morf and T. Kailath, *Square-root algorithms for least-squares estimation*, IEEE Transactions on Automatic Control, AC-20(4): 487-497 (1975).

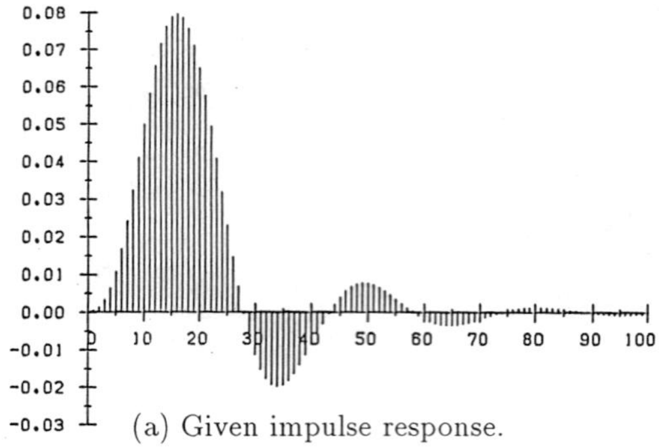

(a) Given impulse response.

Table I Coefficient sensitivity S .

Realization	Sensitivity S
Canonical	20379.96
Optimal	105.718184
Suboptimal	105.718186

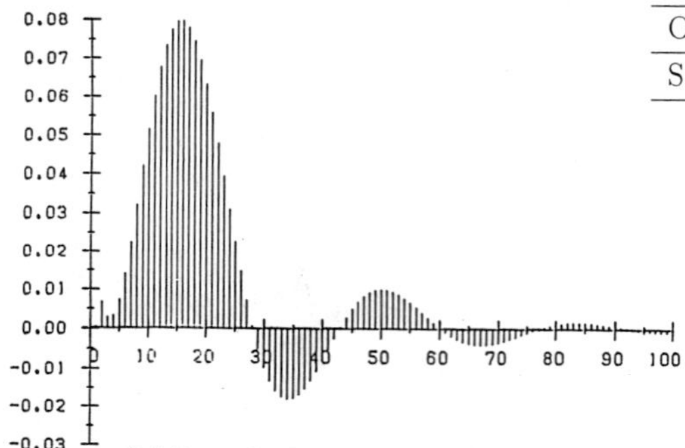

(b) Impulse response of the designed suboptimal realization.

FIGURE 2 Impulse responses of 1-D digital filters.

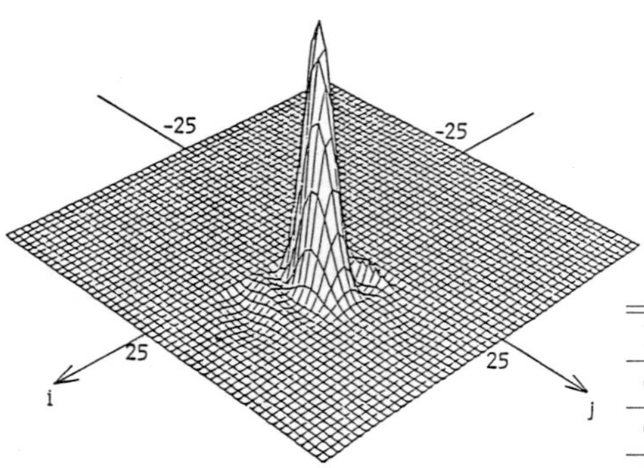

Table II Coefficient sensitivity S .

Realization	Sensitivity S
Canonical	113888.7069
Optimal	46.1158
Suboptimal	46.1603

FIGURE 3 Impulse response of the designed 2-D suboptimal realization.

Identification of Parameters and Noise Variances for Linear Systems with Plant Input Noise

Tohru Katayama

Abstract

This paper deals with the identification problem for a discrete–time linear system in the presence of plant input noise as well as measurement noise, where it is assumed that the noises are Gaussian white and mutually uncorrelated. Based on the innovation representation, or the steady state Kalman filter, we apply the maximum likelihood method to the identification of the plant parameters and the variance of the innovation process. The noise variances are then obtained either by utilizing the correlations of residuals, or by solving a nonlinear optimization problem, where the constraint is the Riccati equation satisfied by the error covariance matrix of the steady state Kalman filter. Numerical results are also included.

1. Introduction

There have been developed various identification techniques for time-invariant linear discrete-time systems [5], [8]. In most cases, the input signal to the plant is known exactly; namely, although the output measurement noise is allowed, no input noise is included except for only few papers. For example, Söderström [13] has studied when and how the dynamic system can be identified based on noise-corrupted input-output data. More recently, Eguchi, Wada and Sagara [4] have developed a recursive least-squares (LS) identification algorithm with bias correction for the linear system subject to both input and output measurement noises. In these papers, however, the control input itself is fed into the plant without noise, so that models therein are called errors-in-variables models, for which some fundamental studies have been presented in [1], [6].

This paper considers the identification problem for discrete–time linear system in the presence of plant input noise as well as measurement noise. The plant input noise naturally arises in many industrial plants such as a fluidized-bed type refuse incineration plant for which the feed of waste is incurred by various noises. In Section 2, we present the SISO plant to be identified, together with discussion from the point of view of the least–squares (LS) method. Section 3 develops an identification algorithm based on the maximum likelihood (ML) method and the state–space model of the plant. In Section 4, we show that the noise variances are obtained either by utilizing the correlations of residuals, or by solving a nonlinear optimization problem, in which the algebraic Riccati equation (ARE) satisfied by the error covariance matrix associated with the steady state Kalman filter is used as a constraint. Numerical results are also included in Section 5.

2. Statement of the problem

Consider an SISO linear system shown in Fig.1, where $u(t)$ is an input variable, $y(t)$ is an observable output, $x(t)$ is a plant output that is not directly accessible, $w(t)$ is a plant noise, and $v(t)$ is an observation noise. It is assumed that $w(t)$ and $v(t)$ are mutually uncorrelated white noises with means zero and variances q and r, respectively. Let the plant transfer function be given by

$$G(z^{-1}) = \frac{b(z^{-1})}{a(z^{-1})} = \frac{b_1 z^{-1} + \cdots + b_n z^{-n}}{1 + a_1 z^{-1} + \cdots + a_n z^{-n}} \tag{2.1}$$

It is assumed that $a(z^{-1})$ is stable and that $a(z^{-1})$ and $b(z^{-1})$ are coprime. From (2.1) and Fig.1, we have

$$a(z^{-1})x(t) = b(z^{-1})[u(t) + w(t)] \tag{2.2a}$$

$$y(t) = x(t) + v(t) \tag{2.2b}$$

Hence, a state-space model for (2.2) is given by an observer form

$$z(t+1) = Az(t) + bu(t) + bw(t)$$

$$y(t) = x(t) + v(t), \quad x(t) = hz(t) \tag{2.3}$$

where

$$A=\begin{bmatrix} -a_1 & 1 & & \\ \vdots & & \ddots & \\ & & & 1 \\ -a_n & & & 0 \end{bmatrix}, \quad b=\begin{bmatrix} b_1 \\ \vdots \\ b_n \end{bmatrix}, \quad h=[1\ 0\ \cdots\ 0] \tag{2.4}$$

The problem to be considered in this paper is to identify the system parameters and noise variances $\{a_1, \cdots, a_n, b_1, \cdots, b_n, q, r\}$ based on the input-output observations $\{u(t), y(t), t = 1, \cdots, N\}$.

First we consider the problem from the point of the LS method. It follows from (2.2) that

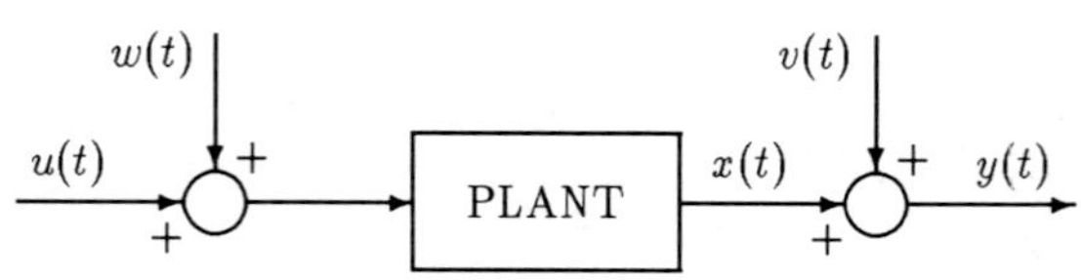

FIGURE 1 The plant.

$$a(z^{-1})y(t) = b(z^{-1})u(t) + \xi(t) \tag{2.5}$$

where

$$\xi(t) = a(z^{-1})v(t) + b(z^{-1})w(t) \tag{2.6}$$

Define $\theta = (a_1, \cdots, a_n,\ b_1, \cdots, b_n)^T$ and $\phi(t) = [-y(t-1) \cdots - y(t-n) u(t-1) \cdots u(t-n)]^T$. Then we have a linear regression model

$$y(t) = \phi^T(t)\theta + \xi(t) \tag{2.7}$$

We see that the application of the LS method to (2.7) yields the biased estimate of θ, since $\xi(t)$ of (2.6) is correlated. In fact, it is not difficult to derive an explicit expression for the bias under the assumption that the underlying processes are stationary and ergodic. Bias corrected LS methods [11], [12], [14] could be applied to the present situation, but they simultaneously require the estimation scheme of noise variances, thus complicating the structure of overall algorithms.

Therefore, possible approaches to the present identification problem will be to apply the instrumental variable (IV) method or the ML method [5], [8]. Since a preliminary numerical study based on the IV method was not favorable, we develop an identification algorithm based on the state–space model (2.3) and the ML method [8].

3. Identification by the ML method

Let $U^{t-1} = \{u(\tau), \tau = t-1, t-2, \cdots\}$, $Y^{t-1} = \{y(\tau), \tau = t-1, t-2, \cdots\}$ be the past input–output observations and $\hat{z}(t)$ be the one-step predicted estimate of $z(t)$, namely, $\hat{z}(t) = E\{z(t)|U^{t-1}, Y^{t-1}\}$. Then, the innovation representation, or the steady state Kalman filter, for (2.3) is given by [2]

$$\begin{aligned} \hat{z}(t+1) &= A\hat{z}(t) + bu(t) + \Gamma e(t) \\ y(t) &= h\hat{z}(t) + e(t) \end{aligned} \tag{3.1}$$

where $e(t) := y(t) - h\hat{z}(t)$ is the innovation process for $y(t)$, and Γ is the $n \times 1$ steady state Kalman gain defined by

$$\Gamma = APh[hPh^T + r]^{-1} \tag{3.2}$$

It is well known that $P = E\{[z(t) - \hat{z}(t)][z(t) - \hat{z}(t)]^T\}$ is the $n \times n$ steady state error covariance matrix satisfying the ARE

$$P = APA^T - APh^T[hPh^T + r]^{-1}hPA^T + qbb^T \tag{3.3}$$

Moreover, the variance of the innovation process is given by

$$\sigma^2 = hPh^T + r \tag{3.4}$$

Eliminating $\hat{z}(t)$ from (3.1) by using (2.4) yields an ARMAX model

$$a(z^{-1})y(t) = b(z^{-1})u(t) + c(z^{-1})e(t) \tag{3.5}$$

where $c(z^{-1}) = 1 + c_1 z^{-1} + \cdots + c_n z^{-n}$, and $c_i := a_i + \gamma_i$, $i = 1, \cdots, n$, and where $\Gamma = (\gamma_1, \cdots, \gamma_n)^T$. It is assumed that $a(z^{-1})$ has no common zeros with $b(z^{-1})$ and $c(z^{-1})$ and that $c(z^{-1})$ is stable.

Moreover, it is well known [8] that the ML method under the assumption that the noise processes are Gaussian leads to the prediction error method that minimizes the variance of prediction error $\varepsilon(t, \theta)$:

$$E_N(\theta) = \frac{1}{N} \sum_{t=1}^{N} \varepsilon^2(t, \theta) \tag{3.6}$$

where $\varepsilon(t, \theta)$ is generated by solving (3.5) with respect to $e(t)$ for fixed $\theta = (a_1, \cdots, a_n, b_1, \cdots, b_n, c_1, \cdots, c_n)^T$.

For the numerical optimization, we apply the Davidon–Fletcher– Powell (DFP) method [10]. The gradients needed are computed as [7]

$$\frac{\partial E_N(\theta)}{\partial a_i} = -\frac{1}{N} \sum_{t=1}^{N} \lambda(t) y(t-i) \tag{3.7a}$$

$$\frac{\partial E_N(\theta)}{\partial b_i} = \frac{1}{N} \sum_{t=1}^{N} \lambda(t) u(t-i) \tag{3.7b}$$

$$\frac{\partial E_N(\theta)}{\partial c_i} = \frac{1}{N} \sum_{t=1}^{N} \lambda(t) \varepsilon(t-i, \theta) \quad (i = 1, \cdots, n) \tag{3.7c}$$

where $\lambda(t)$ is the adjoint variable that satisfies

$$\lambda(t) + \sum_{i=1}^{n} c_i \lambda(t+i) + 2\varepsilon(t, \theta) = 0 \tag{3.8}$$

with the boundary conditions $\lambda(t) = 0, t = N+1, \cdots, N+n$.

4. Identification of noise variances

There have been presented several methods for identifying the noise variances for discrete-time stochastic systems [3], [9]. Here we present two identification algorithms for noise variances.

First, we note from (2.5) and (3.5) that $\xi(t)$ of (2.6) has two expressions:

$$\xi(t) = a(z^{-1})v(t) + b(z^{-1})w(t) = c(z^{-1})e(t) \tag{4.1}$$

Thus, assuming that $w(t)$, $v(t)$, $e(t)$ are stationary and ergodic, we get a set of linear equations

$$\left(\sum_{j=\ell}^{n} a_j a_{j-\ell}\right) r + \left(\sum_{j=\ell+1}^{n} b_j b_{j-\ell}\right) q = \left(\sum_{j=\ell}^{n} c_j c_{j-\ell}\right) \sigma^2, \quad \ell = 0, 1, \cdots, n \tag{4.2}$$

by computing the correlations of $\xi(t)$, where $a_0 = c_0 = 1$. Thus, replacing a_j, b_j, c_j and σ^2 by their ML estimates, and solving the overdetermined equations (4.2) by the LS method, we get the estimates $\hat{q}$, $\hat{r}$ of noise variances.

Next, we present a different method for estimating the variances q, r based on the state-space model. Rewriting the ARE of (3.3),

$$P = \tilde{A}P\tilde{A}^T + r\Gamma\Gamma^T + qbb^T \tag{4.3}$$

where $\tilde{A}$ is given by

$$\tilde{A} := A - \Gamma h = \begin{bmatrix} -c_1 & 1 & & & \\ -c_2 & & 1 & & \\ \vdots & & & \ddots & \\ & & & & 1 \\ -c_n & & & & 0 \end{bmatrix} \tag{4.4}$$

Thus, since P is symmetric, we have $(n+1)(n+2)/2$ constraint equations from (3.2), (3.4) and (4.3). But since quantities q, r and P consist of $(n^2+n+4)/2$ unknowns, we see that for $n \geq 2$ the number of constraint equations is larger than that of unknown quantities to be determined.

Therefore, we wish to minimize

$$J(\mu) = \| \Gamma(hPh^T + r) - APh^T \|^2 + \delta \| hPh^T + r - \hat{\sigma}^2 \|^2, \quad \delta > 0 \tag{4.5}$$

with respect to $\mu := (q, r)$ subject to (4.3). Let Λ be the $n \times n$ symmetric adjoint matrix. Then combining the constraint equation (4.3) with (4.5), we get

$$\bar{J} = J + tr[\Lambda(P - \tilde{A}P\tilde{A}^T - r\Gamma\Gamma^T - qbb^T)] \tag{4.6}$$

Thus the gradients necessary for applying the DFP method are given by

$$\frac{\partial \bar{J}}{\partial q} = -b^T\Lambda b \tag{4.7}$$

$$\begin{aligned} \frac{\partial \bar{J}}{\partial r} &= 2r(\| \Gamma \|^2 + \delta) - \Gamma^T\Lambda\Gamma + \Gamma^T(\Gamma h - A)Ph^T \\ &\quad + hP(h^T\Gamma^T - A^T)\Gamma + 2\delta(hPh^T - \hat{\sigma}^2) \end{aligned} \tag{4.8}$$

where Λ satisfies the Lyapunov equation

$$\begin{aligned} \frac{\partial \bar{J}}{\partial P} &= \Lambda - \tilde{A}^T\Lambda\tilde{A} + \tilde{A}^T\tilde{A}Ph^Th + h^ThP\tilde{A}^T\tilde{A} \\ &\quad - r(h^T\Gamma^T\tilde{A} + \tilde{A}^T\Gamma h) + 2\delta h^Th[hPh^T + r - \hat{\sigma}^2] = 0 \end{aligned} \tag{4.9}$$

It should be noted that the identification method of noise variances developed above is valid even if there is no control input, namely, $u(t) \equiv 0$. In this case, we may assume that $b_1 = 1$ without loss of generality, so that we have $n-1$ additional unknowns in (4.3). Therefore, the number of unknowns becomes $(n+1)(n+2)/2$, which equals that of constraint equations of (3.2), (3.4) and (4.3). Since it is, however, very difficult to solve these nonlinear equations, the approach based on the DPF method may also be used for the numerical solution, where the gradient with respect to b is given by $\dfrac{\partial \bar{J}}{\partial b} = -2q\Lambda b$ where b_1 fixed as unity.

5. Numerical results

Simulation studies are carried out for the second and third order plants:

$$G_1(z^{-1}) = \frac{z^{-1} + 0.5z^{-2}}{1 - 1.5z^{-1} + 0.7z^{-2}} \tag{5.1}$$

$$G_2(z^{-1}) = \frac{0.2(z^{-1} + 0.5z^{-2})}{(1 - 0.8z^{-1})(1 - 1.5z^{-1} + 0.7z^{-2})} \tag{5.2}$$

It is assumed that the control input is a binary random signal with $u(t) = \pm 1$, and that $q = 0.04$, $r = 1.0$. As shown in TABLE 1, the ML approach gives good results for the system parameter identification. TABLE 2 shows the identification results for noise variances for $N = 1000$, where COR and RIC respectively denote the results obtained by solving (4.2) and by solving the constrained optimization problem, for which the initial values are: $q^{(0)} = 0.02$, $r^{(0)} = 0.8$. Based on several numerical studies, we have observed that although COR often results in negative values of $\hat{q}$, RIC approach always gives reasonable results. This is achieved at the cost of computational effort associated with the nonlinear optimization problem. Further study remains to be done to clarify the difference between COR and RIC methods.

TABLE 1 Results of parameter identification for $G_1(z^{-1})$

	$N = 500$		$N = 1000$		
	LS	ML	LS	ML	True
a_1	-1.117	-1.504	-1.057	-1.502	-1.5
a_2	0.350	0.708	0.292	0.705	0.7
b_1	1.119	0.976	0.963	0.976	1.0
b_2	0.935	0.518	1.010	0.525	0.5
c_1	(-0.5)	-1.176	(-0.5)	-1.176	-1.188
c_2	(0.8)	0.502	(0.8)	0.485	0.490

Numbers in parenthesis denote the initial values for ML estimation.

TABLE 2 Identification of noise variances with $N = 1000$

	$G_1(z^{-1})$		$G_2(z^{-1})$	
	COR	RIC	COR	RIC
q	0.0401	0.0456	-0.270	0.0545
r	0.930	0.926	0.916	0.938

6. Conclusion

We have developed identification techniques for a discrete-time stochastic system, paying much attention to the identification of noise variances. Simulation studies show that the feasibility of the present algorithm based on the ML method and the state–space model. Further studies are needed to extend the present approach to the case where the plant is multivariable and is subjected to colored input noise.

ACKNOWLEDGMENT

The author would like to express this sincere thanks to Dr. K. Wada for his useful comments and to T. Tsuchiya for his assistance in simulation studies.

References

[1] B. D. O. Anderson, Identification of scalar errors-in-variables models with dynamics. *Automatica,* vol. 21, no. 6, pp. 709–716, 1985.

[2] B. D. O. Anderson and J. B. Moore, *Optimal Filtering.* Prentice–Hall, 1979.

[3] P. R. Belanger, Estimation of noise covariance matrices for a linear time-varying stochastic processes. *Automatica*, vol. 10, no. 3, pp. 267–275, 1974.

[4] M. Eguchi, K. Wada and S. Sagara, Identification of pulse transfer function in the presence of input–output noise. *Preprints 19th JAACE Symp. Stochastic Systems Theory and its Applications*, pp. 201–204, 1987.

[5] G. C. Goodwin and R. L. Payne, *Dynamic System Identification: Experiment Design and Data Analysis.* Academic, 1977.

[6] R. E. Kalman, Identifiability and modeling in econometrics. In Krishnaiah, P. R. (Ed.), *Developments in Statistics*, vol. 4, pp. 97–136, Academic, 1983.

[7] R. L. Kashyap, Maximum likelihood identification of stochastic linear systems. *IEEE Trans. Automat. Control*, vol. AC–15, no. 1, pp. 25–34, 1970.

[8] L. Ljung, *System Identification: Theory for the User.* Prentice–Hall, 1987.

[9] R. K. Mehra, On the identification of variances and adaptive Kalman filtering. *IEEE Trans. Automat. Control*, vol. AC–15, no. 2, pp. 175–184, 1970.

[10] D. G. Luenberger, *Introduction to Linear and Nonlinear Programming.* Addison–Wesley, 1973.

[11] S. Sagara and K.Wada, On–line modified least squares parameter estimation of linear discrete dynamic systems. *Int. J. Control*, vol. 25, no. 3, pp. 329–343, 1977.

[12] H. Sakai and M. Kawase, Recursive parameter estimation of an autoregressive process disturbed by white noise. *Int. J. Control*, vol. 30, no. 6, pp. 949–966, 1979.

[13] T. Söderström, Identification of stochastic linear system in presence of input noise. *Automatica,* vol. 21, no. 6, pp. 709–716, 1981.

[14] P. Stoica and T. Söderström, Bias correction in least-squares identification. *Int. J. Control*, vol. 35, no. 3, pp. 449–457, 1982.

System Balancing Approach to a Unified Design of 1-D and 2-D State-Space Digital Filters

Masayuki Kawamata and Masami Iwatsuki

Abstract

This paper proposes a new design methodology of 1-D and 2-D state-space digital filters, called the unified design method. The unified design method simultaneously performs approximation and synthesis for given time or spatial domain specifications, and yields suboptimal state-space filter structures with respect to coefficient sensitivity. It is given in closed form, and needs no complicated calculation but the singular value decomposition of Hankel matrices. Approximation and synthesis in the unified design is combined by system balancing concept of state-space digital filters.

1. Introduction

It is desired in the design of one-dimensional (1-D) and two-dimensional (2-D) digital filters that digital filters have minimum approximation error and minimum quantization effects. Thus, the problem of designing 1-D and 2-D digital filters has been divided into two important stages as approximation and synthesis as shown in FIGURE 1 (a). Hitherto, approximation of digital filters has been studied mainly by means of the transfer function, whereas synthesis of filter structures with minimum quantization effects has been studied mainly by means of the state-space representation. Therefore, the conventional design methods have the following two problems: (a) The design procedures are complicated, because they involve different mathematical models. (b) Redundant computations are needed in the design, because the information in approximation stage is not handed to synthesis stage.

This paper proposes a very simple and unified design method of 1-D and 2-D state-space digital filters using system balancing concept. The unified design method of digital filters is one stage algorithm both for approximation and synthesis. FIGURE 1 (b) shows the concept of the unified design. This paper also emphasizes the importance of system balancing concept or the controllability and observability Gramians in the design of 1-D and 2-D digital filters.

This paper is organized as follows. Section 2 considers the unified design of 1-D state-space digital filters. First, 1-D state-space digital filters and balanced realizations are introduced. Coefficient sensitivity of 1-D state-space digital filters is analyzed from a statistical point of view, and is shown to be equal to the round-off noise. Next, the minimum coefficient sensitivity realization is synthesized by applying a simple scaling and rotation to the balanced realization. Thus, if there exists a state-space approximation method such that the resultant filter structure is balanced, then we can obtain a unified design method for state-space digital filters. Kung's balanced approximation is one candidate of approximation for our unified design method. Thus, we will propose our unified design method by combining Kung's balanced approximation and the synthesis method of minimum sensitivity realizations. Section 3 is a natural extension of Section 2 to the 2-D case. It deals with a unified design of 2-D state-space digital filters with separable denominators. Section 4 gives numerical design examples to illustrate the effectiveness of the proposed unified design method.

2. Unified design of 1-D digital filters

2.1 state-space model

In this subsection, we will give a brief review of 1-D state-space digital filters. Consider the transfer function $H(z) = N(z)/D(z)$ of an n-th order stable digital filter. A 1-D digital filter with the transfer function $H(z)$ can be represented by the following state equations:

$$x(k+1) = Ax(k) + bu(k) \quad (1)$$

$$y(k) = cx(k) + du(k) \quad (2)$$

where $x(k)$ is an n-th order state-vector, $u(k)$ is the scalar input, $y(k)$ is the scalar output, and A, b, c and d are real constant coefficient matrices with proper sizes. We call the system described by (1) and (2) the state-space digital filter, and denote it as $DF(A, b, c, d)$.

The controllability Gramian K and observability Gramian W are important matrices to analyze state-space digital filters from a system theoretical point of view. They are defined for $DF(A, b, c, d)$ as the solutions of the following Liapunov equations:

$$K = AKA^t + bb^t, \quad W = A^tWA + c^tc \quad (3)$$

We can get a new realization $DF(A', b', c', d) = DF(T^{-1}AT, T^{-1}b, cT, d)$, which has the same input-output behavior as $DF(A, b, c, d)$, by an equivalent transformation T defined as $x'(k) = T^{-1}x(k)$. This equivalent transformation is used to represent a variety of realization structures of 1-D state-space digital filters. A transformed realization $DF(A', b', c', d)$ by T has the following Gramians:

$$K' = T^{-1}KT^{-t}, \quad W' = T^tWT. \quad (4)$$

We see from (4) that although the eigenvalues of both K' and W' change under equivalent transformation, the eigenvalues $\theta_i^2, i = 1, 2, ..., n$ of the product $K'W'$ do not. These invariants $\theta_i > 0, i = 1, 2, ..., n$ are called the 2-nd order modes of the corresponding filter [10].

Realization structures of 1-D state-space digital filters are classified by their Gramians. If both of the Gramians K and W of a filter are equal to the diagonal matrix given by

$$\Theta = diag(\theta_1, \theta_2, ..., \theta_n) \tag{5}$$

then the filter is called the balanced realization [8], [9].

2.2 Coefficient sensitivity

• ℓ_2 scaling

The state variables of state-space digital filters should be scaled such that the dynamic range is fully utilized without causing excessive overflow. The ℓ_2 norm scaling is convenient and it is adopted in this paper. Assume that the input of $DF(A, b, c, d)$ is a white Gaussian process with zero mean and unit variance. Then, the variances of the state variables $E[x_i^2], i = 1, 2, ..., n$ are represented by the i-th diagonal elements of the following covariance matrix

$$E[xx^t] = K \tag{6}$$

where the covariance matrix K of $DF(A, b, c, d)$ is seen to be the controllability Gramian of the filter. In ℓ_2 norm scaling, the transformed and/ or scaled filter $DF(T^{-1}AT, T^{-1}b, cT, d)$ must satisfy

$$(T^{-1}KT^{-t})_{ii} = 1, \ i = 1, 2, ..., n \tag{7}$$

where $(\ )_{ii}$ expresses the i-th diagonal element of the matrix [2]. In the following analysis of coefficient sensitivity, a state space digital filter $DF(A, b, c, d)$ is assumed, without loss of generality, to be scaled.

• Analysis of coefficient sensitivity

Suppose that a scaled $DF(A, b, c, d)$ is implemented with fixed-point arithmetic, of which coefficient wordlength is ℓ bits. To analyze coefficient sensitivity of the filter, we consider the following state-space model:

$$\tilde{x}(k+1) = [A + \Delta A(k)]\tilde{x}(k) + [b + \Delta b(k)]u(k) \tag{8}$$
$$\tilde{y}(k) = [c + \Delta c(k)]\tilde{x}(k) + [d + \Delta d(k)]u(k) \tag{9}$$

where $\Delta A(k), \Delta b(k), \Delta c(k)$, and $\Delta d(k)$ represent quantization errors of coefficients. For the purpose of simplicity of analysis, we assume that the errors of coefficient quantization $\Delta A_{ij}(k), (i, j = 1, ..., n)$ are white and uniformly distributed in the range $[-2^{-\ell}/2, 2^{-\ell}/2]$, and have no correlations with each other and with the input. The errors of coefficient quantization $\Delta b_i(k), \Delta c_j(k), \Delta d(k)$ are also assumed to be random variables with the same statistical property as $\Delta A_{ij}(k)$.

The generation of the output error due to coefficient quantization is expressed by subtracting (1) from (8) and (2) from (9) as

$$e_x(k+1) = Ae_x(k) + \Delta A(k)x(k) + \Delta b(k)u(k) \tag{10}$$
$$\Delta y(k) = ce_x(k) + \Delta c(k)x(k) + \Delta d(k)u(k) \tag{11}$$

where $\Delta y = \tilde{y} - y, e_x = \tilde{x} - x$, and second-order quantities are ignored. The output error variance $E[\Delta y^2]$ in the steady state under a white Gaussian input with zero mean and unit variance is obtained as [3]

$$E[\Delta y^2] = \sigma^2(n+1)\sum_{i=1}^{n} W_{ii} + \sigma^2(n+1) \tag{12}$$

where the noise matrix W is seen to be the observability Gramian of $DF(A,b,c,d)$, and $\sigma^2 = 2^{-2\ell}/12$ is the variance of the errors of coefficient quantization. We define the coefficient sensitivity S of $DF(A,b,c,d)$ by $E[\Delta y^2]/\sigma^2$.

• Minimization of coefficient sensitivity

The coefficient sensitivity $S = E[\Delta y^2]/\sigma^2$ in (12) is dependent highly on realization structures of the filter. The structures with minimum sensitivity are called optimal realizations in this paper. From (12), we readily note that the coefficient sensitivity of $DF(A,b,c,d)$ is equivalent to the output error variance of the filter due to roundoff noise after multiplication. Thus, we can apply the minimization for roundoff noise proposed in [2] and [10] to the minimization of coefficient sensitivity. Therefore, from any initial realization $DF(A,b,c,d)$ with the associated Gramians K and W, the optimal realization $DF(A_o,b_o,c_o,d)$ can be synthesized by the following equivalent transformation T:

$$T = T_0 R_1 \Lambda R_0^t \tag{13}$$

where

$$\Lambda = diag(\lambda_1, \lambda_2, ..., \lambda_n), \quad \lambda_i = (\sum_{p=1}^{n} \theta_p)^{1/2}/n\theta_i, i = 1,2,...,n, \tag{14}$$

T_0 is given by

$$T_0 T_0^t = K, \tag{15}$$

and R_0 and R_1 are orthogonal matrices such that

$$(R_0 \Lambda^{-2} R_0^t)_{ii} = 1, i = 1,2,...,n \tag{16}$$

$$R_1^t(T_0^t W T_0)R_1 = diag(\theta_1^2, \theta_2^2, ..., \theta_n^2). \tag{17}$$

Thus, the minimum coefficient sensitivity of the filter is

$$S_{min} = (n+1)(\sum_{i=1}^{n} \theta_i)^2/n + (n+1). \tag{18}$$

2.3 Unified design

• Equivalent Transformations between optimal realizations and balanced realizations

Until now, although much attention has been paid to the synthesis of the optimal realization, few has been paid to the study of its structural properties. Considering the structural properties from system balancing concept, we can reveal that the optimal realization is a scaled and rotated balanced realization. The result is given briefly as follows:

Substituting (13) into (4), we get the Gramians of $DF(A_o, b_o, c_o, d)$ as follows:

$$K_o = (\rho^{1/2}R_0^t)^{-1}\Theta(\rho^{1/2}R_0^t)^{-t}, \quad W_o = (\rho^{1/2}R_0^t)^t\Theta(\rho^{1/2}R_0^t) \tag{19}$$

where

$$\rho = \sum_{i=1}^{n} \theta_i/n. \tag{20}$$

Noting that Θ is just the Gramians of the balanced realization, we see from (19) that the optimal realization is related to the balanced realization by the simple scaling and rotation as

$$T = \rho^{1/2}R_0^t. \tag{21}$$

• Design procedure

In the above discussion, we have shown that the optimal realization can be synthesized by applying the equivalent transformation $T = \rho^{1/2}R_0^t$ to the balanced realization. This implies that approximation and synthesis can be unified if we utilize an approximation method which can produce balanced realizations. To the author's best knowledge, such an approximation method has not been proposed. However, Kung's balanced approximation method [5] can approximate a given impulse response and produce a nearly balanced realization. Thus, combining the above transformation with Kung's approximation method, we get our unified design method which can result in a suboptimal (nearly optimal) realization. Therefore, our unified design method becomes as follows:

Suppose that a design specification is given by the impulse response $h_0, h_1, .., h_{N-1}$. The Hankel matrix is formed as

$$H = \begin{bmatrix} h_1 & h_2 & h_3 & \dots & h_{N-1} \\ h_2 & h_3 & \dots & h_{N-1} & 0 \\ h_3 & \vdots & & 0 & \vdots \\ \vdots & h_{N-1} & 0 & & \vdots \\ h_{N-1} & 0 & \dots & \dots & 0 \end{bmatrix} \tag{22}$$

Suppose that the singular values σ_i's of H satisfy $\sigma_i > \sigma_{i+1}$ and $\sigma_n \gg \sigma_{n+1}$ for some n. Then we can write the singular value decomposition (SVD) of H as

$$H = U_1\Sigma_1V_1 + U_2\Sigma_2V_2 \tag{23}$$

where

$$\Sigma_1 = diag(\sigma_1, ..., \sigma_n), \; \Sigma_2 = diag(\sigma_{n+1}, ..., \sigma_{N-1}) \tag{24}$$

and where U_1, U_2, V_1 and V_2 are orthogonal matrices. Using (21) and Kung's approximation method, we have the following suboptimal realization $DF(A_s, b_s, c_s, d)$ which approximates the given impulse response and has the nearly minimum coefficient sensitivity:

$$A_s = R_0(\Sigma_1^{-1/2}U_1^t)(U_1\Sigma_1^{1/2})^{\uparrow}R_0^t \tag{25}$$

$$b_s = \rho^{-1/2} R_0 (\text{the first column of } \Sigma_1^{1/2} V_1) \tag{26}$$

$$c_s = (\text{the first row of } U_1 \Sigma_1^{1/2}) \rho^{1/2} R_0^t \tag{27}$$

$$d = h_0 \tag{28}$$

where ρ and R_0 are found from (20) and (16) by substituting θ_i with σ_i , and $(\ )^\dagger$ denotes the operation of one row shift upward with the last row filled by zeros.

3. Unified design of 2-D digital filters

We restrict ourselves to a class of 2-D digital filters with separable denominators. This is because this class of 2-D digital filters share the fundamental theory with 1-D digital filters, such as realizations, stability, etc. with 1-D digital filters, and thus they are mathematically tractable.

Consider a stable and causal 2-D digital filter with the transfer function $H(z_1, z_2) = N(z_1, z_2)/D_1(z_1)D_2(z_2)$, where $D_1(z_1)$ and $D_2(z_2)$ are polynomials of degrees n in z_1^{-1} and m in z_2^{-1},respectively. A 2-D digital filter with the transfer function $H(z_1, z_2)$ can be described by Roesser's local state-space model as follows [11], [1]:

$$\begin{bmatrix} x^h(i+1,j) \\ x^v(i,j+1) \end{bmatrix} = \begin{bmatrix} A_1 & A_2 \\ 0 & A_4 \end{bmatrix} \begin{bmatrix} x^h(i,j) \\ x^v(i,j) \end{bmatrix} + \begin{bmatrix} b_1 \\ b_2 \end{bmatrix} u(i,j) \tag{29}$$

$$y(i,j) = \begin{bmatrix} c_1 & c_2 \end{bmatrix} \begin{bmatrix} x^h(i,j) \\ x^v(i,j) \end{bmatrix} + du(i,j) \tag{30}$$

where $u(i,j)$ is the input, $y(i,j)$ is the output, $x^h(i,j)$ and $x^v(i,j)$ are the n-th order horizontal and the m-th order vertical state vectors, respectively, and A_1, A_2, A_4, b_1, b_2, c_1, c_2, and d are real coefficient matrices with appropriate sizes.

The 2-D digital filter represented by Eqs. (29) and (30) is called the 2-D state-space digital filter with separable denominators , and denoted as $2DDF\ (A_1,$ A_2, A_4, b_1, b_2, c_1, c_2, $d)$. We can get a new realization $2DDF(T_1^{-1}A_1T_1$, $T_1^{-1}A_2T_4$, $T_4^{-1}A_4T_4$, $T_1^{-1}b_1$, $T_4^{-1}b_2$, c_1T_1, c_2T_4, $d)$, which has the same input-output behavior as $2DDF(A_1$, A_2, A_4, b_1, b_2, c_1, c_2, $d)$, by an equivalent transformation matrix T_1 and T_4 as $x^{h'} = T_1^{-1}x^h$ and $x^{v'} = T_4^{-1}x^v$. The controllability and observability Gramians also play an important role in analysis of 2-D state-space digital filters. They are obtained as the solutions of the following Liapunov equations [6],[4]:

$$K^h = A_1K^hA_1^t + A_2K^vA_2^t + b_1b_1^t,\ W^h = A_1^tW^hA_1 + c_1^tc_1 \tag{31}$$

$$K^v = A_4K^vA_4^t + b_2b_2^t,\ W^v = A_4^tW^vA_4 + A_2^tW^hA_2 + c_2^tc_2 \tag{32}$$

The matrices K^h and K^v are called the horizontal and vertical controllability Gramians, respectively of the 2-D state-space digital filter. The matrices W^h and W^v are called the horizontal and vertical observability Gramians,respectively of the 2-D state-space digital filter. The Gramians of a transformed $2DDF(T_1^{-1}A_1T_1$, $T_1^{-1}A_2T_4$, $T_4^{-1}A_4T_4$, $T_1^{-1}b_1$, $T_4^{-1}b_2$, c_1T_1, c_2T_4, $d)$ become

$$K^{h'} = T_1^{-1}K^hT_1^{-t},\ K^{v'} = T_4^{-1}K^vT_4^{-t} \tag{33}$$

$$W^{h'} = T_1^t W^h T_1, \; W^{v'} = T_4^t W^v T_4 \tag{34}$$

2-D state-space digital filters with separable denominators share the fundamental results for balanced realizations and coefficient sensitivity with 1-D state-space digital filters, that is, balanced realizations are well defined and the minimum coefficient sensitivity realization is proved to be a scaled and rotated balanced realization. Therefore, we can obtain a unified design of 2-D state-space digital filters with separable denominators as the 1-D case. The approximation method proposed by Lashgari, et al. [6] is one candidate as 2-D approximation for our unified design [12].

4. Numerical Examples

4.1 1-D filter

The specification of our example is given by the impulse response h_k shown in FIGURE 2(a). The singular values of the Hankel matrix H of this example satisfy $\sigma_3 \gg \sigma_4$, and thus we can write the SVD of H as (23), substituting 3 for n. The coefficient matrices of the suboptimal realization obtained by (25),(26), (27), and (28) are omitted here. The impulse response of the filter designed is shown in FIGURE 2(b). The approximation error is

$$max|h'_k - h_k| = 0.005413 \tag{35}$$

where h'_k is the impulse response of the designed suboptimal realization. The coefficient sensitivity is shown in Table I together with those of the other two realizations for comparison. Note from Table I that the coefficient sensitivity of the suboptimal realization is the same as that of the optimal realization to 5 decimal places. Thus, we can consider it as optimal in a practical sense.

4.2 2-D filter

The 2-D impulse response h_{ij} as the design specification is given by

$$h_{ij} = \begin{cases} 0.256322exp[-0.103203(i-5)^2 + (j-i)^2] & (0,0) \leq (i,j) \leq (12,15) \\ 0 & \text{otherwise.} \end{cases} \tag{36}$$

Coefficient matrices of the suboptimal realization of order $(n, m) = (4, 4)$ obtained by the unified design method proposed here is omitted in this paper. FIGURE 3 shows the impulse response of the designed suboptimal realization. The approximation error is

$$max|h'_{ij} - h_{ij}| = 0.017517 \tag{37}$$

where h'_{ij} is the impulse response of the designed suboptimal realization. This result shows that the impulse response of the designed suboptimal realization results in a very good approximation of the given specification h_{ij}.

Table II shows coefficient sensitivity of three realizations, where the optimal realization is given by the synthesis method in [7]. From this table, we can observe that the coefficient sensitivity of the suboptimal realization obtained by the unified design method is extremely close to that of the optimal realization.

5. Concluding remarks

This paper has proposed the unified design method of 1-D state-space digital filters and 2-D state-space digital filters with separable denominators. The unified design can approximate given time or spatial domain specifications and yield sub-optimal realization structures with respect to coefficient sensitivity. In the unified design, system balancing concept of state-space digital filters plays the very important role, i.e., time or spatial domain approximation is performed by "balanced approximation method" and optimal realization structures are a kind of balanced realizations.

Multi-Dimensional (M-D) digital filters with *separable denominators* share the fundamental properties with 1-D digital filters, such as realizations, and stability, etc. and thus system balancing concept is well defined for these digital filters. Therefore, our unified design methodology can be extended to higher-dimensional digital filters with *separable denominators*. As for 2-D and M-D digital filters with *non-separable denominators*, system balancing concept is not well defined. In order to extend the unified design to M-D digital filters with *non-separable denominators*, fundamental research on system balancing is necessary.

References

[1] T. Hinamoto, "Realization of a state-space model from two-dimensional input-output map," IEEE Trans. vol. CAS-27, No. 1 (1980), 36-44.

[2] S. Y. Hwang, "Minimum uncorrelated unit noise in state-space digital filtering," IEEE Trans. vol. ASSP-25, No. 4 (1977), 273-281.

[3] M. Iwatsuki, M. Kawamata, and T. Higuchi, "Synthesis of minimum sensitivity structures in linear systems using controllability and observability measures," Proc. of Int. Symp. on ASSP, Tokyo (1986) 501-504.

[4] M. Kawamata and T. Higuchi, "Synthesis of 2-D separable denominator digital filters with minimum roundoff noise and no overflow oscillations," IEEE Trans. , vol. CAS-33, No. 4 (1986) 365-372.

[5] S. Kung, "A new identification and model reduction algorithm via singular value decompositions," Proc. 12th Annual Ascilomar Conf. on Circuits, Systems, and Computers, Pacific Grove, California (1978) 705-714.

[6] B. Lashgari, L. M. Silverman, and J-F. Abramatic, "Approximation of 2-D separable in denominator digital filters," IEEE Trans. vol. CAS-30, No. 2 (1983) 107-121.

[7] T. Lin, M. Kawamata, and T. Higuchi, "Minimization of sensitivity of 2-D systems and its relation to 2-D balanced realizations," Trans. of IEICE, vol. E 70, No. 10 (1987) 938-944.

[8] B. C. Moore, "Principal component analysis in linear systems: controllability, observability, and model reduction," IEEE Trans. vol. AC-26, No. 1 (1981) 17-32.

[9] L. Pernebo and L. M. Silverman, "Model reduction via balanced state space representations," IEEE Trans. AC-27, No. 2 (1982) 382-287.

[10] R. A. Roberts and C. T. Mullis, Digital Signal Processing, Addison Wesley, 1987.

[11] R. P. Roesser, "A discrete state-space model for linear image processing," IEEE Trans. vol. AC-20, No. 2 (1975) 1-10.

[12] Z. Zhao, M. Kawamata, and T. Higuchi, "A new property of optimal realizations of CRSD 2-D digital filters and its application to the direct spatial-domain design," Trans. of IEICE, vol. E 69, No. 10 (1986) 1084-1092.

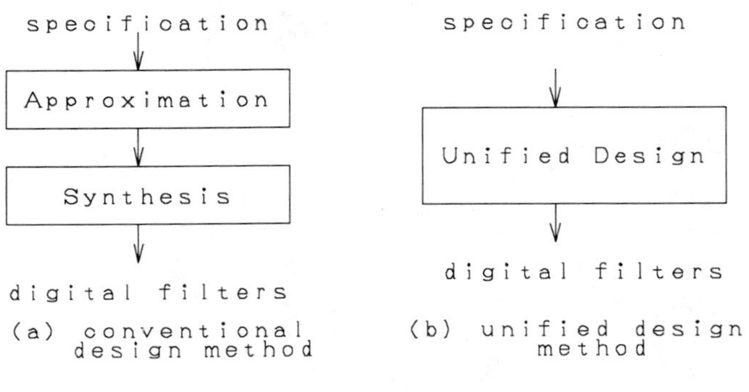

FIGURE 1 Design method of digital filters.

STATE-SPACE REALIZATION OF ORTHOGONAL FILTERS

Hidenori Kimura, Ko-ichi Ichikawa

ABSTRACT

A new method of state space realization of orthogonal filters is given based on the factorization of J-orthogonal matrices. This factorization leads to the cascade construction of filters using the two types of rotors. The resulting realization is simple and numerically superior.

1. INTRODUCTION

The orthogonal realization of digital filters, which was introduced in [1], is a method of utilizing the numerical stability of orthogonal transformations for the filter operation. It naturally leads to the cascade structure containing rotors which is suitable for VLSI implementation and results in the large throughput rate.

In the frequency domain, the orthogonal system is regarded as a lossless system. Therefore, the synthesis of orthogonal filters is closely related to the synthesis of lossless or passive circuits [2][3] , as in the case of wave digital filters [4]. An extensive exposition of the theory of orthogonal filters in the frequency domain in the framework of classical circuit theory or classical function theory is found in [5]. An advanced treatment of orthogonal filters from the viewpoint of lossless inverse scattering is found in [6].

In the state space, the orthogonal filter implies the orthogonality of the system map from (state, input) to (next state, output). This is a straightforward definition with rich structure. Many interesting system-theoretic properties of orthogonal filters can be extracted.

In this paper, we propose a synthesis method of orthogonal filters described purely in the state space. Cascade structure of J-lossless system is fully exploited in the state space, which is related to the new interpolation theory in the state space [7]. It is shown that any J-unitary system map can be represented as the product of matrices, each of which contains only one rotor or one "J-rotor". The synthesis method is straightforward and simple.

2. LOSSLESS EMBEDDING

Let H(z) be the transfer function of a stable digital filter which is to be constructed. After an appropriate scaling, we can assume that H(z) is bounded, i.e.,

$$|H(z)|^2 \leq 1, \qquad |z| = 1. \tag{1}$$

The lossless embedding is a method of realization of H(z) through a lossless filter $\Sigma(z)$ with 2 port described as in Fig. 1. Here, a filter $\Sigma(z)$ with 2 ports described by

$$\Sigma(z) = \begin{bmatrix} \Sigma_{11}(z) & \Sigma_{12}(z) \\ \Sigma_{21}(z) & \Sigma_{22}(z) \end{bmatrix} \quad (2)$$

is said to be lossless, if

$$\Sigma(z)\Sigma^*(z) = I, \quad |z| = 1, \quad (3a)$$

$$\Sigma(z)\Sigma^*(z) \leq I, \quad |z| \geq 1. \quad (3b)$$

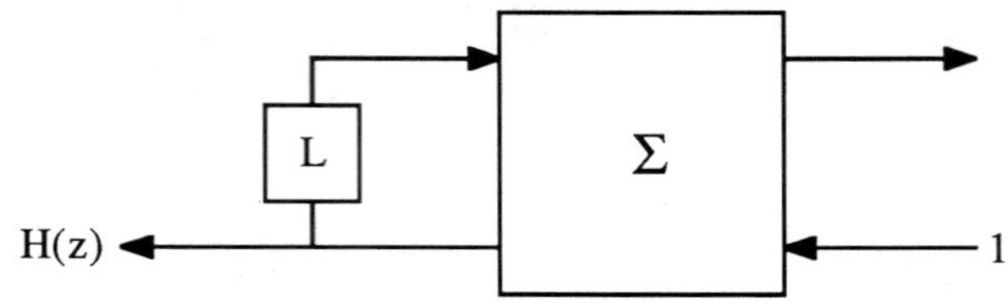

Fig. 1 Lossless Embedding

According to Fig. 1, H(z) is given by

$$H(z) = (1 - \Sigma_{21}(z)L(z))^{-1}\Sigma_{21}(z). \quad (4)$$

We are concerned with the construction of a lossless $\Sigma(z)$ satisfying (4) for a given H(z), as well as the cascade realization of $\Sigma(z)$ as described in Fig. 2.

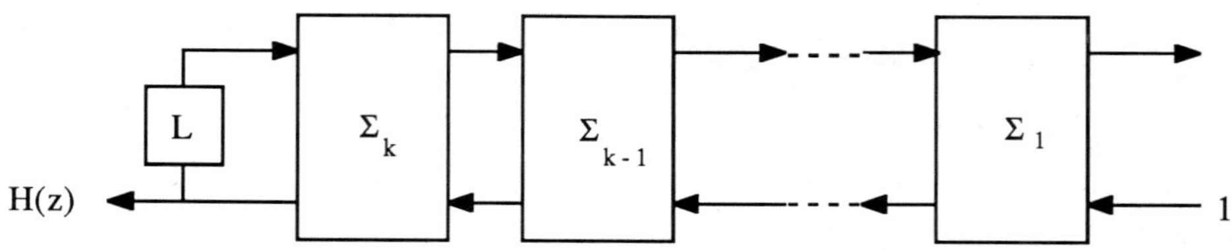

Fig. 2 Cascade Realization

In the synthesis of a cascade structure like Fig. 2, it should be avoided to create a delay-free loop which causes computational deadlocks. For that purpose, it is sufficient to make the (2,1) element of the direct feedthrough of $\Sigma_i(z)$ zero, i.e.,

$$\Sigma_{21}(\infty) = 0 \quad (5)$$

for each section. If a lossless system satisfies (5), we call it *computable*.

The input-output description of a system in Fig. 3 is given by

$$\begin{pmatrix} a_2(z) \\ b_1(z) \end{pmatrix} = \Sigma(z) \begin{pmatrix} a_1(z) \\ b_2(z) \end{pmatrix}. \tag{6}$$

It is sometimes convenient to use an alternative description of Fig. 3 called a *chain-scattering representation*, which is described as

$$\begin{pmatrix} a_2(z) \\ b_2(z) \end{pmatrix} = \Theta(z) \begin{pmatrix} a_1(z) \\ b_1(z) \end{pmatrix}. \tag{7}$$

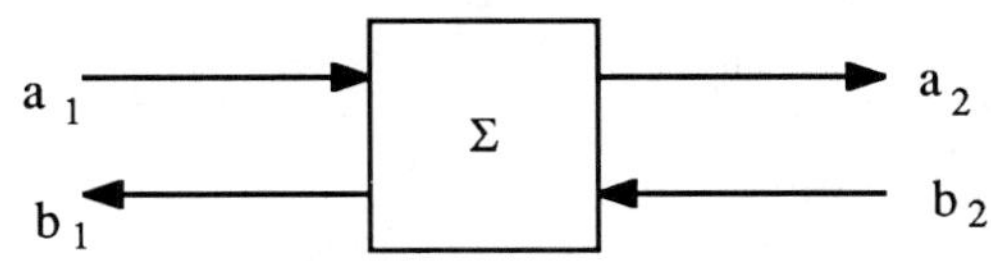

Fig. 3 Input-output Description

It is easily seen that $\Theta(z)$ is given in term of $\Sigma(z)$ by

$$\Theta(z) = \begin{bmatrix} \Sigma_{11}(z) - \Sigma_{12}(z)\Sigma_{22}^{-1}(z)\Sigma_{21}(z) & \Sigma_{12}(z)\Sigma_{22}^{-1}(z) \\ -\Sigma_{22}^{-1}(z)\Sigma_{21}(z) & \Sigma_{22}^{-1}(z) \end{bmatrix}.$$

It is well-known that $\Sigma(z)$ is lossless, if and only if $\Theta(z)$ is *J-lossless,* i.e.,

$$\Theta(z)\, J\, \Theta^*(z) = J, \quad |z| = 1, \tag{8a}$$

$$\Theta(z)\, J\, \Theta^*(z) \le J, \quad |z| \ge 1. \tag{8b}$$

$$J = \begin{bmatrix} 1 & 0 \\ 0 & -1 \end{bmatrix}$$

The chain scattering formulation is particularly convenient for the construction of the cascade structure like Fig. 2, because the chain scattering matrix of the cascade connection of two systems is given by the product of each chain scattering matrix.

It is well-known that a system with state-space form $\Sigma(z) = D + C(zI - A)^{-1}B$ is lossless, if and only if

$$\begin{bmatrix} A & B \\ C & D \end{bmatrix} \begin{bmatrix} P & 0 \\ 0 & I \end{bmatrix} \begin{bmatrix} A & B \\ C & D \end{bmatrix}^T = \begin{bmatrix} P & 0 \\ 0 & I \end{bmatrix} \tag{9}$$

for some $P > 0$. For a lossless system, we can always choose $P = I$ by an appropriate similarity transformation. Such realization is called *an orthogonal realization*. Similarly, a state-space form $\Theta(z) = D + C(zI - A)^{-1}B$ is J-lossless, if and only if

$$\begin{bmatrix} A & B \\ C & D \end{bmatrix}\begin{bmatrix} P & 0 \\ 0 & J \end{bmatrix}\begin{bmatrix} A & B \\ C & D \end{bmatrix}^T = \begin{bmatrix} P & 0 \\ 0 & J \end{bmatrix}. \tag{10}$$

If we choose P = I, such a realization is called *a J-orthogonal realization.*

Now, we derive a state-space form of a lossless system, or its J-lossless counterpart, satisfying (4) or Fig. 1. Let

$$H(z) = \frac{P(z)}{Q(z)}, \tag{11}$$

where Q(z) is a Hurwitz polynominal of degree n. We assume that (1) holds. A lossless system V(z) is given by

$$V(z) = \frac{z^{-n}Q(z^{-1})}{Q(z)}.$$

Let

$$\begin{bmatrix} H(z) \\ V(z) \end{bmatrix} = \begin{bmatrix} J_1 \\ J_2 \end{bmatrix} + \begin{bmatrix} H_1 \\ H_2 \end{bmatrix}(zI - F)^{-1}G, \tag{12}$$

be a state-space realization, such that

$$\begin{bmatrix} F & G \\ H_2 & J_2 \end{bmatrix}\begin{bmatrix} F & G \\ H_2 & J_2 \end{bmatrix}^T = I. \tag{13}$$

Since V(z) is lossless, we can always find its orthogonal realization satisfying (13). A state-space form of the lossless embedding is given by the following lemma, the proof of which is straightforward and omitted.

LEMMA 1. If (12) and (13) hold, then, a computable lossless system which embeds H(z) in the sense of Fig. 1 for L = 1 is given in the form of J-orthogonal realization as

$$\Theta(z) = D + C(zI - A)^{-1}B \tag{14}$$

where

$$A = F - GJ_1^{-1}H_1, \quad B = [K(FH_1^T + GJ_1) \quad GJ_1^{-1}K^{-1}]\begin{bmatrix} I & -I \\ 0 & I \end{bmatrix},$$

$$C = \begin{bmatrix} I & -J_2 \\ 0 & I \end{bmatrix}\begin{bmatrix} H_2 \\ -J_1^{-1}H_1 \end{bmatrix}, \quad D = \begin{bmatrix} K(H_2H_1^T + J_2J_1) & -KC_1H_1^T \\ 0 & J_1^{-1}K^{-1} \end{bmatrix},$$

$$K = (H_1H_1^T + J_1^2)^{-1/2}.$$

3. FACTORIZATION OF J-ORTHOGONAL SYSTEMS

In the preceding section, we found a state-space form of a J-lossless embedding $\Theta(s)$ of $H(s)$ in (14). The next thing to do is to represent $\Theta(s)$ in the cascade form

$$\Theta(s) = \Theta_1(s)\,\Theta_2(s)\cdots\Theta_k(s), \tag{15}$$

in order to construct a cascade structure of Fig. 2.

Without loss of generality, we can assume that $A = F - GJ_1^{-1}H_1$ in (14) is of upper triangular form

$$A = \begin{bmatrix} A_1 & * & \cdots & * \\ 0 & A_2 & \cdots & * \\ \vdots & \vdots & & \\ 0 & 0 & \cdots & A_k \end{bmatrix}, \tag{16}$$

where A_i is at most 2×2. We can transform each A in the upper triangular form (16) by an orthogonal similarity transformation. Let

$$R = \begin{bmatrix} A & B \\ C & D \end{bmatrix} \tag{17}$$

for (14). The cascade structure (15) is represented in the state-space form as

$$R = \begin{bmatrix} A_1 & 0 & B_1 \\ 0 & I & 0 \\ C_1 & 0 & D_1 \end{bmatrix} \begin{bmatrix} I & 0 & 0 & 0 \\ 0 & A_2 & 0 & B_2 \\ 0 & 0 & I & 0 \\ 0 & C_2 & 0 & D_2 \end{bmatrix} \cdots\cdots \begin{bmatrix} I & 0 & 0 \\ 0 & A_k & B_k \\ 0 & C_k & D_k \end{bmatrix}, \tag{18}$$

where A is of the form (16) and

$$C = [\, C_1 \;\; C_2 \;\cdots\; C_k \,]. \tag{19}$$

The factored form (18) represents itself the cascade realization of the filter having R as its system map. In order to get (18), it is necessary to find B_i, D_i for a given A_i, such that

$$\begin{bmatrix} A_i & B_i \\ C_i & D_i \end{bmatrix} \begin{bmatrix} I & 0 \\ 0 & J \end{bmatrix} \begin{bmatrix} A_i & B_i \\ C_i & D_i \end{bmatrix}^T = \begin{bmatrix} I & 0 \\ 0 & J \end{bmatrix}. \tag{20}$$

If A_i is of order 1, we can show that under the computability condition, the representation

$$\begin{bmatrix} A_i & B_i \\ C_i & D_i \end{bmatrix} = \Delta N_2(\phi_1) M_3(\phi_2) \tag{21}$$

holds, where

$$\Delta = \mathrm{diag}[\,\delta_1\,\delta_2\,\delta_3\,], \qquad \delta_i = \pm 1$$

$$N_2(\phi) = \begin{bmatrix} \cos\phi & -\sin\phi & 0 \\ \sin\phi & \cos\phi & 0 \\ 0 & 0 & 1 \end{bmatrix}, \quad M_3(\phi) = \begin{bmatrix} \cosh\phi & 0 & \sinh\phi \\ 0 & 1 & 0 \\ \sinh\phi & 0 & \cosh\phi \end{bmatrix}.$$

Thus, ϕ_1 and ϕ_2 can be computed from the relations

$$A_i = \delta_1 \cos\phi_1 \cosh\phi_2, \quad C_i = \begin{bmatrix} \delta_2 \sin\phi_1 \cosh\phi_2 \\ \delta_3 \sinh\phi_2 \end{bmatrix}.$$

The matrices B_i and D_i are computed from (21).

If A_i is of order 2, the factorization corresponding to (21) is given by

$$\begin{bmatrix} A_i & B_i \\ C_i & D_i \end{bmatrix} = \Delta N_2(\phi_1) N_3(\phi_2) N_{32}(\phi_3) M_4(\phi_4), \tag{22}$$

where

$$\Delta = \mathrm{diag}[\,\delta_1\,\delta_2\,\delta_3\,\delta_4], \qquad \delta_i = \pm 1$$

$$N_2(\phi) = \begin{bmatrix} \cos\phi & -\sin\phi & 0 \\ \sin\phi & \cos\phi & 0 \\ 0 & 0 & I_2 \end{bmatrix}, \quad N_3(\phi) = \begin{bmatrix} \cosh\phi & 0 & \sinh\phi & 0 \\ 0 & 1 & 0 & 0 \\ \sinh\phi & 0 & \cosh\phi & 0 \\ 0 & 0 & 0 & 1 \end{bmatrix}$$

$$N_{32}(\phi) = \begin{bmatrix} 1 & 0 & 0 & 0 \\ 0 & \cos\phi & -\sin\phi & 0 \\ 0 & \sin\phi & \cos\phi & 0 \\ 0 & 0 & 0 & 1 \end{bmatrix}, \quad M_4(\phi) = \begin{bmatrix} \cosh\phi & 0 & \sinh\phi \\ 0 & I_2 & 0 \\ \sinh\phi & 0 & \cosh\phi \end{bmatrix}$$

Therefore, ϕ_1, ϕ_2, ϕ_3 and ϕ_4 are calculated from the relations

$$A_i = \begin{bmatrix} \delta_1\cos\phi_1\cos\phi_2\cosh\phi_4 & \delta_1(-\sin\phi_1\cos\phi_3 + \cos\phi_1\sin\phi_2\sin\phi_3) \\ \delta_2\cos\phi_1\cos\phi_2\cosh\phi_4 & \delta_2(\cos\phi_1\cos\phi_3 - \sin\phi_1\sin\phi_2\sin\phi_3) \end{bmatrix}$$

$$C_i = \begin{bmatrix} \delta_3\sin\phi_3\cosh\phi_4 & \delta_3\cos\phi_2\sin\phi_3 \\ \delta_4\sinh\phi_4 & 0 \end{bmatrix}.$$

It should be noted that, in this case, we need an additional orthogonal transformation which makes the (2,2)-block of C_i zero.

From the above results, we can factorize R in (17) in the form (18). A. similar factorization method is found in [8].

Example. The transfer function is given by

$$H(z) = \frac{0.064 + 0.128z^{-1} + 0.064z^{-2}}{1.0 - 1.1682z^{-1} + 0.04241z^{-2}}.$$

A state-space realization (12) is given by

$$\begin{bmatrix} F & G \\ H & J \end{bmatrix} = \begin{bmatrix} 0.8203 & 0.5719 & 0.0 \\ -0.2426 & 0.3479 & 0.9056 \\ 0.3923 & 0.2239 & 0.064 \\ 0.5179 & -0.7429 & 0.4241 \end{bmatrix}.$$

Then, the realization (14) is given by

$$\begin{bmatrix} A & B \\ C & D \end{bmatrix} = \begin{bmatrix} 0.8203 & 0.5719 & 0.9861 & -0.9861 \\ -5.7937 & -2.8203 & 0.0892 & 6.3662 \\ -2.0816 & -2.2266 & 0.1403 & 2.8828 \\ -6.1296 & -3.4948 & 0.0 & 7.1282 \end{bmatrix}$$

with $K = 2.192$. An orthogonal transformation gives

$$\begin{bmatrix} A & B \\ C & D \end{bmatrix} = \begin{bmatrix} -1.0 & 6.3656 & -0.2105 & 6.3691 \\ 0.0 & -1.0 & -0.9675 & -0.9675 \\ -1.500 & 2.6533 & 0.1403 & 2.8828 \\ -1.500 & 6.8964 & 0.0 & 7.1282 \end{bmatrix}$$

The J-orthogonal factor corresponding to

$$\begin{bmatrix} A_1 \\ C_1 \end{bmatrix} = \begin{bmatrix} -1.0 \\ -1.5 \\ -1.5 \end{bmatrix}$$

gives rise to $(\phi_1, \phi_2) = (-0.9829,\ -1.1949)$ for the choice $(\delta_1\ \delta_2\ \delta_3) = (-1, 1, 1)$. Also,

$$\begin{bmatrix} A_2 \\ C_2 \end{bmatrix} = \begin{bmatrix} -1.0 \\ 2.6533 \\ 6.8964 \end{bmatrix}$$

gives rise to $(\phi_1, \phi_2) = (-1.3151,\ 2.0514)$ for the choice $(\delta_1\ \delta_2\ \delta_3) = (-1, 1, 1)$. The block diagram of the realization of H(z) is illustrated in Fig. 4.

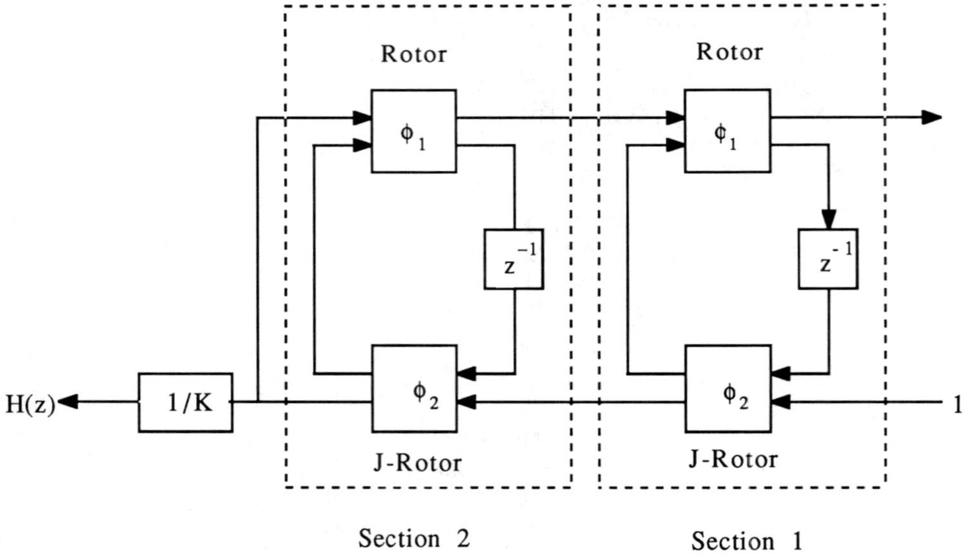

Fig. 4 A Realization of H(z)

4.CONCLUSION

A synthesis method of orthogonal filters has been proposed in the state space. The method is based on the factorization of the J-unitary system map into the product of system maps with simpler structure. The algorithm is described entirely in the state space and is straightforward and simple.

REFERENCE

[1] E. Deprettere and P. Dewilde, "Orthogonal cascade realization of real multiport digital filters," *Circuit Theory and Appl.*, vol. 8, pp. 245-272 (1980).

[2] V. Belevitch, *Classical Network Theory*, Holden-Day, San Francisco, 1968.

[3] J. D. Rhodes, P. C. Marston and D. C. Youla, "Explicit solutions for the synthesis of two-variable transmission-line networks," *IEEE Trans. Circuit Th.*, vol. CT-20, pp. 504-511, 1973.

[4] A. Fettweis, "Digital filter structures related to classical filter networks," A.E.U. band 25, pp. 79-89, 1981.

[5] P. Dewilde, "Advanced digital filters," in *Modern Signal Processing*, T. Kailath (ed.) Hemisphere Publishing Co., New York, 1985.

[6] C. V. K. Prabhakara Rao, *A Systems Approach to the Multiport Lossless Inverse Scattering Problem*, Dept. of Electrical Engineering, The Delft Univ. of Technology, 1985.

[7] H. Kimura, "Conjugation, interpolation and model-matching in H^∞," *Int. J. Control*, vol. 49, pp. 269-307, 1989.

[8] D. Alpay and I. Gohberg, "Unitary rational matrix functions," *Operator Theory: Adavances and Applications*, vol. 33, pp. 75-222, Birkhäuser, Basel, 1988.

Implementation of Arbitrary Realizations in Orthogonal Systems with Applications to Principle Axis Realizations

C. V. K. Prabhakara Rao
G. J. Maskarinec

Abstract:

From a numerical point of view, it is attractive to embed a digital system F(z) as a partial transfer of a lossless 2-port system with an orthogonal realization matrix $\mathcal{T}$ = [A B; C D]. Existing theory allows the *transfer function* F(z) to be embedded, but without control over the resulting realization of the partial transfer, say $\mathcal{T}_{pt}$. In this paper, we begin with an *arbitrary realization* $\mathcal{T}_F$ of a *scalar* F(z) and attempt to embed it in an orthogonal system, such that the resulting realization of the partial transfer $\mathcal{T}_{pt} = \mathcal{T}_F$. In general, this equality is achieved at the expense of having to use a number of additional signal channels $q>1$. We derive an algorithm to perform the embedding when the original realization $\mathcal{T}_F$ of F(z) is a contractive matrix. We apply this result to embed the balanced realization and more generally, any principle axis realization associated with a given F(z) in an orthogonal 2-port system. We first show that the principle axis realizations of a contractive transfer function F(z) will be contractive. Consequently, we will always be able to orthogonally embed the principle axis realizations corresponding to a contractive transfer function. We conclude with a numerical example.

0. INTRODUCTION

Motivation: In the implementation of digital systems, one finds in the literature two significant approaches to achieve numerical robustness. In one approach, one attempts to reduce all numerical computations in the system into an sequence of elementary orthogonal transformations, since orthogonal transformations are known to have the best numerical conditioning. The resulting signal processing structures are known as Orthogonal filters. In order to obtain orthogonal implementations of arbitrary digital filters, it is necessary to embed the given filter into an all-pass filter, which then is implemented as an orthogonal structure. It must be clarified at once that this embedding can be done without increasing the dynamical complexity (state space dimension), although the number of inputs and outputs is increased. The original digital filter is now realized as a partial transfer of the embedding orthogonal system by applying appropriate boundary conditions. Although the embedding orthogonal system is numerically ideally conditioned, it does not follow that the partial transfer under the specified boundary conditions is also well conditioned numerically. To the best of our knowledge, there are no hard results on this matter, although many examples as well as intuition suggest good conditioning of the partial transfer as well.

In the other approach, one states the problem as a state space synthesis problem, and attempts to find those realizations which optimize the filter's numerical performance, assuming standard arithmetic based on multiplications and additions. It turns out that the ideal candidates here form a class of structures called collectively as principle axis realizations. When these strucures process random input signals, the internal state variable nodes exhibit uncorrelated signal fluctuations and can have equal variances too. In a sense, these structures are internally balanced and 'orthogonally decompose' the signal processing involved. However, these structures are based on classical multiplications and do not exploit the numerical robustness of orthogonal transformations. It would be interesting to study such internally balanced structures implemented via orthogonal transformations.

Led by the above observations, we explored signal processing structures that simultaneously exploit the structrural balancedness properties as well as orthogonal numerical implementation. As such, the problem can be viewed either as designing orthogonal embedding systems with structurally balanced partial transfers, or as implementing principle axis realizations in orthogonal structures. We shall take the first viewpoint and show in this paper that it is possible to design orthogonal filters with specified principle axis realizations for the partial transfers under specified boundary conditions. This paper restricts its results to one particular type of boundary condition although other interesting boundary conditions exist as well. Furthermore, this paper deals with embedding only systems with scalar transfer functions.

Literature Review: Orthogonal filters, as we use the term, were first defined by Dewilde and Deprettere [Dp],[DpDe]. The synthesis of Orthogonal filters is based on principles of Lossless Inverse Scattering [DeDy], [Pr],[RaKa]. (There have been many more works published on this subject, but the ones quoted are a sufficient sample for our purpose.)

The problem of synthesizing numerically optimal filter structures in state space, leading to principle axis realizations, has been solved by Mullis and Roberts [MuRo]. Independent of numerical robustness, principle axis realizations may be viewed as desirable structures for implementing dynamical systems, and also find applications in control and model reduction [Mo].

Organization of Paper: In section I, we review results from existing literature about Orthogonal Implementations, Principle Axis Realizations and Abstract Realization Theory, which shall form the main tool for solving the problem addressed in this paper. Section II deals with the norms of principle axis realizations of transfer functions. We shall show that the principle axis realizations of contractive transfer functions are contractive matrices. In Section III, we shall show that the principle axis realizations of contractive transfer functions can be embedded in Orthogonal structures with a particular type of boundary condition and give a numerical example. We conclude with closing remarks in Section IV.

Notation: $\mathbb{C}$ shall denote the extended complex plane $\mathbb{C} = \{z\}$, whereas $\mathbb{D}, \mathbb{E}, \mathbb{T}$ shall denote the open unit disc, the unit circle and the complement of the closed unit disc as: $\mathbb{D} = \{z: |z| < 1\}$, $\mathbb{T} = \{z: |z| = 1\}$ and $\mathbb{E} = \{z: |z| > 1\}$.

The superscript * stands for conjugate transpose of a matrix or simply the conjugate for a complex scalar. The symbol 'I_m' has been used for an mxm identity matrix, with the subscript being dropped when the dimension is unambigous from the context. When A and B are square matrices of same dimensions, A>B stands for (A-B) being a positive definite matrix. Similarly, A≥B stands for (A-B) being positive semidefinite. Matrices are sometimes defined by writing its rows separated by a semicolon, for example, T = [A B; C D]. Although no special notation is reserved for indicating matrices, the operators corresponding to matrices are usually outlined, as in operator $\mathbb{A}$ and matrix A.

L^2 stands for the Hilbert space of square integrable functions on the unit circle $\mathbb{T}$; H^2 stands for the Hardy space of analytic functions in the unit disc $\mathbb{D}$; K^2 stands for the Hardy space of analytic functions in $\mathbb{E}$. Note that H^2 and K^2 are also Hilbert spaces with respect to the L^2 innerproduct. Norms in Hilbert spaces are indicated by $\|\bullet\|$. A subscript 'L^2' or 'op' or 'E' is used sometimes for clarity to distinguish 'L^2' norm or operator norm in the appropriate Hilbert space or Euclidean norm respectively.

The symbols $\mathcal{T}$, $\mathcal{T}_F$, $\mathcal{T}_\Sigma$ are used to denote the system matrix [A B; C D] of a general transfer function or F(z) or $\Sigma(z)$ respectively. Finally, we define the z-transform of a discrete time sequence $\{u(k)\}$ as $U(z) = \Sigma_k u(k)z^k$. Thus stable transfer functions are analytic in $\mathbb{D}$, with poles outside the unit circle. Finally, '#EOP#' denotes 'End Of Proof'.

I. BASICS

I. 1 Orthogonal 2-Port Systems:

Consider a causal, stable, scalar, rational transfer function F(z) which is analytic in the unit disc $\mathbb{D}$ and of Smith-McMillan degree N. Let $\mathcal{T}_F$ be a minimal (N+1)x(N+1) realization matrix:

$$\mathcal{T}_F = [A_F \;\; B_F \, ; \; C_F \;\; D_F] \tag{1a}$$

F(z) is related to the realization as:

$$F(z) = D_F + C_F\,[\,I - z\,A_F\,]^{-1}\;B_F \tag{1b}$$

$\mathcal{T}_F$ is said to be an orthogonal realization of F(z) or an orthogonal system if, by definition, [Dp], [DeDp]

$$\mathcal{T}_F^{*}\,\mathcal{T}_F = I_{(N+1)} \tag{2}$$

Such orthogonal systems are too restrictive as it is easily shown that if $\mathcal{T}_F$ is orthogonal, then F(z) is an all-pass (inner) function.

This restriction can be overcome by embedding [Dp], [DeDp]. Specifically, suppose F(z) is also a contractive function, i.e. $|\,F(z)\,| \le 1$ in $\mathbb{D}$. (If F(z) is not contractive, it can be scaled appropriately to satisfy this condition.) Referring to Figure 1, F(z) can be realized as the partial transfer from A_1-input to A_2-output of a "lossless 2-port system" with the Y_2-input set to zero.

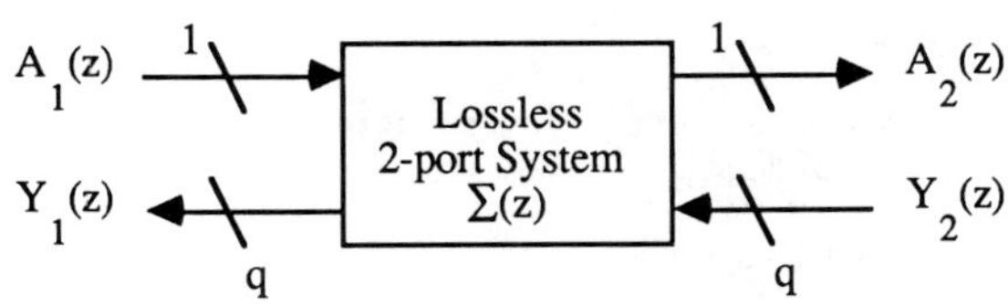

Figure 1 - Lossless 2-port System

The lossless 2-port system can be described by a scattering matrix $\Sigma(z)$ defined as:

$$\begin{bmatrix} A_2(z) \\ Y_1(z) \end{bmatrix} = \Sigma(z) \begin{bmatrix} A_1(z) \\ Y_2(z) \end{bmatrix}; \qquad \Sigma(z) = \begin{bmatrix} \Sigma_{11}(z) & \Sigma_{12}(z) \\ \Sigma_{21}(z) & \Sigma_{22}(z) \end{bmatrix} \tag{3a}$$

where $\Sigma(z)$ satisfies the following "losslessness" condition:

$$I - \Sigma^*(z)\,\Sigma(z) \; \begin{cases} = 0 & |z| = 1 \\ \geq 0 & |z| < 1 \end{cases} \tag{3b}$$

It is easy to see that $\Sigma_{11}(z) = F(z)$. Furthermore, $\Sigma(z)$ being a matrix all-pass function admits an orthogonal realization τ_Σ:

$$\begin{bmatrix} x(k+1) \\ a_2(k) \\ y_1(k) \end{bmatrix} = \tau_\Sigma \begin{bmatrix} x(k) \\ a_1(k) \\ y_2(k) \end{bmatrix} \qquad \text{where } \tau_\Sigma = \begin{bmatrix} A & B_1 & B_2 \\ C_1 & D_{11} & D_{12} \\ C_2 & D_{21} & D_{22} \end{bmatrix} \tag{4}$$

where $x(\bullet)$ is the Nx1 state vector and $a_1(k)$, $y_2(k)$ and $a_2(k)$, $y_1(k)$ are the time domain inputs and outputs respectively. τ_Σ may be found either recursively or globally using Lossless Inverse Scattering techniques [DeDy], [Pr]. For a scalar F(z), we obtain a τ_Σ corresponding to a $\Sigma(z)$, with q=1. In other words, a scalar F(z) can be embedded in a 2-input 2-output orthogonal system. The realization of F(z) is given by $\tau_F = [A \;\; B_1;\; C_1 \;\; D_{11}]$, a submatrix of τ_Σ. Note that there is no control over τ_F and hence, it cannot be prespecified.

I. 2 Principle Axis Realizations

Let F(z) have a realization matrix τ_F as given in (1). The controllability and observability Grammians, $W_{F,C}$ and $W_{F,O}$ respectively, are defined by:

$$W_{F,C} = A_F\, W_{F,C}\, A_F^* + B_F\, B_F^* \tag{5a}$$

$$W_{F,O} = A_F^*\, W_{F,O}\, A_F + C_F^*\, C_F \tag{5b}$$

A realization is said to be a principle axis realization if both $W_{F,C}$ and $W_{F,O}$ are diagonal. Specifically, there are three types of principle axis realizations [MuRo], [Mo]:

1) Internally Balanced: $\tau_{F,b} = [A_b \;\; B_b\,;\, C_b \;\; D_b]$ where $W_{F,C} = W_{F,O} = \Sigma$ (6a)

2) Input Normal: $\tau_{F,i} = [A_i \;\; B_i\,;\, C_i \;\; D_i]$ where $W_{F,C} = I$, $W_{F,O} = \Sigma^2$ (6b)

3) Output Normal: $\tau_{F,o} = [A_o \;\; B_o\,;\, C_o \;\; D_o]$ where $W_{F,C} = \Sigma^2$, $W_{F,O} = I$ (6c)

where Σ is a diagonal matrix.

The following facts concerning principle axis realizations are either known or are easily derived [PeSi], [WiKu], [Mo]:

1) A_b is a contractive matrix, i.e. $\| A_b \| \leq 1$

2) A balanced realization can be found such that the following J- symmetry properties hold:

$$(J\,A_b)^* = (J\,A_b) \tag{7}$$
$$B_b = J\,C_b^* \tag{8}$$

where J is a signature matrix, i.e. $J = \mathrm{diag}\,(\pm 1, \;\ldots\;, \pm 1)$

3) The three principle axis realizations are related by the following diagonal similarity transformations:

$$\mathcal{T}_{F,i} = [\, \Sigma^{-1/2} \oplus 1 \,]\, \mathcal{T}_{F,b} \,[\, \Sigma^{1/2} \oplus 1 \,] \tag{9a}$$

$$\mathcal{T}_{F,o} = [\, \Sigma^{1/2} \oplus 1 \,]\, \mathcal{T}_{F,b} \,[\, \Sigma^{-1/2} \oplus 1 \,] \tag{9b}$$

where Σ is defined in (6).

4) The Input Normal and Output Normal realizations can be found such that the following symmetry property holds:

$$\mathcal{T}_{F,o} = J\, \mathcal{T}_{F,i}^{*}\, J \tag{10}$$

Although A_b has been shown to be a contractive matrix, to our knowledge there are no results concerning the norm of the realization matrix $\mathcal{T}_F$. In the following, we will show that if F(z) is contractive, then all three principle axis realizations are contractive.

I. 3 Abstract Realization Theory [Fu], [PrDe]:

Define the space of finite energy signals L^2, the space of finite energy signals in the strict past K^2, and the space of finite energy signals in the future H^2 as follows:

$$L^2 = \text{span}\,\{\, e^{ik\theta},\ k = 0, \pm1, \pm2, \ldots \,\} \tag{11}$$
$$K^2 = \text{span}\,\{\, z^k,\ k = -1, -2, \ldots \,\} \tag{12}$$
$$H^2 = \text{span}\,\{\, z^k,\ k = 0, 1, 2, \ldots \,\} \tag{13}$$

Note that H^2 and K^2 are Hardy spaces of analytic functions in the open unit disc $\mathbb{D}$ and the complement of the closed unit disc $\mathbb{E}$ respectively.

Let a given causal, stable, linear shift invariant (LSI) system with transfer function F(z), analytic and bounded in $\mathbb{D}$ be considered as a LSI Input/Output operator from L^2 to L^2. The so-called Restricted Input/Output operator H is the Hankel operator defined as the map from K^2 to H^2 as follows:

$$u(z) \in K^2 \longrightarrow H\, u(z) = P\, \{F(z) \cdot u(z)\} \tag{14}$$

where P denotes orthogonal projection onto H^2. The state space $\mathbb{H}$ of the system is a subspace of H^2, defined in terms of the Hankel operator H, as follows:

$$\mathbb{H} = \text{Range}\, H \tag{15}$$

A realization for F(z) is now obtained as:

$$\mathcal{T} = \begin{bmatrix} \mathbb{A} & \mathbb{B} \\ \mathbb{C} & \mathbb{D} \end{bmatrix} = \begin{bmatrix} R & R\,F(z)\cdot \\ \Pi & F(0)\cdot \end{bmatrix} \tag{16}$$

where R is the restricted shift operator and Π is the projection operator onto constants. Specifically, the $\mathbb{A}, \mathbb{B}, \mathbb{C}, \mathbb{D}$ operators are defined as follows:

$$\mathbb{A}:\ x(z) \in \mathbb{H} \longrightarrow \mathbb{H} \ni A\, x(z) = R\, x(z) = [x(z) - x(0)]\, /\, z \tag{17}$$
$$\mathbb{B}:\ u \in \mathbb{C} \longrightarrow \mathbb{H} \ni B\, u = R\, F(z)\cdot u = u \cdot [F(z) - F(0)]\, /\, z \tag{18}$$
$$\mathbb{C}:\ x(z) \in \mathbb{H} \longrightarrow \mathbb{C} \ni C\, x(z) = \Pi\, x(z) = x(0) \tag{19}$$
$$\mathbb{D}:\ u \in \mathbb{C} \longrightarrow \mathbb{C} \ni D\, u = F(0)\cdot u \tag{20}$$

Define $\mathbb{S}$ to be the space of 2x1 vectors whose top entry comes from $\mathbb{H}$ and the bottom entry comes from $\mathbb{C}$.

$$\mathbb{S} = \left\{ \begin{bmatrix} x \\ y \end{bmatrix} : \ x \in \mathbb{H}, \ y \in \mathbb{C} \right\}$$

Thus, $\mathfrak{T}$ is an operator mapping $\mathbb{S}$ into itself. Choosing a set of basis functions in $\mathbb{S}$, the operators $\mathbb{A}, \mathbb{B}, \mathbb{C}, \mathbb{D}$ and $\mathfrak{T}$ become constant matrices A,B,C,D and $\mathcal{T}$. Moreover, if the basis is orthonormal, then the operator norm of $\mathfrak{T}$ equals the Euclidean norm of $\mathcal{T}$:

$$\| \mathfrak{T} \|_{op} = \| \mathcal{T} \|_E \tag{21}$$

The Euclidean norm of $\mathcal{T}$ affects the numerical behavior of the system and will be the subject of Section II.

II. NORMS OF PRINCIPLE AXIS REALIZATIONS

In this section, we shall show that the principal axis realizations $\mathcal{T}_o$, $\mathcal{T}_i$, $\mathcal{T}_b$ of a contractive transfer function F(z) are contractive matrices (i.e. their Euclidean norms are bounded by 1). We shall demonstrate this fact in three steps. First, we show that the Output Normal Realization $\mathcal{T}_0$ is obtained by choosing an orthonormal basis in $\mathbb{S}$. Next, we show that for a contractive F(z), the operator norm of $\mathfrak{T}_0$ is bounded by 1. From these two facts, it follows that the Euclidean norm of $\mathcal{T}_o$ is bounded by 1. Finally, we extend this result to deduce the contractivity of $\mathcal{T}_i$ and $\mathcal{T}_b$.

Lemma 1: Let F(z) be the transfer function and $\mathfrak{T}$ the system operator of a system. Choose a set of orthonormal basis in $\mathbb{S}$, resulting in a realization matrix $\mathcal{T}$ = [A B; C D]. Then the observability Grammian $W_o = I$, i.e. from (5)

$$A^*A + C^*C = I \tag{22}$$

Proof: Establishing (22) is equivalent to establishing

$$\| Ax \|^2 + \| Cx \|^2 = \| x \|^2 \tag{23}$$

where x is an arbitrary constant vector of appropriate dimensions, and the indicated norms are Euclidean vector norms. The assumption of orthonormal basis in $\mathbb{S}$ implies that (23) can be written equivalently as:

$$\| \mathbb{A}\, x(z) \|^2 + \| \mathbb{C}\, x(z) \|^2 = \| x(z) \|^2 \tag{24}$$

where x(z) is an arbitrary element in the state space $\mathbb{H}$ and the indicated norms are L^2 norms. Writing $[x(z) = x_0 + x_1.z + x_2.z^2 + ..]$ and substituting for $\mathbb{A}$ and $\mathbb{C}$, the left hand side of (24) becomes

$$\| R\, x(z) \|^2 + \| \Pi\, x(z) \|^2 = \| \sum_1^{\infty} x_k z^{k-1} \|^2 + |x_0|^2$$

$$= \sum_1^{\infty} |x_k|^2 + |x_0|^2 = \| \sum_0^{\infty} x_k z^k \|^2$$

$$= \| x(z) \|^2 \tag{25}$$

Note that we used Parseval's identity in going from line 1 to line 2.

#EOP#

Lemma 2: Assume the conditions stipulated in Lemma 1. There exists a unitary state transformation matrix U such that the Output Normal Realization is given by:

$$\mathcal{T}_o = \begin{bmatrix} U & 0 \\ 0 & 1 \end{bmatrix} \mathcal{T} \begin{bmatrix} U^* & 0 \\ 0 & 1 \end{bmatrix}$$

Proof: According to Lemma 1, an orthonormal basis in $\mathbb{S}$ results in a realization matrix $\mathcal{T}$ with observability Grammian $W_o = I$, although the controllability Grammian W_c may not be diagonal. However, since all minimal realizations are related by similarity (state) transformations, there exists a nonsingular matrix T such that

$$\mathcal{T}_o = \begin{bmatrix} T^{-1} & 0 \\ 0 & 1 \end{bmatrix} \mathcal{T} \begin{bmatrix} T & 0 \\ 0 & 1 \end{bmatrix}$$

The above transformation changes W_o of $\mathcal{T}$ into $T^* W_o T = T^* T$, which must again be the identity matrix. This means that T is a unitary matrix, corresponding to a simple rotation of the original orthonormal basis in $\mathbb{H}$.

#EOP#

Theorem 1: Let F(z) be a stable causal transfer function. There exists an orthonormal basis in the state space $\mathbb{H}$, such that the resulting realization matrix $\mathcal{T}$, given by (16), is the Output Normal Realization.

Proof: Follows at once from Lemmas 1 and 2.

#EOP#

Theorem 2: Let F(z) be a contractive transfer function, i.e. $|F(z)| \le 1$ for $|z| \le 1$. Let $\mathcal{T}$ be the corresponding system operator given by (16). Then

$$\| \mathcal{T} \|_{op} \le 1 \tag{26}$$

Proof: The proof essentially rests on the fact that for a contractive transfer function, the energy of the input signal has to be greater than or equal to the energy of the output signal. That is, if u(z) is an input signal in L^2,

$$\| F(z).u(z) \|_{L2} \le \| F(z) \|_\infty \|u(z)\|_{L2} \le \|u(z)\|_{L2} \tag{27}$$

Now the details: Let $x(z) \in \mathbb{H}$ and $u \in \mathbb{C}$. We have

$$\| \mathcal{T} \begin{bmatrix} x(z) \\ u \end{bmatrix} \|_{L2}^2 = \| \begin{bmatrix} R x(z) + R F(z) u \\ x(0) + F(0) u \end{bmatrix} \|_{L2}^2$$

$$= \| R \{x(z) + F(z)\, u\} \|_{L2}^2 + | \{x(0) + F(0)\, u\} |_{L2}^2$$

$$= \| x(z) + F(z)\, u \|_{L2}^2 \le \| x(z) \|_{L2}^2 + \| F(z)\, u \|_{L2}^2$$

$$\leq \| x(z) \|_{L2}^2 + | u |^2 \qquad \text{from (27)}$$

$$= \left\| \begin{bmatrix} x(z) \\ u \end{bmatrix} \right\|_{L2}^2$$

This shows that the norm of the operator $\mathfrak{T}$ is less than or equal to 1.

#EOP#

Theorem 3: Let F(z) be a contractive transfer function. The Euclidean norm of its Output Normal Realization matrix $\mathcal{T}_o$ is less than or equal to 1:

$$\| \mathcal{T}_o \|_E \leq 1 \tag{28}$$

Proof: Follows at once from Theorems 1 and 2 and equation 21.

#EOP#

Theorem 4: Let F(z) be a contractive transfer function. The Euclidean norm of its Input Normal Realization matrix $\mathcal{T}_i$ is less than or equal to 1:

$$\| \mathcal{T}_i \|_E \leq 1 \tag{29}$$

Proof: Recall from (10) that the Input Normal Realization is related to the Output Normal Realization as follows:

$$\mathcal{T}_i = J \, \mathcal{T}_o^* \, J$$

where J is a diagonal signature matrix. Thus the Euclidean norm of $\mathcal{T}_i$ equals that of $\mathcal{T}_o$, which has been shown to be less than or equal to 1.

#EOP#

Lemma 3: Let F(z) be the system transfer function (not necessarily contractive) and let $\mathcal{T}_i$, $\mathcal{T}_o$ and $\mathcal{T}_b$ its Input Normal, Output Normal and Balanced Realizations respectively. Then

$$\| \mathcal{T}_b \|_E \leq \| \mathcal{T}_i \|_E = \| \mathcal{T}_o \|_E \tag{30}$$

Proof: Recall that $\mathcal{T}_b$ is J-symmetric and is related to $\mathcal{T}_o$ by a diagonal similarity transformation, (See equations 7,9). (7) implies that the singular values and eigenvalues of [J $\mathcal{T}_b$] are equal in magnitude, whereas (9) implies that the eigenvalues of [J $\mathcal{T}_b$] and [J$\mathcal{T}_o$] are equal as seen below:

$$J \, \mathcal{T}_b = J \begin{bmatrix} \Sigma^{-1/2} & 0 \\ 0 & 1 \end{bmatrix} \mathcal{T}_o \begin{bmatrix} \Sigma^{1/2} & 0 \\ 0 & 1 \end{bmatrix} = \begin{bmatrix} \Sigma^{-1/2} & 0 \\ 0 & 1 \end{bmatrix} J \, \mathcal{T}_o \begin{bmatrix} \Sigma^{1/2} & 0 \\ 0 & 1 \end{bmatrix} \tag{31}$$

= a similarity transformation of [J $\mathcal{T}_o$]

Now, $\| \mathcal{T}_b \| = \| J \, \mathcal{T}_b \|$ = max singular value of J $\mathcal{T}_b$

= max | {eigenvalue of J $\mathcal{T}_b$} | due to symmetry

= max | {eigen value of J $\mathcal{T}_o$} | due to (31)

$\leq \| J \, \mathcal{T}_o \| = \| \mathcal{T}_o \|$

$= \| J \, \mathcal{T}_i^* J \| = \| \mathcal{T}_i \|$ due to (10)

#EOP#

Theorem 5: Let F(z) be a contractive transfer function. The Euclidean norm of its

Balanced Realization matrix $\mathcal{T}_b$ is less than or equal to 1:

$$\| \mathcal{T}_b \|_E \leq 1 \tag{32}$$

Proof: Follows at once from Theorem 3 and Lemma 3.

#EOP#

III.Principle Axis Realizations & Orthogonal Embeddings

Recall that the existing theory allows a scalar transfer function F(z) to be embedded in a lossless 2-port with q = 1, but without control over the resulting realization of the $\Sigma_{11}(z)$ partial transfer. We begin with an arbitrary realization $\mathcal{T}_F$ of a scalar F(z) and attempt to embed it an an orthogonal system, such that the resulting realization of the $\Sigma_{11}(z)$ partial transfer is $\mathcal{T}_F$. In general, this equality is achieved at the expense of having to use a number of signal channels q > 1. Recall that $\mathcal{T}_F$ is a submatrix of $\mathcal{T}_\Sigma$. Thus, the realization embedding problem is to embed a given realization matrix $\mathcal{T}_F$ as a submatrix in $\mathcal{T}_\Sigma$, an orthogonal matrix of larger dimension. Note, however, that the dynamical complexity (given by the dimension of the state space) of $\Sigma(z)$ is not any greater. The embedding result is summarized in the Theorem 6 and applied to principle axis realizations in the Corollary 1.

Theorem 6: Let $\mathcal{T}_F$ be an nxn realization matrix of a transfer function F(z). Then $\mathcal{T}_F$ can be embedded in an (n+m)x(n+m) orthogonal matrix $\mathcal{T}_\Sigma = [\ \mathcal{T}_F \ \ \mathcal{T}_{12}\,;\ \mathcal{T}_{21} \ \ \mathcal{T}_{22}\ ]$ with $m \leq n$ iff:

(i) $\mathcal{T}_F$ is a contrative matrix
(ii) $m \geq \text{rank}\,(I - \mathcal{T}_F^{\,*}\, \mathcal{T}_F)$

Furthermore, the entries of $\mathcal{T}_\Sigma$ are given by:

$$\mathcal{T}_{12} = U\, M^*\, Q_1 \qquad \mathcal{T}_{21} = Q_2\, M\, V^* \qquad \mathcal{T}_{22} = Q_3\, \Sigma_m\, Q_1 \tag{33a}$$

$$\text{where } \mathcal{T}_F = U\, [\ \Sigma_m \oplus I_{n-m}\ ]\, V^* \quad \text{(SVD)} \tag{33b}$$

$$M = [\ (I_m - \Sigma_m{}^2)^{1/2} \qquad \mathbf{0}_{mx(n-m)}\ ] \tag{33c}$$

Q_1, Q_2 and Q_3 are arbitrary mxm orthogonal matrices such that $Q_3 = -\,Q_1$.

Proof: Assume $\mathcal{T}_\Sigma$ is orthogonal, i.e. $I - \mathcal{T}_\Sigma^{\,*}\, \mathcal{T}_\Sigma = 0$.

We have $I - \mathcal{T}_F^{\,*}\, \mathcal{T}_F = \mathcal{T}_{21}^{\,*}\, \mathcal{T}_{21} \geq 0$, which proves that $\mathcal{T}_F$ is contractive. Furthermore, $\text{rank}\,(\,I - \mathcal{T}_F^{\,*}\, \mathcal{T}_F\,) = \text{rank}\,(\,\mathcal{T}_{21}^{\,*}\, \mathcal{T}_{21}\,) \leq m$.

To prove the converse, assume $\mathcal{T}_F$ is contractive and $m \geq \text{rank}(I - \mathcal{T}_F^{\,*}\, \mathcal{T}_F)$. The contractivity of $\mathcal{T}_F$ guarentees that $(I_m - \Sigma_m{}^2)^{1/2}$ in (33c) exists. $m \geq \text{rank}\,(I - \mathcal{T}_F^{\,*}\mathcal{T}_F)$ implies that the number of singular values of $\mathcal{T}_F$ less than one is at least m. Substituting the values in (33a-c), it is straight forward to show that $\mathcal{T}_F$ is an orthogonal matrix.

#EOP#

Corollary 1: If F(z) is a contractive transfer function then it can always be embedded as a principle axis realization in an orthogonal system.

Proof: The contractivity of F(z) implies that the principle axis realizations are all contractive matrices. (See Theorems 3,4, &5). Theorem 6 ensures that they can be in turn embedded in an orthogonal system.

#EOP#

Numerical Example: Consider the following transfer function:

$$F(z) = \frac{0.165\,(z+2)}{z^2 + 0.25\,z - 0.125}$$

A J-symmetric balanced realization of F(z) is given by:

$$\tau_{F,b} = \begin{bmatrix} .4346 & -.4154 & .5003 \\ .4154 & -.6846 & -.2922 \\ .5003 & .2922 & 0 \end{bmatrix}$$

Expressing $\tau_{F,b}$ after singular value deomposition as $\tau_{F,b} = U\,\Sigma\,V^*$ yields:

$$U = \begin{bmatrix} -.5536 & -.5253 & .6464 \\ .3722 & .5380 & .7563 \\ .7449 & -.6593 & .1023 \end{bmatrix};\ \Sigma = \begin{bmatrix} .5178 & 0 & 0 \\ 0 & .6370 & 0 \\ 0 & 0 & 1.0 \end{bmatrix};\ V = \begin{bmatrix} .5536 & .3722 & -.7449 \\ -.5253 & -.5380 & -.6593 \\ .6464 & -.7563 & .1023 \end{bmatrix}$$

Here m = number of singular values less than 1 is 2. Σ_2 and M in equation 33 can be now computed as:

$$\Sigma_2 = \begin{bmatrix} .5178 & 0 \\ 0 & .6370 \end{bmatrix} \quad \text{and} \quad M = \begin{bmatrix} .8555 & 0 & 0 \\ 0 & .7708 & 0 \end{bmatrix}$$

$\tau_{12}, \tau_{21}, \tau_{22}$ then become:

$$\tau_{12} = \begin{bmatrix} -.4736 & -.4049 \\ .3185 & .4147 \\ .6373 & -.5082 \end{bmatrix};\quad \tau_{21} = \begin{bmatrix} .4736 & .3185 & -.6373 \\ -.4049 & -.4147 & -.5082 \end{bmatrix};\quad \tau_{22} = \begin{bmatrix} -.5178 & 0 \\ 0 & -.6370 \end{bmatrix}$$

τ_Σ is given by $\tau_\Sigma = [\ \tau_F\ \ \tau_{12}\ ;\ \tau_{21}\ \ \tau_{22}\ ]$, whose orthogonality is easily checked.

IV. CONCLUSION

In this paper, we addressed the problem of embedding principle axis realizations in orthogonal systems. The embedding orthgonal system will, in general, have additional inputs and outputs, although the dynamical complexity or the state space dimension is not increased. The principle axis realization is retrieved by a terminating the extra inputs with a zero and ignoring the other outputs. We showed that for this type of embedding to be possible, it is necessary and sufficient that the transfer function of the original system be a contractive transfer function, which will guarantee that its principle axis realizations are contractive matrices (proved in this paper). A construction and numerical example for the embedding orthogonal system is also given. An important finding here is that, in general, one requires more than one additional input and output in the embedding system.

The results in this paper are limited to the case when the original system is an SISO system. Furthermore, we only considered one type of boundary condition (termination by a zero) for the embedding orthogonal system.

V. REFERENCES

[DeDy] P. Dewilde, H. Dym, "Lossless Inverse Scattering, Estimation Theory and Digital Filters", IEEE Trans. Info. Theory, vol. IT-30, no. 4, pp. 644-662, July 1984.

[Dp] E. Deprettere, "Orthgonal Filters", Ph.D. Thesis, Delft University of Technology, Delft, The Netherlands, 1981.

[DpDe] E. Deprettere, P. Dewilde, "Orthogonal Cascade Realization of Real Multiport Digital Filters", Intl. J. Circuit Theory and Applns., vol. 8, pp. 245-272, 1980.

[Fu] P. Fuhrmann, Linear Systems and Operators in Hilbert Space, McGraw Hill Inc., 1981.

[Mo] B.C. Moore, "Principal Component Analysis in Linear Systems: Controllability, Observability, and Model Reduction", IEEE Trans. Auto. Control, vol. AC-26, no. 1, pp.17-32, Feb. 1981.

[MuRo] C.T. Mullis, R.A. Roberts, "Synthesis of Minimum Roundoff Noise Fixed Point Digital Filters", IEEE Trans. Circuits and Systems, vol. CAS-23, no. 9, pp. 551-562, Sept. 1976.

[Pr] C.V.K. Prabhakara Rao, "A Systems Approach to the Multiport Lossless Inverse Scattering Problem", Ph.D. Thesis, Delft University of Technology, Delft, The Netherlands, 1985.

[PrDe] C.V.K. Prabhakara Rao, P. Dewilde, "System Theory for Lossless Wave Scattering", in Operator Theory: Advances and Applications, vol. 19, pp. 333-358, Birkhauser Verlag, Basel, 1986.

[PeSi] L. Pernebo, L. M. Silverman, "Model Reduction via Balanced State Space Representations", IEEE Trans. on Auto. Control, vol. AC-27, no. 2, pp. 382-387, Apr. 1982.

[RaKa] S.K. Rao, T. Kailath, "Orthogonal Digital Filters for VLSI Implementation", IEEE Trans. Circuits and Systems, vol. CAS-31, no.11, pp. 933-945, Nov 1984.

[WiKu] D.A. Wilson, A. Kumar, "Symmetry Properties of Balanced Systems", IEEE Trans. on Auto. Control, vol. AC-28, no. 9, pp. 927-929, Sept. 1983.

ADAPTIVE SINGULAR VALUE DECOMPOSITION AND ITS USE FOR ADAPTIVE SIGNAL SEPARATION

Dirk Callaerts and Joos Vandewalle

Many problems in signal processing deal with the extraction of a signal from a set of combined signals. Very recently some SVD-based methods have been developed for this signal separation problem. At the conceptual level the Singular Value Decomposition (SVD), a numerically reliable matrix decomposition technique known from linear algebra [6], provides a unifying framework and a numerically robust approach for the formulation and computation of a solution. However, the classical algorithms to compute the SVD implement a typical off-line strategy, while for real-time multichannel signal processing on-line and adaptive updating techniques, that incorporate a priori knowledge, are required. In this contribution an on-line adaptive algorithm for SVD-based signal separation is presented. The method has already been applied with success to the extraction of the fetal electrocardiogram from potential recordings on the maternal skin [1], [2], [10].

1 Introduction

Recent work showed that the Singular Value Decomposition of the data matrix [7], constructed from simultaneously sampled signals, constitutes a main tool in signal separation problems [3], [4], [11]. But also in linear least squares problems (adaptive filtering, linear prediction), in methods using principal component reduction (image and speech processing, modal analysis), in rank-determination problems (identification, model reduction) etc. SVD is used with growing interest. The wide acceptance of the SVD as a reliable method in numerical algebra is due to the existence of a stable and accurate algorithm for its computation : the Golub-Reinsch algorithm [5]. There are a lot of subroutines from well established libraries available, using this basic Golub-Reinsch algorithm. These subroutines return the three matrices U, Σ and V^t from the **complete** decomposition. In most cases, however, only the U-matrix is required at any time, e.g. for adaptive AR modelling.

The Golub-Reinsch algorithm furthermore implements a typical **off-line** strategy of data analysis : all data should be available at once before the computational work can start. In signal processing applications, this off-line technique has a number of important drawbacks. First, the dimensions $(p \times q)$ of the data matrix A can be very large. Here p is the number of signals and q the number of time instants at which these signals are sampled.

In most applications q is very large, which results in an inadmissible computation time $O(qp^2)$. Big efforts have been made to reduce this objection e.g. by introducing strategies for parallel computing, by applying fast Givens rotations, by computing the partial SVD, by inserting acceleration steps in the iterative part of the algorithm, by using very powerful computers, ... Furthermore, even if these remedies could accelerate the computational work, an off-line strategy is still not optimal with regard to time variations within the time interval q under consideration. This requires an **adaptive** technique, such that the method is able to "follow" the most recent events, occurring in the data, and to "forget" events, that happened in the far past. It is therefore also important that **a priori** knowledge can be incorporated, for instance the SVD of a previous step. This is elementary in signal processing applications, where usually the dimensions of the data matrix are growing with time, such that updating of the computed parameters is required, each time some new information is available.

This clearly stresses the importance of and the need for an **adaptive on-line computation strategy** for SVD-based adaptive signal separation. This new technique should provide U (and Σ) at any instant, by processing data and using these for updating, as soon as they become available. In this paper such an algorithm is presented. Since the Generalized SVD of a matrix pair can be computed as two consecutive SVD's, it will be shown that the presented algorithm can also be used for a GSVD-based signal separation method [3], [4].

2 Development of the algorithmic scheme

The algorithm that will be outlined in this section is based on some features of the SVD of a matrix A and its relation with the eigenvalue decomposition of $A\,A^t$. The algorithmic scheme is founded on the following very important theorem :

Theorem 1 : Updating the SVD after adding a column :

Given $p \times k$ matrix $A_k = [\mathbf{a}_1\,\mathbf{a}_2\,\ldots\,\mathbf{a}_k]$ with known SVD, $A_k = U_k\,\Sigma_k\,V_k^t$. Add one column $\mathbf{a}_{k+1}$ to A_k such that

$$A_{k+1} = [A_k\ \mathbf{a}_{k+1}] = [\mathbf{a}_1\,\mathbf{a}_2\,\ldots\,\mathbf{a}_{k+1}] \tag{1}$$

If the SVD of A_{k+1} is written as $A_{k+1} = U_{k+1}\,\Sigma_{k+1}\,V_{k+1}^t$, then U_{k+1} and Σ_{k+1} can be computed from U_k, Σ_k and $\mathbf{a}_{k+1}$ as follows

$$\Sigma_{k+1}^2 = Q^t\,(\Sigma_k^2 + U_k^t\,\mathbf{a}_{k+1}\,\mathbf{a}_{k+1}^t\,U_k)\,Q \tag{2}$$

$$U_{k+1} = U_k\,Q \tag{3}$$

where Q is an orthogonal $p \times p$ matrix.

Proof

Since the matrix V_k in the SVD of A_k has the property that $V_k^t\,V_k = I$, it is clear that

$$A_k\,A_k^t = (U_k\,\Sigma_k\,V_k^t)\,(V_k\,\Sigma_k\,U_k^t) = U_k\,\Sigma_k^2\,U_k^t \tag{4}$$

Due to the fact that U_k is an orthogonal matrix $(U_k^t\,U_k = U_k\,U_k^t = I)$, and because of the associativity of matrix multiplications, one can prove from Eq. (4) that

$$U_k^t\,(A_k\,A_k^t)\,U_k \;=\; (U_k^t\,A_k)\,(A_k^t\,U_k) \;=\; \Sigma_k^2 \tag{5}$$

where Σ_k^2 is a square diagonal matrix with nonnegative elements on the diagonal. After adding a column $\mathbf{a}_{k+1}$ to A_k, a symmetric matrix C_{k+1} can be defined as the product

$$\begin{aligned} C_{k+1} &= (U_k^t\,A_{k+1})\,(A_{k+1}^t\,U_k) \\ &= \begin{bmatrix} U_k^t\,A_k & U_k^t\,\mathbf{a}_{k+1} \end{bmatrix} \begin{bmatrix} A_k^t\,U_k \\ \mathbf{a}_{k+1}^t\,U_k \end{bmatrix} \\ &= (U_k^t\,A_k)\,(A_k^t\,U_k) \;+\; (U_k^t\,\mathbf{a}_{k+1})\,(\mathbf{a}_{k+1}^t\,U_k) \\ &= \Sigma_k^2 \;+\; (U_k^t\,\mathbf{a}_{k+1})\,(\mathbf{a}_{k+1}^t\,U_k) \end{aligned} \tag{6}$$

The extra term in Eq. (6) is a rank-one perturbation of the diagonal matrix Σ_k^2. Due to the existence of stable algorithms to diagonalize C_{k+1} (symmetric QR-iteration, Jacobi methods, Lanczos diagonalisation, ...) an orthogonal matrix Q can be found such that

$$B_{k+1} \;=\; Q^t\,C_{k+1}\,Q \tag{7}$$

is diagonal. With Eq. (6) this can be written as

$$B_{k+1} \;=\; (Q^t\,U_k^t)\,(A_{k+1}\,A_{k+1}^t)\,(U_k\,Q) \tag{8}$$

Since $U_k\,Q$ is orthogonal, since B_{k+1} is diagonal and since the SVD of A_{k+1} is unique, it follows from Eq. (8) that

$$\Sigma_{k+1}^2 \;=\; Q^t\,(\Sigma_k^2 \;+\; U_k^t\,\mathbf{a}_{k+1}\,\mathbf{a}_{k+1}^t\,U_k)\,Q \tag{9}$$

$$U_{k+1} \;=\; U_k\,Q \tag{10}$$

□

Based on the previously defined theorem, an on-line algorithm can be outlined. This algorithm assumes that the matrix A is built up column by column. Each time a column $\mathbf{a}_i$ is added to $A_{i-1} = [\mathbf{a}_1 \ldots \mathbf{a}_{i-1}]$ it computes the U_i- and Σ_i-matrices in the SVD of A_i by updating the matrices from the previous step.

The algorithmic scheme can then be presented as follows :

Initialisation $\quad U_0 = I_{p,p} \quad B_0 = O_{p,p}$

Loop $\quad$ for $i = 1$ to q

> **generate output** $\quad \mathbf{y}_i = U_{i-1}^t\,\mathbf{a}_i$
>
> **compute** C_i $\quad C_i = B_{i-1} + \mathbf{y}_i\,\mathbf{y}_i^t$
>
> **update** B_i **by diagonalizing** C_i $\quad B_i = Q_i^t\,C_i\,Q_i$
>
> **update** U_i $\quad U_i = U_{i-1}\,Q_i$

end loop

At the end of each loop, the matrices U_i and $\Sigma_i^2 = B_i$ in the SVD of A_i are exactly known. The memory requirements for this on-line method are restricted to p^2 locations for the elements of U_i and $p(p+1)/2$ locations for the upper triangle of C_i. For the p

elements of B_i, one can use p locations on the diagonal of C_i. In many multichannel signal processing applications the output is a set of linear combinations or weighted sums of the input signals. The coefficients in these linear combinations are found as the columns of the U-matrix. Since the input is assumed to be gradually available (column per column), the algorithm can generate immediately an output column for each available input column and knowing the U-matrix at that instant. This is exactly the first step in the algorithm.

3 Comparison with the method of normal equations

The attentive reader may have noticed that the matrix product AA^t appears in the outline of the new algorithm. It is a well-known problem in numerical algebra that the "normal" matrix AA^t is frequently ill-conditioned and influenced greatly by round-off errors [5]. Especially the small singular values are sensitive to the implicit squaring [7]. Attempts to improve the condition of AA^t are useless, once this normal matrix has been formed. Osborne [8], however, proves that the loss of information, resulting from forming AA^t, is minimal in case the normal matrix is diagonal.

Therefore the matrix A is first premultiplied by a $p \times p$ matrix Q to form a new matrix $B = QA$, such that $B\,B^t$ is diagonal. This is almost exactly what is done in the presented algorithm. The matrix Q in question is in this case the orthogonal matrix U_{i-1} from the previous update. The formation of C_i is implicitely not performed as $C_i = U_{i-1}^t\,(A_i\,A_i^t)\,U_{i-1}$ but as $C_i = (U_{i-1}^t\,A_i)\,(U_{i-1}^t\,A_i)^t$ making C_i almost diagonal, since U_{i-1} is an estimate of U_i. Using double precision arithmetic on the diagonal of C_i only, further reduces ill-conditioning of the problem.

4 Adaptivity of the algorithm

It should be stressed at this point that the algorithm, outlined in this section, is **on-line**, because data are processed as soon as they are available, but not **adaptive**. Adaptivity is the possibility of the method to follow time variations, to recover from sudden changes and to track new events. The matrices B_i and U_i at the end of each loop in the presented algorithm constitute the knowledge that the system has built up from the past. This knowledge determines the way of processing and it is quite evident that this knowledge should be as actual as possible. Therefore, it is desirable that events that happened in the remote past don't have any influence on the present processing anymore, while on the other hand, the very recent events will have to correct the processing severely.

Another problem that occurs while the on-line method proceeds, is the increase of the norm of the diagonal elements of matrix B_i. A result of this is that the contribution of $\mathbf{y}_i\,\mathbf{y}_i^t$ to the matrix C_i will be of less importance for the disturbance of diagonality of B_{i-1}. This means that the orthogonal transformation Q_i, will be almost equal to the unity matrix. New information will not have any influence on the present processing anymore. These problems can be solved by windowing or exponentially weighting the data signals.

1. **Windowing** : The most straightforward way to restrict the "knowledge" that the system builts up to the recent past is accomplished by windowing : while a new contribution updates the present knowledge, a contribution from the past is

removed simultaneously. If w is the width of the window, then the second step in the algorithm would change into

$$C_i = B_{i-1} + \mathbf{y}_i \mathbf{y}_i^t - U_{i-1}^t \mathbf{a}_{i-w} \mathbf{a}_{i-w}^t U_{i-1} \tag{11}$$

This equation implements the ideal moving window-function. In practice, however, this is not the most elegant solution. It requires the storage of w input columns, in order to eliminate the "oldest" contribution. Moreover, the width of the window has to be considerably big, because the a priori knowledge must be based on enough information.

2. **Exponential weighting :** Another solution is found by incorporating some "forgetting function" in an extremely simple way. In each pass of the loop in the presented algorithm, the singular values are reduced slightly by multiplying them by some factor α lying between 0 and 1, such that step two of the algorithm changes into

$$C_i = \alpha B_{i-1} + \mathbf{y}_i \mathbf{y}_i^t \tag{12}$$

This is equivalent to an exponential weighting of the input data. The closer α lies to 1, the slower the system forgets data from the past and the more influence data from the further past can still have. The smaller α, the faster some changes can be recovered by the method. However α should be large enough in order to limit sensitivity of the method to detailed short interval signal characteristics.

5 On-line method for GSVD-based signal separation

Recently a new signal separation method has been developed [3], [4], based on the computation of the Generalized SVD of two matrices $A\,(p \times q)$ and $B\,(p \times k)$, given by

$$\begin{aligned} A &= X^{-t} D_A Q_A^t \\ B &= X^{-t} D_B Q_B^t \end{aligned} \tag{13}$$

where Q_A and Q_B are matrices with p orthonormal columns, X is a non-singular $p \times p$ matrix and where $D_A = \text{diag}(\alpha_1, \ldots, \alpha_p)$ and $D_B = \text{diag}(\beta_1, \ldots, \beta_p)$, $(\alpha_i, \beta_i \geq 0)$, are square diagonal matrices. In fact the signal separation method only needs the computation of the X matrix in this GSVD. The following theorem shows that this can be obtained by two consecutive SVD's.

Theorem 2 : Four step method for GSVD computation

The computation of the X-matrix in the GSVD of a matrix pair (A, B) can be performed by the following four-step method

- *Step 1 : compute the SVD of B* $\quad B = U_B \Sigma_B V_B^t$
- *Step 2 : define a transformation $Q = \Sigma_B^{-1} U_B^t$ on the matrix A such that*

$$Q\,A = A^* = \Sigma_B^{-1} U_B^t A$$

- *Step 3 : compute the SVD of A^** $\quad A^* = (U_A^*)(\Sigma_A^*)(V_A^*)^t$
- *Step 4 : compute the X-matrix in the GSVD of (A, B) as*

$$X = Q^t U_A^* = U_B \Sigma_B^{-1} U_A^*$$

Proof : see [9] and [3].

Suppose matrix B is time-invariant, such that its SVD has to be computed only once, as a kind of initialisation, using a classical off-line method. If on the other hand, A is the data matrix, growing with time, an algorithm for the on-line computation of X can be outlined as follows :

Initialisation $X_0 = U_B \Sigma_B^{-1}$ $D_0 = O_{p,p}$

Loop for $i = 1$ to q

generate output $\mathbf{y}_i = X_{i-1}^t \mathbf{a}_i$

normalize the columns of X_i

compute C_i $C_i = \alpha D_{i-1} + \mathbf{y}_i \mathbf{y}_i^t$

update D_i **by diagonalizing** C_i $D_i = Q_i^t C_i Q_i$

update X_i $X_i = X_{i-1} Q_i$

end loop

6 On-line adaptive FECG-extraction

In [2] and [10] an SVD-based technique has been developed to eliminate the maternal electrocardiogram (MECG) from potential measurements taken from electrodes on the skin of a pregnant subject, such that the fetal electrocardiogram (FECG) is obtained.

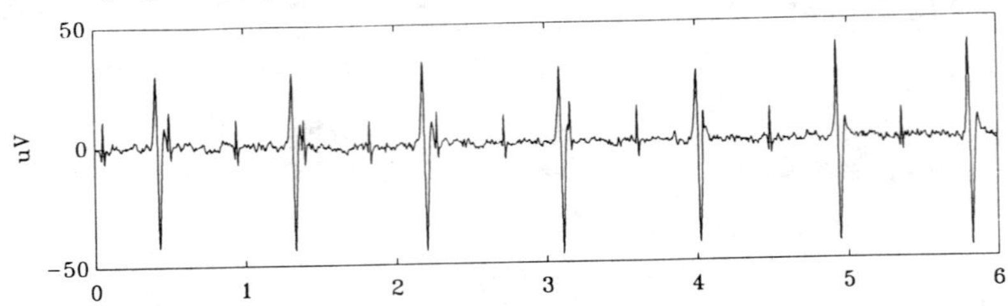

Figure 1: *A typical abdominal signal containing MECG and FECG*

Figure 1 shows a typical signal recorded on the abdomen of the mother and containing both the strong maternal and weak fetal ECG. Suppose six such signals are recorded simultaneously ($p = 6$) and the on-line adaptive algorithm (with $\alpha = 0.997$ and with only one Jacobi rotation per cycle for the diagonalization) is applied to the corresponding data matrix, then one of the output signals is as shown in Figure 2.

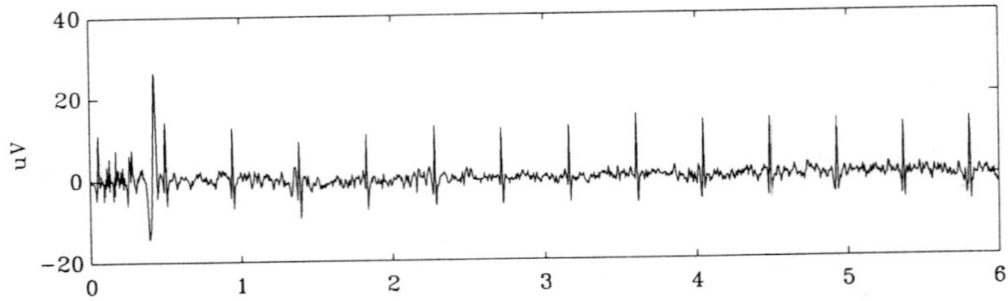

Figure 2: *One of the signals resulting after application of the on-line adaptive algorithm for the SVD-based signal separation method*

It is clear from this figure that maternal and fetal ECG are completely separated after only one maternal QRS-complex, which contains enough information for the signal separation method to discriminate between MECG and FECG.

If on the other hand the GSVD-based signal separation technique is applied, the signal shown in Figure 3 is obtained. For this method matrix B is constructed as a concatenation of a number of isolated maternal QRS-complexes, selected from the recorded signals, while matrix A is the original data matrix.

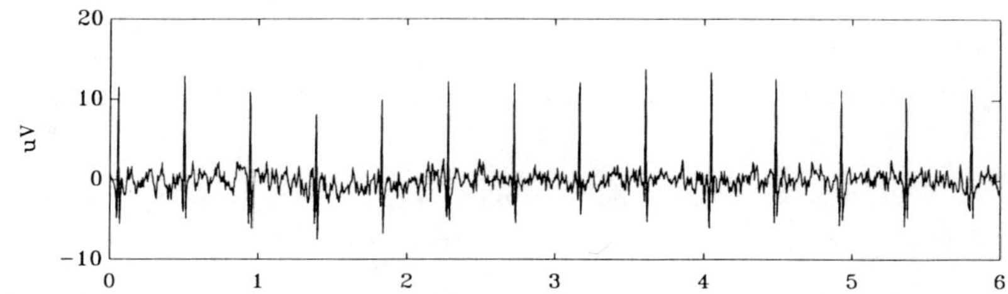

Figure 3: *One of the signals resulting after application of the on-line adaptive algorithm for the GSVD-based signal separation method*

The SVD of B is computed once as a kind of initialisation. From then on the X matrix is updated each time with the on-line algorithm outlined in section 5, this time with $\alpha = 0.9995$ and also with only one Jacobi rotation per cycle for the diagonalization. The forgetting factor α has to be closer to 1 in this case because otherwise the method forgets to quick the initializing SVD of B.

7 Conclusion

In this paper an on-line adaptive algorithm has been outlined for real-time SVD- and GSVD-based signal separation. The algorithm has proven to be very useful for the on-line extraction of the fetal ECG from signals recorded on the maternal skin. It is at the moment also used for principal component reduction in topographic mapping in EEG analysis.

References

[1] CALLAERTS, D., VANDERSCHOOT, J., VANDEWALLE, J., SANSEN, W., "An on-line adaptive algorithm for signal processing using SVD", in *Signal Processing III*, Amsterdam, The Netherlands : Elsevier Science Publ. (North Holland), EURASIP 86, The Hague, pp. 953-956, Sept. 1986.

[2] CALLAERTS, D., VANDEWALLE, J., SANSEN, W., MOONEN, M., " On-line Algorithm for Signal Separation based on SVD", in *SVD and Signal Processing : Algorithms, Applications and Architectures*, Deprettere, E., Editor, North-Holland, pp. 269-276, Sept. 1987.

[3] CALLAERTS, D., DE MOOR, B., VANDEWALLE, J., SANSEN, W., "Comparison of SVD-methods to extract the Fetal Electrocardiogram from cutaneous electrode recordings", Submitted to Medical & Biological Eng. & Comp., Jan. 1989.

[4] CALLAERTS, D., VANDEWALLE, J., SANSEN, W., "Comparison of signal separation methods based on SVD and GSVD", *Proc. of 12ieme Coll. sur le Traitement du Signal et des Images GRETSI*, Juan-les-Pins (Fr), pp. 177-180, June 1989.

[5] GOLUB, G.H., REINSCH, C., "Singular Value Decomposition and Least Squares Solutions", *Numer. Math.*, vol. 14, pp. 403-430, 1970.

[6] GOLUB, G.H., VAN LOAN, C.F., "Matrix Computations", North Oxford Academy, 1983.

[7] KLEMA, V.C., LAUB, A.J., "The singular value decompostion : its computation and some applications", *IEEE Trans. Automatic Control*, vol. AC-25, no. 2, pp. 164-167, 1980.

[8] OSBORNE, E.E., "On Least Squares Solutions of Linear Equations", *Journal ACM*, 8, pp. 628-636, (1961).

[9] PARLETT, B.N., "The Symmetric Eigenvalue Problem", *Prentice Hall*, Englewood Cliffs, N.J., 1980.

[10] VANDERSCHOOT, J., CALLAERTS, D., SANSEN, W., VANDEWALLE, J., VANTRAPPEN, G., JANSSENS, J., "Two Methods for Optimal MECG Elimination and FECG Detection from Skin Electrode Signals ", *IEEE Trans. Biomed. Eng.*, vol. BME-34, No. 3, pp. 233-243, March 1987.

[11] VANDEWALLE, J., CALLAERTS, D., (1988) "Singular Value Decomposition : a powerful concept and tool in signal processing", *Proc. of Conf. on Mathematics in Signal Processing*, Warwick, Dec. 1988.

QRD-BASED LATTICE-LADDER ALGORITHM FOR ADAPTIVE FILTERING

I K Proudler, J G McWhirter & T J Shepherd

Abstract

A new lattice-ladder algorithm for adaptive filtering is presented. In common with other lattice-ladder algorithms for adaptive filtering, this algorithm only requires $O(p)$ operations in the solution of a p-th order problem. The algorithm is, however, derived from the QR-decomposition (QRD) based recursive least squares minimisation algorithm and hence does not suffer from the numerical instabilities found in the some other algorithms. Some early computer simulation results are presented that indicate that the output residuals produced by the new, fast algorithm are in good agreement with those from the more established, $O(p^2)$ QRD recursive least squares minimisation algorithm.

1. Introduction

Least squares minimisation can be applied to a wide range of digital signal processing problems including that of adaptive filtering. The adaptive filtering problem is, however, special in the sense that the data matrix has a Toeplitz structure. Various algorithms [3] have been devised that take advantage of this redundancy to reduce the computational load, for a p-th order filter, from $O(p^2)$ to $O(p)$. Unfortunately these fast algorithms are not explicitly well conditioned and some tend to be numerically unstable.

It is also possible, however, to solve a least squares minimisation problem using the technique of QR decomposition. This technique has the advantage that all operations involved are orthogonal rotations and thus the resulting algorithm is well-conditioned and hence, for most practical purposes, numerical stable. The recursive QRD algorithm can be implemented on the triangular systolic array devised by Gentleman and Kung [2] and subsequently modified by McWhirter [6]. This algorithm solves the general least squares minimisation problem and will require $O(p^2)$ operations to generate the solution to a p-th order adaptive filter problem.

Recently Cioffi [1] showed how to take advantage of the time-shift redundancy that is present in the problem of the linear prediction of time series data and produced a QRD-based Fast Kalman algorithm (see Proudler, McWhirter & Shepherd [5] for a simpler derivation). This algorithm can recursively update the solution to a

Copyright © Controller HMSO London 1989

p-th order linear prediction problem in *O(p)* operations per time instant. In this paper we show how to derive a QRD-based lattice algorithm that solves the linear prediction problem, recursively in time and order, again in *O(p)* operations. The extension of this latter algorithm to the solution of the adaptive filtering problem is also given. The resulting algorithm has a lattice-ladder structure. Ling [4] appears to have derived a similar algorithm by exploiting the relationship between Givens' rotation based algorithms and the recursive modified Gram-Schmidt method. In common with Cioffi's original formulation of the QRD-based fast Kalman algorithm, the lattice-ladder algorithm presented here operates on pre-windowed data (i.e. all data before the first time instant is assumed to be zero).

The key to the fast lattice-ladder adaptive filtering algorithm is the development of a fast (lattice) algorithm for the solution of an associated linear prediction problem. The time series involved in this linear prediction problem is the "data" sequence that is to be filtered to match the "reference" sequence. In section 2 the connection between these two, related, problems is outlined and the solution to the adaptive filtering problem is derived from that of the linear prediction problem. Section 3 contains the mathematics of the transformation between successive orders of the linear prediction problem that is the basis of the fast linear prediction algorithm. Having derived the formal solution to the problem at a given time instant the time recursive update is developed, in section 4, which leads, finally, to the fast algorithm. The results of some preliminary computer simulations are also presented.

2. Adaptive Filtering via Linear Prediction

Consider a (p+1)st order adaptive filter with reference signal y(n) and data signal x(n), both assumed to be real sequences. The vector of filter coefficients, at time n, $\underline{\tilde{\omega}}(n) = |\tilde{\omega}_0(n), \ldots, \tilde{\omega}_p(n)|^t$ is determined by minimising the total adaptation error $\tilde{E}(n) = \|B(n)\ \underline{\tilde{e}}(n)\|$ where

$$(1) \qquad \underline{\tilde{e}}(n) = \tilde{X}_{p+1}(n)\,\underline{\tilde{\omega}}(n) + \underline{y}(n) \qquad \tilde{X}_{p+1}(n) = \begin{vmatrix} x(1) & \ldots & x(1-p) \\ \vdots & & \vdots \\ x(n) & \ldots & x(n-p) \end{vmatrix} \qquad \underline{y}(n) = |y(1), \ldots, y(n)|^t$$

The matrix $B(n) = \mathrm{DIAG}(\beta^{n-1}, \beta^{n-2}, \ldots, 1)$, where $0<\beta\leq 1$, represents an exponential forgetting factor that allows the algorithm to work with quasi-stationary signals. The solution of least squares minimisation problems via QR decomposition is now well established [2][6] and, in this case, requires the determination of an orthogonal matrix $\tilde{Q}_{p+1}(n)$ that transforms the matrix $B(n)\ \tilde{X}_{p+1}(n)$ into an upper triangular form. To complete the solution, various other quantities are also required to be transformed by multiplication by $\tilde{Q}_{p+1}(n)$. With $\underline{1} = |0, \ldots, 0, 1|^t$, let

$$(2) \qquad \tilde{Q}_{p+1}(n)\, B(n) \left| \tilde{X}_{p+1}(n) \;\; \underline{1} \;\; \underline{y}(n) \right| = \left| \begin{matrix} \tilde{R}_{p+1}(n) & \tilde{\underline{a}}_{p+1}(n) & \tilde{\underline{u}}_{p+1}(n) \\ \mathbf{0} & \tilde{\underline{g}}_{p+1}(n) & \tilde{\underline{v}}_{p+1}(n) \end{matrix} \right|$$

where $\tilde{\underline{g}}_{p+1}(n) = |0, \ldots ,0, \tilde{\gamma}_{p+1}(n)|^t$ and $\tilde{R}_{p+1}(n)$ is upper triangular (as indicated by the shading). It can be shown [6] that the adaptation residual can be obtained directly from the product of $\tilde{\gamma}_{p+1}(n)$ and $\tilde{\alpha}_{p+1}(n)$, the last element of the vector $\tilde{\underline{v}}_{p+1}(n)$. The utility of this lies in the fact that the a-posteriori residual $\tilde{e}(n,n) = \tilde{e}_n(n)$ $= y(n) + \sum_{i=0}^{p} \tilde{\omega}_i(n)\, x(n-i)$ (see equation (1)) and hence represents that part of the reference signal uncorrelated with x(n). This is the data that is most usually required in an adaptive filtering problem.

The matrix $\tilde{X}_{p+1}(n)$ depends entirely on the sequence x(n) and has a Toeplitz structure. This means that a certain amount of redundancy exists in the problem that could be exploited to reduce the computational requirements. The means by which this is achieved is via the problem of the linear prediction of the data sequence x(n). Define the p-th order (forward) linear prediction problem, at time n, as the estimation of x(n) based upon the data {x(n-1), ..., x(n-p)}. This involves the determination of the vector of filter coefficients $\underline{\omega}(n) = |\omega_0(n), \ldots, \omega_{p-1}(n)|^t$ that minimises the total prediction error $E(n) = \|B(n)\ \underline{e}(n)\|$ where

$$(3) \qquad \underline{e}(n) = X_p(n)\, \underline{\omega}(n) + \underline{y}_{f,p}(n) \qquad X_p(n) = \left| \begin{matrix} x(0) & \ldots x(1-p) \\ \vdots & \vdots \\ x(n-1) & \ldots x(n-p) \end{matrix} \right| \qquad \underline{y}_{f,p}(n) = |x(1), \ldots , x(n)|^t$$

Notice that $|\underline{y}_{f,p}(n)\ X_p(n)| = \tilde{X}_{p+1}(n)$ so that having converted the matrix $B(n)\ X_p(n)$ into a triangular form, for the solution of the linear prediction problem, then the matrix $B(n)\ \tilde{X}_{p+1}(n)$ itself has very nearly been triangularised. This fact, and a fast linear prediction algorithm, is the key to the development of an algorithm for solving the adaptive filter problem in *O(p)* operations.

Following [6], introduce an n×n orthogonal matrix $Q_p(n)$ that converts the matrix $B(n)\ X_p(n)$ in to an upper-triangular matrix. Let

$$(4) \qquad Q_p(n)\, B(n) \left| \underline{y}_{f,p}(n)\ X_p(n) \;\; \underline{1} \;\; \underline{y}(n) \right| = \left| \begin{matrix} \underline{u}_{f,p}(n) & R_p(n) & \underline{a}_p(n) & \underline{u}_{J,p}(n) \\ \underline{v}_{f,p}(n) & \mathbf{0} & \underline{g}_p(n) & \underline{v}_{J,p}(n) \end{matrix} \right|$$

where $g_p(n) = |0, \ldots ,0, \gamma_p(n)|^t$. As in the adaptive filter case, the prediction residual is the product of $\gamma_p(n)$ and $\alpha_{f,p}(n)$, the last element of the vector $\underline{v}_{f,p}(n)$. Bearing in mind the decomposition for $\tilde{X}_{p+1}(n)$ given above, we may now generate the solution to the adaptive filter problem, from that to the linear prediction problem, by applying two further rotations:

(5) $$Q^{v}_{J,p+1}(n)\begin{vmatrix} \underline{u}_{f,p}(n) & R_p(n) & \underline{a}_p(n) & \underline{u}_{J,p}(n) \\ \underline{v}_{f,p}(n) & \mathbf{0} & \underline{g}_p(n) & \underline{v}_{J,p}(n)\end{vmatrix} = \begin{vmatrix} \underline{u}_{f,p}(n) & R_p(n) & \underline{a}_p(n) & \underline{u}_{J,p}(n) \\ \lambda_{f,p}(n) & \underline{0}^t & \theta_p(n) & \varphi_{J,p}(n) \\ \underline{0} & \mathbf{0} & \underline{\tilde{g}}_{p+1}(n) & \underline{\tilde{v}}_{p+1}(n)\end{vmatrix}$$

(6) $$Q^{u}_{J,p+1}(n)\begin{vmatrix} \underline{u}_{f,p}(n) & R_p(n) & \underline{a}_p(n) & \underline{u}_{J,p}(n) \\ \lambda_{f,p}(n) & \underline{0}^t & \theta_p(n) & \varphi_{J,p}(n) \\ \underline{0} & \mathbf{0} & \underline{\tilde{g}}_{p+1}(n) & \underline{\tilde{v}}_{p+1}(n)\end{vmatrix} = \begin{vmatrix} \tilde{R}_{p+1}(n) & \underline{\tilde{a}}_{p+1}(n) & \underline{\tilde{u}}_{p+1}(n) \\ \mathbf{0} & \underline{\tilde{g}}_{p+1}(n) & \underline{\tilde{v}}_{p+1}(n)\end{vmatrix}$$

This completes the derivation of the formal connection between the solution to the (p+1)st order adaptive filtering problem and that to the p-th order linear prediction problem. Note, however, that if the quantity of interest is the adaptation residual then the step outlined in equation (6) is not required. This is because both $\tilde{\gamma}_{p+1}(n)$ and $\tilde{\alpha}_{p+1}(n)$ are contained within the matrix on the right hand side of equation (5).

3. Linear Prediction by Order Iteration

In order to be able to take advantage of the above connection (equations (5) and (6)) a fast linear prediction algorithm has to be available. The basis of the fast (lattice) algorithm for linear prediction is the fact that the data required for the solution of a p-th order problem is essentially the same as that used in solving a (p-1)st order problem. Thus by iterating in order, savings in computational requirements can be made. In the following calculation we carry through the extra variables that will be required for the ultimate goal of solving the adaptive filtering problem. Note that

(7) $$\begin{vmatrix} \underline{v}_{f,p}(n) & X_p(n) \end{vmatrix} = \begin{vmatrix} \underline{v}_{f,p-1}(n) & X_{p-1}(n) & \underline{v}_{b,p-1}(n-1) \end{vmatrix}$$

where $\underline{v}_{b,p-1}(n-1) = |x(1-p), \ldots , x(n-p)|^t$ can be considered to be the reference vector of an auxiliary linear prediction problem: the so-called backward linear prediction problem. In the q-th order problem, an estimate, at time n, of x(n-q) is formed from a linear combination of the data {x(n), ..., x(n-q+1)}. The solution to this problem depends on the triangularisation of the matrix $B(n+1)\,X_q(n+1)$. Hence let (cf. equation (4))

$$(8) \qquad Q_q(n+1)\,B(n+1)\left|X_q(n+1)\;\underline{y}_{b,q}(n)\;\underline{\iota}\right| = \left|\begin{matrix} R_q(n+1) & \underline{u}_{b,q}(n) & \underline{a}_q(n+1) \\ \mathbf{0} & \underline{v}_{b,q}(n) & \underline{g}_q(n+1)\end{matrix}\right|$$

From equations (4), (7) and (8) we have

$$(9) \qquad Q_{p-1}(n)\,B(n)\left|\underline{y}_{f,p}(n)\;X_p(n)\;\underline{\iota}\;\underline{y}(n)\right| =$$
$$\left|\begin{matrix} \underline{u}_{f,p-1}(n) & R_{p-1}(n) & \underline{u}_{b,p-1}(n-1) & \underline{a}_{p-1}(n) & \underline{u}_{J,p-1}(n) \\ \underline{v}_{f,p-1}(n) & \mathbf{0} & \underline{v}_{b,p-1}(n-1) & \underline{g}_{p-1}(n) & \underline{v}_{J,p-1}(n)\end{matrix}\right|$$

The key to the solution to the p-th order linear prediction problem will have been found if $X_p(n)$ has been triangularised (equation (4)). From equation (9), it can be seen that this has very nearly been achieved. Define an orthogonal matrix $Q_{f,p-1}(n)$ to be that which rotates the components of the vector $\underline{v}_{b,p-1}(n-1)$ into one consisting of a non-zero top element with zeros elsewhere:

$$(10) \qquad Q_{f,p-1}(n)\left|\begin{matrix} \underline{u}_{f,p-1}(n) & R_{p-1}(n) & \underline{u}_{b,p-1}(n-1) & \underline{a}_{p-1}(n) & \underline{u}_{J,p-1}(n) \\ \underline{v}_{f,p-1}(n) & \mathbf{0} & \underline{v}_{b,p-1}(n-1) & \underline{g}_{p-1}(n) & \underline{v}_{J,p-1}(n)\end{matrix}\right|$$
$$= \left|\begin{matrix} \underline{u}_{f,p-1}(n) & R_{p-1}(n) & \underline{u}_{b,p-1}(n-1) & \underline{a}_{p-1}(n) & \underline{u}_{J,p-1}(n) \\ \mu_{f,p-1}(n) & \underline{0}^t & \varepsilon_{b,p-1}(n-1) & \delta_{p-1}(n) & \mu_{J,p-1}(n) \\ \underline{v}_{f,p}(n) & \mathbf{0} & \underline{0} & \underline{g}_p(n) & \underline{v}_{J,p}(n)\end{matrix}\right|$$
$$= \left|\begin{matrix} \underline{u}_{f,p}(n) & R_p(n) & \underline{a}_p(n) & \underline{u}_{J,p}(n) \\ \underline{v}_{f,p}(n) & \mathbf{0} & \underline{g}_p(n) & \underline{v}_{J,p}(n)\end{matrix}\right|$$

where the last step follows by definition. Equation (10), ignoring the right hand columns, shows that it is possible to generate, entirely by orthogonal rotations, the solution to the p-th order linear prediction problem if we have to hand to solutions to the (p-1)st order linear prediction and auxiliary problems. In order to complete the order-recursive algorithm for the linear prediction problem, the order update equation for the auxiliary (backward) problem must be derived. This can be done along similar line to the above except that it proves necessary to factor the matrix analogous to $Q_{f,p}(n)$ into two parts (cf. equations (5) and (6)). For completeness, the mathematics are reproduced in the appendix but, due to lack of space, in a somewhat terse form (see [5] for more details).

4. *O(p)* Algorithm

As it stands equation (10) implies that it would take *O(n-p)* operations to generate the p-th order linear prediction solution from the (p-1)st ones. Indeed consideration of equation (5) shows that it would require a further $O(n\text{-}p)$ operations to convert the linear prediction solution into the adaptive filtering one. A fast algorithm can, however, be developed by utilising the time recursive properties of the problem. At time (n+1) we have to operate on the matrix (cf. equations (2) and (4))):

$$\text{(11)}\quad \left| \tilde{X}_{p+1}(n+1)\ \underline{1}\ \underline{y}(n+1) \right| = \left| \underline{y}_{f,p}(n+1)\ X_p(n+1)\ \underline{1}\ \underline{y}(n+1) \right|$$
$$= \left| \begin{array}{cccccc} \underline{y}_{f,p-1}(n) & & X_{p-1}(n) & \underline{y}_{b,p-1}(n-1) & \underline{0} & \underline{y}(n) \\ x(n+1) & x(n) & \ldots\ x(n-p+2) & x(n-p+1) & 1 & y(n+1) \end{array} \right|$$

in order to convert the sub-matrix $X_p(n+1)$ (respectively $\tilde{X}_{p+1}(n+1)$) into an upper triangular form so as to solve the linear prediction (adaptive filtering) problem. As a result of the partitioning indicated in equation (11) and the rotations $Q_{p-1}(n)$ and $Q_{f,p-1}(n)$ defined in equations (9) and (10) respectively we have that

$$\text{(12)}\quad \left| \begin{array}{cc|c} Q_{f,p-1}(n)Q_{p-1}(n) & \underline{0} & B(n+1) \\ \underline{0}^t & 1 & \end{array} \right| \left| \begin{array}{cccccc} \underline{y}_{f,p-1}(n) & & X_{p-1}(n) & \underline{y}_{b,p-1}(n-1) & \underline{0} & \underline{y}(n) \\ x(n+1) & x(n) & \ldots\ x(n-p+2) & x(n-p+1) & 1 & y(n+1) \end{array} \right|$$
$$= \left| \begin{array}{ccccc} \beta\underline{u}_{f,p-1}(n) & \beta R_{p-1}(n) & \beta\underline{u}_{b,p-1}(n-1) & \underline{0} & \beta\underline{u}_{J,p-1}(n) \\ \beta\mu_{f,p-1}(n) & \underline{0}^t & \beta\varepsilon_{b,p-1}(n-1) & \underline{0} & \beta\mu_{J,p-1}(n) \\ \beta\underline{v}_{f,p}(n) & \mathbf{0} & \underline{0} & \underline{0} & \beta\underline{v}_{J,p}(n) \\ x(n+1) & x(n)\ \ldots\ x(n-p+2) & x(n-p+1) & 1 & y(n+1) \end{array} \right|$$

Following [6] introduce the "time-update" rotation matrix $\hat{Q}_{p-1}(n+1)$ that annihilates the vector $|x(n), \ldots, x(n-p+2)|$ by rotating it against the matrix $\beta R_{p-1}(n)$:

$$\text{(13)}\quad \hat{Q}_{p-1}(n+1) \left| \begin{array}{ccccc} \beta\underline{u}_{f,p-1}(n) & \beta R_{p-1}(n) & \beta\underline{u}_{b,p-1}(n-1) & \underline{0} & \beta\underline{u}_{J,p-1}(n) \\ \beta\mu_{f,p-1}(n) & \underline{0}^t & \beta\varepsilon_{b,p-1}(n-1) & \underline{0} & \beta\mu_{J,p-1}(n) \\ \beta\underline{v}_{f,p}(n) & \mathbf{0} & \underline{0} & \underline{0} & \beta\underline{v}_{J,p}(n) \\ x(n+1) & x(n)\ \ldots\ x(n-p+2) & x(n-p+1) & 1 & y(n+1) \end{array} \right|$$
$$= \left| \begin{array}{ccccc} \underline{u}_{f,p-1}(n+1) & R_{p-1}(n+1) & \underline{u}_{b,p-1}(n) & \underline{a}_{p-1}(n+1) & \underline{u}_{J,p-1}(n+1) \\ \beta\mu_{f,p-1}(n) & \underline{0}^t & \beta\varepsilon_{b,p-1}(n-1) & \underline{0} & \beta\mu_{J,p-1}(n) \\ \beta\underline{v}_{f,p}(n) & \mathbf{0} & \underline{0} & \underline{0} & \beta\underline{v}_{J,p}(n) \\ \alpha_{f,p-1}(n+1) & \underline{0}^t & \alpha_{b,p-1}(n) & \gamma_{p-1}(n+1) & \alpha_{J,p-1}(n+1) \end{array} \right|$$

where $\alpha_{b,p-1}(n)$ is the normalised (p-1)st order backward prediction residual (cf. $\alpha_{f,p-1}(n)$). The time-update of the p-th order forward linear prediction problem is complete once the triangular matrix $R_p(n+1)$ has been formed. Hence define a rotation matrix $\hat{Q}_{f,p-1}(n+1)$ such that

$$(14)\quad \hat{Q}_{f,p-1}(n+1)\begin{vmatrix} \underline{u}_{f,p-1}(n+1) & R_{p-1}(n+1) & \underline{u}_{b,p-1}(n) & \underline{a}_{p-1}(n+1) & \underline{u}_{J,p-1}(n+1) \\ \beta\mu_{f,p-1}(n) & \underline{0}^t & \beta\varepsilon_{b,p-1}(n-1) & \underline{0} & \beta\mu_{J,p-1}(n) \\ \beta\underline{v}_{f,p}(n) & \mathbf{0} & \underline{0} & \underline{0} & \beta\underline{v}_{J,p}(n) \\ \alpha_{f,p-1}(n+1) & \underline{0}^t & \alpha_{b,p-1}(n) & \gamma_{p-1}(n+1) & \alpha_{J,p-1}(n+1) \end{vmatrix}$$

$$= \begin{vmatrix} \underline{u}_{f,p-1}(n+1) & R_{p-1}(n+1) & \underline{u}_{b,p-1}(n) & \underline{a}_{p-1}(n+1) & \underline{u}_{J,p-1}(n+1) \\ \mu_{f,p-1}(n+1) & \underline{0}^t & \varepsilon_{b,p-1}(n) & \delta_{p-1}(n+1) & \mu_{J,p-1}(n+1) \\ \beta\underline{v}_{f,p}(n) & \mathbf{0} & \underline{0} & \underline{0} & \beta\underline{v}_{J,p}(n) \\ \alpha_{f,p}(n+1) & \underline{0}^t & 0 & \gamma_p(n+1) & \alpha_{J,p}(n+1) \end{vmatrix}$$

$$= \begin{vmatrix} \underline{u}_{f,p}(n+1) & R_p(n+1) & \underline{a}_p(n+1) & \underline{u}_{J,p}(n+1) \\ \underline{v}_{f,p}(n+1) & \mathbf{0} & \underline{g}_p(n+1) & \underline{v}_{J,p}(n+1) \end{vmatrix}$$

This competes the solution to the linear prediction problem ($\alpha_{f,p}(n+1)$ and $\gamma_p(n+1)$ have been calculated). Note that all of the quantities in the matrix on the left hand side of equation (14) are available if the solution to the p-th order linear prediction problem at time n, the (p-1)st order linear prediction problem at time n+1 and the (p-1)st order auxiliary (backward) problem at time n are known. Thus only the rotations summarised in equation (14) need actually be performed explicitly. This requires a fixed amount of computation because only eight elements, one of which is zero, of the left hand matrix in equation (14) are affected by the required rotations. In order to complete the lattice algorithm for the linear prediction problem, the fast update for the auxiliary (backward) problem must be derived. This can be done along similar line to the above (see appendix).

The solution to the adaptive filtering problem could then be generated by multiplying the matrix on the right hand side of equation (14) by $Q^v_{J,p+1}(n+1)$ (see equation (5)). Note, however, that by virtue of equation (5), we have

(15)
$$\begin{vmatrix} Q^{v}_{J,p+1}(n) & \underline{0} \\ \underline{0}^{t} & 1 \end{vmatrix} \begin{vmatrix} \underline{u}_{f,p}(n+1) & R_p(n+1) & \underline{a}_p(n+1) & \underline{u}_{J,p}(n+1) \\ \beta \underline{v}_{f,p}(n) & \mathbf{0} & \underline{0} & \beta \underline{v}_{J,p}(n) \\ \alpha_{f,p}(n+1) & \underline{0}^{t} & \gamma_p(n+1) & \alpha_{J,p}(n+1) \end{vmatrix}$$

$$= \begin{vmatrix} \underline{u}_{f,p}(n+1) & R_p(n+1) & \underline{a}_p(n+1) & \underline{u}_{J,p}(n+1) \\ \beta\lambda_{f,p}(n) & \underline{0}^{t} & 0 & \beta\varphi_{J,p}(n) \\ \underline{0} & \mathbf{0} & \underline{0} & \beta\, \underline{\tilde{v}}_{p+1}(n) \\ \alpha_{f,p}(n+1) & \underline{0}^{t} & \gamma_p(n+1) & \alpha_{J,p}(n+1) \end{vmatrix}$$

Now define a rotation matrix $\hat{Q}^{v}_{J,p+1}(n+1)$ that annihilates the element $\alpha_{f,p}(n+1)$:

(16)
$$\hat{Q}^{v}_{J,p+1}(n+1) \begin{vmatrix} \underline{u}_{f,p}(n+1) & R_p(n+1) & \underline{a}_p(n+1) & \underline{u}_{J,p}(n+1) \\ \beta\lambda_{f,p}(n) & \underline{0}^{t} & 0 & \beta\varphi_{J,p}(n) \\ \underline{0} & \mathbf{0} & \underline{0} & \beta\, \underline{\tilde{v}}_{p+1}(n) \\ \alpha_{f,p}(n+1) & \underline{0}^{t} & \gamma_p(n+1) & \alpha_{J,p}(n+1) \end{vmatrix}$$

$$= \begin{vmatrix} \underline{u}_{f,p}(n+1) & R_p(n+1) & \underline{a}_p(n+1) & \underline{u}_{J,p}(n+1) \\ \lambda_{f,p}(n+1) & \underline{0}^{t} & \theta_p(n+1) & \varphi_{J,p}(n+1) \\ \underline{0} & \mathbf{0} & \underline{0} & \beta\, \underline{\tilde{v}}_{p+1}(n) \\ 0 & \underline{0}^{t} & \tilde{\gamma}_{p+1}(n+1) & \tilde{\alpha}_{p+1}(n+1) \end{vmatrix}$$

$$= \begin{vmatrix} \underline{u}_{f,p}(n+1) & R_p(n+1) & \underline{a}_p(n+1) & \underline{u}_{J,p}(n+1) \\ \lambda_{f,p}(n+1) & \underline{0}^{t} & \theta_p(n+1) & \varphi_{J,p}(n+1) \\ \underline{0} & \mathbf{0} & \underline{\tilde{g}}_{p+1}(n+1) & \underline{\tilde{v}}_{p+1}(n+1) \end{vmatrix}$$

To complete the triangularisation of $\tilde{X}_{p+1}(n+1)$ (see equation (11)) and hence the solution of the (p+1)st order adaptive filter problem at time (n+1) all that would be required is the application of the rotation matrix $Q^{u}_{J,p+1}(n+1)$ (see equation (6)). However it can be seen from equation (16) that all quantities of interest (viz. $\tilde{\alpha}_{p+1}(n+1)$ and $\tilde{\gamma}_{p+1}(n+1)$) have already been generated. All of the quantities on the left hand side of equation (16) are available if the adaptive filter problem has been solved at time n and the linear prediction problem at time (n+1). Hence only the rotations defined by $\hat{Q}^{v}_{J,p+1}(n+1)$ need be implemented and indeed only six elements, one of which is zero, of the matrix in equation (16) are actually changed by these rotations. Thus the algorithm requires only a fixed number of rotations per stage and hence is *O(p)* for a p-th order problem. A "pseudo-code" listing of the algorithm is given in the appendix.

This algorithm has been simulated on a computer and compared with the full $O(p^2)$ QRD triangular array algorithm. Initial results show that the two algorithms produce virtually identical results. Both algorithms were used, independently, to simulate a 6-th order filter attempting to remove sinusoidal interference from a square wave signal. The adaptation error sequences produced by the two algorithms, using 32-bit floating-point arithmetic, were identical to within at least five decimal places, for sequences several hundred time samples long.

5. Conclusion

We have presented a fast lattice-ladder algorithm for solving the adaptive filter problem (with pre-windowed data). The algorithm has as its root the QRD recursive least squares minimisation algorithm and therefore is expected to be numerically stable. Preliminary computer simulations would seem to confirm this expectation. More demanding, and realistic, simulations would never-the-less be highly desirable.

6. References

[1] Cioffi, JM, Fast QR/Frequency Domain RLS Adaptive Filters, I.E.E.E. trans. A.S.S.P. - to be published (see Cioffi, JM, High Speed Systolic Implementation of Fast QR Adaptive Filters, Proc. I.E.E.E. International Conference on A.S.S.P., vol. DIII, 1584-1587, New York, 11-14th April 1988).

[2] Gentleman, WM; Kung, HT, Matrix Triangularisation by Systolic Array, Proc. S.P.I.E., vol. 298, Real Time Signal Processing IV, 1981.

[3] Haykin, S, Adaptive Filter Theory, Prentice-Hall, Englewood Cliffs, New Jersey, USA, 1981.

[4] Ling, F, Efficient Least-Squares Lattice Algorithms based on Givens Rotation with Systolic Array Implementations, proc. I.E.E.E. International Conference on A.S.S.P., Glasgow, UK, May 1989, paper no. 40.D9.4.

[5] Proudler, IK; McWhirter, JG; Shepherd, TJ, Fast QRD-based Algorithms for Least Squares Linear Prediction, Proc. I.M.A. Conference on Mathematics in Signal Processing, Warwick, England, 13-15th December 1988.

[6] McWhirter, JG, Recursive Least Squares Minimisation using a Systolic Array, Proc. S.P.I.E., vol. 431, Real Time Signal Processing IV, 105-112, 1983.

7. Appendix

7.1. Algorithm

This algorithm, written in pseudo-ALGOL, calculates the adaptive filter residual for a (p+1)st order system fed with a pre-windowed data sequence x(n) and a reference sequence y(n).

START

INITIALISE {all variables} := 0;

FOR n FROM 1 DO

LET $\alpha_{f,0}(n) := x(n)$; $\alpha_{b,0}(n) := x(n-1)$; $\alpha_{J,0}(n) := y(n)$; $\gamma_0(n) := 1$;

FOR q FROM 1 TO p DO

LET $\varepsilon_{b,q-1}(n-1) := \sqrt{(\beta\varepsilon_{b,q-1}(n-2))^2 + (\alpha_{b,q-1}(n-1))^2}$;

IF $\varepsilon_{b,q-1}(n-1) = 0$ THEN LET $\cos\Theta_f := 1$; $\sin\Theta_f := 0$

ELSE LET $\cos\Theta_f := \beta\varepsilon_{b,q-1}(n-2) / \varepsilon_{b,q-1}(n-1)$;

$\sin\Theta_f := \alpha_{b,q-1}(n-1) / \varepsilon_{b,q-1}(n-1)$ END_IF;

LET $\mu_{f,q-1}(n) := \cos\Theta_f\, \beta\mu_{f,q-1}(n-1) + \sin\Theta_f\, \alpha_{f,q-1}(n)$;

$\alpha_{f,q}(n) := \cos\Theta_f\, \alpha_{f,q-1}(n) - \sin\Theta_f\, \beta\mu_{f,q-1}(n-1)$;

$\mu_{J,q-1}(n) := \cos\Theta_f\, \beta\mu_{J,q-1}(n-1) + \sin\Theta_f\, \alpha_{J,q-1}(n)$;

$\alpha_{J,q}(n) := \cos\Theta_f\, \alpha_{J,q-1}(n) - \sin\Theta_f\, \beta\mu_{J,q-1}(n-1)$; $\gamma_q(n) := \cos\Theta_f\, \gamma_{q-1}(n)$;

COMMENT prediction residual = $\gamma_q(n)\, \alpha_{f,q}(n)$ COMMENT

LET $\varepsilon_{f,q-1}(n) := \sqrt{(\beta\varepsilon_{f,q-1}(n-1))^2 + (\alpha_{f,q-1}(n))^2}$;

IF $\varepsilon_{f,q-1}(n) = 0$ THEN LET $\cos\Theta_b := 1$; $\sin\Theta_b := 0$

ELSE LET $\cos\Theta_b := \beta\varepsilon_{f,q-1}(n-1) / \varepsilon_{f,q-1}(n)$; $\sin\Theta_b := \alpha_{f,q-1}(n) / \varepsilon_{f,q-1}(n)$

END_IF;

LET $\mu_{b,q-1}(n-1) := \cos\Theta_b\, \beta\mu_{b,q-1}(n-2) + \sin\Theta_b\, \alpha_{b,q-1}(n-1)$;

$\alpha_{b,q}(n) := \cos\Theta_b\, \alpha_{b,q-1}(n-1) - \sin\Theta_b\, \beta\mu_{b,q-1}(n-2)$;

$\gamma_q(n+1) := \cos\Theta_b\, \gamma_{q-1}(n)$;

COMMENT backward prediction residual = $\gamma_q(n+1)\, \alpha_{b,q}(n)$ COMMENT

LET $\lambda_{f,q}(n) := \sqrt{(\beta\lambda_{f,q}(n-1))^2 + (\alpha_{f,q}(n))^2}$;

IF $\lambda_{f,q}(n) = 0$ THEN LET $\cos\Theta_J := 1$; $\sin\Theta_J := 0$

ELSE LET $\cos\Theta_J := \beta\lambda_{f,q}(n-1) / \lambda_{f,q}(n)$; $\sin\Theta_J := \alpha_{f,q}(n) / \lambda_{f,q}(n)$

END_IF;

LET $\varphi_{J,q}(n) := \cos\Theta_J\, \beta\varphi_{J,q}(n-1) + \sin\Theta_J\, \alpha_{J,q}(n)$;

$\tilde{\alpha}_{q+1}(n) := \cos\Theta_J\, \alpha_{J,q}(n) - \sin\Theta_J\, \beta\varphi_{J,q}(n-1);\ \tilde{\gamma}_{q+1}(n) := \cos\Theta_J\, \gamma_q(n)$

END_DO

filtered_residual $:= \tilde{\gamma}_{p+1}(n)\tilde{\alpha}_{p+1}(n)$

END_DO

FINISH

7.2. Auxiliary Linear Prediction Problem

Consider data required for the solution of the p-th order auxiliary problem (equation (8)). It is possible to identify within this matrix the data required for the (p-1)st order problems (cf. equation (7)). If the data sequence x(n) is pre-windowed (i.e. $x(n) = 0$ for $n \leq 0$) then from equation (8) we have

$$(17)\quad \begin{vmatrix} 1 & \underline{0}^t \\ \underline{0} & Q_{p-1}(n) \end{vmatrix} B(n+1) \begin{vmatrix} X_p(n+1) & \underline{y}_{b,p}(n) & \underline{1} \end{vmatrix}$$
$$= \begin{vmatrix} 0 & \underline{0}^t & 0 & 0 \\ \underline{u}_{f,p-1}(n) & R_{p-1}(n) & \underline{u}_{b,p-1}(n-1) & \underline{a}_{p-1}(n) \\ \underline{v}_{f,p-1}(n) & \mathbf{0} & \underline{v}_{b,p-1}(n-1) & \underline{g}_{p-1}(n) \end{vmatrix}$$

Define an orthogonal matrix $Q^{v}_{b,p-1}(n)$ as follows:

$$(18)\quad Q^{v}_{b,p-1}(n) \begin{vmatrix} 0 & \underline{0}^t & 0 & 0 \\ \underline{u}_{f,p-1}(n) & R_{p-1}(n) & \underline{u}_{b,p-1}(n-1) & \underline{a}_{p-1}(n) \\ \underline{v}_{f,p-1}(n) & \mathbf{0} & \underline{v}_{b,p-1}(n-1) & \underline{g}_{p-1}(n) \end{vmatrix}$$
$$= \begin{vmatrix} \varepsilon_{f,p-1}(n) & \underline{0}^t & \mu_{b,p-1}(n-1) & \eta_{p-1}(n) \\ \underline{u}_{f,p-1}(n) & R_{p-1}(n) & \underline{u}_{b,p-1}(n-1) & \underline{a}_{p-1}(n) \\ \underline{0} & \mathbf{0} & \underline{v}_{b,p}(n) & \underline{g}_{p}(n+1) \end{vmatrix}$$

Equation (18) shows that it is possible to generate, recursively in order, all quantities related to the auxiliary problem that are required in equation (14). The fast time update algorithm is developed as follows: using a decomposition of the matrix $\begin{vmatrix} X_p(n+2) & \underline{y}_{b,p}(n+1) & \underline{1} \end{vmatrix}$ similar to that in equation (11) and by virtue of equations (17) and (18) and the definition of $\hat{Q}_{p-1}(n+1)$ (cf. equation (13)) we have

$$(19)\quad \begin{vmatrix} 1 & \underline{0}^t \\ \underline{0} & \hat{Q}_{p-1}(n+1) \end{vmatrix} \begin{vmatrix} Q^v_{b,p-1}(n) & \underline{0} \\ \underline{0}^t & 1 \end{vmatrix} \begin{vmatrix} 1 & \underline{0}^t & 0 \\ \underline{0} & Q_{p-1}(n) & \underline{0} \\ 0 & \underline{0}^t & 1 \end{vmatrix} B(n+2)$$

$$\begin{vmatrix} X_p(n+2) & \underline{y}_{b,p}(n+1) & \underline{1} \end{vmatrix} = \begin{vmatrix} \beta\epsilon_{f,p-1}(n) & \underline{0}^t & \beta\mu_{b,p-1}(n-1) & 0 \\ \underline{u}_{f,p-1}(n+1) & R_{p-1}(n+1) & \underline{u}_{b,p-1}(n) & \underline{a}_{p-1}(n+1) \\ \underline{0} & \mathbf{0} & \beta\underline{v}_{b,p}(n) & \underline{0} \\ \alpha_{f,p-1}(n+1) & \underline{0}^t & \alpha_{b,p-1}(n) & \gamma_{p-1}(n+1) \end{vmatrix}$$

Finally define a rotation matrix $\hat{Q}^v_{b,p-1}(n+1)$ such that

$$(20)\quad \hat{Q}^v_{b,p-1}(n+1) \begin{vmatrix} \beta\epsilon_{f,p-1}(n) & \underline{0}^t & \beta\mu_{b,p-1}(n-1) & 0 \\ \underline{u}_{f,p-1}(n+1) & R_{p-1}(n+1) & \underline{u}_{b,p-1}(n) & \underline{a}_{p-1}(n+1) \\ \underline{0} & \mathbf{0} & \beta\underline{v}_{b,p}(n) & \underline{0} \\ \alpha_{f,p-1}(n+1) & \underline{0}^t & \alpha_{b,p-1}(n) & \gamma_{p-1}(n+1) \end{vmatrix}$$

$$= \begin{vmatrix} \epsilon_{f,p-1}(n+1) & \underline{0}^t & \mu_{b,p-1}(n) & \eta_{p-1}(n+1) \\ \underline{u}_{f,p-1}(n+1) & R_{p-1}(n+1) & \underline{u}_{b,p-1}(n) & \underline{a}_{p-1}(n+1) \\ \underline{0} & \mathbf{0} & \beta\underline{v}_{b,p}(n) & \underline{0} \\ 0 & \underline{0}^t & \alpha_{b,p}(n+1) & \gamma_p(n+2) \end{vmatrix}$$

This completes the algorithm as all the quantities of interest, as regards direct residual extraction, have now been found. Again note that all of the quantities in the matrix on the left hand side of equation (20) are available at the correct moment. Thus only the rotations summarised in equation (20) need actually be performed explicitly.

Computationally Efficient, Numerically Stable Adaptive IIR Filters

Phillip A. Regalia

Abstract

The normalized recursive lattice filter is well suited for adaptive IIR filtering since the stability properties of the structure can be ensured theoretically in time varying environments. Previously proposed lattice algorithms, however, have met with prohibitive gradient computational complexity of $O(N^2)$, where N is the filter order. To overcome this computational burden, two new adaptive algorithms are proposed for the lattice filter, corresponding to the Steiglitz-McBride method and a maximum likelihood approach. The algorithms give consistent parameter estimates and are computationally competitive with existing schemes, but have the distinct advantage of remaining stable during the adaptation process. This stability property implies more robust convergence behavior.

I Introduction

The use of recursive filters in adaptive applications has attracted the interest of many researchers due to the superior system modelling abilities of IIR filters compared to their FIR counterparts, particularly in applications where the physical system to be modelled is not well approximated by a finite impulse response description. Adaptive IIR filters however introduce some intrinsic difficulties not encountered in the FIR schemes, such as the possibility of local minima in the error surface, and potentially unstable behavior if an eigenvalue (i.e., "pole") of the feedback portion migrates outside the unit circle.

The problem of local minima can be averted by using adaptive schemes distinct from the output error method, such as the Steiglitz-McBride method [14] or the hyperstable model reference approach [5, 7]. The problem of stability, on the other hand, depends heavily on the filter structure used in the adaptive scheme. One of the better studied structures in adaptive IIR filtering is the direct form, in which the input sequence $\{u(n)\}$ and the output sequence $\{y(n)\}$ of the filter are related as

$$\sum_{k=0}^{N} a_k(n)\, y(n-k) = \sum_{k=0}^{N} h_k(n)\, u(n-k), \qquad a_0 = 1. \tag{1}$$

Unfortunately, this does not represent an efficient parameterization, as the structure remains quite susceptible to instability problems in time varying environments. In practice, the update algorithm must be complemented with an eigenvalue test applied to the instantaneous characteristic equation to guard against instability. Even if the eigenvalues of the feedback portion remain strictly contained within the unit circle, the structure of (1) exhibits numerical problems in the presence of finite precision arithmetic, such as high susceptibility to quantization limit cycles, a high accumulation of roundoff noise, and generally poor internal signal scaling. The severity of these problems varies with the eigenvalue locations, which in the adaptive filtering context are assumed unknown. We should point out that even a hyperstable adaptive filter can suffer from certain numerical instability problems, if constructed around a direct form filter [6].

These problems have motivated researchers to study adaptive IIR filters composed of the parallel or cascade connection of lower order sections [4, 9]. An intrinsic problem encountered with this strategy is that a given transfer function no longer retains a unique mapping into the coefficient space of the structure, since the coefficients of one section can be interchanged with those of another without altering the overall transfer function. Simulations indicate that this lack of uniqueness tends to retard the convergence of the filter by introducing unstable manifolds into the error surface [9, 12].

A superior strategy from the stability point of view is to develop an adaptive algorithm for a time varying structure whose stability properties can be ensured theoretically. The present paper will consider the normalized lattice filter, as sketched in Fig. 1. We shall show in Section II.A that provided

$$\cos \phi_k(n) \neq 0, \qquad \text{for all } n, k, \tag{2}$$

the time varying filter is exponentially stable. The stability condition (2) is trivial to check in the lattice structure. It is also interesting to note that (2) is equivalent to the eigenvalues of the filter lying strictly inside the unit circle at all sample instants, and hence (2) provides a physically meaningful constraint.

Previous attempts at applying IIR lattice structures to adaptive scenarios have resulted in a gradient complexity of $O(N^2)$ computations per sample instant [3, 10], where N is the filter order. A simplified gradient scheme was subsequently proposed in [1], but one can show that the gradient vector so obtained at any stationary point is not persistently exciting of sufficient order. As such the concept of a stable stationary point does not apply, leading to parameter drift, although this explanation was not recognized in [1].

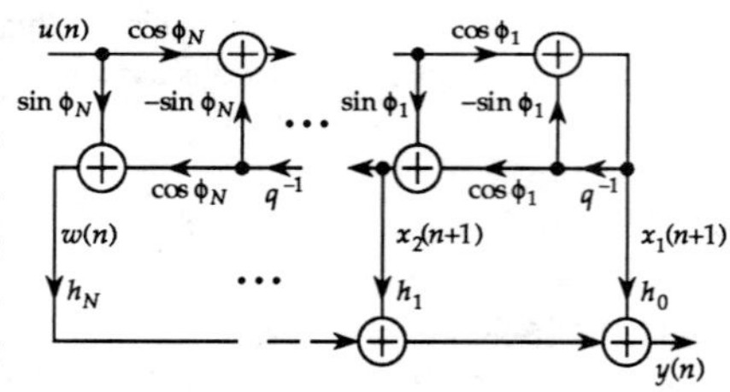

FIGURE 1: The normalized lattice filter.

The present paper derives two new lattice based algorithms for adaptive IIR filtering and system identification. Both algorithms possess a gradient complexity of $O(N)$, consisting of two normalized lattice filters to compute all the gradient and error signals necessary. The first algorithm is a reinterpretation of the Steiglitz-McBride identification method, while the second is a variation on the output error method. The set of possible stationary points is shown to be consistent with those of the direct form Steiglitz-McBride and output error methods, respectively, whose properties are well studied [13, 15]. These algorithms thus represent for the first time a class of computationally efficient, numerically stable adaptive IIR filters.

Notation

Matrices are set in upper case bold font, while vectors appear as lower case bold font. Superscript "t" denotes matrix transposition. The identity matrix is denoted by **I**, with a subscript if necessary to indicate its dimension.

II The Steiglitz-McBride Method Revisited

The Steiglitz-McBride identification scheme is sketched in Fig. 2, where $u(n)$ is the observed input to the plant, and $g(n)$ is the observed output, including the measurement noise $v(n)$ which is assumed uncorrelated with the input $u(n)$. At a heuristic level, an initial estimate $D_{n-1}(q)$

of the denominator of the plant is used as a prefilter for both the input and output sequences relating the plant transfer function. From the prefiltered signals, the appropriate polynomials $N_n(q)$ and $D_n(q)$ are found to minimize some measure of the error sequence. The prefilters are then updated to $1/D_n(q)$, and the procedure seeks $D_{n+1}(q)$ and $N_{n+1}(q)$. A stationary point corresponds to $D_{n+1}(q) = D_n(q)$, indicating no further improvement in the choice of denominator polynomial; the filter structure at convergence then appears as Fig. 3.

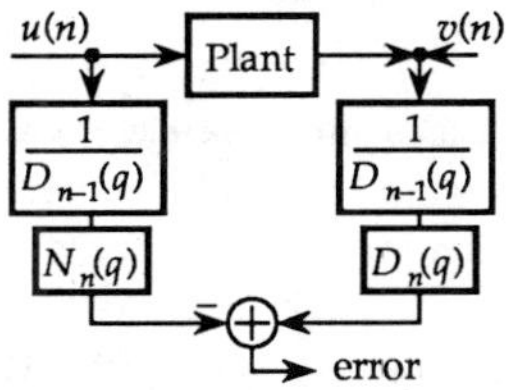

FIGURE 2: Schematic of Steiglitz-McBride identification technique

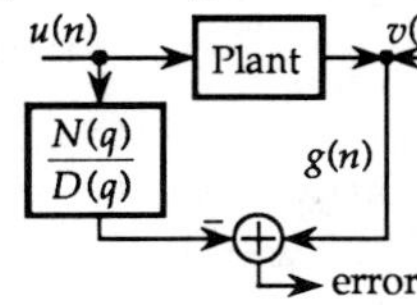

FIGURE 3: Equivalent system of Fig. 2 at any stationary point.

A. Choice of Model Structure

To develop a structural interpretation, let us introduce the (modified) state space description

$$y(n) = \mathbf{h}^t(n)\begin{bmatrix}\mathbf{x}(n+1)\\ w(n)\end{bmatrix}, \qquad \begin{bmatrix}\mathbf{x}(n+1)\\ w(n)\end{bmatrix} = \mathbf{Q}(n)\begin{bmatrix}\mathbf{x}(n)\\ u(n)\end{bmatrix}, \tag{3}$$

where $\mathbf{h}^t = [h_0, \ldots, h_N]$ is a tap vector, $\mathbf{x} = [x_1, \ldots, x_N]^t$ is the state vector, $w(n)$ is an intermediate signal, and $\mathbf{Q}$ is an $(N+1) \times (N+1)$ matrix which varies parametrically with N parameters, such that it is generically invertible. For example, the choice

$$\mathbf{Q} = \begin{bmatrix}-\mathbf{a}^t & 1\\ \mathbf{I}_N & \mathbf{0}\end{bmatrix}, \qquad \mathbf{a}^t = [a_1 \ \cdots \ a_N] \tag{4}$$

results in a direct form filter, where the coefficients $\{a_k\}$ and $\{h_k\}$ are the direct coefficients of the difference equation in (1). In this case, one has

$$\mathbf{Q}^{-1} = \begin{bmatrix}\mathbf{0} & \mathbf{I}_N\\ 1 & \mathbf{a}^t\end{bmatrix}. \tag{5}$$

If instead $\mathbf{Q}$ is chosen as

$$\mathbf{Q} = \mathbf{U}_1\,\mathbf{U}_2 \cdots \mathbf{U}_N, \qquad \mathbf{U}_k = \begin{bmatrix}\mathbf{I}_{k-1} & & & \\ & -\sin\phi_k & \cos\phi_k & \\ & \cos\phi_k & \sin\phi_k & \\ & & & \mathbf{I}_{N-k}\end{bmatrix}, \tag{6}$$

then (3) describes the normalized lattice filter of Fig. 1. Since $\mathbf{Q}$ is orthogonal, one easily obtains $\mathbf{Q}^{-1} = \mathbf{Q}^t$. Note that the Givens rotations of (6) can be implemented using CORDIC techniques [16, 17], in which case the angle ϕ_k of each planar rotator is controlled by a single register. A tunable realization, using $\{\phi_k\}_{k=1}^N$ as the adjustable parameter set, then becomes quite practicable.

The orthogonal nature of $\mathbf{Q}$ leads to very robust numerical behavior of the structure in time varying environments. Specifically, if we set $u(n) = 0$ for all n to obtain a homogeneous recursive system, then the orthogonality of $\mathbf{Q}(n)$ gives

$$\mathbf{x}^t(n+1)\,\mathbf{x}(n+1) - \mathbf{x}^t(n)\,\mathbf{x}(n) = -w^2(n), \qquad \text{for all } n, \tag{7}$$

i.e., the state energy of the homogeneous system decreases according to the energy delivered to $w(n)$, which suggests a candidate Lyapunov function of the form $V[\mathbf{x}(n)] = \mathbf{x}^t(n)\,\mathbf{x}(n)$. Partition now $\mathbf{Q}(n)$ as

$$\mathbf{Q}(n) = \begin{matrix} & N & 1 \\ N & \mathbf{A}(n) & \mathbf{b}(n) \\ 1 & \mathbf{c}^t(n) & d(n)\end{matrix}. \tag{8}$$

According to the results of [2], if the system is uniformly observable from $w(n)$, meaning that there exists a positive integer M such that

$$\sum_{n=m}^{m+M} \Phi^t(n,m)\,\mathbf{c}(m)\,\mathbf{c}^t(m)\,\Phi(n,m) > \mathbf{0}, \qquad \text{for all } n, \qquad \Phi(m,n) = \mathbf{A}(m-1)\cdots\mathbf{A}(n), \tag{9}$$

then

$$V[\mathbf{x}(n+M+1)] - V[\mathbf{x}(n)] < 0, \qquad \text{for all } n, \tag{10}$$

and the homogeneous portion of the time varying system is exponentially stable. Now, the matrix $\mathbf{Q}(n)$ of (6) turns out to be an upper Hessenburg matrix with the subdiagonal elements given by $\cos\phi_k(n)$. For a Hessenburg system description, one can readily show that (9) holds with $M = N$ (the state vector dimension) if and only if the subdiagonal elements of the Hessenburg matrix are nonzero for all time. Since these subdiagonal elements are $\cos\phi_k(n)$, the condition (2) implies $V[\mathbf{x}(n+N+1)] - V[\mathbf{x}(n)] < 0$ for all n, i.e., the N-step state transition matrix is contractive. This gives the desired exponential stability of the homogeneous system, and also ensures that roundoff errors in the state vector computation are not unduly amplified by the feedback loop of the filter.

B. A Generalized Interpretation

Using the model structure (3) and (4) for the direct form filters of Fig. 2, we arrive at the structure of Fig. 4, where for now the switch is set to position "1". The matrix $\mathbf{Q}^{-1}(n)$ need not be implemented in full, as only one output from this matrix is required. The gradient signals with respect to the $\mathbf{h}(n)$ vector are obtained from the state signals of the left side prefilter, while the gradients with respect to the parameters of $\mathbf{Q}^{-1}(n)$ will in general be some function of the state signals of the right side prefilter, the exact form depending on the structure of $\mathbf{Q}^{-1}(n)$. The updated parameters are then used in $\mathbf{Q}(n)$ in the prefilters for the next sample instant.

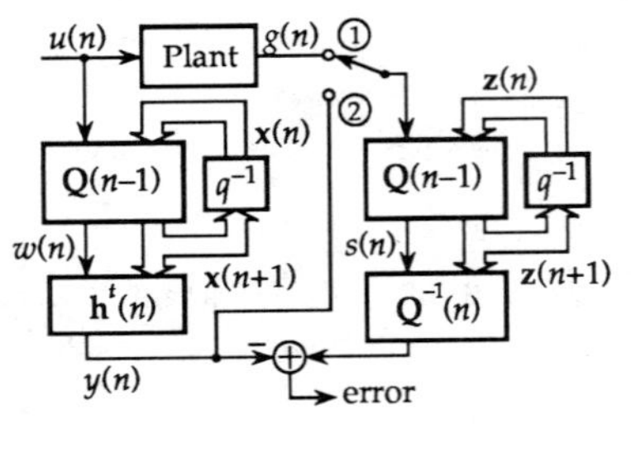

FIGURE 4: Structural interpretation of the adaptive algorithms.

If the previous estimates $\mathbf{h}^t(n-1)$ and $\mathbf{Q}^{-1}(n-1)$ are tentatively used in place of the (as yet undetermined) estimates $\mathbf{c}^t(n)$ and $\mathbf{Q}^{-1}(n)$, we can define an a priori error $\alpha(n)$ in the direct form version as

$$\alpha(n) = [1 \quad \mathbf{a}^t(n-1)] \begin{bmatrix} \mathbf{z}(n+1) \\ s(n) \end{bmatrix} - \mathbf{h}^t(n-1) \begin{bmatrix} \mathbf{x}(n+1) \\ w(n) \end{bmatrix} \tag{11}$$

where $\mathbf{x} = [x_1, \ldots, x_N]^t$ is the state vector of the left side prefilter in Fig. 4, while $\mathbf{z} = [z_1, \ldots, z_N]^t$ is the state vector of the right side prefilter. Note that $[1 \quad \mathbf{a}^t(n-1)]$ is simply the bottom row of $\mathbf{Q}^{-1}(n-1)$ according to (5), so that the first term on the right-hand side of (11) is simply the plant output $g(n)$. Likewise, the second term is the output from the adaptive filter using the most recently available parameter estimates. A set of gradient signals can now be obtained as

$$-\frac{\partial\alpha(n)}{\partial h_k} = \begin{cases} x_{k+1}(n+1), & i = 0, \ldots, N-1; \\ w(n), & i = N. \end{cases} \tag{12a}$$

$$-\frac{\partial\alpha(n)}{\partial a_k} = \begin{cases} -z_{k+1}(n+1), & i = 1, \ldots, N-1; \\ -s(n), & i = N. \end{cases} \tag{12b}$$

One can verify that these gradient signals agree with those presented in [15].

The realization of Fig. 4 suggest an immediate generalization of the Steiglitz-McBride technique to structures other than the direct form by substituting the appropriate $\mathbf{Q}(n)$ matrix and $\mathbf{h}(n)$ vector. Using the model (6) for the normalized lattice filter, we obtain for the a priori error

$$\alpha(n) = [0 \ \cdots \ 0 \quad 1]\, \mathbf{Q}^t(n-1) \begin{bmatrix} \mathbf{z}(n+1) \\ s(n) \end{bmatrix} - \mathbf{h}^t(n-1) \begin{bmatrix} \mathbf{x}(n+1) \\ w(n) \end{bmatrix} \tag{13}$$

where the states $\mathbf{x}(n+1)$ and $\mathbf{z}(n+1)$ are now computed using two normalized lattice structures. As above, the first term on the right-hand side of (13) is the plant output, and the second term is the output from the adaptive filter using the most recent parameter estimates. Although the gradients with respect to the rotation parameters $\{\phi_k\}$ in the matrix $\mathbf{Q}$ [cf. (6)] may appear complicated due to the nonlinear mapping from these rotation parameters to the error $\alpha(n)$, the gradient signals can be shown using (6) and some arithmetic to be

$$-\frac{\partial \alpha(n)}{\partial h_k} = \begin{cases} x_{k+1}(n+1), & k = 0, 1, \ldots, N-1; \\ w(n) & k = N. \end{cases} \tag{14a}$$

$$-\frac{\partial \alpha(n)}{\partial \phi_k} = -z_k(n) \prod_{i=k+1}^{N} \cos \phi_i(n), \qquad k = 1, 2, \ldots, N. \tag{14b}$$

The filter implementation is seen to consist of two normalized lattice structures: one to compute the a priori error $\alpha(n)$ in (13) and the gradient signals (14*a*) with respect to the elements of $\mathbf{h}$, the other to compute to gradients signals (14*b*) with respect to the rotation parameters. The computational load is essentially the same as that found in the direct form implementation, with the added benefit of providing the numerical and time varying stability not found in the direct form version.

III A Modified Output Error Method

A simple variation results if the second lattice filter is connected to the adaptive filter output rather than the plant output, as shown by changing the switch to position "2" in Fig. 4. The a priori error then becomes

$$\begin{aligned} \alpha(n) &= g(n) - [0 \ \cdots \ 0 \quad 1]\, \mathbf{Q}^t(n-1) \begin{bmatrix} \mathbf{z}(n+1) \\ s(n) \end{bmatrix} \\ &= g(n) - \mathbf{h}^t(n-1) \begin{bmatrix} \mathbf{x}(n+1) \\ w(n) \end{bmatrix} \end{aligned} \tag{15}$$

Again $\alpha(n)$ becomes the output from the plant minus the output from the adaptive filter. The gradient signals are those of (14) upon interpreting $\mathbf{z}(n)$ as the state signals of the right-hand side filter in Fig. 4. If the system parameters are set to the true values, it is seen that the gradient signals coincide with those obtained from the Steiglitz-McBride scheme in the absence of any measurement noise. Hence the true parameter values correspond to a stable stationary point of the corresponding update algorithm. In this alternate arrangement though, the measurement noise $v(n)$ no longer enters into the gradient computations. Thus if this measurement noise is uncorrelated with the input sequence, its presence does not introduce a bias into the parameter estimates. We shall call this algorithm the Modified Output Error Method (MOEM), as the next section shows that the method is closely related to maximum likelihood identification.

IV Possible Convergent Points

To examine possible stationary points of the algorithms described above, let us assume that the parameters of the filters under question are fixed. If the parameters are updated using a steepest descent or Gauss-Newton algorithm, a necessary condition to be satisfied at any stationary point is [8]

$$E[\psi(n)\,\alpha(n)] = \mathbf{0}, \tag{16}$$

where $\psi(n)$ is the vector of gradient signals, $\alpha(n)$ is the error signal, and $E[\cdot]$ denotes mathematical expectation. Although the sequence $\{\alpha(n)\}$ depends only on the external transfer function of the (stationary) adaptive filter, the gradient sequence $\{\psi(n)\}$ depends on the specific filter structure as well.

The possible solutions to (16) are well studied for the direct form parameterization [8, 13, 15]. It is shown in [11] that these results remain valid for the lattice parameterization developed here. The demonstration in [11] uses the fact that, if a lattice filter and a direct form filter are configured to provide the same external transfer function, their gradient signals are related as

$$\psi_l(n) = \mathbf{T}\,\psi_d(n), \tag{17}$$

where $\psi_d(n)$ is the gradient vector computed from a direct form realization, $\psi_l(n)$ is the gradient vector from a lattice realization, and $\mathbf{T}$ is an invertible matrix. Note that if (17) is satisfied, then clearly

$$E[\psi_l(n)\,\alpha(n)] = \mathbf{T}\,E[\psi_d(n)\,\alpha(n)], \tag{18}$$

and since $\mathbf{T}$ is invertible,

$$E[\psi_l(n)\,\alpha(n)] = \mathbf{0} \quad \Leftrightarrow \quad E[\psi_d(n)\,\alpha(n)] = \mathbf{0}, \tag{19}$$

so that the possible stationary points are independent of the parameterization provided $\mathbf{T}$ has full rank. The specific form for $\mathbf{T}$ is developed in [11]. Note that for parallel and cascade forms $\mathbf{T}$ is not in general invertible, so that (19) does not apply for these structures and additional saddle points can be introduced into the error surface [9].

Since $\mathbf{T}$ is invertible with the proposed lattice realization, the characterization of stationary points developed in [15] for the direct form Steiglitz-McBride method applies to the lattice parameterization as well. Consider next the MOEM. When applied to the direct form parameterization, the gradient vector at any stationary point becomes

$$\psi_d(n) = \left[\frac{1}{A(q)}u(n) \quad \cdots \quad \frac{q^{-N}}{A(q)}u(n) \quad -\frac{q^{-1}H(q)}{A^2(q)}u(n) \quad \cdots \quad -\frac{q^{-N}H(q)}{A^2(q)}u(n)\right]^t, \tag{20}$$

where $H(q)/A(q)$ is the transfer function of the filter at convergence. Careful inspection of (20) shows that these are precisely the same signals computed by the direct form output error method, and the set of possible stationary points is thus consistent with those derived in [13]. Again by invertibility of $\mathbf{T}$ in (17), the stationary points of the lattice version of the MOEM are also consistent with those derived in [13]. We should point out that the lattice implementation of the MOEM does *not* compute gradients with respect to the output error. Nonetheless, the set of possible stationary points remains consistent with those computed by the output error method. The advantage here is that all gradient signals are available using only two lattice structures. By comparison, the gradient signals with respect to the output error require N+1 lattice structures; see [10].

V Simulation Example

The lattice version of the Steiglitz-McBride algorithm was used to track a time varying plant, which for this example is configured as a time varying all-pass function, as might be used to model a time varying delay. For the simulation the plant output was generated according to

$$g(n) = w_p(n) + v(n), \qquad \begin{bmatrix} \mathbf{x}_p(n+1) \\ w_p(n) \end{bmatrix} = \mathbf{Q}_p(n) \begin{bmatrix} \mathbf{x}_p(n) \\ u(n) \end{bmatrix} \tag{21}$$

where subscript "p" denotes plant, $v(n)$ is additive white noise, and $\mathbf{Q}_p(n)$ is orthogonal for all n. If the input sequence $u(n)$ is white noise, then the output $g(n)$ can be shown to be asymptotically white noise, and the cross correlation between the white noise processes $\{u(n)\}$ and $\{g(n)\}$ can be shown to be the time varying impulse response of the plant.

The coefficients of the lattice filter were adjusted using the normalized steepest descent algorithm:

$$\theta(n) = \theta(n-1) + \frac{\mu}{\|\psi(n)\|^2}\psi(n)\,\alpha(n), \tag{22}$$

where the vector θ contains the parameters $\{\phi_k\}_{k=1}^N$ and $\{h_k\}_{k=0}^N$, and ψ contains the corresponding gradient components from (14). The plant measurement noise $v(n)$ was adjusted to achieve an output signal to noise ratio of 40 dB, and the stepsize was set at $\mu = 0.32$. Figure 5 shows the coefficient trajectories for the rotation parameters $\{\phi_k(n)\}$ averaged over ten independent runs, for an example in which the plant admitted a fourth order model. The dashed lines indicate the parameter trajectories corresponding to a perfect identification. The tap parameters $\{h_i(n)\}$ were also adapted, converging on the average to their desired values $h_4 = 1.0$, $h_i = 0.0$, $i = 0, \ldots, 3$. The error signal evolution is shown in Fig. 6, and is seen to exhibit bursts when the parameter ϕ_3 changes direction, but otherwise the tracking is seen to be quite faithful.

VI Concluding Remarks

Two lattice algorithms have been presented for adaptive IIR filtering and system identification. These algorithms demonstrate that a gradient complexity of $O(N)$ computations is quite feasible for the lattice realization, and that convergence need not be sacrificed as in [1]. The fundamental advantage of the lattice filter over the direct form structure is that stability can be ensured theoretically during the adaptation process, thus overcoming a significant drawback associated with direct form adaptive filters. Moreover, the parameterization makes immediately available a set of reflection coefficients as $k_m = \sin\phi_m$, which can be exploited for modelling and/or signal analysis purposes.

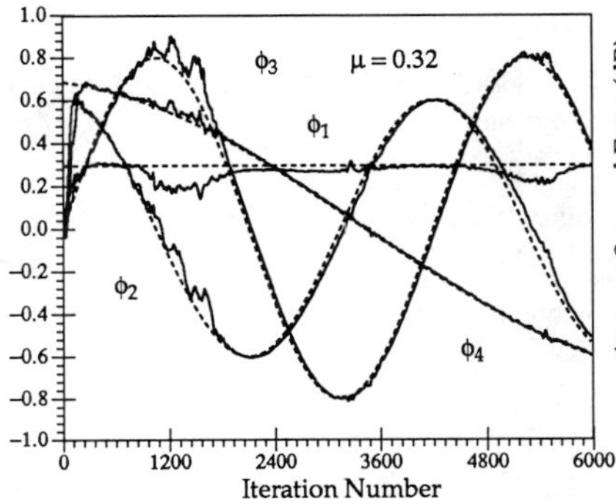

FIGURE 5: Parameter trajectories in tracking experiment.

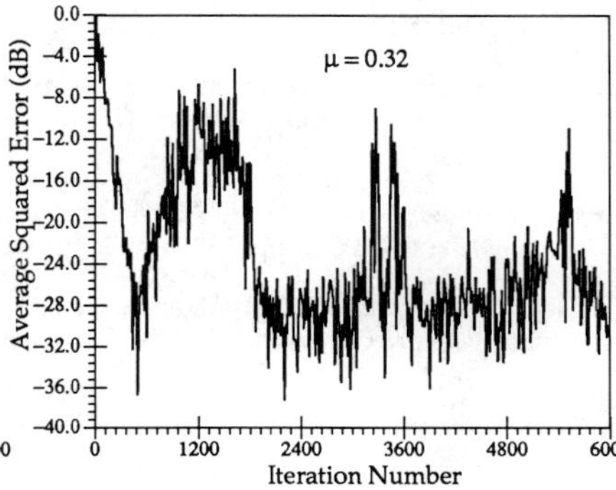

FIGURE 6: Error signal evolution.

References

[1] I. L. Ayala, "On a New Adaptive Lattice Algorithm for Recursive Digital Filters," *IEEE Trans. Acoustics, Speech, and Signal Processing*, vol. ASSP-30, pp. 316–319, 1982.

[2] R. R. Bitmead and B. D. O. Anderson, "Lyapunov Techniques for the Exponential Stability of Linear Difference Equations with Random Coefficients," *IEEE Trans. Automatic Control*, vol. AC-25, pp. 782–787, August 1980.

[3] S. Horvath, Jr., "Lattice Form Adaptive Recursive Digital Filters: Algorithms and Applications," *Proc. IEEE Int. Symp. Circuits and Systems*, pp. 128–133, 1980.

[4] W. K. Jenkins and M. Nayeri, "Adaptive Filters Realized with Second Order Sections," *Proc. IEEE Int. Conf. Acoustics, Speech, and Signal Processing*, Tokyo, April 1986, pp. 2103–2106.

[5] C. R. Johnson, Jr., "A Convergence Proof for a Hyperstable Adaptive Recursive Filter," *IEEE Trans. Information Theory*, vol. IT-25, pp. 745–749, November 1979.

[6] C. R. Johnson, Jr., and T. Taylor, "Failure of a Parallel Adaptive Identifier with Adaptive Error Filtering," *IEEE Trans. Automatic Control*, vol. AC-25, pp. 1248–1250, December 1980.

[7] I. D. Landau, "Unbiased Recursive Identification Using Model Reference Adaptive Techniques," *IEEE Trans. Automatic Control*, vol. AC-21, pp. 194–202, April 1976.

[8] L. Ljung and T. Söderström, *Theory and Practice of Recursive Identification*, MIT Press: Cambridge, MA, 1983.

[9] M. Nayeri and W. K. Jenkins, "Alternate Realizations of Adaptive IIR Filters and Properties of Their Performance Surfaces," *IEEE Trans. Circuits and Systems*, vol. CAS-36, pp. 485–496, April 1989.

[10] D. Parikh, N. Ahmed, and S. D. Stearns, "An Adaptive Lattice Algorithm for Recursive Filters," *IEEE Trans. Acoustics, Speech, and Signal Processing*, vol. ASSP-28, pp. 110–112, 1980.

[11] P. A. Regalia, "Stable and Efficient Lattice Algorithms for Adaptive IIR Filtering," Technical Report, Institut National des Télécommuncations: Evry, France, May 1989.

[12] J. J. Shynk, "Performance of Alternate Adaptive IIR Filter Realizations," *Proc. Asilomar Conf. Circuits, Systems, and Computers*, Pacific Grove, CA, November 1987.

[13] T. Söderström and P. Stoica, "Some Properties of the Output Error Method," *Automatica*, vol. 18, pp. 93–99, 1982.

[14] K. Steigltz and L. E. McBride, "A Technique for the Identification of Linear Systems," *IEEE Trans. Automatic Control*, vol. AC-10, pp. 461-464, October 1964.

[15] P. Stoica and T. Söderström, "The Steiglitz-McBride Algorithm Revisited—Convergence Analysis and Accuracy Aspects," *IEEE Trans. Automatic Control*, vol. AC-26, pp. 712–717, June 1981.

[16] J. E. Volder, "The CORDIC Trigonometric Computing Technique," *IEEE Trans. Electronic Computers*, vol. EC-8, pp. 330–334, September 1959.

[17] J. S. Walther, "A Unified Algorithm for Elementary Functions," *Proc. 1971 AFIPS Spring Joint Computer Conference*, pp. 379–385, Atlantic City, NJ, May 1971.

On the Orthogonal Projection method for Adaptive Filters

P.C.W. Sommen

1 Abstract

Convergence characteristics of adaptive filters are influenced by the statistical properties (such as correlation) of the input signal. In order to remove this dependency, decorrelation can be applied. This can be achieved by multiplying the update part of the adaptive scheme with the inverse autocorrelation matrix $\mathbf{R}$ *of the input signal. This leads to the Recursive Least Squares (RLS) algorithm.*
Using this algorithm, both the dimension of the autocorrelation matrix $\mathbf{R}$ *and the length of the adaptive filter are equal to N. However, the relevant dimension of* $\mathbf{R}$ *is determined by statistical properties of the input signal and needs not to be equal to N. Orthogonal Projection (OP) algorithms can decouple the dimension of the autocorrelation matrix and the length of the adaptive filter. To derive the OP method for adaptive filters the echo canceller problem is used as a vehicle.*

2 Introduction

The echo canceller, which uses an adaptive filter for the coefficient update, is a well known structure in many communication applications. Here we will use this echo canceller as a vehicle to describe the OP algorithm. A sketch of this application is given in Fig. 1. The echo path impulse response h is assumed to have a finite

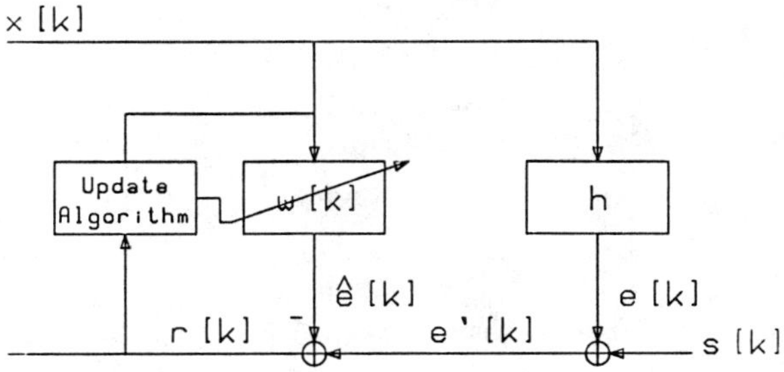

Figure 1: *Basic echo canceller structure*

impulse response length of order N. The input signal $x[k]$ produces via this echo path h an echo signal $e[k]$. The near–end signal $s[k]$ is added to this echo $e[k]$. The adaptive filter uses a transversal model of the echo path and makes a replica $\hat{e}[k]$ of the echo signal $e[k]$. Theoretically the residual signal $r[k] = s[k] + e[k] - \hat{e}[k]$ in steady state will in average be equal to the signal $s[k]$. For simplicity reasons we will assume in this paper the signal $s[k] = 0$.

We define the inner product $< \underline{\mathbf{x}}[k], \underline{\mathbf{h}} >$ of an N dimensional vector $\underline{\mathbf{x}}[k]$ with an N dimensional vector $\underline{\mathbf{h}}$ as the next convolution sum:

$$< \underline{\mathbf{x}}[k], \underline{\mathbf{h}} > \doteq \sum_{i=0}^{N-1} x[k-i]h_i = \underline{\mathbf{x}}^T[k]\underline{\mathbf{h}}. \tag{1}$$

Furthermore two vectors are orthogonal if their inner product is zero. Likewise the norm (length) of a vector is given by:

$$||\underline{\mathbf{x}}[k]|| \doteq \sqrt{< \underline{\mathbf{x}}[k], \underline{\mathbf{x}}[k] >}. \tag{2}$$

Now we will give a geometric interpretation of the echo cancellation problem, which will lead to the well known Normalized Least Mean Square (NLMS) algorithm [1]. The update procedure, as given in Fig.2, is to make a projection of the difference

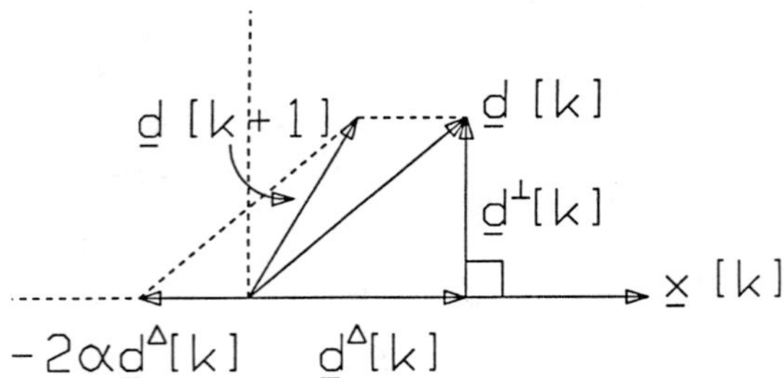

Figure 2: *Geometric interpretation of the NLMS algorithm*

vector $\underline{\mathbf{d}}[k] = \underline{\mathbf{h}} - \underline{\mathbf{w}}[k]$ on the vector $\underline{\mathbf{x}}[k]$. By doing this we decompose $\underline{\mathbf{d}}[k]$ as:

$$\underline{\mathbf{d}}[k] = \underline{\mathbf{d}}^{\perp}[k] + \underline{\mathbf{d}}^{\Delta}[k] \tag{3}$$

with $\underline{\mathbf{d}}^{\perp}[k]$ orthogonal and $\underline{\mathbf{d}}^{\Delta}[k]$ parallel to $\underline{\mathbf{x}}[k]$. This implies that:

$$\begin{aligned} < \underline{\mathbf{x}}[k], \underline{\mathbf{d}}^{\perp}[k] > &= 0 \\ \underline{\mathbf{d}}^{\Delta}[k] &= c \cdot \underline{\mathbf{x}}[k] \end{aligned} \tag{4}$$

where c is some scalar. Assuming $s[k] = 0$ we see from Fig. 1 that we can write the residual signal $r[k]$ as:

$$\begin{aligned} r[k] &= \underline{\mathbf{x}}^T[k]\underline{\mathbf{d}}[k] = < \underline{\mathbf{x}}[k], \underline{\mathbf{d}}[k] > = \\ &= < \underline{\mathbf{x}}[k], \underline{\mathbf{d}}^{\Delta}[k] > = c||\underline{\mathbf{x}}[k]||^2. \end{aligned} \tag{5}$$

With this result we are able to calculate the scalar c and we can write the parallel component $\underline{\mathbf{d}}^{\Delta}[k]$ as:

$$\underline{\mathbf{d}}^{\Delta}[k] = c \cdot \underline{\mathbf{x}}[k] = \frac{r[k]}{||\underline{\mathbf{x}}[k]||^2} \cdot \underline{\mathbf{x}}[k]. \tag{6}$$

We are interested in both reducing the length and rotating vector $\underline{\mathbf{d}}[k]$ in such a way that it becomes "more orthogonal" to $\underline{\mathbf{x}}[k]$. This can be achieved by subtracting a small part of the vector $\underline{\mathbf{d}}^{\Delta}[k]$ from the vector $\underline{\mathbf{d}}[k]$, as shown in Fig.2, thus:

$$\begin{aligned}\underline{\mathbf{d}}[k+1] &= \underline{\mathbf{d}}[k] - 2\alpha\underline{\mathbf{d}}^{\Delta}[k] \\ &= \underline{\mathbf{d}}[k] - 2\alpha\frac{\underline{\mathbf{x}}[k]r[k]}{||\underline{\mathbf{x}}[k]||^2}\end{aligned} \tag{7}$$

which leads, by using $\underline{\mathbf{d}}[k] = \underline{\mathbf{h}} - \underline{\mathbf{w}}[k]$, to the NLMS algorithm:

$$\underline{\mathbf{w}}[k+1] = \underline{\mathbf{w}}[k] + 2\alpha\frac{\underline{\mathbf{x}}[k]r[k]}{||\underline{\mathbf{x}}[k]||^2}. \tag{8}$$

Heuristically the above geometric approach can be explained as follows:
The update algorithm rotates the vector $\underline{\mathbf{d}}[k]$ in such a way that the inner product $< \underline{\mathbf{d}}[k+1], \underline{\mathbf{x}}[k+1] >$ becomes smaller then the previous inner product. This inner product is a measure for the final misadjustment. The reduction can be accomplished in two ways: by reducing the length of $\underline{\mathbf{d}}[k]$ or by rotating $\underline{\mathbf{d}}[k]$ in such a way that it comes more orthogonal to $\underline{\mathbf{x}}[k]$. The above algorithm achieves both. With a large adaptation constant ($\alpha \approx 1/2$) the algorithm rotates $\underline{\mathbf{d}}[k]$ in one step orthogonal to $\underline{\mathbf{x}}[k]$, which implies an inner product that is zero. The length however is not very small. In practical situations however the impulse response $\underline{\mathbf{h}}$ may change, which implies that $\underline{\mathbf{d}}[k+1]$ will rotate and will not be orthogonal to $\underline{\mathbf{x}}[k+1]$ any more. The result is that the inner product will not be zero any more. For this reason the inner product has large fluctuations, which implies a large final misadjustment. On the other hand choosing a small adaptation constant α leads to a slow algorithm, where the final difference vector will be both orthogonal and small, leading to a small final misadjustment.
It is known from literature [1] that the convergence behaviour of the NLMS algorithm is influenced when the input signal $x[k]$ is not a white noise process. The geometric interpretation of such a white noise process (flat spectrum) is that the vectors $\underline{\mathbf{x}}[k]$ and $\underline{\mathbf{x}}[k-i]$ are orthogonal to each other (for all $i \neq 0$) which implies that their inner product is zero. For a non white input signal the vectors $\underline{\mathbf{x}}[k]$ and $\underline{\mathbf{x}}[k-i]$ are in general not orthogonal. For this reason this implies that the new update is not orthogonal to the previous update in the above mentioned method. Thus an update made in one iteration may be counteracted in the next iteration. By this we can see that the convergence behaviour is influenced.

3 L–step Orthogonal Projection Algorithm

In [3,2] an L–step Orthogonal Projection (OP) method is introduced, where the above geometric approach is extended to L dimensions. With this OP method we make a projection on an L dimensional plane rather then on a line as in the previous section. In general there are two variants of this idea:

1. The "sliding" procedure using every iteration only one new input signal sample. This is the L–step Orthogonal Projection method (OP) as described in [3].

2. The "block" approach which uses L new samples and performs only one update every L samples. This is the Block Orthogonal Projection (BOP) algorithm. For this we introduce the block index m denoting the mth block of length L.

We will describe here the second approach. At the end of this section we give the results of the first one.
The procedure of the BOP method is as follows (see Fig. 3): make a projection of $\underline{\mathbf{d}}[m]$ on an L dimensional plane, rather then on a line, spanned by the vectors $\underline{\mathbf{x}}_0[m], \cdots, \underline{\mathbf{x}}_{L-1}[m]$, with

$$\underline{\mathbf{x}}_i[m] = \underline{\mathbf{x}}[mL-i] = (x[mL-i], \cdots, x[mL-i-N+1])^T . \tag{9}$$

Thus we have:

$$\underline{\mathbf{d}}[m] = \underline{\mathbf{d}}^{\Delta}[m] + \underline{\mathbf{d}}^{\perp}[m] \tag{10}$$

with

$$\begin{aligned} < \underline{\mathbf{d}}^{\perp}[m], \underline{\mathbf{x}}_i[m] > &= 0 \quad \text{for all } i = 0, \cdots L-1 \\ \underline{\mathbf{d}}^{\Delta}[m] &= \sum_{i=0}^{L-1} c_i[m]\underline{\mathbf{x}}_i[m] = \chi[m]\underline{\mathbf{c}}[m] \end{aligned} \tag{11}$$

where we used

$$\begin{aligned} \underline{\mathbf{c}}[m] &\doteq (c_0[m], \cdots, c_{L-1}[m])^T \\ \chi[m] &\doteq (\underline{\mathbf{x}}_0[m], \cdots, \underline{\mathbf{x}}_{L-1}[m])^T. \end{aligned} \tag{12}$$

The residual signal on block basis, assuming $\underline{\mathbf{s}}[k] = \underline{0}$, can be written as:

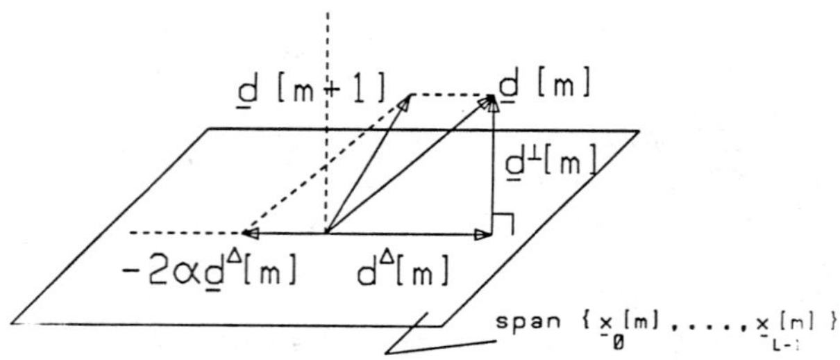

Figure 3: *The BOP method*

$$\underline{\mathbf{r}}[m] \doteq (r_0[m], \cdots, r_{L-1}[m])^T = \chi^T[m]\underline{\mathbf{d}}[m]. \tag{13}$$

Note that this is a typical block expression: the same adaptive weights are only updated once every L samples. Now an equivalent procedure as in the previous section can be used to calculate the vector $\underline{\mathbf{c}}[m]$. For this we note that $\underline{\mathbf{d}}^{\perp}[m]$ is perpendicular to all vectors $\underline{\mathbf{x}}_i[m]$ for $i = 0, 1, \cdots, L-1$. In vector matrix notation this leads to:

$$\begin{aligned} \underline{\mathbf{r}}[m] &= \chi^T[m]\underline{\mathbf{d}}[m] \\ &= \chi^T[m]\left(\underline{\mathbf{d}}^{\Delta}[m] + \underline{\mathbf{d}}^{\perp}[m]\right) \\ &= \chi^T[m]\underline{\mathbf{d}}^{\Delta}[m] \\ &= \chi^T[m]\chi[m]\underline{\mathbf{c}}[m]. \end{aligned} \tag{14}$$

The solution of these equations is given by:

$$\underline{\mathbf{c}}[m] = \hat{\mathbf{R}}^{-1}[m]\underline{\mathbf{r}}[m] \tag{15}$$

where the $L \times L$ estimate of autocorrelation matrix is defined as:

$$\hat{\mathbf{R}}[m] \doteq \chi^T[m]\chi[m] \tag{16}$$

from which the $(p,q)^{th}$ element is equal to:

$$(\hat{\mathbf{R}}[m])_{p,q} = < \underline{\mathbf{x}}_p[m], \underline{\mathbf{x}}_q[m] > = \underline{\mathbf{x}}_p^T[m]\underline{\mathbf{x}}_q[m]. \tag{17}$$

Now we can reduce the length of vector $\underline{\mathbf{d}}[m]$ with some amount of the vector $\underline{\mathbf{d}}^\Delta[m]$ according to the next update equation:

$$\begin{aligned} \underline{\mathbf{d}}[m+1] &= \underline{\mathbf{d}}[m] - 2\alpha\underline{\mathbf{d}}^\Delta[m] & (18)\\ &= \underline{\mathbf{d}}[m] - 2\alpha\chi[m]\hat{\mathbf{R}}^{-1}[m]\underline{\mathbf{r}}[m]. & (19) \end{aligned}$$

This leads to the BOP update equation:

$$\underline{\mathbf{w}}[m+1] = \underline{\mathbf{w}}[m] + 2\alpha\chi[m]\hat{\mathbf{R}}^{-1}[m]\underline{\mathbf{r}}[m]. \tag{20}$$

This update equation generates after every L input samples one new adaptive weight vector. In the meanwhile it calculates L "new" residual signals, which implies L convolutions, with the "old" adaptation vector $\underline{\mathbf{w}}[m]$
This is in contrast to the sliding approach. The derivation is the same as above which results in the next L–step OP algorithm:

$$\underline{\mathbf{w}}[k+1] = \underline{\mathbf{w}}[k] + 2\alpha\chi[k]\hat{\mathbf{R}}^{-1}[k]\underline{\mathbf{r}}[k] \tag{21}$$

with:

$$\begin{aligned} \chi[k] &\doteq (\underline{\mathbf{x}}_0[k], \cdots, \underline{\mathbf{x}}_{L-1}[k]) = (\underline{\mathbf{x}}[k], \cdots, \underline{\mathbf{x}}[k-L+1]) \\ \underline{\mathbf{r}}[k] &\doteq (r_0[k], \cdots, r_{L-1}[k])^T \neq (r[k], \cdots, r[k-L+1])^T. & (22) \end{aligned}$$

This algorithm generates after every new input sample a new adaptive weight vector. For each update we need the vectors $\underline{\mathbf{x}}_0[k], \cdots, \underline{\mathbf{x}}_{L-1}[k]$. Furthermore we need each iteration k, L "new" residual signals $r_i[k]$. This will result in a large complexity increasement in comparison to the BOP algorithm with block length L. In [6] an efficient implementation is given of this algorithm when the input signal can be modelled with an Auto Regressive (AR) process.
It is clear that the OP algorithm (21) is an uncoupled version of the well–known RLS algorithm [1] which in its most simple form is given by:

$$\underline{\mathbf{w}}[k+1] = \underline{\mathbf{w}}[k] + 2\alpha\mathbf{R}_N^{-1}\underline{\mathbf{x}}[k]r[k] \tag{23}$$

where the inverse autocorrelation matrix $\mathbf{R}_N^{-1}$ is updated by a "simple" recursive scheme. In this equation we use an autocorrelation matrix $\mathbf{R}$ of dimension $N \times N$. As a final comment to this section we note that in [5] (p69) the suggestion is made that the inverse autocorrelation matrix $\hat{\mathbf{R}}^{-1}[m]$ used in the OP algorithm (21), can be calculated by using the same "recursive" techniques as used in the RLS scheme [1].

4 BOP using the Gram–Schmidt procedure

The BOP concept of the previous section was to make a projection of the vector $\underline{\mathbf{d}}[m]$ on the space which was spanned by the basis:

$$\{\underline{\mathbf{x}}_0[m], \cdots, \underline{\mathbf{x}}_{L-1}[m]\}. \tag{24}$$

In this section we will use the same BOP concept, but we will first derive an orthogonal basis by applying the Gram–Schmidt procedure.
For $i = 0$ to $L-1$ this procedure leads to the next set of orthogonal vectors:

$$\tilde{\underline{\mathbf{x}}}_i[m] = \underline{\mathbf{x}}_i[m] - \sum_{q=0}^{i-1} \frac{< \underline{\mathbf{x}}_i[m], \tilde{\underline{\mathbf{x}}}_q[m] >}{||\tilde{\underline{\mathbf{x}}}_q[m]||^2} \tilde{\underline{\mathbf{x}}}_q[m] \tag{25}$$

where the sum $\sum_{q=0}^{-1}$ is defined as zero.
Note here that in general these orthogonal vectors are not shifted versions of each other, in contrast to the input signal vectors. Thus, with exception of the borders, we have for the input signal:

$$(\underline{\mathbf{x}}_i[m])_t = x[mL - i - t] = (\underline{\mathbf{x}}_0[m])_{t+i} \tag{26}$$

while:

$$(\tilde{\underline{\mathbf{x}}}_i[m])_t \neq (\tilde{\underline{\mathbf{x}}}_0[m])_{t+i}. \tag{27}$$

Following the Gram–Schmidt procedure (25) it is obvious that we can rotate the original basis of input vectors to an orthogonal basis, spanning the same space. This leads for $i = 0, \cdots, L-1$ to:

$$\tilde{\underline{\mathbf{x}}}_i[m] = \sum_{q=0}^{i} \gamma_q^i[m] \underline{\mathbf{x}}_q[m]. \tag{28}$$

with $\gamma_i^i[m] = 1$ for all $i = 0, 1, \cdots, L-1$. In matrix notation we can write these equations as follows:

$$\tilde{\chi}[m] = \chi[m] \cdot \mathbf{\Gamma}[m] \tag{29}$$

where we used the $L \times L$ upper triangular matrix

$$\mathbf{\Gamma}[mL] \doteq \begin{pmatrix} \gamma_0^0[m] & \gamma_0^1[m] & \cdot & \cdot & \cdot & \gamma_0^{L-1}[m] \\ 0 & \gamma_1^1[m] & \cdot & \cdot & \cdot & \gamma_1^{L-1}[m] \\ \cdot & \cdot & \cdot & \cdot & \cdot & \cdot \\ \cdot & \cdot & \cdot & \cdot & \cdot & \cdot \\ \cdot & \cdot & \cdot & \cdot & \cdot & \cdot \\ 0 & 0 & \cdot & \cdot & 0 & \gamma_{L-1}^{L-1}[m] \end{pmatrix}. \tag{30}$$

Now the BOP concept is to make a projection of the vector $\underline{\mathbf{d}}[m]$ on the orthogonal basis. This leads to:

$$\begin{aligned} < \underline{\mathbf{d}}^{\perp}[m], \tilde{\underline{\mathbf{x}}}_i[m] > &= 0 \quad \text{for } i = 0, \cdots, L-1 \\ \underline{\mathbf{d}}^{\Delta}[m] &= \tilde{\chi}[m]\underline{\mathbf{c}}[m] \end{aligned} \tag{31}$$

(32)

By defining a rotated (transformed) residual signal vector as:

$$\tilde{\underline{\mathbf{r}}}[m] = \mathbf{\Gamma}^T[m]\underline{\mathbf{r}}[m] \tag{33}$$

we can derive the next Gram–Schmidt BOP update algorithm (GS–BOP):

$$\underline{\mathbf{w}}[m+1] = \underline{\mathbf{w}}[m] + 2\alpha\tilde{\chi}[m]\tilde{\mathbf{R}}^{-1}[m]\tilde{\underline{\mathbf{r}}}[m] \tag{34}$$

where the $L \times L$ rotated (transformed) "autocorrelation" matrix $\tilde{\mathbf{R}}[m]$ is diagonal because of the orthogonality property of the vectors $\tilde{\underline{\mathbf{x}}}_i[m]$. Thus:

$$\tilde{\mathbf{R}}^{-1}[m] = diag\{||\tilde{\underline{\mathbf{x}}}_0[m]||^2, \cdots, ||\tilde{\underline{\mathbf{x}}}_{L-1}[m]||^2\} \tag{35}$$

From the above definitions it can be shown that the $L \times L$ autocorrelation matrix $\hat{\mathbf{R}}[m]$ can be diagonalized as:

$$\mathbf{\Gamma}^T[m]\hat{\mathbf{R}}[m]\mathbf{\Gamma}[m] = \tilde{\mathbf{R}}[m]. \tag{36}$$

Using this equation it can be shown by straightforward substitution that (20) and (34) are equivalent equations from which the adaptive weights converge, in average, to the same values. The main difference is that the matrix $\tilde{\mathbf{R}}[m]$ is diagonal and $\hat{\mathbf{R}}[m]$ not.
In [4] (p99) some efficient alternatives are given to compute the matrix $\mathbf{\Gamma}[m]$. One of them is using the Householder transform [7].
Furthermore it is straightforward to rewrite the GS–BOP algorithm (34) as:

$$\underline{\mathbf{w}}[m+1] = \underline{\mathbf{w}}[m] + 2\alpha\sum_{i=0}^{L-1}\frac{\tilde{\underline{\mathbf{x}}}_i[m]\tilde{r}_i[m]}{||\tilde{\underline{\mathbf{x}}}_i[m]||^2}. \tag{37}$$

From this update equation we see directly that at every iteration we make a (NLMS) correction of the difference vector in L orthogonal directions. Thus a correction in one direction is not influenced by a correction in another direction. On the other hand one of the drawbacks of the GS–BOP algorithm is that at every new iteration we are dealing with a new space, which need not to be orthogonal to the previous space. This "non–orthogonality" between separate iterations has not been investigated in this paper.
As a final comment of this section we note that it is obvious that this GS–BOP algorithm can also be implemented in a sliding way, leading to the L step GS–OP algorithm:

$$\underline{\mathbf{w}}[k+1] \;=\; \underline{\mathbf{w}}[k] + 2\alpha\tilde{\chi}[k]\tilde{\mathbf{R}}^{-1}[k]\tilde{\underline{\mathbf{r}}}[k] \tag{38}$$

with

$$\begin{aligned} \tilde{\chi}[k] &= \chi[k]\mathbf{\Gamma}[k] \\ \tilde{\mathbf{R}}[k] &= diag\{||\tilde{\underline{\mathbf{x}}}_0[k]||^2, \cdots, ||\tilde{\underline{\mathbf{x}}}_{L-1}[k]||^2\} \\ \tilde{\underline{\mathbf{r}}}[k] &= \mathbf{\Gamma}^T[k]\underline{\mathbf{r}}[k]. \end{aligned} \tag{39}$$

As before this is equivalent to:

$$\underline{\mathbf{w}}[k] = \underline{\mathbf{w}}[k] + 2\alpha\sum_{i=0}^{L-1}\frac{\tilde{\underline{\mathbf{x}}}_i[k]\tilde{r}_i[k]}{||\tilde{\underline{\mathbf{x}}}_i[k]||^2}. \tag{40}$$

5 Conclusions

The main goal of this paper was to give algorithms which decouple the length (N) of the adaptive filter and the dimension (L) of the autocorrelation matrix from the input signal of the adaptive filter. Two algorithms are given. The first one makes a projection on a space S rather then on a line as in the NLMS algorithm. The second approach is to apply first a Gram–Schmidt procedure to construct an orthogonal basis for S, and after that apply a projection on this basis. Both algorithms can be applied using "sliding" or "block" processing techniques.
Points of further research are the complexity of the algorithms and the problem of "non–orthogonality" between two separate iterations as described in the paper.

6 Acknowledgement

The author wish to thank Philips Research Laboratories Eindhoven for their agreement with this publication. Furthermore C.J. van Valburg is acknowledged for his cooperation.

References

[1] S.T. Alexander; "Adaptive Signal Processing, Theory and Applications", Springer–Verlag, 1986

[2] Kaoru Furosawa, Takuji Furusawa; "A geometric interpretation of adaptive algorithms"; Globecom conference, Tokyo, Japan, nov. 15–18, 1987, pp 49.7.1–49.7.5

[3] Kazuhiko Ozeki, Tetsuo Umeda; "An adaptive Filtering Algorithm Using an Orthogonal Projection to an Affine Subspace and its properties"; Electronics and Communications in Japan, vol. 67-A, no.5, 1984, pp 19–27
Translated from Denshi Tsushin Gakkai Ronbunshi, vol. 67A, no. 2, Feb. 1984, pp 126–132

[4] Beresford N. Parlett; "The Symmetric Eigenvalue Problem", Prentice–Hall, Inc., Englewood Cliffs, N.J. 07632

[5] U. Schültheiss; "Über die Adaption eines Kompensators für akustische Echos", Phd Report 1988, Darmstadt, Fortschritt–berichte, VDI Verlag Düsseldorf, Reihe 10, Informatik Kommunikationstechnik, Nr.90

[6] P.C.W. Sommen and C.J. van Valburg; "Efficient Realisation of Adaptive Filter using an Orthogonal Projection Method", Proceedings ICASSP 1989, Glasgow Scotland.

[7] Allan O. Steinhardt; "Householder Transforms in Signal Processing", IEEE Magazine, July 1988, pp 4–12

REALIZATIONS OF PRODUCTS AND WIENER-HOPF FACTORS FOR A CLASS OF MATRIX FUNCTIONS ANALYTIC IN A STRIP

JOSEPH A. BALL AND LONNIE E. CARPENTER

We give explicit realization formulas for the factors in a right canonical Wiener-Hopf factorization in terms of realizations of the factors in a left factorization for a matrix function $W(\lambda)$ of the form $W(\lambda) = I + \hat{k}(\lambda)$, where $\hat{k}$ is the (bilateral) Laplace transform of a function k from the class $e^{\omega|t|}L_1^{m\times m}(\mathbb{R})$, $\omega < 0$. Here we work with realizations $W(\lambda) = I + C(\lambda I - A)^{-1}B$ which are admissible in the sense of Bart, Gohberg and Kaashoek. We also give a direct proof that the product of two admissible realizations is again admissible.

1. Introduction.

As is well known (see e.g. [2]) any proper rational matrix function $W(\lambda)$ can be represented in terms of four matrices A, B, C, D as the transfer function

$$W(\lambda) = D + C(\lambda I - A)^{-1}B \tag{1.1}$$

of a linear system

$$\begin{aligned} \dot{x}(t) &= Ax(t) + Bu(t) \\ u(t) &= Cx(t) + Du(t). \end{aligned} \tag{1.2}$$

This has proved to be an important tool, both conceptually and computationally, in many problems concerning rational matrix functions, e.g. factorization, interpolation and applications to control theory (see e.g. [1], [3], [10]). Extensions of these ideas to nonrational matrix functions have not developed as quickly. The theory usually goes through as in the rational case if A, B, C, D can be taken to be bounded linear operators on a Hilbert (or more generally a Banach) space. Unfortunately this assumption eliminates many important examples, especially for continuous time systems theory problems. In general one must take A, B, C unbounded, or equivalently introduce a delicately rigged Hilbert space structure; there have been a number of approaches to do this with successively better and better results (see eg. [8], [11], [12], [15]). In this paper we focus on a class of realizations (A, B, C) studied by Bart, Gohberg and Kaashoek [4-7] which we shall term BGK-admissible We give a proof directly in terms of realizations that the product of two admissible realizations is again admissible and we verify formulas for the factors in a right canonical Wiener-Hopf factorization of a function with an admissible realization in terms of the realizations of the factors in a left canonical factorization. This last result extends work of Ball and Ran [2] done for the rational case, and the result of Ran [14] for J-symmetric Wiener-Hopf factorization. We also mention that Curtain and Ran [9] recently have extended the symmetric version to a larger class of infinite dimensional systems. Whereas the work in [14] and [9] hinged on solutions of operator Lyapunov equations with unbounded coefficients, i.e., an equation of the form

$$AQx + QA^*x = \Phi x \text{ for } x \in D(A)$$

for the unknown bounded operator Q, where A is the (unbounded) generator for a C_0 semigroup and Φ is A^* bounded, a key role in this paper is played by operator Sylvester equations

$$A_+Qx - QA_-x = \Phi x \text{ for } x \in D(A_-)$$

where A_+ and $-A_-$ are generators of C_0 semigroups.

2. Admissible realization triples.

Let m be a positive integer and $\mathbb{C}$ represent the complex numbers. By $\mathcal{D}_1^m(\mathbb{R})$ we denote the linear subspace of $L_1^m(\mathbb{R})(L_1(\mathbb{R})$ functions with values in $\mathbb{C}^m)$ consisting of all f in $L_1^m(\mathbb{R})$ for which there exists a g in $L_1^m(\mathbb{R})$ such that

$$f(t) = \begin{cases} \int_{-\infty}^t g(s)ds & a.e. \text{ on } (-\infty, 0] \\ -\int_t^\infty g(s)ds & a.e. \text{ on } [0, \infty). \end{cases}$$

We shall need the notion of exponential dichotomy for a densely defined linear operator on a Banach space X.

<u>Definition 2.1</u> We say that the densely defined operator A on the Banach space X is exponentially dichotomous if X admits a topological direct sum decomposition $X = X_- \oplus X_+$ with the following properties:

1. The decomposition reduces A, i.e., the domain $D(A)$ has a direct sum decomposition $D(A) = [D(A) \cap X_-] \oplus [D(A) \cap X_+]$ and A maps $D(A) \cap X_-$ (respectively $D(A) \cap X_+$) into X_- (respectively X_+).

2. The restriction of $-A$ to X_- is the infinitesimal generator of an exponentially decaying strongly continuous semigroup.

3. The restriction of A to X_+ is the infinitesimal generator of an exponentially decaying strongly continuous semigroup.

We define the bisemigroup $E(t,A)$ generated by the exponentially dichotomous operator A by

$$E(t;A)x = \begin{cases} -e^{tA_-}Px, & t<0 \\ e^{tA_+}(I-P)x, & t>0 \end{cases}$$

where the exponential notation represents the strongly continuous semigroups generated by A_+ and $-A_-$ (see [13]). Here A_+ is the restriction of A to X_+, A_- is the restriction of A to X_-, and P is the projection onto X_- along X_+ (called the *separating projection*) for the exponential dichotomy of A. The exponential decay condition in Definition 2.1 implies that there is number $\omega < 0$ such that

$$\sup_{t\neq 0} e^{-\omega|t|}\|E(t;S)\| < \infty.$$

We shall refer to any such ω as an exponential type for A.

We next define the class of realizations (A,B,C) considered in [4-7]. Here we assume $A: D(A) \to X,\ B: \mathbb{C}^m \to X,\ C: D(C) \to \mathbb{C}^m$ where $D(A)$ and $D(C)$ are dense in X.

Definition 2.2 The triple $\theta = (A,B,C)$ is *BGK admissible* if the following conditions are satisfied:

1. A is exponentially dichotomous of exponential type ω.

2. $D(C) \supset D(A)$ and C is A-bounded,

3. There exists a linear operator $\Lambda_\theta : X \to L_1^m(\mathbb{R})$ such that

(a) $$\sup_{\|x\|\leq 1} \int_{-\infty}^{\infty} e^{-\omega|t|}\|\Lambda_\theta x(t)\|dt < \infty, \tag{2.1}$$

(b) Λ_θ maps $D(A)$ into $D_1^m(\mathbb{R})$ and $\Lambda_\theta x = CE(t,A)x$ for $x \in D(A)$. (2.2)

Note that the content of condition (3) is that the realization triple θ has a bounded *observability operator* $\Lambda_\theta : X \to L_1^m(\mathbb{R})$ despite the fact that the output map C itself is unbounded.

Note that B, being a linear operator from C^m into X is automatically bounded. Also 3a implies that Λ_θ is bounded and maps X into $L_{1,\omega}^m(\mathbb{R})$, where $L_{1,\omega}^m(\mathbb{R}) = \{f \in L_1^m(\mathcal{R}) | e^{-\omega|t|}f(t)$

$\in L_1^m(\mathbb{R})\}$. Using 3b and the fact that $D(A)$ is dense in X, one sees that Λ_θ is determined uniquely. Since ω is negative, $L_{1,\omega}^m(\mathbb{R})$ is a linear subspace of $L_1^m(\mathbb{R})$. The spaces X and $\mathbb{C}^m$ are respectively called the state space and input/output space of the triple θ.

Next we introduce an operator $\Gamma_\theta : L_1^m(\mathbb{R}) \to X$ which depends only on A and B. Γ_θ is defined by

$$\Gamma_\theta\phi = \int_{-\infty}^{\infty} E(t;A)B\phi(t)dt. \tag{2.3}$$

Note that Γ_θ is well-defined, linear and bounded, and can be thought of as a controllability operator.

For $\theta = (A, B, C)$ a BGK-admissible realization triple for each fixed y in $\mathbb{C}^m$ the function $\Lambda_\theta By$ is in $L_{1,\omega}^m(\mathbb{R})$. This enables us to define a matrix valued function $k_\theta \in L_{1,\omega}^{m\times m}(\mathbb{R})$ by

$$k_\theta(t)y = [\Lambda_\theta By](t).$$

We call k_θ the *kernel associated with the realization triple* θ. Moreover, since $k_\theta \in L_{1,\omega}^{m\times m}(\mathbb{R})$, the bilateral Laplace transform $\hat{k}_\theta$ of k_θ is an $m \times m$ matrix function analytic on the strip $\{\lambda : |Re\lambda| < -\omega\}$. Since ω is also the exponential type for A, this same strip is in the resolvent set of A and it can be shown that $\hat{k}_\theta(\lambda) = C(\lambda I - A)^{-1}B$. We also have the converse.

THEOREM 2.3. *(see [7]). An $m \times m$ matrix function $W(\lambda)$ analytic on a vertical strip $\{\lambda : |Re\lambda| < -\omega\}$ can be represented as*

$$W(\lambda) = W_\theta(\lambda) = I + C(\lambda I - A)^{-1}B \tag{2.4}$$

for an admissible realization triple $\theta = (A, B, C)$ if and only if W has the form

$$W(\lambda) = I + \hat{k}(\lambda) \tag{2.5}$$

where $\hat{k}$ is the bilateral Laplace transform of a function k in $L_{1,\omega}^{m\times m}(\mathbb{R})$.

Since $L_{1,\omega}^{m\times m}$ is a Banach algebra (without a unit) under convolution (see [13]), it follows that the class of matrix functions $W(\lambda)$ of the form (2.5) with k in $L_{1,\omega}^{m\times m}(\mathbb{R})$ is closed under multiplication. On the other hand, if W_j are two matrix functions associated with realization triples $\theta_j = (A_j, B_j, C_j)$ as in (2.4), then the product $W = W_1W_2$ formally is associated with the realization triple $\theta = (A, B, C)$ given by

$$\theta = \left(\begin{bmatrix} A_1 & B_1C_2 \\ 0 & A_2 \end{bmatrix}, \begin{bmatrix} B_1 \\ B_2 \end{bmatrix}, [C_1 \ C_2] \right). \tag{2.6}$$

The next result is that the product $\theta = \theta_1 * \theta_2$ defined by (2.6) of two admissible realization triples is again admissible, and indeed provides a realization representation for the product function $W(\lambda) = W_1(\lambda)W_2(\lambda)$.

THEOREM 2.4. *If $\theta_j = (A_j, B_j, C_j)$ are admissible realization triples for $j = 1, 2$, then the product $\theta = \theta_1 * \theta_2$ is also admissible and*

$$W_\theta(\lambda) = W_{\theta_1}(\lambda)W_{\theta_2}(\lambda).$$

Sketch of the proof Since $\theta_j = (A_j, B_j, C_j)$ is an admissible realization triple, there exists a decomposition $X_j = X_{j+} \oplus X_{j-}$ of the state space X_j for θ_j so that with respect to this decomposition we have

$$A_j = \begin{bmatrix} A_{j+} & 0 \\ 0 & A_{j-} \end{bmatrix}, \ B_j = \begin{bmatrix} B_{j+} \\ B_{j-} \end{bmatrix}, \ C_j = [C_{j+} \ C_{j-}] \tag{2.7}$$

for $j = 1, 2$. Note that the resulting stable triples $\theta_{j+} = (A_{j+}, B_{j+}, C_{j+})$ and antistable triples $\theta_{j-} = (A_{j-}, B_{j-}, C_{j-})$ are also admissible for $j = 1, 2$.

LEMMA 2.5. *The operator $\begin{bmatrix} A_{1+} & B_{1+}C_{2+} \\ 0 & A_{2+} \end{bmatrix}$ with domain equal to $Dom(A_{1+}) \oplus Dom(A_{2+})$ is the generator of a C_0-semigroup.*

PROOF: The proof relies on a perturbation result of Hille and Phillips (see p. 400 of [13]); we omit the details for lack of space. ∎

We also need the following result concerning Sylvester equations with unbounded operator coefficients. We omit the proof.

LEMMA 2.6. *If (A_+, B_+, C_+) is a stable admissible realization and (A_-, B_-, C_-) is an antistable admissible realization, then there is a unique bounded operator $Q : X_- \to X_+$ given by*

$$Qx = \int_0^\infty e^{tA_+} B_+ C_- e^{-tA_-} x \, dt$$

for $x \in D(A_-)$, such that

(i) $Q : Dom A_- \to Dom A_+$

and (ii) $A_+Qx - QA_-x = -B_+C_-x$ for $x \in Dom(A_-)$.

If the realization triples θ_j have the form (2.7) for $j = 1, 2$, then the product $\theta = \theta_1 * \theta_2$ has the form (A, B, C) with

$$A = \begin{bmatrix} A_{1+} & 0 & B_{1+}C_{2+} & B_{1+}C_{2-} \\ 0 & A_{1-} & B_{1-}C_{2+} & B_{1-}C_{2-} \\ 0 & 0 & A_{2+} & 0 \\ 0 & 0 & 0 & A_{2-} \end{bmatrix}$$

$$B = \begin{bmatrix} B_{1+} \\ B_{1-} \\ B_{2+} \\ B_{2-} \end{bmatrix}, \quad C = [C_{1+}, C_{1-}, C_{2+}, C_{2-}] \tag{2.8}$$

By Lemma 2.6 there exist bounded solutions $Q_1 : X_{2-} \to X_{1+}$ and $Q_2 : X_{2+} \to X_{1-}$ of the Sylvester equations

$$A_{1+}Q_1x - Q_1A_{2-}x = -B_{1+}C_{2-}x \text{ for } x \in Dom(A_{2-}) \tag{2.9}$$

$$A_{1-}Q_2x - Q_2A_{2+}x = -B_{1-}C_{2+}x \text{ for } x \in Dom(A_{2+}). \tag{2.10}$$

Then, since Q_1 and Q_2 are bounded, it is easily checked that S defined by

$$S = \begin{bmatrix} I & 0 & 0 & Q_1 \\ 0 & Q_2 & I & 0 \\ 0 & I & 0 & 0 \\ 0 & 0 & 0 & I \end{bmatrix}$$

is invertible, and hence the state space $X := X_{1+} \oplus X_{1-} \oplus X_{2+} \oplus X_{2-}$ for $\theta_1 * \theta_2$ has the direct sum decomposition

$$X = X_\alpha \oplus X_\beta$$

where

$$X_\alpha = Im \begin{bmatrix} I & 0 \\ 0 & Q_2 \\ 0 & I \\ 0 & 0 \end{bmatrix}, \quad X_\beta = Im \begin{bmatrix} 0 & Q_1 \\ I & 0 \\ 0 & 0 \\ 0 & I \end{bmatrix}. \tag{2.11}$$

One now must check that $X_\alpha \oplus X_\beta$ forms the needed exponential dichotomy for the operator A defined in (2.8). The appropriate domain for A works out to be $D(A) = D(A_{1+}) \oplus D(A_{1-}) \oplus D(A_{2+}) \oplus D(A_{2-})$. The verification that $A(X_\alpha \cap D(A)) \subset X_\alpha$ and $A(X_\beta \cap D(A)) \subset X_\beta$ uses the Sylvester equations (2.9) and (2.10). The verification that $A|X_\alpha$ and $(-A)|X_\beta$ are generators for exponentially decaying C_0-semigroups uses Lemma 2.5.

To check that θ is an admissible realization triple it remains to check conditions (2) and

(3) in Definition 2.2. Verification of the A-boundedness of C is straightforward from the definitions. For condition (3), after some computations one can verify that Λ is defined by

$$\Lambda = [\Lambda_{1+}, \Lambda_{1-}, k_{1+} * \Lambda_{2+} + k_{1-} * \Lambda_{2+} + \Lambda_{2+}, k_{1+} * \Lambda_{2-} + k_{1-} * \Lambda_{2-} + \Lambda_{2-}]$$

has all the requisite properties. Here $*$ denotes convolution.

3. Left versus right Wiener-Hopf factorization.

In this section we assume that we are given matrix functions Y_+ and $Y-$ analytic and invertible on the closed right (respectively left) half planes with realizations

$$Y_+(\lambda) = I + C_+(\lambda I - A_+)^{-1}B_+ \tag{3.1}$$

and

$$Y_-(\lambda) = I + C_-(\lambda I - A_-)^{-1}B_- \tag{3.2}$$

associated with admissible realization triples. Since Y_+ is analytic on the whole right half plane, without loss of generality we may assume that A_+ itself generates an exponentially decaying C_0 semigroup (i.e. the exponential dichotomy is trivial and involves only the forward time semigroup). Similarly we may assume that $-A_-$ is the generator of an exponentially decaying semigroup. By a result of [7], then the inverses also have stable (respectively antistable) admissible realizations

$$Y_+(\lambda)^{-1} = I - C_+(\lambda I - A_+^\times)^{-1}B_+ \tag{3.3}$$

$$Y_-(\lambda)^{-1} = I - C_-(\lambda I - A_-^\times)^{-1}B_-. \tag{3.4}$$

where $A_+^\times = A_+ - B_+C_+$, $A_-^\times = A_- - B_-C_-$. If a matrix function $W(\lambda)$ has a factorization

$$W(\lambda) = Y_+(\lambda)Y_-(\lambda)$$

with factors Y_+, Y_- analytic and invertible on the closed right (respectively left) half plane, W is said to have a *left* canonical Wiener-Hopf factorization.

A *right canonical* Wiener-Hopf factorization is one of the form

$$W(\lambda) = W_-(\lambda)W_+(\lambda)$$

where W_+ and W_- are similarly analytic and invertible on the closed right (respectively left) half plane. In the following we indicate how one can compute realizations for the factors of a right canonical factorization in terms of realizations for the factors of a left canonical factorization. The following result extends the main result of [2] to a class of nonrational matrix functions. The proof relies on the geometric factorization principle of Bart, Gohberg Kaashoek and van Dooren (see [3]) as adapted to Wiener-Hopf factorization for functions with exponential dichotomy in [7].

THEOREM 3.1. *Suppose Y_+ (respectively Y_-) is an $m \times m$ matrix function with BGK-admissible stable (respectively antistable) realization (3.1) (respectively (3.2)) such that $det Y_+(\lambda)$ (respectively $det Y_-(\lambda)$) does not vanish on the closed right (respectively left) half plane. Let P and Q denote the unique solutions of the operator Sylvester equations*

$$A_+Qx - QA_-x = -B_+C_-x \text{ for } x \in D(A_-)$$

$$A_-^\times Px - PA_+^\times x = B_-C_+x \text{ for } x \in D(A_+)$$

given by Lemma 2.6.

Then $W := Y_+Y_-$ has a right canonical Wiener-Hopf factorization

$$W(\lambda) = W_-(\lambda)W_+(\lambda)$$

if and only if $(I - QP)$ or equivalently $(I - PQ)$ is invertible. In this case the factors $W_-(\lambda)$ and $W_+(\lambda)$ have admissible realizations given by

$$W_-(\lambda) = I + (C_+Q + C_-)(\lambda I - A_-)^{-1}(I - PQ)^{-1}(-PB_+ + B_-)$$

and

$$W_+(\lambda) = I + (C_+ + C_-P)(I - QP)^{-1}(\lambda I - A_+)^{-1}(B_+ - QB_-)$$

with inverses given by

$$W_-(\lambda)^{-1} = I - (C_+Q + C_-)(I - PQ)^{-1}(\lambda I - A_- + B_-C_-)^{-1}(-PB_+ + B_-)$$

and

$$W_+(\lambda)^{-1} = I - (C_+ + C_-P)(\lambda I - A_+ + B_+C_+)^{-1}(I - QP)^{-1}(B_+ - QB_-).$$

Both authors were partially supported by National Science Foundation grant DMS-8701625.

References

[1] J.A. Ball, I. Gohberg and L. Rodman, *Interpolation Problems for Rational Matrix Functions*, monograph in preparation.

[2] J.A. Ball and A.C.M. Ran, Left versus right canonical Wiener-Hopf factorization, in *Constructive Methods of Wiener-Hopf Factorization* 0T21, Birkhauser (Basel), 1986, 9-38.

[3] H. Bart, I. Gohberg and M.A. Kaashoek, *Minimal Factorization of Matrix and Operator Functions.* Operator Theory: Advances and Applications, 0T1, Birkhäuser Verlag, (Basel), 1979.

[4] H. Bart, I. Gohberg and M.A. Kaashoek, Exponentially dichotomous operators and inverse Fourier transforms, Report 8511/m, Econometric Institute, Erasumus University Rotterdam, The Netherlands, 1985.

[5] H. Bart, I. Gohberg and M.A. Kaashoek, Fredholm theory of Wiener-Hopf equations in terms of realization of their symbols, *Integral Equations and Operator Theory* 8, (1985), 590-613.

[6] H. Bart, I. Gohberg and M.A. Kaashoek, Wiener-Hopf factorization, inverse fourier transforms and exponentially dichotomous operators, *J. Functional Analysis*, (1986), 1-41.

[7] H. Bart, I. Gohberg and M.A. Kaashoek, Wiener-Hopf equations with symbols analytic in a strip, in *Constructive Methods of Wiener-Hopf Factorization* 0T21, Birkhauser (Basel), 1986, pp. 39-75.

[8] R. Curtain, A synthesis of time and frequency domain methods for the control of infinite-dimensional systems: a system theoretic approach, to appear in the series SIAM Frontiers in Applied Mathematics.

[9] R. Curtain and A.C.M. Ran, Explicit formulas for Hankel norm approximations of infinite-dimensional systems, *Integral Equations and Operator Theory* 12 (1989), 455-469.

[10] B. Francis, *A Course in H_∞ Control Theory*, Springer-Verlag (Berlin-Heidelberg-New York), 1987.

[11] P. Fuhrmann, *Linear Systems and Operators in Hilbert Space*, McGraw-Hill (New York), 1981.

[12] J.W. Helton, Systems with infinite-dimensional state space: the Hilbert space approach, Proc. IEEE (1976), 145-160.

[13] E. Hille and R.S. Phillips: *Functional Analysis and Semigroups.* American Mathematics Society, Providence, R.I., 1957.

[14] A.C.M. Ran, Hankel Norm approximation for infinite dimensional systems and Wiener-Hopf factorization, *Modelling, Robustness and Sensitivity Reduction in Control Systems*, NATO ASI Series, F34, Springer-Verlag Berlin Heidelberg, 1987, pp. 57-68.

[15] G. Weiss, The representation of regular linear systems on Hilbert spaces, Proceedings of the Conference on Distributed Parameter Systems, Vorau, 1988, to appear.

FACTORIZATION OF MATRIX POLYNOMIALS IN THE SOLUTION OF NONMONIC MATRIX DIFFERENTIAL EQUATIONS

V. Hernández M. Gassó

Abstract

The solution of the nonmonic matrix differential equation $A_n x^{(n)}(t) + \dots + A_o x(t) = b(t)$, is related to the linear factorization of the matrix polynomial $L(\lambda) = A_n \lambda^n + \dots + A_1 \lambda + A_o$ given by $L(\lambda) = (A_n \lambda - T_1)(I\lambda - T_2) \dots (I\lambda - T_n)$, by using the strict equivalence between the companion pencil of $L(\lambda)$ and an upper bidiagonal pencil defined by matrices A_n, T_1, T_2, ..., T_n.

1. Introduction

Consider the nonmonic matrix differential equation

$$(1) \qquad \sum_{j=0}^{n} A_j x^{(j)}(t) = b(t)$$

where $A_j \in C^{mxm}$, $j = 0, 1, \dots, n$, $b(t)$ is a C^m-valued function of the real variable t and $L(\lambda) = A_n \lambda^n + \dots + A_1 \lambda + A_o$ is a regular matrix polynomial, that is $\det L(\lambda) \neq 0$. It is well known [4], [5], that the general solution of (1) can be obtained from a resolvent triple of $L(\lambda)$. Let (X,Y,Z) be a resolvent triple for $L(\lambda)$, where $X = [X_1, X_2]$, $Y = Y_1 \oplus Y_2$ and $Z = col[Z_1, Z_2]$. If the size of Y_1 is pxp, where $p = degree[\det L(\lambda)]$, and there exists a positive integer q such that $Y_2^q = 0$, then for every (q+n-1)-times continuously differentiable C^m-valued function b(t), the general solution of (1) is given by

$$x(t) = X_1 e^{Y_1 t} c + \int_{t_o}^{t} X_1 e^{Y_1 (t-s)} Z_1 b(s) ds - \sum_{i=0}^{q-1} X_2 Y_2^i Z_2 b^{(i)}(t)$$

where $c \in C^p$ is an arbitrary vector.

In this paper we relate the problem (1) with the associated algebraic matrix equation

$$(2) \qquad \sum_{j=0}^{n} A_j X^j = 0$$

and the concept of solvents of (2) [3]. The set of matrices $T_1, T_2, \ldots, T_n \in C^{mxm}$ is said to form a chain of solvents of (2) if T_i is a solution of the algebraic matrix equation $L_i(X)=0$, $i = 2, 3, \ldots, n$, where $L_n(\lambda) = L(\lambda)$, $L_1(\lambda) = A_n\lambda - T_1$ and $L_i(\lambda) = L_{i-1}(\lambda)(I\lambda - T_i)$, $i = n, n-1, \ldots, 2$. Then it follows that

$$(3) \qquad L(\lambda) = (A_n\lambda - T_1)(I\lambda - T_2) \ldots (I\lambda - T_n).$$

Reciprocally, if $L(\lambda)$ admits a descomposition in linear factors of the form (3), then $T_1, T_2, \ldots, T_n$ is a chain of solvents of (2). From [6, proposition 1, page 781], it can be deduced that the factorization given by (3) is related to the strict equivalence between the pencils $C_L(\lambda)$ and $J(\lambda)$, where $C_L(\lambda)$ is the companion pencil of $L(\lambda)$

$$C_L(\lambda) = \mathrm{diag}(I, \ldots, I, A_n)\ \lambda - C_L$$

$$(4) \qquad C_L = \begin{bmatrix} 0 & I & 0 & \cdot & \cdot & \cdot & 0 \\ 0 & 0 & I & \cdot & \cdot & \cdot & 0 \\ \cdot & \cdot & \cdot & \cdot & \cdot & \cdot & \cdot \\ \cdot & \cdot & \cdot & \cdot & \cdot & \cdot & \cdot \\ 0 & 0 & 0 & \cdot & \cdot & \cdot & I \\ -A_o & -A_1 & -A_2 & \cdot & \cdot & \cdot & -A_{n-1} \end{bmatrix}$$

and $J(\lambda)$ is the Jordan pencil defined by

$$J(\lambda) = \mathrm{diag}(I, \ldots, I, A_n)\ \lambda - J$$

$$(5) \qquad J = \begin{bmatrix} T_n & I & & & & \\ & T_{n-1} & I & & & \\ & & \cdot & & \cdot & \\ & & & & T_2 & I \\ & & & & & T_1 \end{bmatrix}.$$

In [1] it has been proved that this strict equivalence holds when $L(\lambda)$ is a matrix polynomial not necessarily regular.

In the next section we obtain explicit expressions of the factors of equivalence, which we need in the solution of the matrix differential equation given by (1).

2. The main result.

Given the matrix polynomial $L(\lambda) = A_n\lambda^n + \ldots + A_1\lambda + A_o$, $A_j \in C^{mxm}$, $j = 0, 1, \ldots, n$, and matrices $T_1, T_2, \ldots, T_n \in C^{mxm}$, we define unit lower block-triangular matrices $Q(T_i)$, $P(T_i)$, $i = 1, 2, \ldots, n$, by the expressions

(6) $$Q(T_i) = \begin{bmatrix} I & & & & & \\ T_i & I & & & & \\ \cdot & T_i & \cdot & & & \\ \cdot & \cdot & \cdot & \cdot & & \\ \cdot & \cdot & \cdot & \cdot & I & \\ T_i^{i-1} & T_i^{i-2} & \cdot & \cdot & T_i & I \end{bmatrix} \in C^{mixmi}$$

(7) $$P(T_i) = \begin{bmatrix} I & & & & & \\ -T_i & I & & & & \\ & -T_i & \cdot & & & \\ & & \cdot & I & & \\ & & & -T_i & I & \\ & & & & -A_nT_i & I \end{bmatrix} \in C^{mixmi}.$$

We call these matrices unit lower block-triangular matrices because their diagonal blocks are equal to the identity matrix. Note that

$P(T_i) = \mathrm{diag}(I, \ldots, I, A_n)Q(T_i)^{-1} + \mathrm{diag}(0, \ldots, 0, I-A_n)$

and $P(T_i) = Q(T_i)^{-1}$ when $A_n = I$. Then we obtain the following result.

Lemma

If $T_n \in C^{mxm}$ is a solution of the algebraic matrix equation $A_nX^n + \ldots + A_1X + A_o = 0$, then

$$P(T_n)\ \mathrm{diag}(I, \ldots, I, A_n)Q(T_n) = \mathrm{diag}(I, \ldots, I, A_n)$$

(8) $$P(T_n)C_LQ(T_n) = \left[\begin{array}{c|ccccc} T_n & I & 0 & \cdot & \cdot & 0 \\ \hline 0 & & & & & \\ \cdot & & & & & \\ \cdot & & & C_M & & \\ \cdot & & & & & \\ 0 & & & & & \end{array}\right]$$

where C_L, C_M are the matrices of the form (4) corresponding to $L(\lambda)$, $M(\lambda)$, with $M(\lambda)$ given by

$$L(\lambda) = M(\lambda)(I\lambda - T_n).$$

Proof: By direct substitution in (8), of matrices $P(T_n)$, $Q(T_n)$, given by (6), (7).

Proposition

If $T_1, T_2, \ldots, T_n$ is a chain of solvents of the algebraic matrix equation (2), then there exist unit lower block-triangular matrices P, $Q \in C^{mnxmn}$, such that

$$\text{(9)} \qquad \begin{aligned} &P\,\mathrm{diag}(I, \ldots, I, A_n)Q = \mathrm{diag}(I, \ldots, I, A_n) \\ &PC_LQ = J \end{aligned}$$

where C_L is the companion matrix of $L(\lambda)$ and J is the Jordan block matrix defined by (5).

Proof: The matrices P, Q are obtained from the previous Lemma,

$$\text{(10)} \qquad \begin{aligned} &P = P_2P_3\ldots P_n \\ &Q = Q_nQ_{n-1}\ldots Q_2 \end{aligned}$$

where Q_i, P_i, i = 2, 3, ..., n , are defined by $Q_n = Q(T_n)$,$Q_{n-1} = \mathrm{diag}(I,Q(T_{n-1})),\ldots,Q_2 = \mathrm{diag}(I,\ldots,I,Q(T_2))$; $P_n = P(T_n)$, $P_{n-1} = \mathrm{diag}(I,P(T_{n-1})),\ldots,P_2 = \mathrm{diag}(I,\ldots,I,P(T_2))$ with $Q(T_i)$, $P(T_i)$ given by (6), (7), respectively, i = 2, 3, ..., n.

From this result we can see that if $L(\lambda)$ is a matrix polynomial, not necessarily regular, which can be decomposed in linear factors, $L(\lambda) = (A_n\lambda-T_1)(I\lambda-T_2)\ldots(I\lambda-T_n)$, then the companion pencil of $L(\lambda)$, $C_L(\lambda)$, is strictly equivalent to the Jordan pencil $J(\lambda)$ defined by (5).

3. Application to differential equations.

We consider the initial value problem

$$\text{(11)} \qquad \begin{aligned} &\sum_{j=0}^{n} A_j x^{(j)}(t) = b(t) \\ &x^{(j-1)}(t_o) = e_j \ , \quad j = 1, 2, \ldots, n \end{aligned}$$

where $e_j \epsilon C^m$, j= 1, 2, ..., n , and $t_o \epsilon R$. By introducing the change of variables $x^{(j-1)}(t) = y_j(t)$, j = 1,2,...,n by performing the coordinate transformation $\mathrm{col}[y_1(t),\ldots,y_n(t)] = Q\ \mathrm{col}[z_1(t),\ldots,z_n(t)]$ and premultiplying by matrix P, where P and Q are given by (10), we obtain, from (9), the following system

$$(12)\quad \begin{aligned} z_1^{(1)}(t) &= T_n z_1(t) + z_2(t), & z_1(t_o) &= f_1 \\ &\;\;\vdots \\ z_{n-1}^{(1)}(t) &= T_2 z_{n-1}(t) + z_n(t), & z_{n-1}(t_o) &= f_{n-1} \\ A_n z_n^{(1)}(t) &= T_1 z_n(t) + b(t), & z_n(t_o) &= f_n \end{aligned}$$

where

$$(13)\quad \begin{aligned} f_i &= E_i Q^{-1} \mathrm{col}[e_1,\dots,e_n], \quad i = 1, 2, \dots, n \\ E_i &= [0,\dots,0,\underset{(i)}{I},0,\dots,0] \in C^{mxnm} \end{aligned}$$

If there exists a $\lambda \in C$ such that $A_n\lambda - T_1$ is invertible then $A_n z_n^{(1)}(t) = T_1 z_n(t)$ is tractable [2]. Let $\hat{A}_n = (A_n\lambda - T_1)^{-1}A_n$ be such that $\mathrm{Ind}(\hat{A}_n) = k$ (the smallest nonnegative integer k such that $\mathrm{rank}(\hat{A}_n^{\,k}) = \mathrm{rank}(\hat{A}_n^{\,k+1})$. If $b(t)$ is a k-times continuously differentiable C^m-valued function around t_o then the vector f_n, given by (13), is a consistent initial vector associated with t_o for the nonhomogeneous equation $A_n z_n^{(1)}(t) = T_1 z_n(t) + b(t)$ if and only if

$$(14)\quad f_n \in \hat{w}_n + R(\hat{A}_n^{\,k})$$

with

$$\begin{aligned} \hat{w}_n &= -(I-\hat{A}_n\hat{A}_n^{\,D}) \sum_{i=0}^{k-1} (\hat{A}_n\hat{T}_1^{\,D})^i \hat{T}_1^{\,D} \hat{b}^{(i)}(t_o) \\ \hat{T}_1 &= (A_n\lambda - T_1)^{-1} T_1 \\ \hat{b}(t) &= (A_n\lambda - T_1)^{-1} b(t) \end{aligned}$$

where $(\ )^D$ denotes the Drazin inverse of a matrix and $R(\ )$ the subspace generated by the columns of a matrix. Then the unique solution of the initial value problem

$$A_n z_n^{(1)}(t) = T_1 z_n(t) + b(t)\ , \quad z_n(t_o) = f_n$$

is given by [2],

$$(15)\quad \begin{aligned} z_n(t) = {} & e^{\hat{A}_n^{\,D}\hat{T}_1(t-t_o)} \hat{A}_n\hat{A}_n^{\,D} f_n + \int_{t_o}^{t} e^{\hat{A}_n^{\,D}\hat{T}_1(t-s)} \hat{A}_n^{\,D}\hat{b}(s)\,ds - \\ & - (I-\hat{A}_n\hat{A}_n^{\,D}) \sum_{i=0}^{k-1} (\hat{A}_n\hat{T}_1^{\,D})^i \hat{T}_1^{\,D} \hat{b}^{(i)}(t). \end{aligned}$$

By backward substitution in (12) we obtain the following result.

Theorem

Let us consider the initial value problem (11), and we suppose that there exists a $\lambda \in C$ such that $\det(A_n\lambda^n+\ldots+A_1\lambda+A_o) \neq 0$. Let $T_1, T_2, \ldots, T_n$ be a chain of solvents of the matrix equation (2), and let $\hat{A}_n = (A_n\lambda-T_1)^{-1}A_n$ be such that $\mathrm{Ind}(\hat{A}_n) = k$. If $b(t)$ is a k-times continuously differentiable C^m-valued function around t_o, then we obtain the following results:

(i) If the vector f_n, given by (13), satisfies the condition (14) then $e_1, e_2, \ldots, e_n$ are consistent initial vectors for the problem (11).

(ii) The unique solution of (11) is given by

$$x(t) = e^{T_n(t-t_o)} f_1 + \int_{t_o}^{t} e^{T_n(t-s_1)} e^{T_{n-1}s_1} f_2 ds_1 + \ldots +$$

$$+\int_{t_o}^{t} \ldots \int_{t_o}^{s_{n-3}} e^{T_n(t-s_1)} \ldots e^{T_2 s_{n-2}} f_{n-1} ds_{n-2} \ldots ds_1 +$$

$$+\int_{t_o}^{t} \ldots \int_{t_o}^{s_{n-2}} e^{T_n(t-s_1)} \ldots e^{T_2(s_{n-2}-s_{n-1})} z_n(s_{n-1}) ds_{n-1} \ldots ds_1$$

with $z_n(t)$ given by (15).

This result is an extension of [7, proposition 1, page 252] where the solution of monic matrix differential equations was considered.

References

[1] M.A. Beitia and I. Zaballa, "Factorization of the matrix polynomial $A(\ )=A_o\lambda^t+\ldots+A_t$", Linear Algebra Appl. 121(1989), 423-432.

[2] S.L. Campbell, "Singular Systems of Differential Equations", Pitman Publishing Ltd., 1980.

[3] J.E. Dennis, J.F. Traub and R.P. Weber, "The algebraic theory of matrix polynomials", SIAM J. Numer. Anal. 13(1976), 831-845.

[4] I.Gohberg, M.A. Kaashoek, L. Lerer and L. Rodman, "Common multiples and common divisors of matrix polynomials, I. Spectral method", Indiana J. of Math., 30(1981), 321-357.

[5] I. Gohberg, M.A. Kaashoek, L. Lerer and L. Rodman, "Common multiples and common divisors of matrix polynomials, II. Vandermonde and resultant matrices", Linear and Multilinear Algebra, 12(1982), 159-203.

[6] I. Gohberg, M.A. Kaashoek and P. Lancaster, "General theory of regular matrix polynomials and band Toepliz operators", Integral Equations and Operator Theory 11(1988), 776-882.
[7] V. Hernández and F. Incertis, "A block bidiagonal form for block companion matrices", Linear Algebra Appl. 75(1986), 241-256.

PROPERTIES OF MINIMAL FIRST DEGREE FACTORIZATIONS FOR ARBITRARY RATIONAL MATRICES

JOOS VANDEWALLE and SHAOHUA TAN

Abstract

In [1] a state space technique is used in order to prove that any complex square or rectangular rational matrix $G(s)$ of Smith-McMillan degree n can always be factorized as a product of n first degree rational matrices, the so-called minimal first degree factorization. The procedure is constructive in the sense that the extraction of a first degree factor can be obtained by performing some linear operations on the system matrix of a state space description of $G(s)$. However there is still much freedom in the factor which is being extracted, e.g., the number of intermediate interconnections may vary depending on the rank of a submatrix of the system matrix; and certain classes of rational matrices may have special factorizations.

It is the aim of this paper to study the properties of minimal first degree factorizations with the emphasis on the reduction of the number of interconnections, since this reduction issue is not only structurally interesting, but also important for implementations (VLSI). It is shown that the number of intermediate connections does not increase when the A-matrix has the simple structure (no Jordan-blocks). In the general case, some upper bounds are derived on the number of interconnections. By exploring some aspects of similarity transformations, we are able to present a modified factorization algorithm which will be shown to produce a minimal factorization with smaller number of interconnections.

1 Introduction

In a recent publication [1], it is shown that any rational matrix $G(s)$ $(m \times p)$ of degree n always admits a *minimal first degree factorization* of the form

$$G(s) = G_1(s) \cdots G_n(s) \tag{1}$$

where each $G_i(s)$ $(i = 1, \ldots, n)$ is of degree 1. While the theory developed there has mainly focussed on splitting the degree of $G(s)$ into the first degree factors, no attention has been paid to the dimensions of the factors generated via the scheme. In this note, we continue the investigation by showing that these dimensions are always upper-bounded by the number $(n + m + p)/2$. When the condition $m - p \geq n - 1$ $(m \geq p)$ holds, the factorization procedure presented in [1] will generate a minimal factorization (1) with the dimensions of the factors less than $m \times m$. It is generally known that the dimensions of the factors also depend upon choices of similarity transformations. Therefore we shall provide a modification using this principle to the original procedure in order to reduce the dimensions of the factors. We show, among other things, that with this modification the dimensions of the factors will generically not exceed m.

2 Factorization of rational matrices via the state space description

To facilitate the ensuing discussion we first lay down a few conventions, and then restate the factorization procedure presented in [1] for easy reference.

Let an $m \times p$ rational matrix $G(s)$ of degree n be given, we shall always assume, throughout this note, that $m \geq p$. If this is not the case, its transpose will be considered. When $G(s)$ is factorized into (1), the dimensions of the factors can be viewed as a sequence of $n-1$ integers, each representing the number of interconnections between two adjacent factors. This sequence is called the connection sequence, and all the results concerning the dimensions of the factors will be stated in terms of such a sequence.

To factorize an arbitrary $G(s)$, let us first realize it in its minimal state-space form, then apply a similarity transformation to bring the system matrix of $G(s)$ into the following form

$$P(s) = \begin{bmatrix} s-\alpha & -h^T & \bar{b}^T \\ 0 & I_{n-1}s - A_r & B_r \\ -c_1 & -\bar{c} & D \end{bmatrix} \tag{2}$$

where the superscript T denotes transpose. Note in particular that c_1, h, b, are vectors. Construct from (2) the following $(m+1) \times (n+p-1)$ submatrix

$$\pi = \begin{bmatrix} -h^T & \bar{b}^T \\ -\bar{c} & D \end{bmatrix}. \tag{3}$$

Then π can be decomposed into

$$\pi = \begin{bmatrix} b_1^T \\ D_1 \end{bmatrix} [-C_r \;\; D_r] \tag{4}$$

where the matrices on the right hand side of (4) have dimensions $(m+1) \times f, f \times (n+p-1)$ respectively, and $f = rank\ \pi$.

Thus a first degree factor of $G(s)$ can be extracted from the left based on (4), i.e.

$$G(s) = G_1(s)G_r(s) \tag{5}$$

and the system matrices for $G_1(s), G_r(s)$ are given respectively by

$$P_1(s) = \begin{bmatrix} s-\alpha & b_1^T \\ -c_1 & D_1 \end{bmatrix}, \quad P_r(s) = \begin{bmatrix} I_{n-1}s - A_r & B_r \\ C_r & D_r \end{bmatrix}$$

The same procedure will be applied to $P_r(s)$ and so on in order to achieve a complete first degree factorization.

It is obvious from the above factorization procedure that the dimensions of the factors are determined by the rank of the matrix π, which is influenced by choices of similarity transformation and its actual structure as depicted in (3). Our intention, therefore, is to explore the structure of π so as to reveal the tendency and the bound of the connection sequence generated by the factorization procedure. Moreover, we shall explore the properties of similarity transformations in relation to the rank of the matrix π, and shall provide two tricks to reduce the dimensions of the factors. These tricks constitute a modification to the original factorization scheme.

3 Simple bounds for the minimal first degree factorization procedure

It is clear from the preceding discussion that the connection sequence is generated as ranks of a sequence of matrices of the form (3). We observe from (3) that the dimension of this matrix is simply $(m_i + 1) \times (n + p - (i + 1))$ at each stage of the factorization. Here $m_0 = m$, and $\{m_i\}$, $i = 1, 2, \ldots, n - 1$ is the connection sequence. If this matrix is such that the number of its rows exceeds the number of its columns right at the start of the factorization, the connection sequence generated by the factorization procedure will never be greater than m, or more precisely we have

Observation 1 *For a given $G(s)$, if $m - p \geq n - 1$ (recall that $m \geq p$) then the connection sequence generated by the factorization procedure will be upperbounded by a descending sequence*

$$\{m - (i - 1)\} \quad (i = 1, \ldots, n - 1)$$

Proof:
To show this, simply notice that if $m - p \geq n - 1$, then $f = rank\ \pi \leq n + p - 1 \leq m$, moreover, at each next factorization stage the number of the columns of the matrix π is reduced by 1.

If the condition $m - p \geq n - 1$ is not met, then the factorization procedure may produce a connection sequence in which some of the elements will exceed m. However, we can derive an upper-bound in this case.

Observation 2 *For any $G(s)$ with $m - p < n - 1$, the connection sequence produced by a factorization scheme is upper-bounded by $(n + m + p)/2$*

Proof:
It is clear that the number of rows in (3) after the ith factorization is given by $1 + m_i \leq 1 + m_{i-1} + 1 \leq 1 + m + i$, whereas the number of columns decreases, being given by $n + p - i - 1$. Now define

$$r := min\{i \mid m + i \geq n + p - i - 1\}. \tag{6}$$

Then $m_i \leq m + r$ for each i: If $i > r$, then $m_i \leq n + p - i \leq n + p - r - 1 \leq m + r$, and for $i \leq r$ one has $m_i \leq m + i \leq m + r$. Now $m + r = n + p - r - 1$ or $m + r = n + p - r$. In both cases one has

$$r \leq [n - (m - p)]/2, \tag{7}$$

and hence $m + r \leq (m + n + p)/2$. This proves the observation.

Let us consider an example in order to illustrate the implication of the above observation. Consider a 3×2 matrix $G(s)$ with degree 5. Then the factorization scheme will produce five first degree factors with the dimensions no greater than 5×5. One should realize that this is simply an upper-bound; hence the factors actually obtained may have far less dimensions than 5×5.

4 Modifications on the factorization procedure in order to reduce the interconnectivity

The upper bounds given above may not seem to be satisfactory if the dimensions of factors need to be seriously limited. This happens, for example, in the VLSI filter implementation in which an increase in dimensions of the factors would mean a considerable increase in the complexity. Recall that the connection sequence is determined as ranks of the matrix (3) constructed at each stage of the factorization and the rank of this matrix depends on the applied similarity transformation. The main idea is therefore to apply suitably chosen similarity transformations to (2) in such a way that the rank of (3) will not increase at all. Applying these additional similarity transformations to $P(s)$ to induce certain structures before going from (2) to (3) is a modification to the original procedure. Though the general implication of this modification may yield further results, some useful results have already been obtained.

Theorem 1 *If the matrix A in a minimal state-space realization of $G(s)$ has the simple structure (no Jordan-blocks), then the connection sequence produced by the modified scheme will not exceed m.*

Proof:
For square $(m \times m)$ rational matrix $G(s)$, the theorem has been shown in, for example, [2, 3]. Here we only consider the case $m > p$.

Start with (2) and assume, without loss of generality, that the first row of D is zero. If this is not the case, we shall apply a nonsingular K to $G(s)$ such that $KG(s)$ has this property, this will have no effect on the factorizability of $G(s)$. Now consider the first row of $[-c_1 \;\; -\bar{c}]$. Obviously, if A is first put in the diagonal form, we can always find a similarity transformation such that the system matrix $P(s)$ is in the form (2) and the first row of $[-c_1 \;\; -\bar{c}]$ will be of the form $[\lambda \; 0 \cdots 0]$. Consequently, the rank of the matrix (3) can at most be m. The same argument can be applied to $G_r(s)$ till the desired factorization is obtained. The proof is complete.

The above result does not seem to carry over to the general case, at least we know of examples which can not be factorized this way [2]. In spite of this fact, we can still come up with a few local tricks to be applied at each stage of the factorization. Consequently, the connection sequence obtained using these tricks may ensure that at least locally the sequence is upper-bounded by m.

Trick 1. Start with the system matrix (2), if it is possible to induce a $1 \times (1+n-1+p)$ row of the form $[\lambda \; e^T \; 0]$ $(\lambda \neq 0)$ in the matrix $[-c_1 \;\; -\bar{c} \; D]$ by using the row operations only, then the dimension of the factorization at this stage will not exceed m. If this is not possible for a pole α, we can shuffle all the other poles of $G(s)$ to check if it is possible.

To prove the assertion, we note that there exists a similarity transformation in this case so as to bring this row into $[\lambda \; 0 \cdots 0]$ while keeping the structure of A

unchanged. As a result, the matrix (3) will not have rank greater than m. To implement this trick, one simply row-compresses D first, then check if the elements in C, corresponding to the zero rows of D are non-zero. If so, pick up one of these rows.

Trick 2. Start with the system matrix of the following form

$$\begin{bmatrix} I_{n-1}s - A_1 & u & B_1 \\ 0 & s-\alpha & b_2^T \\ c_1 & c_2 & D \end{bmatrix} \tag{8}$$

If it is possible to induce a $(n-1+1+p) \times 1$ column of the form $[e^T \; \lambda \; 0]^T$ in the matrix $[B_1^T \; b_2 \; D^T]^T$ by using the column operations only, then the dimension of the factorization at this stage will not exceed p. If this is not possible for a pole α, all the poles of $G(s)$ can be shuffled to check if it is possible.

Note that Trick 2 is not a simple algebraic dual of Trick 1. It is a property of the matrix $[B_1^T \; b_2 \; D^T]^T$, and works independently from Trick 1.

Combining these two tricks with our original factorization procedure, we arrive at the following modified version of the scheme, which will produce a minimal first degree factorization with limited number of interconnections:

Modified Algorithm

- Step 1 : minimally realized $G(s)$ into $P(s)$;
- Step 2 : Apply a similarity transformation to bring A into upper triangular.
- Step 3 : Apply Trick 1 to the transformed system matrix. If successful, go to step 5; otherwise go to the next step;
- Step 4 : Apply Trick 2. If successful go to step 5; otherwise conclude that at this stage the dimension of the factorization may be greater than m, and go to the next step;
- Step 5 : Factorize the matrix (3) so as to obtain the factors;
- Step 6 : If n is 1, then stop; otherwise replace $G(s)$ by $G_r(s)$ and go to Step 2.

Our computation experiences with the above algorithm have shown that it almost always gives a limited interconnection factorization. It especially works well for rectangular and singular square rational matrices.

5 Conclusions

By analysing in detail the factorization procedure presented in [1], we have derived an upper-bound for the connection sequence. We have also shown that under various conditions the connection sequence will actually be less than m.

Based on the observation that similarity transformations will change the rank of the matrix (3), a modification to the original procedure is also devised. By applying this modified scheme in dealing with factorization problems, one may find it a very attractive algorithm.

Acknowledgement We thank the anonymous referee for his suggestions which has improved the presentation considerably.

References

[1] Tan S.H. and Vandewalle J., "On factorizations of rational matrices", *IEEE Trans. on Circuits and Systems*, Sept. 1988, pp. 1179-1182.

[2] Dewilde P. and Vandewalle J., "On the factorization of a nonsingular rational matrix", *IEEE Trans. Circuits Syst.*, Vol. CAS-22, pp. 637-645, 1975.

[3] Cohen N, "On minimal factorizations of rational matrix functions", *Integr. Eq. Oper. Theory*, Vol. 6, pp. 647-671, 1983.

[4] Bart H., Gohberg I., Kaashoek M. and Van Dooren P., "Factorizations of transfer functions", *SIAM J. Control Opt.*, Vol. 18, pp. 675-696, 1980.

Polynomial matrix factorizations via pole-zero cancelation

P. Van Dooren

Abstract

In this paper we give an algorithm for obtaining factorizations $P(\lambda) = P_1(\lambda) \cdot P_2(\lambda)$, where $P_1(\lambda)$ and/or $P_2(\lambda)$ are polynomial matrices and $P_1(\lambda)$ is regular. Moreover the factors $P_1(\lambda)$ and $P_2(\lambda)$ are such that either the *poles* of $[P_1(\lambda)]^{-1}$ and $P_2(\lambda)$ are in a prescribed set Γ of the extended complex plane, or their *zeros*. Such factorizations cover e.g. the specific cases of coprime factorization, polynomial factor extraction and GCD extraction. The algorithm works on the state space (or generalized state space) realization of $P(\lambda)$ and derives the corresponding realizations of the factors.

1 Introduction

Several problems occurring in the literature of linear systems theory can be rephrased as a factorization problem of some rational matrix $R(\lambda)$. We consider a certain class of such factorizations, namely where the $p \times m$ rational matrix $R(\lambda)$ is factored into a product of two rational matrices :

$$R(\lambda) = R_1(\lambda) \cdot R_2(\lambda), \tag{1}$$

where $R_1(\lambda)$ is $p \times p$ regular and of minimal degree and where, given a set Γ of the extended complex plane, one of the following two conditions is satisfied :

a) the *poles* of $R_1^{-1}(\lambda)$ and $R_2(\lambda)$ lie in Γ (2.a)

b) the *zeros* of $R_1^{-1}(\lambda)$ and $R_2(\lambda)$ lie in Γ (2.b)

We give a general algorithm for obtaining such factorizations using the state space (or generalized state space) realization of $R(\lambda)$ and yielding the corresponding realizations of the factors. Applying this to problems involving polynomial matrices, one solves the following standard polynomial factorization problems :

1) Coprime factorization

Given a $p \times m$ rational matrix $R(\lambda)$, one wants to find polynomial matrices $D(\lambda)$ and $N(\lambda)$ with $D(\lambda)$ regular, such that $R(\lambda) = D^{-1}(\lambda) \cdot N(\lambda)$. This fits into the above formulation with condition (2.a), where $\Gamma = \{\infty\}$.

2) GCD extraction

Let $P_i(\lambda)$, $i, 1, \cdots, k$ be a set of polynomial matrices of dimensions $p \times m_i$, then their greatest common (left) divisor (GCD) is defined as the *regular* polynomial matrix $D(\lambda)$ such that : $P(\lambda) = D(\lambda) \cdot Q(\lambda)$ where $P(\lambda) = [P_1(\lambda), \cdots, P_k(\lambda)]$ and $Q(\lambda) = [Q_1(\lambda), \cdots, Q_k(\lambda)]$ and where the *quotients* $Q_i(\lambda)$ form together a polynomial matrix $Q(\lambda)$ with *no* Smith zeros anymore. Hence, both $D^{-1}(\lambda)$ and $Q(\lambda)$ have all their McMillan zeros at ∞, which is thus condition (2.b) with $\Gamma = \{\infty\}$.

3) Polynomial factor extraction

Given a $p \times m$ polynomial matrix $P(\lambda)$, one wants to find a $p \times p$ *regular divisor* $P_1(\lambda)$ which contains all the zeros of $P(\lambda)$ inside a set Γ_f *not* containing the point at infinity : $P(\lambda) = P_1(\lambda) \cdot P_2(\lambda)$. The factors $P_1^{-1}(\lambda)$ and $P_2(\lambda)$ have thus their zeros in the complement $\Gamma = \Gamma_f^c$ and this factorization then satisfies (2.b).

We show in the sequel that there always exist such factorizations for any set Γ and that they are in fact far from unique unless some additional conditions are imposed. We typically require the (McMillan) degree δ of $R_1(\lambda)$ to be minimal, which implies that

a) $\delta(R_1) = \#$*poles* of $R(\lambda)$ outside Γ (3.a)

b) $\delta(R_1) = \#$*zeros* of $R(\lambda)$ outside Γ (3.b)

depending on the choice (2.a) or (2.b) for the factorization (1). Notice also that to each factorization of the above type there corresponds a "dual" factorization where the role of $R_1(\lambda)$ and $R_2(\lambda)$ is interchanged. These are easily obtained by working as above on the transpose of $R(\lambda)$ and transposing back the obtained factors. These dual factorizations are often called a "left" and "right" factorization, respectively.

Although the above theory holds for proper rational matrices (since we suppose $R(\lambda)$ has a state space realization), we can apply it here to polynomial matrices. One can indeed always perform a 1st degree conformal mapping $\lambda = (a\mu + b)/(c\mu + d)$ which does not affect the degree of rational matrices and which turns $P(\lambda)$ into a *proper* rational matrix $R(\mu) \doteq P(\frac{a\mu+b}{c\mu+d})$. The set $\Gamma(\lambda)$ of course has to be transformed accordingly to $\Gamma(\mu)$. Another way of by-passing this difficulty is to use *generalized state space realizations* [14], but since the theory is slightly more involved then, we prefer to stick to state space models. Below, we therefore talk about rational matrices and their state space representations and we only comment in the last section on problems involving polynomial matrices.

2 A recursive approach

For the development of our algorithm we were strongly inspired by [1], [4], [9]. There it is shown that it is always possible to find a regular $p \times p$ rational transfer function $C(\lambda)$ of degree 1, i.e. with one pole γ and one zero δ, such that in the product $R_2(\lambda) = C(\lambda) \cdot R(\lambda)$, *either* δ cancels with a pole (say α_1) of $R(\lambda)$, *or* γ cancels with a zero (say β_1) of $R(\lambda)$. For such a cancelation to occur, one of course needs $\delta = \alpha_1$, respectively, $\gamma = \beta_1$, but in the matrix case some additional vector conditions are required and can always be satisfied as shown in [4]. Let us represent the rational matrix $R(\lambda)$ via its poles $\{\alpha_i \mid i = 1, \cdots, \ell\}$ and zeros $\{\beta_j \mid j = 1, \cdots, k\}$, then the product $R_2(\lambda) = C(\lambda) \cdot R(\lambda)$ can be represented as either of the following two :

$$\left[\frac{\not\delta}{\gamma}\right]\left(\frac{\beta_1,\cdots,\beta_k}{\not\alpha_1,\cdots,\alpha_\ell}\right) = \left(\frac{\beta_1,\beta_2,\cdots,\beta_k}{\gamma,\alpha_2,\cdots,\alpha_\ell}\right), \quad \left[\frac{\delta}{\not\gamma}\right]\left(\frac{\not\beta_1,\cdots,\beta_k}{\alpha_1,\cdots,\alpha_\ell}\right) = \left(\frac{\delta,\beta_2,\cdots,\beta_k}{\alpha_1,\alpha_2,\cdots,\alpha_\ell}\right). \quad (4,5)$$

Here we use square brackets for *regular* rational matrices and round ones for (possibly) *singular* rational matrices. The number of poles is by definition the McMillan degree of the (regular or singular) rational matrix. Notice that for regular rational matrices, this also equals the number of zeros, while for the singular case, the number of zeros can be less than the McMillan degree [14]. Moreover, the inverse $C^{-1}(\lambda)$ of a regular rational matrix has its poles and zeros interchanged. Therefore, from (4-5) one obtains for $R(\lambda) = C^{-1}(\lambda).R_2(\lambda) = R_1(\lambda) \cdot R_2(\lambda)$:

$$\left(\frac{\beta_1,\cdots,\beta_k}{\alpha_1,\cdots,\alpha_\ell}\right) = \left[\frac{\gamma}{\alpha_1}\right]\cdot\left(\frac{\beta_1,\cdots,\beta_k}{\gamma,\alpha_2,\cdots,\alpha_\ell}\right) \text{ or } \left(\frac{\beta_1,\cdots,\beta_k}{\alpha_1,\cdots,\alpha_\ell}\right) = \left[\frac{\beta_1}{\delta}\right]\cdot\left(\frac{\delta,\beta_2,\cdots,\beta_k}{\alpha_1,\cdots,\alpha_\ell}\right). \quad (6,7)$$

It is shown in [4] that in the construction of $C(\lambda)$ in (4) the pole γ can be chosen *arbitrarily* while δ is fixed since it must cancel with α_1. If α_1 would have been the only *pole* of $R(\lambda)$ outside a given set Γ, then the factorization (6) with $R_1(\lambda) = C^{-1}(\lambda)$ would correspond to a factorization as described in section 1. All conditions would indeed be satisfied iff γ was chosen inside Γ as well, which is always possible as indicated above. The same holds for (5) where $\gamma = \beta_1$ cancels a zero β_1 of $R(\lambda)$. If β_1 was the only *zero* outside Γ, then (7) would satisfy all conditions of the factorization described in section 1, provided δ was chosen inside Γ. The factor $C(\lambda)$ in (4-5) has thus "dislocated" one pole $\alpha_1 = \delta$ (resp. zero

$\beta_1 = \gamma$) outside Γ to a pole γ (resp. zero δ) inside Γ, this using a factor $R_1^{-1}(\lambda) = C(\lambda)$ which itself has a pole γ (resp. zero δ) inside Γ. When several poles $\{\alpha_i \mid i = 1, \cdots, \ell_0\}$ (resp. zeros $\{\beta_j \mid j = 1, \cdots, k_0\}$) are outside Γ, they can be dislocated recursively one after the other by such first degree sections $C_i(\lambda)$ as described in (4-7). This then yields :

$$\left[\frac{\cancel{\delta}_{\ell_0}}{\gamma_{\ell_0}}\right] \cdots \left[\frac{\cancel{\delta}_2}{\gamma_2}\right] \cdot \left[\frac{\cancel{\delta}_1}{\gamma_1}\right] \cdot \left(\frac{\beta_1, \beta_2, \cdots, \beta_k}{\cancel{\alpha}_1, \cdots, \cancel{\alpha}_{\ell_0}, \alpha_{\ell_0+1}, \cdots, \alpha_\ell}\right) = \left(\frac{\beta_1, \beta_2, \cdots, \beta_k}{\gamma_1, \cdots, \gamma_{\ell_0}, \alpha_{\ell_0+1}, \cdots, \alpha_\ell}\right), \tag{8}$$

$$\left[\frac{\delta_{\ell_0}}{\cancel{\gamma}_{\ell_0}}\right] \cdots \left[\frac{\delta_2}{\cancel{\gamma}_2}\right] \cdot \left[\frac{\delta_1}{\cancel{\gamma}_1}\right] \cdot \left(\frac{\cancel{\beta}_1, \cdots, \beta_{k_0}, \beta_{k_0+1}, \cdots, \beta_k}{\alpha_1, \alpha_2, \cdots, \alpha_\ell}\right) = \left(\frac{\delta_1, \cdots, \delta_{k_0}, \beta_{k_0+1}, \cdots, \beta_k}{\alpha_1, \alpha_2, \cdots, \alpha_\ell}\right). \tag{9}$$

Considering the product of the regular factors $C_i(\lambda)$ as $R_1^{-1}(\lambda)$ and the right hand side as $R_2(\lambda)$, this certainly satisfies the imposed conditions since (1) $R_1(\lambda)$ is constructed to be regular, (2) the poles (resp. zeros) of $R_2(\lambda)$ and $R_1^{-1}(\lambda)$ are in Γ by construction and (3) the degree of $R_1(\lambda)$ equals the number of poles (resp. zeros) to be moved inside Γ. Notice that the latter also sheds some light on condition (3) of *minimal* degree. Indeed, any solution $R_1(\lambda)$ to (1-2) has degree *at least* equal to (3) as is easily seen from the above discussion. The above recursive scheme thus allows us to generate factorizations of the type described in section 1, and this for any given set Γ.

The major disadvantage of the above "transfer function"-approach is its complexity. First one computes the poles (and zeros) and the partial fraction expansion of $R(\lambda)$. From the coefficient matrices of this expansion one constructs the $C_i(\lambda)$ factors and after each pole/zero cancelation with a factor $C_i(\lambda)$, the expansion has to be updated. The most appealing methods for calculating poles and zeros use state space models [5]. One could then as well try to solve the problem using this parametrization, which is now done in the next two sections.

3 Dislocating poles in state space

Since we assumed (without loss of generality) that $R(\lambda)$ has no infinite poles, it has a realization quadruple $\{A, B, C, D\}$, i.e. $R(\lambda) = C(\lambda I - A)^{-1}B + D$, which we denote as :

$$R(\lambda) \sim \left[\begin{array}{c|c} A & B \\ \hline C & D \end{array}\right]. \tag{10}$$

Let also

$$R_1^{-1}(\lambda) \sim \left[\begin{array}{c|c} F & G \\ \hline H & J \end{array}\right]. \tag{11}$$

In order to construct a quadruple for the product $R_2(\lambda) = R_1^{-1}(\lambda).R(\lambda)$ we make use of the following lemma, which can be found implicitly in the literature ([2], see also [12]).

Lemma 1 : *Let* $\left[\begin{array}{c|c} F & G \\ \hline H & J \end{array}\right]$ *and* $\left[\begin{array}{c|c} A & B \\ \hline C & D \end{array}\right]$ *be realizations of two transfer matrices, then a realization of the product of the two corresponding transfer matrices (in that order) is given by the constant matrix product :*

$$\left[\begin{array}{cc|c} F & 0 & G \\ 0 & I & 0 \\ \hline H & 0 & J \end{array}\right] \cdot \left[\begin{array}{cc|c} I & 0 & 0 \\ 0 & A & B \\ \hline 0 & C & D \end{array}\right] = \left[\begin{array}{cc|c} F & GC & GD \\ 0 & A & B \\ \hline H & JC & JD \end{array}\right]. \tag{12}$$

We now derive necessary and sufficient conditions for canceling the poles of a transfer function $R(\lambda)$. Let us choose a *minimal* realization (10) for $R(\lambda)$ where A is in Schur form and where the eigenvalues of A outside Γ are all grouped in the top left corner A_{11}. Let us assume that the blocks A_{ii} have dimensions $n_i \times n_i$ for $i = 1, 2$ (where $n = n_1 + n_2$) :

$$R(\lambda) \sim \left[\begin{array}{cc|c} A_{11} & A_{12} & B_1 \\ 0 & A_{22} & B_2 \\ \hline C_1 & C_2 & D \end{array}\right] \tag{13}$$

(such a realization can always be obtained by updating (10) with a *unitary* state space transformation) [13]. Then the canceling factor $R_1^{-1}(\lambda)$ will thus have degree n_1 according to condition (3). Let its *minimal* realization be given by (11) where F has order n_1. Following Lemma 1, a (non-minimal) realization for $R_2(\lambda)$ is then given by :

$$\left[\begin{array}{ccc|c} F & & & G \\ & I_{n_1} & 0 & \\ & 0 & I_{n_2} & \\ \hline H & & & J \end{array}\right] \cdot \left[\begin{array}{ccc|c} I_{n_1} & & & \\ & A_{11} & A_{12} & B_1 \\ & 0 & A_{22} & B_2 \\ \hline & C_1 & C_2 & D \end{array}\right] = \left[\begin{array}{ccc|c} F & GC_1 & GC_2 & GD \\ & A_{11} & A_{12} & B_1 \\ & & A_{22} & B_2 \\ \hline H & JC_1 & JC_2 & JD \end{array}\right]. \tag{14}$$

This has all its poles inside Γ iff the eigenvalues of A_{11} are either unobservable or uncontrollable, since A_{22} is Γ-stable by assumption and F is chosen to be Γ-stable. Since we assumed (13) to be a minimal realization, the eigenvalues of A_{11} are controllable in (13) and hence also in (14). Let $\mathcal{X}$ be the invariant subspace of (14) corresponding to the eigenvalues of A_{11}. This space is uniquely defined since the spectrum of A_{11} is disjoint from the rest of the poles of (14). A basis for $\mathcal{X}$ is easily seen to be :

$$\mathcal{X} = < \begin{bmatrix} X \\ I_{n_1} \\ 0 \end{bmatrix} >, \tag{15}$$

where X is the (unique) solution of the Sylvester equation :

$$XA_{11} - FX = GC_1. \tag{16}$$

Since this space must be unobservable one has $[H, JC_1, JC_2]\mathcal{X} = 0$ or :

$$HX + JC_1 = 0. \tag{17}$$

A reduced realization for $R_2(\lambda)$ is then obtained using the state space transformation

$$T^{-1} = \begin{bmatrix} I_{n_1} & -X & 0 \\ & I_{n_1} & 0 \\ & & I_{n_2} \end{bmatrix}, \; T = \begin{bmatrix} I_{n_1} & X & 0 \\ & I_{n_1} & 0 \\ & & I_{n_2} \end{bmatrix}. \tag{18}$$

Performed on (14) this yields, because of (16)(17) :

$$R_2(\lambda) \sim \left[\begin{array}{ccc|c} F & 0 & GC_2 - XA_{12} & GD - XB_1 \\ 0 & A_{11} & A_{12} & B_1 \\ 0 & 0 & A_{22} & B_2 \\ \hline H & 0 & JC_2 & JD \end{array}\right] \sim \left[\begin{array}{cc|c} F & GC_2 - XA_{12} & GD - XB_1 \\ 0 & A_{22} & B_2 \\ \hline H & JC_2 & JD \end{array}\right], \tag{19}$$

which again has state space dimension $n_1 + n_2$. Summarizing we thus proved the following theorem (see [12] for more details).

Theorem 1 : *Let $R(\lambda)$ be a $p \times m$ rational matrix with a minimal realization of order $n = n_1 + n_2$ as given in (13), where $\Lambda(A_{22}) \subset \Gamma$ and $\Lambda(A_{11}) \subset \Gamma_c$, the complement of Γ. Then $R_1(\lambda)$ and $R_2(\lambda)$, realized by*

$$R_1^{-1}(\lambda) \sim \left[\begin{array}{c|c} F & G \\ \hline H & J \end{array}\right], \; R_2(\lambda) \sim \left[\begin{array}{cc|c} F & GC_2 - XA_{12} & GD - XB_1 \\ 0 & A_{22} & B_2 \\ \hline H & JC_2 & JD \end{array}\right], \tag{20}$$

satisfy the requested conditions of the factorization (1-3) iff (i) $\{F,G,H,J\}$ *represents a regular transfer matrix, (ii)* $\Lambda(F) \subset \Gamma$, *and (iii) the following equation is satisfied :*

$$\left[\begin{array}{c|c} F & G \\ \hline H & J \end{array}\right] \cdot \left[\begin{array}{c} X \\ \hline C_1 \end{array}\right] = \left[\begin{array}{c} X \\ \hline 0 \end{array}\right] \cdot A_{11}. \tag{21}$$

The question now of course remains how to find matrices F, G, H, J and X satisfying the conditions of Theorem 1. We know from section 2 that such a solution must exist, but we would like a *constructive* proof *not* relying on this section. For this, one shows that the regularity of the system $\{F,G,H,J\}$ implies X to be invertible. Indeed, from (21) it follows that $\left[\begin{array}{c} X \\ \hline C_1 \end{array}\right]$ and X must have the same rank and hence the same kernel, say $\mathcal{N}$. But then (21) also implies that $A_{11}\mathcal{N} \subset \mathcal{N}$, and hence $\mathcal{N}$ must be an unobservable subspace of (A_{11}, C_1) which contradicts the assumptions (for a more detailed proof, we refer to [12]).

Lemma 2 : *Let* (A_{11}, C_1) *be observable and let* $\{F,G,H,J\}$ *satisfy (21). Then* $\{F,G,H,J\}$ *represents a regular system only if* X *in (21) is regular. Moreover, the eigenvalues of* A_{11} *– i.e. the poles of* $R(\lambda)$ *– to be canceled are the zeros of the regular system* $\{F,G,H,J\}$.

Since X is invertible it can be "absorbed" into the quadruple $\{F,G,H,J\}$ as a state space transformation. Putting

$$\{\hat{F},\hat{G},\hat{H},\hat{J}\} = \{X^{-1}FX,\ X^{-1}G,\ HX,\ J\}, \tag{22}$$

it follows indeed from (21) that we are looking for a system $\{\hat{F},\hat{G},\hat{H},\hat{J}\}$ satisfying :

$$\left[\begin{array}{c|c} \hat{F} & \hat{G} \\ \hline \hat{H} & \hat{J} \end{array}\right] \left[\begin{array}{c} I \\ \hline C_1 \end{array}\right] = \left[\begin{array}{c} A_{11} \\ \hline 0 \end{array}\right], \tag{23}$$

and this is easily solved via the following procedure :

Algorithm 1

Step 1 : Determine $\hat{F}, \hat{G}$, with $\Lambda(\hat{F}) \subset \Gamma$ by solving the pole placement problem : $\hat{F} = A_{11} - \hat{G}C_1$. This has always a solution since (A_{11}, C_1) is observable.

Step 2 : Determine $[\ \hat{H} \mid \hat{J}\ ]^T$ as any basis for the null space of $[\ I_{n_1} \mid C_1^T\ ]$, i.e. $[\ \hat{H} \mid \hat{J}\ ] = M[-C_1 \mid I\ ]$ for an arbitrary *invertible* M.

Since the regularity of the system $\{\hat{F},\hat{G},\hat{H},\hat{J}\}$ follows from the invertibility of $\hat{J} = M$, all conditions of Theorem 1 are satisfied. We have thus derived here a constructive proof that the undesired *poles* of a transfer function $R(\lambda)$ can be canceled by a regular transfer function $R_1^{-1}(\lambda)$ whose degree equals the number of poles to be canceled (here n_1), whose zeros will be those unwanted poles and whose poles can be chosen arbitrarily in Γ.

4 Dislocating zeros in state space

Here we consider the case of canceling the undesired zeros of $R(\lambda)$ that are outside a specified set Γ by the poles of $R_1^{-1}(\lambda)$. As before, let (11) be a *minimal* realization of the factor $R_1^{-1}(\lambda)$ and let n_1 be its state space dimension. Then according to Lemma 1, the product $R_2(\lambda) = R_1^{-1}(\lambda) \cdot R(\lambda)$ is realized by :

$$R_2(\lambda) \sim \left[\begin{array}{cc|c} F & GC & GD \\ 0 & A & B \\ \hline H & JC & JD \end{array}\right]. \tag{24}$$

We would like the poles of $R_1^{-1}(\lambda)$ – i.e. the spectrum of F – to cancel with the undesired zeros of $R(\lambda)$, which are assumed to be n_1 in number. Since the (F,H)-pair is observable in

(11) it is also observable in (24) and the spectrum of F in (24) must thus be uncontrollable. Since (A,B) is controllable in (24) the controllable subspace $\mathcal{X}$ of the realization (24) must be of the form [3] :

$$\mathcal{X} = < \begin{bmatrix} X \\ I_n \end{bmatrix} > , \tag{25}$$

where X satisfies [3] :

$$XA - FX = GC \ ; \quad XB = GD. \tag{26}$$

The state space transformation $T = \begin{bmatrix} I_{n_1} & X \\ 0 & I_n \end{bmatrix}$ applied to (24) then yields :

$$R_2(\lambda) \sim \left[\begin{array}{cc|c} F & 0 & 0 \\ 0 & A & B \\ \hline H & JC+HX & JD \end{array}\right], \tag{27}$$

where we used (26) for the zero blocks in the top row of (27). Condition (26) is only a *necessary* condition for the pole-zero cancelation to occur (see [12] for an example where (26) is met and no pole-zero cancelation occurs). Additional conditions have to be imposed on X in order to enforce a zero dislocation. In order to do so one has to explicitate first what zeros of $R(\lambda)$ have to be dislocated as was e.g. done in (13) for the poles. A similar decomposition for the zeros is now derived using the following lemmas, proven in [11] :

Lemma 3 : *Let U be any invertible transformation such that*

$$\left[\begin{array}{c|c} \lambda I - A & B \\ \hline -C & D \end{array}\right] \cdot U = \left[\begin{array}{c|c} \lambda\hat{E} - \hat{A} & \lambda\hat{B} - \hat{F} \\ \hline 0 & \hat{D} \end{array}\right], \tag{28}$$

where has linearly independent columns. Then the generalized eigenvalues of $\lambda\hat{E} - \hat{A}$ are the zeros of $R(\lambda) = C(\lambda I - A)^{-1}B + D$.

Notice that, since $[-C \mid D\,]\cdot U = [\,0 \mid \hat{D}\,]$, $rank.\hat{D}$ equals $rank.([-C \mid D\,])$ and the number of columns $\hat{m}$ of $\hat{D}$ and m of D may thus differ. We also remark that in practice one uses unitary matrices U, which yields a numerically stable construction of what was called the *zero pencil* $(\lambda\hat{E} - \hat{A})$ in [11]. The use of unitary matrices is also maintained in the following lemma, leading to the separation between two parts of the spectrum of $\lambda\hat{E} - \hat{A}$.

Lemma 4 : *Let $\lambda\hat{E} - \hat{A}$ be an arbitrary singular pencil with spectrum $\Lambda(\hat{E},\hat{A})$ – i.e. $\Lambda(\hat{E},\hat{A})$ is the set of generalized eigenvalues of $(\lambda\hat{E} - \hat{A})$. Then for any complementary sets Γ and Γ_c of the extended complex plane separating $\Lambda(\hat{E},\hat{A})$ in two disjoint parts Λ_Γ and Λ_{Γ_c}, there exist unitary transformations Q and Z such that :*

$$Q^*(\lambda\hat{E} - \hat{A})Z = \begin{bmatrix} \lambda\hat{E}_{11} - \hat{A}_{11} & \lambda\hat{E}_{12} - \hat{A}_{12} & \lambda\hat{E}_{13} - \hat{A}_{13} \\ 0 & \lambda\hat{E}_{22} - \hat{A}_{22} & \lambda\hat{E}_{23} - \hat{A}_{23} \\ 0 & 0 & \lambda\hat{E}_{33} - \hat{A}_{33} \end{bmatrix}, \tag{29}$$

whereby (i) $\Lambda(\hat{E}_{11},\hat{A}_{11}) = \Lambda_\Gamma$; $\Lambda(\hat{E}_{22},\hat{A}_{22}) = \Lambda_{\Gamma_c}$; $\Lambda(\hat{E}_{33},\hat{A}_{33}) = \emptyset$, (ii) $\lambda\hat{E}_{11} - \hat{A}_{11}$ is right invertible for $\lambda \in \Gamma_c$, (iii) $\lambda\hat{E}_{22} - \hat{A}_{22}$ is invertible for $\lambda \in \Gamma$, and (iv) $\lambda\hat{E}_{33} - \hat{A}_{33}$ is left invertible.

This essentially says that there exists a (generalized) block Schur decomposition (29) with the generalized eigenvalues of $\lambda\hat{E} - \hat{A}$ inside Γ gathered in $\lambda\hat{E}_{11} - \hat{A}_{11}$, and the remaining ones gathered in $\lambda\hat{E}_{22} - \hat{A}_{22}$. This leads to the following theorem (proven in [12]).

Theorem 2 : *One can always update a minimal realization (10) by a unitary state space transformation Q such that its zero pencil (28) is automatically in generalized Schur form (29), i.e. :*

$$\left[\begin{array}{c|c} Q^*(\lambda I - A)Q & Q^*B \\ \hline -CQ & D \end{array}\right] = \left[\begin{array}{ccc|c} \lambda I_{n_1} - A_{11} & -A_{12} & -A_{13} & B_1 \\ -A_{21} & \lambda I_{n_2} - A_{22} & -A_{23} & B_2 \\ -A_{31} & -A_{32} & \lambda I_{n_3} - A_{33} & B_3 \\ \hline -C_1 & -C_2 & -C_3 & D \end{array}\right] \tag{30}$$

and

$$\left[\begin{array}{ccc|c} \lambda I_{n_1}-A_{11} & -A_{12} & -A_{13} & B_1 \\ -A_{21} & \lambda I_{n_2}-A_{22} & -A_{23} & B_2 \\ -A_{31} & -A_{32} & \lambda I_{n_3}-A_{33} & B_3 \\ \hline -C_1 & -C_2 & -C_3 & D \end{array}\right] V$$

$$= \underbrace{\left[\begin{array}{ccc|c} \lambda \hat{E}_{11}-\hat{A}_{11} & \lambda \hat{E}_{12}-\hat{A}_{12} & \lambda \hat{E}_{13}-\hat{A}_{13} & \lambda \hat{F}_1-\hat{B}_1 \\ 0 & \lambda \hat{E}_{22}-\hat{A}_{22} & \lambda \hat{E}_{23}-\hat{A}_{23} & \lambda \hat{F}_2-\hat{B}_2 \\ 0 & 0 & \lambda \hat{E}_{33}-\hat{A}_{33} & \lambda \hat{F}_3-\hat{B}_3 \\ \hline 0 & 0 & 0 & \hat{D} \end{array}\right]}_{\hat{n}_1 \quad \hat{n}_2 \quad \hat{n}_3 \quad \hat{m}} \begin{array}{l} \}n_1 \\ \}n_2 \\ \}n_3 \\ \}p \end{array} \tag{31}$$

where $\lambda\hat{E}_{11}-\hat{A}_{11}$, $\lambda\hat{E}_{22}-\hat{A}_{22}$ and $\lambda\hat{E}_{33}-\hat{A}_{33}$ satisfy conditions (i)(ii)(iii)(iv) of Lemma 4 (hence $\hat{n}_1 \geq n_1$, $\hat{n}_2 = n_2$, $\hat{n}_3 < n_3$).

Without loss of generality we can thus assume that our state space realization is in a form satisfying (31) where $\lambda\hat{E}_{22}-\hat{A}_{22}$ contains the zeros of $R(\lambda)$ to be dislocated. This form is to be considered an analogue to (13), now isolating the *zeros* to be canceled in a separate block $\lambda\hat{E}_{22}-\hat{A}_{22}$. The fact that now there is a third block $\lambda\hat{E}_{33}-\hat{A}_{33}$ is due to the possible singularity of the pencil (30-31) – and hence of $R(\lambda)$ (see [11]). Indeed, when $R(\lambda)$ happens to be right invertible, so will (30-31), and the block $\lambda\hat{E}_{33}-\hat{A}_{33}$ vanishes [11]. We are now ready to formulate a theorem on zero dislocation analogously to Theorem 1.

Theorem 3 : *Let $R(\lambda)$ be a $p \times m$ rational matrix with a minimal realization of order $n = n_1 + n_2 + n_3$ as given in (30-31), where $\Lambda(\hat{E}_{11}, \hat{A}_{11}) \subset \Gamma$ and $\Lambda(\hat{E}_{22}, \hat{A}_{22}) \subset \Gamma_c$, the complement of Γ and $\Lambda(\hat{E}_{33}, \hat{A}_{33}) = \emptyset$. Then $R_1(\lambda)$ and $R_2(\lambda)$, realized by*

$$R_1^{-1}(\lambda) \sim \left[\begin{array}{c|c} F & G \\ \hline H & J \end{array}\right], \quad R_2(\lambda) \sim \left[\begin{array}{c|c} \lambda I - A & B \\ \hline -JC - HX & JD \end{array}\right], \tag{32}$$

satisfy the requested conditions of the factorization (1-3) iff (i) $\{F, G, H, J\}$ represents a regular transfer matrix, (ii) the zeros of $\{F, G, H, J\}$ lie in Γ, and (iii) the following equation is satisfied for the $n_2 \times (n_1 + n_2 + n_3)$ matrix $X = [\,0 \;\; X_2 \;\; X_3\,]$:

$$\left[\begin{array}{cc} X & G \end{array}\right] \cdot \left[\begin{array}{c|c} A & B \\ \hline C & D \end{array}\right] = F \cdot \left[\begin{array}{cc} X & 0 \end{array}\right], \tag{33}$$

with X_2 invertible

For the construction of a solution to the above theorem, we can again "absorb" the invertible factor X_2 into the quadruple $\{F, G, H, J\}$ as a state space transformation. Putting

$$\{\hat{F}, \hat{G}, \hat{H}, \hat{J}\} = \{X_2^{-1}FX_2,\; X_2^{-1}G,\; HX_2,\; J\}, \tag{34}$$

it then follows that we are looking for a system $\{\hat{F}, \hat{G}, \hat{H}, \hat{J}\}$ satisfying :

$$\left[\begin{array}{ccc} I & Y_3 & \hat{G} \end{array}\right] \cdot \left[\begin{array}{cc|c} \lambda\hat{E}_{22}-\hat{A}_{22} & \lambda\hat{E}_{23}-\hat{A}_{23} & \lambda\hat{F}_2-\hat{B}_2 \\ 0 & \lambda\hat{E}_{33}-\hat{A}_{33} & \lambda\hat{F}_3-\hat{B}_3 \\ \hline 0 & 0 & \hat{D} \end{array}\right] = [\,\lambda I - \hat{F}\,] \cdot \left[\begin{array}{ccc} \hat{E}_{22} & \hat{Y}_3 & \hat{Y}_4 \end{array}\right] \tag{35}$$

and this is easily solved via the following procedure :

Algorithm 2

Step 1 : Put $\hat{F} = \hat{A}_{22}\hat{E}_{22}^{-1}$

Step 2 : Solve $Y_3(\lambda\hat{E}_{33}-\hat{A}_{33}) - (\lambda I - \hat{F})\hat{Y}_3 = -(\lambda\hat{E}_{23}-\hat{A}_{23})$ for Y_3 and $\hat{Y}_3$ (see [10][8])

Step 3 : Put $\hat{Y}_4 = \hat{F}_2 + Y_3\hat{F}_3$ and solve $\hat{G}$ from $\hat{G}\hat{D} = \hat{B}_2 + Y_3\hat{B}_3 - \hat{F}\hat{Y}_4$

Step 4 : Choose $\hat{H}$, $\hat{J}$ such that the zeros of $\{\hat{F}, \hat{G}, \hat{H}, \hat{J}\}$ lie in Γ. For *finite* zeros this is a pole placement problem since then $\hat{J}$ is invertible. Choose then K such that $\Lambda(\hat{F} + \hat{G}K) \subset \Gamma$ and then solve for $[\,\hat{H}\;\hat{J}\,]^T$ as any basis for the null space of $[\,I_{n_2} \mid K^T\,]$, i.e. $[\,\hat{H} \mid \hat{J}\,] = M[-K \mid I\,]$ for an arbitrary *invertible* M.

5 Conclusion

In this paper a method was presented to perform factorizations involving polynomial matrices, with constraints on the poles and/or zeros of the factors. The method uses a state-space approach in contrast to recursive solutions using transfer function techniques [1], [4], [9]. The main advantage of the present method is its algorithmic simplicity : first the poles/zeros of the original matrix are computed via a (generalized) Schur decomposition of the corresponding state space model. From thereon, one merely has to solve a pole placement problem in order to find the realizations of the two factors (this is just a set of linear equations to solve !). A similar "block"-approach was also recently used in [6] for a related problem. In contrast to the recursive transfer function approach, all poles/zeros can be dislocated simultaneously, which is useful when *real* factorizations are requested (this is more involved in the transfer function approach [9]). On the other hand, it is shown in [12] that the state space method can also be implemented in a recursive fashion, which is useful for other types of *rational* factorizations (inner-outer factorization, all-pass extraction, ...).

Notice that the strength of the present approach is the use of the Schur decomposition as a starting point for the construction of the factors. For problems as coprime factorization and GCD extraction, it could be argued that no eigenvalue computation is required since these problems can be solved in a finite number of operations. But the use of eigenvalue methods here can be compared with the solution of Sylvester and Lyapunov equations (these are also sets of linear equations !) where again Schur methods prove to be attractive from a computational point of view (see [7], [13], [8]).

References

[1] V. Belevitch, *Classical Network Theory*, San Francisco, Holden Day, 1968

[2] C. Chen, *Introduction to Linear System Theory*, New York, Bolt Rineh. Winston, 1970

[3] C. Desoer, *Notes for a second course on linear systems*, V. Nostrand, New York, 1970

[4] P. Dewilde, J. Vandewalle, "On the factorization of a nonsingular rational matrix", *IEEE Trans. Circ. Syst.*, **CAS-22**, 637-645, 1975

[5] E. Emami, P. Van Dooren, "Computation of zeros of linear multivariable systems", *Automatica*, **18**, 415-430, 1982

[6] I. Gohberg, M. Kaashoek, A. Ran, "Partial pole and zero displacement by cascade connection", *SIAM Matr. Anal. & Appl.*, **10**, 316-325, 1989

[7] G. Golub, S. Nash, C. Van Loan, "A Hessenberg-Schur method for the problem AX+XB =C", *IEEE Trans. Aut. Contr.*, **AC-24**, 909-913, 1979

[8] B. Kågström, L. Westin, "Generalized Schur methods with condition estimators for solving the generalized Sylvster equation $(AR - LB, DR - LE) = (C, F)$", *IEEE Trans. Aut. Contr.*, **AC-34**, July 1989

[9] J. Vandewalle, P. Dewilde, "On the irreducible cascade synthesis of a system with a real rational transfer matrix", *IEEE Trans. Circ. Syst.*, **CAS-24**, 481-494, 1977

[10] P. Van Dooren, "Reducing subspaces : definitions, properties, algorithms", in *Matrix Pencils*, Eds. Ruhe–Kågström, Lect. Notes Math, **973**, 58-73, Springer, New York, 1983

[11] P. Van Dooren, "Factorization of a rational matrix : the singular case", *Int. Eq. Op. Th.*, **7**, 704-741, 1984

[12] P. Van Dooren, "Rational and polynomial matrix factorizations via recursive pole-zero cancelation", Internal Rept. Philips Research Laboratory, Belgium, 1989

[13] P. Van Dooren, M. Verhaegen, "On the use of unitary state space transformations", in *Contemporary Mathematics*, **47**, AMS, 1985

[14] G. Verghese, P. Van Dooren, T. Kailath, "Properties of the system matrix of a generalized state space system", *Int. J. Contr.*, **30**, 235-243, 1979

A SPECTRAL FACTORIZATION APPROACH TO LQ-OPTIMAL CONTROL OF INFINITE-DIMENSIONAL STATE-SPACE SYSTEMS

F.M. Callier and J. Winkin

Abstract

We consider the standard LQ-optimal control of a Hilbert state-space system $\dot{x} = Ax + Bu$, $y = Cx$, where 1) A is the generator of a C_0-semigroup, 2) B and C are bounded operators, 3) the pair (A, B) is exponentially stabilizable and the pair (C, A) is exponentially detectable, and 4) the pair (A, B) is approximately reachable. Such a system has a transfer matrix with entries in the algebra $\hat{\mathcal{B}}$ of fractions over $\hat{\mathcal{A}}_-$, [1]; moreover the LQ-optimal state feedback operator K_0 is obtained from the algebraic operator Riccati equation, [5], [6]. We report, [4], a result similar to [7, Theorem 2], namely that this feedback operator can be recovered alternatively 1) by solving a spectral factorization problem delivering a bistable spectral factor with entries in $\hat{\mathcal{A}}_-$, and 2) by obtaining the constant solution of an operator diophantine equation over $\hat{\mathcal{A}}_-$. This procedure is implicitly present in [8].

Notations. $\delta(\cdot - t_1)$ is the Dirac pulse at t_1 ; $\hat{f}$ is the Laplace transform of f ; Mat(A) is the set of matrices with entries in the set A ; $\mathcal{L}(X, Y)$, ($\mathcal{L}(X)$), is the space of linear bounded operators from X into Y , (from X into X). $D(A)$ denotes the domain of the operator A .

1. Introduction

In transfer function feedback system design diophantine equations are used for controller determination given the closed loop dynamics, e.g. [9, Ch. 3], [10, Eq. (6.2.32)]. In this paper in Theorem 3 below we report an infinite-dimensional generalization of this philosophy, viz. we generalize [7, Theorem 2] : Kucera's procedure to solve an LQ-optimal control problem by spectral factorization (closed loop dynamics) and a diophantine equation : 1) the lumped plant is generalized to a distributed one described by a standard semigroup state-space system, 2) the matrix fraction transfer functions use "distributed proper-stable" algebra instead of polynomial algebra. The objective is to create a relatively simple learning model for alternatively solving the LQ-problem.

Section 2 below describes our distributed system transfer function and state-space framework connected through Theorem 1. Section 3 reports the classical Riccati-solution of the LQ-problem (Theorem 2) and the alternative solution (Theorem 3). Concluding remarks are made in Section 4.

2. Transfer Function and State-Space System Framework

The following classes of distributed system transfer functions are described in [1], [2].

Let $\sigma \le 0$. An impulse response f is said to be in $\mathcal{A}(\sigma)$ if for $t < 0$, $f(t) = 0$, and for $t \ge 0$, $f(t) = f_a(t) + f_{sa}(t)$ where the regular functional part $f_a \in L_{1,\sigma}$ i.e. $\int_0^\infty |f_a(t)| \exp(-\sigma t)\, dt < \infty$, and the singular (atomic) part $f_{sa} := \sum_{i=0}^{\infty} f_i \delta(\cdot - t_i)$, where $t_0 = 0$, $t_i > 0$ for $i = 1, 2, \cdots$ and $f_i \in \mathbb{C}$ for $i = 0, 1, \cdots$ with $\sum_{i=0}^{\infty} |f_i| \exp(-\sigma t_i) < \infty$. An impulse response f is said to be in $\mathcal{A}_-$ iff $f \in \mathcal{A}(\sigma)$ for some $\sigma < 0$. $\mathcal{A}(\sigma)$ and $\mathcal{A}_-$ are convolution algebras. $\hat{\mathcal{A}}_-$ (class of Laplace transforms of elements of $\mathcal{A}_-$) is our selected class of distributed proper-stable transfer functions. It contains the multiplicative subset $\hat{\mathcal{A}}_-^\infty$ i.e. of transfer functions in $\hat{\mathcal{A}}_-$ that are bounded away from zero at infinity in Re $s \ge 0$, i.e. that are biproper-stable. Possibly open-loop unstable transfer functions are selected to be in $\hat{\mathcal{B}}$, where $\hat{f} \in \hat{\mathcal{B}}$ iff $\hat{f} = \hat{n}\hat{d}^{-1}$ where $\hat{n} \in \hat{\mathcal{A}}_-$ and $\hat{d} \in \hat{\mathcal{A}}_-^\infty$, (where $\hat{d}$ may be chosen biproper-stable *rational*). Multivariable plants have transfer matrices $\hat{P}$ in Mat($\hat{\mathcal{B}}$) described by a right matrix fraction, $\hat{P} = \hat{N}\hat{D}^{-1}$ where $\hat{N}$ and $\hat{D}$ are in Mat($\hat{\mathcal{A}}_-$) and det $\hat{D}$ is in $\hat{\mathcal{A}}_-^\infty$; if this holds and there exist $\hat{U}$ and $\hat{V}$ in Mat($\hat{\mathcal{A}}_-$) such that $\hat{U}\hat{N} + \hat{V}\hat{D} = I$ (Bezout identity) (or equivalently $[\hat{D}^T(s)\ \hat{N}^T(s)]^T$ has full column rank in Re $s \ge 0$), then we say that $\hat{P} \in$ Mat($\hat{\mathcal{B}}$) has a r.c.f. (right coprime fraction) $(\hat{N}, \hat{D})$ in Mat($\hat{\mathcal{A}}_-$) . Without loss of generality one may always choose $\hat{D}(s)$ biproper-stable *rational* such that $\hat{D}(\infty) = I$, [2, Proof of Theorem 2.1].

State-space systems below are of class SGB (i.e. *semigroup systems with bounded sensing and control*); see [6] for the details.

Definition 1. An m-input p-output state-space system SGB is described by the equations

$$\begin{cases} \dot{x}(t) = Ax(t) + Bu(t) & \text{for } x(0) = x_0 \in D(A) \\ y(t) = Cx(t) & \end{cases} \quad t \ge 0 , \tag{1}$$

where a) $x(t) \in X$, a Hilbert space with inner product $< \cdot, \cdot >$, $u(t) \in \mathbb{R}^m$, $y(t) \in \mathbb{R}^p$;
b) $A : D(A) \subset X \to X$ is the infinitesimal generator of a C_0-semigroup $(e^{At})_{t \ge 0}$ of bounded linear operators in X , i.e. $e^{At} \in \mathcal{L}(X)\ \ \forall\, t \ge 0$, and
c) B and C are bounded linear operators, i.e. $B \in \mathcal{L}(\mathbb{R}^m, X)$ and $C \in \mathcal{L}(X, \mathbb{R}^p)$.■

In the sequel a C_0-semigroup $(e^{At})_{t \ge 0}$ will be said to be *exp.* (exponentially) *stable* iff there exist $\sigma < 0$ and $M > 0$ such that $\|e^{At}\| \le M \exp(\sigma t)$ on $t \ge 0$, ($\|\cdot\|$ is the uniform operator norm).

Moreover, for a given system SGB we say that
a) the operator pair (A, B) is *exp. stabilizable* iff there exists a *stabilizing feedback* $K \in \mathcal{L}(X, \mathbb{R}^m)$, i.e. such that the C_0-semigroup $(e^{(A+BK)t})_{t \ge 0}$ is exp. stable.
b) the operator pair (C, A) is *exp. detectable* iff there exists a *stabilizing injection* $F \in \mathcal{L}(\mathbb{R}^p, X)$, i.e. such that the C_0-semigroup $(e^{(A+FC)t})_{t \ge 0}$ is exp. stable.

Now, any system SGB has a bounded linear *reachability operator* $G_{t_1} \in \mathcal{L}(L^2[0, t_1; \mathbb{R}^m], X)$ given for any $t_1 > 0$ by

$$G_{t_1} u = \int_0^{t_1} e^{A(t_1 - s)} Bu(s)\, ds , \tag{2}$$

and we say that (A, B) is *approximately reachable* (in finite time) iff

$$X = \overline{\bigcup_{t_1>0} \Re a[G_{t_1}]} \tag{3}$$

or equivalently [6, Proof of Theorem 3.15]

$$B^\star e^{A^\star t} x = 0 \text{ on } t \geq 0 \quad \text{implies} \quad x = 0 \tag{4}$$

(i.e. $(B^\star, A^\star)$ is *observable* on $t \geq 0$),
or equivalently, [6, Proof of Theorem 3.15], with $K \in \mathcal{L}(X, \mathbb{R}^m)$ any possible stabilizing feedback,

$$B^\star[(A + BK)^\star]^{-(n+1)} x = 0 \text{ for all } n = 0, 1, 2, \cdots \text{ implies } x = 0 \ . \tag{5}$$

Finally, the transfer function $\hat{P}$ of any system SGB is given by

$$\hat{P}(s) = C(sI - A)^{-1} B \tag{6}$$

as a well-defined $p \times m$-matrix valued function holomorphic in some right-half plane. Furthermore [11, Theorem 3.2] [12, Theorem 2, pp. 25-30]

Theorem 1. *For any system SGB with transfer function $\hat{P}$ given by (6), let (A, B) be exp. stabilizable and (C, A) be exp. detectable.*
Then, a) $\hat{P} \in Mat(\hat{\mathcal{B}})$, where for any stabilizing feedback $K \in \mathcal{L}(X, \mathbb{R}^m)$, the matrix pair $(\hat{N}, \hat{D})$ in $Mat(\hat{\mathcal{A}}_-)$ given by

$$(\hat{N}, \hat{D}) := (C(sI - A - BK)^{-1} B, I + K(sI - A - BK)^{-1} B) \tag{7}$$

is a r.c.f. of $\hat{P}$.
b) $\quad (e^{At})_{t \geq 0}$ *is exp. stable*
iff

$$\hat{P} \in \mathrm{Mat}(\hat{\mathcal{A}}_-) \ .\blacksquare$$

Comment 1. Recall the stable-unstable sum decomposition of [2, Proof of Theorem 2.1] or [6, p. 75 et seq.] with (A_u, B_u, C_u) a *matrix* minimal realization of the unstable *rational* part of $\hat{P}$ above : one can always reduce the feedback K in (7) to a finite-dimensional state feedback matrix K_u applied to the unstable part : (7) reads then

$$(\hat{N}, \hat{D}) = (\hat{P}\hat{D}, \hat{D}) \tag{7a}$$

where

$$\hat{D}(s) = [I - K_u(sI - A_u)^{-1} B_u]^{-1} \tag{7b}$$

is biproper-stable *rational* with $\hat{D}(\infty) = I$, ($\hat{D}(s)^{-1}$ is the control-loop return difference).

3. Spectral Factorization Based LQ-Optimal Control

For a system SGB of Theorem 1 one considers the *LQ-optimal control problem* : for any initial state $x_0 \in X$, find a square-integrable control u_0 which minimizes the cost functional

$$J(x_0, u) = \int_0^\infty (< C^\star Cx(t), x(t) > + < u(t), u(t) >) \, dt \, . \tag{8}$$

The solution of this problem is obtained by finding the nonnegative self-adjoint operator $Q \in \mathcal{L}(X)$ wich solves the (ORE) (operator Riccati equation), viz.

$$Q(D(A)) \subset D(A^\star) \tag{9a}$$

and

$$[A^\star Q + QA + C^\star C - QBB^\star Q]x = 0 \qquad \forall\, x \in D(A) \, . \tag{9b}$$

Indeed [5], [6, Ch. 4]

Theorem 2. *For any system SGB with (A, B) exp. stabilizable and (C, A) exp. detectable, the (ORE) has a unique nonnegative self-adjoint solution $Q_0 \in \mathcal{L}(X)$ and for any initial state $x_0 \in X$ the quadratic cost (8) is minimized by the optimal state feedback law*

$$u_0(t) = K_0 x(t) := -B^\star Q_0 x(t) \quad \text{on } t \geq 0 \tag{10}$$

where the optimal feedback operator $K_0 = -B^\star Q_0 \in \mathcal{L}(X, I\!R^m)$ is stabilizing, i.e. the feedback semigroup

$$(e^{(A+BK_0)t})_{t\geq 0} \text{ is exp. stable.} \blacksquare \tag{11}$$

We are now ready to report our main result [3, Theorem 2] [4, Theorems 2 and 3].

Theorem 3 (LQ-Optimal control by spectral factorization). *Consider any system SGB as in Definition 1 such that (A, B) is exp. stabilizable and (C, A) is exp. detectable and assume that*

$$(A, B) \text{ is approximately reachable.} \tag{12}$$

Let $K \in \mathcal{L}(X, I\!R^m)$ be such that $(e^{(A+BK)t})_{t\geq 0}$ is exp. stable and let

$$(\hat{N}(s), \hat{D}(s)) := (C(sI - A - BK)^{-1}B, I + K(sI - A - BK)^{-1}B) \tag{7}$$

such that the pair $(\hat{N}, \hat{D})$ in Mat($\hat{\mathcal{A}}_-$) is a right-coprime fraction of the transfer function $\hat{P}$ in Mat($\hat{\mathcal{B}}$) of SGB. Define also

$$\hat{\mathcal{N}}(s) := (sI - A - BK)^{-1}B \tag{13}$$

Consider now any solution of the spectral factorization problem

$$\hat{N}_\star \hat{N} + \hat{D}_\star \hat{D} = \hat{R}_\star \hat{R} \quad \text{for } s = j\omega \tag{14}$$

where $\hat{F}_\star(s) := \hat{F}(-\bar{s})^\star$ *and* $\hat{R} \in Mat(\hat{\mathcal{A}}_-)$ *(together with its inverse) is a* spectral factor.

Consider finally the LQ-optimal feedback $K_0 = -B^\star Q_0 \in \mathcal{L}(X, \mathbb{R}^m)$ *of Theorem 2.*

Then

a) Any spectral factor of (14) is given by

$$\hat{R}(s) = U[I - K_0(sI - A)^{-1}B]\hat{D}(s) \tag{15}$$

where U *is a unitary constant matrix and* $I - K_0(sI - A)^{-1}B$ *in* $Mat(\hat{B})$ *is the LQ-optimal control-loop return difference*

b) The LQ-optimal feedback $K_0 \in \mathcal{L}(X, \mathbb{R}^m)$ *is given by*

$$K_0 = -\mathcal{U}^{-1}\mathcal{V} = -\mathcal{U}^\star\mathcal{V} \tag{16}$$

where $(\mathcal{U}, \mathcal{V}) \in \mathcal{L}(\mathbb{R}^m) \times \mathcal{L}(X, \mathbb{R}^m)$ *, with* $\mathcal{U}$ *unitary, is the unique constant solution of the operator diophantine equation*

$$\mathcal{U}\hat{D}(s) + \mathcal{V}\hat{\mathcal{N}}(s) = \hat{R}(s) \in \mathrm{Mat}(\hat{\mathcal{A}}_-) \tag{17}$$

induced by the operator right coprime fraction over $\hat{\mathcal{A}}_-$ *of the input-state transfer function described by*

$$(sI - A)^{-1}B = \hat{\mathcal{N}}(s)\hat{D}(s)^{-1} \tag{18}$$

with operator Bezout identity over $\hat{\mathcal{A}}_-$ *given by*

$$\hat{D}(s) - K\hat{\mathcal{N}}(s) = I \;.\blacksquare \tag{19}$$

Comments 2. a) As usual (15) describes the spectral factor $\hat{R}(s)$ (closed loop dynamics) as the product of the (control-loop) return-difference times the open-loop dynamics $(\hat{D}(s))$. For a proof see Theorem 2 of [3], [4]. See also [13, Chapter 4].
b) [4, Theorem 3]. (18)-(19) follows by (7) and (13), where 1) $\hat{\mathcal{N}}(s)$ is exp. stable in the time-domain, and so an $\mathcal{L}(\mathbb{R}^m, X)$-valued $\hat{\mathcal{A}}_-$ function, 2) $\hat{D}(s) \in \mathrm{Mat}(\hat{\mathcal{A}}_-)$ and 3) $K \in \mathcal{L}(X, \mathbb{R}^m)$: hence $(\hat{\mathcal{N}}, \hat{\mathcal{D}})$ is an $\hat{\mathcal{A}}_-$ operator r.c.f. of $(sI - A)^{-1}B$: compare with [7, Eq. (16)]. Hence Eq. (17) is a diophantine equation over the distributed proper-stable algebra $\hat{\mathcal{A}}_-$ which for a given spectral factor $\hat{R}$ (closed-loop dynamics) determines the controller K_0 because $(\hat{\mathcal{N}}, \hat{\mathcal{D}})$ is $\hat{\mathcal{A}}_-$-coprime and (A, B) is approximately reachable : compare with [7, Eq. (19)]. Note that because of (18)-(19), (15) reads

$$\hat{R}(s) = U\hat{D}(s) - UK_0\hat{\mathcal{N}}(s) \tag{20}$$

whence, modulo reachability, (12) i.e. (4), Eq. (17) has the unique constant solution $(\mathcal{U}, \mathcal{V}) = (U, -UK_0)$ leading to (16) : the same philosophy leads to similar results in [7, Theorem 2] and Theorem 3, modulo the exchange of polynomial algebra for $\hat{\mathcal{A}}_-$-algebra.
c) The unique *standard* spectral factor of (14) is such that $\hat{R}(\infty) = I$, whence without loss of generality $\mathcal{U} = U = I$. Hence without loss of generality (16)-(17) and (19) imply

$$\hat{R}_a(s) := \hat{R}(s) - I = (K - K_0)\hat{\mathcal{N}}(s) \in \mathrm{Mat}(\hat{\mathcal{A}}_-) \;, \tag{21}$$

where, [1, Fact 2.3], the left and right hand sides are holomorphic in a neighborhood of $s = 0$. Hence the following Taylor series matrix-coefficient identification scheme

$$[n!]^{-1}\hat{R}_a^{(n)}(0) = (K_0 - K)(A + BK)^{-(n+1)}B \quad \text{for } n = 0, 1, 2, \cdots \tag{22}$$

which determines $K_0 - K$ and thus K_0 uniquely by the reachability condition (5). This is an analogue of the polynomial matrix coefficient identification scheme defining K_0 in [7, Theorem 2].

d) Eq. (22) is probably of theoretical interest since with $\mathcal{U} = U = I$, Eq. (15) gives rise to

$$\hat{R}(s)\hat{D}(s)^{-1} = I - K_0(sI - A)^{-1}B \in \text{Mat}(\hat{\mathcal{B}}) \tag{23}$$

with in many cases both sides meromorphic in $\mathbb{C}$. $K_0 \in \mathcal{L}(X, \mathbb{R}^m)$ is then recovered simply by residue calculus at the open-loop poles. Of course this idea also applies at the "*stabilized* open-loop poles" (eigenvalues of $A + BK$) in (21).

4. Concluding Remarks.

The most salient feature of this paper is the similarity in learning content of Theorem 3 and [7, Theorem 2], about solving the LQ-problem by spectral factorization and a diophantine equation. Practical implementation has been studied by one author (F.M.C.) on a simple heat equation example : it turns out that in this (monovariable) case spectral factorization is realizable by searching for the closed loop spectrum (pole placement of a few dominant modes) and feedback operator determination by simple almost finite-dimensional residue calculus. However more needs to be done.

References

[1] F.M. Callier and C.A. Desoer, "An Algebra of Transfer Functions for Distributed Linear Time-Invariant Systems", IEEE Trans. Circ. and Sys., Vol. 25, pp. 651-662, 1978; Ibidem, Vol. 26, p. 360, 1979; Vol. 27, pp. 320-323, 1980.

[2] F.M. Callier and C.A. Desoer, "Stabilization, Tracking and Disturbance Rejection in Multivariable Convolution Systems", Annales de la Société Scientifique de Bruxelles, T. 94, I, pp. 7-51, 1980.

[3] F.M. Callier and J. Winkin, "On Spectral Factorization for Multivariable Distributed Systems", Report No. 88/5, University of Namur (Belgium), Dept. of Math., 1988; presented at the 27th CDC Conference, Austin (Texas), December 1988.

[4] F.M. Callier and J. Winkin, "On Spectral Factorization and LQ-Optimal Regulation for Multivariable Distributed Systems", Report No. 88/9, University of Namur (Belgium), Dept. of Math., 1988; submitted.

[5] J. Zabczyk, "Remarks on the Algebraic Riccati Equation in Hilbert Space", Appl. Math. Opt., Vol. 2, pp. 251-258, 1976.

[6] R.F. Curtain and A.J. Pritchard, "Infinite-Dimensional Linear System Theory", Springer-Verlag, New-York, 1978.

[7] V. Kucera, "New Results in State Estimation and Regulation", Automatica, Vol. 17, No. 5, pp. 745-748, 1981.

[8] J.H. Davis and B.M. Barry, "A Distributed Model for Stress Control in Multiple Locomotive Trains", Appl. Math. Opt. Vol. 3, pp. 163-190, 1977.

[9] V. Kucera, "Discrete Linear Control : The Polynomial Equation Approach", Wiley, Chichester, 1979.

[10] M. Vidyasagar, "Control System Synthesis : A Factorization Approach", MIT Press, Cambridge (MA), 1985.

[11] C.A. Jacobson and C.N. Nett, "Linear State-Space Systems in Infinite-Dimensional Space : The Role and Characterization of Joint Stabilizability/ Detectability", IEEE Trans. Aut. Control, Vol. 33, No. 6, pp. 541-549, 1988.

[12] C.A. Jacobson, "The Structure of Exponential Stabilization of a Class of Linear Distributed Parameter Systems", Ph.D. Thesis, Rensselaer Polytechnic Institute, Troy (NY), 1986.

[13] J. Winkin, "Spectral Factorization and Feedback Control for Infinite-Dimensional Systems", Doctoral Thesis, University of Namur (Belgium), Dept. of Math., May 1989.

SOME CONTROL PROBLEMS FOR ABSTRACT VOLTERRA FUNCTIONAL-DIFFERENTIAL EQUATIONS

C. Corduneanu

Abstract

This paper is an introduction to the study of Control Problems, with special regard to Optimal Control, for functional differential equations involving abstract Volterra (non-anticipative or causal) operators. The goal of this investigation is to unify and generalize Control Theory for equations that allow as special cases the ordinary differential equations, the equations with delays, and the integrodifferential equations. Under adequate conditions on the operator, the relationship between control and state variables, at least in the linear case, can be always described by means of an integral operator of Volterra type, in the classical sense. This feature facilitates the discussion of various control problems, sometimes by reducing them to similar ones, attached to systems described by classical Volterra equations.

An LQ Optimal Control Problem

Let us consider the linear control system described by the functional-differential equations

(1) $\dot{x}(t) = (Lx)(t) + u(t), \ t \in [0,T],$

where $x : [0,T] \to \mathbf{R}^n$, L stands for a linear abstract Volterra operator acting on the space $L^2\left([0,T], \mathbf{R}^n\right)$, and $u : [0,T] \to \mathbf{R}^n$ stands for the controller. Of course, it is not necessary to assume that x and u have the same dimension. In the right hand side of (1) one could substitute (Bu)(t) for u(t), where B is an operator taking a convenient function space whose elements are m-dimensional vectors, into a function space whose elements are n-dimensional vectors. For simplicity, we will assume here that m = n, and deal with equation (1).

As usual, an initial condition

(2) $x(0) = x^0 \in \mathbf{R}^n$

will be attached to (1), which means that for every control u, there is a unique trajectory x(t;u) of (1), (2) (see [3]).

We will also consider a functional (cost functional) to be minimized:

(3) $$C(x;u) = \int_0^T \left\{ < P(t)\,x(t),\ x(t) > + < Q(t)\,u(t),\ u(t) > \right\} dt,$$

where x(t) stands for x(t;u), and P(t), Q(t) are square matrices of order n, satisfying further properties to be specified in the sequel.

As far as the controller u is concerned, we will assume that

(4) $u \in C \subset L^2\big([0,T], \mathbf{R}^n\big)$,

where C is a closed convex set.

In order to carry out the discussion of this simple control problem, we need a representation formula for the solution of the equation (1), under initial condition (2). In other words, we need a "variation of constants" formula for functional-differential equations involving abstract Volterra operators. This formula has been obtained in [3], and it looks

(5) $$x(t;u) = X(t,0)x^0 + \int_0^t X(t,s)\, u(s)ds,$$

where the resolvent kernel X(t,s) can be constructed in terms of the operator L, occurring in equation (1). The kernel X(t,s) is defined for $0 \le s \le t \le T$, and satisfies certain properties that will be precised in the sequel.

The formula (5) reduces to the simpler form

(6) $$x(t;u) = \int_0^t X(t,s)\, u(s)ds,\ t \in [0,T]\ ,$$

in case we choose the initial value $x^0 = \theta \in \mathbf{R}^n$. It is useful to notice that choosing $x^0 = \theta$, the generality of the problem is not really diminished. Indeed, if we want to reduce the case of condition (2) to the special case $x^0 = \theta$, we have to operate the substitution

(7) $$y(t) = x(t) - X(t,0)x^0,$$

which leads to the equation

(8) $$\dot{y}(t) = (Ly)(t) + u(t),\ \ t \in [0,T],$$

for which $y(0) = \theta \in \mathbf{R}^n$.

Therefore, without loss of generality, we can deal with the problem:

(9) $$\min C(x;u)$$

where x is expressed in terms of u by the formula (6), while u is subject to the constraint (4).

Under natural assumptions on the equation (1), and matrices P(t) and Q(t), the above optimal control problem can be reduced to the well-known proposition that states: in a Hilbert space, any closed convex set contains a unique element with minimum norm.

Let us now formulate the precise hypotheses under which the existence of the optimal control is obtained from the general proposition stated above.

With respect to the operator L occurring in equation (1), we will assume:

a) L is a linear continuous Volterra type operator on the space $L^2\left([0,T], \mathbb{R}^n\right)$.

The matrices P(t) and Q(t) occurring in (3) are assumed to satisfy:

b) P(t) and Q(t) have square integrable entries on [0,T], $P(t) = P^*(t) \geq O$, and $Q(t) = Q^*(t) > O$.

Both equalities in b) are supposed to hold a.e. on [0,T], while the condition on Q(t) means there exists $\lambda > 0$, such that

$$(10) \quad \int_0^T < Q(t)\, u(t),\ u(t) > dt \geq \lambda \int_0^T |u(t)|^2\, dt,$$

for all $u \in L^2\left([0,T], \mathbb{R}^n\right)$.

The following result, related to the linear-quadratic optimal control problem formulated above, can be proven.

<u>Theorem 1</u>. The optimal control problem (1), (2), (3), (4) has a unique solution $u \in C$, provided conditions a) and b) stated above are satisfied.

<u>Proof</u>. Under assumption a) the unique solution (a. e.) of the equation (1), satisfying the initial condition $x(0) = \theta \in \mathbb{R}^n$, is given by formula (5). As mentioned above, restricting the initial condition to $x(0) = \theta$ does not imply a loss of generality.

We shall now prove (see also [4]) that there exist two positive constants α and β, such that

$$(11) \quad \alpha \int_0^T |u(t)|^2\, dt \leq C(x;u) \leq \beta \int_0^T |u(t)|^2\, dt,$$

for any $u \in L^2\left([0,T], \mathbb{R}^n\right)$. In (11), x stands for the x(t;u) given by formula (6).

Taking into account the conditions imposed on P(t) and Q(t) in b), it is obvious that the first inequality in (11) is verified with $\alpha = \lambda$. As far as the second inequality in (11) is concerned, it can be shown by applying Schwarz's inequality that

$$(12) \quad C(x;u) \leq K \int_0^T \left\{ |x(t)|^2 + |u(t)|^2 \right\} dt,$$

for some positive K, such that

$$(13) \quad \max\left\{ \int_0^T \|P(t)\|^2\, dt,\ \int_0^T \|Q(t)\|^2\, dt \right\} \leq K^2.$$

The matrix norm can be chosen to be the Euclidean norm. In order to derive the second inequality (11) from (12), it suffices to show that for some positive M one has

$$(14)\quad \int_0^T |x(t)|^2\, dt \le M \int_0^T |u(t)|^2 dt,$$

where x(t) stands for the x(t;u) given by (6). It is obvious that (14) means nothing else but the continuity of the Volterra operator defined by (6), on the space $L^2\big([0,T],\ \mathbf{R}^n\big)$. But it can be easily seen (see, for instance, [2]) that this operator is completely continuous if considered from $L^2\big([0,T],\ \mathbf{R}^n\big)$ into $C\big([0,T],\ \mathbf{R}^n\big)$. Therefore, the existence of a positive M such that (14) holds true for any $u \in L^2\big([0,T],\ \mathbf{R}^n\big)$ is proven. Combining (12) and (14), we obtain the second inequality (11) with $\beta = KM$.

Let us now define another inner product on $L^2\big([0,T],\ \mathbf{R}^n\big)$, by letting

$$(15)\quad < u, v >_1 = \int_0^T \Big\{ < P(t)\, x(t), y(t) > + < Q(t)\, u(t), v(t) > \Big\}\, dt,$$

where $x(t) = x(t;u)$ is given by (6), and $y(t) = x(t;v)$. It is easily checked that (15) defines on $L^2\big([0,T],\ \mathbf{R}^n\big)$ an inner product. Because of (11), the usual norm of $L^2\big([0,T],\ \mathbf{R}^n\big)$ is equivalent to the norm derived from the inner product defined by (15). Hence, $L^2\big([0,T],\ \mathbf{R}^n\big)$ is again a Hilbert space if we endow it with the norm

$$(16)\quad \| u \|_1 = \big\{C(x;u)\big\}^{1/2},$$

with x given by (6).

Since in the new topology the set C is still closed, while the convexity has not been changed by changing the topology, the proof of the Theorem 1 is complete.

Remark. The result of Theorem 1 can be applied to the case when the linear plant is described by the linear differential–delay equation

$$(17)\quad \dot{x}(t) = L\big(t, x_t\big) + u(t),$$

under initial conditions

$$(18)\quad x(0) = x^0 \in \mathbf{R}^n,\ \ x_0 = \phi \in S,$$

where $x_t(s) = x(t+s)$, $-h \leq s < 0$, and S stands for a convenient function space $S\big([-h,0], \mathbb{R}^n\big)$, the case $h = \infty$ being acceptable. Under suitable conditions, the operator $x \to L\big(t, x_t\big)$ is a continuous linear operator of Volterra type, and the application of Theorem 1 is immediate.

A Maximum Principle Result

Instead of equation (1), we will consider in this section a nonlinear perturbed equation

$$(19) \quad \dot{x}(t) = (Lx)(t) + f\big(t, x(t); u(t)\big), \quad t \in [0,T],$$

where L has the same meaning as in the preceding section, while f(t,x;u) is a map from $[0,T] \times \mathbb{R}^n \times U$ into $\mathbb{R}^n$, with $U \subset R^m$ a set providing the constraints for the control variable u, which will be assumed to be closed.

If we associate with (19) an initial condition of the form (2), then under suitable conditions the existence and uniqueness of solution of (19) is assured on the whole interval [0,T]. More precisely, using the formula (6), we can reduce the problem (19), (2) to the Volterra integral equation

$$(20) \quad x(t) = X(t,0)\, x^0 + \int_0^t X(t,s)\, f\big(s, x(s); u(s)\big)\, ds.$$

Equation (20) is, in general, a nonlinear equation of Volterra–Hammerstein type, which is equivalent to (19), (2). While for (20) we look for continuous solutions, it is interesting to notice that such a solution will be absolutely continuous, verifying (19) a. e. on [0,T].

In other words, the relationship in between the control and state variables is described by the Volterra equation (20), which has the classical form.

If we consider now a cost functional

$$(21) \quad C(x;u) = g^0\big(x(T)\big),$$

and impose the terminal constraints

$$(22) \quad g_i\big(x(T)\big) \leq 0, \quad i = 1,2,\ldots, p,$$

$$(23) \quad g_i\big(x(T)\big) = 0, \quad i = p+1,\ldots, q,$$

then the optimal problem associated with (19), (2), (21)–(23) becomes of a less abstract nature when (19), (2) are substituted by (20) alone.

The choice of C(x;u) in the form (21), instead of

$$C(x;u) = \int_0^T f^0\Big(t,\ x(t);\ u(t)\Big)\, dt$$

does not reduce the generality of the problem. Indeed, it is well known that introducing a new state variable

$$y(t) = \int_0^t f^0\Big(s,\ x(s);\ u(s)\Big)\, ds,$$

one has $C(x;u) = y(T)$, and therefore we have to solve the problem of minimizing y(T), which turns out to be a problem in $n + 1$ variables, but of the form (21).

Since optimal control problems of the type (20)–(23) have been considered in the literature by many authors (see, for instance, [3], Ch. VI, where references are provided), it appears possible to discuss the problem initially formulated for the equation (19), with abstract Volterra operator, by reducing it to the equation (20), in which only classical operators appear.

We shall choose, for illustration, a result concerning the validity of the Maximum Principle for control problems governed by Volterra equations, due to D. A. Carlson [1]. See also [3].

The following conditions must be imposed on our problem (19), (2), (21)–(23) in order to be able to reduce it to the problem (20)–(23), and then apply the available result mentioned above.

a) The same as in Theorem 1.

b) The matrix-valued function X(t,s), appearing in (20), is continuous in (t,s), for $0 \leq s \leq t \leq T$.

Let us point out that X(t,s) is not, generally speaking, a continuous function. The proof of its existence (see [3]) is not entirely constructive and while certain regularity properties can be obtained for X(t,s), such as the absolute continuity of X(t,0), it is not clear whether the continuity of X(t,s) can be obtained under the sole assumption of continuity for the Volterra operator L. However, an important special case when the continuity of X(t,s) is secured is when L is a time-invariant operator, in which case $X(t,s) = X(t-s,\ 0)$, and therefore the absolute continuity of X(t,0) implies the continuity of X(t,s).

c) The map $f\colon\ [0,T] \times \mathbf{R}^n \times U \to \mathbf{R}^n$ is continuous, together with its Jacobian matrix $(\partial f/\partial x)$, where $U \subset R^m$ is a closed set.

d) f and $\partial f/\partial x$ are locally bounded on $[0,T] \times \mathbf{R}^n \times U$, in the sense that for every compact $K \subset \mathbf{R}^n$, there exist constants $M = M(K) > 0$ and $N = N(K) > 0$, such that

(24) $$|f(t,x;u)| \leq M, \quad \left|\frac{\partial}{\partial x} f(t,x;u)\right| \leq N,$$

as soon as $(t,x\ u) \in [0,T] \times K \times U$.

e) The maps g^0: $\mathbb{R}^n \to \mathbb{R}$ and g_i: $\mathbb{R}^n \to \mathbb{R}$, $i = 1,2,\ldots, q$, are continuously differentiable.

Under these assumptions, the following result can be stated.

Theorem 2. Let us consider the optimal control problem (19), (2), (21)–(23), i.e., to minimize $g^0\big(x(T)\big)$ with respect to all measurable controllers u: $[0,T] \to U$, such that constraints (19), (2), (22), (23) are satisfied. Assume that the conditions a)–e) stated above are verified, and let (x,u) be an optimal pair. Then there exist constant multipliers λ_j, $j = 0,1,\ldots, q$, and a map ψ: $[0,T] \to \mathbb{R}^n$, such that the following conditions hold true:

(i) $$\sum_{j=0}^{q} |\lambda_j| = 1, \quad \lambda_j \le 0, \quad 0 \le j \le p, \quad \lambda_j g_j\big(x(T)\big) = 0, \quad 1 \le j \le p;$$

(ii) $$\psi(t) = \left[-d(\lambda)\, X(T,t) + \int_t^T \psi(s)\, X(s,t) ds \right] \cdot \frac{\partial}{\partial x} f\big(t,x(t);u(t)\big);$$

where

$$-d(\lambda) = \lambda_0 \frac{\partial}{\partial x} g^0\big(x(T)\big) + \sum_{j=1}^{q} \lambda_j \frac{\partial}{\partial x} g_j\big(x(T)\big);$$

(iii) for almost all $t \in [0,T]$, one has

$$\left[d(\lambda)\, X(T,t) - \int_t^T \psi(s)\, X(s,t) ds \right] \cdot f\big(t,x(t);u(t)\big)$$

$$= \max_{v \in U} \left\{ \left[d(\lambda)\, X(T,t) - \int_t^T \psi(s)\, X(s,t) ds \right] \cdot f\big(t,x(t);v\big) \right\}.$$

Proof. If we concentrate on the integral equation (20), whose solution is defined on the whole interval $[0,T]$ for given $u(t)$, the optimal control problem of minimizing $g^0\big(x(T)\big)$ under constraints (22), (23), implies we have only to check the validity of the hypotheses made in [1] (see also [3], where the formulation is closer to the exposition adopted in the present paper). This is a routine task, and we leave it to the reader.

Remark 1. It is obvious that condition (iii) provides the Maximum Principle corresponding to the problem (19), (2), (21)–(23). Instead of the classical Hamiltonian, we have a more complicated function. The fact that $X(t,s)$ does appear in the expression of the new "Hamiltonian", makes more difficult the application of such procedures. It is clear that the construction of $X(t,s)$ represents a basic aspect for making the method effectively applicable.

Remark 2. As expected — to some extent — the adjoint function $\psi(t)$ must satisfy a linear Volterra integral equation. Under the hypotheses adopted in this paper, this adjoint equation has always a unique locally integrable solution on the interval $[0,T]$.

Acknowledgement: The author gratefully acknowledges partial support from the ONR (Contract N00014-88-K0080).

References

[1] D. A. Carlson, "An elementary proof of the Maximum Principle for optimal control problems governed by a Volterra integral equation," J. Opt. Th. Appl 54 (1987), 43-61.

[2] C. Corduneanu, Integral Equations and Stability of Feedback Systems. Academic Press, New York, 1973.

[3] C. Corduneanu, Integral Equations and Applications (in print).

[4] C. Corduneanu, "Control problems for abstract Volterra functional-differential equations," in the volume Modern Optimal Control, E. Roxin (Ed.), M. Dekker, New York, 1989, 41-48.

Discrete Observability of Linear Parabolic Systems

D.S. Gilliam, B.A. Mair, C.F. Martin

Abstract

In this paper we investigate the problem of recovering the initial states of distributed parameter systems governed by linear parabolic partial differential equations, from discrete data.

1 Introduction

The observability of linear systems of the form

$$\dot{w}(t) = Aw(t), \;\; w(0) = f \tag{1.1}$$

with observation

$$y(t) = Cw(t) \tag{1.2}$$

is of significant practical and theoretical importance especially as it is a first step in the problem of actually determining the initial state f from the observations y (cf. [1], [8]). Although the continuous observability (i.e. $y(t) = 0, \forall\, 0 < t < T \Rightarrow f = 0$) of these systems has been studied by various authors (cf. [1], [6], [10]), it is important for practical purposes to investigate discrete observability (cf. [3], [5], [7]). For this reason, this paper seeks to develop a stable technique for determining f from a discrete set of values of $y(t)$.

Concerning the recovery of the initial state of the observed system (1.1), (1.2), it is interesting to note that observability ensures the existence of a corresponding unique initial state. Now, if (1.1), (1.2) is a finite dimensional system, the dependence of the initial state on the data is continuous, however, this is not the case for infinite dimensional systems (cf. [8], [6], [4]).

Various techniques to overcome the lack of continuity have been investigated (cf. [6], [8]). The Tychonoff regularization technique consists of restricting the domain of the initial states and introducing a regularizing parameter either in the minimization of a certain quadratic functional or in the solution of a perturbed linear system (cf. [6], [8]). To discuss this regularization as applied in this paper and to clarify the preceding comments on observability, we introduce the following terminology.

We assume the operator A is self-adjoint on a Hilbert space H with simple, decreasing eigenvalues $\{-\lambda_j : j \in J, \ \ \lambda_j > 0\}$ and corresponding orthonormal eigenfunctions $\{\varphi_j : j \in J\}$, where $J = \{1, 2, \ldots, n\}$ or $\{1, 2, 3, \ldots\}$ according as A is $n \times n$ or infinite dimensional.

Definition 1.1 *The system (1.1), (1.2) is said to be eigen-observable if*

$$C\varphi_j \neq 0, \ \forall_j \in J.$$

Definition 1.2 *The system (1.1), (1.2) is said to be discretely observable if for any set $\{t_j : j \in J\}$ of distinct times,*

$$y(t_j) = 0, \forall_j \in J \Rightarrow f = 0$$

In this paper, we consider scalar output, i.e. C densely defined in H with values in $\mathbf{R}$, and assume there is an "admissible" set of sequences $\{t_j\}$ such that $\{y(t_j)\} \in l^2$.

Definition 1.3 *The system (1.1), (1.2) is said to be discretely recovery observable if for any admissible sequence $\{t_j : j \in J\}$ of discrete times, the map*

$$P : f \in H \to \{y(t_j); j \in J\} \in l^2$$

has a continuous inverse on its range.

Under these assumptions on A (given before Definition 1.1), the solution of (1.1) is given in terms of a semigroup as

$$w(t) = \sum_{j \in J} \exp(-\lambda_j t) f_j \varphi_j \tag{1.3}$$

where f_j is the Fourier coefficient $< f, \varphi_j >$ of f. Recovering f is equivalent to determining each f_j.

From (1.3), the discrete observed data gives rise to an infinite system

$$LF = Y \tag{1.4}$$

where F and Y are column vectors in $\mathbb{R}^J$ having j-th entry $f_j C\varphi_j$ and $y(t_j)$ respectively, and L is the matrix with (k,j)-th entry $\exp(-\lambda_j t_k)$.

If H has finite dimension n, the matrix L is invertible for any set $\{t_1, t_2, \ldots, t_n\}$ of distinct times ([7]). A proof of this fact is obtained from Polya's property W, [9], a condition which is based on the non-vanishing of Wronskian determinants for the functions $\{\exp(-\lambda_i t)$. The system (1.4) is uniquely solvable for F, hence all the Fourier coefficients are uniquely determined if and only if $C\varphi_j \neq 0$, for all $j = 1, 2, \ldots, n$. In the finite dimensional case, eigen-observability, discrete observability and continuous discrete recoverability all coincide.

Rather than discuss the problem of observability in an abstract setting, we present a study of the problem of recovering initial data for parabolic equations determined by regular Sturm-Liouville spatial operators on a finite interval with separated boundary conditions.

We consider discrete observations $y(t_k) = Cw(t_k)$ where C satisfies a general condition which includes the case of idealized point sensors and spatial averages over some effective sensing region. As usual (cf [8]), (1.4) is equivalent to

$$L^* L F = L^* Y \tag{1.5}$$

where $L^* L$ is now self-adjoint, non negative and compact. Thus, it does not have a continuous inverse. This leads to the introduction of a regularizing parameter $\alpha > 0$ (see (3.3)) in a perturbation of (1.5).

In this paper, we show that one can obtain an approximation to the initial data f by solving a finite truncation of the perturbed equation even in the presence of noisy finite dimensional data $\tilde{y}(t_1), \ldots, \tilde{y}(t_n)$. The continuous dependence of this approximate solution on the observations is obtained for all initial states in a suitable compact set D of our Hilbert space H.

2 Statement of the Problem

In this paper we consider the distributed system

$$\begin{aligned} w_t(x,t) &= Aw(x,t), a < x < b, t > 0 \\ w(x,0) &= f(x), a < x < b \\ y(t) &= Cw(\cdot,t) \end{aligned} \tag{2.1}$$

with boundary conditions

$$B_1 w(\cdot,t) = 0, t > 0$$

$$B_2 w(\cdot,t) = 0, t > 0$$

where the spatial operator A is given by

$$A = \frac{1}{w(x)} \left\{ \frac{d}{dx} \left(p(x) \frac{d}{dx} \right) - q(x) \right\}$$

and,

$$B_1 u = h_1 u(a) - k_1 u'(a)$$

$$B_2 u = h_2 u(b) + k_2 u'(b)$$

where prime denotes differentiation with respect to x, $p \in \mathcal{C}^1[a,b]$; $q, w \in \mathcal{C}[a,b]$; p, w are strictly positive on $[a,b]$; and h_1, k_1, h_2, k_2 are non-negative constants such that $h_1 + k_1 \neq 0$ and $h_2 + k_2 \neq 0$.

The operator A defines a self-adjoint operator on the Hilbert space $H = L^2_w[a,b]$, with the "weighted" inner product

$$< u, v >= \int_a^b u(x)\bar{v}(x)w(x)dx$$

and dense domain

$$\mathcal{D}\ (A) \ = \ \{u \in L^2_w[a,b] : Au \in L^2_w[a,b], B_1 u = B_2 u = 0\}.$$

In the present setting, (1.3) is obtained by considering the regular Sturm-Liouville eigenvalue problem:

$$\begin{array}{lcl} A\varphi(x) & = & -\lambda\varphi(x), a < x < b \\ B_1\varphi & = & 0 \\ B_2\varphi & = & 0 \end{array} \tag{2.2}$$

This system has countably many distinct eigenvalues $\{-\lambda_j : j \in J,\ 0 < \lambda_i < \lambda_{i+1}\}$. with corresponding eigenfuctions $\{\varphi_j : j \in J\}$.

Hence the solution w of (2.1) is given by

$$w(x,t) = \sum_{j=1}^{\infty} f_j \varphi_j(x) \exp(-\lambda_j t) \tag{2.3}$$

where

$$f = \sum_{j=1}^{\infty} f_j \varphi_j \tag{2.4}$$

In what follows we assume that

$$\{C\varphi_j\} \text{ is a bounded sequence.} \tag{2.5}$$

The main aim of this paper is to develop a computationally feasible method to approximate the initial state f from a finite set of approximate outputs $\tilde{y}(t_1), \ldots, \tilde{y}(t_n)$ and to demonstrate the continuous dependence of the resulting approximation on the observations. It seems reasonable to assume that the system is discretely observable. Hence we impose the additional condition:

$$C\varphi_j \neq 0, \text{ for } j = 1, 2, 3, \ldots. \tag{2.6}$$

Now, let $\{t_k\}$ be an increasing, unbounded sequence of times, then for each k,

$$y(t_k) = Cw(t_k) = \sum_{j=1}^{\infty} \exp(-\lambda_j t_k) f_j C\varphi_j \tag{2.7}$$

which can be expressed in matrix form as

$$LF = Y \tag{2.8}$$

where, $F = (f, C\varphi_1, f_2 C\varphi_2 \ldots)^t, Y = (y(t_1), y(t_2), \ldots)^t$, and L is the infinite matrix with (k,j)-th the entry $\exp(-\lambda_j t_k)$.

Since $\{f_j\} \in l^2$ and $\{C\varphi_j\} \in l^\infty$, then $F \in l^2$, hence we consider L as an operator on the sequence space l^2. Although L does not map l^2 into itself for arbitrary $\{t_k\}$, by using (2.7) and the Cauchy-Schwartz inequality, it is easily seen that if:

$$\sum_{j,k=1}^{\infty} \exp(-2\lambda_j t_k) < \infty \tag{2.9}$$

then the "direct problem " of determining the sequence $\{y(t_k)\}$ from f is well-posed. We therefore require the admissible times $\{t_k\}$ to satisfy (2.9).

3 Regularization

To compensate for the ill-posedness of this inverse problem we restrict the domain of L to a compact subset of l^2.

Definition 3.1 *Fix $M > 0$ and let*

$$D = \{f \in \mathcal{C}^2[a,b] : \ max\ \{||f||_\infty, ||f'||_\infty, ||f''||_\infty\} \leq M\},$$

$T : H \to l^2$ defined by $T(f) = (f_1 C\varphi_1, f_2 C\varphi_2, \ldots)^t$ and $\tilde{D} = T(D)$.
Note that the conditions on D imply that

$$|f_j| \leq \ (constant)\ /\lambda_j \ for\ j = 1, 2, 3, \ldots \tag{3.1}$$

Now, from the characterization of compact subsets of Hilbert spaces in [2], it follows that D is a compact subset of H. Also from (2.5) and (2.6), T is continuous and one-to-one. Hence, $\tilde{D}$ is compact and by Tikhonov's theorem (cf. [8]), we obtain

Theorem 3.1 *The restriction of T to D has a continuous inverse on $\tilde{D}$.*

It therefore suffices to investigate the continuous dependence of $F = T(f) \in \tilde{D}$ in (2.8) on the output Y. Equation (2.8) is equivalent to:

$$SF = \hat{Y} \tag{3.2}$$

where $\hat{Y} = L^*Y$ and $S = L^*L$ is compact, one-to-one, self-adjoint and non-negative-definite. Thus, for every $\alpha > 0$, the resolvent operator $(S + \alpha I)^{-1}$ is continuous on H and $||(S + \alpha I)^{-1}|| \leq 1/\alpha$ (cf [8]). Hence we consider the equation

$$(S + \alpha I)F_\alpha = \hat{Y} \tag{3.3}$$

By using the resolution of the identity for S, we obtain (cf [8], [4]),

Theorem 3.2 $\lim_{\alpha \to 0+} ||F_\alpha - F|| = 0.$

Now, for any infinite matrix P with (k, j)-entry p_{kj} we introduce the following infinite matrices.

$$P_{n1} = \begin{bmatrix} p_{11} & p_{12} & \cdots & p_{1n} & \\ \vdots & \vdots & & \vdots & 0_{n\times\infty} \\ p_{n1} & p_{n2} & \cdots & p_{nn} & \\ & 0_{\infty\times n} & & & 0_{\infty\times\infty} \end{bmatrix}$$

$$P_{n2} = \begin{bmatrix} & p_{1(n+1)}, & p_{1(n+2)}, & \cdots \\ 0_{n\times\infty} & \vdots & \vdots & \\ & p_{n(n+1)}, & p_{n(n+2)}, & \cdots \\ 0_{\infty\times n} & & 0_{\infty\times\infty} & \end{bmatrix}$$

$$P_{n3} = \begin{bmatrix} & 0_{n\times n} & & & 0_{n\times\infty} \\ p_{(n+1)1}, & p_{(n+1)2}, & \cdots & p_{(n+1)n} & \\ p_{(n+2)1}, & p_{(n+2)2}, & \cdots & p_{(n+2)n} & 0_{\infty\times\infty} \\ \vdots & \vdots & \vdots & & \end{bmatrix}$$

$$P_{n4} = \begin{bmatrix} 0_{n\times n} & & 0_{n\times\infty} & \\ & p_{(n+1)(n+1)}, & p_{(n+1)(n+2)} & \cdots \\ 0_{\infty\times\infty} & p_{(n+2)(n+1)}, & p_{(n+2)(n+2)} & \cdots \\ & \vdots & \vdots & \vdots \end{bmatrix}$$

Also, for a vector $Z = (z_1, z_2, \ldots)^t \in l^2$, define $Z_{n1} = (z_1, \cdots, z_n, \vec{0})^t$ and $Z_{n2} = Z - Z_{n1} = (0, \ldots, 0, z_{(n+1)}, z_{(n+2)}, \ldots)^t$. Define the corresponding subspaces of l^2 as follows

$$l^2_{n1} = \{Z \in l_2 : Z = Z_{n1}\}, \quad l^2_{n2} = \{Z \in l^2 : Z = Z_{n2}\}.$$

Then l^2_{n2} is the orthogonal complement of l^2_{n1}.

With this notation, S_{nj} is a self-adjoint, one-to-one, non-negative definite (or positive), continuous operator on l^2_{nj} for $j = 1, 4$. The equations (3.2), (3.3) now become

$$S_{n1}F_{n1} + S_{n2}F_{n2} = \hat{Y}_{n1} \tag{3.4}$$

$$S_{n3}f_{n1} + S_{n4}F_{n2} = \hat{Y}_{n2} \tag{3.5}$$

$$(S_{n1} + \alpha I_{n1})F_{\alpha,n1} + S_{n2}F_{\alpha,n2} = \hat{Y}_{n1} \tag{3.6}$$

$$S_{n3}F_{\alpha,n1} + (S_{n4} + \alpha I_{n4})F_{\alpha,n2} = \hat{Y}_{n2} \tag{3.7}$$

Elementary calculations now give

$$F_{\alpha,n1} = (S_{n1} + \alpha I_{n1})^{-1}\hat{Y}_{n1} - E_{\alpha,n}, \tag{3.8}$$

where

$$E_{\alpha,n} = (S_{n1} + \alpha I_{n1})^{-1}S_{n2}(S_{n4} + \alpha I_{n4})^{-1}[S_{n3}(F_{n1} - F_{\alpha,n1}) + S_{n4}F_{n2}]$$

Now, suppose $Z = Z_{n1} \in l^2_{n1}$. Define $\tilde{Z}_{n1} = (L_{n1})^* Z$ and consider the finite system:

$$(S_{n1} + \alpha I_{n1})G_{\alpha,n1} = \tilde{Z}_{n1} \tag{3.9}$$

Theorem 3.3 *For each $F \in \tilde{D}$, corresponding output Y given by (2.8) and $\varepsilon > 0$, there exist $\alpha(\varepsilon) > 0$ and $N(\varepsilon)$ such that $||G_{\alpha(\varepsilon),n1} - F|| < \varepsilon$ if $||Z_{n1} - Y_{n1}|| < \frac{\varepsilon\alpha(\varepsilon)}{\tau}$ and $n \geq N(\varepsilon)$, where*

$$\tau = \text{(constant)} \left\{ \sum_{j,k=1}^{\infty} \exp(-2\lambda_j t_k) \right\}^{1/2}$$

For a proof of this, see [4]

This gives rise to the following method of recovering the initial state. Given approximate values $z_1, z_2, \ldots, z_n$ of the observations $y(t_1), \ldots, y(t_n)$, respectively, form the column vector

$$Z^{(n)} = (z_1, \ldots, z_n, 0, \ldots),$$

and solve the finite system

$$(B_{n1} + \alpha I_{n1})G_{\alpha,n1} = (L_{n1})^* Z^{(n)}. \tag{3.10}$$

Let $G_{\alpha,n1} = (g_1, g_2, \ldots, g_n, 0 \ldots)$. From Theorems 3.3 and 3.1, it follows that

$$\tilde{f}(x) = \sum_{j=1}^{n} \left(\frac{g_j}{C\varphi_j} \right) \varphi_j(x)$$

is an approximation to f which depends continuously on the data $Z^{(n)}$ if n is sufficiently large and α is sufficiently small.

Acknowledgements: The authors wish to acknowledge support in the following agencies: D.S. Gilliam partially supported by Texas Advanced Research Program grant # 0219-44-5195 and AFOSR grant. B.A. Mair partially supported by Texas Advanced Research Program grant # 0219-44-5195. C.F. Martin partially supported by NSA grant # MDA 904-85 H0009.

References

[1] R.F. Curtain, A.J. Pritchard, Infinite Dimensional Linear systems Theory, LNCS 8, Springer-Verlag, 1978.

[2] N. Dunford, J. Schwartz, Linear Operators I, Interscience, 1963.

[3] D.S. Gilliam, Zhu Li and C. Martin, Discrete observability of the heat equation on bounded domains, Int. J. control, 48, 2 (1988), 755-780.

[4] D.S. Gilliam, B.A. Mair and C. Martin, Determination of initial states of parabolic systems from discrete data, submitted.

[5] D.S. Gilliam, C. Martin, Discrete observability and Dirichlet series, Systems & Control letters 9 (1987) 345-348.

[6] T. Kobayashi, Initial state determination for distributed parameter systems, SIAM J. Control and Optimization, 14, 5(1976), 934-944.

[7] C. Martin, J. Smith, Approximation, interpolation and sampling, Contemporary Math. 68(1987), 227-252.

[8] M.M. Lavrent'ev, V.G. Romanov and S.P. Shishatskii, Ill-posed Problems of Mathematical Physics and Analysis, Translations of Mathematical Monographs, Vol. 64, American Math. Soc., 1986.

[9] G. Polya, On the mean-value theorem corresponding to a given linear homogeneous differential equation, Bull. Amer. Math. Soc., 24(1922), 312-324.

[10] Y. Sakawa, Observability and related problems for partial differential equations of parabolic type, SIAM J. Control 13(1975), 15-27.

VIRTUAL ZEROS AND CUT-OFF FREQUENCES FOR DISTRIBUTED PARAMETER SYSTEMS

LUCIANO PANDOLFI

Abstract. *Frequency domain and time domain description of linear systems are well estabilished methods for lumped parameter systems ([1]) and a great deal of results are now available in the distributed parameter case too ([2, 3]). Distributed systems are in fact a very rich field and it may well be possible that a general theory that covers all the important features does not exist.*

In recent times large classes of systems were studied, mostly with the use of semigroup theory. In this case, the relationships between time and frequency domain descriptions were investigated ([4]). But, one variable semigroup theory hides some of the key features of distributed parameter systems. In this paper we investigate one of them.

1. Introduction

Linear finite dimensional systems are described, in the frequency domain, by *rational* transfer functions. The key properties of rational transfer functions are zeros and poles, and their multiplicities (more correctly, their "structure"). For nonrational transfer functions, the properties of poles and zeros can be studied and related to the properties of the internal description of the system, very much as in the lumped parameter case; but, new possibilities arise, which only in part can be detected from the transfer function, if input and output spaces are finite dimensional. Moreover, a transfer function may now have non-isolated or essential singularities which contain important informations on the properties of the system: for example, in the case of delay systems the structure of the essential (not isolated in general) singularity at $z = \infty$ contains informations on the small solutions ([5]). The zeros (which may even coincide with poles or more general type of singularities) give informations on the frequences which are "blocked" by the system: i.e. if z_0 is a zero, then there are input signals $u(t) = e^{z_0 t}p(t)$ ($p(.)$ being a polynomial) such that the output is identically zero, provided that a suitable initial datum is assigned to the system. Hence, in the concrete cases of stable systems, the output which corresponds to this input function will exponentially decay, even if $Re z_0 \geq 0$. In this paper we are going to present a new possibility which virtually corresponds to a zero (a special input gives "practically" zero output) and which is not encountered in the finite dimensional case.

The organization of this paper is as follows: in the next section a special example is presented, which justifies the arguments in sect. 3, where "virtual zeros" of a transfer function are related to the internal description of the class of systems presented in that same section. Further considerations (on virtually invariant subspaces and on relations with different descriptions) may be found in sections 4, 5.

2. An Example

In the finite dimensional theory, only one independent variable is considered, which usually represents time. in the case of distributed parameters, more independent variables are often (not always) present, but this fact is hidden by the usual semigroup approach. In fact, one of the variables (the time variable) is singled out and the system is then represented formally by an ordinary differential equation:

$$\frac{d}{dt}x(t) = Ax + Bu.$$

The function $x(\cdot)$ takes values in a suitable Banach space. Unfortunately, this approach, which has so many merits, is not always suited or convenient, since many important features of a distributed parameter system may be lost, as the following example shows. We consider a system described by:

$$(1a) \qquad E_{tt} = E_{zz} + E_{xx} \quad 0 < x < 1,\ t > 0,\ z > 0.$$

Eq. (1a) describes the electric field $E(\cdot)$ between two infinite, parallel plates in the direction of the z-axis, between $x = 0$ and $x = 1$ and this is the simplest example of a wave guide. In order to solve Eq. (1a) some boundary conditions must be imposed. We consider the case of perfectly conducting plates, so that we impose the conditions:

$$(1b) \qquad E(t,z,0) = 0,\ E(t,z,1) = 0$$

and we must know $E(t,0,x)$.

In fact, the boundary value $E(t,0,x)$ is under the control of an operator, who sends a message through the guide, described by a function $t \to u(t)$. The action of this signal on the system is represented by:

$$(1c) \qquad E(t,0,x) = \Phi(x)u(t)$$

where $\Phi(\cdot)$ depends on the physical structure of the system.

The data assigned till now do not specify a well posed problem (for example, we did not specify the initial value of the electric field), but this fact is of no concern to us, since the following arguments apply equally well even to ill posed problems. We shall however add the prescription that the function $z \to E(t,x,z)$ must be bounded, since no amplification can be performed by the guide.

The signal injected into the guide is extracted at some position z_0, $z_0 \gg 0$, which needs not be the end of the guide. *We note that we cannot assume that z_0 is $+\infty$, since in general the limit: $lim_{z\to+\infty} E(t,x,z)$ will not exist. So, it is not convenient to hide the variable z by making use of the semigroup approach.*

The extracted signal (i.e. the output of the system) $y(.)$ will be an average of the electric field at the position z_0,

$$(2) \qquad y(t,z_0) = \int_0^1 \psi(x)E(t,z_0,x)dx$$

(the dependence of y on z_0 is explicitly indicated, since the point z_0 at which the signal is extracted could be changed by the operator in some application).

Let us assume now that the operator injects the signal $u(t) = e^{i\tau t}u_0$ (τ is any complex number; if $\tau = \alpha + i\beta$ we intend that the injected signal is $u(t) = e^{-\beta t}[cos(\alpha t) + i\,sin(\alpha t)]u_0$.) We look for solutions to Eq.s (1a-c) of the form:

$$(3) \qquad E(t,z,x) = e^{i\tau t}e^{i\zeta z}E(x).$$

A simple substitution shows that such a solution exists and $E(x) = \Phi(x)u_0$ if the function $x \to E(x)$ and the numbers ζ, τ are related by:

$$(4) \qquad (\zeta^2 - \tau^2)E = E_{xx} \qquad E(0) = E(1) = 0.$$

Hence, a non zero solution of the form (3) exists iff

$$(5) \qquad \tau^2 - \zeta^2 = n^2\pi^2,\ n \in N$$

Eq. (5) may be satisfied by reals as well as complex values of ζ, τ. For any $\tau \in C$ we find $\zeta \in C$ which solves Eq. (5); but, we cannot allow any complex value for ζ. In fact if, for example, ζ has zero real part, then we have the following cases:

a. if $\zeta = i\delta, \delta < 0$ the signal would increase without bound while crossing the guide. As we said, this case is physically impossible.

b. if $\zeta = i\delta, \delta > 0$ then the signal exponentially decreases while crossing the guide. The output, although not zero, will be undetectable if $z_0 \gg 0$.

The same cases as a. b. happen if $\zeta = \gamma + i\delta, \delta < 0$, or $\delta > 0$. So, the only realistic (i.e. bounded) solutions to Eq. (5) which produces a detectable output are those for which $Im\zeta = 0$.

The pairs (τ, ζ), $Im\zeta > 0$ identify solutions to Eq. (1) which are physically meaningful, but which do not produce a usable output (signals that "do not propagate through the guide" in the electrician jargoon). Such pairs will be called *virtual zeros.*

Now, we have this case, that for a given real value of τ only those values of n can be considered in Eq. (5) for which a real solution ζ exists. Hence, $\tau^2 - n^2\pi^2$ must be real and non negative; i.e., if τ is given, the number n, which depends on some adjustable properties of the system, must be chosen such that: $n^2\pi^2 \leq \tau^2$. The number n can be changed, provided that a new choice is made for the function $x \longrightarrow \Phi(x)$ and, in concrete cases, only few choices are possible. Conversely, once that a choice of $\Phi(\cdot)$ is made, the value n_0 for n is given and only input signals with $\tau > n_0\pi$ may propagate trough the guide. The number $n_0\pi$ is called *the cut off frequency* of the guide. It is prescribed by the generator of the function $x \longrightarrow \Phi(x)$. The previous example shows that the pair (ζ, τ) may be considered as a zero if ζ belongs to some set (the set $Im\zeta > 0$ in the previous example).

Now, we can abstract the following general observations from the previous special example:

1. We call *characteristic pair* of the system a pair (τ, ζ) such that a signal $e^{\tau t}e^{\zeta z}E(x)$ exists (i.e., in the previous example, those pairs with $Im\, i\zeta \geq 0$, and $(i\tau, i\zeta)$ solution to Eq. (5). Note that we replaced $i\tau$ and $i\zeta$ with τ and ζ);
2. A set $N \subseteq C$ is given, and any characteristic pair (τ, ζ) such that ζ belongs to N is called a *virtual zero.*
3. If τ is such that ζ belongs to the boundary of N it is called a *cut off frequency.*

Of course, analogous arguments can be repeated in the case that more then two privileged variables are present.

In the next sections we try to develop a formal setting which will contain the previous example. Before doing this, we remark that neither causality nor well posedness were used.

3. Virtual zeros for a class of linear systems

We assume for simplicity that only two privileged variables are present, that we call t, z and we consider the differential operator:

$$DE = [\sum_{i=0}^{r}\sum_{j=0}^{i} a_{ij}\frac{\partial^i}{\partial t^j\, \partial z^{i-j}}]E \tag{6}$$

Where E is a function of t, z which takes values in the domain of a linear operator A from a complex linear space X to itself. Usually, X is a Banach or a Hilbert space.

We consider the differential equation

$$DE = AE \tag{7a}$$

togheter with an additional condition

$$E(t, 0) = \Phi u(t). \tag{7b}$$

where $u \in dom\Phi \subset U$ (U being a second complex linear space) and Φ is a linear operator from U to X. We search for solutions to Eq. (7) of the form:

$$E(t, z) = P(t, z)e^{\tau t}e^{\zeta z} \tag{8a}$$

$P(t, z)$ being a polynomial:

$$P(t, z) = \sum_{\nu=0}^{n}\sum_{\mu=0}^{m} t^{\nu} z^{\mu} E_{\nu\mu} \tag{8b}$$

We assume that a solution of the form (8a) exists, with $P(.)$ given by (8b) and $E_{n,r} \neq 0$ for some value of r. We want to show that solutions of the form (8) come into chains; i.e., we want to show that if a solution exists with $n > 0$ then a solution exists of the same form as (8a), but with the polynomial $P(.)$ of degree $n - 1$ with respect to the time variable t. This is reminiscent of the finite dimensional theory.

Clearly, initial and boundary conditions cannot be interchanged: in the example of the previous section, the boundary condition at $z = 0$ represents the input to the system, while the datum for

$t = 0$ is not under the control of the operator. So, in general, it is not possible to exchange the role of the variables t, z in the previous result; i.e., it will not be possible in general to reduce the degree of $P(\cdot)$ with respect to the z variable.

So, let us assume that problem (7) has a solution of the form (8) with $n > 0$. Then, from condition (7b), we know that there exists at least one function $u(.)$ such that:

$$(9) \qquad \Phi u(t) = e^{\tau t} \sum_{\xi=0}^{n} t^{\xi} E_{\xi 0}.$$

Moreover, we note the following equalities, which are easily tested by recurrence ($E(t, z)$ is defined in (8)):

$$\frac{\partial^k}{\partial t^k} e^{\tau t} t^{\nu} = e^{\tau t} \sum_{s=0}^{k} \binom{k}{s} \frac{\nu!}{(\nu - s)!} \tau^{k-s} t^{\nu - s}$$

so that, after few manipulations, the function $(t, z) \to DE(t, z)$ is given by:

$$e^{\tau t} e^{\zeta z} \sum_{\nu=0}^{n} \{ \sum_{s=0}^{r} \chi(n - s - \nu) \sum_{i=s}^{r} \sum_{j=s}^{i} a_{ij} \binom{j}{s} \frac{(\nu + s)!}{\nu!} \tau^{j-s} H(i - j, \nu + s, \zeta, z) \} t^{\nu}$$

where $\chi(.)$ is the characteristic function of $[0, +\infty)$ and

$H(\rho, \nu, \zeta, z) = \sum_{\mu=0}^{m} \sum_{\xi=0}^{\rho} \binom{\rho}{\xi} \frac{\mu!}{(\mu-\xi)!} \zeta^{\rho-\xi} z^{\mu-\xi} E_{\nu\mu}$ (in this and the following formulas we intend that $(1/s!) = 0$ if s is negative and, consequently, $\binom{j}{s} = 0$ if $s < 0$ or $s > j$; moreover, we put $E_{rk} = 0$ if $r > n$, $k > m$). Consequently, $E(t, z)$ is a solution to Eq. (7a) if and only if $A \sum_{\xi=0}^{m} E_{\nu\xi} z^{\xi} = \sum_{s=0}^{r} \chi(n - s - \nu) \sum_{i=s}^{r} \sum_{j=s}^{i} a_{ij} \binom{j}{s} \frac{(\nu+s)!}{\nu!} H(i - j, \nu + s, \zeta, z) \tau^{j-s}$. Equating the coefficients of z^{ξ} on both sides we get:

Lemma 1. *$E(t, z)$ is a solution to the equation $DE = AE$ if and only if:*

$$AE_{\nu\xi} = \sum_{\mu=\xi}^{\xi+r} \sum_{s=\nu}^{\nu+r} \chi(n - s) \frac{s!}{\nu!} \frac{\mu!}{\xi!} Q(\mu - \xi, s - \nu) E_{s\mu}$$

where:

$$Q(\rho, s) = \sum_{i=s}^{r} \sum_{j=s}^{i} a_{ij} \binom{j}{s} \binom{i - j}{\rho} \tau^{i-s} \zeta^{i-j-\rho}.$$

◇

Now, let us assume that $E(t, z)$ is the solution of both the following equations,

$$(10) \qquad DE = AE \qquad E(t, 0) = \Phi u(t).$$

Then

Theorem 2. *There exists a solution to the problem (7), which has degree (n-1) with respect to the time variable t.*

Proof: We consider the function

$F(t, z) = e^{\tau t} e^{\zeta z} \sum_{\nu=0}^{n-1} \sum_{\mu=0}^{m} t^{\nu} z^{\mu} F_{\nu\mu}$ where: $F_{s\xi} = (s + 1) E_{s+1,\xi}, \quad 0 \leq s \leq n - 1$. It is an easy matter to verify that $AF_{\nu\xi} = \sum_{\mu=\xi}^{\xi+r} \sum_{s=\nu}^{\nu+r} \chi(n - 1 - s) \frac{s!}{\nu!} \frac{\mu!}{\xi!} Q(\mu - \xi, s - \nu) F_{s\mu}$ so that $F(s, z)$ is a solution to $DF = AF$. Moreover,

$$F(t, 0) = e^{\tau t} \sum_{\nu=0}^{n-1} t^{\nu} F_{\nu 0} = e^{\tau t} \sum_{\nu=0}^{n-1} t^{\nu} (\nu + 1) E_{\nu+1,0} = \Phi e^{\tau t} \sum_{\nu=0}^{n-1} t^{\nu} (\nu + 1) u_{\nu+1}$$

since we assumed that $E(t, z)$ satisfies condition (7b).

◇

Remark. ***We note the analogy with the zeros of finite dimensional systems: if τ_0 is a zero of a transfer function then we can find a sequence $(x_k(\cdot), u_k(\cdot))$ of suitable input and state functions, which give zero output. The poles behave in fact differently: if τ_0 is a pole of a transfer function then an input function whose transform has a pole at τ_0 of a certain order, produces an output which has a pole at τ_0, of higher order.***

◇

We note again that the previous construction cannot be performed in order to reduce the degree with respect to z, since the roles of the two variables t and z are not the same, because of the second condition in (7b). But, of course, this procedure can be iterated (always with respect to the t variable) and after k iterations we find a solution whose polynomial part has degree $n-k$ with respect to t, if $k \leq n$. Let us denote by $E_n(t,z)$ the solution to the problem (7) given by (8) and by $E_{n-1}(t,z)$, $E_{n-2}(t,z)$, ..., $E_0(t,z)$ the functions that are obtained by the previous iterations. The sequence $\{E_n(t,z), ..., E_0(t,z)\}$ will be called a *Jordan chain* of lenght $(n+1)$, which corresponds to the pair (τ, ζ). It may be that the previous Jordan chain is the "final section" of a longer chain. If this is not the case, then we say that the Jordan chain is *maximal.*

Now, let us write $E_k(t,z)$ in the form $E_k(t,z) = e^{\tau t} e^{\zeta z} \sum_{s=0}^{k} t^s \Psi_{sk}(z)$, $E_k(t,z)$ being the $(n-k+1)$-th element of the Jordan chain of lenght $(n+1)$. From the previous calculations we know that the functions $\Psi_{s,k}(z)$ are not independent. In fact, let us assume that we know $E_n(t,z)$,

$$E_n(t,z) = \sum_{\nu=0}^{n} t^{\nu} \{\sum_{\mu=0}^{m} z^{\mu} E_{\nu\mu}\}.$$

Then:

$$E_{n-k}(t,z) = e^{\tau t} e^{\zeta z} \sum_{\nu=0}^{n-k} t^{\nu} \Psi_{\nu,n-k}(z) = \sum_{\nu=0}^{n-k} t^{\nu} \{\sum_{\mu=0}^{m} z^{\mu} (\nu+1) E_{\nu+1,\xi}\};$$

$$\Psi_{\nu,n-k}(z) = \frac{(\nu+1)!}{(\nu+k-1)!} \Psi_{\nu+k,n}(z).$$

We consider the linear space

$$N = span\{\Psi_{sk}(z)\}_{0 \leq k \leq n, 0 \leq s \leq n}$$

which is a subspace of the space of the X–valued functions on $0 \leq z \leq \infty$.

The space N is associated to the pair (ν, ξ) which, from now on, is a virtual zero. Consequently, a solution which takes values in N for each t gives virtually zero output if z_0 is large, as it is usually the case. Now:

Lemma 3. *We have: $N = span\{E_n(t,z), t \geq 0\}$.*

Proof: It is clear that for each t the function $z \to E_n(t,z)$ is an element of N. Conversely, we show that $\Psi_{sk}(z)$ can be obtained from $z \to E_n(t,z)$. As we noted, $\Psi_{s,k}(z)$ is a multiple of $\Psi_{s+k,n}(z)$, so that we confine ourselves to consider the elements $\Psi_{s,n}(z)$, $0 \leq s \leq n$. Now, $\Psi_{0,n}(z) = E_n(0,z)$; the element $\Psi_{1,n}(z)$ is obtained as follows: $\Psi_{1,n}(z)$ is the value, at $t=0$, of $(1/z)[E_n(t,z) - \Psi_{0,n}(z)]$. This procedure can be iterated and this proves the lemma.

◇

Now, a given initial signal $u(\cdot) = \sum_{\nu=0}^{n} u_\nu t^\nu$ is not detected at the output if the corresponding solutions $E_n(t,z)$ evolve in the subspace N. This observation is preliminary to any consideration on decoupling subspaces which resembles the known finite dimensional case, as we show in the next section. We note that it may be important to identify subspaces of signals which do not propagate trhoughout a given line in some applications, for example, when several different channels are connected to the same signal generator.

4. Output Decoupled Subspaces

We consider again the subspace N of the previous section. Any initial datum $E_0(z)$ in N is kept in N by the input function $\Phi u(t) = e^{\tau t}\sum_{\nu=0}^{n} t^{\nu} E_{\nu 0}$ (and, the control function exists, may be not uniquely identified). Let us assume that $u(.)$ takes values in the finite dimensional space C^m, so that we can assume that it takes values in $[Ker\Phi]^{\perp}$, i.e. that it is a linear function of $e^{\tau t}\sum_{\nu=0}^{n} t^{\nu} E_{\nu 0}$; consequently, the value of $u(t)$ at time t is a linear function of $z \rightarrow E(t,z)$. This result parallels the well known fact that a system invariant subspace is kept invariant by a feedback control.

Let us recall the following fact from finite dimensional theory: if V is a subspace which is constructed from a Jordan chain relative to a zero λ then λ is an eigenvalue of the closed loop obtained by using a feedback control which keeps V invariant. We show an analogous statement in the present context. We assume for simplicity that Φ is invertible (this is not restrictive) and we call L the operator such that $LE(t,z) = E(t,0)$. Then, $u = \Phi^{-1}LE(t,z)$. We consider the last element $E_0(t,z)$ of the Jordan chain: $E_0(t,z) = e^{\tau t}e^{\zeta z}\sum_{\mu=0}^{m} E_{0,\mu} z^{\mu}$. We can apply Lemma 1 to this element and we see that:

$$AE_{00} = \sum_{\mu=0}^{r}\sum_{i=s}^{r}\sum_{j=s}^{i} a_{ij}\frac{(i-j)!}{(i-j-\mu)!}\tau^{i}\zeta^{i-j-\mu}E_{0\mu}.$$

Let us call $D_\mu(\tau,\zeta)$ the polynomial

$$D_\mu(\tau,\zeta) = \sum_{i=s}^{r}\sum_{j=s}^{i} a_{i,j}\frac{(i-j)!}{(i-j-\mu)!}\tau^{i}\zeta^{i-j-\mu},$$

so that: $AE_{00} = \sum_{\mu=0}^{r} D_\mu(\tau,\zeta)E_{0\mu}$.

In the special, but important, case that $E_0(t,z) = e^{\tau t}e^{\zeta z}E_{00}$, (i.e. that $\mu = 0$) $E_{00} \neq 0$ we find that:

$$AE_{00} = D_0(\tau,\zeta)E_{00}$$

(since we put $E_{0\mu} = 0$ if $\mu > m$.) Moreover, we recall that there exists a vector u_0 such that $E_{00} = \Phi u_0$ (so that u_0 is different from zero). Consequently, we have:

Theorem 1. *Let (τ,ζ) be a virtual zero. Then, there exists a vector u_0 such that:*

$$e^{\tau t}\Phi u_0 = L\,E(t,z) \neq 0$$

$$\textbf{(11)} \qquad \Phi u_0 \in Ker\{D_0(\tau,\zeta)I - A\}.$$

◇

Signals of the special form $E(t,z) = e^{\tau t}e^{\zeta z}E_{00}$ are of preminent interest for the applications and the previous result show that they depend upon the zeros of the 2–D operator valued polynomial $D_0(\tau,\zeta)I - A$.

2–D matrix valued polynomials were studied by systems theorists mostly for the applications to 2–D discrete time systems and "zeros" for this class of systems were investigated only by few authors ([5]). In particular, a signal of the form $E(t,z) = e^{\tau t}e^{\zeta z}E_{00}$ which does not propagate through the guide corresponds to the "zeros" of $\{D_0(\tau,\zeta)I - A\}$ provided that ζ belongs to the subset N of the complex plane. In the very important case that $X = C$ and A is the multiplication by a complex number then the virtual zeros are the points of an algebric curve which belong to a prescribed region of the complex plane.

5. Comparaison with different methods

We compare now the previous approach with the methods which are more common in the control community. We confine ourselves to the simple example of sect. 2, that we write again for more clarity:

$$E_{tt} = E_{xx} + E_{zz} \quad E(t,z,0) = E(t,z,1) = 0 \quad E(t,0,x) = \Phi(x)u(t)$$

the output being:

$$y(t, z_0) = \int_0^1 \eta(x) E(t, x, z_0) dx.$$

We insist that we are not assuming well posedness and, in fact, this problem is not well posed. From a physical point of view, we did not specify whether the guide ends at z_0, or, if it continues; in the firs case we did not specify whether it is open or it is closed; in the second case we did not specify the end of the guide or, if it is infinite, the expected behaviour of the electric field for large values of z. And, in fact, we could even think that the lenght of the guide is not fixed, the output being sensed at different positions. However, let us assume for the moment that z_0 is fixed and that we add enough conditions so to have a well posed problem. In this case it maight be convenient to study the problem *via* the semigroup approach with respect to the time variable t. In this case, a zero is a complex number τ with this property: that it is possible to find non-zero functions $u(t) = e^{\tau t} u_0$, $E(t, x, z) = e^{\tau t} E_0(x, z)$ which solves the problem, and such that $y(t) \equiv 0$. From a physical point of view, if the lenght of the guide is large, so that $y(t) \sim E(t, x, z_0)$, this means that z_0 is a *node* of the electric field, which is quite a different problem from the one considered previously: a node of the electric field is encountered if the field is read at a precise location along the $z-$axis, while whether a point (τ, ζ) is a virtual zero does not depend on z_0, if $z_0 \gg 0$.

A second classical approach to the previous problem is *via* Laplace transformation in the time variable and finite Laplace transformation in the $z-$variable. We get the following equalities, where $\widehat{}$ means Laplace transformation in both t and z variables and $\widetilde{}$ means transformation with respect to the time variable (the argument is explicitly indicated only when this may avoid ambiguities):

(12a) $$\tau^2 \widehat{E} = \zeta^2 E + \widehat{E}_{xx} + e^{-\zeta z_0} E_z(z_0) - E_z(0) +$$
$$+ \zeta e^{-\zeta z_0} E(z_0) - \zeta E(0) + \widetilde{E}_t(0) + \tau \widetilde{E}(0)$$

(12b) $$(\tau^2 - \zeta^2)\widehat{E} = \widehat{E}_{xx} + \{ e^{-\zeta z_0} [\widetilde{E}_z(\tau, z_0, x) +$$
$$+ \zeta \widetilde{E}(\tau, z_0, x)] - \zeta \widetilde{E}(\tau, 0, x) - \widetilde{E}_z(\tau, 0, x) + E_t(\tau, 0, x) + \tau E(\tau, 0, x)\}$$

The previous calculation is giustified only for $Re\, \tau > 0$, since $(t, z, x) \longrightarrow E(t, z, x)$ is bounded but, in general, not decaying for $t \longrightarrow +\infty$. while a virtual zero may even correspond to an unbounded input signal.

A second difficulty with this approach is that both the input and the output depend only on the time variable, so that we should add the quite involved conditions:

$$\widetilde{E}(\tau, 0, x) = \Phi \widetilde{u}(\tau) \widetilde{y}(\tau) = \int_0^1 \eta(x) \widetilde{E}(\tau, x, z_0) dx.$$

The first expression may be replaced in (12a), while the second one contains $\widetilde{E}(\tau, x, z_0)$, which should be calculated from (12b) (and suitable additional conditions). The expression which is obtained is quite involved, since $\widetilde{E}(\tau, z_0, x)$ must be expressed through an inversion formula and, moreover, the difficulties are as in the previous approach, that $\widehat{y}(\tau) = 0$ identifies the nodes of the system, and there is no simple way to identify the virtual zeros.

Finally, let us consider the fact that z_0, in some applications, may not be fixed: so, it may be convenient in this case to consider it as a variable, and to use Laplace transformations in t and z. We find:

$$\int_0^{+\infty} \int_0^{+\infty} e^{-\tau t} e^{-\zeta z} E_{zz}(t, z) dz dt =$$

$$\int_0^{+\infty} e^{-\tau t} E_z(t, 0) dt - \zeta \int_0^{+\infty} e^{-\tau t} E(t, 0) dt + \zeta^2 \widetilde{E}(\tau, \zeta),$$

if we assume that $E_z(t,0) \equiv 0$, which means that the end of the guide is closed with an insulating material. Consequently, $(\tau^2 - \zeta^2 - D_{xx})\widehat{E}(\tau,\zeta,x) = -\zeta\Phi(x)\widehat{u}(\tau)$ (we have not distinguished any more between transformations with respect to both the variables t, z, or with respect to only one of them, since this appears clearly from the formula). Consequently:

$$\widehat{y}(\tau,\zeta) = -\zeta\int_0^1 \Psi(s)[\tau^2 - \zeta^2 - D_{xx}]^{-1}\Phi(s)ds\widehat{u}(\tau)$$

This expression defines a transfer function:

$$T(\tau,\zeta) = -\zeta\int_0^1 \Psi(s)[(\tau^2 - \zeta^2)I - D_{xx}]^{-1}\Phi(s)ds,$$

on $Re\zeta > 0$. We saw already that $\Phi(x) = \sin n_0\pi x$ for some value n_0 of n. For $\tau^2 - \zeta^2 \neq n_0^2\pi^2$ the transfer function now depends on the same 2–D operator in (12):

$$T(\tau,\zeta) = -\zeta\int_0^1 \Psi(r)[\frac{1}{\nu}\frac{\sin h\nu r}{\sin h\nu}\int_0^1 (\sin h\nu(r-s))\Phi(s)\,ds -$$

$$\int_0^r \frac{1}{\nu}[\sin h\nu(r-s)]\Phi(s)\,ds]dr,$$

where $\nu = (\tau^2 - \zeta^2)^{1/2}$. Moreover, $\widehat{y}(\tau,\zeta) = T(\tau,\zeta)\widehat{u}(\tau)$. A direct computation shows that the input $e^{\tau t}u_0$ produces, for suitable initial conditions, the output $y(t,z) = e^{\tau t}e^{\zeta z}[T(\tau,\zeta)u_0]$, with $\tau^2 - \zeta^2 = n^2\pi^2$ (n being integer): the characteristic pairs belong to the sets over which the transfer function *is not* defined. The virtual zeros are those characteristic pairs such that ζ belongs to a prescribed subset of C: the set where $Re\,\zeta < 0$ in the example we have described. This observation suggests that in some applications a weaker definition could be used: If ϵ is a prescribed tolerance, then we can consider the pairs (τ,ζ) such that $||T(\tau,\zeta)|| < \epsilon$.

Acknowledgment: *Paper written with financial support of the Italian "Ministero della Ricerca Scientifica".*

REFERENCES

[1] Conte, G., Perdon, R.M., *A Geometric approach to the theory of 2–D systems,* IEEE Trans. Automatic Control, **AC–33**, 946-950.

[2] Curtain, R.F., *A Synthesis of Time and Frequency Domain Methods for the Control of Infinite-Dimensional Systems: a System Theoretic Approach* to appear in "SIAM Series: Frontiers in Applied Mathematics".

[3] Pandolfi, L. *Some Properties of the Frequency domain Description of boundary Control Systems,* J. Math. Analysis Appl., 142, 219-241, 1989.

[4] Pandolfi, L. *The Structure of Singularities of Holomorphic Matrices,* Proc. Royal Soc. of Edinburgh, 107A, 1-14, 1987.

[5] Rosenbrock, H.H., *State Space and Multivariable Theory*, Nelson, London, 1970.

"ON THE WELL–POSEDNESS AND RELAXABILITY OF NONLINEAR DISTRIBUTED PARAMETER SYSTEMS"

N.S. Papageorgiou

Abstract

In this work we examine the relation existing between well–posedness (sensitivity) and relaxability of nonlinear distributed parameter systems. We introduce the notion of "strong calmness" which describes the dependence of the value of the problem on perturbations of the state constraints and we show that it is equivalent to "relaxability". We also present an equivalent, control free description of the relaxed problem and we prove a density result.

1. Introduction.

It is well known that if we want to derive necessary conditions for optimality, things do simplify if we have some convexity hypothesis at our disposal and this partly motivates the introduction of the relaxed system, wherein the original dynamical equations are replaced by their convexified versions. Another important reason to consider the relaxed system, is that it always has a solution under very mild hypotheses. But then we need to know if and when the relaxation process introduces new better solutions or leaves the value of the problem unchanged and an original optimal control, optimal for the relaxed system too (relaxability). On the other hand, given the optimal control problem, it is important–especially when state constraints are present – to have a precise mathematical formulation to express the fact that the original problem is well posed in the sense that arbitrarily small perturbations of the data, do not drastically change the value of the problem. The first to formalize this stability concept was Clarke [1], who for this purpose introduced the notion of calmness. He then proved that calmness implies relaxability for finite dimensional systems with no state constraints (see theorem 2 in [1]). In this paper we consider nonlinear distributed parameter systems with state constraints. We introduce a stronger notion of calmness and we show that it is in fact equivalent to relaxability. We present the results without proof. A detailed exposition will appear elsewhere.

2. Strong calmness.

Let Y be any Banach space. We will be using the following notations:

$P_{f(c)}(Y) = \{A \subseteq Y$: nonempty, closed, (convex)$\}$ and $P_{(w)k(c)}(Y)$ $= \{A \subseteq Y$: nonempty, (w–) compact, (convex)$\}$.

Now let H be a separable Hilbert space and X a dense linear subspace carrying the structure of a separable, reflexive Banach space and with the embedding $X \hookrightarrow H$ compact. Identifying H with its dual (pivot space), we have $X \hookrightarrow H \hookrightarrow X^*$, where all embeddings are continuous, dense and compact. So (X, H, X^*) is a Gelfand triple. By $\|\cdot\|$ (resp. $|\cdot|$, $\|\cdot\|_*$), we will denote the norm of X (resp. of H, X^*), while by $<\cdot,\cdot>$ we will denote the duality brackets for the pair (X, X^*) and by $(\cdot,\cdot)$ we will denote the inner product of H. Recall that $<\cdot,\cdot>\Big|_{X\times H} = (\cdot,\cdot)$. We will model the control space using a separable Banach space Y.

Consider the following nonlinear, distributed parameter optimal control problem of Lagrange type:

$$\left\{\begin{array}{l} J(x,u) = \int_0^b L(t,x(t),\ u(t))\,dt \to \inf = p(0) \\ \text{s.t. } \dot{x}(t) + A(t,x(t)) = f(t,x(t),\ u(t)) \text{ a.e.} \\ x(0) = x_0,\ x(t) \in C(t),\ u(t) \in U(t) \text{ a.e.} \end{array}\right\} (*)$$

We will need the following hypotheses on the data of (*).

H(A): $A: T \times X \to X^*$ is a map s.t. (1) $t \to A(t,x)$ is measurable, (2) $x \to A(t,x)$ is sequentially continuous from X_w into X^*_w (where X_w (resp. X^*_w) denotes the space X (resp. X^*) with the w–topology), (3) $x \to A(t,x)$ is monotone, (4) $<A(t,x), x> \geq c_1 \|x\|^2$ a.e. $c_1 > 0$ and (5) $\|A(t,x)\|_* \leq g(t) + c_2 \|x\|^2$ a.e. with $g(\cdot) \in L^\infty_+$, $c_2 > 0$,

H(f): $f: T \times H \times Y \to H$ is a map s.t. (1) $t \to f(t,x,u)$ is measurable, (2) $(x,u) \to f(t,x,u)$ is sequentially continuous from $H \times Y_w$ into H_w and (4) $|f(t,x,u)| \leq a(t) + b(|x| + \|u\|)$ a.e. with $a(\cdot) \in L^2_+$, $b > 0$,

H(U): $U: T \to P_{fc}(Y)$ is an L^2–integrably bounded multifunction,

H(C): $C: T \to P_f(H)$ is an L^1–integrably bounded multifunction with $x_0 \in C(0) \cap X$,

H(L): $L: T \times H \times Y \to \overline{\mathbb{R}} = \mathbb{R} \cup \{+\infty\}$ is a proper measurable integrand s.t. $\phi(t) \leq L(t,x,u)$ a.e. with $\phi(\cdot) \in L^1$,

H_a: There exist $(x,u) \in W(T) \times L^2(Y)$ satisfying the constraints of (*) s.t. $J(x,u) < \infty$.

Recall that $W(T) = \{x(\cdot) \in L^2(X): \dot{x}(\cdot) \in L^2(X^*)\}$. We know that $W(T) \subseteq C(T,H)$ (see for example Lions [4]). To problem (*) we associate the following perturbed problem

$$J(x,u) = \int_0^b L(t,x(t), u(t))\, dt \to \inf = P(\epsilon)$$

$$\left.\begin{cases} \text{s.t. } \dot{x}(t) + A(t,x(t)) = f(t,x(t), u(t)) \text{ a.e.} \\ x(0) = x_0, \int_0^b d_H(x(t), C(t))\, dt \leq \epsilon, \ u(t) \in U(t) \text{ a.e.} \end{cases}\right\} (*)_\epsilon$$

Let $V = \{m: \mathbb{R}_+ \to \overline{\mathbb{R}}_+$ s.t. $m(\cdot)$ is nondecreasing and $\lim_{\epsilon \downarrow 0} m(\epsilon) = m(0) = 0\}$. We will say that (*) is "strongly calm" if and only if there exists $m(\cdot) \in V$ s.t. $\varliminf_{\epsilon \downarrow 0} \frac{P(\epsilon) - P(0)}{m(\epsilon)} > -\infty$. Note that Clarke [1] defined "calmness" using $m(\epsilon) = \epsilon$.

Using our stronger notion of calmness we can easily check that:

Proposition 2.1: If hypotheses H(A), H(f), H(U), H(C), H(L) and H_a hold, then $P(\cdot)$ is right continuous at zero iff (*) is strongly calm.

So strong calmness is equivalent to well–posedness.

Let $m(\cdot) \in V$ and set $K(m) = \inf\{J(x,u) + m(\int_0^b d_H(x(t), C(t))\, dt)$ s.t. $\dot{x}(t) + A(t,x(t)) = f(t,x(t), u(t))$ a.e., $x(0) = x_0$, $u(t) \in U(t)$ a.e.$\}$. We have:

Proposition 2.2: If hypotheses H(A), H(f), H(U), H(C), H(L) and H_a hold, then $P(\cdot)$ is right continuous iff there exists $m \in V$ s.t. $P(0) = K(m)$ iff (*) is strongly calm.

Also to problem (*) we can associate the following penalized version, in which we perturb the cost criterion instead of the constraints.

$$\left.\begin{cases} J(x,u) + \frac{1}{\epsilon} \int_0^b d_H(x(t),\ C(t))\, dt \to \inf = Q(\epsilon) \\ \text{s.t. } \dot{x}(t) + A(t, x(t)) = f(t, x(t), u(t)) \text{ a.e.} \\ x(0) = x_0,\ u(t) \in U(t) \text{ a.e.} \end{cases}\right\} (*)'_\epsilon$$

Observe that for $\epsilon = 0$, $(*)'_\epsilon$ reduces to (*) and so $Q(0) = P(0)$. The next proposition tells us that the previous analysis is also valid for the above "penalized" problem.

Proposition 2.3: If hypotheses $H(A)$, $H(f)$, $H(U)$, $H(C)$, $H(L)$ and H_a hold, then $Q(\cdot)$ is right continuous at zero iff $P(\cdot)$ is.

3. Relaxability and strong calmness.

In this section we will show that well–posedness in the sense of right continuity of $P(\cdot)$ at zero and thus strong calmness of the original problem is equivalent to the "relaxability" of the system. To this end we introduce the following relaxed system:

$$\left.\begin{cases} J_r(x,\lambda) = \int_0^b \int_Y L(t, x(t),\ z)\, \lambda(t)\,(dz)\, dt \to \inf = P_r(0) \\ \text{s.t. } \dot{x}(t) + A(t, x(t)) = \int_Y f(t, x(t),\ z)\, \lambda(t)(dz) \text{ a.e.} \\ x(0) = x_0,\ \lambda(\cdot) \in S_\Sigma,\ x(t) \in C(t) \end{cases}\right\} (*)_r$$

where $\Sigma(t) = \{\lambda \in M_+^1(Y): \lambda(U(t)) = 1\}$, with $M_+^1(Y)$ being the space of probability measures on $M_+^1(Y)$. We will say that (*) is "relaxable" iff $P(0) = P_r(0)$.

We will need the following stronger hypotheses on the data

$H(f)_1$: $f: T \times H \times Y \to H$ is a map s.t. (1) $t \to f(t,x,u)$ is measurable, (2) $(x,u) \to f(t,x,u)$ is sequentially continuous from $H \times Y_w$ into H_w, (3) $|f(t,x,u) - f(t,y,u)| \le k_M(t)\ |x-y|$ a.e. for all $\|u\| \le M$ with $k_M(\cdot) \in L_+^1$ and (4) $|f(t,x,u)| \le a(t) + b(t)\ (|x| + \|u\|)$ a.e. with $a(\cdot), b(\cdot) \in L_+^2$.

$H(U)_1$: $U: T \to P_{fc}(Y)$ is a measurable multifunction s.t. $U(t) \subseteq W$ a.e. with $W \in P_{wkc}(Y)$.

$\underline{H(L)_1}$: L: $T \times H \times Y \to \mathbb{R}$ is an integrand s.t. (1) $(t,x,u) \to L(t,x,u)$ is measurable, (2) $(x,u) \to L(t,x,u)$ is continuous from $X \times W_w$ into $\mathbb{R}$, where W_w denotes the set W with its relative w–topology, and (3) for every $B \subseteq H$ compact, $t \to \inf\,[L(t,x,u): x \in B, u \in W]$, belongs in L^1.

With a weaker hypothesis on L (namely joint measurability, lower semicontinuity on $H \times W_w$ and for each $B \subseteq H$ compact $-\infty < \int_0^b \inf$ $[L(t,x,u): x \in B, u \in W]\,dt$, we can show that $(x,\lambda) \to J_r(x,\lambda)$ is l.s.c. on $C_B \times (L^\infty(T, M(W_w)), w^*)$, where $C_B = \{x(\cdot) \in C(T, X_w): x(t) \in B, t \in T\}$ and $M(W_w)$ is the space of Radon measures on W_w. Using this fact, we can prove the following theorem relating well–posedness and relaxability.

Theorem 3.1: If hypotheses $H(A)$, $H(f)_1$, $H(U)_1$, $H(C)$, $H(L)_1$ and H_a hold, then $P(\cdot)$ is right continuous iff (*) is relaxable.

So combining theorem 3.1 with the results of section 2, we get the following complete characterization of relaxability:

Theorem 3.2: If hypotheses $H(A)$, $H(f)_1$, $H(U)_1$, $H(C)$, $H(L)_1$ and H_a hold, then the following statements are equivalent: (1) problem (*) is relaxable, (2) $P(\cdot)$ is right continuous at zero, (3) $Q(\cdot)$ is right continuous at zero, (4) problem (*) is strongly calm.

The proofs of these results are based on some density results concerning the trajectories of the controlled evolution equations proved by the author in [5] and [6].

4. An alternative form of the relaxed problem.

Let $p: T \times H \times X^* \to \overline{\mathbb{R}} = \mathbb{R} \cup \{+\infty\}$ be defined by $p(t,x,v) = \inf\{L(t,x,u): v + A(t,x) = f(t,x,u), u \in U(t)\}$, with $\inf \emptyset = +\infty$. So $p(t,x,v)$ represents the minimum cost of producing velocity v at time t using admissible controls and given that the state of the system is x. Using $p(\cdot,\cdot,\cdot)$ we can have the following control free formulation of the relaxed problem:

$$\hat{J}_r(x) = \int_0^b p^{**}(t,x(t), \dot{x}(t))\, dt \to \inf = \hat{P}_r(0)$$

$$\left.\begin{cases} \text{s.t. } \dot{x}(t) + A(t,x(t)) \in \overline{\text{conv}}\ F(t,x(t)) \text{ a.e.} \\ x(0) = x_0,\ u(t) \in U(t) \text{ a.e., } x(t) \in C(t) \end{cases}\right\} (\hat{*})_r$$

with $F(t,x) = \cup\{f(t,x,u): u \in U(t)\}$ and $p^{**}(\cdot,\cdot,\cdot)$ is the second convex conjugate of $p(t,\cdot,\cdot)$. The next theorem tells us that problems $(*)_r$ and $(\hat{*})_r$ have the same value.

<u>Theorem 4.1</u>: <u>If</u> hypotheses $H(A)$, $H(f)$, $H(U)_1$, $H(C)$, $H(L)_1$ and H_a hold, <u>then</u> $P_r(0) = \hat{P}_r(0)$.

The proof of this result is rather involved and is based on properties of Radon measures, of measurable multifunctions and of the convex conjugates. Also an interesting byproduct of the proof is that $(*)_r$ and $(\hat{*})_r$ have equivalent dynamics, hence the same trajectory sets. Furthermore from theorem 4.1 we deduce that system (*) is relaxable <u>iff</u> $P(0) = \hat{P}_r(0)$.

5. <u>A density result</u>.

The problem of whether the original trajectories are dense in the relaxed ones, can not be answered using the results of [5] because of the presence of state constraints. In fact it is a nontrivial problem and in this section we present a solution to it.

We need the following lemmata, which are also interesting in their own as general results about multifunctions and convex sets.

<u>Lemma I</u>: <u>If</u> Z is a Banach space, $C: T \to P_{fc}(Z)$ is a Hausdorff continuous multifunction with bounded values and for all $t \in T$ int $C(t) \neq \emptyset$, <u>then</u> the set of continuous selectors of C, denoted by $CS(C)$, is nonempty and int $CS(C) = CS(\text{int } C)$.

The first conclusion of the lemma follows from Michael's selection theorem, while the proof of the second conclusion is based on the fact that $t \to \text{bd } C(t)$ is Hausdorff continuous too (see DeBlasi–Pianigiani [2]).

Lemma II: If Z is a Banach space, $A, B \subseteq Z$ are nonempty, $\overline{A}$ is convex, B is closed convex with $\text{int}\, B \neq \emptyset$ and $A \cap \text{int}\, B \neq \emptyset$, then $\overline{A \cap B} = \overline{A} \cap B$.

The proof of this lemma is based on some simple convex analytic arguments.

Now we are ready for the density result in the presence of state constraints. We were able to prove it for systems with linear dynamics $\dot{x}(t) + A(t)\, x(t) = B(t)\, u(t)$ a.e., $x(0) = x_0$, $u \in S^1_U = \{$integrable selectors of $U(\cdot)\}$. We will need the following hypotheses:

$H(A)_1$: $A: T \times X \to X^*$ is a map s.t. (1) $A(t)(\cdot)$ is linear, monotone, (2) $\|A(t)x\|_* \leq k(t)\, \|x\|$ a.e. with $k(\cdot) \in L^1_+$ (i.e. $A(t)(\cdot) \in \mathscr{L}(X, X^*)$), (3) $\langle A(t)\, x, x\rangle \geq c_1\, \|x\|^2$ $c_1 > 0$, (4) $\|A(t')x - A(t)\, x\|_* \leq m|t'-t|\ \|x\|$ $m > 0$.

$H(B)$: $B \in L^2(T, \mathscr{L}(Y, X^*))$, $H(U)_2$: $U(\cdot)$ is L^2–integrably bounded,

$H(C)_1$: $C: T \to P_{fc}(H)$ is h–continuous.

From Tanabe [7] (section 5.4) we know that under $H(A)_1$ $\{A(t)\}_{t \in T}$ generates a strongly continuous evolution operator $\Phi(t,s) \in \mathscr{L}(H)$ $0 \leq s \leq t \leq b$ and a trajectory $x(\cdot) \in W(T)$ can be written as

$x(t) = \Phi(t,0)\, x_0 + \int_0^t \Phi(t,s)\, B(s)\, u(s)\, ds$, $t \in T$. We will assume the following about $\Phi(\cdot,\cdot)$:

H_c: $\Phi(t,s)$ is compact for $t - s > 0$.

Let $\hat{S}(x_0)$ be the set of trajectories of the original problem and $\hat{S}_r(x_0)$ the set of the relaxed ones.

Theorem 5.1: If hypotheses $H(A)_1$, $H(f)_1$, $H(U)_1$, $H(C)_1$, H_c hold and there exists $x(\cdot) \in S(x_0)$ s.t. $x(t) \in \text{int}\, C(t)$ $t \in T$, then $\hat{S}_r(x_0) = \overline{\hat{S}(x_0)}$, the closure in $C(T,H)$.

The proof is based on the two lemmata above and the unconstrained density results proved in [5].

Our work extends to distributed parameter systems those of Dontchev–Morduhovic [3] and Zolezzi [8].

ACKNOWLEDGEMENT: This work was supported by NSF–Grant D.M.S.–8802688.

REFERENCES

[1] F.H. Clarke: "Admissible relaxation and variational control problems" J. Math. Anal. Appl. 51 (1975), pp. 557–576.

[2] F. DeBlasi–G. Pianigiani: "Remarks on Hausdorff continuous multifunctions and selections" Comm. Math. Univ. Carol 24 (1983), pp. 553–562.

[3] A. Dontchev – B. Morduhovic: "Relaxation and well–posedness of nonlinear optimal control processes" Systems and Control Letters 3 (1983), pp. 177–179.

[4] J.–L. Lions: "Optimal Control of Systems Governed by P.D.E.'s Springer, New York (1970).

[5] N.S. Papageorgiou: "Properties of the relaxed trajectories of evolution equations and optimal control" SIAM J. Control and Optim. 27 (1989), pp. 267–289.

[6] N.S. Papageorgiou: "Relaxation of infinite dimensional variational and control problems with state constraints" Kodai Math. Jour – to appear.

[7] H. Tanabe: "Equations of Evolution" Pitman, London (1979).

[8] T. Zolezzi: "Well posedness and stability analysis in optimization" Proceedings of the "Fermat Days", ed. J.–B. Hiriart–Urruty, North Holland, New York, pp. 305–320.

On the Solution of DDP in Infinite-Dimensional Systems

Hans Zwart

Abstract

Under mild conditions on the infinite dimensional system, Σ, we shall prove the following. If there exists a second system Σ_f which solves the Disturbance Decoupling Problem (DDP), then there exists also a linear, time-invariant, memoryless system, thus a feedback law, which solves it. The proof will be given entirely in geometric terms. However this feedback is not necessarily bounded.

1 Introduction

In this paper we will investigate under which conditions the disturbances entering an infinite dimensional, linear, time invariant system can be rejected by a suitable compensator. We will show that this problem is equivalent to a geometric problem. By which we mean that the disturbance decoupling problem can be formulated in terms of subspace inclusions. Let Σ denote the infinite dimensional, linear, time-invariant system, which is given by the following equations.

$$(1)\qquad \begin{aligned} \dot{x}(t) &= Ax(t) + Bu(t) + Eq(t); \\ z(t) &= Dx(t); \qquad\qquad t \geq 0 \\ x(0) &= x_0 \end{aligned}$$

where $x(.), u(.), q(.)$ and $z(.)$ are time trajectories in respectively $\mathcal{X}, I\!R^m, \mathcal{Q}$ and $\mathcal{Z}$, where $\mathcal{X}, \mathcal{Q}$ and $\mathcal{Z}$ are general Banach spaces, B, D and E are bounded linear operators, and A is a generator of a C_0-semigroup $T_A(t)$ on $\mathcal{X}$. This system is schematized in figure 1 by a signal flow graph.

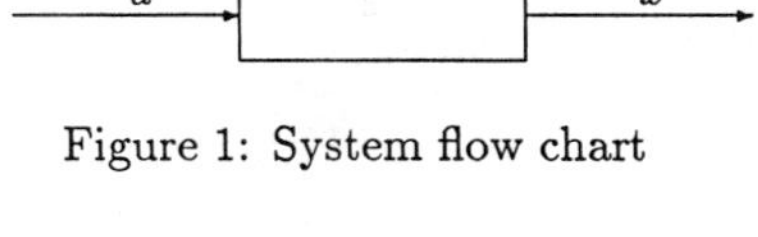

Figure 1: System flow chart

In this system we regard the input $q(.)$ as an undesired influence (disturbance) on the system. In general this input will influence the output $z(.)$. The dependence of the solution $z(.)$ on $q(.)$ and $u(.)$ will be represented in the notation $z(t; x_0, u, q)$, which denotes the solution of (1) with input function u, disturbance function q and initial state $x(0) = x_0$. In the last twenty years a lot of research has been done on the so called problem of disturbance decoupling, see e.g. [1,6] for finite dimensional systems and [2,4,5,7,9] for the infinite dimensional case. That is the problem to give necessary and sufficient conditions for the existence of a (bounded) linear feedback law F such that, by applying the input $u = Fx$, the disturbances has no influence on the output $z(.)$. Thus for all disturbance functions q we have $z(t, x_0, Fx, q) = z(t, x_0, Fx, 0)$, and so the output is decoupled from the disturbances. In figure 2 we give this configuration pictorially.

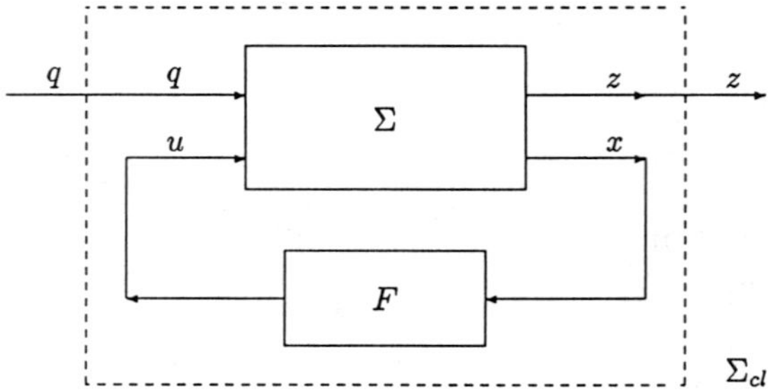

Figure 2: Feedback system flow chart with static feedback

In this paper we will consider a weaker version of this problem. Instead of looking for a bounded linear feedback law, we will be looking for a (causal) compensator such that, in closed loop, the disturbances have no influence on the output z. Before we state the formal definition of this generalized problem we will investigate the (old) problem more closely.

2 Disturbance Decoupling Problems

In this section we will investigate the problem as schematized in figure 2 more closely. Suppose it is possible to find a bounded feedback law F such that no disturbance enters the output z. Then we have that for every disturbance input q

$$0 = \int_0^t DT_{A+BF}(t-s)Eq(s)ds$$

This is only possible if $DT_{A+BF}(t)E \equiv 0$, see e.g. [9]. Let $\mathcal{R}$ denote the reachable subspace of the system $(A + BF, E)$. This subspace is defined as

$$\mathcal{R} = \overline{span}\{T_{A+BF}(t)Eq; q \in \mathcal{Q}, t \leq 0\}$$

By the solvability of the disturbance decoupling problem it is clear that

(2) $\quad ImE \subset \mathcal{R} \subset \ker D$

and furthermore the semigroup property of T_{A+BF} implies that $\mathcal{R}$ satisfies

(3) $\quad T_{A+BF}(t)\mathcal{R} \subset \mathcal{R}$

On the other hand, suppose that there exists a subspace $V \subset \mathcal{X}$, which for some bounded F satisfies (2) and (3). Then the DDP is solvable with this F. So we see that subspaces, which are invariant under $T_{A+BF}(t)$ for some F, play a key role in the Disturbance Decoupling Problem. Therefore this invariant property has a special name.

Definition 2.1 Let V be a linear subspace of $\mathcal{X}$. This subspace is closed loop invariant if there exists a bounded feedback law such that $T_{A+BF}(t)V \subset V$ for all $t \geq 0$.

From the DDP we see that we are looking for a closed loop invariant subspace contained in the kernel of D. If the state space is finite dimensional simple dimension arguments imply that there exists always a largest closed loop invariant subspace contained in the kernel of D. In the general case we may always define this subspace.

Definition 2.2 $\mathcal{V}^*(\ker D)$ will denote the largest closed loop invariant subspace contained in the kernel of D, i.e if V is a closed loop invariant subspace contained in the kernel of D, then $V \subset \mathcal{V}^*(\ker D)$.

In [4,9] it is shown that in the infinite dimensional case $\mathcal{V}^*(\ker D)$ need not exist necessarily. By the boundedness of $T_{A+BF}(t)$ for every t it is easy to show that $\mathcal{V}^*(\ker D)$ must be closed. Furthermore if it exists then we have the following result.

Theorem 2.3 *If $\mathcal{V}^*(\ker D)$ exists, then DDP is solvable if and only if $ImE \subset \mathcal{V}^*(\ker D)$. Furthermore the feedback that solves the DDP can be any bounded feedback F that satisfies $T_{A+BF}(t)\mathcal{V}^*(\ker D) \subset \mathcal{V}^*(\ker D)$.*

These results can all be found in [2]. As mentioned in the introduction we will consider a generalized version of the disturbance decoupling problem. Instead of looking for a bounded linear feedback law, we will be looking for a (causal) compensator such that, in closed loop, the disturbances have no influence on the output z. We will investigate the (old) problem more closely to see what is a suitable definition for the generalized problem. As we have seen in the beginning of this section DDP is solvable if and only if $DT_{A+BF}(t)Eq \equiv 0$ for every $q \in \mathcal{Q}$, see e.g. [9]. By standard C_0-semigroup theory we have that, see e.g. [3, theorem 2.31].

$$(4) \quad T_{A+BF}(t)Eq = T_A(t)Eq + \int_0^t T_A(t-s)BFT_{A+BF}(s)Eqds$$

Since F is bounded we have that $FT_{A+BF}(t)Eq$ is a continuous function on $[0,\infty)$. Equation 4 is the same as

$$T_{A+BF}(t)Eq = \int_0^t T_A(t-s)Eq\delta(s)ds + \int_0^t T_A(t-s)BFT_{A+BF}(s)Eqds$$

This is exactly the impulse response of system 1 to the disturbance impulse $q\delta(t)$. So we can reformulate the (old) disturbance decoupling problem as finding a feedback F such that if $x_0 = 0$, then for all disturbances of the form $q * \delta$, there exists a continuous input u making $z \equiv 0$. In this paper we will generalize this definition of DDP. Note that this definition could also be interpreted for a system with no disturbance input, as knowing the initial state up to an element in ImE and making the output z independent of this lack of information.
Now our DDP amounts to constructing a second system Σ_f, which takes as input $x(.)$ and gives as output a control input $u(.)$, such that $z(.)$ has become independent of the disturbance input $q(.)$, see figure 3.
In a formal definition this is:

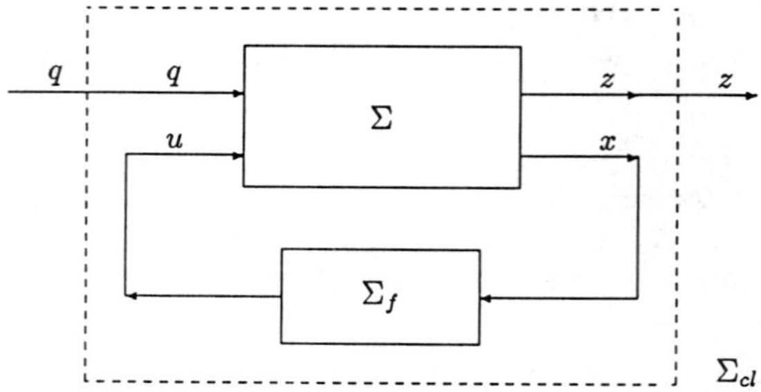

Figure 3: Feedback system flow chart

Definition 2.4 The Disturbance Decoupling Problem with Dynamic state feedback (DDPD):
Find if possible a causal compensator Σ_f such that for initial state $x_0 = 0$ and disturbance impulse $\delta * q$ there exists a continuous input u, which is the output of the compensator, such that the impulse response from the disturbances to the measurement $z(.)$ is zero. Thus the disturbance has no influence on the measurement.

In this definition we assumed as little as posible about the compensator. This enables us to allow every compensator, which may be non-linear, time dependent, as long as the closed loop system has a unique and well defined solution. The formulation in definition 2.4 is close to an open loop formulation, see also theorem 3.5. In the formal definition of this disturbance decoupling problem we assumed that the output of the second system, which is the input of the to-be-controlled system, is continuous. This will simplify the proofs of the theorems considerably. Similar theorems can be obtained by looking at piecewise continuous functions, but presently it is unknown which results one would obtain if the functions were (locally) integrable.
The addition "D" in DDPD represents the fact that we allow general compensators as generator of the input functions. For finite dimensional systems it is well known that DDPD is solvable if and only if DDP is solvable. Under very mild conditions on Σ we shall prove a similar result for infinite dimensional systems. If there exists a second system Σ_f which solves the DDPD, then there exists also a linear, time-invariant, memoryless system which solves it. The proof will be given in geometric terms.

3 Open Loop Invariance

In this section we will investigate the behavior of system (1) with no disturbance and furthermore we assume that there is one input thus $u \in I\!R$. So we regard the equation

$$(5) \qquad \begin{aligned} \dot{x}(t) &= Ax(t) + bu(t); \\ z(t) &= Dx(t); \qquad\qquad t \geq 0 \\ x(0) &= x_0 \end{aligned}$$

where $x(.), u(.)$ and $z(.)$ are time trajectories in respectively $\mathcal{X}, \mathbb{R}$ and $\mathcal{Z}$, $\mathcal{X}$ and $\mathcal{Z}$ are general Banach spaces, $b \in \mathcal{X}$, D is a bounded linear operator, and A is a generator of a C_0-semigroup $T_A(t)$ on $\mathcal{X}$. Furthermore we assume that the transfer function from u to z is not identically zero.
Here we shall make precise what we mean if we say that the solution of (5) remains in V. Furthermore we shall give some interesting properties of this concept, that are useful in the sequel. By the solution of (5) for $u(.) : [0, \infty) \mapsto \mathbb{R}$ we mean the mild solution:

$$(6) \quad x(t) = T_A(t)x_0 + \int_0^t T_A(t-s)bu(s)ds$$

In the solution of DDP the notion of closed loop invariance played a key role. For the DDPD this role will be played by the concept of open loop invariance.

Definition 3.1 Open Loop Invariance
A subspace $V \subset \mathcal{X}$ is said to be open loop invariant if for every $x_0 \in V$ there exists a continuous $u(.) : [0, \infty) \mapsto \mathbb{R}$ such that the solution of (5) remains in V.

Remark:
If the state space is finite dimensional, then it is standard to allow arbitrary measurable functions in the definition of open loop invariance. One obtains there as a result that the input function could be chosen to be continuous. However if the state space is infinite dimensional, then this result no longer holds, see [9]. So we have to restrict the class of input functions in this definition and here we have restricted ourselves to continuous functions. A result that was believed to hold, is that this definition is the same as if one were to use piecewise continuous functions. The next example shows that this conjecture is false.

Example 3.2 In this example we shall study the delay system

$$(7) \quad \dot{y}(t) = y(t-1) + u(t); \; t \geq 0 \; \textit{and} \; y(t) = y_0(t); \; -1 \leq t \leq 0$$

This system can be rewritten in the form (5) if we make the following definitions.

$$x(t) = (y(t), y(t-\tau); 0 \leq \tau \leq 1)$$

with $x \in M^2((-1,0); \mathbb{R})$ where

$$M^2((-1,0); \mathbb{R}) = \mathbb{R} \oplus L^2((-1,0); \mathbb{R})$$

$$\begin{aligned} &A : D(A) \mapsto M^2((-1,0); \mathbb{R}) \; \textit{with} \\ &D(A) = (\varphi_0, \varphi) \in M^2((-1,0); \mathbb{R}) | \varphi_0 \in \mathbb{R}, \; \varphi \; \textit{such that} \; \dot{\varphi} \in L^2((-1,0); \mathbb{R}) \\ &A(\varphi_0, \varphi) = (\varphi(-1), \dot{\varphi}) \end{aligned}$$

$$b : \mathbb{R} \mapsto M^2((-1,0); \mathbb{R}) \; \textit{with} \; bu = (u, 0).$$

$$D : M^2((-1,0); \mathbb{R}) \mapsto \mathbb{R} \; \textit{with} \; D(r, \varphi) = r$$

It can be shown that A generates a semigroup on $M^2((-1,0); \mathbb{R})$, see [3, page 48]. We shall show that the subspace

$$(8) \quad V := \{0\} \oplus \{\varphi \in L^2((-1,0); \mathbb{R}) | \varphi \; \textit{is continuous on} \; [-1,0] \; \textit{and} \; \varphi(0) = 0\}$$

is an open loop invariant subspace contained in $\ker D$. Let $(0,\varphi)$ be an element of V, then by defining $u(t) = -\varphi(t-1)$ for $0 \leq t \leq 1$ we have that

$$\dot{y}(t) = y(t-1) + u(t) = 0$$

for $0 \leq t \leq 1$ and $y(0) = 0$. This implies that $y(t) = 0$ for $0 \leq t \leq 1$. Now we define $u(t)$ to be zero for $t \geq 1$, and with (7) this gives that $y(t) \equiv 0$ for $t \geq 0$. The state at time t is $(y(t), y(t-\tau);\ 0 \leq \tau \leq 1)$ and this is in V for all $t \geq 0$. Furthermore by the continuity of $y(.)$ for $t \geq -1$ we have that $u(.)$ is continuous. So V is open loop invariant. If we extend the subspace V to

$$\tilde{V} := \{0\} \oplus \{\varphi \in L^2((-1,0); I\!R) | \varphi \text{ is piecewise continuous on } [-1,0]\},$$

then $\tilde{V}$ is a subspace contained in $\ker D$. It is easily seen that $\tilde{V}$ is not open loop invariant. However if we would weaken the definition of open loop invariance by allowing piecewise continuous function then $\tilde{V}$ becomes open loop invariant. So a subspace which is open loop invariant for piecewise continuous input functions need no longer be open loop invariant for continuous inputs. □

In the example above one can also show that the subspace defined by (8) is the largest open loop invariant subspace contained in the kernel of D. The largest open loop invariant subspace contained in a given closed subspace exists always, and is defined by

Definition 3.3 Let K be a closed linear subspace of $\mathcal{X}$. Then $\mathcal{V}^*_{ol}(K)$ denotes the subspace containing all $x_0 \in \mathcal{X}$ such that there exists a continuous input $u(t)$ such that the solution of (5) remains in K.

Lemma 3.4 *$\mathcal{V}^*_{ol}(K)$ is the largest open loop invariant subspace contained in K.*

Proof: See [9, theorem 3.12].

Note that if V is a closed subspace which is open loop invariant, then $\mathcal{V}^*_{ol}(V) = V$. Now we will give the relation between $\mathcal{V}^*_{ol}(\ker D)$ and the solvability of DDPD. In contrast with $\mathcal{V}^*(\ker D)$ we have that $\mathcal{V}^*_{ol}(\ker D)$ always exists. This together with the next theorem makes this subspace very useful in the study of disturbance decoupling problems.

Theorem 3.5 *DDPD is solvable if and only if ImE is contained in $\mathcal{V}^*_{ol}(\ker D)$.*

Proof: By definition we have that DDPD is solvable if and only if for every $q(t) = \delta(t) * q_0$ there exists a continuous input $u(.)$ such that $z(.) \equiv 0$, where $z(.)$ is the solution of (1). So $0 = z(t) = \int_0^t DT_A(t-s)bu(s)ds + \int_0^t DT_A(t-s)E\delta(s)q_0 ds$. Rewriting this solution gives

$$(9) \quad 0 = DT_A(t)Eq_0 + \int_0^t DT_A(t-s)bu(s)ds.$$

So ImE is contained in $\mathcal{V}_{ol}^*(\ker D)$. On the otherhand, if ImE is contained in $\mathcal{V}_{ol}^*(\ker D)$, then (9) is solvable for every $q_0 \in \mathcal{Q}$. Thus DDPD is solvable then. □

One of the main results of Zwart [9] is presented next. This theorem gives the relation between $\mathcal{V}_{ol}^*(\ker D)$ and $\mathcal{V}^*(\ker D)$. Furthermore it gives a condition under which DDP and DDPD are equivalent problems.

Theorem 3.6 *Consider the system as described in equation (5).* □

- *If* $\mathcal{V}_{ol}^*(\ker D)$ *is a closed subspace, then* $\mathcal{V}^*(\ker D)$ *exists and is equal to* $\mathcal{V}_{ol}^*(\ker D)$.
- *Suppose that* $\mathcal{V}_{ol}^*(\ker D)$ *is a closed subspace.*

  Then DDPD is solvable if and only if DDP is solvable.

Proof: See [9, theorem III.12], and theorems 3.5 and 2.3.

In the next section we will prove that if DDPD is solvable, then there exists always a linear and time invariant feedback law which solves the problem too. However this feedback law need not be a bounded operator.

4 Existence of a Feedback Law

As mentioned in the title in this section we will show that if DDPD is solvable, then it is also solvable with a linear and time-invariant feedback law, so by $u(t) = Fx(t)$ for some F. Therefore we will prove first that for an open loop invariant subspace the input u can be chosen in feedback form i.e. $u(t) = Fx(t)$ for some feedback.

Theorem 4.1 *Consider the single input system (5) with the largest open loop invariant subspace* $\mathcal{V}_{ol}^*(\ker D)$. *Then the following assertions hold*

1. *For every* $x_0 \in \mathcal{V}_{ol}^*(\ker D)$ *there exists one and only one* $u(x_0(.) \in C([0,\infty); I\!R)$ *such that the solution of*

   (10) $\dot{x}(t) = Ax(t) + bu_{x_0}(t) \qquad x(0) = x_0$

   remains in $\mathcal{V}_{ol}^*(\ker D)$.

2. *There exists a linear operator* $F : \mathcal{V}_{ol}^*(\ker D) \mapsto I\!R$ *such that for every* $x_0 \in \mathcal{V}_{ol}^*(\ker D)$ *there exists a solution of*

   (11) $\dot{x}(t) = (A + bF)x(t) \qquad x(0) = x_0$

   which remains in $\mathcal{V}_{ol}^*(\ker D)$.

Proof: Let x_0 be an element of $\mathcal{V}_{ol}^*(\ker D)$. From the definition of $\mathcal{V}_{ol}^*(\ker D)$ we have the existence of a continuous input function u_{x_0} such that the solution of (10) remains in $\mathcal{V}_{ol}^*(\ker D)$. We will show that this input is unique. Let x_0 be an element of $\mathcal{V}_{ol}^*(\ker D)$, then there exists a u_{x_0} such that

$$x(t) = T_A(t)x_0 + \int_0^t T_A(t-s)bu_{x_0}(s)ds$$

is contained in $\ker D$. Suppose there exists another input function $\tilde{u}$ such that the solution of (10) is also contained in $\ker D$. Then $x(t)-\tilde{x}(t) = \int_0^t T_A(t-s)b(u_{x_0}(s)-\tilde{u}(s))ds$ is contained in $\ker D$. Standard convolution theory now gives that $DT_A(t)b \equiv 0$, but this implies that the transfer from the input u to the output z is identically zero. This is in contradiction with our assumption, see the assumptions after equation (5). Thus the input function is unique.
Let x_0 be an element of $\mathcal{V}^*_{ol}(\ker D)$ and let u_{x_0} be the unique input function as defined in the first part of this proof. Define

$$(12) \quad F : \mathcal{V}^*_{ol}(\ker D) \mapsto I\!R \qquad\qquad Fx_0 = u_{x_0}(0)$$

By the continuity of u_{x0} this is well defined. By the uniqueness of the input functions and the linearity of the system (5) we have that F is linear. We will show that if $x(t)$ is the solution of (10) with input $u_{x_0}(t)$, then $u_{x_0}(t) = Fx(t)$ for all $t \geq 0$. Let a positive time instant t_0 be given. Then $x(t_0)$ is an element of $\mathcal{V}^*_{ol}(\ker D)$, since $u_{x(t_0)}(t) := u_{x_0}(t+t_0)$ keeps the solution of (10) in $\ker D$. This input is unique, thus we have that $u_{x_0}(t_0) = u_{x(t_0)}(0) = Fx(t_0)$. □

Remark:
If the transfer function from u to z is identically zero, then the largest open loop invariant subspace is the largest closed loop invariant equals the unobservable subspace. Thus the feedback law can be chosen zero, see [9].

We have from theorem 4.1 that for every $x_0 \in \mathcal{V}^*_{ol}(\ker D)$ there exists a solution to (11). In general it will not be possible to define a solution of this equation for all initial values outside the largest open loop invariant subspace. So in particular $A + bF$ will not generate a C_0-semigroup in general, see example 4.3.

Example 4.2 In this example we will consider example 3.2 again. The open loop invariant subspace V defined by (8) is equal to the largest open loop invariant subspace contained in $\ker D$. Since if $x_0 = (r_0, \varphi_0)$ is an element of $\mathcal{V}^*_{ol}(\ker D)$, then from the fact that $x_0 \in \ker D$, it follows tha $r_0 = 0$. Furthermore for $0 \leq t < 1$ we must have that

$$\dot{y}(t) = \varphi(t-1) + u(t) = 0$$

for some continuous u. So φ is continuous on $(-1, 0)$. Since u is continuous on $[0, \infty)$ we get that $\varphi(0) = 0$. Thus $x_0 \in V$, which proves the assertion. Now we can define the feedback law F from theorem 4.1. Recall that the state at time t is $x(t) = (y(t), y(t-\tau); 0 \leq \tau \leq 1)$. It is not hard to see that $F : V \mapsto I\!R$ with $Fx = F(0, \varphi) = \varphi(-1)$ is the feedback we are looking for.

In [4] an example was given of an open loop invariant subspace for which there would exists no feedback law that makes the subspace invariant. Pandolfi used this subspace to conclude that the largest closed loop invariant subspace, $\mathcal{V}^*(\ker D)$, need not exist always. The non existence of a feedback law seems in contraction with theorem 4.1. In the next example we will reproduce this subspace, we will calculate the feedback law according to 4.1 and show why there is no conflict with the result of Pandolfi [4].

Example 4.3 Let the system be described by the equation

$$(13)\quad \begin{aligned} \dot{x}_1(t) &= x_2(t) - x_2(t-1) \\ \dot{x}_2(t) &= u(t) \\ z(t) &= x_1(t) \end{aligned}$$

As state space we take $I\!R \oplus I\!R \oplus L^2((-1,0); I\!R)$ and the state at time t is $(x_1(t), x_2(t), x_2(t-\tau); 0 < \tau < 1)$. In order to derive $\mathcal{V}^*_{ol}(\ker D)$ we calculate all initial values for which there exists a continuous u making $z \equiv 0$. Let (r, x, h) be an initial value. Since in particular $0 = y(0) = x_1(0) = r$ we have that $r = 0$. Furthermore since $y = 0$ we have that $\dot{y} = 0$, and thus from equation (13) we obtain that

$$(14)\quad x_2(t) = x_2(t-1)\ \forall t \geq 0$$

For $t \geq 0$ we have that

$$(15)\quad x_2(t) = \int_0^t \dot{x}_2(s)ds + x_2(0) = \int_0^t u(s)ds + x_2(0)$$

Since u is continuous we have that x_2 is continuous differentialble, and using $x_2(t-1) = h(t-1)$ for $0 < t < 1$ together with (14) gives that $h \in C^1((-1,0); I\!R)$. Using (14), (15) again we see that

$$(16)\quad u(t) = \dot{x}_2(t-1)$$

keeps $z \equiv 0$. Thus we obtain that $\mathcal{V}^*_{ol}$ is $\{(0, x, h); x \in I\!R$ and $h \in C^1((-1,0); I\!R)$ with $h(-1) = h(0) = x\}$.
In (16) the feedback law is also given, i.e.

$$F(0, x, h) = \dot{h}(-1)$$

Thus there exists a feedback that keeps $\mathcal{V}^*_{ol}$ invariant, but this feedback is a highly unbounded operator on the state space, and these operators were excluded by Pandolfi in [4].

Now we will use the results of theorem 4.1 to formulate and prove the main theorem of this paper.

Theorem 4.4 *Consider the system (1), where we assume that the system is single input, thus $m = 1$. Furthermore let $\mathcal{V}^*_{ol}(\ker D)$ denote the largest open loop invariant subspace contained in the kernel of D.*
If DDPD is solvable, then there exists a linear feedback law which solves the problem too. Furthermore the feedback law can be chosen to be the operator F from theorem 4.1.

Proof: From theorem 3.5 we have that if DDPD is solvable, then ImE is contained in $\mathcal{V}^*_{ol}(\ker D)$. Let Eq be an element of ImE, then we have from theorem 4.1 that the solution, $x_q(t)$ of (11) with $x_0 = Eq$ remains in $\mathcal{V}^*_{ol}(\ker D)$. Define $u_q(t)$ as $Fx_q(t)$, and so $x_q(t) = T_A(t)Eq + \int_0^t T_A(t-s)bu_q(s)ds = T_A(t)Eq + \int_0^t T_A(t-s)bFx_q(s)ds$. The last equation is exactly the response on $\delta * q$ of the system Σ_{cl} with compensator $\Sigma_f = F$. Since $\mathcal{V}^*_{ol}(\ker D)$ is contained in $\ker D$ we have that $z_q(t) = Dx_q(t) = 0$. Thus DDPD is solvable with the linear feedback F from theorem 4.1. □

Remark:
The feedback law can not always be chosen to be a bounded feedback law, see for example 4.2 or [9, example E.9].

Acknowledgement
I would like to thank Ruth Curtain and Henk Nijmeijer for their useful comments.

References

[1] G. BASILE and G. MARRO; Controlled and Conditioned Invariant Subspaces in Linear System Theory. Journal of Optimization Theory and Applications, Vol. 3, no. 5, pp. 306-315, 1969.

[2] R.F. CURTAIN; Invariance Concepts in Infinite Dimensions. SIAM Journal of Control and Optimiz, Vol. 24, no. 5, pp. 1009-1031, Sept. 1986.

[3] R.F. CURTAIN and A.J. PRITCHARD; Infinite Dimensional Linear Systems Theory. Lecture Notes in Control and Information Sciences 8, Springer Verlag, Berlin, 1978.

[4] L. PANDOLFI; Disturbance Decoupling and Invariant Subspaces for Delay Systems. Applied Mathematics and Optimization 14, pp. 55-72, 1986.

[5] E.J.P.G. SCHMIDT and R.J. STERN; Invariance Theory for Infinite Dimensional Linear Control Systems. Applied Mathematics and Optimization 6, pp. 113-122, 1980.

[6] W.M. WONHAM; Linear Multivariable Control: A Geometric Approach. Springer Verlag, A.M.S. 10, 1978.

[7] H.J. ZWART; Characterization of all Controlled Invariant Subspaces for Spectral Systems. SIAM Journal of Control and Optimization, Vol. 26, no. 2, pp. 369-387, 1988.

[8] H.J. ZWART; Equivalence between Open Loop and Closed Loop Invariance for Infinite Dimensional Systems: A Frequency Domain Approach. SIAM Journal of Control and Optimization, Vol. 26, no. 5, 1988.

[9] H.J. ZWART; Geometric Theory for Infinite Dimensional Systems. Lecture Note in Control and Information Science, vol. 115, Springer Verlag, Berlin, 1989.

The Parallel Solution of the Matrix Eigenproblem with Applications in Control Theory

Daniel Boley
Robert Maier
Haesun Park
P.J. Eberlein

ABSTRACT

We present some of the techniques we used to parallelize two different algorithms for the non-symmetric matrix eigenvalue problem. The non-symmetric matrix eigenvalue problem plays a central role in the analysis of control systems. We discuss the QR algorithm and a non-symmetric Jacobi algorithm, how they were converted to parallel implementations, and present some computational results illustrating the success of the parallel implementations.

1. INTRODUCTION

The rapid solution of very large non-symmetric matrix eigenvalue problems plays an essential role in many fields. In particular in control, it is well known that computation of the modes of a dynamical system is basic to the analysis of the behavior of such systems. In this paper, we discuss some of the techniques that we have used to obtain a parallel implementation of some algorithms for this problem. Because space is short, we limit our discussion to the parallelization itself. Many interesting questions arise regarding the convergence, the specific choice of heuristics among the many available, etc. These problems, though critical for a successful algorithm, are not directly related to the problem of converting the methods to a parallel implementation. For this reason, we defer the discussion.

The two main algorithms used to obtain all the eigenvalues of a non-symmetric matrix are the QR algorithm [2, 11] and the non-symmetric Jacobi algorithm [5, 10, 12, 13]. These methods use different techniques, but in most cases the result is the Schur decomposition of the given matrix A:

$$A = U T U^H \tag{1}$$

where U is a unitary matrix, and T is an upper triangular matrix. Other versions of the Jacobi algorithm actually diagonalize A [3], but we do not discuss that version, since many of the techniques are similar.

We assume the target parallel environment is a network of relatively small processors, each with its own local memory, communicating exclusively by messages. An example is a hypercube, on which these algorithms have actually been implemented.

2. QR ALGORITHM

The QR algorithm, as implemented in EISPACK, has been one of the most successful serial algorithms for the computation of eigenvalues and eigenvectors. The process begins with a preliminary reduction to upper Hessenberg form before proceeding to the iterative part, in order to reduce the cost of each pass through the iterative part from $O(n^3)$ to $O(n^2)$. We will show how we can implement this on a linear array of $O(n)$ processors, obtaining a speed-up of $O(n)$.

The basic tools [7] are the Householder transformation, which is used to annihilate the elements below the subdiagonal, and the plane (Givens) rotation, which is used to annihilate a single element. To illustrate the techniques we use, we concentrate on the iterative part of the method, which is in Figure 1, part II. We omit the details of the origin shift and deflation for the simplicity of the explanation [2]. In steps 1-5, the matrix H is factored by zeroing out the subdiagonal element in each column k by using the appropriate plane rotation in plane $k,k+1$. The plane rotation in the p,q plane $R(p,q)$ is a matrix identical to an identity matrix, except that the elements r_{pp}, r_{pq}, r_{qp}, r_{qq} are as follows (where c and s satisfy $c^2+s\bar{s}=1$ and c is real):

$$\begin{bmatrix} r_{pp} & r_{qp} \\ r_{pq} & r_{qq} \end{bmatrix} \equiv \begin{bmatrix} c & s \\ -\bar{s} & c \end{bmatrix} \tag{2}$$

At the k-th stage, a rotation is created to annihilate the subdiagonal in column k, and then this rotation must be applied across the entire rows k and $k+1$. In steps 6-9, the same rotations must be applied on the right, so that the net effect is a similarity transformation to the matrix. In this way the eigenvalues are preserved.

I. Reduce A to upper Hessenberg H by unitary similarity transformations

II. Apply QR algorithm repeatedly to H: until H converges to upper triangular

1. Factor $H \to QR$
    2. For $k=1, \ldots, n-1$ do: zero $h_{k+1,k}$:
        3. Generate and apply rotation which maps

$$\begin{bmatrix} h_{kk} \\ h_{k+1,k} \end{bmatrix} \to R(k,k+1) \cdot \begin{bmatrix} h_{kk} \\ h_{k+1,k} \end{bmatrix} \equiv \begin{bmatrix} * \\ 0 \end{bmatrix}$$

        4. For $j=k+1, \ldots, n$ do
            5. Apply rotation to kth and $k+1$-th rows:

$$\begin{bmatrix} h_{kj} \\ h_{k+1,j} \end{bmatrix} \to R(k,k+1) \cdot \begin{bmatrix} h_{kj} \\ h_{k+1,j} \end{bmatrix}$$

6. Re-form $H^{\text{new}} \leftarrow RQ$
    7. For $j=1, \ldots, n-1$ do:
        8. For $i=1, \ldots, j+1$ do
            9. Apply rotation to jth and $j+1$th columns:

$$\begin{bmatrix} h_{ij} & h_{i,j+1} \end{bmatrix} \to \begin{bmatrix} h_{ij} & h_{i,j+1} \end{bmatrix} \cdot R(j,j+1)$$

FIGURE 1. One Explicit QR Iteration applied to the upper Hessenberg matrix H.

We show how this can be parallelized by applying some transformations to the algorithm itself. For the implementation, we assume that the matrix is distributed among the processors by column (i.e. one column per processor in the simplest case), and that the processors are in a linear array and can communicate with their two immediate neighbors.

First, we change the loop in steps 2-5 to the form of nested for loops:

For k=1, . . . , n−1 do:
    For j=k, . . . , n do:
        If j=k then
            Generate and apply rotation as in step 3 of Figure 1.
        Else Apply rotation as in step 5 of Figure 1.

Then we swap the nesting of the loops to put the j loop on the outside, since j is the number of the column being operated on:

For j=1, . . . , n do:
    For k=1, . . . , $min(j,n-1)$ do:
        If j=k then
            Generate and apply rotation as in step 3 of Figure 1.
        Else Apply rotation as in step 5 of Figure 1.

Next, we distribute this algorithm to the processors, as is outlined in Figure 2. Since we have distributed the matrix, so that processor P_j holds column j, it is natural to have processor P_j carry out the j-th pass of each loop. Then we obtain the steps 2-5 and 9-10, 15 of Figure 2.

At each stage the rotations must be transmitted to the remaining processors in order to apply them to the entire rows (step 5), and elements from the neighboring column j+1 are needed in step 15. The rotation $R(j,j+1)$ is generated in processor P_j, and must be passed to the processors to its right. Each processor P_i, $i > j$ must receive the rotation from its left neighbor, use it, and pass it on to its right, until it reaches the last processor P_n. Hence we need a "receive" and "send" as in steps 5 and 6, respectively.

In steps 9-10 and 15, we need to obtain entries from the neighboring processor P_{j+1}, which are in steps 14 and 16. However, for each "receive" on the local processor P_j, there must be a corresponding "send" in the remote processor, and vice versa, so we need steps 12 and 13. Note that in the loop at step 9, the j is increasing, hence, each element h_{ij} in the j-th column is operated on with $h_{i,j-1}$, before $h_{i,j+1}$. That is, each element in the j-th column must be loaned temporarily to processor P_{j-1} *before* it is used in its home processor P_j. In order to cut the time the left neighbor must wait for the arrival of h_{ij}, we can move up the "send h_{ij} to left" to just after its previous use, i.e. at step 7 in Figure 2 (identifying i with k). Note that the processors P_n and P_1 work in slightly different ways than other processors - P_n does not generate any rotation and P_1 does not send its elements to the neighbors to get updated. These considerations are omitted from Figure 2 for simplicity.

Several issues related to the QR algorithm still remain to be addressed. The most critical one is how to handle the shifts. In the simplest scheme, the shift is based on the n,n element, and thus can be computed by processor P_n. In order to transmit this shift to all the processors, we assume there is a wrap-around connection from processor P_n to processor P_1, on which the shift can be sent. Then all that is needed is to transmit the shift ahead or with the first rotation $R(1,2)$ to all the other processors. Convergence and deflation are handled as outlined in [1].

matrix size	number of processors	time (sec) (parallel)	ratio wrt line above	speed-up over serial code
64	2	63		1.2
128	4	267	4.2	2.1
256	8	1077	4.0	3.8
512	16	4531	4.2	7.8
1024	32	18191	4.0	15.6

TABLE 1. Typical Parallel QR times on NCUBE/seven

1. Factor
    2. For $j=1, \ldots, n$ do on processor P_j:
        3. For $k=1, \ldots, j$ do:
            4. If $j=k$ then Generate and apply rotation (G_k in Table 1)

$$\begin{bmatrix} h_{kk} \\ h_{k+1,k} \end{bmatrix} \rightarrow R(k,k+1) \cdot \begin{bmatrix} h_{kk} \\ h_{k+1,k} \end{bmatrix} \equiv \begin{bmatrix} * \\ 0 \end{bmatrix}$$

            5. Else Receive $R(k,k+1)$ from left & Apply it: (L_k in Table 1)

$$\begin{bmatrix} h_{kj} \\ h_{k+1,j} \end{bmatrix} \rightarrow R(k,k+1) \cdot \begin{bmatrix} h_{kj} \\ h_{k+1,j} \end{bmatrix}$$

            6. Send $R(k,k+1)$ to right.
            7. (see text).

8. Re-form
    9. For $j=1, \ldots, n-1$ do on processor P_j:
        10. For $i=1, \ldots, j+1$ do
            11. If $i \neq j+1$ then
                12. Send h_{ij} to left.
                13. Receive h_{ij} from left.
            14. Receive $h_{i,j+1}$ from right.
            15. Apply rotation to j-th column: (R_j in Table 1)

$$\begin{bmatrix} h_{ij} & h_{i,j+1} \end{bmatrix} \rightarrow \begin{bmatrix} h_{ij} & h_{i,j+1} \end{bmatrix} \cdot R(j,j+1)$$

            16. Send $h_{i,j+1}$ to right.

FIGURE 2. Outline of Parallel QR Algorithm.

We illustrate in Table 1 the results of the clustered algorithm when run on the NCUBE/seven using up to 32 processors on matrices as large as 1024×1024. Instead of embedding a linear array of processors onto a hypercube [8], we used the Gray code [8] properties of the hypercube interconnection network since it reduces the total communication cost [2]. We put $1024/32=32$ columns on each processor, grouping the columns into clusters of $w=4$ consecutive columns per cluster, and then distributing the clusters in wrap-around fashion among all the $p=n/32$ processors. In Table 1, column 4 is the ratio of the parallel time over the parallel time for the previous line. The fact that these entries are all approximately equal to 4 as the problem size and number of processors are doubled simultaneously indicates that the method is indeed behaving as an $O(n^2)$ method, compared to the serial algorithm which is $O(n^3)$. Column 5 is the speed-up over the serial EISPACK code. We would expect that as the number of processors doubles, the speed-up would also double, and this is in fact the case, approximately. Finally, note that the speed-ups (column 5) are approximately equal to half the number of processors, indicating an efficiency of approximately 50%.

3. JACOBI ALGORITHM

An alternate algorithm for the eigenproblem is the Jacobi algorithm. This algorithm, in its original form for hermitian matrices, dates back to Jacobi, 1846, and is described in e.g. [7]. It is generally slower than the QR algorithm as a serial algorithm, but as we shall illustrate, it possesses much more parallelism, so that in a parallel environment, there is no clear winner between these two algorithms.

In this paper, we concentrate on the techniques needed to parallelize this algorithm, once the serial scheme has been defined. There are several variations of the Jacobi algorithm for non-symmetric matrices. We first describe the particular variation [5, 10] we have used, since it gives the better results in terms of accuracy and convergence, and then mention some of the others.

In the Jacobi algorithm, as we have used it, the goal is to reduce a matrix A to upper triangular form by applying a sequence of unitary plane rotations as similarity transformations. Specifically, the rotation is constructed as in (2) and then applied to A to obtain $A^{(k+1)} \equiv R^H A^{(k)} R$. In this application, only rows and columns p and q of A are changed. In the Jacobi method, the values c and s are computed so that $a_{qp}^{(k)}$ becomes zero. That is, we compute a 2×2 Schur decomposition

$$\begin{bmatrix} c & -s \\ \bar{s} & c \end{bmatrix} \begin{bmatrix} a_{pp}^{(k)} & a_{pq}^{(k)} \\ a_{qp}^{(k)} & a_{qq}^{(k)} \end{bmatrix} \begin{bmatrix} c & s \\ -\bar{s} & c \end{bmatrix} = \begin{bmatrix} a_{pp}^{(k+1)} & a_{pq}^{(k+1)} \\ 0 & a_{qq}^{(k+1)} \end{bmatrix}. \tag{3}$$

Then, a complete Jacobi sweep consists of applying such rotations R using all possible index pairs $1 \le p < q \le n$ in some order. At each step we zero out the q,p element, which is then filled in in the next step. If the transformations are performed in all possible planes, then the sequence $\{A^{(k)}\}$ should approach the desired form. In the sequential algorithm, the order used was typically cyclic by rows, or cyclic by columns, but these orderings do not lead to direct parallelism. In contrast to the symmetric case [9], no Jacobi algorithm for non-symmetric eigenvalue problems is proved to converge.

In summary, the procedure Jacobi can be described as in Figure 3. In other variations of the Jacobi method for non-symmetric matrices, it is possible to use only index pairs (p,q) such that $q = p+1$ [12], or to diagonalize A by relaxing the condition that R be unitary [3].

Jacobi Method

1. Choose an Ordering of the index pairs (p,q)
2. Set $V^{(0)} \equiv I$, and set k=0
3. Repeat until Convergence
    4. Loop through all the index pairs (p,q) in the order chosen
        5. Construct rotation R in (p,q) plane as in (2) satisfying (3)
        6. Apply R: $A^{(k+1)} \equiv R^H A^{(k)} R$
        7. Accumulate R: $V^{(k+1)} \equiv V^{(k)} R$

FIGURE 3. Sketch of Jacobi Algorithm.

In order to parallelize this algorithm, we first consider a two-dimensional mesh of processors. Assuming that we have $n/2 \times n/2$ processors, each processor can hold a 2×2 block of the matrix $A^{(k)}$. We start by using the index pairs (1,2), (3,4), etc. Each of these rotations can be computed and applied simultaneously. Each diagonal processor carries out step 5 in Figure 3, and then sends the resulting rotation to the other processors in the same row and column. Then all the processors can carry out step 6. Thus $O(n)$ rotations can be generated and applied in $O(n)$ time; that is, $O(n^2)$ arithmetic can be carried out in constant time on $O(n^2)$ processors. We can embed the two-dimensional mesh on a hypercube to implement the algorithm. But it is more efficient to embed a linear array of processors and distribute the matrix by columns as shown in Figure 4a.I for n=8 and p=4, where the numbers represent the indices of the columns, the rectangles the processors. Note that when the rotations in the (1,2), (3,4), (5,6), and (7,8) planes are performed, no communication is needed for column updates, but row updates do require message passing.

To reduce the communication cost, we use an ordering in which message exchanges occur only between the neighboring processors, while minimizing the message traffic. Previous orderings have satisfied the requirements of nearest neighbor communication, but they either involve at least two sends and receives per processor for the column interchanges or do not achieve good load balancing [6]. However, we have used an ordering that has the property that each processor sends and receives only one column per stage, cutting the message traffic by about half. We illustrate the ordering we have used in Figures 4a and 4b. The two schemes I and II are applied alternately, then the pattern returns to its original state after $2(n-1)$ stages. Implementation of this ordering requires a ring connection. Thus, we view the hypercube as a ring connected linear array of processors. For other equivalent orderings, see [6].

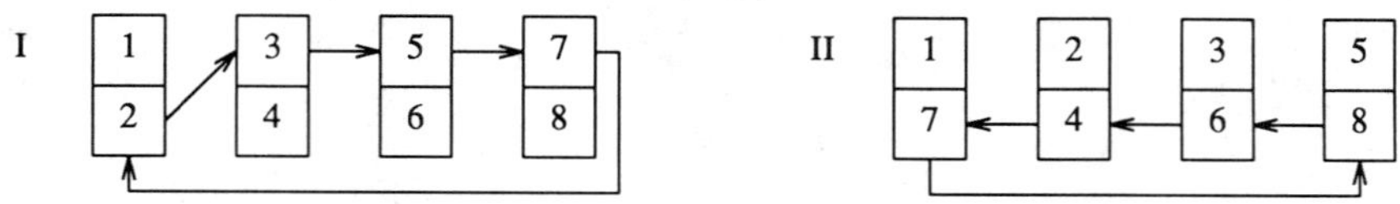

FIGURE 4a. Ring ordering scheme for $n = 8$ and $p = n/2$

stage 1. (1, 2) (3, 4) (5, 6) (7, 8)
stage 2. (1, 7) (2, 4) (3, 6) (5, 8)
stage 3. (1, 4) (2, 6) (3, 8) (5, 7)
stage 4. (1, 5) (4, 6) (2, 8) (3, 7)
stage 5. (1, 6) (4, 8) (2, 7) (3, 5)
stage 6. (1, 3) (6, 8) (4, 7) (2, 5)
stage 7. (1, 8) (6, 7) (4, 5) (2, 3)

FIGURE 4b. One sweep of the ring ordering for $n = 8$

In order to cut the message traffic further, thus to improve the efficiency, we have used a one-sided variation of the algorithm in Figure 3, in which only columns need to be updated and interchanged, at a cost of some more arithmetic. In step 6, instead of applying R from both sides, we compute $B^{(k+1)} \equiv B^{(k)}R$, where $B^{(0)}$ is initialized to be A. At every step, we can always recover the latest iterate $A^{(k)}$ from $V^{(k)^H}B^{(k)}$. But $A^{(k)}$ needs to be computed only to construct a new R in step 5, and in this situation, each processor needs to compute only the appropriate 2×2 strategic elements of $A^{(k)}$. For example, at a certain stage in the method, a processor that would have contained two pth and qth columns of the matrix $A^{(k)}$ in the two-sided method, will contain the corresponding columns of $B^{(k)}$ and $V^{(k)}$. To carry out step 5, we need only the 2×2 submatrix at the (p,q) plane, that is, $a_{pp}^{(k)}$, $a_{pq}^{(k)}$, $a_{qp}^{(k)}$, and $a_{qq}^{(k)}$. Since the relation $A^{(k)} = V^{(k)^H}B^{(k)}$ is satisfied, these four necessary elements can be recovered at a cost of four inner products [10]. After the rotation is computed, all processors may simultaneously update their vectors. Note that rotations need not be transmitted to other processors, thus the communication cost is reduced substantially.

When blocks of more than two columns are distributed to each processor, all possible index pairs within each processor are chosen in the first stage of every sweep; thereafter, index pairs are formed between two blocks inside each processor as shown in Figure 5 for n=8 and p=2.

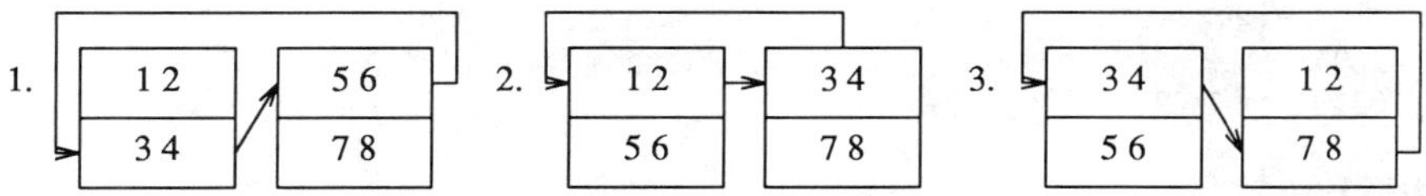

stage	processor 1	processor 2
1	(1,2) (3,4) (1,3) (1,4) (2,3) (2,4)	(5,6) (7,8) (5,7) (5,8) (6,7) (6,8)
2	(1,5) (1,6) (2,5) (2,6)	(3,7) (3,8) (4,7) (4,8)
3	(3,5) (3,6) (4,5) (4,6)	(1,7) (1,8) (2,7) (2,8)

FIGURE 5. One sweep of the ring ordering on blocks of columns

There are several different methods to check for convergence [10]. We counted the number of rotations performed each sweep. Iterations are halted when one complete sweep with no rotations has taken place.

We have run the one-sided algorithm on an Intel iPSC/2 hypercube. We illustrate some typical timings in Table 2. These timings show that the parallelism is so evident in the Jacobi algorithm that high efficiencies can be obtained even on matrices of relatively small size. But unlike the Jacobi algorithms for the symmetric matrices, the convergence for the non-symmetric case seems relatively slow.

matrix size	number of processors	time (sec)	speed-up over serial code	efficiency in %
16	8	4.963	3.837	47.97
	4	6.680	2.851	71.28
	2	11.260	1.691	84.57
	1	19.045	-	-
32	16	23.192	11.197	69.98
	8	43.778	5.932	74.15
	4	85.213	3.047	76.18
	2	150.036	1.731	86.54
	1	259.674	-	-
64	32	140.600	23.010	71.90
	16	238.400	13.570	84.48
	8	457.413	7.073	88.41
	4	843.649	3.835	95.87
	2	1674.537	1.932	96.60
	1	3235.205	-	-

TABLE 2. Parallel Jacobi times on iPSC/2 hypercube

4. REMARKS

We have described the basic techniques we have used to convert two popular and effective algorithms for the non-symmetric eigenvalue problem to parallel forms, suitable for an environment such as the hypercube. For both algorithms, the QR and the Jacobi, we have shown with some representative timings that the parallel implementations did indeed exhibit a parallel efficiency in excess of 50%.

Although we have presented the actual implementation timings of both algorithms, we would like to point out that they are not appropriate for comparing performance of the two algorithms. Obviously two different architectures would make difference in the actual timing. More important is that while the QR algorithm did not accumulate the Schur vectors, the Jacobi algorithm accumulated Schur vectors, which is in fact necessary for implementing the one-sided Jacobi method. The QR algorithm was implemented with single precision and the Jacobi with double precision, and accordingly, they used different tolerances for the stopping criterion. All of the above can make a large difference in the actual speed of the algorithms. We have yet to compare the effectiveness of these two methods; this will be reserved for future work.

The work of Boley, Park, and Eberlein was partially supported by NSF grants DCR-8420935, DCR-8519029, and CCC-8813493, NSF grant CCR-8813493, and NSF grants ASC-8705104 and CCR-8812521, respectively.

REFERENCES

[1] D. Boley, "Solving the generalized eigenvalue problem on a synchronous linear processor array," *Parallel Computing* vol. 3 no. 2, pp 153-66, May 1986.

[2] D.Boley and R. Maier, "A parallel QR algorithm for the nonsymmetric eigenvalue problem," Univ. of Minn. Dept. of Comput. Sci. report TR-88-12, Jan., 1988.

[3] P.J. Eberlein, "Solution to the complex eigenproblem by a norm reducing Jacobi type method," *Numer. Math.*, 14, pp. 232-245, 1970.

[4] P.J. Eberlein, "On using the Jacobi method on the hypercube," *SIAM Proc. of the 2nd Conf. on Hypercube Multiprocessors,* Ed. M.T. Heath, pp. 605-611, 1987.

[5] P.J. Eberlein, "On the Schur decomposition of a matrix for parallel computation," *IEEE Transactions on Computers,* C-36, pp 167-174, 1987.

[6] P.J. Eberlein and H. Park, "Efficient implementation of Jacobi algorithm and Jacobi sets on distributed memory architectures," *Tech Report TR-89-10,* Univ. of Minn., Dept. of Comput. Sci., Mar., 1989.

[7] G. H. Golub and C. Van Loan, Matrix Computations; Johns Hopkins, 1983.

[8] S. Lennart Johnsson, "Communication efficient basic linear algebra computations on hypercube architectures," Yale Univ. Dept. of Comput. Science Research Report YALEU/DCS/RR-361, Sep., 1985.

[9] F.T. Luk and H. Park, "A proof of convergence for two parallel Jacobi SVD algorithms," *IEEE Trans. on Comput.*, 1989, to appear.

[10] H. Park and P.J. Eberlein, "Eigensystem computation on hypercube architectures," *Proc. for the Fourth Conference on Hypercube Concurrent Computers and Application,* 1989, to appear.

[11] B.T. Smith, J.M. Boyle, Y. Ikebe, V.C. Klema, C.B. Moler, Matrix eigensystem routines, EISPACK Guide, 2nd edition, Springer, 1970.

[12] G.W. Stewart, "A Jacobi-like algorithm for computing the Schur decomposition of a non- Hermitian matrix," *SIAM J. Sci. and Stat. Comp.*, pp. 853-864, 1985.

[13] R. Van de Geijn, "Storage schemes for efficient parallel methods for solving linear eigenvalue problems," pre-print, Univ. of Texas - Austin, 1988.

ON THE INVERSION OF CAUCHY MATRICES

I. Gohberg and I. Koltracht

Abstract

It was observed in [GKKL] that one of the proposed fast algorithms gives a much higher accuracy than standard numerically stable algorithms do on the inversion of the Hilbert matrix, $A = \{1/(i+j-1)\}_{i,j=1}^{n}$, which is known to be very ill conditioned. In this paper we explain this phenomenon. The explanation is based on an analysis of the notion of the condition number and also on the fact that the effect of round-off errors in the mentioned fast algorithm is negligible.

Introduction

One of the fast algorithms for matrices with parallel recursive structure in [GKKL] was designed for Hilbert type matrices A which satisfy the Sylvester matrix equation

$$D_t A - A D_s = \sum_{i=1}^{\alpha} g_i h_i^T \qquad (1)$$

where $D_t = \mathrm{diag}\{t_1,\cdots,t_n\}$, $D_s = \mathrm{diag}\{s_1,\cdots,s_n\}$ and $g_i,h_i \in \mathbb{R}^n$, $i=1,\cdots,\alpha$. One well known example of a Hilbert type matrix is the Hilbert matrix itself, namely,

$$A = \left\{\frac{1}{i+j-1}\right\}_{i,j=1}^{n},$$

which satisfies (1) with $t_i=i, s_i=-i+1, i=1,\cdots,n$, $\alpha=1$, $g_1=h_1=(1,\cdots,1)^T$.

The fast algorithm for Hilbert type matrices was compared in [GKKL] with standard numerically stable algorithms, the Gaussian elimination with partial pivoting and the QR algorithm with column pivoting (as well as with some other fast algorithms), on the example of the Hilbert matrix. For n=10 the relevant data is reproduced in the following table.

Algorithm	$(A^{-1})_{10,10}$
Gaussian Elimination	33125049
QR algorithm	40498954
Fast algorithm [GKKL]	44914172416
Exact value	44914183600

TABLE 1

The exact value is taken from [GK]. The inaccurate results of the Gaussian elimination and the QR algorithms are consistent with the ill conditioning of the Hilbert matrix, namely, $k(A) = \|A\|\|A^{-1}\| \sim e^{3.5n}$, (see [GK]).

The main objective of this paper is to explain the unusual accuracy of the fast algorithm from [GKKL]. (Similarly high accuracy was also achieved by this algorithm on randomly generated Cauchy matrices, namely, matrices of the form

$$A = \left\{\frac{1}{t_i - s_j}\right\}_{i,j=1}^{n},$$

where $t_1,\cdots,t_n,s_1,\cdots,s_n$ are 2n different real numbers. Such matrices satisfy (1) with $\alpha = 1$, $g_1 = h_1 = (1,\cdots,1)^T$.)

For this purpose we start with the analysis of the notion of the condition number of a map. Let F be a continuous map from one finite dimensional normed space to another, $F : \mathbb{R}^p \to \mathbb{R}^q$. Assuming that the zero vector belong neither to the domain nor to the range of F the condition number $k(F,a)$ of the problem of computing Fa is defined as follows (see [R], [BL]):

$$k(F,a) = \lim_{\epsilon\to 0} \sup_{\substack{\|x-a\|\le\epsilon \\ x\ne a}} \left\{\frac{\|Fx-Fa\|}{\|Fa\|} : \frac{\|x-a\|}{\|a\|}\right\}, \qquad (2)$$

where ϵ is so small that the ball $\{x \in \mathbb{R}^p |\ \|x-a\|\le\epsilon\}$ belongs to the domain of F. Although $\mathbb{R}^p$ and $\mathbb{R}^q$ may be equipped with different norms we will use the same notation for both of them as long as it does not cause any confusion. If F is differentiable at a then

$$k(F,a) = \frac{\|F'(a)\|\ \|a\|}{\|Fa\|} \qquad (3)$$

where $\|F'(a)\|$ is the norm of the linear map $F'(a) : \mathbb{R}^p \to \mathbb{R}^q$. Since

$$\frac{\|Fx-Fa\|}{\|Fa\|} \leq k(F,a) \frac{\|x-a\|}{\|a\|} + o(\|x-a\|) \qquad (4)$$

the condition number $k(F,a)$ shows by how much the relative error in a can be magnified on computing Fa.

For illustration consider two examples. Let $F : \mathbb{R}^p \to \mathbb{R}$ be defined by $Fa=a^Tf$ where f is a fixed vector. Then for $a\neq 0$ such that $a^Tf\neq 0$

$$k(F,a) = \frac{\|f\| \|a\|}{|a^Tf|}$$

where, clearly, $F'(a)x = x^Tf$. This number can be large if the sizes of f and a are large relative to a^Tf.

Let $F : \mathbb{R}^{n\times n} \to \mathbb{R}^{n\times n}$ be the map of matrix inversion on all nonsingular $n\times n$ matrices. Then

$$k(F,A) = \|A\| \|A^{-1}\|.$$

In this case $F'(A)X = -A^{-1}XA^{-1}$.

In general, the condition number $k(F,a)$ has the following significance. Suppose Fa is to be computed in finite arithmetic with a unit round-off error u. Then the representation $fl(a)$ of a in this arithmetic satisfies the inequality

$$\frac{\|fl(a) - a\|_\infty}{\|a\|_\infty} \leq u$$

Hence it follows from (4) that

$$\frac{\|Ffl(a) - Fa\|}{\|Fa\|} \leq k(F,a)u + o(u).$$

Thus even without the effect of round-off errors in the performed calculation the relative error in the computed value of Fa can be of size $k(F,a)u$.

The problem of computing Fa is said to be well or ill conditioned depending on how large $k(F,a)$ is relative to u and on the desired accuracy in the computed value of Fa. (In the context of matrix inversion see [W], [GV], [B] for more details).

The explanation of numerical results from Table 1 is based on two observations. First, we observe that the condition number as defined in (2) depends on the map with its natural domain of definition and the point a in this domain, rather than on the point a only.

The second observation is concerned with the general definition of the relative error. If in the infinity norm

$$\frac{\|x-a\|_\infty}{\|a\|_\infty} = \frac{\max\limits_{i=1,\cdots,p} |x_i-a_i|}{\max\limits_{i=1,\cdots,p} |a_i|} = \gamma$$

then for $|a_{io}| = \|a\|_\infty$ we have

$$\left|\frac{x_{io} - a_{io}}{|a_{io}|}\right| \leq \gamma$$

while the relative error in other coordinates can be arbitrary. As a simple example consider $b = (1.000,\ 0.001)$ and $\tilde{b} = (1.010,\ 0.010)$. Then $\|b - \tilde{b}\|_\infty/\|b\|_\infty = 10^{-2}$ while $|b_2-\tilde{b}_2|/|b_2| = 9$.

We will make use of these two observations by taking into account the map F <u>with its domain</u> when estimating the condition number associated with computation of Fa. We will also modify, whenever possible, the definition of the condition number by replacing relative errors in norm by relative errors in each component. Such a modification is possible when neither a nor Fa have zero coordinates. Let

$$F(a_1,\cdots,a_p) = (f_1(a_1,\cdots,a_p),\cdots,f_q(a_1,\cdots,a_p)).$$

If $a_i \neq 0$, $i=1,\cdots,p$, $f_j(a_1,\cdots,a_p) \neq 0$, $j=1,\cdots,q$, then define

$$c(F,a) = \max_{i=1,\cdots,q} \lim_{\epsilon\to 0} \sup_{\substack{\delta(x-a)\leq\epsilon \\ x\neq a}} \left\{\frac{|f_i(x) - f_i(a)|}{|f_i(a)|} : \delta(x-a)\right\} \quad (5)$$

where

$$\delta(x-a) = \max_{i=1,\cdots,p} \frac{|x_i - a_i|}{|a_i|} .$$

If F is differentiable at a then it is shown in Section 1 that in the infinity norm

$$c(F,a) = \|D_{Fa}^{-1} F'(a)\ D_a\|_\infty, \quad (6)$$

where $D_{Fa} = \text{diag}\{f_1(a),\cdots,f_q(a)\}$, $D_a = \text{diag}\{a_1,\cdots,a_p\}$ and $D_{Fa}^{-1} F'(a)D_a$ is considered as a linear map from $\mathbb{R}^p$ to $\mathbb{R}^q$ both equipped with the infinity norms. For this condition number we have

$$\delta(Fx-Fa) \leq c(F,a)\delta(x-a) + o(\delta(x-a)), \quad (7)$$

which gives more accurate information (than (4) does) about the effect of errors in a on the computed Fa.

To illustrate the difference between $k(F,a)$ and $c(F,a)$ consider the following simple example. Let F be the matrix inversion map and $A = \mathrm{diag}\{a_1,a_2\}$ where $a_1>a_2>o$. Then, obviously, $k(F,a) = a_1a_2^{-1}$, which can be arbitrarily large. It is clear, however, that A can be inverted accurately. On the other hand if we consider $F : \mathbb{R}^2 \to \mathbb{R}^2$, $F(a_1,a_2) = (a_1^{-1}, a_2^{-2})$, that is the inversion map restricted to the set of all 2×2 diagonal matrices only, we have

$$c(F,a) = \left\| \begin{bmatrix} a_1 & 0 \\ 0 & a_2 \end{bmatrix} \begin{bmatrix} -1/a_1^2 & 0 \\ 0 & -1/a_2^2 \end{bmatrix} \begin{bmatrix} a_1 & 0 \\ 0 & a_2 \end{bmatrix} \right\|_\infty = 1.$$

In Section 1 the formulas (6) – (7) are obtained. In Section 2 a bound on $c(F,a)$ is obtained for F being the matrix inversion map on the set of Cauchy matrices, $F : \mathbb{R}^{2n} \to \mathbb{R}^{n\times n}$

$$F(t_1,\cdots,t_n,s_1,\cdots,s_n) = \left[\left\{ \frac{1}{t_i-s_j} \right\}_{i,j=1}^{n} \right]^{-1}.$$

It is shown that

$$c(F,a) \le \frac{8n-6}{\Delta}\|a\|_\infty \tag{8}$$

where

$$a = (t_1,\cdots,t_n,s_1,\cdots,s_n) = (a_1,\cdots,a_{2n})$$

and

$$\Delta = \min_{\substack{i\neq j \\ i,j=1,\cdots,2n}} \{|a_i-a_j|\}.$$

For the Hilbert matrix with n=10 we have $\|a\|_\infty = 10$, $\Delta = 1$ and hence $c(F,a) \le 760$ which is, of course, small comparative to the general condition number of the 10×10 Hilbert matrix $k(F,A) \sim 10^{15}$.

In Section 3 it is shown that the fast algorithm for Cauchy matrices yields a negligible round–off error accumulation, namely, if $\tilde{A}_{ij}^{-1}$ is the exact value of an entry of the inverse of the Cauchy matrix defined by $(fl(t_1),\cdots,fl(t_n),fl(s_1),\cdots,fl(s_n))$ and $fl(\tilde{A}_{ij}^{-1})$ is the corresponding value computed with round–off errors, then

$$\max_{i,j=1,\cdots,n} \frac{\left|fl(\tilde{A}_{ij}^{-1}) - \tilde{A}_{ij}^{-1}\right|}{\left|\tilde{A}_{ij}^{-1}\right|} \le 4nu + o(u). \tag{9}$$

The estimates (8) and (9) explain now accurate results of the numerical experiments in [GKKL], as these results are obtained by a numerically stable algorithm applied to a well conditioned problem.

Finally we remark that developed techniques can also be applied to the explanation of other numerical examples in [GKKL] concerning Vandermonde and Toeplitz matrices. This will be published elsewhere.

1. The Condition Number of a Map

In this section we consider the condition number of a continuous map from one finite dimensional normed space to another. First we give the well-known definition and properties of the condition number of such a map and then modify the definition to suite our purpose.

Let $F : \mathbb{R}^p \to \mathbb{R}^q$ be a continuous map defined on an open set $D_F \subset \mathbb{R}^p$ such that $0 \notin D_F$. For a given $a \in D_F$ such that $Fa \neq 0$ and $\epsilon > 0$ small enough such that

$$B(a,\epsilon) = \{x \in \mathbb{R}^p \mid \|x-a\| \leq \epsilon\} \subset D_F$$

the ϵ-condition number $k(F,a,\epsilon)$ is defined by (see [R], [BL])

$$k(F,a,\epsilon) = \sup_{\substack{x \in B(a,\epsilon) \\ x \neq a}} \left\{ \frac{\|Fx-Fa\|}{\|Fa\|} : \frac{\|x-a\|}{\|a\|} \right\} . \tag{10}$$

This number gives the worst possible magnification of the relative error in a (of size at most ϵ) when a is transformed into Fa. Since $k(F,a,\epsilon)$ is monotone decreasing with ϵ there exists

$$\lim_{\epsilon \to 0} k(F,a,\epsilon) = k(F,a)$$

By definition the number $k(F,a)$ is called the condition number of the map F at the point a. If F is differentiable at a then it is easy to see that

$$k(F,a) = \frac{\|F'(a)\| \, \|a\|}{\|Fa\|}$$

where $\|F'(a)\|$ is the norm of the linear map $F'(a) : \mathbb{R}^p \to \mathbb{R}^q$. In this case for x sufficiently close to a

$$\frac{\|Fx-Fa\|}{\|Fa\|} \leq k(F,a) \frac{\|x-a\|}{\|a\|} + o(\|x-a\|)$$

Suppose now that the vectors a and Fa have no zero coordinates, that is

$$F(a_1,\cdots,a_p) = (f_1(a_1,\cdots,a_p),\cdots,f_q(a_1,\cdots,a_p))$$

where $a_i \neq 0$, $i=1,\cdots,p$ and $f_j(a) \neq 0$, $j=1,\cdots,q$. In this case we can modify the definition of the condition number of computing Fa as follows.

Definition

Let $F : \mathbb{R}^p \to \mathbb{R}^q$ *be continuous in a neighborhood of* $a = (a_1,\cdots,a_p)$, $a_i \neq 0$, $i=1,\cdots,p$, *such that* $Fa = (f_1(a),\cdots,f_q(a))$, $f_j(a) \neq 0$, $j=1,\cdots,q$. *Then the condition number of* F *at* a *is*

$$c(F,a) = \max_{i=1,\cdots,q} \lim_{\epsilon\to 0} \sup_{\substack{\delta(x-a)\le\epsilon \\ x\neq a}} \left\{ \frac{|f_j(x)-f_j(a)|}{|f_j(a)|} : \delta(x-a) \right\} \quad (11)$$

where by definition

$$\delta(x-a) = \max_{i=1,\cdots,p} \frac{|x_i-a_i|}{|a_i|}. \quad (12)$$

Suppose now that F is differentiable at a. Since for $j=1,\cdots,q$

$$|f_j(x) - f_j(a)| \le |\nabla f_j(a)(x-a)| + o(\|x-a\|)$$

where, as usual, $\nabla f_j(a) = [\partial f_j/\partial a_1 \cdots \partial f_j/\partial a_p]$, we can write

$$\frac{|f_j(x) - f_j(a)|}{|f_j(a)|} \le \frac{|\nabla f_j(a) D_a D_a^{-1}(x-a)|}{|f_j(a)|} + o(\|x-a\|)$$

where $D_a = \text{diag}\{a_1,\cdots,a_p\}$. Hence

$$\frac{|f_j(x)-f_j(a)|}{|f_j(a)|} \le \left\|\frac{1}{f_j(a)} \nabla f_j(a) D_a\right\|_1 \delta(x-a) + o(\|x-a\|)$$

as it is clear that $\|D_a^{-1}(x-a)\|_\infty = \delta(x-a)$. Finally we get

$$\max_{j=1,\cdots,q} \frac{|f_j(x) - f_j(a)|}{|f_j(a)|} \le \|D_{Fa}^{-1} F'(a) D_a\|_\infty \delta(x-a) + o(\|x-a\|)$$

where $D_{Fa} = \text{diag}\left\{f_1(a),\cdots, f_q(a)\right\}$ and $\|D_{Fa}^{-1} F'(a) D_a\|_\infty$ is the norm of the linear map $D_{Fa}^{-1} F'(a) D_a$ from $\mathbb{R}^p$ to $\mathbb{R}^q$ both equipped with the infinity norm. Comparing with (11) we see that in this case

$$c(F,a) = \|D_{Fa}^{-1} F'(a) D_a\|_\infty . \quad (13)$$

As $\delta(x-a) \le \|D_a^{-1}\|_\infty \|x-a\|_\infty$ and $\|x-a\|_\infty \le \|D_a\|_\infty \delta(x-a)$ we also have

$$\delta(Fx-Fa) \le \|D_{Fa}^{-1} F'(a) D_a\|_\infty \delta(x-a) + o(\delta(x-a)). \quad (14)$$

In particular, if $fl(a)$ is the computer representation of a then $\delta(fl(a)-a) \le u$, where $u = \beta^{-t}$ is the unit round-off error, and hence

$$\delta(Ffl(a) - Fa) \le c(F,a)u + o(u). \quad (15)$$

This inequality means that one cannot guarantee more than

$$-\log_\beta (c(F,a)u) = t - \log_\beta c(F,a)$$

correct significant figures in each coordinate of the computed Fa.

2. The Condition Number of Inverting a Cauchy Matrix

Consider a Cauchy matrix

$$A = \left\{\frac{1}{t_i - s_j}\right\}_{i,j=1}^{n}$$

where $t_1,\cdots,t_n,s_1,\cdots,s_n$ are all different numbers. Without any loss of generality we may assume that these numbers are different from zero. Indeed, $t_1+\alpha,\cdots,t_n+\alpha,s_1+\alpha,\cdots,s_n+\alpha$ define the same matrix, and we can always chose α such that none of them is equal to zero.

It is well known (see [K], for example) that A^{-1} exists and

$$A_{ij}^{-1} = \frac{1}{t_j - s_i}\,\frac{\prod_{k=1}^{n}(t_k-s_i)(t_j-s_k)}{\prod_{\substack{k=1\\k\neq i}}^{n}(s_k-s_i)\prod_{\substack{k=1\\k\neq j}}^{n}(t_j-t_k)}\,. \tag{16}$$

It follows from (16) that $A_{ij}^{-1}\neq 0$, $i,j=1,\cdots,n$. We are interested in estimating the condition number of the matrix inversion map on the set of Cauchy matrices, $F:\mathbb{R}^{2n}\to\mathbb{R}^{n\times n}$,

$$F(t_1,\cdots,t_n,s_1,\cdots,s_n) = \left[\left\{\frac{1}{t_i-s_j}\right\}_{i,j=1}^{n}\right]^{-1}. \tag{17}$$

Theorem 1

$$c(F,a) \le \frac{(8n-6)}{\Delta}\|a\|_\infty$$

where

$$a = (t_1,\cdots,t_n,s_1,\cdots,s_n) = (a_1,\cdots,a_{2n})$$

and

$$\Delta = \min_{\substack{i,j=1,\cdots,2n\\ i\neq j}} |a_i - a_j|.$$

Proof

The $2n\times n^2$ Jacobian matrix $F'(a)$ is of the form

$$F'(a) = \left\{\frac{\partial F_{ij}(a)}{\partial a_u}\right\}_{u=1,\ i,j=1}^{2n\qquad n}.$$

Let us estimate the size of the entries of the matrix $D_{Fa}^{-1}F'(a)$. Using the formula (16) for $F_{ij}(a)$ we have for $a_u = t_u$, $u\neq i,j$

$$\frac{1}{F_{ij}(a)}\,\frac{\partial F_{ij}(a)}{\partial t_u} = \frac{\partial}{\partial\lambda}\left[\log\left[\frac{(\lambda-s_i)}{(t_j-\lambda)}\cdot\text{const}\right]\right]_{\lambda=t_u}$$

$$= \frac{\partial}{\partial\lambda}\,[\log(\lambda-s_i)-\log(t_j-\lambda)+\text{const}]_{\lambda=t_u} = \frac{1}{t_u-s_i} + \frac{1}{t_j-t_u}.$$

Similarly

$$\frac{1}{F_{ij}(a)}\frac{\partial F_{ij}(a)}{\partial s_u} = \frac{1}{s_u-t_j} + \frac{1}{s_i-s_u},\quad u \neq i,j,$$

$$\frac{1}{F_{ij}(a)}\frac{\partial F_{ij}(a)}{\partial t_i} = \frac{1}{t_i-s_i} + \frac{1}{t_j-t_i},$$

$$\frac{1}{F_{ij}(a)}\frac{\partial F_{ij}(a)}{\partial s_j} = \frac{1}{s_j-t_j} + \frac{1}{s_i-s_j},$$

$$\frac{1}{F_{ij}(a)}\frac{\partial F_{ij}(a)}{\partial s_i} = \sum_{\substack{k=1\\k\neq i}}^{n} \frac{1}{s_k-s_i} - \sum_{k=1}^{n} \frac{1}{t_k-s_i}$$

and

$$\frac{1}{F_{ij}(a)}\frac{\partial F_{ij}(a)}{\partial t_j} = \sum_{k=1}^{n} \frac{1}{t_j-s_k} - \sum_{\substack{k=1\\k\neq j}}^{n} \frac{1}{t_j-t_k}.$$

The row i,j of the matrix $D_{Fa}^{-1}F'(a)$ has the form

$$\left[\frac{1}{F_{ij}(a)}\frac{\partial F_{ij}(a)}{\partial a_1},\ldots,\frac{1}{F_{ij}(a)}\frac{\partial F_{ij}(a)}{\partial a_{2n}}\right].$$

Using the obtained expressions for its entries we get the following bound on the 1–norm of this row

$$(2n-1)/\Delta + (2n-1)/\Delta + 2(n-1)/\Delta + 2(n-1)/\Delta = (8n-6)/\Delta.$$

Thus the ∞-norm of $D_{Fa}^{-1}F'(a)D_a$ is bounded by $(8n-6)\|D_a\|_\infty/\Delta$

$= (8n-6)\|a\|_\infty/\Delta$. □

For the n×n Hilbert matrix A we have $\Delta = 1$, $\|a\|_\infty = n$ and hence

$$c(F,a) < (8n-6)n.$$

In particular, for $n = 10$, $c(F,a) < 740$.

We remark that $(8n-6)\|a\|_\infty/\Delta$ is only a bound on the condition number of inverting the Cauchy matrix, in reality $c(F,a)$ can be even smaller.

3. The Effect of Round–off Errors

Let us consider now the question of computing the inverse of a Cauchy matrix in floating point arithmetic with the unit round–off error $u = \beta^{-t}$ where β is the machine base and t is the precision (see [W], [GV]).

The Gaussian elimination and the QR algorithm compute the value of the general matrix inversion map on the Cauchy matrix A. Therefore the general condition number $k(F,A)$ applies, and in this case one can not guarantee more than $t-\log_\beta k(F,A)$ correct significant digits in the computed inverse (in fact, in its largest components only). In the experiment described in Table 1 $\beta=10$ and $t=-8$. Since for the 10×10 Hilbert matrix A $k(F,A)\sim10^{15}$ it follows that no correct significant figures can be guaranteed, which explains the inaccurate results obtained by the Gaussian elimination and the QR algorithm. (This does not, of course, contradict the fact that both algorithms are numerically stable.)

Consider now the fast algorithm from [GKKL], which in the case of Cauchy matrices simplifies to the following.

Algorithm

1. Start with

$$\varphi_1(1) = t_1-s_1$$

2. For $j=1,\cdots,n-1$

For $i=1,\cdots,j$

$$\varphi_{j+1}(i) = \varphi_j(i)\,\frac{t_{j+1}-s_i}{s_{j+1}-s_i}.$$

For $i=j+1$

$$\varphi_{j+1}(j+1) = \frac{\prod_{k=1}^{j+1}(t_k-s_{j+1})}{\prod_{k=1}^{j}(s_k-s_{j+1})}.$$

3. For $i=1,\cdots,n$

$$\gamma_n(i) = \frac{\prod_{k=1,\,k\neq i}^{n}(t_n-s_k)}{\prod_{k=1}^{n-1}(t_n-t_k)}\,\varphi_n(i)\ .$$

4. In particular, for $i=1,\cdots,n$

$$A^{-1}_{in} = \gamma_n(i).$$

This algorithm computes values of the map (17) of inverting Cauchy matrices. Therefore the relevant condition number is $c(F,a)$ estimated in (8). For the 10×10 Hilbert matrix we have $c(F,a)\leq740$, which means that in the experiment represented in Table 1 there are $8-\log_{10}740\geq5$ guaranteed correct significant figures in each of the components of the inverse, provided there are no round-off errors in the computations.

Let us consider now the effect of round-off errors in performing elementary arithmetic operations. Such errors are given by

$$\frac{|fl(x\otimes y) - x\otimes y|}{|x\otimes y|} \leq u$$

where $\otimes$ stands for $+$, $-$, $\cdot$, $/$ (see [W], [GV]). We remark that the algorithm for Cauchy matrices presented above amounts to the recursive estimation of the last column of $\tilde{A}^{-1}=\{\tilde{A}^{-1}_{ij}\}^{n}_{i,j=1}$ according to the formulas (16), where $\tilde{A}$ is the Cauchy matrix corresponding to $(fl(t_1),\cdots,fl(t_n),fl(s_1),\cdots,fl(s_n)) = (\tilde{t}_1,\cdots,\tilde{t}_n,\tilde{s}_1,\cdots,\tilde{s}_n)$. Therefore we will estimate the effect of round-off errors in evaluating (16).

Theorem 2

Let $fl(\tilde{A}^{-1}_{ij})$, $i,j=1,\cdots,n$ *denote values computed with round-off errors according to (16) from the data* $(\tilde{t}_1,\cdots,\tilde{t}_n,\tilde{s}_1,\cdots,\tilde{s}_n)$ *and let* $\tilde{A}^{-1}_{ij}$ *denote the exact values computed from this data according to (16). Then*

$$\max_{i,j=1,\cdots,n} \frac{|fl(\tilde{A}^{-1}_{ij})-\tilde{A}^{-1}_{ij}|}{|\tilde{A}^{-1}_{ij}|} \leq 4nu + o(u) \tag{18}$$

Proof

It follows from standard estimates for round-off errors ([W], 24.2) that

$$fl\left[\frac{\prod\limits_{\substack{k=1\\k\neq j}}^{n} fl(\tilde{t}_k-\tilde{s}_i)\cdot \prod\limits_{k=1}^{n} fl(\tilde{t}_j-\tilde{s}_k)}{\prod\limits_{\substack{k=1\\k\neq i}}^{n} fl(\tilde{s}_k-\tilde{s}_i)\cdot \prod\limits_{\substack{k=1\\k\neq j}}^{n} fl(\tilde{t}_j-\tilde{t}_k)}\right] = \tilde{A}^{-1}_{ij}(1 + 4nu_{ij}) + o(u)$$

where $|u_{ij}| \leq u$. Thus

$$|fl(\tilde{A}^{-1}_{ij}) - \tilde{A}^{-1}_{ij}|\leq|\tilde{A}^{-1}_{ij}|4nu + o(u).$$

□

Since

$$fl(\tilde{A}^{-1}_{ij}) - \tilde{A}^{-1}_{ij} = \tilde{A}^{-1}_{ij}4nu + o(u)$$

and

$$\tilde{A}^{-1}_{ij} - A^{-1}_{ij} = A^{-1}_{ij}\,\frac{8n-6}{\Delta}\|a\|_{\infty}u + o(u)$$

it follows that

$$fl(\tilde{A}^{-1}_{ij}) - A^{-1}_{ij} = A^{-1}_{ij}(4n + \frac{8n-6}{\Delta} \|a\|_{\infty})u + o(u).$$

Hence there are least $t-\log_{\beta}(4n+\frac{8n-6}{\Delta}\|a\|_{\infty})$ correct significant figures in computed $fl(\tilde{A}^{-1}_{ij})$.

In the case of the 10×10 Hilbert matrix one can guarantee $8-\log_{10}780 \geq 5$ correct significant figures which agrees with the data presented in Table 1.

This paper was supported by the NSF grant DMS–8801961.

References

[BL] G.R. Belitskii and Y.I. Lyubich, "Matrix Norms and Their Applications", OT Vol. 36, Birkhäuser, 1988.

[B] J.R. Bunch, "The Week and Strong Stability of Algorithms in Numerical Linear Algebra", Lin. Alg. Appl., 88/89, 1987. pp. 49–66.

[GKKL] I. Gohberg, T. Kailath, I. Koltracht and P. Lancaster, "Linear Complexity Parallel Algorithms for Linear Systems of Equations with Recursive Structure", Lin. Alg. Appl., 88/89, 1987, pp. 271–315.

[GV] G.H. Golub and C.F. Van Loan, "Matrix Computations", The Johns Hopkins University Press, 1983.

[GK] R.T. Gregory and D.I. Karney, "A Collection of Matrices for Testing Computational Algorithms", Wiley–Interscience, 1969.

[K] D.E. Knuth, "The Art of Computer Programming", Addison–Wesley, 1968.

[R] J.R. Rice, "A Theory of Condition", J. SIAM Numer. Anal., 3, 1966, pp. 287–310.

[W] J.H. Wilkinson, "Rounding Errors in Algebraic Processes", Prentice–Hall, 1963.

THE LANCZOS ALGORITHM WITH INDEFINITE INNER PRODUCT

B. N. Parlett and H. C. Chen

Abstract A quadratic eigenvalue problem with symmetric positive definite coefficient matrices may be reduced to linear form while retaining symmetry in the new coefficients but neither of them will be positive definite. Formally the symmetric Lanczos algorithm and subspace iteration may be used to compute some eigenpairs of the linear problem. The trouble is that the basis vectors are orthogonal with respect to an indefinite inner product so there is no assurance that they will be linearly independent. Nevertheless this is an attractive way to solve the original problem and we discuss how to implement it and how it relates to the unsymmetric Lanczos procedures. We discuss complex origin shifts, the Look-Ahead version, reorthogonalization, and error bounds.

1. INTRODUCTION

In the study of Mechanics, when one analyzes how small displacements from an equilibrium state evolve in time one is led to the familiar equations of motion

$$\mathbf{M}\,\ddot{\mathbf{q}}(t)+\mathbf{C}\,\dot{\mathbf{q}}(t)+\mathbf{K}\,\mathbf{q}(t)=\mathbf{f}(t) \tag{1.1}$$

where **M**, **C**, and **K** are, respectively, the $n \times n$ mass, damping, and stiffness matrices and $\ddot{\mathbf{q}}(t)$, $\dot{\mathbf{q}}(t)$, and $\mathbf{q}(t)$ are the $n \times 1$ acceleration, velocity, and displacement vectors.

For many solid structures attached to the earth the damping coefficients are small compared with those of mass and stiffness. For structures in space damping plays a bigger role and the undamped model may not be a good approximation. To understand the response of the system to a variety of external forces $\mathbf{f}(t)$ it is still useful to know the system's dominant natural modes of vibration. This leads to the quadratic eigenvalue problem

$$(\lambda_i^2\mathbf{M}+\lambda_i\,\mathbf{C}+\mathbf{K})\,\mathbf{w}_i=\mathbf{0} \qquad i=1,\ldots,2n \tag{1.2}$$

In general λ_i will be complex and the associated modal shapes are given by the real and imaginary parts of $\mathbf{w}_i$. In the problems we consider **M**, **C**, **K** are real, symmetric, and positive definite, although we can permit **K** to be semi-definite without difficulty. In this note we describe an efficient implementation of the Look-Ahead Lanczos algorithm for computing a few eigenvalues (and vectors) near any point in the complex plane.

A Lanczos program similar to ours has been developed independently (see [Nour-Omid & Regelbrugge, 1988]) and used effectively to solve the equations of motion (1.1) directly without computing eigensolutions.

We follow standard notational conventions in the field of matrix computations. In particular, capital letters denote matrices, small Roman letters denote column vectors, and small Greek letters denote scalars. Also v^t denotes the transpose of v and $\bar{\alpha}$ denotes the conjugate of a complex number α. Unless the contrary is indicated $\|\mathbf{x}\|$ denotes the Euclidean norm of **x**, $\|\mathbf{B}\|$ denotes the spectral norm of **B**.

2. REDUCTION TO LINEAR FORM

There are several ways of rewriting (1.2) as a linear eigenvalue problem. When **K** and **M** are positive definite then 0 is not an eigenvalue and one reduction is

$$\lambda_i \begin{bmatrix} \mathbf{C} & \mathbf{M} \\ \mathbf{M} & \mathbf{0} \end{bmatrix} \begin{bmatrix} \mathbf{w}_i \\ \lambda_i \mathbf{w}_i \end{bmatrix} - \begin{bmatrix} -\mathbf{K} & \mathbf{0} \\ \mathbf{0} & \mathbf{M} \end{bmatrix} \begin{bmatrix} \mathbf{w}_i \\ \lambda_i \mathbf{w}_i \end{bmatrix} = \begin{bmatrix} \mathbf{0} \\ \mathbf{0} \end{bmatrix} \tag{2.1}$$

Recall that a symmetric matrix is indefinite if it has eigenvalues of both signs.

Definition 1. Let $\mathbf{Y} \in \mathbb{R}^{n \times n}$ be symmetric but indefinite. The bilinear form defined by

$$(\mathbf{u}, \mathbf{v})_{\mathbf{Y}} := \mathbf{v}^t \, \mathbf{Y} \, \mathbf{u}$$

is called a pseudo (or indefinite or improper) inner product. It obeys all the axioms of an inner product except positivity.

Definition 2. Let $\mathbf{F} \in \mathbb{R}^{n \times n}$. **F** is symmetric with respect to **Y**, if $\mathbf{F}^t \, \mathbf{Y} = (\mathbf{Y}\,\mathbf{F})^t = \mathbf{Y}\,\mathbf{F}$.

Equation (2.1) is a special case of the problem $(\lambda \mathbf{A} - \mathbf{B})\,\mathbf{z} = \mathbf{0}$ where **A** and **B** are symmetric but indefinite. A matrix pair (**A**, **B**) is sometimes called a matrix pencil. See [Gantmacher, 1959] for example. In our case we have a symmetric indefinite pencil. Excellent references for further study of these pencils see [Gohberg, Lancaster, and Rodman, 1983 and 1986].

Lemma. $\mathbf{B}^{-1}\mathbf{A}$ is symmetric with respect to both **A** and **B**. $\mathbf{A}\,\mathbf{B}^{-1}$ is symmetric with respect to both $\mathbf{A}^{-1}$ and $\mathbf{B}^{-1}$.

It turns out that the Lanczos algorithm may be invoked with the operator $\mathbf{B}^{-1}\mathbf{A}$ using a pseudo inner product defined by **A** or **B**. The three term recurrence still holds in this more general situation. There is more on the Lanczos algorithm in Section 4.

Subspace Iteration [Bathe and Wilson, 1976] or [Rutishauser, 1970] (also called the method of Simultaneous Iterations) also extends formally to the indefinite case but the Rayleigh Ritz approximations produced at each step are no longer optimal in any meaningful sense.

In contrast to the definite case both algorithms can break down or become unstable when close to breakdown. Alarming things can happen with an indefinite (or improper) inner product : a set of *orthogonal* vectors might be linearly dependent. The geometry associated with such an improper inner product is the geometry of relativity theory.

Next we wish to point out that it is possible to work with $\mathbf{A}\,\mathbf{B}^{-1}$ instead of $\mathbf{B}^{-1}\mathbf{A}$ but we see no advantage to this formulation in our problem.

The need for shifts of origin (see next section) suggests the use of an operator $\mathbf{H}^{-1}\mathbf{A}$ and the pseudo inner product defined by **A**. Here $\mathbf{H}^{-1}$ may be $\mathbf{B}^{-1}$ or $(\mathbf{B} - \sigma\,\mathbf{A})^{-1}$ or $\mathrm{Re}\,[\,(\mathbf{B} - \sigma\,\mathbf{A})^{-1}\,]$ or $\mathrm{Im}\,[\,(\mathbf{B} - \sigma\,\mathbf{A})^{-1}\,]$.

The form (2.1) is well known. In a recent paper [Borri & Mantegazza 1977] form (2.1) is explicitly proposed when **M**, **C**, and **K** are all symmetric. However, no mention is made of the fact that the Rayleigh quotients used in that paper may overflow or yield 0 / 0. A conventional alternative to the pseudo-symmetric form (2.1) is to apply the two sided Lanczos algorithm to the unsymmetric form obtained by changing the sign in row 2 of (2.1).

We recall that the attraction of the Lanczos algorithm is that a few of the largest eigenvalues may be found by stopping the algorithm long before the full n steps. It is usual practice to give the algorithm an operator such as $\mathbf{H}^{-1}\mathbf{A}$ whose largest eigenvalues are shifted reciprocals of the ones we really want. We can expect to find the p largest eigenvalues within max $\{p+8, 2p\}$ steps.

3. ORIGIN SHIFTS

There are applications (e.g. space structures) in which the stiffness matrix **K** is singular. This makes $\mathbf{B} = diag(-\mathbf{K}, \mathbf{M})$ singular too. In this situation, we need to solve a shifted problem

$$(\lambda-\sigma)\,\mathbf{A}\,\mathbf{z} = (\mathbf{B} - \sigma\,\mathbf{A})\,\mathbf{z}$$

where $\sigma \neq 0$ is a shift. To preserve the block diagonal form, we note that the $\mathbf{B} - \sigma\,\mathbf{A}$ can be factored into

$$\begin{bmatrix} \mathbf{I} & -\sigma\mathbf{I} \\ \mathbf{0} & \mathbf{I} \end{bmatrix} \begin{bmatrix} (-\mathbf{K}-\sigma\mathbf{C}-\sigma^2\mathbf{M}) & \mathbf{0} \\ \mathbf{0} & \mathbf{M} \end{bmatrix} \begin{bmatrix} \mathbf{I} & \mathbf{0} \\ -\sigma\mathbf{I} & \mathbf{I} \end{bmatrix} \tag{3.1}$$

We need to factor only **M** and the shifted stiffness matrix $\mathbf{K} + \sigma\,\mathbf{C} + \sigma^2\,\mathbf{M}$, which has the same banded structure as **K**, in order to solve $(\mathbf{B} - \sigma\,\mathbf{A})\mathbf{u} = \mathbf{A}\mathbf{v}$ for **u** given **v**. Thus we may use the problem $(\mathbf{B} - \sigma\mathbf{A})^{-1}\mathbf{A}\mathbf{z} = \frac{1}{(\lambda - \sigma)}\mathbf{z}$ to obtain eigenvalues close to σ.

In several applications the user would like to explore part of the complex plane and see how certain eigenvalues vary with changes in the matrix elements. Unfortunately, if σ is complex then $\mathbf{B} - \sigma\,\mathbf{A}$ is not real and so $(\mathbf{B} - \sigma\,\mathbf{A})^{-1}\,\mathbf{A}$ would force the use of complex arithmetic in Lanczos.

A way out of this quandary was proposed in [Parlett and Saad, 1987]. The operator given to the Lanczos algorithm is either

$$\mathrm{Re}\,[\,(\mathbf{B} - \sigma\,\mathbf{A})^{-1}\,]\,\mathbf{A} \quad \text{or} \quad \mathrm{Im}\,[\,(\mathbf{B} - \sigma\,\mathbf{A})^{-1}\,]\,\mathbf{A}\ . \tag{3.2}$$

Here $\mathrm{Re}\,\mathbf{G} := \frac{1}{2}\,(\mathbf{G} + \overline{\mathbf{G}})$, $\mathrm{Im}\,\mathbf{G} := \frac{1}{2i^*}\,(\mathbf{G} - \overline{\mathbf{G}})$, and $\overline{\mathbf{G}}$ is the conjugate of **G**. Both operators are real and have complex conjugate pairs of eigenvalues.

Let λ, **z** be an eigenpair of (**B**, **A**), i.e. they satisfy $\lambda\,\mathbf{A}\,\mathbf{z} = \mathbf{B}\,\mathbf{z}$. By subtracting $\sigma\,\mathbf{A}\,\mathbf{z}$ and $\overline{\sigma}\,\mathbf{A}\,\mathbf{z}$ from each side and inverting one finds that

$$\mathrm{Re}\,[\,(\mathbf{B} - \sigma\,\mathbf{A})^{-1}\,]\,\mathbf{A}\,\mathbf{z} = \frac{1}{2}\,\left(\frac{1}{\lambda-\sigma} + \frac{1}{\lambda-\overline{\sigma}}\right)\mathbf{z}\ , \tag{3.3a}$$

$$\mathrm{Im}\,[\,(\mathbf{B} - \sigma\,\mathbf{A})^{-1}\,]\,\mathbf{A}\,\mathbf{z} = \frac{1}{2\,i^*}\,\left(\frac{1}{\lambda-\sigma} - \frac{1}{\lambda-\overline{\sigma}}\right)\mathbf{z}\ , \tag{3.3b}$$

To recover λ from the computed eigenvalues of either of these operators requires merely solving a quadratic equation for the root that is closer to σ. However λ may be recovered directly, with extra work, from the Rayleigh quotient

$$\lambda = \mathbf{z}'\,\mathbf{B}\,\mathbf{z} / \mathbf{z}'\,\mathbf{A}\,\mathbf{z}\ .$$

In order to exploit narrow bandwidth it is convenient to use complex arithmetic in (3.1) to factor $\mathbf{B} - \sigma\,\mathbf{A}$ and thus to solve , for $\mathbf{w} \in \mathbb{C}^n$, $(\mathbf{B} - \sigma\,\mathbf{A})\,\mathbf{w} = \mathbf{A}\,\mathbf{v}$ for any given $\mathbf{v} \in \mathbb{R}^n$. However, the subroutine returns either $\mathrm{Re}\,\mathbf{w}$ or $\mathrm{Im}\,\mathbf{w}$ and so there is no complex arithmetic in the Lanczos algorithm. The major computation is in factoring $\mathbf{K}+\sigma\,\mathbf{C}+\sigma^2\,\mathbf{M}$. This is done once. Note that the pseudo-inner product is defined by **A** and is independent of σ.

Both the real and imaginary part of $(\mathbf{B} - \sigma\, \mathbf{A})^{-1}$ provide good enhancement of the eigenvalues close to σ. It is shown in [Parlett and Saad, 1987] that the imaginary part suppresses the unwanted eigenvalues far from σ more strongly than does the real part. However common sense suggests that if $\mathrm{Im}(\sigma) > \mathrm{Re}(\sigma)$ then one should use the imaginary part, otherwise stick with the real part.

4. CONNECTION BETWEEN LANCZOS ALGORITHMS

The original formulations given by Lanczos (in [Lanczos, 1950]) considered the standard symmetric eigenvalue problem $\mathbf{A}\,\mathbf{z} = \lambda\,\mathbf{z}$, i.e. $\mathbf{B} = \mathbf{I}$. The Lanczos vectors $\{\mathbf{q}_k\}_{k=1}^{n}$ for symmetric $n \times n$ $\mathbf{A}$ obey the well-known 3-term recurrence relation

$$\mathbf{A}\,\mathbf{q}_k = \beta_k \mathbf{q}_{k-1} + \alpha_k \mathbf{q}_k + \beta_{k+1}\mathbf{q}_{k+1} \qquad k = 1, \ldots, n \tag{4.1}$$

with $\mathbf{q}_0 = \mathbf{0}$. The $\{\mathbf{q}_k\}_{1}^{n}$ form an orthonormal set. The attractive feature of the Lanczos algorithm for the generalized problem $\mathbf{A} - \lambda\,\mathbf{B}$, where $\mathbf{B}$ is symmetric, positive definite is that, as in (4.1)

$$\mathbf{B}^{-1}\,\mathbf{A}\,\mathbf{q}_k = \beta_k \mathbf{q}_{k-1} + \alpha_k \mathbf{q}_k + \beta_{k+1}\mathbf{q}_{k+1}$$

The Lanczos vectors are still orthonormal but in the inner product defined by $\mathbf{B}$ or by $\mathbf{A}$. Usually they will not be orthogonal in the ordinary Euclidean sense.

In our pseudo-symmetric eigenvalue problem $\mathbf{H}^{-1}\mathbf{A} - \lambda\,\mathbf{I}$ we face the possibility that the Lanczos process may break down and may be unstable when close to breakdown. It is important to know when instability occurs and so we prefer to normalize the $\{q_k\}$ to satisfy

$$\|\mathbf{q}_k\| = 1\ , \qquad \omega_k := (\mathbf{q}_k, \mathbf{q}_k)_{\mathbf{A}}\ , \qquad \text{for all } k\ . \tag{4.2}$$

A tiny value of ω_k is a sign of danger. The three term recurrence, in matrix form, is

$$\begin{aligned} \mathbf{H}^{-1}\,\mathbf{A}\,\mathbf{Q}_n - \mathbf{Q}_n\,\Omega_n^{-1}\,\mathbf{T}_n &= \mathbf{0}\ , \\ \mathbf{T}_n &= \mathbf{Q}_n^t\,\mathbf{A}\,\mathbf{H}^{-1}\,\mathbf{A}\,\mathbf{Q}_n\ , \\ \Omega_n &= \mathbf{Q}_n^t\,\mathbf{A}\,\mathbf{Q}_n\ . \end{aligned} \tag{4.3}$$

provided that breakdown does not occur. Note that $\mathbf{T}_n$ is symmetric. Equating the kth column on each side of (4.3) yields the three term recurrence for our pseudo-symmetric formulation

$$\mathbf{H}^{-1}\,\mathbf{A}\,\mathbf{q}_k = (\beta_k/\omega_{k-1})\,\mathbf{q}_{k-1} + (\alpha_k/\omega_k)\,\mathbf{q}_k + (\beta_{k+1}/\omega_{k+1})\,\mathbf{q}_{k+1}\ , \qquad k < n\ . \tag{4.4}$$

It can be shown that the pseudo symmetric algorithm is a disguised form of the familiar two-sided algorithm given by Lanczos for the nonsymmetric eigenproblem. It turns out that the ωs defined above are directly related to the quality of the $\mathbf{q}$s as a basis for the Krylov subspace.

Lemma:

$$\sigma_j(Q_j) \geq \min_{1 \le i \le j} \frac{|\,\omega_i\,|}{\sqrt{j}\ \|\mathbf{A}\|}\ . \tag{4.5}$$

Here σ_j denotes the smallest singular value. In our case it can be shown that

$$\|\mathbf{A}\| \leq \|\mathbf{C}\| + \|\mathbf{M}\|\ .$$

5. HYPERBOLIC PAIRS AND LOOK-AHEAD LANCZOS

The breakdown of the (generalized) form of the Lanczos algorithm may be interpreted as breakdown in the triangular factorization of a certain matrix. This connection is well known (see [Householder, 1964] and gives valuable insight but there is insufficient space to review it here.

When $\mathbf{A}$ is indefinite then there exist vectors $\mathbf{w} \neq \mathbf{0}$ such that $\mathbf{w}^t \mathbf{A} \mathbf{w} \leq \mathbf{0}$ and, except in trivial cases, there will be starting vectors $\mathbf{q}_1$ and j-vectors $\mathbf{v}$ such that the A-norm of $\mathbf{Q}_j \mathbf{v}$ vanishes for large enough j.

Definition: $\mathbf{v}$ is isotropic if $\mathbf{v}^t \mathbf{A} \mathbf{v} = (\mathbf{v}, \mathbf{v})_{\mathbf{A}} = 0$.

It is helpful to regard the occurrence of breakdown (i.e. when $\mathbf{q}_j$ is isotropic) not as disaster but as a reminder that $\mathbf{A}$ is indefinite. There is a natural way to proceed. The idea goes back to Lagrange but was put to use by D. G. Luenberger to extend the conjugate gradient algorithm [Luenberger, 1969] and by [Bunch, Kauffman, and Parlett, 1976] to stablize triangular factorization.

In our context the idea is to modify our generalized Lanczos algorithm as follows. If $\mathbf{q}_j$ is nearly isotropic then choose the value of α_{j+1} in (4.3) so that $\mathbf{q}_{j+1}$ is isotropic. Provided that $\mathbf{q}_j \mathbf{A} \mathbf{q}_{j+1}$ is not too small these Lanczos vectors still provide a good basis. They are no longer strictly $\mathbf{A}$-orthogonal. Moreover $\mathbf{T}$ is no longer strictly tridiagonal. Nevertheless $\mathbf{Q}_{j+1}^t \mathbf{A} \mathbf{Q}_{j+1}$ is block diagonal with 2×2 diagonal blocks of the form

$$\begin{bmatrix} 0 & \omega_{j,j+1} \\ \omega_{j,j+1} & 0 \end{bmatrix}$$

whenever the modification is used. We follow Luenberger [Luenberger, 1969] in calling $\mathbf{q}_j$, $\mathbf{q}_{j+1}$ a hyperbolic pair.

In practice there is no need to wait until $\mathbf{q}_j^t \mathbf{A} \mathbf{q}_j$ is negligible before switching to a hyperbolic pair for $\mathbf{q}_j$ and $\mathbf{q}_{j+1}$. By forsaking strict $\mathbf{A}$-orthogonality we gain a better conditioned basis. Similar thinking for the two sided algorithm yields the Look-Ahead algorithm [Parlett, Taylor, and Liu, 1985] but the procedure there is much more complicated. Our experience is limited but in our comparisons the Look-Ahead version has always produced good bases even when simple Lanczos would yield values for $\sigma_j(Q_j)$ like 10^{-6}.

6. LOSS OF ORTHOGONALITY

Known techniques for counteracting the loss of orthogonality due to finite precision arithmetic extend in a natural way to the A-orthogonal case.

7. HOW TO SOLVE $\mathbf{T} - \lambda \boldsymbol{\Omega}$

Space does not permit a survey of the various ways that have been proposed for solving this problem in an efficient way. Suffice it to say here that there is active research on this topic.

8. ERROR BOUNDS

Let

$$(\mathbf{T}_j - \theta \boldsymbol{\Omega}_j)\, \mathbf{s} = 0 \ , \tag{8.1}$$

with $\|\mathbf{s}\| = 1$, define a typical eigenpair $(\theta, \mathbf{s})$ of the reduced problem. Let $\mathbf{F}$ be the operator given to the Lanczos algorithm and let $\mathbf{F}$ be symmetric with respect to $\mathbf{A}$. After j steps, in exact arithmetic,

$$\mathbf{F}\,\mathbf{Q}_j - \mathbf{Q}_j\,\mathbf{\Omega}_j^{-1}\,\mathbf{T}_j = \mathbf{q}_{j+1}\,(\beta_{j+1} / \omega_{j+1})\,\mathbf{e}_j^t \tag{8.2}$$

The 'Ritz vector' corresponding to θ in (8.1) is defined by $\mathbf{y} := \mathbf{Q}_j\ \mathbf{s}$. It follows that

$$\|\mathbf{F}\,\mathbf{y} - \mathbf{y}\,\theta\| = (\beta_{j+1} / \omega_{j+1})\ |s(j)| \tag{8.3}$$

since $\|\mathbf{q}_{j+1}\| = 1$. Unfortunately, it is not possible to evaluate $\|\mathbf{y}\|$ without computing $\mathbf{y}$. Note that $\bar{\mathbf{y}}^t\ \mathbf{A}\,\mathbf{y} = \bar{\mathbf{s}}^t\ \mathbf{\Omega}_j\ \mathbf{s}$ but this quantity might be 0 or negative in general.

For indefinite $\mathbf{A}$ we may invoke the results presented in [Kahan, Parlett, and Jiang, 1982] and adapt them to our problem. Two residual norms are required. Let

$$\mathbf{u} := \mathbf{F}\,\mathbf{y} - \mathbf{y}\,\theta \quad , \ \bar{\mathbf{v}}^t := \bar{\mathbf{x}}^t\ \mathbf{F} - \theta\,\bar{\mathbf{x}}^t \ , \tag{8.4}$$

where $\bar{\mathbf{y}}^t\ \mathbf{x} \neq 0$, $\|\mathbf{y}\| = \|\mathbf{x}\| = 1$. Now we are obliged to work in $\mathbb{C}^n$ rather than $\mathbb{R}^n$.

Theorem With the notation given above $(\theta, \mathbf{y}, \bar{\mathbf{x}}^t)$ is an eigentriple of $\mathbf{F} - \mathbf{E}$ for some $\mathbf{E}$ satisfying

$$\|\mathbf{E}\| \le \max\{\ \|\mathbf{u}\|,\ \|\bar{\mathbf{v}}^t\|\ \}\ .$$

Note that $\|\bar{\mathbf{v}}^t\| = \|\mathbf{v}\|$. In other words, if $\|\mathbf{u}\|$ and $\|\mathbf{v}\|$ are both tiny then $(\theta, \mathbf{y}, \bar{\mathbf{x}}^t)$ is an eigentriple for an operator almost indistinguishable from the given one.

In our case, if $\mathbf{F} = \mathbf{H}^{-1}\ \mathbf{A}$ then $\mathbf{F}' = \mathbf{A}\ \mathbf{H}^{-1}$ and we have from (8.2),

$$\mathbf{H}^{-1}\ \mathbf{A}\,\mathbf{y} - \mathbf{y}\,\theta = \mathbf{q}_{j+1}\,(\beta_{j+1} / \omega_{j+1})\ s(j)\ .$$

The second residual is obtained easily by premultiplying by $\mathbf{A}$.

$$(\mathbf{A}\,\mathbf{H}^{-1})\,(\mathbf{A}\,\mathbf{y}) - (\mathbf{A}\,\mathbf{y})\,\theta = \mathbf{A}\ \mathbf{q}_{j+1}\,(\beta_{j+1} / \omega_{j+1})\ s(j)\ .$$

Let $\beta(j) = (\beta_{j+1} / \omega_{j+1})\ |s(j)|$ then $(\theta, \mathbf{y}, \mathbf{A}\,\mathbf{y})$ is an eigentriple of $\mathbf{F} - \mathbf{E}$ for some $\mathbf{E}$ satisfying

$$\|\mathbf{E}\| \le \beta(j)\ \max\left\{ \frac{1}{\|\mathbf{y}\|},\ \frac{\|\mathbf{A}\ \mathbf{q}_{j+1}\|}{\|\mathbf{A}\ \mathbf{y}\|} \right\}\ .$$

When the computation has proceeded enough that $|\beta(j)|$ is very small then first order perturbation theory may be invoked to obtain an accurate error estimate for θ by regarding $\mathbf{F}$ as a change to $\mathbf{F} - \mathbf{E}$. If λ is the eigenvalue of $\mathbf{F}$ closest to θ we have, in general,

$$|\lambda - \theta| = \frac{\|\mathbf{y}\| \cdot \|\mathbf{x}\|}{|\bar{\mathbf{y}}^t\ \mathbf{x}|} \cdot \frac{|\bar{\mathbf{y}}^t\ \mathbf{E}\,\mathbf{x}|}{\|\mathbf{y}\| \cdot \|\mathbf{x}\|} + O\,(\|\mathbf{E}\|^2)\ ,$$

The first term on the left is independent of $\mathbf{E}$ and is the condition number of θ as an eigenvalue of $\mathbf{F} - \mathbf{E}$:

$$\mathrm{cond}(\theta)\ :=\ \|\mathbf{y}\| \cdot \|\mathbf{x}\|\ /\ |\bar{\mathbf{y}}^t\ \mathbf{x}|\ .$$

We now adapt these results to our problem. Note that

$$\|\mathbf{y}\|^2 = |\bar{\mathbf{s}}^t\ \mathbf{Q}_j^t\,\mathbf{Q}_j\ \mathbf{s}| \le j\ |\bar{\mathbf{s}}^t\ \mathbf{s}| = j \quad , \quad \|\bar{\mathbf{y}}^t\ \mathbf{A}\,\mathbf{y}\| = |\bar{\mathbf{s}}^t\ \mathbf{Q}_j^t\,\mathbf{A}\,\mathbf{Q}_j\ \mathbf{s}| = |\bar{\mathbf{s}}^t\ \mathbf{\Omega}_j\ \mathbf{s}|\ .$$

Now we can obtain an error bound. As $\|\mathbf{E}\| \to 0$,

$$|\lambda - \theta| \le \mathrm{cond}(\theta) \cdot \|\mathbf{E}\| + O\,(\|\mathbf{E}\|^2)\ ,$$

$$\le \frac{\|\mathbf{y}\| \cdot \|\mathbf{A}\ \mathbf{y}\|}{|\bar{\mathbf{y}}^t\ \mathbf{A}\,\mathbf{y}|}\ \beta(j)\ \max\left\{ \frac{1}{\|\mathbf{y}\|},\ \frac{\|\mathbf{A}\ \mathbf{q}_{j+1}\|}{\|\mathbf{A}\ \mathbf{y}\|} \right\} + O\,(\|\mathbf{E}\|^2)\ ,$$

$$\leq \frac{\beta(j)}{|\mathbf{s}^t\, \boldsymbol{\Omega}_j\, \mathbf{s}|} \max\{\, \|\mathbf{A}\,\mathbf{y}\|\,, \|\mathbf{A}\,\mathbf{q}_{j+1}\| \cdot \|\mathbf{y}\|\,\} + O(\|\mathbf{E}\|^2) \quad ,$$

$$\leq \frac{\beta(j)\sqrt{j}}{|\mathbf{s}^t\, \boldsymbol{\Omega}_j\, \mathbf{s}|} \|\mathbf{A}\| + O(\|\mathbf{E}\|^2) \quad .$$

The attraction of the last inequality is that the dominant term can be calculated at step j, without recourse to n-vectors, provided that $\|\mathbf{A}\|$ is provided along with the subroutine that multiplies vectors by **A**. We use $\beta(j) / |\mathbf{s}^t\, \boldsymbol{\Omega}_j\, \mathbf{s}|$ as a provisional error estimate. When the required number of Ritz values θ have passed this test then, and not before, the Ritz vector **y** may be computed. At that point the more precise factor

$$\max\{\, \|\mathbf{A}\,\mathbf{y}\|, \|\mathbf{A}\,\mathbf{q}_{j+1}\| \cdot \|\mathbf{y}\|\,\}$$

may be computed at the cost of forming **A y**. At the end of a Lanczos run a multiple of $\mathbf{A}\,\mathbf{q}_{j+1}$ is available and its Euclidean norm can be computed at the cost of one dot product.

9. NUMERICAL EXAMPLES

Our numerical results are encouraging and will be reported in future communications. Our codes have been run on realistic examples with n up to 192.

The authors gratefully acknowledge partial support from contract AFOSR-88-0048.

REFERENCES

[1] K. J. Bathe and E. L. Wilson, *Numerical Methods in Finite Element Method*, Prentice Hall, Englewood Cliffs, N. J. (1976).

[2] M. Borri and P. Mantegazza, "Efficient solution of quadratic eigenproblems arising in dynamic analysis of structures," *Comput. Meths. Appl. Mech. Engrg.* **12**, 19-31 (1977).

[3] M. Brebner and J. Grad, "Eigenvalues of $Ax=\lambda Bx$ for real symmetric matrices A and B computed by reduction to pseudo-symmetric form and the HR process," *J. Lin. Alg. & Appls.*, **43**, 59-118 (1982).

[4] Bunch, Kauffman, and Parlett, "Decomposition of a Symmetric Matrix," *Numer. Math.*, **27**, 95-109 (1976).

[5] A. Bunse-Gerstner, "An analysis of the HR algorithm for computing eigenvalues of a matrix," *J. Lin. Alg. & Appls.*, **35**, 155-173 (1981).

[6] H. C. Chen and R. L. Taylor, "Properties and Solutions of the Eigensystem of Nonproportionally Damped Linear Dynamic Systems," *Report UCB/SEMM-86/10*, Department of Civil Engineering, University of California at Berkeley (1986).

[7] EISPACK - The Eigensystem Subroutine Package, Argonne National Laboratories, 9700 S. Cass Ave., Argonne, IL.

[8] F. R. Gantmacher, *The Theory of Matrices*, Vol. 2, Chelsea Pub. Co., New York (1959).

[9] I. Gohberg, P. Lancaster, and L. Rodman, *Matrices and Indefinite Scalar Products*, Birkhauser, Basel, (1983).

[10] I. Gohberg, P. Lancaster, and L. Rodman, "Quadratic Matrix Polynomials with a Parameter", *Advances Appl. Math.*, **7**, 253-281 (1986).

[11] Householder, *The Theory of Matrices in Numerical Analysis,* Blaisdell, New York (1964).

[12] W. Kahan, B. N. Parlett, and E. Jiang, "Residual Bounds on Approximate Eigensystems of Nonnormal Matrices," *SIAM J. Appl. Math.*, **19**, 470-484 (1982).

[13] C. Lanczos, "An Iterative Method for the Solution of the Eigenvalues Problem of Linear Differential and Integral Operator," *J. Res. Natl. Bur. Standards*, **45**, 255-282, (1950).

[14] D. G. Luenberger, "Hyperbolic Pairs and the method of Conjugate Gradients," *SIAM J. Appl. Math.*, **17**, 1263-1267 (1969).

[15] B. Nour-Omid and Regelbrugge, "Lanczos Method for Dynamic Analysis of Damped Structural Systems," (1988).

[16] B. Nour-Omid, B. N. Parlett, T. Ericsson, and P. S. Jensen, "How to Implement the Spectral Transformation," *Mathematics of Computation,* **48**, 663-673, (1987).

[17] B. N. Parlett, *The symmetric Eigenvalue Problem,* Prentice-Hall, Englewood Cliffs, NJ, (1980).

[18] B. N. Parlett, "Two Monitoring Schemes for the Lanczos Algorithm," *Computing Methods in Applied Sciences and Engineering, V*, ed. by R. Glowinski and J. L. Lions, North-Holland Press (1982).

[19] B. N. Parlett, B. Nour-Omid, and Z. S. Liu, "How to maintain Semi-orthogonality among Lanczos Vectors," *Report PAM-74*, Center for Pure and Appl. Math., University of California at Berkeley (1988).

[20] B. N. Parlett and Y. Saad, "Complex Shift and Invert Strategies," *J. Lin. Alg. & Appls.*, **88/89**, 623-650 (1987).

[21] B. N. Parlett and D. S. Scott, "The Lanczos Algorithm with Selective Orthogonalization," *Math. Comp.*, **33**, 217-238, (1979).

[22] B. N. Parlett, D. R. Taylor, and Z. S. Liu, "The Look-Ahead Lanczos Algorithm for Large Unsymmetric Eigenproblems," J. L. Lions and R. Glowinski ed., *Proceedings of the 6th International Conference on Computing Methods in Applied Sciences and Engineering* Versailles, France, Dec. 12-16, 1984, INRIA, North-Holland (1985).

[23] H. Rutishauser, "Simultaneous Iteration Methods for Symmetric Matrices," *Numer. Math.*, **16**, 205-223 (1970).

[24] H. Simon, "The Lanczos Algorithm with Partial Reorthogonalization." *Math. Comp.*, **42**, 115-142 (1984).

[25] O. C. Zienkiewicz, *The Finite Element Method,* 3rd edn., McGraw-Hill, London, (1977).

Overview of Krylov subspace methods with applications to control problems

Youcef Saad

Abstract

This paper gives an overview of projection methods based on Krylov subspaces with emphasis on their application to solving matrix equations that arise in control problems. The main idea of Krylov subspace methods is to generate a basis of the Krylov subspace $\text{Span}\{v, Av, \ldots, A^{m-1}v\}$, and seek an approximate solution to the original problem from this subspace. Thus, the original matrix problem of size N is approximated by one of dimension m, typically much smaller than N. Krylov subspace methods have been very successful in solving linear systems (Conjugate Gradients, GMRES,..) and eigenvalue problems (Lanczos, Arnoldi,..) and are now just becoming popular for solving nonlinear equations. We will show how they can be used to solve partial pole placement problems, Sylvester's and Lyapunov's equations.

1 Introduction

Krylov subspace methods have become a very useful and popular tool to solve large sets of linear and nonlinear equations, and large eigenvalue problems. One of the reason for their popularity is their simplicity and their generality. These methods have been increasingly accepted as efficient and reliable alternative to the more expensive methods that are usually employed for solving dense problems. This trend is likely to accelerate as models are becoming more complex and give rise to larger and larger matrix problems.

It is interesting to observe that, surprisingly, there has been very little done in developing algorithms that can solve the very large matrix problems that arise in control. Yet, there are now applications in this area which lead to equations involving very large sparse matrices. This is the case for example in models that involve partial differential equations in several space dimensions, or in applications related to large space structures [4]. Another typical example is that of electrical networks [5].

In [24] we proposed a method for partial pole placement, which consists of placing a few of the poles of the matrix, namely only those that are unstable. The methods proposed are based on projecting the problem onto a small invariant subspace of A

associated with the unstable eigenvalues. Datta and Saad [9] considered several ways of solving special Sylvester equations and some related problems. Recently, we have considered a few numerical methods for solving large Lyapunov equations [25].

The purpose of this paper is to describe the general concepts used in Krylov subspace methods and to give an overview of the different ways in which they are used. As will be seen the method is fairly universal in that it can be used in various forms to provide solution methods for virtually any linear problem. However, the success of the method depends critically on the nature of the matrices at hand. For example conjugate gradient type methods are very successful in solving symmetric positive definite or nonsymmetric positive real linear systems but have been rather unsuccessful with highly indefinite problems.

The next section is a brief introduction to Krylov subspaces. Section 3 discusses the application of the method to linear systems, while section 4 is on eigenvalue problems. Section 5 will be on evaluating exponentials of A times a vector and some applications to Lyapunov equations.

2 Krylov subspaces

Given a square matrix A and a nonzero vector v, the subspace defined by

$$K_m \equiv span\ \{v, Av, A^2v, ...A^{m-1}v\} \tag{1}$$

is referred to as a the m-th Krylov subspace associated with the pair (A, v) and is denoted by $K_m(A, v)$ or simply by K_m if there is no ambiguity. We start by stating a few elementary properties of Krylov subspaces. Recall that the minimal polynomial of a vector v is the nonzero monic polynomial p of lowest degree such that $p(A)v = 0$. Clearly, the Krylov subspace K_m is the subspace of all vectors in $\mathbb{C}^N$ which can be written as $x = p(A)v$, where p is a polynomial of degree not exceeding $m - 1$.

Proposition 2.1 *The Krylov subspace K_m is of dimension m if and only if the degree of the minimal polynomial of v with respect to A is not less than m.*

In practice it is uncommon that the degree of the minimal polynomial is less than N, even in exact arithmetic. If this were to happen then it is usually helpful rather than harmful because of the following proposition.

Proposition 2.2 *Let μ be the degree of the minimal polynomial of v. Then K_μ is invariant under A and $K_m = K_\mu$ for all $m \geq \mu$.*

Thus, in case μ is small we can work work in subspace of dimension μ and be able to solve the problem exactly in this small subspace.

Working directly with the basis $\{A^jv\}_{j=0,\ldots,m-1}$ is likely to lead to serious numerical difficulties. Most Krylov subspace methods utilize either orthogonal or bi-orthogonal bases of K_m. Thus, the procedure introduced by Arnoldi [1] builds an orthogonal basis of the Krylov subspace K_m by the following algorithm.

Arnoldi's algorithm:

1. Start: Choose a vector v_1 of norm 1.

2. Iterate: for $j = 1, 2, \ldots, m$ compute,

$$h_{ij} = (Av_i, v_j) \quad i = 1, 2, \ldots, j \tag{2}$$

$$w = Av_j - \sum_{i=1}^{j} h_{ij} v_i \tag{3}$$

$$h_{j+1,j} = \|w\|_2 \tag{4}$$

$$v_{j+1} = w / h_{j+1,j} \tag{5}$$

This algorithm is mathematically equivalent to a Gram-Schmidt process applied to the power sequence $v, Av, \ldots., A^{m-1}v$, in that it would deliver the same sequence of v_i's *in exact arithmetic.* The algorithm will stop if the vector w computed in (4) vanishes which happens if the degree of the minimal polynomial for v is j. This is referred to a 'lucky' breakdown since as was seen above it means that the original problem (linear system, eigenvalue problem) can be solved exactly in a j-th dimensional subspace.

The following are a few simple but important properties satisfied by the algorithm.

Proposition 2.3 *The vectors $v_1, v_2, \ldots, v_m$ form an orthonormal basis of the subspace $K_m = span\{v_1, Av_1, \ldots, A^{m-1}v_1\}$.*

Proposition 2.4 *Denote by V_m the $N \times m$ matrix with column vectors $v_1, \ldots, v_m$ and by H_m the $m \times m$ Hessenberg matrix whose nonzero entries are defined by the algorithm. Then the following relations hold:*

$$AV_m = V_m H_m + h_{m+1,m} v_{m+1} e_m^T \tag{6}$$

$$V_m^T A V_m = H_m \tag{7}$$

Note that when A is symmetric then (7) implies that the matrix H_m is tridiagonal symmetric and as a result Arnoldi's algorithm simplifies into an algorithm which involves only three consecutive vectors at each step. The corresponding algorithm is the well-known Lanczos algorithm.

The second of the relations in the proposition indicates that the Hessenberg matrix H_m is nothing but the matrix representation of the projection of A onto K_m, with respect to the orthogonal basis V_m. Analysis of various projection methods based on Krylov subspaces, indicate that, loosely speaking, K_m contains the most significant information of A, in that the outermost eigenvalues of A are well represented by those of its projection onto K_m, for large enough m. The main idea of Krylov subspace methods is to project the original problem into K_m. In the next sections we will see how this is done via simple Galerkin type procedures, for standard linear algebra problems.

The relation (6) has been exploited in [9] for solving special Sylvester's equations that arise in the design of reduced-dimensional state estimator. The Arnoldi and block-Arnoldi algorithms have been used in [6] to compute numerically the controllability of a linear system.

3 Krylov subspace methods for solving linear systems

Given an initial guess x_0 to the linear system

$$Ax = b, \tag{8}$$

a general *projection method* seeks an approximate solution x_m from an affine subspace $x_0 + K_m$ of dimension m by imposing the Petrov-Galerkin condition

$$b - Ax_m \perp L_m \tag{9}$$

where L_m is another subspace of dimension m. A Krylov subspace method is a method for which the subspace K_m is the Krylov subspace

$$K_m(A, r_0) = span\{r_0, Ar_0, A^2r_0, \ldots, A^{m-1}r_0\}, \tag{10}$$

in which $r_0 = b - Ax_0$. The different versions of Krylov subspace methods arise from different choices of the subspaces K_m and L_m and from the ways in which the system is preconditioned. The most common choices of K_m and L_m are the following.

1. $L_m = K_m = K_m(A, r_0)$. The conjugate gradient method is a particular instance of this method when the matrix is symmetric positive definite. Another method in this class is the Full Orthogonalization Method (FOM) [21] which is closely related to Arnoldi's method for solving eigenvalue problems [1]. Also in this class is ORTHORES [14], a method that is mathematically equivalent to FOM. Axelsson [2] also derived a similar algorithm for general nonsymmetric matrices.

As an example we outline here the FOM method for solving linear systems. Assume that we take $v_1 = r_0/\|r_0\|_2$ and run m steps of Arnoldi's method described in the previous section. Then, the approximate solution is of the form $x_0 + V_m y_m$ where y_m is some m-vector. The Galerkin condition (9) with $L_m = K_m$ gives immediately that $y_m = H_m^{-1}\|r_0\|_2 e_1$.

2. $L_m = AK_m; K_m = K_m(A, r_0)$. With this choice of L_m, it can be shown, see e.g., [26] that the approximate solution x_m minimizes the residual norm $\|b - Ax\|_2$ over all candidate vectors in $x_0 + K_m$. In contrast, there is no similar optimality property known for methods of the first class when A in nonsymmetric. Because of this, many methods of this type have been derived for the nonsymmetric case [3, 14, 11, 27]. The Conjugate Residual method [7] is the analogue of conjugate gradient method that is in this class. The GMRES algorithm [27] is an extension of the Conjugate Residual method to nonsymmetric problems.

3. $L_m = K_m(A^T, r_0); K_m = K_m(A, r_0)$. Clearly, in the symmetric case this class of methods reduces to the first one. In the nonsymmetric case, the biconjugate gradient method (BCG) due to Lanczos [15] and Fletcher [12] is a good representative of this class. There are various mathematically equivalent formulations of the biconjugate gradient method [22], some of which are more numerically viable than others. An efficient variation on this method, called CGS (Conjugate gradient squared) was proposed by Sonneveld [28].

Apart from the above three basic methods there are a number of techniques for nonsymmetric problems that are mathematically equivalent to solving the normal equations $A^TAx = A^Tb$ or $AA^Ty = b$ by the conjugate gradient method. We will comment that these methods are often too quickly dismissed as inferior because of the fact that the condition number of the original problem is squared. For problems that are strongly indefinite they do represent, however, the only viable alternative, since none of the above three types of methods would work in this situation.

One of the possible applications of the methods described here is in the frequency response calculations in input-output analysis. For example, in the single input single output case, one needs to compute $c(A - j\omega I)^{-1}b$ for many values of ω. We observe that the Krylov subspaces are invariant under arbitrary shifts to the matrix A, i.e., $K_m(A, v) = K_m(A - sI, v)$ for any s. This suggests using the same Krylov subspace $K_m(A, v)$ to get approximations to all of the solution vectors $(A - j\omega I)^{-1}b$ via the formula

$$x_m(\omega) = \beta V_m(H_m - j\omega I)^{-1}e_1 \tag{11}$$

where we have set $\beta = \|b\|$ and where we assume that the Arnoldi algorithm is started with $v_1 = b/\|b\|_2$, i.e., $x_0 = 0$. This technique was suggested in [10] and can be regarded as an extension of the earlier technique proposed by Laub in [16] for dense problems. Numerical experiments are currently being performed.

An important factor in the success of conjugate gradient-like methods is the preconditioning technique. This typically consists of replacing the original linear system (8) by, for example, the equivalent system

$$M^{-1}Ax = M^{-1}b \tag{12}$$

In the classical case of the incomplete LU preconditionings, the matrix M is of the form $M = LU$ where L is a lower triangular matrix and U is an upper triangular matrix such that L and U have the same structure as the lower and upper triangular parts of A respectively. In the general sparse case, the incomplete factorization is obtained by performing the standard LU factorization of A and dropping all fill-in elements that are generated during the process. This is referred to as ILU(0), or IC(0) in the symmetric case.

4 Krylov subspace methods for eigenvalue problems

An idea that is basic to sparse eigenvalue calculations is that of projection processes [23]. Given a subspace K spanned by a system of m orthonormal vectors $V \equiv [v_1, \ldots, v_m]$ a projection process onto $K \equiv span\ \{V\}$ computes an approximate eigenpair $\tilde{\lambda} \in \mathbb{C}, \tilde{u} \in K$ that satisfy the Galerkin condition,

$$(A - \tilde{\lambda} I)\tilde{u} \perp K \tag{13}$$

The approximate eigenvalues $\tilde{\lambda}$ are the eigenvalues of the $m \times m$ matrix $C = V^T A V$. The corresponding approximate eigenvectors are the vectors $\tilde{u}_i = V y_i$ where y_i are the eigenvectors of C. Similarly, the approximate Schur vectors are the vector columns of VU, where $U = [u_1, u_2, \ldots, u_m]$ are the Schur vectors of C, i.e., $U^T C U$ is quasi-upper triangular. Thus, one possible method for computing eigenvalues/ eigenvectors of large sparse matrices is to use the Arnoldi process [1, 20] which is a projection process onto $K_m = span\{v_1, Av_1, \ldots, A^{m-1}v_1\}$. Once the Arnoldi vectors $v_1, \ldots, v_m$ have been generated we can use V_m for a projection process onto K_m. The matrix $V_m^T A V_m$ which is needed for this purpose is nothing but the upper Hessenberg matrix H_m generated by the algorithm.

Note that the Arnoldi algorithm utilizes the matrix A only to compute successive matrix by vector products $w = Av$, so sparsity can be exploited. As m increases, the eigenvalues of H_m that are located in the outermost part of the spectrum start converging towards corresponding eigenvalues of A. However, the difficulty with the above algorithm is that as m increases cost and storage increase rapidly. One solution is to use the method iteratively: m is fixed and the initial vector v_1 is taken at each new iteration as a linear combination of some of the approximate eigenvectors. Moreover, there are several ways of accelerating convergence by preprocessing v_1 by a Chebyshev iteration before restarting, i.e., by taking $v_1 = t_k(A)z$ where z is again a linear combination of eigenvectors.

A technique related to Arnoldi's method is the nonsymmetric Lanczos algorithm [18, 8] which produces a nonsymmetric tridiagonal matrix instead of a Hessenberg matrix. Unlike Arnoldi's process, this method requires multiplications by both A and A^T at every step. On the other hand it has the big advantage of requiring little storage (5 vectors). Although no comparisons of the performances of the Lanczos and the Arnoldi type algorithms have been made, the Lanczos methods are usually recommended whenever the number of eigenvalues to be computed is large.

Finally, if the matrix is banded an efficient solution is the shift and invert strategy which consists of using one of the above iterative methods (subspace iteration, Arnoldi, or Lanczos) for the matrix $(A - \sigma I)^{-1}$, where σ is some shift chosen say at the center of some small region of the complex plane where eigenvalues are sought. The matrix $(A - \sigma I)^{-1}$, need not be explicitly computed: all we need is to factor $(A - \sigma I)$ into LU and subsequently at each step of the iterative method solve two triangular systems one with L and the other with U. Thus band structure can be fully exploited. In [19] several implementations of the shift and invert strategy are considered and the problem of avoiding complex arithmetic when A is real is addressed.

An important application of the eigenvalue algorithms is to provide a small invariant subspace that will represent the critical modes of the system. For example unstable modes, assuming that there are just a few of them, can then be displaced by a technique described in [24]. The main idea in [24] is now summarized.

Let A be an N x N real nonsymmetric matrix whose eigenvalues are

$$\lambda_1, \lambda_2, \ldots, \lambda_k, \lambda_{k+1}, \ldots, \lambda_N$$

and b be a given real vector. The problem considered is to find a vector f so that the modified matrix

$$M = A - bf^T$$

has the given eigenvalues

$$\mu_1, \mu_2, \ldots, \mu_k, \lambda_{k+1}, \ldots, \lambda_N.$$

In other words we would like to assign the eigenvalues $\lambda_1, \lambda_2, \ldots, \lambda_k$ of A into $\mu_1, \mu_2, \ldots, \mu_k$, while leaving the rest of the spectrum of A unchanged, with the rank one perturbation $-bf^T$. This is referred to as the *partial pole assignment problem.*

To solve this problem we assume that we have computed the partial Schur factorization for A^T:

$$A^T Q = QR,$$

where Q is an N x k matrix whose columns form an orthonormal basis of the left invariant subspace associated with $\lambda_i, i = 1, \ldots k$ and R is a k x k upper quasi-triangular matrix. We then seek a solution f in the form $f = Qg$. Denoting by s the vector $s = Q^T b$, the matrix $M^T Q$ is such that,

$$M^T Q = \left[A^T - fb^T\right] Q = QR - Qgb^T Q = Q\left[R - gs^T\right] = Q\left[R^T - sg^T\right]^T$$

The above equation means that the choice $f = Qg$ makes the subspace spanned by Q also invariant under M^T. Moreover, the eigenvalues of the matrix M associated with this invariant subspace are the eigenvalues of the k x k matrix $C_k = R^T - sg^T$. Thus it suffices to assign the eigenvalues of this small matrix to be $\mu_i, i = 1, \ldots k$, by an appropriate choice of the vector g. It was shown in [24] that this solves the problem and that the solution is feasible under some mild conditions on b and the μ_i's. Nichols [17] proposed improvements on this scheme to provide robust partial pole placement techniques. In [9] alternative techniques for partial pole placement were derived from a special technique to solve Sylvester's equation.

5 Approximations to $e^A v$ and applications

Computing approximations to the exponential of a matrix is usually not too hard a problem for small dense matrices. For large matrices, this can be rather difficult to do because of the fact that e^A will in general be a dense matrix even when A is very sparse. However, it is often not the exponential of the matrix that is sought but its product with some vector v. The question of approximating $e^A v$ for any given vector v was considered in [13] where polynomial and rational approximations to the exponential were used. Here we summarize only the method proposed in [13] that is based on polynomial approximation to $e^A v$:

$$e^A v \approx p_{m-1}(A)v \tag{14}$$

where p_{m-1} is a polynomial of degree $m - 1$. Thus, the above approximation is an element of the Krylov subspace (1) and it is convenient to express it in the orthonormal basis $V_m = [v_1, v_2, v_3, \ldots, v_m]$ generated by Arnoldi's algorithm seen earlier. We can write the desired approximation to $x = e^A v$ as $x_m = p_m(A)v$ or equivalently $x_m = V_m y$ where y is an m-vector. In [13], the choice $y = \beta e^{H_m} e_1$ with $\beta = \|v\|_2$ was suggested, leading to the following formula for arbitrary t,

$$e^A v \approx \beta V_m e^{H_m} e_1 \tag{15}$$

The quality of this approximation was also analyzed in [13] and the following result was shown.

Theorem 5.1 *Let A be any square matrix and let $\rho = \|A\|_2$. Then the error of the approximation (15) is such that*

$$\|e^A v - \beta V_m e^{H_m} e_1\|_2 \leq 2\beta \frac{\rho^m e^\rho}{m!}. \tag{16}$$

Experiments reported in [13], reveal that this approximation can be very accurate even for moderate values of the degree m. The theorem shows convergence of the approximation (15) for fixed t, as m increases to ∞. However, note that the above approximation is exact when $m = N$, see [13].

One application of the above formulas is that one can approximate $e^{tA} v$ for all t as

$$e^{tA}v \approx \beta V_m e^{tH_m} e_1. \tag{17}$$

This provides a way of solving ordinary differential equations of the type $\dot{x} = Ax + b$ which was the original problem considered in [13]. Moderate degrees can provide reasonably high accuracy in the solutions. Another direct application is described in [25], where the controllability Grammian,

$$X = \int_0^\infty e^{\tau A} bb^T e^{\tau A^T} d\tau. \tag{18}$$

was approximated by replacing the function $e^{\tau A}b$ by its approximation (17). Note that X is known to be the solution of the Lyapunov equation

$$AX + XA^T + bb^T = 0. \tag{19}$$

A rather unexpected result shown in [25] is that the approximation provided by the above integration process is nothing but a Galerkin method applied to (19) over the subspace of matrices of the form $V_m G V_m^T$, where V_m is fixed and G runs over the set of $m \times m$ matrices.

6 Conclusion

The techniques described in the previous section, and other recent developments elsewhere, suggest that in many of the problems in control, one can work in a Krylov subspace of reasonably small dimension. For example although the theory is not established for this case, the above approach for Lyapunov equations can be extended to Riccati's equation using a Galerkin point of view. Many of the optimization techniques in control can be carried out by replacing the full n-dimensional variable $x(t)$ by an approximation derived from replacing x by its m dimensional approximation (17). This may open up interesting new approaches and theoretical questions as to the accuracy of the corresponding approximations.

References

[1] W. E. Arnoldi. The principle of minimized iteration in the solution of the matrix eigenvalue problem. *Quart. Appl. Math.*, 9:17–29, 1951.

[2] O. Axelsson. Conjugate gradient type-methods for unsymmetric and inconsistent systems of linear equations. *Linear Algebra Appl.*, 29:1–16, 1980.

[3] O. Axelsson. A generalized conjugate gradient, least squares method. *Num. Math.*, 51:209–227, 1987.

[4] M.J. Balas. Trends in large space structure control theory: fondest dreams, wildest hopes. *IEEE Trans. Aut. Contr.*, AC-2:522–535, 1982.

[5] P. Kokotovic B.H. Krogh. Feedback control of overloaded networks. *IEEE Trans. Aut. Cont.*, AC-29:704–711, 1984.

[6] D. Boley and G. H. Golub. The lanczos-arnoldi algorithm and controllability. *Systems and Control Letters*, 4:317–324, 1987.

[7] R. Chandra. *Conjugate Gradient Methods for Partial Differential Equations.* PhD thesis, Yale University, Computer Science Dept., New Haven, CT. 06520, 1978.

[8] J. Cullum and R. Willoughby. A Lanczos procedure for the modal analysis of very large nonsymmetric matrices. In *Proceedings of the 23rd Conference on Decision and Control, Las Vegas*, 1984.

[9] B. N. Datta and Y. Saad. Arnoldi methods for large Sylvester-like observer matrix equations. in preparation.

[10] B. N. Datta and Y. Saad. Solution of large linear algebra problems in control. Presentation at the 1986 SIAM meeting on Linear Algebra and its Applications, Boston Mass., Aug 1986.

[11] H. C. Elman. *Iterative Methods for Large Sparse Nonsymmetric Systems of Linear Equations.* PhD thesis, Yale University, Computer Science Dept., New Haven, CT., 1982.

[12] R. Fletcher. Conjugate gradient methods for indefinite systems. In G.A. Watson, editor, *Proceedings of the Dundee Biennal Conference on Numerical Analysis 1974*, pages 73–89, New York, 1975. University of Dundee,Scotland, Springer Verlag.

[13] E. Gallopoulos and Y. Saad. On the parallel solution of parabolic equations. In R. De Groot, editor, *Proceedings of the International Conference on Supercomputing 1989, Heraklion, Crete, June 5-9, 1989.* ACM press, 1989.

[14] K. C. Jea and D. M. Young. Generalized conjugate gradient acceleration of nonsymmetrizable iterative methods. *Linear Algebra Appl.*, 34:159–194, 1980.

[15] C. Lanczos. Solution of systems of linear equations by minimized iterations. *J. Res. Nat. Bur. Standards*, 49:33–53, 1952.

[16] A. J. Laub. Robust stability of linear systems- some computational considerations. In C. Wilde et al., editor, *Information linkages between applied mathematics and industry.* Academic Press, 1980.

[17] N. K. Nichols. Robustness in partial pole placement. *IEEE Trans. Aut. Cont.*, 32:728–732, 1987.

[18] B. N. Parlett and D. R. Taylor andf Z. S. Liu. A look-ahead Lanczos algorithm for nonsymmetric matrices. *Mathematics of Computation*, 44:105–124, 1985.

[19] B.N. Parlett and Y. Saad. Complex shift and invert strategies for real matrices. Technical Report YALEU/ DCS-RR-424, Yale University, Computer Science Dept., New-Haven, Connecticut, 1985.

[20] Y. Saad. Variations on Arnoldi's method for computing eigenelements of large unsymmetric matrices. *Linear Algebra Appl.*, 34:269–295, 1980.

[21] Y. Saad. Krylov subspace methods for solving large unsymmetric linear systems. *Mathematics of Computation*, 37:105–126, 1981.

[22] Y. Saad. The Lanczos biorthogonalization algorithm and other oblique projection methods for solving large unsymmetric systems. *SIAM J. Numer. Anal.*, 19:470–484, 1982.

[23] Y. Saad. Projection methods for solving large sparse eigenvalue problems. In B. Kagstrom and A. Ruhe, editors, *Matrix Pencils, proceedings, Pitea Havsbad*, pages 121–144, Berlin, 1982. University of Umea, Sweden, Springer Verlag. Lecture notes in Math. Series, Number= 973.

[24] Y. Saad. Projection and deflation methods for partial pole assignment in linear state feedback. *IEEE Trans. Aut. Cont.*, 33:290–287, 1988.

[25] Y. Saad. Numerical solution of large lyapunov equations. Technical Report 89-20, RIACS, Ms 230-5, NASA Ames, Moffett Field, CA 94035, 1989. Also in these proceedings.

[26] Y. Saad and M. H. Schultz. Conjugate gradient-like algorithms for solving nonsymmetric linear systems. *Mathematics of Computation*, 44(170):417–424, 1985.

[27] Y. Saad and M. H. Schultz. GMRES: a generalized minimal residual algorithm for solving nonsymmetric linear systems. *SIAM J. Sci. Statist. Comput.*, 7:856–869, 1986.

[28] P. Sonneveld. CGS, a fast Lanczos-type solver for nonsymmetric linear systems. *SIAM J. Scient. Statist. Comput.*, 10(1):36–52, 1989.

Numerical Linear Algebra on Shared Memory Multiprocessors

A.H. Sameh and H.A.G. Wijshoff

Abstract

The development of parallel computers has had a significant impact on the design of algorithms in numerical linear algebra. For efficient utilization of these machines one needs to exploit parallelism and the vector capability of each processor. Most importantly one needs to manage the memory system properly. In particular, for shared memory multiprocessors in which memory is often hierarchically structured, algorithms have to be carefully designed in order to profit from the faster access times of caches, local, or cluster memories compared to that of the global memory. In this paper we present some new and important techniques for designing algorithms which allow both parallelism/vectorization as well as efficient utilization of hierarchical memory systems.

1 Introduction

With the introduction of the first vector processors (Cray 1 and CDC 205) it became apparent that new approaches in the design of algorithms had to be adopted in order to exploit the architectural features of these machines. These approaches consisted mainly of algorithms which exhibit a large degree of regularity, so that in their implementation one need only employ simple DO-loop structures. These DO-loop structures directly translate into vector operations. In order to overcome the vector start-up overhead, the length of these operations had to be made sufficiently large. It is not surprising, therefore, that algorithms for dense matrices and regularly structured sparse matrices can be successfully implemented on these machines so as to attain high computational rates.

As the architectures of these early supercomputers evolved into multiprocessor architectures with complex memory systems (Cray-2,XMP,YMP, the Alliant FX series, and Cedar), algorithm design that relies only on simple DO-loop structures ceases to be effective. In fact, increasing the vector length destroys the data locality of the algorithm. This means that data cannot be kept locally to a single processor or a processor complex, in which case the access times significantly increase. Because of this, the design of algorithms for such architectures involves three, possibly contradictory, goals: enhancing vectorization, parallelization, and data locality.

For an architecture like Cedar, which is a cluster-based organization, different levels of parallelization and data locality can be distinguished. Large grain parallelism is realized by dividing a computation across clusters. Within each cluster medium grain parallelism can be obtained by utilizing each cluster's CE's, and fine grain parallelism is realized from the vector units in each CE. Further, data can reside in the global memory, cluster memory, cluster cache, or vector registers. Thus, for efficient utilization of Cedar, it is necessary that one not only use these three levels of parallelism but also manage its four levels of memory.

In this paper we describe some new and important techniques to realize these goals in the areas of dense and sparse matrix computations. In the next section, we briefly describe the target architecture: Cedar. We present algorithms for dense matrix computations in section 3 as well as strategies used to design and analyze them. Sparse matrix algorithms are treated in section 4. Specifically we present iterative algorithms for structured sparse matrices, and direct solvers for unstructured sparse systems.

2 Cedar Architecture

The Cedar multiprocessor [15] has a hierarchically structured architecture with shared memory at two levels. At the lowest level of the hierarchy are the Computational Elements (CE's) which are pipelined vector processors. The next level of the hierarchy consists of several clusters of CE's. The highest level of the Cedar architecture is formed by coupling several clusters together via a global interconnection network and a global shared memory.

At present, a single Cedar cluster is a slightly modified Alliant FX/8. The Alliant FX/8 combines vector and parallel capabilities with a two-level memory structure (cache and main memory). Each CE (up to eight are possible) is a register Cray-1-like vector processor. A separate concurrency control bus shared by the eight CE's facilitates rapid synchronization. An 8 CE cluster allows a processing rate of up to 68 Mflops including vector startup. The CE's share (through a crossbar switch) a 128(512)-Kbyte write-back cache which allows up to eight simultaneous 64-bit word accesses per 170 ns cycle. The shared cache is connected to the main memory via a bus with a maximum bandwidth of 23 Mwords(64-bit) per second.

All clusters have access to a global shared memory. This global memory is accessed through two unidirectional switching networks. These are packet-switching omega networks built from 8×8 cross-bar switches. They are pipelined and use wide data paths for packet transmission. Buffers are provided in each switching element to increase memory bandwidth.

3 Dense Matrix Computations

When implementing algorithms on parallel computers distinction has to be made between implicit and explicit parallelism. The first form of parallelism can be viewed as the parallelization obtained by program transformations automatically done by a restructuring (autotasking) compiler. As these transformations often do not require any interaction with the user, this kind of parallelism seems to be ideal. However, although restructuring compilers are becoming more and more powerful, their program transforming capabilities are limited. Also, some forms of parallelism will never be detected by restructuring compilers as they require major redesign of the algorithm. A second form of implicit parallelism is obtained by imbedding most vectorization and parallelism in certain primitives which are implemented carefully on a given parallel computer (usually in assembler language). These primitives, in turn, may be called by higher level routines to realize efficient utilization of the machine. Examples of such primitives are the Basic Linear Algebra Subroutines (BLAS) [8,12] which used frequently in scientific computations. Explicit parallelism is obtained by the user and invariably involves a major redesign of the algorithm.

3.1 BLAS3

The performance of BLAS1(vector-vector operations) [6] and BLAS2 primitives(matrix-vector operations) [8] on hierarchical structured memory architectures, specifically the

Cedar architecture, is limited to only a fraction of the peak performance (on one cluster of the Cedar architecture only up to 10-25 Mflops is achieved). This is caused by the fact that cache/local memory cannot be exploited by these primitives. Higher level BLAS routines had to be defined which would allow better data locality. These higher level BLAS routines (BLAS3) were found to be matrix-matrix operations [12]. These routines are now being part of the CSRD scientific library. Recently in the design of LAPACK [7,9] these higher BLAS3 routines are being employed to obtain a better implementation of LINPACK and EISPACK.

In order to illustrate the BLAS3 primitives we consider the basic matrix multiplication and add algorithm: $C = C + A * B$, where A is an $n_1 \times n_2$ matrix, B is an $n_2 \times n_3$ matrix, and C is an $n_1 \times n_3$ matrix. In order to exploit data locality the matrices A, B and C have to be partitioned into blocks. Let A, B and C be partitioned into submatrices of dimensions m_1 by m_2, m_2 by m_3, and m_1 by m_3, respectively. Then the algorithm for obtaining C is basically:

do $i = 1, k_1$
  do $k = 1, k_2$
    do $j = 1, k_3$
      $C_{ij} = C_{ij} + A_{ik} * B_{kj}$
end do

where $n_1 = k_1 m_1, n_2 = k_2 m_2$ and $n_3 = k_3 m_3$ with k_1, k_2 and k_3 being integers greater than 1. The block operations in the innermost loop can be done in parallel constituting large grain parallelism. Each of these block operations, in turn, contains a reasonable amount of potential parallelism. This is achieved by slicing each block into horizontal slices such that each resulting column in each slice of the C block can be accumulated by using an m_2-adic instruction. Hence the optimization of this primitive on one Cedar cluster requires that m_1 be a multiple of the length of the vector register (32), m_2 be large enough so that a significant fraction of the peak performance of the m_2-adic instruction is achieved, and m_3 be either large or a multiple of the number of CE's.

The choice of the actual parameters m_1, m_2 and m_3 also has to take into account the exploitation of the cache/local memory. This leads to an optimization problem. In [12] a methodology for obtaining the actual parameters is described. It is shown that, under reasonable assumptions including $n_1 > \sqrt{CS}$ (CS is the cache size in 64-bit words, respectively the size of local memory), the parameters have to fulfill the following conditions:
(1) If $n_2(p + n_1) \leq CS$ then $m_1 = n_1$ and $m_2 = n_2$.
(2) If $n_2(p + n_1) \geq CS$ then $m_2 = \min(n_2, CS(\sqrt{CS} + p)^{-1})$ and $m_1 = (CS/m_2) - p$.

For the Alliant FX/8, this primitive can achieve a performance of around 55 Mflops which is a dramatic improvement over the performance of BLAS1 and BLAS2. For more details the reader is referred to [12]. In the remainder of this section we describe how BLAS3 primitives can be used to obtain high performance in two examples of dense matrix computations: a block LU decomposition and a block orthogonal factorization scheme. Other algorithms where BLAS3 are useful are reductions to upper-Hessenberg and tridiagonals forms via similarity orthogonal transformations.

3.2 Block LU Decomposition

A linear system of equations $Ax = b$ can be solved by decomposing the $n \times n$ matrix A into a product of a lower triangular matrix L and an upper triangular matrix U. In order to utilize BLAS3 primitives a block decomposition has to be used instead of the classical LU decomposition which only requires BLAS1 or BLAS2 routines [6,8]. This algorithm decomposes A into the product of a block lower triangular matrix L_k and a block upper triangular matrix U_k. Let A be diagonally dominant then A can be written as the following product:

$$A = \begin{pmatrix} A_{11} & A_{12} \\ A_{21} & A_{22} \end{pmatrix} = \begin{pmatrix} I & 0 \\ L_{21} & I \end{pmatrix} \begin{pmatrix} U_{11} & U_{12} \\ 0 & B \end{pmatrix},$$

where A is square and of the order of k. The block LU algorithm is given by:
(1) $A_{11} \leftarrow A_{11}^{-1}$
(2) $A_{21} \leftarrow L_{21} = A_{21}A_{11}$
(3) $A_{22} \leftarrow B = A_{22} - L_{21}A_{12}$
(4) Proceed recursively on the matrix B.
If A is positive definite the resulting decomposition is a block LDL^T. Since A is assumed to be diagonally dominant, explicit inversion of the diagonal blocks can be obtained via the Gauss-Jordan algorithm [17] or an LU decomposition without pivoting. In case that A is not diagonally dominant, partial pivoting is easily introduced in a manner similar to that used in the block algorithms which produce the standard LU factorization, e.g. [3]. This block LU decomposition achieves around 45 Mflops, i.e a speed improvement over the classical LU decomposition by a factor of 3-5.5 (depending on the BLAS2 primitives used for the classical LU) on the Alliant FX/8.

3.3 A Block Householder Scheme

In solving the linear least squares problem:

$$\min ||f - Ax||_2,$$

where A is an $m \times n$ matrix of rank n, $(m > n)$, one needs to obtain the factorization,

$$QA = \begin{bmatrix} R \\ 0 \end{bmatrix},$$

in which Q is an orthogonal matrix and R is a nonsingular upper triangular matrix of order n. This factorization can be obtained by elementary reflectors see [19]. The Householder transformation is an orthogonal matrix P of the form $P = I - uu^T$, where $u \in \mathbf{R}^m$ and $u^T u = 2$. If $x \in \mathbf{R}^m$, then Px is a vector in the orthogonal complement of the set $\{u\}$. In particular P can be constructed so that Px is a multiple of e_1, the first column of the identity matrix. Bischof and Van Loan [1] proposed a block scheme which depends on the fact that the product of k elementary reflectors $Q_k = (P_k, \ldots, P_2, P_1)$, with $P_i = I_m - u_i u_i^T$, can be expressed as a rank-k update of the identity of order m, i.e.

$$Q_k = I_m - V_k U_k^T,$$

where $V_1 = U_1 = u_1, V_j = (P_j V_{j-1}, u_j)$ and $U_j = (U_{j-1}, u_j)$, for $j = 2, 3, \ldots, k$.

The block algorithm may be described as follows. Write the matrix A as $[A_1, B]$, where A_1 consists of the first k columns of A. Proceed with the usual Householder reduction scheme by generating the k elementary reflectors P_1 through P_k, such that

$$(P_k \ldots P_2 P_1)A_1 = \begin{bmatrix} R_1 \\ 0 \end{bmatrix},$$

where R_1 is upper triangular of order k, without modifying the matrix B. If the product $Q_k = P_k \ldots P_2 P_1 = I - V_k U_k^T$ is accumulated as each P_i is generated, the matrix B is updated via

$$B \leftarrow (I - V_k U_k^T)B.$$

Note that this later operation is just a BLAS3 operation which can be done very efficiently. The process is then repeated on the modified B with another "well-chosen" blocksize, and so on until the factorization is completed. In [1] it is shown that this block scheme is as numerically stable as the classical Householder scheme. The block scheme, however, requires roughly $(1+2/p)$ times the arithmetic operations needed by the classical sequential algorithm, where p is the number of blocks. This extra operation count is nullified by the performance achieved. On the Alliant FX/8 the block scheme yields a performance of up to 38 Mflops for matrices of order 1024×1000, whereas the classical Householder scheme achieves a maximum computational rate of onlyo 10 Mflops.

4 Sparse Matrix Computations

Sparse matrix computations are characterized by several features: (i) complex data handling, (ii) irregular data streams, (iii) indirect addressing, and (iv) a low ratio of arithmetic operations to data element references. The first feature is caused by the fact that sparse matrices are stored in a condensed format in order to minimize the storage requirements, which leads to substantial overhead. The second leads to lower effective memory bandwidth, the third complicates vectorization and parallelization of the sparse matrix codes, and the fourth can lead to excessive data movement and loss of data locality. All of the above can lead to much lower achievable computational rates compared to those for dense matrix computations.

4.1 Structured Sparse Problems

Structured sparse matrices arise mainly from the numerical handling of partial differential equations by either finite element or finite differences techniques. These matrices differ from general sparse matrices in the sense that mostly a diagonal storage scheme can be used, which can be exploited to obtain longer vector operations. Also irregular data streams do not occur as frequently as for general sparse matrices which alleviates the constraints on the effective memory bandwidth. Further, complex data manipulations are not needed, thus minimizing overhead. The exploitation of data locality, however, still remains as a major problem.

4.1.1 A Preconditioned Conjugate Gradient Scheme

The preconditioned conjugate gradient method [21] has become one of the most effective algorithms for handling linear systems which arise from the 5-point finite difference discretization of second-order self-adjoint problems. Consider the problem

$$\frac{-\partial}{\partial x}(a(x,y)\frac{\partial x}{\partial y}) - \frac{\partial}{\partial y}(b(x,y)\frac{\partial x}{\partial y}) + c(x,y)u = g(x,y) \tag{1}$$

on the unit square with Dirichlet boundary conditions, where $a, b \geq 0$, and $c \geq 0$. Using a uniform mesh of width $h = 1/(n+1)$, n-even, with point red-black ordering we obtain (after diagonal scaling) a linear system of the form:

$$\begin{bmatrix} I & H \\ H^T & I \end{bmatrix} \begin{bmatrix} x_R \\ x_B \end{bmatrix} = \begin{bmatrix} f_R \\ f_B \end{bmatrix}, \tag{2}$$

where $H = [\Gamma_i, H_i, \Omega_i]$ is block tridiagonal of order $n^2/2$ in which Γ_i and Ω_i are diagonal matrices, and H_i is upper bidiagonal for i-even and lower bidiagonal for i-odd. The vector x_B can now be obtained by solving the reduced system,

$$(I - H^T H)x_B = (f_B - H^T f_R), \tag{3}$$

using the conjugate gradient method with the diagonal of $(I - H^T H)$ as preconditioner. So, the basic computational task for solving this reduced system are:

1. inner-products: $\alpha = x^T y$,
2. scalar*vector + vector: $z = \alpha x + y$,
3. matrix*vector: $q = (I - H^T H)p$,
4. solving $Ms = r$, where M is the preconditioner.

Since M is the diagonal of $(I - H^T H)$, step 3 consumes the highest percentage of time per iterations. On the Alliant FX/8 this reduced system approach achieves a performance of 18 Mflops for n around 32 but drops sharply to approximately 8 Mflops for larger values of n. This is caused by the fact that data locality cannot be exploited for the matrix-vector multiply, so that for larger values of n the matrix $(I - H^T H)$ does not reside in cache but in memory. In spite of this behavior, this PCG scheme requires the least time on the Allinat FX/8 for achieving a given level of the residual norm compared to other preconditioned schemes such as point [21] or block [4] incomplete factorization, or polynomial preconditioners, [16] and [18], on systems resulting from natural ordering of the grid points.

4.1.2 The Block-Kaczmarz Algorithm

The accelerated block-SSOR scheme as described in [2] solves large block-tridiagonal nonsymmetric linear systems of equations. These linear systems arise from the finite element or finite difference discretization of elliptic partial differential equations [14]. This method is ideally suited for multiprocessors and can be used when the the matrix of coefficients has unknown eigenvalue distribution, its symmetric part is indefinite, or when its eigenvalues are known to lie on both sides of the imaginary axis. The block-SSOR scheme is a projection method that is obtained by applying the SSOR method to the normal equations, with either block row or block column partitioning of the matrix. By a suitable permutation of the block rows the system may be partitioned as

$$\begin{bmatrix} B_1^T \\ B_2^T \end{bmatrix} x = \begin{bmatrix} f_1 \\ f_2 \end{bmatrix},$$

so that the k-th iteration of the block-row SSOR method is given by

$$\begin{aligned} z_1 &= x_k \\ z_2 &= z_1 + \omega(g_1 - P_1 z_1) \\ z_3 &= z_2 + \omega(g_2 - P_2 z_2) \\ z_4 &= z_3 + \omega(g_2 - P_2 z_3) \\ z_5 &= z_4 + \omega(g_1 - P_1 z_4) \\ x_{k+1} &= z_5 \end{aligned}$$

where $g_i = B_i(B_i^T B_i)^{-1} f_i = B^+ f_i, i = 1, 2$ and $P_i, i = 1, 2$ is the projector onto the space spanned by the columns of B_i. Here the permutation is chosen such that each B_i is a block diagonal matrix. Thus the linear least squares problem in the iteration above actually consists of several smaller independent linear least squares problems. This iteration process converges for $0 < \omega < 2$, irrespective of the eigenvalue distribution of the matrix. For this particular partitioning it can be shown that the optimal value for ω is 1.0, which simplifies considerably the iterative scheme. This results in solving the linear system

$$(I - Q) = Rf,$$

where $Q = Q_1Q_2Q_1, Q_i = (I - P_i), i = 1, 2$, and $R = [(I + Q_1Q_2)(B_1^+)^T, Q_1(Q_2 + I)(B_2^+)^T]$ via the iteration

$$x_{k+1} = Qx_k + Rf.$$

The matrix $(I - Q)$ is symmetric positive definite and thus ideally suited for acceleration by the conjugate gradient method. Numerous experimental results indicate that this scheme is more robust than Krylov-based subspace methods and even more efficient on multiprocessors especially for three-dimensional problems.

4.2 Direct Solvers for Unstructured Sparse Systems

Implementations of direct solvers for general sparse matrices on vector/parallel architectures mostly perform very poorly. This is mainly caused by the fact that the exploitation of parallelism and vectorization in such codes is strongly dependent on the sparsity structure of the matrix A. In the case that the sparsity pattern of the matrix A is arbitrary, it is almost impossible to exploit any parallelism/vectorization a priori. There are essentially three different techniques for generating parallelism in direct solvers for general linear systems of equations. The first technique exploits parallelism at the level of the rank-1 update. Secondly multiple rank-1 updates can be performed in parallel. This mostly requires a search for a set of diagonal or triangular pivots. Thirdly a global ordering (tearing technique) can be used to decompose the sparse matrix into blocks which can be factored in parallel. In this section we describe two solvers, D2 (based on a parallel pivot search), and PSSS (based on a global ordering), in more detail.

4.2.1 The D2 Solver

This solver [5] is based on a parallel pivot search that finds at each stage of the factorization of the sparse matrix A a compatible pivot set S of size m, followed by a parallel rank-m update. A set of pivots is called compatible at a certain stage of the factorization if the pivots are not connected with each other in the associated graph of the matrix. Secondly the pivots have to satisfy a Markowitz cost criterion.

The algorithm for finding a set of compatible pivots essentially works as follows. First, a single stable pivot with the lowest Markowitz cost (LM) is found. All rows with nonzeros in the pivot column and all columns with nonzeros in the pivot row are then disabled. No other pivots can be considered from disabled rows or columns. Thereafter each partially independent task searches a single row for a stable pivot with lowest cost in the row. Elements in disabled columns are ignored. As an additional fill-in heuristic, the tasks are executed roughly in order of increasing length.

If the best pivot in the row has Markowitz cost $\leq k.LM$, for some factor k, then this is a potential pivot. At this point, however, two (or more) tasks could have potential pivots that are compatible with current set but incompatible with each other. To exclude one or the other, the pivot set is locked. If the pivot row and column are not disabled, the pivot is added to S and all rows with nonzeros in the pivot column and all columns with nonzeros in the pivot row are disabled. S is then unlocked. If the task is successful, it stores the pivot row in U and the structure of the pivot column in L.

The rank-m update is divided into two phases, each consisting of a set of completely independent tasks. The first phase updates the columns of A and the second updates the rows of A. A row update task first finds all the nonzero elements in the row that are in pivot columns. It then scatters all corresponding pivot rows and forms their sum, stores the multipliers in L, gathers the updates into the row, and then gathers the fill-in caused by each pivot row. Memory is allocated from a global free list, divided into length vl records. The column update task is similar but simpler since no numerical computations are done.

4.2.2 The PSSS Solver

This solver is based on finding large grain parallelism such that the factorization of a general sparse matrix A can be divided up into partitions which allow the clusters or processors to work in parallel on the problem. (On Cedar, finer granularity parallelism can be exploited within a cluster.) Large grain parallelism is obtained by ordering the matrix by a hybrid ordering **H*** which is related to tearing techniques for nonsymmetric matrices [10]. Tearing techniques, however, are based mostly on nonsymmetric orderings (column and row interchanges). The **H*** algorithm differs in that it uses an initial nonsymmetric ordering (**H0**) and subsequent symmetric orderings to achieve the desired form. The **H*** ordering produces a bordered block upper-triangular form. The ordering is a hybrid form combining *Tarjan's algorithm* [20], the **H1** ordering which is derived from *Tarjan's algorithm*, and the **H2** ordering which is based on a modified form of *Nested Dissection* [11] that exploits the nonsymmetric structure of the matrix. Additionally, **H*** tries to enhance the stability of the pivoting strategies used in the factorization algorithm. This is accomplished via three techniques. First, the **H0** phase generates a transversal such that the element on the diagonal for each column is within a bound of the largest element within the column. In the **H1** and **H2** phases, two techniques are used which monitor the size of the elements placed in the border.

The factorization and solution of the partitioned system can exploit the above ordering in several ways. These include modifications to: a standard sparse LU decomposition, a block LU decomposition, a decomposition combined with a low rank update such as the Woodbury formula [13], and iterative methods which may use the decomposition as a preconditioner. For the moment only attention is paid to the modifications of the standard sparse LU decomposition to exploit the form effectively and generate an accurate solution.

The multicluster implementation of the solver first performs the **H*** ordering on a single cluster and then distributes the matrix to the other clusters and places certain portions of the border in the global memory to be shared by the clusters. The clusters then compute the LU decomposition of each diagonal block. This factorization can be done using dense techniques suitable for the cluster's architecture (BLAS3-based partial pivoting) or a parallel sparse factorization routine which exploits finer grain parallelism. Of course, the diagonal blocks which result from **H*** may not be nonsingular or well-conditioned. This is handled by either artificially forcing nonsingularity or by dynamically redefining the border during the factorization. In the latter technique the unknowns which cause difficulty are cast into the border and eliminated at a later stage. After these factorizations are completed the off-diagonal entries in the upper triangular part of the matrix are updated. The entries in the border are then updated. This update may exploit the hierarchical structure of the border depending upon its size. Finally, the diagonal block coupling the border elements is factored. This is typically dense but may be performed on a single cluster or multiple clusters depending on its size. The forward solve is partitioned naturally as a result of the factorization. However, during the factorization phase the U matrix is redistributed to allow an efficient backward solve (which may be done repeatedly if coupled with an iterative method).

This work was supported in part by the National Science Foundation under Grants No. US NSF DCR84-10110 and US NSF DCR85-00970, by the US Department of Energy under Grant No. US DOE DEFG01-85ER25001, the Air Force Office of Scientific Research under Grant No. AFOSR-85-0211, and the IBM Donation.

References

[1] C. Bischof and C. Van Loan. The WY representation for products of Householder matrices. *SIAM J. Sci. Stat. Comput.*, 8(2):s2–s13, 1987.

[2] R. Bramley and A.H. Sameh. A robust parallel solver for block tridiagonal systems. In *Proc. 1988 Int. Conf. on Supercomp.*, ACM Press, pp. 39-54 1988.

[3] D. Calahan. Block-oriented local-memory-based linear equation solution on the Cray-2: uniprocessor algorithms. In *Proc. Intl. Conf. Par. Processing*, pp. 375–378, IEEE Computer Society Press, 1986.

[4] P. Concus, G.H. Golub, and G. Meurant. Block preconditioning for the conjugate gradient method. *SIAM J. Sci. Stat. Comput.*, 6:220–252, 1985.

[5] T. Davis. Ph.D. thesis (in preparation). University of Illinois at Urbana-Champaign.

[6] J. Dongarra, J. Bunch, C. Moler, and G. W. Stewart. *LINPACK User's Guide.* SIAM Publications, Philadelphia, 1979.

[7] J. Dongarra, J. Du Croz, I. Duff, and S. Hammarling. *A proposal for a set of Level 3 basic linear algebra subprograms.* Technical Report TM-88, Math. and Comp. Sci. Div., Argonne National Laboratory, Argonne, IL, 1987.

[8] J. Dongarra, J. Du Croz, S. Hammarling, and R. Hanson. *A proposal for an extended set of Fortran basic linear algebra subprograms.* Technical Report TM-41, Math. and Comp. Sci. Div., Argonne National Laboratory, Argonne, IL, 1984.

[9] J. Demmel, J. Dongarra, J. Du Croz, A. Greenbaum, S. Hammarling, and D. Sorensen. *Prospectus for the development of a linear algebra library for high-performance computers.* Technical Report TM-97, Math. and Comp. Sci. Div., Argonne National Laboratory, Argonne, IL, 1987.

[10] I. Duff, A. Erisman, J. Reid. *Direct Methods for Sparse Matrices.* Clarendon Press, Oxford, 1986.

[11] A. George Nested dissection of a regular finite-element mesh. *SIAM J. Numer. Anal.*, 10:345–363, 1973.

[12] K. Gallivan, W. Jalby, U. Meier, and A. Sameh. Impact of hierarchical memory systems on linear algebra algorithm design. *Intl. J. Supercomputer Appl.*, 2(1):12–48, 1988. Presented at the Level 3 BLAS Workshop, Argonne National Laboratory, January 1987.

[13] G. Golub and C. Van Loan. *Matrix Computations.* The Johns Hopkins University Press, Baltimore, 1983.

[14] C. Kamath and A. Sameh. *A projection method for solving nonsymmetric linear systems on multiprocessors. Parallel Computing*, 9(3):291–312, 1989.

[15] D. Kuck, E. Davidson, D. Lawrie, A. Sameh. *Parallel supercomputing today and the Cedar approach* in Experimental Parallel Computing Architectures, J. Dongarra ed. pp. 1-20, North-Holland, New York, 1987.

[16] G. Meurant. The block preconditioned conjugate gradient method on vector computers. *BIT*, 24:623–633, 1984.

[17] G. Peters and J. Wilkinson. On the stability of Gauss-Jordan elimination with pivoting. *CACM*, 18:20–24, 1975.

[18] Y. Saad. Practical use of polynomial preconditioning for the conjugate gradient method. *SIAM J. Sci. Stat. Comput.*, 6:865–881, 1985.

[19] A. Sameh. Solving the linear least squares problem on a linear array of processors. In *Algorithmically-Specialized Parallel Computers*, pp. 191–200, Academic Press, West Lafayette, IN., 1985. Proc. 1982 Purdue Univ. Workshop.

[20] R.E. Tarjan. Depth first search and linear graph algorithms. *SIAM J. Comput.*, 1:146–160, 1972.

[21] H. van der Vorst. A vectorizable variant of some ICCG methods. *SIAM J. Sci. Stat. Comput.*, 3:350–356, 1982.

New Decompositions of the Inverse of a Toeplitz Matrix

Gregory Ammar Paul Gader

Abstract.

The Gohberg-Semencul formula expresses the inverse of a Toeplitz matrix as the difference of products of lower triangular and upper triangular Toeplitz matrices. After this decomposition is computed, the solution of the n-by-n Toeplitz system of equations $Mx = b$ can be obtained in $O(n \log n)$ arithmetic operations using fast Fourier transform techniques. In this paper the idea of cyclic displacement structure is used to show that the upper triangular factors in the Gohberg-Semencul formula can be replaced by circulant matrices. This factorization of the inverse of M leads to a computational savings of more than 35 per cent of the work required by the Gohberg-Semencul formula in computing $M^{-1}b$. This formula also leads to another formula that provides a computational savings of over 42 per cent.

1. Introduction.

In 1972, Gohberg and Semencul [7] presented formulas that provide a decomposition of the inverse of a Toeplitz matrix into the difference of products of lower triangular and upper triangular Toeplitz matrices (see, e.g., [6],[9]). These formulas have been and continue to be useful in both theoretical and computational issues involving Toeplitz matrices. In particular, these formulas play an important role in several algorithms for the solution of Toeplitz systems of equations.

Ideas involving the displacement representation of a matrix can be used to generalize the Gohberg-Semencul formula [11], [4]; in particular, any square matrix can be written as the sum of products of lower triangular and upper triangular Toeplitz matrices. If the number of terms in this sum is small, efficient algorithms, analogous with algorithms for Toeplitz matrices, can be developed.

In this paper the idea of cyclic displacement is used to derive formulas analogous with the Gohberg-Semencul formula. These new formulas involve circulant and skew-circulant matrices, and can provide for computationally more efficient methods for the calculation of matrix-vector products. Toeplitz inversion formulas involving circulant matrices have also been presented by Lerer and Tismenetsky [13].

Let

$$M = M_n = \begin{bmatrix} \mu_0 & \mu_{-1} & \mu_{-2} & \cdots & \mu_{1-n} \\ \mu_1 & \mu_0 & \mu_{-1} & & \mu_{2-n} \\ \mu_2 & \mu_1 & \mu_0 & \ddots & \vdots \\ \vdots & \vdots & \ddots & \ddots & \mu_{-1} \\ \mu_{n-1} & \mu_{n-2} & \cdots & \mu_1 & \mu_0 \end{bmatrix} = [\mu_{i-j}]_{i,j=0}^{n-1}$$

be a Hermitian ($\mu_{-j} \equiv \overline{\mu}_j$) positive definite Toeplitz matrix of order n. Several efficient algorithms have been presented for solving the Toeplitz system of equations $M x = b$ (see, e.g., [14], [8], [2] and the references therein). Many of these algorithms rely on the fact that M^{-1} is determined by its last column. In fact, several algorithms for the solution of $M x = b$ can be described as a two-phase process:

Phase one: The computation of the last column of M^{-1}.

Phase two: The computation of $M^{-1} b$.

Algorithms for phase one include $O(n^2)$ algorithms such as the Levinson-Durbin algorithm, as well as $O(n \log^2 n)$ methods such as the generalized Schur algorithm (see [2], [3]).

The Gohberg-Semencul formula provides an explicit representation of M^{-1} in terms of its last column. Specifically, let $e_0, e_1, \ldots, e_{n-1}$ be the columns of the identity matrix I of order n, and define $\delta_{n-1} > 0$ and $r = [\rho_j]_{j=0}^{n-1}$ by $\rho_{n-1} = 1$ and $\delta_{n-1} M^{-1} e_{n-1} = r$. Let $L(x)$ denote the lower triangular Toeplitz matrix whose first column is $x \in \mathbb{C}^n$, and let $Z = L(e_1)$ and $J = [e_{n-1}, e_{n-2}, \ldots, e_0]$ denote the *downshift matrix* and the *reversal matrix*, respectively. The *Gohberg-Semencul formula* is then given as follows.

Proposition 1. $\delta_{n-1} M^{-1} = L(r_1) L(r_1)^* - L(r_0) L(r_0)^*$ *, where* $r_0 = Z r$ *and* $r_1 = J \overline{r}$.
□

This formula is, in fact, a manifestation of the Christoffel-Darboux formula for Szegö polynomials. See, for example, [12], [2] for more details.

A major computational use of the Gohberg-Semencul formula is in the solution phase of algorithms for solving Toeplitz systems of equations. For sufficiently large n, the matrix-vector product $M^{-1} b$ can be efficiently computed from the Gohberg-Semencul formula using fast Fourier transform (FFT) techniques [10].

2. Cyclic Displacements.

Unless stated otherwise, all matrices are in $\mathbb{C}^{n\times n}$. We let A^* and x^* denote the complex conjugate transpose of the matrix A and the (column) vector x, respectively.

The *displacement* of a matrix A is given by $\mathbf{D}_0(A) = A - Z A Z^T$, and $\alpha = \text{rank}(\mathbf{D}_0(A))$ is the *displacement rank* of A. Proposition 1 summarizes some fundamental aspects of displacement.

Proposition 2. [11] *Any square matrix A is uniquely determined by its displacement. Furthermore,*

$$\mathbf{D}_0(A) = \sum_{m=1}^{\alpha} x_m y_m^* \quad \textit{if and only if} \quad A = \sum_{m=1}^{\alpha} L(x_m) L(y_m)^* ,$$

where $x_m, y_m \in \mathbb{C}^n$. □

Thus, any square matrix A can be written as the sum of products of lower triangular and upper triangular Toeplitz matrices, and the displacement rank of A is the minimum number of terms needed in the summation. Toeplitz matrices fit nicely into this more general framework. In particular, a Toeplitz matrix and its inverse have displacement rank $\alpha \le 2$, and the Gohberg-Semencul formula is equivalent to the identity

$$\mathbf{D}_0(M^{-1}) = \frac{1}{\delta_{n-1}}(r_1 r_1^* - r_0 r_0^*) . \tag{1}$$

It is noted in [11] that other displacement techniques can be considered by replacing the matrix Z with another matrix. For example, let $C(x)$ denote the circulant matrix whose first column is $x \in \mathbb{C}^n$, and let $E = C(e_1)$ denote the *cyclic downshift matrix*. The *cyclic displacement* of the matrix A is then defined by $\mathbf{D}_1(A) = A - E A E^*$. Aspects of cyclic displacement and displacements based on other group matrices are developed in [5]. One important difference between these displacements and the standard displacement is that a matrix is not uniquely determined by a finite-group-based displacement. In the case of cyclic displacements, we have the following proposition.

Proposition 3.

(1) $\mathbf{D}_1(A) = \mathbf{D}_1(B)$ *if and only if* $A - B$ *is a circulant matrix.*

(2) A square matrix A is uniquely determined by its cyclic displacement and any one of its rows or columns.

Proof. Since $\mathbf{D}_1(A) = [\alpha_{jk} - \alpha_{j-1,k-1}]_{j,k=0}^{n-1}$ (where the indices are taken modulo n),

$\mathbf{D}_1(A)$ provides a homogeneous relationship between the entries of each cyclic diagonal of A. (The kth cyclic diagonal of A has entries $\alpha_{j+k,k}$, $0 \le k < n$.) Statement (1) then follows from the fact that $\mathbf{D}_1(A) = 0$ if and only if the entries of A are constant along the cyclic diagonals; that is, if and only if A is a circulant matrix. Also note that A is determined by $\mathbf{D}_1(A)$ together with one element from each cyclic diagonal of A, and (2) follows from the fact that any row or column of A contains exactly one element from each cyclic diagonal of A. □

More specifically, we have the following result, which is a slight generalization of some results in [5].

Proposition 4. *If the cyclic displacement of A is given as the sum $\mathbf{D}_1(A) = \sum_{m=1}^{\alpha} x_m y_m^*$, where $\alpha = \mathrm{rank}(\mathbf{D}_1(A)) > 0$ then:*

(1) The vectors x_m, y_m satisfy the homogeneity relation $\sum_{m=1}^{\alpha} C(x_m) C^*(y_m) = 0$.

(2) *$A = C_l + \sum_{m=1}^{\alpha} L(x_m) C(y_m)^*$, where C_l is the circulant matrix with the same last row as that of A.*

(3) $A = C_f + \sum_{m=1}^{\alpha} C(x_m) L(y_m - e_0^* y_m e_0)^*$, where C_f is the circulant matrix with the same first column as that of A.

Proof. Since $E^n = I$, we have

$$0 = \sum_{l=0}^{n-1} E^l (A - EAE^*)(E^l)^* = \sum_{l=0}^{n-1} \sum_{m=1}^{\alpha} (E^l x_m)(E^l y_m).$$

Let $B = [\beta_{jk}] = \sum_{l=0}^{n-1} (E^l x)(E^l y)^*$. Then for $0 \le j,k < n$, we have $\beta_{j,k} = \sum_{l=0}^{n-1} \xi_{j-l} \eta_{k-l}$ (The indices are taken modulo n throughout this proof.) Consequently, $B = C(x) C^*(y)$, and the homogeneity relation follows.

Now let $B = [\beta_{j,k}]_{j,k=0}^{n-1} = \mathbf{D}_1(A)$. Then for any $k > 0$, we have

$$\alpha_{j,k} = \alpha_{j-k,0} + \sum_{l=1}^{k} \beta_{j-k+l,l} .$$

If $B = xy^*$, then $\sum_{l=1}^{k} \beta_{j-k+l,l} = \sum_{l=1}^{k} \xi_{j-k+l} \bar{\eta}_l$, so that

$$B = \begin{bmatrix} \xi_0 & \xi_{n-1} & \cdots & \xi_1 \\ \xi_1 & \xi_0 & \ddots & \vdots \\ \vdots & \ddots & \ddots & \xi_{n-1} \\ \xi_{n-1} & \cdots & \xi_1 & \xi_0 \end{bmatrix} \begin{bmatrix} 0 & \bar{y}_1 & \cdots & \bar{y}_{n-1} \\ & 0 & \ddots & \vdots \\ & & \ddots & \bar{y}_1 \\ & & & 0 \end{bmatrix};$$

that is, $C(x) L(y - e_0^* y e_0)^*$. Statement (3) then follows from a similar argument applied to $B = \sum_m x_m y_m^*$. The proof of (2) is analogous. □

Observe that the homogeneity relation implies that $\sum_m C(x_m)L(y_m) = -\sum_m C(x_m)L(y_m')^*$, where $y' = [\eta_j']_{j=0}^{n-1}$ with $\eta_0' = 0$ and $\eta_j' = \overline{\eta}_{n-j}$ for $0 < j < n$. Thus, the lower triangular Toeplitz matrices in Proposition 4 can be replaced by upper triangular Toeplitz matrices.

Since $E = Z + e_0 e_{n-1}^*$, there is a direct relationship between standard displacements and cyclic displacements. The following proposition, which is easily verified, provides one description of this relationship.

Proposition 5. *Let $x = A\, e_{n-1}$ and $y^* = e_{n-1}^* A$ be the last column and last row of the square matrix A, respectively. Then*

$$\mathbf{D_1}(A) = \mathbf{D_0}(A) - e_0(E\, y)^* - Z\, x\, e_0^* \; . \; \square$$

It is well known that the matrix-vector product $C(x)b$ can be computed using three fast Fourier transforms (FFTs) of order n. The product $L(x)b$ is computed using three FFTs of order $2n$ by embedding $L(x)$ into a circulant matrix of order $2n$. Since an FFT requires $O(n \log n)$ arithmetic operations, we see that the circulant-vector multiplication requires roughly half the computation of a Toeplitz matrix-vector multiplication. Consequently, cyclic displacement-based decompositions can be more efficient than displacement-based decompositions in the computation of matrix-vector products of the form $A\, b$ for sufficiently large n. (For small values of n direct multiplication of $L(x)b$ can be more efficient than using FFT techniques.) Since the efficient matrix-vector multiplication provided by a displacement representation often plays an important role in displacement-based algorithms, cyclic displacement formulas could be used to improve the efficiency of such algorithms. In the next section we consider the use of cyclic displacement representations in the derivation of Toeplitz inversion formulas.

3. New Toeplitz Inversion Formulas.

Let M be a Hermitian positive definite Toeplitz matrix, define r, r_0, r_1 and δ_{n-1} as in Proposition 1, and let $A = \delta_{n-1} M^{-1}$. Then $ZAe_{n-1} = r_0$ and $EAe_{n-1} = r_0 + e_0$ so (1) and Proposition 5 imply

$$\mathbf{D_1}(A) = r_1 r_1^* - r_0 r_0^* - e_0(e_0 + r_0)^* - r_0 e_0^*$$

Statement (2) of Proposition 4 then yields

$$A = L(r_1)C(r_1)^* - L(r_0)C(r_0)^* - C(r_1) - L(r_0) + C(r_1)$$

because $C(e_0+r_0)^* = C(r_1)$, $L(e_0) = C(e_0) = I$, and $C(r_1)$ is the circulant matrix with the same last row as that of A. Hence,

$$A = L(r_1)C(r_1)^* - L(r_0)(C(r_0)+I)^*$$

and so we have the following decomposition of A.

Proposition 6. *Let r_0 and r_1 be defined as in Proposition 1. Then*

$$\delta_{n-1}M^{-1} = L(r_1)C(r_1)^* - L(r_0)C(r_1) \quad \Box$$

This formula represents the circulant variant of the Gohberg-Semencul formula presented in [1].

Proposition 6 leads to other formulas for the inverse of a Toeplitz matrix as follows. A *factor circulant matrix with factor* τ, or more simply, a τ-circulant is a matrix of the form

$$C_\tau(x) = \begin{bmatrix} \xi_0 & \tau\xi_n & \cdots & \tau\xi_2 & \tau\xi_1 \\ \xi_1 & \xi_0 & \tau\xi_n & & \tau\xi_2 \\ \xi_2 & \xi_1 & \xi_0 & \ddots & \vdots \\ \vdots & \ddots & \ddots & \ddots & \tau\xi_n \\ \xi_n & \cdots & \xi_2 & \xi_1 & \xi_0 \end{bmatrix}$$

Observe that $L(x) = C_0(x)$ and $C(x) = C_1(x)$. Moreover, it can be shown that if $\tau \neq 0$ then the matrix-vector product $C_\tau(x)b$ can be computed using three FFTs or order n plus $O(n)$ additional computation.

More Toeplitz inversion formulas can be obtained by using Proposition 6 together with the fact that $2L(x) = C_\tau(x) + C_{-\tau}(x)$. In particular, letting $\tau = 1$, and observing the fact that $C_1(x)$ and $C_1^*(y)$ commute, we obtain the following decomposition of $A = \delta_{n-1}M^{-1}$.

Proposition 7.

$$2A = C_{-1}(r_1)C_1^*(r_1) + C_{-1}^*(r_1)C_1(r_1) \quad \Box$$

Let $\phi(n)$ denote the amount of computation required to perform one FFT of order n. If M is a real matrix, then we let $\phi(n)$ denote the the work required by a real FFT, which is roughly half of that of a complex FFT. In [10] a method is described for computing $M^{-1}b$ in $16\phi(n)$ operations (neglecting $O(n)$ terms) using the Gohberg-Semencul formula. It is shown in [1] that the computation using the Gohberg-Semencul formula can be performed using $14\phi(n)$ operations. It is further shown in [1] that $M^{-1}b$ can be obtained using Proposition 6 in $9\phi(n)+O(n)$ operations. Thus, over 35% computational savings is achieved in the second phase of a Toeplitz solver. Similar consideration shows that the decomposition of Proposition 7 provides a method for the computation of $M^{-1}b$ in $8\phi(n)$ operations. Thus, Proposition 7 can be used to obtain a computational savings of over 42 per cent relative to the use of the Gohberg-Semencul formula in the second phase of a Toeplitz solver.

It can be seen that, although the first phase of a Toeplitz solver requires a higher order amount of computation than the second phase, the second phase requires a substantial amount of effort relative to the first phase for moderately sized n [1]. Moreover, in situations in which $M x_j = b_j$ is to be solved for several vectors b_j, the factorization phase of a Toeplitz solver is performed once, and the second phase is performed repeatedly. The value of the improved efficiency of the solution phase afforded by the decompositions of Propositions 6 and 7 will increase dramatically in these situations.

4. Concluding Remarks.

Displacement-based techniques have been valuable in the development of algorithms for Toeplitz matrices and in extending these algorithms to efficient algorithms for matrices of small displacement rank. The crucial idea is the decomposition of a matrix into sums of products of triangular Toeplitz matrices. This provides a condensed representation of the matrix, and provides for efficient matrix-vector multiplication using FFT techniques. The above discussion shows that displacement representations can be used to obtain cyclic displacement representations. The resulting decompositions will then involve circulant matrices, and will allow for the more efficient computation of the matrix-vector product. Such products are used extensively in many algorithms, and in filtering applications, so we expect that cyclic displacement decompositions will lead to improved efficiency in certain aspects of these algorithms.

This research was supported in part by the National Science Foundation under grant DMS-8704196 and by the Institute for Mathematics and Its Applications with funds provided by the National Science Foundation.

References

[1] G. S. Ammar and P. Gader, *A Variant of the Gohberg-Semencul Formula Involving*

Circulant Matrices, SIAM J. Matrix Anal. Appl., submitted for publication.

[2] G. S. Ammar and W. B. Gragg, *The Generalized Schur Algorithm for the Superfast Solution of Toeplitz Systems,* in Rational Approximation and its Applications in Mathematics and Physics (J. Gilewicz, M. Pindor and W. Siemaszko, editors), Lecture Notes in Mathematics 1237, Springer-Verlag, Berlin, 1987, 315-330.

[3] G. S. Ammar and W. B. Gragg, *Superfast Solution of Real Positive Definite Toeplitz Systems,* SIAM J. Matrix Anal. Appl., 9 (1988), 61-76.

[4] B. Friedlander, M. Morf, T. Kailath and L. Ljung, *New Inversion Formulas for Matrices Classified in Terms of Their Distance from Toeplitz Matrices,* Lin. Alg. and Appl., 27 (1979), 31-60.

[5] P. Gader, *Displacement Operator Based Decompositions of Matrices Using Circulants or Other Group Matrices,* Lin. Alg. and Appl., submitted for publication.

[6] I. C. Gohberg and I. A. Fel′dman, Convolution Equations and Projection Methods for their Solution, American Mathematical Society, Providence, 1974.

[7] I. C. Gohberg and A. A. Semencul, *On the Inversion of Finite Toeplitz Matrices and their Continuous Analogs* (in Russian), Mat. Issled 7 (1972), 201-223.

[8] G. H. Golub and C. F. Van Loan, Matrix Computations, Johns Hopkins University Press, Baltimore, 1984.

[9] I. S. Iohvidov, Hankel and Toeplitz Matrices and Forms, Algebraic Theory, Birkhäuser, Boston, 1982.

[10] J. R. Jain, *An Efficient Algorithm for a Large Toeplitz Set of Linear Equations,* IEEE Trans. Acoust. Speech Signal Process., 27 (1979), 612-615.

[11] T. Kailath, S. Y. Kung and M. Morf, *Displacement Ranks of Matrices and Linear Equations,* J. Math. Anal. Appl., 68 (1979), 395-407.

[12] T. Kailath, A. Vieira and M. Morf, *Inverses of Toeplitz Operators, Innovations, and Orthogonal Polynomials,* SIAM Rev., 20 (1978), 106-119.

[13] L. Lerer and M. Tismenetsky, *Generalized Bezoutian and the Inversion Problem for Block Matrices, I. General Scheme,* Integral Equations Operator Theory, 9 (1986), 790-819.

[14] W. Trench, *An Algorithm for the Inversion of Finite Toeplitz Matrices,* SIAM J. Appl. Math., 12 (1964), 515-522.

Estimating Distance to Uncontrollability, Upper & Lower Bounds

Daniel Boley

We apply a Bauer-Fike eigenvalue bound to develop computational methods to obtain upper and lower bounds on the sensitivity to perturbations in the coefficients of the controllability property for linear time-invariant dynamical systems.

1. INTRODUCTION

We develop a numerical upper and lower bound to the distance from a linear time-invariant dynamical system to an uncontrollable system. A note on notation, all norms used in this paper are the vector or matrix 2-norm, as appropriate.

The results in this paper are based on the following result from the perturbation theory of matrix eigenvalues, which we state in the form useful for this paper (see [7] and references therein for a complete statement):

Bauer-Fike Theorem Given a $n\times n$ matrix A with a complete set of eigenvalues $\lambda_1, \ldots, \lambda_n$ and corresponding eigenvectors $\mathbf{v}_1, \ldots, \mathbf{v}_n$. Let $V := (\mathbf{v}_1, \ldots, \mathbf{v}_n)$ be the matrix of eigenvectors. Let Δ be another arbitrary $n\times n$ matrix, and let $\bar{\lambda}$ be any eigenvalue of $A+\Delta$. Then for at least one λ_i, $1\le i\le n$, the following bound holds:

$$|\bar{\lambda}-\lambda_i| \le K(V)\|\Delta\|, \tag{1}$$

where $K(V) \equiv \|V\|\cdot\|V^{-1}\|$ is the condition number of V. ∏

In [4], it was shown that one can replace the term $K(V)$ with n/s_i, with s_i defined by $s_i \equiv \mathbf{w}_i^H\mathbf{v}_i \,/\, \|\mathbf{w}_i\|\cdot\|\mathbf{v}_i\|$ It is not always clear which gives the tighter bound, so in this paper we use the bound

$$|\bar{\lambda}-\lambda_i| \le K_i\|\Delta\|, \text{ where } K_i \equiv \min\left\{K(V)\,,\,\frac{n}{s_i}\right\}. \tag{2}$$

The main goal of this paper is to present some results regarding matrix pencils, of the form $A-\lambda B$, where λ is a free parameter and A,B are $n\times p$ matrices with $n>p$. In the classical theory of matrix pencils [6] [9], it is well known that any pencil is equivalent to its *Kronecker Canonical Form* (KCF), which is a pseudo-diagonal matrix with diagonal blocks of the form L, L^{T}, and/or J, where

$$L = \begin{bmatrix} I \\ (0,\ldots,0) \end{bmatrix} + \lambda \begin{bmatrix} (0,\ldots,0) \\ I \end{bmatrix}$$

is a matrix with one more row than column, and J is a square matrix in Jordan Canonical Form. We call L a "tall-thin" K-block, L^T a "short-fat" K-block, and J the "regular" part.

In this paper, we deal exclusively with tall-thin pencils. Such pencils always have at least $n-p$ tall-thin K-blocks. In [3], we showed that the set of all tall-thin pencils with only tall-thin K-blocks is open and dense in the set of all pencils of the same shape. Hence, given a tall-thin pencil, the logical question to ask is whether it has any other types of K-blocks, and if not, what is the distance to the nearest pencil which does. In [12] and [8], algorithms were proposed that compute the complete KCF for a given pencil guaranteed to be exact for a pencil close to the original given pencil (backward stable). If the KCF computed in this way has only tall-thin K-blocks, then one is still left with determining how far it is from a pencil with other types of K-blocks. In this paper, we attempt to estimate this distance from both above and below. A detailed algebraic analysis for square pencils was given by Waterhouse [13], but beyond that surprisingly little has been found in the literature on this topic.

We use the following characterization of the Kronecker structure of a pencil.

Definition 1. A matrix pencil $A-\lambda B$ is said to be *deficient* if there exists some λ for which it is not full rank, where λ is a complex number or "Infinity". If $A-\lambda B$ is always full rank for any value of λ then it is said to be *non-deficient.* $\square$

All tall-thin pencils have at least one tall-thin K-block. The condition that the pencil be deficient is equivalent to the existence of at least one value λ and vector $\mathbf{x}$ such that $(A-\lambda B)\mathbf{x}=0$, and corresponds to the existence of at least a regular part or a short-fat block. We call such a vector $\mathbf{x}$ a right *annihilating vector* of the pencil associated with the *annihilating value* λ. These are also a generalized eigenvector and value, respectively, if they are associated with the regular part (if any). If there is a short-fat part, then every complex number (including infinity) is an annihilating value, but at most a finite number of these can be also generalized eigenvalues. The eigenvalues, if any, will be exactly those values of λ at which the matrix $A-\lambda B$ has a rank less than the overall maximum rank.

The work in this paper was motivated by the many roles matrix pencils play in Control Systems Theory. As an example, consider a time-invariant linear system

$$\dot{\mathbf{x}} = F\mathbf{x} + G\mathbf{u};\ \mathbf{y} = H\mathbf{x} + D\mathbf{u}. \tag{3}$$

A classical result from Control Theory is the PBH test (see e.g. [9]), which states that the system (3) is controllable if and only if the matrix pencil

$$P^T(\lambda) = \begin{bmatrix} \lambda I - F & | & G \end{bmatrix} = \begin{bmatrix} -F & | & G \end{bmatrix} - \lambda \begin{bmatrix} -I & | & 0 \end{bmatrix} \tag{4}$$

has full rank for any complex value of λ. From a numerical point of view, one may say that if pencil (4) has a small singular value for some value of λ then a small perturbation to the coefficients to (3) can yield an uncontrollable system [10].

The rest of this paper is organized as follows. First we examine a method for computing whether a given pencil is deficient of not. Next we develop an upper bound for the distance to the nearest deficient pencil. Finally we develop a lower bound for this distance, together with a bound on the possible locations of the generalized eigenvalues (if any) of the pencil in the complex plane.

2. FIND PENCIL RANK DEFICIENCY

In this section we address the problem of determining whether a given rectangular pencil is deficient or not. Specifically, given a $n \times p$ pencil $A-\lambda B$, with $n>p$, determine whether or not $A-\lambda B$ loses rank for any λ, including possibly λ infinite. This is equivalent to asking whether or not the pencil has any short-fat K-blocks or regular part. If B has full column rank, and the pencil never loses rank for any finite value of λ, then there are no short-fat blocks nor any regular part.

Consider a $n \times p$ pencil $A-\lambda B$ with $n>p$. Choose arbitrary $n \times (n-p)$ matrices C,D. We can then examine the square $n \times n$ generalized eigenvalue problem

$$(A,C)\mathbf{v} = \lambda(B,D)\mathbf{v} \tag{5}$$

We partition the vector $\mathbf{v}$ as $\mathbf{v} \equiv (\mathbf{x}^T,\mathbf{y}^T)^T$, where $\mathbf{x}$ is a p-vector, and $\mathbf{y}$ is an $(n-p)$-vector. It is a simple matter to derive the following

Proposition 1. Given an $n \times p$ pencil $A-\lambda B$ with $n>p$, and given arbitrary full-rank $n \times (n-p)$ matrices C,D, the following are equivalent:

(a) $A-\lambda B$ is a deficient pencil.

(b) Equation (5) has an annihilating vector $\mathbf{v}_0$ whose last $n-p$ components $\mathbf{y}_0$ are zero. Call the corresponding annihilating value λ_0.

Furthermore, we have

(c) If $B = (I,0)^T$, then all the annihilating vectors $\mathbf{v}_0$ and corresponding values λ_0 are exactly the generalized eigenpairs for the regular part of the pencil.

Proof: Both (a) and (b) are equivalent to the statement

(d) There is an n-vector $\mathbf{x}$ and scalar λ_0 which satisfies $A\mathbf{x} = \lambda_0 B\mathbf{x}$, or else $B\mathbf{x} = 0$. In the latter case, we say $\lambda_0 = \infty$. If B has the form of statement (c), then it has full rank, and there can be no short fat blocks. Hence it follows that there is a regular part, and that the λ_0's are exactly the eigenvalues of that regular part. Otherwise, there is no regular part. ∎

Based on this Proposition, we have a simple procedure for computing the existence of a regular part or short-fat block in a pencil. Given a tall-thin $n \times p$ pencil, choose the $n \times (n-p)$ augmentation matrices C,D to obtain the square eigenvalue problem (5). If there are any annihilating vectors of (5) whose last $n-p$ entries are zero, then there is a regular part or short-fat block, otherwise there is not.

In the special case that $B = (I,0)^T$, choose $D = (0,I)^T$ to turn (5) into an ordinary eigenproblem, and the annihilating vectors above into ordinary eigenvectors. If the last $n-p$ entries of any of those eigenvectors are zero, then the corresponding eigenvalues are exactly the eigenvalues of the regular part of the original pencil. Otherwise, there is no regular part.

However, this method does not give any hint as to the sensitivity of the result to perturbations in the coefficients. Therefore, we develop some bounds that indicate whether a given pencil is "numerically close" to a deficient pencil, in the next sections.

3. UPPER BOUNDS

In this section, we examine the problem of computing an upper bound on the distance to a deficient pencil. Specifically, consider a non-deficient $n \times p$ pencil $A - \lambda B$. In this case, we know that B has full rank. We would like to estimate the size of the perturbation E to the matrix A that is needed to obtain a deficient pencil $A + E - \lambda B$. This perturbed pencil will have a regular part, but no "short fat" blocks, since B has full rank. In this section we develop a simple upper bound for $\|E\|$.

In [5] and [10], it was shown that the smallest perturbation E can be obtained by solving the minimization problem

$$\min_s \sigma_{\min}(A - sB) \tag{6}$$

where $\sigma_{\min}(M)$ denotes the smallest singular value of the matrix M, and s varies over the entire complex plane. If we denote by σ^* and s^* the minimum in (6) and the value of s achieving that minimum, respectively, then $\|E\| = \sigma^*$. In [3], we discussed an expensive descent method that would often converge to the minimum (6). In this section, we would like to address a much simpler scheme that can be used to obtain an upper bound, which often not only provides a good estimate for $\|E\|$, but also provides an estimate for that value of s that yields the minimum in (6).

We start with the $n \times p$ pencil $A - \lambda B$. Choose some arbitrary full-rank $n \times (n-p)$ matrices C, D. And, in the case $B = (I, 0)^T$, choose $D = (0, I)^T$. Let

$$\lambda_i \, , \, \mathbf{v}_i \equiv \begin{pmatrix} \mathbf{x}_i \\ \mathbf{y}_i \end{pmatrix} , \, i = 1, \ldots, k \tag{7}$$

be the generalized eigenvalues and vectors for (5), where $\mathbf{x}_i$ denotes the first p components of $\mathbf{v}_i$. For each i, we have the equation $9A, C)\mathbf{v}_i = \lambda_i (B, D)\mathbf{v}_i$. We can rewrite this as $(\lambda_i B - A)\mathbf{x}_i = (C - \lambda_i D)\mathbf{y}_i$. We define the residual for each i by $\mathbf{r}_i := (A - \lambda_i B)\mathbf{x}_i$, and the perturbation E_i to be

$$E_i := -\frac{\mathbf{r}_i \mathbf{x}_i^T}{\|\mathbf{x}_i\|^2} \equiv (\lambda_i B - A)\frac{\mathbf{x}_i \mathbf{x}_i^T}{\|\mathbf{x}_i\|^2} = (C - \lambda_i D)\frac{\mathbf{y}_i}{\|\mathbf{x}_i\|} \cdot \frac{\mathbf{x}_i^T}{\|\mathbf{x}_i\|} \tag{8}$$

Then $A + E_i - \lambda_i B$ is a deficient pencil, for each i. Let σ_i, $\mathbf{u}_i$, $\mathbf{w}_i$ be, respectively, the smallest singular value and the corresponding left and right singular vectors of $A - \lambda_i B$, for each i. Then $E'_i := -\sigma_i \mathbf{u}_i \mathbf{w}_i^T$ is another smaller perturbation yielding a deficient pencil.

By taking norms of (8), we obtain a bound for these perturbations: $\|E'_i\| \le \|E_i\| \le \|(C - \lambda_i D)\mathbf{y}_i\| \, / \, \|\mathbf{x}_i\|$. If E denotes that perturbation with smallest norm yielding a deficient pencil, then E satisfies

$$\|E\| \equiv \sigma^* \leq \beta_1 \equiv \min_i \frac{\|(C-\lambda_i D)\mathbf{y}_i\|}{\|\mathbf{x}_i\|}. \tag{9}$$

We can summarize the result in

Proposition 2. Let $A-\lambda B$ be a $n\times p$ pencil, with $n>p$. Let C,D be two arbitrary full-rank $n\times(n-p)$ matrices. Then the smallest perturbation E such that $A+E-\lambda B$ is a deficient pencil satisfies the bound (9), where $\lambda_i,\mathbf{v}_i$, $i=1,\ldots,k$ are the eigen-pairs of the generalized eigenproblem (5), and $\mathbf{x}_i$, $\mathbf{y}_i$ are defined by (7).

Proof: This follows from the above discussion. All we must note is that from Proposition 1, if the pencil $A-\lambda B$ is already deficient, then $E=0$ automatically satisfies (9). In fact, if B has full column rank, one of the $\mathbf{y}_i$ should be zero by Proposition 1, so the bound will be hard. ∏

So, given a tall-thin pencil, augment it with two full rank matrices C,D to obtain (5). In the case $B=(I,0)^T$, choose $D=(0,I)^T$ to turn (5) into an ordinary eigenproblem. Then one upper bound is β_1 given by formula (9). Another tighter bound is

$$\beta_2 \equiv \min_i \sigma_{\min}(A-\lambda_i B). \tag{10}$$

One question is how to choose C,D. One goal is to make the augmented square eigenproblem as well-conditioned as possible. So far, the only requirement we have stated is that C,D have full column rank. To keep the condition number as low as possible, it is best to choose C,D to each have orthonormal columns. Two possible choices are (a) orthonormal basis of a random space, and (b) orthonormal basis of the space orthogonal to the columns of A and B. This last choice has the effect of limiting the increase to the condition numbers of (A,C) and (B,D) with respect to inversion, and hence is a heuristic attempt to obtain a reasonably low condition number with respect to the eigenproblem. In any case, the algorithms are intended to provide a-posteriori estimates for a given pencil, and in that context it is easy to check that the condition number of the resulting eigenproblem is reasonably small. The numerical examples below were carried out with choice (b).

4. LOWER BOUNDS

In this section, we show how to extend the results of the previous section for the special case of pencils $A-\lambda B$ such that $B=(I,0)^T$, to obtain a lower bound and to obtain a disk in the complex plane in which the value s which achieves the minimum in (6) must be located.

Let s^* be the complex value achieving the minimum in (6). and let σ^* be the smallest singular value of $A-s^*B$. Augment $A-sB$ with extra columns C and $D=(0,I)^T$ to obtain the square matrix (A,C) as in Section 2, and so that (5) is an ordinary eigenproblem for the matrix (A,C). Then the smallest singular value τ of $(A,C)-s^*I$ satisfies $\tau\leq\sigma^*$, since augmenting with extra columns can only reduce the smallest singular value [7]. So s^* is an exact eigenvalue of $(A,C)+\tau\Delta$, for some matrix Δ such that $\|\Delta\|=1$. Denote the eigenvalues of (A,C) by $\lambda_1,\ldots,\lambda_n$. Then, for at least one such eigenvalue, the modified Bauer-Fike Theorem implies that $|\lambda_i-s^*|\leq\tau K_i$, where K_i is defined in (2).

Next, let α be smallest singular value of $A-\lambda_i B$. Then

$$|\alpha-\sigma^*| \le \|(A-\lambda_i B)-(A-s^*B)\| \le |\lambda_i - s^*| \le \tau K_i \le \sigma^* K_i. \tag{11}$$

From this formula, we can draw two conclusions. One is that $\alpha \le \sigma^*(K_i+1)$, yielding the lower bound on σ^*:

$$\sigma^* \ge \frac{\alpha}{K_i+1} \ge \frac{\beta_2}{K_i+1}, \tag{12}$$

where β_2 is the upper bound defined in (10).

The other conclusion from (11) is, using (10) and (9),

$$|\lambda_i - s^*| \le \sigma^* K_i \le \beta_2 K_i \le \beta_1 K_i. \tag{13}$$

We can summarize this in

Proposition 3. Given a pencil $A-sB$, where $B=(I,0)^T$, and an arbitrary (full-rank) augmentation of this pencil to a square matrix (A,C) as in (5), then

(a) The value of s that achieves the minimum in $\sigma^* \equiv \min_s \sigma_{\min}(A-sB)$ is located within a disk in the complex plane whose center is on an eigenvalue λ_i of (A,C) and whose radius is bounded by (13), for some i.

(b) A lower bound on σ^* is provided by (12). ∎

We note that the backward stability of these methods depends on the backward stability of the methods used to obtain the eigendecompositions. If those are backward stable, then these bounds will be exact for a pencil numerically close to the original pencil, and the residual from the eigenvalue methods indicates how far from that original pencil we have strayed. But in the cases we have tried, the size of this residual was never more than 1E-11, much less than the computed bounds themselves.

5. NUMERICAL EXAMPLES

We illustrate the bounds with the examples taken from [3]. Each example represents a time-invariant linear system of the general form $\dot{\mathbf{x}}=F\mathbf{x}+G\mathbf{u}$, from which we form the pencil (4). The Staircase Algorithm [11] [1] applied to a single input system transforms F into an upper Hessenberg form, with G a multiple of $\mathbf{e}_1$. The pencil (4) has a regular part if and only if a subdiagonal element of the Hessenberg form is zero, so an obvious upper bound is simply the magnitude of the smallest subdiagonal element. This is the second column of Table 1. In the third column are shown upper bounds obtained by the expensive experimental descent method described in [3]. In the last column of Table 1 are lower bounds from the theory in [2] which was based on using the product of the subdiagonal elements.

Example 1 is defined by

$$F = \begin{pmatrix} 0 & 1 \\ -1 & 0 \end{pmatrix} \text{ and } G = \begin{pmatrix} 1 \\ 0 \end{pmatrix}.$$

In Example 2, we start with a single-input system already in Staircase form [11] [12], with $G=(1,0,\ldots,0)^{\mathrm{T}}$, and F equal to the matrix

$$\left|\begin{matrix} -1 & -1 & -1 & -1 & -1 & -1 & 7 \\ 1 & -1 & -1 & -1 & -1 & -1 & 6 \\ 0 & 1 & -1 & -1 & -1 & -1 & 5 \\ 0 & 0 & 1 & -1 & -1 & -1 & 4 \\ 0 & 0 & 0 & 1 & -1 & -1 & 3 \\ 0 & 0 & 0 & 0 & 1 & -1 & 2 \\ 0 & 0 & 0 & 0 & 0 & 1 & 1 \end{matrix}\right|$$

Example 3 is one with a particularly ill-conditioned eigenvalue problem. The system is defined by

$$F = \begin{bmatrix} -149 & 537 & -27 \\ -50 & 180 & -9 \\ -154 & 546 & -25 \end{bmatrix} \text{ and } G = \begin{bmatrix} 1 \\ 1 \\ 1 \end{bmatrix}.$$

The poles (eigenvalues) for this system are 1, 2, and 3.

The bounds reported in Table 2 were obtained using the indicated formulas. We note that the bounds take only $O(n+p)^3$ work to obtain, so the computation is relatively fast. Note that the upper bounds from [3] in Table 1 are tighter, but are much more expensive to obtain.

In Table 3, we give the eigenvalue λ_i of the augmented matrix at which the upper bounds were taken, together with the radius (13) about this value within which the minimum in (6) is located.

Example #	upper bound staircase [1]	upper bound from [3]	lower bound from [2]
1	1.0	6.6144E-01	1.2500E-01
2	1.0	6.7690E-04	4.3654E-08
3	1.1610E-02	4.3715E-03	8.5774E-07

Table 1. Bounds from older methods

Example # equ.	upper bound 1 (9)	upper bound 2 (10)	lower bound (12)
1	7.2561E-01	7.0545E-01	3.1480E-01
2	8.8790E-04	7.3074E-04	7.2095E-05
3	1.1507E-02	4.6607E-03	8.6385E-04

Table 2. Bounds from methods in this paper.

Example #	λ_i achieving min in (10)	radius about given λ_i equ. (13)
1	-1.6899E-01+1.1509E+00i	8.7545E-01
2	1.9998E+00+6.5937E-16i	6.6758E-03
3	2.4534E+00+0.0000E+00i	2.0485E-02

Table 3. Values of λ at which upper bounds in Table 2 were obtained.

When the staircase algorithm is applied to Example 3, we obtain (items in parenthesis are close to the machine epsilon):

*F*new =

2.8300E+02	6.9026E+02	-1.3400E+02
-1.1458E+02	-2.7946E+02	5.4837E+01
(-1.6584E-15)	-1.1610E-02	2.4570E+00

*G*new =

-1.7321E+00
(-2.5339E-16)
(4.2062E-16)

ACKNOWLEDGEMENTS

This research was partially supported by NSF Grants DCR-8420935 and DCR-8519029.

REFERENCES

[1] D. Boley, A. Emami-Naeini, G. F. Franklin: A New Algorithm for Canonical Decomposition of Linear Systems; Proceedings of 19th IEEE Conference on Decision and Control, Albuquerque, N.M., pp 199-200, held Dec. 1980.

[2] D. Boley: A Perturbation Result for Linear Control Problems; SIAM J. Algebraic & Discrete Methods 6 #1, pp 66-72, Jan. 1985.

[3] D. Boley: Computing the rank-deficiency of rectangular matrix pencils; Systems & Contr. Lett. 9, pp207-214, 1987.

[4] J. Demmel: On condition numbers and the distance to the nearest ill-posed problem; Num. Math. vol 51, pp251-289, 1987.

[5] R Eising: Between Controllable and Uncontrollable; Systems & Control Letters vol 4 #5, pp263-264, July 1984.

[6] F R Gantmakher: Theory of Matrices; Celsea Publ., New York, vols 1 and 2, 1959.

[7] G H Golub, C Van Loan: Matrix Computations; Johns Hopkins, 1983.

[8] B Kagstrom: RGSVD - An algorithm for computing the Kronecker structure and reducing subspaces of singular $A-\lambda B$ pencils; SIAM Sci & Stat Comput vol 7 # 1, pp 185-211, Jan 1986.

[9] T Kailath: Linear Systems; Prentice Hall, 1980.

[10] G Miminis: Numerical algorithms for controllability and eigenvalue computation; Masters Thesis, School of Computer Sci, McGill Univ, May 1981.

[11] C C Paige: Properties of numerical algorithms related to controllability; IEEE Trans Auto Contr vol AC-26 #1, pp 130-138, Feb 1981.

[12] P Van Dooren: The computation of Kronecker's canonical form of a singular pencil; Lin Algebra & Appl vol 27, pp 103-141, 1979.

[13] W C Waterhouse: The codimension of singular matrix pairs; Lin Algebra & Appl vol 57, pp 227-245, 1984.

On Derivative and Proportional Feedback Design for Descriptor Systems

Angelika Bunse-Gerstner Volker Mehrmann
Nancy K. Nichols

Abstract

For linear time-invariant singular systems it is customary to use a proportional state or output feedback control in order to achieve a desired closed loop behaviour. Derivative feedback is rarely considered. In this paper we examine how derivative feedback in descriptor systems can be used to alter the structure of the system pencil under various controllability conditions. It is shown that derivative and proportional feedback controls can be constructed such that the closed loop system has a given form and is also regular. This property ensures the solvability of the resulting system of dynamic-algebraic equations. Applications are also presented. For the linear-quadratic optimal control problem, conditions are given under which a derivative feedback transforms a singular problem into a standard L-Q problem, leaving the cost functional positive-definite. We discuss the problem of pole placement with derivative feedback alone and in combination with the usual proportional state feedback.

1 Introduction

We consider linear time-invariant continuous dynamical systems of the form

$$\begin{aligned} E dx/dt &= Ax(t) + Bu(t), \qquad x(t_0) = x_0 && (1)\\ y(t) &= Cx(t), && (2) \end{aligned}$$

where $E, A \in \mathbf{R}^{n\times n}$, $B \in \mathbf{R}^{n\times m}$, $C \in \mathbf{R}^{p\times n}$ and $\operatorname{rank} B = m \leq n$, $\operatorname{rank} C = p \leq n$. Here $x(t) \in \mathbf{R}^n$ is the state, $y(t) \in \mathbf{R}^p$ is the output, and $u(t) \in \mathbf{R}^m$ is the input or control of the system. Such systems are called *descriptor* or *generalized state-space* systems. In the case $E = I$, the identity matrix, we refer to (1)-(2) as a *standard* system.

Descriptor systems arise naturally in a variety of practical circumstances [13,10] and have recently been investigated in a number of papers [3,4,5,8,9,10] [11,12,14,15,16,17,18,19,20]. The response of a descriptor system can be described in terms of the eigenstructure of the matrix pencil

$$\alpha E - \beta A. \qquad (3)$$

In order to alter the behaviour of the system, it is customary to use proportional state or output feedback to modify the matrix A. The closed loop system pencil then becomes

$$\alpha E - \beta(A + BFC), \tag{4}$$

where the control is taken to be $u = Fy + v$. In the theory of matrix pencils, the roles of E and A are interchangeable, but the analogous use of derivative state or output feedback in multivariable systems has received only little attention in the literature. Derivative feedback modifies the matrix E, and the closed loop system pencil then becomes

$$\alpha(E + BGC) - \beta A, \tag{5}$$

where the control is taken to be $u = -G\dot{y} + v$.

A few recent papers [4,12,15,20] have studied aspects of derivative feedback for singular systems with regular matrix pencils. A control of the restricted form $u = F(\alpha x - \dot{x}) + v$ is discussed in [4,15,20]. In [12] state feedback of the form $u = -G\dot{x} + Fx + v$ is studied for the the pole placement problem. In these papers the derivative feedback's main task is to transform E into a nonsingular matrix E + BG.

Derivative feedback has been used as a tool in practical control system design as reported in [3]. A discrete-time observer has been constructed using both current and past information on the output and the state estimation. This leads to a system for the error with a system pencil of the form

$$\alpha(E + GC) - \beta(A + FC). \tag{6}$$

Even for nonsingular E the use of the derivative output feedback has here turned out to be useful. It was shown in [3] that choosing G such that the condition number of $E + GC$ is small gives improved state estimates.

In this paper we consider both derivative and proportional feedback and examine the properties that can be achieved with these types of feedback in various applications. In particular we discuss the pole placement problem and the linear quadratic regulator problem. In Section 2 we introduce notation and some preliminary results. Details of the results presented here can be found in [1], where the corresponding discrete case is also discussed.

2 Definitions and Preliminaries

2.1 Eigenstructure of Descriptor Systems

The system (1) is said to be *solvable* if and only if the system pencil (3) is *regular*, that is

$$\det(\alpha E - \beta A) \not\equiv 0 \quad \forall(\alpha, \beta) \in \mathbf{C}^2 \setminus \{0, 0\} \tag{7}$$

(See [2,19].) The behaviour of the system response $x(t)$ is then governed by the eigenstructure of the system pencil. For a regular pencil the *generalized eigenvalues* are defined to be the pairs $(\alpha_j, \beta_j) \in \mathbf{C}^2$ such that

$$\det(\alpha_j E - \beta_j A) = 0, \quad j = 1, 2, \ldots n. \tag{8}$$

Eigenvalue pairs (α_j, β_j) where $\beta_j \neq 0$ are said to be *finite* and, without loss of generality, can be taken to have the 'value' $\lambda_j = \alpha_j / \beta_j$. Pairs where $\beta_j = 0$ are said to be *infinite* eigenvalues. The maximum number of finite eigenvalues which a pencil can have is less than or equal to the rank of E. (For a pencil which is *not* regular, the generalized eigenvalues can be similarly defined as the pairs (α_j, β_j) such that the pencil loses rank.)

For regular pencils the solution of the system equations can be characterized in terms of the Kronecker Canonical Form (KCF) [7]. In this case there exist non-singular matrices X and Y (representing the right and left generalized eigenvectors and principal vectors of the system pencil, respectively) which transform E and A into the KCF :

$$Y^T E X = \begin{bmatrix} I & 0 \\ 0 & N \end{bmatrix}, \qquad Y^T A X = \begin{bmatrix} J & 0 \\ 0 & I \end{bmatrix}. \tag{9}$$

Here J is a Jordan matrix corresponding to the finite eigenvalues of the pencil and N is a nilpotent matrix such that $N^m = 0, \ \ N^{m-1} \neq 0$, also in Jordan form corresponding to the infinite eigenvalues. The *index* of the system is defined to be equal to the degree m of nilpotency. (For pencils which are not regular, the KCF and the index of the system can be defined similarly. See [1].)

We observe that a descriptor system is regular and index 0 if and only if E is non-singular. In this case the system can be reformulated as a standard system and the usual theory applies. In practice the reduction to standard form can be numerically unstable, however, (if E is ill-conditioned!) and, hence, even for index 0 systems, it is preferable to work directly with the generalized state-space form.

We observe also that a descriptor system is regular and index ≤ 1 if and only if it has exactly $q = \operatorname{rank} E$ finite eigenvalues. Conditions for the system to be regular and index ≤ 1 are given in [8]. For such systems there exists a unique solution for all admissible controls which satisfy certain initial consistency conditions. For higher index systems, impulses can arise in the response of the system if the control is not sufficiently smooth [17]. It is, therefore, desirable to select a feedback which ensures that the closed loop system is regular and index ≤ 1 if possible.

2.2 Controllability and Observablilty of Descriptor Systems

The definitions of controllablility and observability for standard control systems can be extended to descriptor systems. Various types of controllablility/observability can be identified, however [19]. Here we investigate the properties of the generalized state-space system (1)-(2) under the following conditions.

Definition 1 *Let $\alpha E - \beta A$ be a regular pencil. Then the triple (E, A, B) and the corresponding descriptor system are said to be* completely controllable *(C-controllable) if and only if*

$$\mathbf{C0} : \text{rank}[\alpha E - \beta A, B] = n, \quad \forall(\alpha, \beta) \in \mathbf{C}^2 \backslash \{0,0\}. \tag{10}$$

Similarly the triple (E, A, C) and the corresponding descriptor system are said to be completely observable *(C-observable) if and only if*

$$\mathbf{O0} : \text{rank} \begin{bmatrix} \alpha E - \beta A \\ C \end{bmatrix} = n, \quad \forall(\alpha, \beta) \in \mathbf{C}^2 \backslash \{0,0\}. \tag{11}$$

We remark that a system is completely controllable and/or completely observable only if

$$\text{rank}[E, B] = n \quad \text{and/or} \quad \text{rank} \begin{bmatrix} E \\ C \end{bmatrix} = n. \tag{12}$$

Complete controllability ensures that for any given initial and final states $x_0, x_f \in \mathbf{R}^n$ of the system, there exists an admissible control which transfers the system from x_0 to x_f in finite time [19]. Hence descriptor systems which are completely controllable can be expected to have similar properties to standard systems. Analogous remarks hold for completely observable systems.

Weaker definitions of controllability and observability are given by the following.

Definition 2 *Let $\alpha E - \beta A$ be a regular pencil. Then the triple (E, A, B) and the corresponding descriptor system are said to be* strongly controllable *(S-controllable) if and only if*

$$\begin{array}{ll} \mathbf{C1} : \text{rank}[\lambda E - A, B] = n, & \forall \lambda \in \mathbf{C}; \\ \mathbf{C2} : \text{rank}[E, AS_\infty, B] = n, & \textit{where the columns of } S_\infty \textit{ span } \mathcal{N}(E). \end{array} \tag{13}$$

Similarly the triple (E,A,C) and the corresponding descriptor system are said to be strongly observable *(S-observable) if and only if*

$$\begin{array}{ll} \mathbf{O1} : \text{rank} \begin{bmatrix} \lambda E - A \\ C \end{bmatrix} = n, & \forall \lambda \in \mathbf{C}; \\ \mathbf{O2} : \text{rank} \begin{bmatrix} E \\ T_\infty^H A \\ C \end{bmatrix} = n, & \textit{where the columns of } T_\infty \textit{ span } \mathcal{N}(E^H). \end{array} \tag{14}$$

We remark that C-controllablity and C-observability imply S-controllability and S-observability, respectively. In the literature, systems which satisfy **C2** (or **O2**) are often described as "controllable (or observable) at infinity" [5,8,17].

We remark also that the properties of controllability and observability of descriptor systems are preserved under certain transformations. Specifically, conditions **C0**, **C1**, **C2**, **O0**, **O1** and **O2** are all preserved under non-singular transformations of the pencil and under proportional state and output feedback. With the exception of conditions **C2** and **O2**, these same conditions are also preserved under derivative state and output feedback.

3 Derivative and Proportional Feedback for Descriptor Systems

In this section we discuss conditions under which we can alter the structure of system pencil (3) by the use of derivative and/or proportional feedback. We show first that if the triples (E, A, B) and (E, A, C) are C-controllable and C-observable, respectively, then the system (1)- (2) can be transformed into a *standard* system by derivative feedback [1].

Theorem 3 *There exists a feedback control $u = -G\dot{y} + v$ such that the system matrix $E+BGC$ is non-singular and the closed loop system defined by the triples $(E+BGC, A, B)$ and $(E+BGC, A, C)$ is C-controllable and C-observable if and only if the triples (E, A, B) and (E, A, C) are C-controllable and C-observable, respectively.*

We remark that if $S \equiv E + BGC$ is non-singular, then under derivative feedback the corresponding closed loop system is equivalent to the standard system

$$\dot{x} = \tilde{A}x + \tilde{B}v, \tag{15}$$

where $\tilde{A} = S^{-1}A, \quad \tilde{B} = S^{-1}B$. Furthermore, the feedback can be implemented directly in terms of the states and the external inputs of the system. Using the closed loop form (15) we find that

$$u = -G\dot{y} + v = Wx + Vv \tag{16}$$

where

$$W = -GCS^{-1}A \quad \text{and} \quad V = I - GCS^{-1}B. \tag{17}$$

We now show that if the system is S-controllable and S-observable, then a closed loop system pencil which is regular and index ≤ 1 can be obtained by proportional *or* derivative feedback [8] and [1].

Theorem 4 *There exists a feedback control $u = Fy + v$ such that $\alpha E - \beta(A + BFC)$ is a regular pencil of index ≤ 1 and the closed loop system defined by the triples $(E, A + BFC, B)$ and $(E, A + BFC, C)$ is S-controllable and S-observable if and only if the triples (E, A, B) and (E, A, C) are S-controllable and S-observable, respectively.*

Theorem 5 *There exists a feedback control* $u = -G\dot{y} + v$ *such that* $\alpha(E + BGC) - \beta A$ *is a regular pencil of index* ≤ 1 *and the closed loop system defined by the triples* $(E + BGC, A, B)$ *and* $(E + BGC, A, C)$ *is S-controllable and S-observable if the triples* (E, A, B) *and* (E, A, C) *are S-controllable and S-observable, respectively.*

We remark that the converse of Theorem 5 does not hold. An example is given in [1].

In the following sections we consider the application of these results to the problems of pole placement and linear-quadratic optimal control for descriptor systems.

4 Pole Placement for Descriptor Systems

The problem of pole placement by derivative and proportional state feedback can be stated as follows.

Problem 1 *Given triple* (E, A, B) *and set* $\mathcal{L} = \{\lambda_1, \lambda_2, ...\lambda_q\}$ *where* $\lambda_j \in \mathbf{C}$ *and* $\lambda_j \in \mathcal{L} \Rightarrow \bar{\lambda}_j \in \mathcal{L}, j = 1, 2, ...q \leq n$, *find* $F \in \mathbf{R}^{m \times p}, G \in \mathbf{R}^{m \times p}$ *such that for some* $X \in \mathbf{C}^{n \times q}$

$$(A + BF)X = (E + BG)X\Lambda, \qquad \Lambda = diag\{\lambda_j\} \tag{18}$$

and

$$\det(\lambda(E + BG) - (A + BF)) \neq 0, \qquad \textit{for some } \lambda \notin \mathcal{L}. \tag{19}$$

We remark that (18) guarantees that the prescribed poles are assigned by the feedback $u = Fx - G\dot{x}$ and (19) ensures that the system pencil is regular.

For C-controllable systems we can assign a full set of n poles to the closed loop system by using a combination of proportional and derivative feedback. By Theorem 3 we can select G such that $E + BG$ is nonsingular and then select F to assign the prescribed poles to the equivalent standard system (15) [1].

Theorem 6 *For any arbitrary set* $\mathcal{L}$ *of* n *self-conjugate (finite) poles, there exists a pair of matrices* F *and* G *solving the pole placement problem, Problem 1, if and only if the triple* (E, A, B) *is C-controllable.*

We remark that with proportional state feedback alone we can assign at most $q = \operatorname{rank} E$ finite poles to the closed loop system. (The remaining $n - q$ infinite poles cannot be reassigned.) The use of derivative feedback alone allows us to reassign up to $\operatorname{rank} A$ poles, including all the infinite poles (but excluding the null poles $\lambda_j = 0$).

For S-controllable systems we can assign exactly $q \equiv \operatorname{rank} E$ poles with regularity of the system pencil. The closed loop system is then index ≤ 1 and regular. We have the following [8] and [1].

Theorem 7 *For any set $\mathcal{L}$ of q self-conjugate (finite) poles, where $q \equiv \operatorname{rank} E$, there exists a solution to the pole placement problem, Problem 1, if and only if the triple (E, A, B) is S-controllable.*

We remark that for S-controllable systems a closed loop system pencil which is regular and index ≤ 1 and has the prescribed poles can be achieved by proportional state feedback alone. We expect, however, that a more well-conditioned dynamic-algebraic system can be obtained by first using a derivative feedback to obtain a regular pencil of index ≤ 1 and then applying proportional feedback to assign the finite poles to specific positions.

We remark also that for the problem of pole assignment by output feedback in descriptor systems analogous results can be obtained.

5 Linear Quadratic Regulator Problems for Descriptor Systems

The linear quadratic optimal control problem for descriptor systems can be stated as follows

Problem 2 *Given triple (E, A, B) and $Q = Q^T \in \mathbf{R}^{n\times n}$, $R = R^T \in \mathbf{R}^{m\times m}$ and $S \in \mathbf{R}^{n\times m}$ such that*

$$\begin{bmatrix} Q & S \\ S^T & R \end{bmatrix} \geq 0 \qquad \text{and} \qquad R > 0, \tag{20}$$

find $F \in \mathbf{R}^{m\times p}, G \in \mathbf{R}^{m\times p}$ such that $u = Fx - G\dot{x}$ minimizes

$$C = \int_{t_0}^{t_f} \begin{bmatrix} x \\ u \end{bmatrix}^H \begin{bmatrix} Q & S \\ S^T & R \end{bmatrix} \begin{bmatrix} x \\ u \end{bmatrix} dt \tag{21}$$

subject to (1).

If E is nonsingular, then the problem can be solved immediately. If E is singular and (12) is satisfied, that is, the triple (E, A, B) is C-controllable, then by Theorem 3 we can choose a matrix G such that $E + BG$ is nonsingular and such that the corresponding closed loop system is equivalent to the standard system (15) with feedback control u given by (16) and $C = I$. By Theorem 3, G can also be selected to ensure that $V = I - GS^{-1}B$ is non-singular, and hence that

$$\tilde{Q} = \begin{bmatrix} I & 0 \\ W & V \end{bmatrix}^H \begin{bmatrix} Q & S \\ S^T & R \end{bmatrix} \begin{bmatrix} I & 0 \\ W & V \end{bmatrix} \geq 0, \qquad V^H R V > 0, \tag{22}$$

where $W = -GS^{-1}A$. The LQR problem then reduces to finding a feedback F such that $v = Fx$ minimizes

$$C = \int_{t_0}^{t_f} \begin{bmatrix} x \\ v \end{bmatrix}^H \tilde{Q} \begin{bmatrix} x \\ v \end{bmatrix} dt \tag{23}$$

subject to (15). This problem is immediately solvable. We obtain the following [1].

Theorem 8 *A solution to the linear quadratic regulator problem, Problem 2, exists if the triple* (E, A, B) *is C-controllable. Furthermore, the closed loop system corresponding to the triple* $(E + BG, A + BF, B)$ *is then stable.*

By similar arguments, if E is singular and the triple (E, A, B) is S-controllable, then Theorem 4 implies that F can be chosen such that the closed loop system pencil $\alpha E - \beta(A + BF)$ is regular and index ≤ 1, and such that the triple $(E, A + BF, B)$ remains S-controllable. The LQR problem is then in the form required by [11] and a solution to the transformed problem can be found. We thus have the following [11] and [1].

Theorem 9 *A solution to the linear quadratic regulator problem, Problem 2, exists if* (E, A, B) *is S-controllable.*

We remark that the latter problem is solved by proportional feedback alone. We expect, however, that the derivative feedback may also be used here to obtain a more well-conditioned dynamic-algebraic system to which the proportional feedback may be applied.

We remark also that similar constructions apply to the linear quadratic output control problems.

6 Conclusions

We investigate here the use of derivative and proportional feedback in descriptor, or generalized state-space systems. We define various conditions for controllability and observability and demonstrate to what extent the structure of the system pencil can be altered by proportional-derivative feedback under these conditions.

It is established that systems which are C-controllable and C-observable can be transformed into *standard* systems by a combination of derivative and proportional state or output feedback. It is shown that in this case, with state feedback, all of the poles of the system can be assigned to prescribed positions; furthermore, the linear quadratic optimal control problem can be solved explicitly to obtain a stable closed loop system.

It is also established that systems which are S-controllable and S-observable can be transformed by proportional-derivative state or output feedback into closed loop systems which are regular and of index ≤ 1. It is shown that by state feedback the maximum number of finite poles can be assigned and the linear quadratic regulator problem can also be solved.

Details of the proofs of these results and examples are given in [1].

References

[1] A. Bunse-Gerstner, V. Mehrmann, N.K. Nichols, "Derivative Feedback for Descriptor Systems,"Sonderforschungsbereich: "Diskrete Strukturen in der Mathematik", Universität Bielefeld, Bielefeld, F.R.G., *preprint.*

[2] S.L. Campbell," Singular Systems of Differential Equations " *Pitman, San Francisco*, 1980.

[3] M.J. Chapman, D.W. Pearson, D.N. Shields," Partial Singular Value Assignment in the Design of Robust Observers for Discrete Time Descriptor Systems" *to appear in IMA J. Control*, .

[4] M. Christodoulou," Decoupling in the Design and Synthesis of Singular Systems" *Automatica*, Vol. 22, No. 2 , 1986,pp. 245–249.

[5] D.J. Cobb," Feedback and Pole Placement in Descriptor Variable Systems" *Int. J. Control* , Vol. 33, No. 6, 1981,pp. 1135–1146.

[6] L.R. Fletcher, J. Kautsky, N.K. Nichols," Eigenstructure Assignment in Descriptor Systems" *IEEE Trans. Autom. Control*, AC–31, 1986,pp. 1138–.

[7] F.R. Gantmacher," Theory of Matrices," Vol. I, II *Chelsea, New York*, 1959.

[8] J. Kautsky, N.K. Nichols, E.K.-W. Chu ," Robust Pole Assignment in Singular Control Sytems" *to appear in Lin. Alg. Appl.*.

[9] F.L. Lewis," Descriptor Systems: Fundamental Matrix, Reachability and Observability Matrices, Subspaces" *Proceedings of 23rd Conference on Decision and Control, Las Vegas, NV*, December 1984,pp. 293–298.

[10] D.G. Luenberger," Dynamic Equations in Descriptor Form " *IEEE Trans. Autom. Control*, Vol. AC–22, 1977,pp. 312–321.

[11] V. Mehrmann," Existence, Uniqueness and Stability of Solutions to Singular Linear Quadratic Control Problems " *to appear in Lin. Alg. Appl.*

[12] R. Mukundan, W. Dayawansa," Feedback Control of Singular Systems – Proportional and Derivative Feedback of the State" *Int. J. Syst. Sci.*, Vol. 14, 1983,pp. 615–632.

[13] H.H. Rosenbrock," Structural Properties of Linear Dynamic Systems" *Int. J. Control*, Vol. 20, No. 2, 1974,pp. 191–202.

[14] M.A. Shayman," Pole Placement by Dynamic Compensation for Descriptor Systems" *to appear Automatica*, 1988.

[15] M.A. Shayman, Z. Zhou," Feedback Control and Classification of Generalized Linear Systems" *IEEE Trans. Autom. Control*, Vol. AC–32, No. 6, 1987,pp. 483–494.

[16] P. Van Dooren,"The Generalized Eigenstructure Problem in Linear System Theory," *IEEE Transactions on Automatic Control*, Vol. 6, 1981, pp. 111-129.

[17] G.C. Verghese, B.C. Lévy, T. Kailath ," A General State Space for Singular Systems" *IEEE Trans. Autom. Control*, AC–26, No. 4, 1981,pp. 811–831.

[18] G.C. Verghese, P. Van Dooren, T. Kailath," Properties of the System Matrix of a Generalized State Space System" *Int. J. Control*, Vol. 30, 1979,pp. 235–243.

[19] E.L. Yip, R.F. Sincovec," Solvability Controllability and Observability of Continuous Descriptor Systems" *IEEE Trans. Autom. Control*, AC–26, 1981,pp. 702–707.

[20] Z. Zhou, M.A. Shayman, T.-J. Tarn," Singular Systems: A New Approach in the Time Domain" *IEEE Trans. Autom. Control*, Vol. AC–32, No. 1, 1987,pp. 42–50.

DETECTING NEARLY UNCONTROLLABLE PAIRS

RALPH BYERS

Abstract. Given a controllable pair of matrices (A, B), the distance to the set of uncontrollable pairs is obtained as the minimum singular value of the set of augmented matrices $[A - \lambda I, B]$ as λ varies over the complex plane. This study develops one dimensional and two dimensional algorithms to calculate or estimate the distance. The algorithms give guaranteed upper and lower bounds.

1. Introduction. One of the fundamental concepts in linear control theory is that of controllability. The pair $(A, B) \in \mathbf{C}^{n\times n} \times \mathbf{C}^{n\times p}$ is *controllable*, if in the control system $\dot{x} = Ax + Bu$, for any initial state $x(0) = x_o$, any desired final state x_1 and any desired time $t_1 > 0$, there is a continuous control function $u(t)$ such that $x(t_1) = x_1$ [47]. Equivalently [29], a pair (A, B), $A \in \mathbf{C}^{n\times n}$ and $B \in \mathbf{C}^{n\times p}$ is controllable, if and only if for all $\lambda \in \mathbf{C}$,

(1) $\quad \text{rank}[A - \lambda I,\ B] = n.$

Call a pair $(A, B) \in \mathbf{C}^{n\times n} \times \mathbf{C}^{n\times p}$ "nearly uncontrollable" if for some "small" perturbations $(E, F) \in \mathbf{C}^{n\times n} \times \mathbf{C}^{n\times p}$, the pair $(A + E, B + F)$ is uncontrollable. Nearly uncontrollable pairs are associated with ill-conditioned computational problems [9, 15, 17, 37, 38]. Moreover, a "nearly uncontrollable" pair may signal fundamental trouble with the mathematical model or even the underlying physical system itself.

It is desirable, therefore to be able to identify not only uncontrollable pairs (A, B) but "nearly uncontrollable" pairs as well. Checking the rank of the controllability matrix $[B, AB, A^2B, \ldots, A^{n-1}B]$ [34] and other classic tests may give no indication of near uncontrollability. Rounding errors may even make some classic tests utterly fail [3, 42, 43].

A measure of "nearness" to uncontrollability is the magnitude of the smallest perturbation $(E, F) \in \mathbf{C}^{n\times n} \times \mathbf{C}^{n\times p}$ that makes the pair $(A + E, B + F)$ uncontrollable. Specifically, this is the quantity

(2) $\quad \delta(A, B) = \inf_{(A+E, B+F) \text{ uncontrollable}} \|[E, F]\|_2.$

We choose the spectral norm in (2) for mathematical and computational convenience. Early proposals of this and similar measures [3, 40, 42] have now been widely accepted [4, 5, 6, 14, 16, 20, 23, 24, 37, 39].

In a few special cases it is possible to evaluate $\delta(A, B)$ exactly or to give tight bounds [37] . However, in the general case, one must rely on numerical methods. Several authors have investigated numerical methods for calculating $\delta(A, B)$. Paige [42] and Boley [3, 4] proposed a heuristic algorithm that uses staircase form [2, 44, 43] to find "small' to perturbation matrices (E, F) that approximately minimize (2).

The generalized eigenvalues of the regular part of $[A,\ B] - \lambda[I,\ 0]$ are the uncontrollable modes. Thus, (A, B) is controllable if and only if the Kronecker Canonical Form (KCF) of $[A,\ B] - \lambda[I,\ 0]$ has no regular part [18, 21].

Modern methods for calculating Kronecker canonical form with their "eigenvalue clustering" heuristic [1, 18, 21, 19, 20, 33] overestimate $\delta(A, B)$, but, in practice, are often correct to within an order of magnitude [20]. However, there are examples for which such heuristics fail to recognize a "nearly uncontrollable" pair [14, 16].

Eising [23] characterized the minimizer (E, F) of (2) as

$$E = x\left((x^H Ax)x^H - x^H A\right), \qquad F = -xx^H B$$

where $x \in \mathbf{C}^n$, $x^H x = 1$ is a minimizer of

$$f(x) = \left(x^H A(I - xx^H)A^H x + x^H BB^H x\right).$$

The superscript "H" denotes the complex conjugate transpose. So, $\delta(A, B)$ could be calculated by applying a general n complex variable optimization method to $f(x)$. A simpler optimization approach also proposed by Eising [24] is to apply a general 1 complex variable optimization method to

$$h(\lambda) = \sigma_{\min}([A - \lambda I,\ B])$$

where $\lambda \in \mathbf{C}$ and $\sigma_{\min}(M)$ denotes the smallest singular value of $M \in \mathbf{C}^{m\times n}$. Boley [5] and Miminis [39] have developed special purpose optimization methods for this problem.

Optimization methods face several difficulties. There may be many local minima. A numerical method must search for all of them in order to guarantee that it finds the global minimum. For example, $h(\lambda)$ may have as many as n (or more) local minima. The number of local minima varies from problem to problem, so it is difficult to know if all local minima have been found. Most optimization methods require smooth objective functions, so $h(\lambda)$ and perhaps its derivatives must also be evaluated accurately — probably with an expensive singular value decomposition [5, 39].

An ideal algorithm for calculating $\delta(A, B)$ would (i) be provably reliable, (ii) provide inexpensive order of magnitude estimates and (iii) provide rigorous upper and lower bounds on $\delta(A, B)$.

2. Two Dimensional Methods. Consider again the function

$$h(\lambda) = \sigma_{\min}([A - \lambda I_n,\ B]) \tag{3}$$

where $\lambda \in \mathbf{C}$, I_n denotes the n-by-n identity matrix and $\sigma_{\min}(M)$ represents the smallest singular value of M. The method discussed in this section evaluate $h(\lambda)$ many times. So, before presenting the algorithms, we first consider an efficient method of evaluating $h(\lambda)$.

The singular values of a matrix are invariant under multiplication by unitary matrices. In particular, if $Q \in \mathbf{C}^{n\times n}$ is unitary, then

$$\sigma_{\min}([A - \lambda I_n,\ B]) = \sigma_{\min}(Q^H[A - \lambda I_n,\ B])\begin{bmatrix} Q & 0 \\ 0 & I_p \end{bmatrix} = \sigma_{\min}([Q^H AQ - \lambda I_n,\ Q^H B])$$

At the cost of approximately $\frac{5}{3}n^3 + n^2 p$ (complex) operations, we may choose Q to reduce A to Hessenberg form [27, Algorithm 7.4-2]. (The matrix A is said to be in Hessenberg form if for $i > j + 1$, $a_{ij} = 0$.)

Hence, at a one time cost of $\frac{5}{3}n^3 + n^2 p$ work, we may take A to be in Hessenberg form. One might think of this condensed form as a "poor cousin" of staircase form [2, 44, 43]. We shall not need the full generality of staircase form.

If $A \in \mathbf{R}^{n\times n}$ is Hessenberg and $B \in \mathbf{R}^{n\times p}$ is a general rectangular matrix, then for any particular value of λ, it is easy to calculate unitary matrices $U \in \mathbf{C}^{n\times n}$ and $V \in \mathbf{C}^{n\times n+p}$ such that $U[A - \lambda I,\ B]V$ is upper trapezoidal of the form $[R, 0]$. For example, U may be taken to be a sequence of $n - 1$ plane rotations such that UA is upper triangular as set forth in the first half of [27, Algorithm 7.4-1]. Applying U costs $O(n^2 p)$ operations. The matrix V could be chosen to set the last p columns to zero. The matrix V might be obtained as a sequence of $p(n - 1)$ plane rotations from a column oriented version of the updating algorithm in [26] and [22, Section 10.4]. Appling V costs $O(n^2 p)$ operations.

A condition estimator (possibly combined with inverse iteration) applied to the resulting triangular matrix $U[A - \lambda I,\ B]V$ can approximate

$$\sigma_{\min}(U[A - \lambda I,\ B]V) = \sigma_{\min}([A - \lambda I,\ B])$$

in $O(n^2)$ operations [13, 12, 28, 30, 41]. After the initial reduction to Hessenberg form, each value of $h(\lambda)$ costs only $O(n^2p)$ operations. So, we may allow a numerical method to evaluate $h(\lambda)$ many times.

Now, consider the problem of finding the minimum value of $h(\lambda)$. The Wielandt-Hoffman [32] theorem applied to the singular value decomposition of $[A - \lambda I,\ B]$ implies

$$|h(\lambda_2) - h(\lambda_1)| \leq |\lambda_2 - \lambda_1|. \tag{4}$$

This in turn implies that if $\lambda_{jk} = j\sqrt{2}\epsilon + ik\sqrt{2}\epsilon$, for $j,\ k \in \mathbf{Z}$, where $\mathbf{Z}$ denotes the set of integers, then

$$\min_{\lambda_{jk}} h(\lambda_{jk}) \geq \delta(A, B) \geq \min_{\lambda_{jk}} h(\lambda_{jk}) - \epsilon.$$

It suffices to consider only those points

$$|\lambda_{jk}| \leq 2\|A\|_2 + 2\|B\|_2. \tag{5}$$

(Somewhat tighter bounds on the location of a global minimizer of $h(\lambda)$ can be found in [39].)

This suggests a numerical method for estimating $\delta(A, B)$ based on the brute force of evaluating $h(\lambda)$ many times: evaluate $h(\lambda)$ at mesh points $\lambda_{jk} = j\sqrt{2}\epsilon + ik\sqrt{2}\epsilon$ that obey (5). We call this approach the *BFI* method. Even for modest choices of ϵ the BFI method is impractical. If $\|A\|_2 + \|B\|_2 = 1/2$ and $\epsilon = .001$, then it uses 10^6 values of $h(\lambda)$!

Although the BFI is too expensive to be practical, it has some attributes worth noting. The rounding error analysis of the unitary transformations is ideal [27, 46]. Heuristic condition number estimation procedures [13, 12, 30, 41] have been extensively tested and are known to be reliable. Nonheuristic estimation procedures like [28] or inverse iteration are provably reliable. So, the BFI method is provably reliable.

The number of mesh points does not depend on n and p, so the BFI method is $\frac{5}{3}n^3 + O(pn^2)$ algorithm to estimate $\delta(A, B)$. It is only the very large coefficient of the low order pn^2 term in the work count that keeps the method from being practical.

The desirable attributes of BFI justify the effort to rescue the algorithm by finding a way to reduce the work count to a manageable level. We need a way to eliminate most of the function values used by the BFI method. Hinrichsen and Motscha [31] introduced a method to calculate the distance from a stable matrix to the nearest unstable matrix that generalizes to the nearest uncontrollable pair problem. We follow their approach here.

The Statements 1 and 2 of the following Theorem are a slight generalization of Corollary 3.2 in [31]. Statement 3 is a slight generalization of [7, p. 5]. The proofs carry over with a few modifications.

THEOREM 2.1. *Consider the path $\lambda = \lambda(t) = \lambda_0 + t\mu$ where $t \in R$ is a real parameter, $\lambda_0 \in \mathbf{C}$ is an arbitrary complex number and $\mu \in \mathbf{C}$ is a complex number of modulus unity. If $g(t) = h^2(\lambda(t))$, then*

1. *g is analytic except at a discrete set of values of t,*
2. *$g''(t) \leq 2$ whenever the second derivative exists and*
3. *$g(t)$ is twice continuously differentiable at a local minimum.*

Proof. Let $T(t) = [A - \lambda(t)I,\ B][A - \lambda(t)I,\ B]^H$ and observe from (3) that $g(t) = h^2(\lambda(t))$ is the eigenvalue of smallest magnitude of $T(t)$. It is well known [36, p. 141], the eigenvalues of $T(t)$ have representations as real analytic functions $\sigma_1^2(t)$, $\sigma_2^2(t)$, $\sigma_3^2(t)$,...,$\sigma_k^2(t)$. We allow the possibility that $k < n$ to account for eigenvalues with constant multiplicity greater than one. The σ_s^2's intersect at a discrete set of points. The corresponding eigenvectors may also be chosen to be real analytic and orthonormal. Statement 1 now follows from the observation that

(6) $g(t) = \min_{1 \le s \le k} \sigma_s^2(t)$.

By continuity, it suffices to prove Statement 2 at a point t at which there is a unique index s with $g(t) = \sigma_s^2(t)$. Let u_{s1}, u_{s2}, u_{s2}, $\ldots u_{sm_s}$ be a real analytic orthonormal basis of the invariant subspace of $T(t)$ corresponding to eigenvalue σ_s^2. Note that

$$\frac{dT}{dt} = -\frac{d\lambda(t)}{dt}A^H - \frac{d\bar{\lambda}(t)}{dt}A + \left(\frac{d\lambda(t)}{dt}\bar{\lambda}(t) + \lambda(t)\frac{d\bar{\lambda}(t)}{dt}\right) I$$

and

$$\frac{d^2T}{dt^2} = 2I.$$

To establish Statement 2, apply [35, p. 80–81] , to get

$$\begin{aligned}
\frac{d^2\sigma_s}{dt^2} &= \frac{1}{m_s}\sum_{j=1}^{m_s} u_{sj}^H \left(\frac{d^2T}{dt^2}\right) u_{sj} \\
&\quad -\frac{1}{m_s}\sum_{p\neq s}\sum_{l=1}^{m_p}\sum_{j=1}^{m_s}\left(u_{pl}^H\frac{dT}{dt}u_{sj}\right)\left(u_{sj}^H\frac{dT}{dt}u_{pl}\right) + \left(u_{pl}^H\frac{dT}{dt}u_{sj}\right)\left(u_{pl}^H\frac{dT}{dt}u_{sj}\right) \\
&= 2 - \frac{1}{m_s}\sum_{p\neq s}\sum_{l=1}^{m_p}\sum_{j=1}^{m_s}\frac{\left|u_{pl}^H\left(\frac{dT}{dt}\right)u_{sj}\right|^2}{\sigma_p - \sigma_s} \\
&\le 2.
\end{aligned}$$

Now, consider Statement 3. Suppose that t_* is a local minimizer of $g(t)$. If there is a single index j such that $g(t) = \sigma_j^2(t)$ in a neighborhood of t_*, then $g(t)$ is real analytic at t_* and Statement 3 holds. Otherwise, in a neighborhood of t_*,

(7) $$g(t) = \begin{cases} \sigma_j^2(t) & t < t_* \\ \sigma_l^2(t) & t > t_* \end{cases}$$

and $l \neq j$. The eigenvalues $\sigma_s^2(t)$ intersect at discrete points, so (7) involves only 2 σ_s's, and $g(t)$ has right and left derivatives of all orders at t_*.

By definition t_* is a local minimizer of g. Equation (6) implies that t_* must also be a local minimizer of $\sigma_j^2(t)$ and $\sigma_l^2(t)$. In particular, $d\sigma_j^2(t_*)/dt = d\sigma_l^2(t_*)/dt = 0$, i.e., the right and left derivatives of $g(t)$ agree at t_*. Hence, $g(t)$ is continuously differentiable at $t = t_*$.

If $d^2\sigma_j(t_*)/dt^2 < d^2\sigma_l(t_*)/dt^2$, then in some neighborhood of t_*, $\sigma_j(t) < \sigma_l(t)$. But in this case, (6) and (7) contradict each other. Similarly, if $d^2\sigma_j(t_*)/dt^2 > d^2\sigma_l(t_*)/dt^2$, then in a neighborhood of t_* $\sigma_l(t) < \sigma_j(t)$ — is also a contradiction. Therefore, $d^2\sigma_j(t)/dt^2 = d^2\sigma_l(t)/dt^2$, i.e., the right and left second derivatives of $g(t)$ agree at t_*. Hence, $g(t)$ is twice continuously differentiable at t_*. □

Let $\nabla(h^2)$ be the gradient and let $\nabla^2(h^2)$ be the Hessian matrix of h^2. Theorem 2.1 implies that $\|\nabla(h^2)\|_2 \le 1$ and $\nabla^2(h^2) \le 2I_n$ in the sense that $2I_n - \nabla^2$ is positive definite. These observations imply the following corollary.

COROLLARY 2.2. *For $z \in \mathbf{C}$, let $f(z)$ be the paraboloid*

(8) $f(z) = a + b\,\mathrm{Re}(z) + c\,\mathrm{Im}(z) + |z|^2$

with coefficients a, b and c chosen so that f interpolates h at three points z_1, z_2, $z_3 \in \mathbf{C}$. If z lies inside the simplex $S = S(z_1, z_2, z_3)$ with vertices z_1, z_2 and z_3, then $h^2(z) \ge f(z)$. In particular,

$$\min_{z\in S} h(z) \ge \sqrt{\min_{z\in S} f(z)}.$$

If the three points z_1, z_2 and z_3 of the corollary are not co-linear, then there exists a unique paraboloid of form (8). Evaluating $\min_{z\in S} f(z)$ is a trivial quadratic program.

The following algorithm is a two dimensional analogue of [31]. It keeps a list of simplices S_1, S_2, $S_3 \ldots S_K$ each of which may contain a global minimizer of h. An upper bound on $\delta(A, B)$ is the smallest value of $h(\lambda)$ over all vertices of the simplices. For each simplex, S_j, it uses Corollary 2.2 to find a lower bound $l_S \leq h(\lambda)$ for $\lambda \in S_j$. If the lower bound shows that S_j does not contain a global minimizer, then S_j is discarded. If the lower bound shows that S_j may contain a global minimizer, then S_j is subdivided into a few smaller subdivisions. The procedure is repeated until the upper bound and lower bound are within a small tolerance of each other.

In the algorithm, a triple of complex numbers (z_1, z_2, z_3) represents the simplex with z_1, z_2 and z_3 as vertices. The symbol l_S denotes the lower bound of Corollary 2.2 applied to a simplex S.

ALGORITHM 2.1.

- **INPUT:** $A \in \mathbf{C}^{n\times n}$, $B \in \mathbf{C}^{n\times p}$, $\epsilon > 0$
- **OUTPUT:** l, $u \in \mathbf{R}$ such that $u \geq \delta(A,B) \geq l$ and $u - l \leq \epsilon$

1. $z \leftarrow 2\|[A, B]\|_1$; $\mathcal{S} \leftarrow \{(z, zi, -z), (z, -zi, -z)\}$;
2. $l \leftarrow \min\{l_S \mid S \in \mathcal{S}\}$; $u \leftarrow \min(h(z), h(zi), h(-z), h(-zi))$
3. **WHILE** $u - l > \epsilon$
   - 3.1 For $S \in \mathcal{S}$, find the lower bound l_S given by Corollary 2.2.
   - 3.2 $\mathcal{T} \leftarrow \{S \in \mathcal{S} \mid l_S \leq u\}$
   - 3.3 Find the center w_S of each simplex $S \in \mathcal{T}$.
   - 3.4 $v \leftarrow \min_{S\in\mathcal{T}} h(w_S)$; $u \leftarrow \min(u, v)$
   - 3.5 $\mathcal{U} \leftarrow \{S \in \mathcal{T} \mid l_S \leq u\}$
   - 3.6 $l \leftarrow \min\{l_S \mid S \in \mathcal{U}\}$
   - 3.7 $\mathcal{S} \leftarrow \{$barycentric subdivisions of $S \in \mathcal{U}$ about $w_S\}$

Note that the old values of $h(\lambda)$ at the vertices of S_j along with a single new function value in the center of S_j contain all the information needed to use Corollary 2.2 to get lower bounds for each of the three subdivisions. So, a subdivision costs only one function value – $O(n^2p)$ work.

In one sense a barycentric subdivision is optimal, because it requires only one extra function evaluation. However, the choice is somewhat dubious, because it leads to long, thin simplices that may cause numerical problems later.

3. One Dimensional Algorithms. Calculating $\delta(A,B)$ is a one variable minimization problem as well as a two variable minimization problem. Trivially,

$$\delta(A,B) = \min_{\rho\in\mathbf{R}} h_1(\rho)$$

where h_1 is defined by

$$h_1(\rho) = \min_{\phi\in\mathbf{R}} h(\rho + i\phi).$$

The first part of this section presents two algorithms for evaluating $h_1(\rho)$. The second part presents an algorithm that gives upper and lower bounds on the minimum value of h_1.

The following theorem suggests a bisection method of evaluating $h_1(\rho)$.

THEOREM 3.1. *Given $(A,B) \in \mathbf{C}^{n\times n} \times \mathbf{C}^{n\times p}$, define G, K, $L \in \mathbf{C}^{n+p\times n+p}$ by*

$$G = \begin{bmatrix} 0_{nn} & 0_{np} & A^H \\ 0_{pn} & 0_{pp} & B^H \\ A & B & 0_{nn} \end{bmatrix}, \tag{9}$$

$$K = \begin{bmatrix} 0_{nn} & 0_{np} & I_n \\ 0_{pn} & 0_{pp} & 0_{pn} \\ I_n & 0_{np} & 0_{nn} \end{bmatrix}, \tag{10}$$

and

$$L = \begin{bmatrix} 0_{nn} & 0_{np} & -I_n \\ 0_{pn} & 0_{pp} & 0_{pn} \\ I_n & 0_{np} & 0_{nn} \end{bmatrix} \tag{11}$$

where I_n denotes the n-by-n identity matrix and 0_{rs} denotes the r-by-s zero matrix. For all $\sigma \geq 0$, $\sigma \geq \delta(A,B)$ if and only if there are numbers ρ, $\phi \in \mathbf{R}$ such that

(12) $\det(G - \sigma I - \rho K - \phi i L) = 0$

Proof. The function $h(\lambda)$ is continuous, bounded below by zero, and unbounded above as $\lambda \to \infty$, so range$(h) = [\delta(A,B), \infty)$. In particular, $\sigma \in$ range(h) if and only if there exists $\lambda \in \mathbf{C}$ for which σ is a singular value of $[A - \lambda I,\, B]$. Equivalently, $\sigma \in$ range(h) if and only if $\sigma \geq 0$ and

$$(13)\qquad \det\left(\begin{bmatrix} 0_{nn} & 0_{np} & (A-\lambda I)^H \\ 0_{pn} & 0_{pp} & B^H \\ (A-\lambda I) & B & 0_{nn} \end{bmatrix} - \sigma I\right) = 0.$$

Equation (12) is a rearrangement of (13) with $\rho = \text{Re}(\lambda)$ and $\phi = \text{Im}(\lambda)$. □

An immediate corollary to Theorem 3.1 is

COROLLARY 3.2. *For all $\sigma \in \mathbf{R}$, $\sigma > h_1(\rho)$ if and only if $\sigma \geq 0$ and the pencil $(G - \sigma I - \rho K) - \lambda L$ has a generalized eigenvalue λ with zero part.*

The Corollary justifies the following bisection algorithm for approximating $h_1(\lambda)$. Given ρ, the algorithm calculates an upper bound α and a lower bound β on $h_1(\rho)$. The convergence tolerance τ specifies the maximum relative error of α and β as approximations to $h_1(\rho)$. If $h_1(\rho)$ is very small or zero, the relative error requirement becomes expensive or impossible to meet, so there is a second convergence tolerance ϵ. The iteration stops if it determines that $h(\rho) < \epsilon$.

ALGORITHM 3.1.

- **INPUT:** $A \in \mathbf{C}^{n\times n}$, $B \in \mathbf{C}^{n\times p}$, $\rho \in \mathbf{R}$, $\tau > 1$, $\epsilon > 0$
- **OUTPUT:** $\alpha \in \mathbf{R}$, $\beta \in \mathbf{R}$ such that $\alpha < h_1(\rho) < \beta$ and either $\beta < \tau\alpha$ or $\beta < \epsilon$

1 $\alpha \leftarrow 0$; $\beta \leftarrow \|[A - \rho I,\, B]\|_2$

2 **DO WHILE** $\beta > \tau \max(\alpha, \frac{\epsilon}{\tau})$

  2.1 $\sigma \leftarrow \sqrt{\beta \max(\alpha, \frac{\epsilon}{\tau})}$

  2.2 **IF** $(G - \sigma I - \rho K) - \lambda L$ has a zero real part eigenvalue λ,

    THEN $\alpha \leftarrow \sigma$

    ELSE $\beta \leftarrow \sigma$

Similar algorithms have been proposed to approximate the distance of a stable matrix to the unstable matrices [10] and calculate H_∞ norms [8].

If $\epsilon = \tau^{-k+1}\|[A - \rho I,\, B]\|_2$, then the loop in Algorithm 3.1 executes $\lceil \log_2(k) \rceil$ times. A substantial $O(n^3)$ amount of work is needed to solve the generalized eigenvalue problem in Step 2.2 [27], so Algorithm 3.1 is too expensive to be used to get highly accurate values of $h_1(\rho)$. However, for most purposes, an order of magnitude estimate is sufficient. If $\tau = 10$ and $\epsilon = 10^{-7}\|[A,\, B]\|_2$, then the loop executes no more than three times.

At the cost of solving several eigenvalue problems, $h_1(\rho)$ can be evaluated accurately by a quadratically convergent variation of Algorithm 3.1. Similar algorithms have been proposed to evaluate H_∞ norms [7, 11] and our space here is limited, so we will only sketch the algorithm.

Fix a value of $\rho \in \mathbf{R}$ and define $g(\phi)$ by

$$g(\phi) = g_\rho(\phi) = h(\rho + i\phi) = \sigma_{\min}([A - (\rho + i\phi)I,\, B]).$$

Note that $\min_{\phi\in\mathbf{R}} g(\phi) = h_1(\rho)$. It follows from Theorem 2.1 that $g(\phi)$ is twice continuously differentiable in a neighborhood of its global minimum. Suppose ϕ_* is a minimizer of $g(\phi)$. The function $g(\phi)$ is approximately quadratic near its minimum, so if $\sigma - g(\phi_*)$ is small enough and if ϕ_1 and ϕ_2 are the roots of $g(\phi) = \sigma$ closest to ϕ_*, then $\phi_* \approx (\phi_1 + \phi_2)/2$.

The algorithm is based on the observation that the real roots $\phi \in \mathbf{R}$ of the equation $g(\phi) = \sigma$ are also real roots of $\det((G - \sigma I - \rho K) - \phi i L) = 0$.

In broad outline, an algorithm for accurate evaluating $h_1(\rho)$ is as follows. Starting with an initial approximation $\sigma = g(\phi) \approx g(\phi_*)$, find the zero-real-part eigvenvalues of $(G - \sigma I - \rho K) - \lambda I$. Select two adjacent ones and call them ϕ_1 and ϕ_2. Obtain a new approximate minimizer $\phi \leftarrow (\phi_1 + \phi_2)/2$ and a new approximate minimum value $\sigma \leftarrow h_1(\phi_*)$.

ALGORITHM 3.2.

- **INPUT:** $A \in \mathbf{R}^{n\times n}$, $B \in \mathbf{R}^{n\times p}$, $\rho \in \mathbf{R}$, $\epsilon > 0$
- **OUTPUT:** $\sigma \in \mathbf{R}$ such that $\sigma \geq h_1(\rho) \geq \sigma - \epsilon$

1. $\phi \leftarrow 0$
2. **DO WHILE** the pencil $(G - (\sigma - \epsilon)I - \rho K) - \lambda I$ has eigenvalues with zero real part.
   - 2.1 Set $\phi_1 i$ and $\phi_2 i$ equal to the two adjacent zero-real-part eigenvalues of largest magnitude.
   - 2.2 $\phi \leftarrow (\phi_1 + \phi_2)/2$
   - 2.3 $\sigma \leftarrow g(\phi)$

The choice $\phi = 0$ in Step 1 takes advantage of the following observation. If A and B are real, then $g(\phi) = g(-\phi)$, and $\phi = 0$ is a critical point of g. If $\phi_* = 0$, then the loop in Step 2 never executes.

In Step 2.1, the selection of the zero-real-part eigenvalues of largest magnitude is arbitrary. It is not always the best choice. Sometimes it may help to replace Step 1 by few steps of Algorithm 3.1 in order to get a better initial choice of ϕ.

Using $\sigma - \epsilon$ instead of σ in Step 2 not only tests $\sigma < g(\phi_*) + \epsilon$, but it also helps avoid converging to a strictly local minimum.

For more details, see the essentially similar algorithms proposed in [7] and [11] for calculating H_∞ norms. In particular, the proof of quadratic convergence given in [7] carries over with few modifications.

The following lemma justifies a one dimensional analog of the BFI method.

LEMMA 3.3. *For all $\rho_1, \rho_2 \in \mathbf{R}$,*

$$|h_1(\rho_1) - h_1(\rho_2)| \leq |\rho_1 - \rho_2|.$$

Proof. Suppose, without loss of generality that $h_1(\rho_1) \leq h_1(\rho_2)$. Recall that $h_1(\rho) = \min_{\phi\in\mathbf{R}} h(\rho + i\phi)$. The function h is continuous, bounded below by zero, and unbounded above as $\rho + i\phi \to \infty$, so, for some $\phi_1, \phi_2 \in \mathbf{R}$, $h_1(\rho_1) = h(\rho_1 + i\phi_1)$ and $h_1(\rho_2) = h(\rho_2 + i\phi_2)$. Furthermore,

$$\begin{aligned} 0 &\geq h_1(\rho_1) - h_1(\rho_2) \\ &= h(\rho_1 + i\phi_1) - h(\rho_2 + i\phi_2) \\ &= (h(\rho_1 + i\phi_1) - h(\rho_2 + i\phi_1)) + (h(\rho_2 + i\phi_1) - h(\rho_2 + i\phi_2)) \\ &\geq h(\rho_1 + i\phi_1) - h(\rho_2 + i\phi_1). \end{aligned}$$

To get the last inequality recall that $h(\rho_2 + i\phi_2) = \min_{\phi\in\mathbf{R}} h(\rho_2 + i\phi) \leq h(\rho_2 + i\phi_1)$. From (4) it follows that

$$0 \geq (h(\rho_1 + i\phi_1) - h(\rho_2 + i\phi_1)) \geq -|\rho_2 - \rho_1|.$$

This proves the lemma. □

Evaluating h_1 at mesh points $\rho_j = j\epsilon$, $j \in \mathbf{Z}$ gives a one dimensional BFI-like approximation

$$\delta(A, B) + \epsilon \geq \min_{j\in\mathbf{Z}} h_1(\rho) \geq \delta(A, B).$$

It is sufficient to consider only those ρ_j with magnitude less than $2\|[A, B]\|_2$. Although this is vastly more efficient than the two dimensional BFI method, it is still too expensive to be practical even for modest choices of ϵ. If $\|[A, B]\|_2 = 1/2$ and $\epsilon = .001$, and h_1 is evaluated by the more efficient, but less accurate Algorithm 3.1 with $\tau = 10$, then the BFI method would need to solve approximately 3000 generalized eigenvalue problems at a cost of $O(n^3)$ [27] work each! So, here again some auxiliary method is required to cut down the number of function values.

One possibility is to apply a general one dimensional optimization method to h_1. One dimensional optimization is a simpler problem than higher dimensional optimization. The problem of finding good search directions does not arise. Even bisection-like algorithms like golden section search apply [25, Section 8.1], [45]. However, optimization methods require smooth objective functions, so the more expensive Algorithm 3.2 is required. Just as in the two dimensional case, there may be several local minima, all of which must be searched to find a global minimum. However, h_1 has fewer local minima than h.

A more likely approach is a one dimensional analog of Algorithm 2.1. Under mild assumptions, the function $h_1^2(\rho)$ is twice continuously differentiable except at finitely points, $\left|\frac{dh_1^2(\rho)}{d\rho}\right| \leq 1$ and $\frac{d^2h_1^2(\rho)}{d\rho^2} \leq 2$. Suppose that f is the quadratic

(14) $\quad f(x) = a + bx + cx^2$

with coefficients a, b and c chosen so that f interpolates h_1^2 at the two points $x_1 < x_2$. If $x \in [x_1, x_2]$, then then $h_1^2(z) \geq f(z)$. This suggests the following modification of Algorithm 2.1 and [31, Algorithm 3.4].

ALGORITHM 3.3.

- **INPUT:** $A \in \mathbf{C}^{n\times n}$, $B \in \mathbf{C}^{n\times p}$, $\epsilon > 0$
- **OUTPUT:** u, $l \in \mathbf{R}$ such that $l \leq \delta(A, B) \leq u$ and $u - l \leq \epsilon$

1. $z \leftarrow 2\|[A,\ B]\|_2$; $\mathcal{S} \leftarrow \{[-z,\ z]\}$
2. $l \leftarrow 0$; $u \leftarrow \min(h_1(-z), h_1(z))$
3. **WHILE** $u - l > \epsilon$

   3.1 $v \leftarrow \min\limits_{[x_1,x_2]\in\mathcal{S}} \min(h_1(x_1), h_1(x_2))$; $u \leftarrow \min(u, v)$

   3.2 For each $[x_1, x_2] \in \mathcal{S}$, set $l_{x_1x_2}$ to the lower bound $l_{x_1x_2} = \sqrt{\min\limits_{x\in[x_1,x_2]} f(x)}$ where $f(x)$ is the quadratic (14).

   3.3 $l \leftarrow \min\limits_{[x_1,x_2]\in\mathcal{S}} l_{x_1x_2}$

   3.6 $\mathcal{S} \leftarrow \{[x_1, \frac{x_1+x_2}{2}] \mid l_{x_1x_2} < u\} \bigcup \{[\frac{x_1+x_2}{2}, x_2] \mid l_{x_1x_2} < u\}$

4. Conclusions. This is a body of work still in progress. Among the algorithms presented here, Algorithm 2.1 and Algorithm 3.3 are the most viable. Both of them require large computational efforts. It is too early to distinguish which is the better approach.

In definition (2), the perturbation (E, F) may be complex even when the original pair (A, B) is real. Often it is more meaningful to measure the real distance

$$\delta_R(A,B) = \inf_{\substack{(A+E,B+F)\ \text{uncontrollable} \\ (E,F)\in\mathbf{R}^{n\times n}\times\mathbf{R}^{n\times p}}} \|[E,F]\|_2.$$

Clearly, $\delta(A, B) \leq \delta_R(A, B)$. However, there is no known simple relationship between $\delta(A, B)$ and $\delta_R(A, B)$. Usually, $\delta(A, B)$ and $\delta_R(A, B)$ are of the same magnitude, but it is not known that this is always the case. The algorithms presented here calculate δ but not δ_R. Staircase form and Kronecker canonical form algorithms tend to measure δ_R [2, 3, 4, 18, 21, 19, 20, 33, 42, 43, 44].

At this writing, an entirely satisfactory algorithm to estimate or calculate $\delta(A, B)$ is still an open question. Even the best of the preceding algorithms are complicated and expensive. A simpler candidate is the following bisection algorithm suggested by Theorem 3.1.

ALGORITHM 4.1.

- **INPUT:** $A \in \mathbf{C}^{n \times n}$, $B \in \mathbf{C}^{n \times p}$, $\epsilon > 0$, $\tau > 0$
- **OUTPUT:** l, $u \in \mathbf{R}$, such that $l \leq \delta(A, B) \leq u$ and either $u < taul$ or $u < \epsilon$

1. $\alpha \leftarrow 0$; $\beta \leftarrow \|[A, B]\|_1$
2. **DO WHILE** $\beta > \tau \max(\alpha, \frac{\epsilon}{\tau})$

   2.1 $\sigma \leftarrow \sqrt{\beta \max(\alpha, \frac{\epsilon}{\tau})}$

   2.2 Let G, K and L be as in (9), (10), (11). **IF** $f(\rho + \phi i) = \det(G - \sigma I - \rho K - \phi i L)$ has a root $\rho + \phi i \in \mathbf{C}$,
   THEN $u \leftarrow \sigma$
   ELSE $l \leftarrow \sigma$

There is no known way to reliably make the decision in Step 2.2.

Department of Mathematics, University of Kansas, Lawrence, Kansas, 66045, USA. Partial support received from the National Science Foundation under grant CCR-8820882, and University of Kansas General Research Allocations #3758-20-038 and #3692-20-038. Portions of this research were completed in 1988 while visiting Oak Ridge National Laboratory and the University of Tennessee.

REFERENCES

[1] T. BEELEN AND P. M. VAN DOOREN, *An improved algorithm for the computation of Kronecker's canonical form of a singular pencil*, Linear Algebra and Its Applications, 105 (1988), pp. 9–65.

[2] T. BEELEN, P. M. VAN DOOREN, AND M. VERHAEGEN, *A class of fast staircase algorithms for generalized state-space systems*, in Proceedings of the American Control Conference, Seattle, Washington, 1986.

[3] D. L. BOLEY, *Computing the Controllability/Observability of a Linear Time-Invariant Dynamic System, A Numerical Approach*, PhD thesis, Stanford University, Stanford, California, June 1981. Computer Science Report STAN-CS-81-860.

[4] ——, *A perturbation result for linear control problems*, SIAM Journal of Algebraic and Discrete Methods, 6 (1985), pp. 66–72.

[5] ——, *Computing rank-deficiency of rectangular matrix pencils*, Systems and Control Letters, 9 (1987), pp. 207–214.

[6] D. L. BOLEY AND W. S. LU, *Measuring how far a controllable system is from an uncontrollable one*, IEEE Transactions on Automatic Control, AC-31 (1986), pp. 249–252.

[7] S. BOYD AND K. BALAKRISHNAN, *A regularity result for the singular values of a transfer matrix and a quadratically convergent algorithm for computing its L_∞-norm*, Tech. Report L-106-89-1, Information Systems Laboratory, Stanford University, Department of Electrical Engineering, Stanford, California 94305, March 1989.

[8] S. BOYD, V. BALAKRISHNAN, AND P. KABAMBA, *A bisection method for computing the H_∞ norm of a transfer matrix and related problems*, tech. report, Information Systems Laboratory, Stanford University, Department of Electrical Engineering, Stanford, California 94305, August 1987. to appear Mathematics of Control, Signals and Systems.

[9] R. BYERS, *Numerical condition of the algebraic Riccati equation*, Contemporary Mathematics, 47 (1985), pp. 35–39.

[10] ——, *A bisection method for measuring the distance of a stable matrix to the unstable matrices*, SIAM Journal on Scientific and Statistical Computing, 9 (1988), pp. 875–881.

[11] D. J. CLEMENTS AND K. L. TEO, *Evaluation of the H_∞-norm*, tech. report, School of Electrical Engineering and Computer Science, University of New South Wales, P. O. Box 1, Kensington, 2033, Australia, March 1989.

[12] A. K. CLINE, A. R. CONN, AND C. F. VAN LOAN, *Generalizing the LINPACK condition estimator*, in Numerical Analysis, Mexico 1981, J. P. Hennart, ed., Springer Verlag, Lecture Notes in Mathematics, 909, 1982, pp. 73–83.

[13] A. K. CLINE, C. M. MOLER, S. G. W., AND J. H. WILKINSON, *An estimate for the condition number of a matrix*, SIAM Journal of Numerical Analysis, 16 (1979), pp. 368–375.

[14] J. W. DEMMEL, *A lower bound on the distance to the nearest uncontrollable system*, technical report, Computer Science Department, Courant Institute of Mathematical Sciences, 251 Mercer St., New York, N. Y. 10012, January 1987.

[15] ——, *On condition numbers and the distance to the nearest ill-posed problem*, Numerische Mathematik, 51 (1987), pp. 251–289.

[16] ——, *Bounds on the distance to the nearest uncontrollable system*, technical report, Computer Science Department, Courant Institute of Mathematical Sciences, 251 Mercer St., New York, N. Y. 10012, 1988.

[17] ——, *On the conditioning of pole assignment*, Technical Report 150, Computer Science Department, Courant Institute of Mathematical Sciences, 251 Mercer St., New York, N. Y. 10012, 1988.

[18] J. W. DEMMEL AND B. KÅGSTRÖM, *Accurate solutions of ill-posed problems in control theory*, in Proceedings of the 25th IEEE Conference on Decision and Control, Athens, Greece, December 1986, pp. 558–563.

[19] ——, *Stably computing the Kronecker structure and reducing subspaces of singular pencils $A-\lambda B$ for uncertain data*, in Large Scale Eigenvalue Problems, J. Cullum and R. A. Willoughby, eds., Elsevier Science Publishers B. V. (North-Holland), 52 Vanderbilt Avenue, New York, N.Y. 10017, 1986, pp. 283–323.

[20] ——, *Distance to the nearest uncontrollable system*. Informal presentation at the Gatlinburg X meeting, Fairfield Glade, TN, October 1987.

[21] ——, *Accurate solutions of ill-posed problems in control theory*, SIAM Journal of Matrix Analysis and Applications, 9 (1988), pp. 126–145.

[22] J. J. DONGARRA, C. B. MOLER, J. R. BUNCH, AND G. W. STEWART, *LINPACK Users' Guide*, SIAM, Philadelphia, 1979.

[23] R. EISING, *The distance between a system and the set of uncontrollable systems*, in Proceedings of the MTNS, Beer-Sheva, Isreal, P. A. Fuhrmann, ed., Springer-Verlag, June 1983, pp. 303–314.

[24] ——, *Between controllable and uncontrollable*, Systems and Control Letters, 4 (1984), pp. 263–264.

[25] G. E. FORSYTHE, M. A. MALCOLM, AND C. B. MOLER, *Computer Methods for Mathematical Computations*, Prentice-Hall, Inc., Englewood Cliffs, N. J. 07632, 1977.

[26] G. GOLUB, *Numerical methods for solving linear least squares problems*, Numerische Mathematik, 7 (1965), pp. 206–216.

[27] G. GOLUB AND C. VAN LOAN, *Matrix Computations*, The Johns Hopkins University Press, Baltimore, Maryland, 1983.

[28] W. W. HAGAR, *Condition estimators*, SIAM Journal of Scientific and Statistical Computing, 5 (1984), pp. 311–316.

[29] M. L. J. HAUTUS, *Controllability and observability conditions of linear autonomous systems*, Proceedings of the Koninklijke Nederlandse Akademie van Wetenschappen, Series A, 72 (1969), pp. 443–448.

[30] N. J. HIGHAM, *A survey of condition number estimation for triangular matrices*, SIAM Review, 29 (1987), pp. 575–596.

[31] D. HINRICHSEN AND M. MOTSCHA, *Optimization problems in the robustness analysis of linear state space systems*, in Proceedings of the Seminar on Approximations and Optimization, Havanna, LN Mathematics, Springer Verlag, 1987.

[32] A. J. HOFFMAN AND H. W. WIELANDT, *The variation of the spectrum of a normal matrix*, Duke Mathematics Journal, 20 (1953), pp. 37–39.

[33] B. KÅGSTRÖM, *RGSVD — an algorithm for computing the Kronecker structure and reducing subspaces of singular $A - \lambda B$ pencils*, SIAM Journal of Scientific and Statistical Computing, 7 (1986), pp. 185–211.

[34] R. KALMAN, *Mathematical description of linear systems*, SIAM Journal of Control and Optimization, 1 (1963), pp. 152–192.

[35] T. KATO, *Perturbation Theory for Linear Operators*, Springer Verlag, New York, N.Y., 2nd ed., 1976.

[36] ——, *A Short Introduction to Perturbation Theory for Linear Operators*, Springer Verlag, New York, N. Y., 1982.

[37] C. KENNEY AND A. J. LAUB, *Controllability and stability radii for companion form systems*, Mathematics of Control, Signals and Systems, (1988).

[38] A. J. LAUB, *Numerical linear algebra aspects of control design computations*, IEEE Transactions

on Automatic Control, AC-30 (1985), pp. 97–108.

[39] G. S. MIMINIS, *Numerical algorithms for controllability and eigenvalue allocation*, master's thesis, School of Computer Science, McGill University, Montreal, P.Q., Canada, May 1981.

[40] B. C. MOORE, *Singular value analysis of linear systems*, in Proceedings of the IEEE Conference on Decision and Control, January 1979.

[41] D. P. O'LEARY, *Estimating matrix condition numbers*, SIAM Journal of Scientific and Statistical Computing, 1 (1980), pp. 205–209.

[42] C. C. PAIGE, *Properties of numerical algorithms relating to controllability*, IEEE Transactions on Automatic Control, AC-26 (1981), pp. 130–138.

[43] P. M. VAN DOOREN, *The generalized eigenstructure problem in linear systems theory*, IEEE Transactions on Automatic Control, AC-26 (1981), pp. 111–129.

[44] P. M. VAN DOOREN, E. EMAMI-NAEINI, AND L. SILVERMAN, *Stable extraction of the Kronecker structure of pencils*, in Proceedings of the 17th IEEE Conference on Decision and Control, January 1979, pp. 521–524.

[45] J. A. WILDE, *Optimal seeking Methods*, Prentice-Hall, Inc., Englewood Cliffs, N.J., 1964.

[46] J. W. WILKINSON, *The Algebraic Eigenvalue Problem*, Oxford University Press, Oxford, 1965.

[47] W. M. WONHAM, *Linear Multivariable Control: A Geometric Approach*, Applications of Mathematics, Springer Verlag, second ed., 1979.

The Upper Semicontinuity of the Least Squares Problems

Jiu Ding

Abstract

In this paper, we investigate the upper semicontinuity of the set-valued map from the linear least squares problems and for this purpose we obtain the error bounds if the problem is arbitrarily perturbed. These bounds are better than the current ones in that our assumption for the problem and perturbation is most general and we only use the usual spectral norm, not Frobenius norm, to get those estimates.

1. Introduction

Consider the linear least squares problem

$$\text{(1.1)} \qquad \begin{aligned} &\text{minimize } \|x\| \\ &\text{subject to } \|Ax - b\| = \min_{z \in \mathbb{C}^n} \|Az - b\| , \end{aligned}$$

where $A \in L(\mathbb{C}^n,\mathbb{C}^m)$ is an $m \times n$ complex matrix, $b \in \mathbb{C}^m$ and $x \in \mathbb{C}^n$. Here, $\|\cdot\| \equiv \|\cdot\|_2$ stands for the usual Euclidean vector norm and for a matrix A, $\|A\|$ is the induced operator norm defined by $\|A\| = \sup_{\|x\| \le 1} \|Ax\|$.

We call the (unique) solution x_{LS} to (1.1) the minimal norm least squares solution and it is given by the concise formula

$$\text{(1.2)} \qquad x_{LS} = A^+b ,$$

where $A^+ \in L(\mathbb{C}^m,\mathbb{C}^n)$ is the Moore-Penrose generalized inverse of A. Thus a point-to-point map

$$x_{LS} : L(\mathbb{C}^n,\mathbb{C}^m) \times \mathbb{C}^m \to \mathbb{C}^n$$

is defined by the rule

$$\text{(1.3)} \qquad x_{LS}(A,b) = A^+b .$$

If A is of full rank, then the map x_{LS} is continuous at the point (A,b) in the common sense. More generally, if $A_n \to A$ and $b_n \to b$ with $\text{rank}(A_n) = \text{rank}(A)$ for sufficiently large n, then $x_{LS}(A_n,b_n) \to x_{LS}(A,b)$. This is guaranteed by a famous Stewart theorem. However, x_{LS} is not continuous at those (A,b) where A are rank-deficient, that is, $\text{rank}(A) < \min(m,n)$. The cause for such discontinuity is that an arbitrary perturbation can increase the rank of the original matrix A.

Recently, M. Wei ([4]) used the singular value decomposition to prove that, if the linear system $Ax = b$ is consistent, then the minimal norm least squares solution $\bar{x}_{LS}$ to the perturbed system $\bar{A}\bar{x} = \bar{b}$ is close to a solution x of $Ax = b$, even when A is not of full rank and $\mathrm{rank}(\bar{A}) \neq \mathrm{rank}(A)$. The purpose of this paper is to explore the general upper semicontinuity property of least squares problems and to describe how to eliminate the theoretical singularity numerically.

The paper is arranged as follows. In the next section, we prove the main results. The numerical consideration is the content of section 3. In the final section, we conclude with several remarks.

2. The Upper Semicontinuity of Least Squares Problems

Suppose we are given the following least squares problem

$$\text{(2.1)} \qquad \text{minimize } \{\|Ax - b\| : x \in \mathbb{C}^n\}$$

with $A \in L(\mathbb{C}^n,\mathbb{C}^m)$, $b \in \mathbb{C}^m$ and $x \in \mathbb{C}^n$.

It is well known that the solution set to (2.1) is the same as the solution set to the normal equation

$$\text{(2.2)} \qquad A^*Ax = A^*b \ .$$

Define a point-to-set map $F : L(\mathbb{C}^n,\mathbb{C}^m) \times \mathbb{C}^m \to 2^{\mathbb{C}^n}$ by

$$\text{(2.3)} \qquad F(A,b) = \{x \in \mathbb{C}^n : A^*Ax = A^*b\} \ ,$$

where $2^{\mathbb{C}^n}$ denotes the set of all the subsets of $\mathbb{C}^n$.

Note that the point-to-point map x_{LS} defined by (1.3) is the minimal norm selection from F, that is, $x_{LS}(A,b) \in F(A,b)$ and $\|x_{LS}(A,b)\| = \min\{\|x\| : x \in F(A,b)\}$. We've seen that this selection x_{LS} is not continuous.

Let us recall some continuity concept from the set-valued map theory. Let X and Y be Banach spaces. We say that a set-valued map $F : X \to 2^Y$ is upper semicontinuous (u.s.c.) at $x_0 \in X$ if for any open set N containing $F(x_0)$, there exists a neighborhood M of x_0 such that $F(M) \subset N$. Here $F(M) = \bigcup_{x \in M} F(x)$.

Before presenting the main results, we give several useful lemmas.

Lemma 1. For A and $\bar{A} = A + \delta A \in L(\mathbb{C}^n,\mathbb{C}^m)$,

$$\text{(2.4)} \qquad \begin{aligned} \bar{A}^+ - A^+ &= -\bar{A}^+\delta A A^+ + \bar{A}^+(\bar{A}^+)^*(\delta A)^*(I - AA^+) \\ &\quad + (I - \bar{A}^+\bar{A})(\delta A)^*(A^+)^*A^+ \ . \end{aligned}$$

Proof. After direct computation, we have

$$\bar{A}^+ - A^+ = -\bar{A}(\bar{A} - A)A^+ + \bar{A}^+(I - AA^+) - (I - \bar{A}^+\bar{A})A^+ \ .$$

Since

$$\begin{aligned} \bar{A}^+(I - AA^+) &= \bar{A}^+\overline{AA}^+(I - AA^+) \\ &= \bar{A}^+(\overline{AA}^+)^*(I - AA^+) = \bar{A}^+(\bar{A}^+)^*\bar{A}^*(I - AA^+) \end{aligned}$$

$$= \bar{A}^+(\bar{A}^+)^*[(I - AA^+)\bar{A}]^* = \bar{A}^+(\bar{A}^*)^+[(I - AA^+)\delta A]^*$$

$$= \bar{A}^+(\bar{A}^+)^*(\delta A)^*(I - AA^+)$$

and

$$(I - \bar{A}^+\bar{A})A^+ = (I - \bar{A}^+\bar{A})A^+AA^+$$

$$= (I - \bar{A}^+\bar{A})(A^+A)^*A^+ = (I - \bar{A}^+\bar{A})A^*(A^+)^*A^+$$

$$= [A(I - \bar{A}^+\bar{A})]^*(A^+)^*A^+ = [-\delta A(I - \bar{A}^+\bar{A})]^*(A^+)^*A^+$$

$$= -(I - \bar{A}^+\bar{A})(\delta A)^*(A^+)^*A^+ ,$$

we have (2.4). Q.E.D.

Lemma 2. Let A and $\bar{A} = A + \delta A \in L(\mathbb{C}^n,\mathbb{C}^m)$. If $\|\delta A\|\|A^+\| < 1$, then $\operatorname{rank}(\bar{A}) \geq \operatorname{rank}(A)$. In addition, if $\operatorname{rank}(\bar{A}) = \operatorname{rank}(A)$, then

$$\|\bar{A}^+\| \leq \|A^+\|/(1 - \|\delta A\|\|A^+\|) . \tag{2.5}$$

Proof. See for example, [4]. Q.E.D.

In the following, the first theorem is about the perturbation estimate for the point-valued map x_{LS} and the second one is the analysis for the set-valued map F.

Theorem 1. Let A, $\bar{A} = A + \delta A \in L(\mathbb{C}^n,\mathbb{C}^m)$ and b, $\bar{b} = b + \delta b \in \mathbb{C}^m$. Suppose $\|\delta A\| \leq \varepsilon\|A\|$ and $\|\delta b\| \leq \varepsilon\|b\|$. Let $\kappa = \kappa(A) = \|A\|\|A^+\|$. If $\kappa\varepsilon < 1$ and $\operatorname{rank}(\bar{A}) = \operatorname{rank}(A)$, then

$$\frac{\|\bar{x}_{LS} - x_{LS}\|}{\|x_{LS}\|} \leq \kappa\varepsilon\left[\frac{1}{(1 - \kappa\varepsilon)}\frac{\|b\|}{\|Ax_{LS}\|} + \frac{1}{1 - \kappa\varepsilon} + \frac{\kappa}{(1 - \kappa\varepsilon)^2}\frac{\|b - Ax_{LS}\|}{\|Ax_{LS}\|} + 1\right]. \tag{2.6}$$

In addition, if the linear system $Ax = b$ is consistent, then

$$\frac{\|\bar{x}_{LS} - x_{LS}\|}{\|x_{LS}\|} \leq \kappa\varepsilon\left[\frac{2}{1 - \kappa\varepsilon} + 1\right] . \tag{2.7}$$

Here, $x_{LS} = A^+b$ and $\bar{x}_{LS} = \bar{A}^+\bar{b}$.

Proof. From Lemma 1,

$$\|\bar{x}_{LS} - x_{LS}\| = \|\bar{A}^+\bar{b} - A^+b\| = \|\bar{A}^+\delta b + (\bar{A}^+ - A^+)b\|$$

$$= \|\bar{A}^+\delta b - \bar{A}^+\delta AA^+b + \bar{A}^+(\bar{A}^+)^*(\delta A)^*(I - AA^+)b$$

$$+ (I - \bar{A}^+\bar{A})(\delta A)^*(A^+)^*A^+b\|$$

$$\leq \|\bar{A}^+\|\|\delta b\| + \|\bar{A}^+\|\|\delta A\|\|x_{LS}\| + \|\bar{A}^+\|^2\|\delta A\|\|b - Ax_{LS}\|$$

$$+ \|\delta A\|\|A^+\|\|x_{LS}\| .$$

With the help of Lemma 2, we get

$$\frac{\|\bar{x}_{LS} - x_{LS}\|}{\|x_{LS}\|} \le \varepsilon\Big[\frac{\|A^+\|\|A\|\|b\|}{(1 - \|\delta A\|\|A^+\|)\|A\|\|x_{LS}\|} + \frac{\|A^+\|\|A\|}{1 - \|\delta A\|\|A^+\|}$$

$$+ \frac{\|A^+\|^2\|A\|^2\|b - Ax_{LS}\|}{(1 - \|\delta A\|\|A^+\|)^2\|A\|\|x_{LS}\|} + \|A^+\|\|A\|\Big]$$

$$\le \kappa\varepsilon\Big[\frac{1}{(1 - \kappa\varepsilon)}\frac{\|b\|}{\|Ax_{LS}\|} + \frac{1}{1 - k\varepsilon} + \frac{\kappa}{(1 - \kappa\varepsilon)^2}\frac{\|b - Ax_{LS}\|}{\|Ax_{LS}\|} + 1\Big]$$

which is (2.6). In particular, if $b \in R(A)$, then $b - Ax_{LS} = 0$. So

$$\frac{\|\bar{x}_{LS} - x_{LS}\|}{\|x_{LS}\|} \le \kappa\varepsilon\Big[\frac{2}{1 - \kappa\varepsilon} + 1\Big] .$$ Q.E.D.

For the next theorem, we assume $\operatorname{rank}(A) = r$ and without loss of generality assume $A = [B,C]$ where the $m \times r$ submatrix B has full rank. In parallel, we have the following result.

Theorem 2. Given A and $\bar{A} = A + \delta A \in L(\mathbb{C}^n,\mathbb{C}^m)$, b and $\bar{b} = b + \delta b \in \mathbb{C}^m$. Suppose $\|\delta A\| \le \varepsilon\|A\|$ and $\|\delta b\| \le \varepsilon\|b\|$. Let $\tilde{\kappa} = \|A\|\|B^+\| = \kappa\|B^+\|/\|A^+\|$. If $\tilde{\kappa}\varepsilon < 1$, then for any $\bar{x} \in F(\bar{A},\bar{b})$, there is $x = (y,z)^T \in F(A,b)$ such that

$$\frac{\|\bar{x} - x\|}{\|x\|}$$

$$\le \tilde{\kappa}\varepsilon\Big[\frac{1}{1 - \tilde{\kappa}\varepsilon}\frac{\|b\|}{\|Ax\|} + \frac{1}{1 - \tilde{\kappa}\varepsilon} + \frac{\tilde{\kappa}}{(1 - \tilde{\kappa}\varepsilon)^2}\frac{\|b-Ax\|}{\|Ax\|} + 1\Big].$$

(2.8)

In particular, if the linear system $Ax = b$ is consistent, then

(2.9) $$\frac{\|\bar{x} - x\|}{\|x\|} \le \tilde{\kappa}\varepsilon\Big[\frac{2}{1 - \tilde{\kappa}\varepsilon} + 1\Big] .$$

Proof.

$$F(A,b) = \{x \in \mathbb{C}^n : A^*Ax = A^*b\}$$

$$= \left\{x = \begin{pmatrix} y \\ z \end{pmatrix} \in \mathbb{C}^r \times \mathbb{C}^{n-r} : \begin{bmatrix} B^*B & B^*C \\ C^*B & C^*C \end{bmatrix}\begin{bmatrix} y \\ z \end{bmatrix} = \begin{bmatrix} B^* \\ C^* \end{bmatrix} b\right\}$$

$$= \left\{\begin{pmatrix} y \\ z \end{pmatrix} \in \mathbb{C}^n : [B^*B, B^*C]\begin{bmatrix} y \\ z \end{bmatrix} = B^*b\right\}$$

$$= \left\{\begin{pmatrix} y \\ z \end{pmatrix} \in \mathbb{C}^n : y = (B^*B)^{-1}[B^*b - B^*Cz]\right\}$$

$$= \left\{\begin{pmatrix} y \\ z \end{pmatrix} \in \mathbb{C}^n : y = B^+b - B^+Cz\right\}$$

$$= \left\{ \begin{pmatrix} B^+b - B^+Cz \\ z \end{pmatrix} \in \mathbb{C}^n : z \in \mathbb{C}^{n-r} \right\}$$

since $\operatorname{rank}(A^*A) = \operatorname{rank}(A) = \operatorname{rank}(B) = \operatorname{rank}(B^*B) = r$ and $B^+ = (B^*B)^{-1}B^*$ for the full column-ranked matrix B.

Now

$$\bar A = A + \delta A = [B,C] + [\delta B,\delta C]$$
$$= [B + \delta B, C + \delta C] = [\bar B,\bar C] \ .$$

Since $\|\delta B\| \leq \|\delta A\|$ by definition, we have

$$\|\delta B\|\|B^+\| \leq \|\delta A\|\|B^+\| \leq \varepsilon\|A\|\|B^+\| = \kappa\varepsilon\|B^+\|/\|A^+\| = \tilde\kappa\varepsilon < 1 \ .$$

By Lemma 2, $\operatorname{rank}(\bar B) = \operatorname{rank}(B)$ and $\|\bar B^+\| \leq \|B^+\|/(1-\|\delta B\|\|B^+\|)$. It follows that

$$F(\bar A,\bar b) = \{\bar x \in \mathbb{C}^n : \bar A^*\bar A\bar x = \bar A^*\bar b\}$$
$$= \left\{ \bar x = \begin{pmatrix} \bar y \\ \bar x \end{pmatrix} \in \mathbb{C}^r \times \mathbb{C}^{n-r} : \begin{bmatrix} \bar B^*\bar B & \bar B^*\bar C \\ \bar C^*\bar B & \bar C^*\bar C \end{bmatrix} \begin{bmatrix} \bar y \\ \bar z \end{bmatrix} = \begin{bmatrix} \bar B^* \\ \bar C^* \end{bmatrix} \bar b \right\}$$
$$\subset \left\{ \begin{pmatrix} \bar y \\ \bar z \end{pmatrix} \in \mathbb{C}^n : [\bar B^*\bar B, \bar B^*\bar C] \begin{bmatrix} \bar y \\ \bar z \end{bmatrix} = \bar B^*\bar b \right\}$$
$$= \left\{ \begin{pmatrix} \bar y \\ \bar z \end{pmatrix} \in \mathbb{C}^n : \bar y = (\bar B^*\bar B)^{-1}[\bar B^*\bar b - \bar B^*\bar C\bar z] \right\}$$
$$= \left\{ \begin{pmatrix} \bar B^+\bar b - \bar B^+\bar C\bar z \\ \bar z \end{pmatrix} \in \mathbb{C}^n : \bar z \in \mathbb{C}^{n-r} \right\} .$$

For any $\bar x \in F(\bar A,\bar b)$, $\bar x = \begin{pmatrix} \bar y \\ \bar z \end{pmatrix} = \begin{pmatrix} \bar B^+\bar b - \bar B^+\bar C\bar z \\ \bar z \end{pmatrix}$. Let

$$x = \begin{pmatrix} y \\ z \end{pmatrix} = \begin{pmatrix} B^+b - B^+C\bar z \\ \bar z \end{pmatrix} .$$

Then $x \in F(A,b)$. From Lemma 1 and the expression of x,

$$\bar y - y = (\bar B^+\bar b - \bar B^+\bar C\bar z) - (B^+b - B^+C\bar z)$$
$$= (\bar B^+\bar b - B^+b) - (\bar B^+\bar Cz - B^+Cz)$$
$$= [\bar B^+\delta b - \bar B^+\delta BB^+b + \bar B^+(\bar B^+)^*(\delta B)^*(b - BB^+b)$$
$$+ (I - \bar B^+\bar B)(\delta B)^*(B^+)^*B^+b]$$
$$- [\bar B^+\delta Cz - \bar B^+\delta BB^+Cz + \bar B^+(\bar B^+)^*(\delta B)^*(Cz - BB^+Cz)$$
$$+ (I - \bar B^+\bar B)(\delta B)^*(B^+)^*B^+Cz]$$
$$= \bar B^+(\delta b - \delta Ax) + \bar B^+(\bar B^+)^*(\delta B)^*(b - Ax)$$
$$+ (I - \bar B^+\bar B)(\delta B)^*(B^+)^*B^+(b-Cz) .$$

Since $\|x\|^2 = \|B^+(b-Cz)\|^2 + \|z\|^2$, $\|B^+(b-Cz)\| \le \|x\|$. Hence

$$\frac{\|\bar{x} - x\|}{\|x\|} = \frac{\|\bar{y} - y\|}{\|x\|}$$

$$\le \frac{\|B^+\|}{1 - \|\delta B\|\|B^+\|}\left(\frac{\|\delta b\|}{\|x\|} + \|\delta A\|\right)$$

$$+ \frac{\|B^+\|^2}{(1 - \|\delta B\|\|B^+\|)^2}\,\frac{\|\delta B\|\|b - Ax\|}{\|x\|} + \frac{\|\delta B\|\|B^+\|\|B^+(b-Cz)\|}{\|x\|}$$

$$\le \frac{\tilde{\kappa}\varepsilon}{1 - \tilde{\kappa}\varepsilon}\left(\frac{\|b\|}{\|Ax\|} + 1\right) + \frac{\tilde{\kappa}^2\varepsilon}{(1 - \tilde{\kappa}\varepsilon)^2}\,\frac{\|b - Ax\|}{\|Ax\|} + \tilde{\kappa}\varepsilon$$

$$\le \tilde{\kappa}\varepsilon\left[\frac{1}{1 - \tilde{\kappa}\varepsilon}\,\frac{\|b\|}{\|Ax\|} + \frac{1}{1 - \tilde{\kappa}\varepsilon} + \frac{\tilde{\kappa}}{(1 - \tilde{\kappa}\varepsilon)^2}\,\frac{\|b - Ax\|}{\|Ax\|} + 1\right].$$

In particular, if $b \in R(A)$, then $b = Ax$. Consequently,

$$\frac{\|\bar{x} - x\|}{\|x\|} \le \tilde{\kappa}\varepsilon\left[\frac{2}{1 - \tilde{\kappa}\varepsilon} + 1\right] . \qquad \text{Q.E.D.}$$

As a corollary of Theorem 2, we have the following result about upper semicontinuity. Let $B(R) = \{x \in \mathbb{C}^n : \|x\| \le R\}$ and define a set-valued map $F_R : L(\mathbb{C}^n,\mathbb{C}^m) \times \mathbb{C}^m \to 2^{\mathbb{C}^n}$ by

(2.11) $$F_R(A,b) = F(A,b) \cap B(R).$$

Corollary 1. Suppose for some R, $F_R(A,b) \ne \emptyset$, then F_R is upper semicontinuous at (A,b).

Proof. Let N be any open subset of $\mathbb{C}^n$ containing $F_R(A,b)$. Because $F_R(A,b)$ is nonempty and compact, N contains an open neighborhood $B(F_R(A,b),\delta)$ defined by $B(F_R(A,b),\delta) = \{x \in \mathbb{C}^n : \text{dist}(x,F_R(A,b)) < \delta\}$. So we may assume $N = B(F_R(A,b),\delta)$ for some $\delta \in (0,1)$. From the proof of Theorem 2, for $\|\delta A\| + \|\delta b\|$ sufficiently small, if $\bar{x} \in F_R(\bar{A},\bar{b})$, then

$$\|\bar{x} - x\| \le \|\bar{B}^+(\delta b-\delta Ax)\|$$
$$+ \|\bar{B}^+(\bar{B}^+)^*(\delta B)^*(b - Ax)\| + \|(\delta B)^*(B^+)^*B^+(b-Cz)\|$$

for some $x = (y,z)^T \in F(A,b)$. Since $\|z\| \le \|\bar{x}\| \le R$, there exists $\eta > 0$ such that $\|\delta A\| + \|\delta b\| < \eta$ implies

$$\|\bar{x} - x\| < \beta \equiv \delta^2/8(R + 1) .$$

If $\|x\| \le R$, then $x \in F_R(A,b)$ and $\|\bar{x} - x\| < \beta < \delta$ implies that $\bar{x} \in N$.

Suppose now $\|x\| > R$. Let $y \in F_R(A,b)$ be defined by

$$\|x - y\| = \min\{\|x - z\| : z \in F_R(A,b)\} .$$

Since $x \notin B(R)$, in the plane determined by x and y using the Law of cosine, we have

$$\|x - y\|^2 + \|y\|^2 \le \|x\|^2 .$$

Now $\|y\|^2 = R^2$ and $\|x\|^2 \le (\|x - \bar{x}\| + \|\bar{x}\|)^2 \le (\beta + R)^2$, it follows that

$$\|x - y\|^2 \le \|x\|^2 - \|y\|^2 \le \beta^2 + 2\beta R ,$$

and so

$$\begin{aligned} \|\bar{x} - y\| &\le \|\bar{x} - x\| + \|x - y\| \\ &< \beta + \sqrt{\beta^2 + 2\beta R} < \beta + \sqrt{\beta(2R + 1)} \\ &< \frac{\delta}{2} + \frac{\delta}{2\sqrt{2(R + 1)}} \cdot \sqrt{2R + 1} < \frac{\delta}{2} + \frac{\delta}{2} = \delta . \end{aligned}$$

This says that $\bar{x} \in N$.

In summary, for all $(\bar{A},\bar{b})$ satisfying $\|\bar{A} - A\| + \|\bar{b} - b\| < \eta$,

$$F_R(\bar{A},\bar{b}) \subset N = B(F_R(A,b),\delta) .$$

Thus we've proved that $F_R : L(\mathbb{C}^n,\mathbb{C}^m) \times \mathbb{C}^m \to 2^{\mathbb{C}^n}$ is upper semi-continuous in $\mathrm{Dom}(F_R) \equiv \{(A,b) \in L(\mathbb{C}^n,\mathbb{C}^m) \times \mathbb{C}^m : F_R(A,b) \ne \emptyset\}$. Q.E.D.

Remark. It is easy to see that the set-valued map F is not upper semicontinuous. This follows from the fact that given a proper linear subspace S of $\mathbb{C}^n$, there is a neighborhood N of S such that N doesn't contain any other subspaces with the same dimension as S.

3. Numerical Least Squares Problems

The discussion in the previous sections tells us that the (minimal norm) least squares problem is not well-posed when the matrix is not full-ranked. At the same time, however, the argument there also implies an idea to eliminate the intrinsical singularity numerically. This section is for such a purpose.

Suppose we solve the following (minimal norm) least squares problem

$$\text{(3.1)} \qquad \text{minimize}\{\|x\| : \|Ax - b\| = \min_{z \in \mathbb{C}^n} \|Az - b\|\}$$

with $\mathrm{rank}(A) = \mathrm{rank}[B,C] = \mathrm{rank}(B) = r$. Owing to the round-off error, we have to solve the following perturbed one

$$\text{(3.2)} \qquad \text{minimize}\{\|\bar{x}\| : \|\bar{A}\bar{x} - \bar{b}\| = \min_{\bar{z} \in \mathbb{C}^n} \|\bar{A}\bar{z} - \bar{b}\|\} .$$

The corresponding normal equations for (3.2) is

$$\bar{A}^*\bar{A}\bar{x} = \bar{A}^*\bar{b}$$

or

$$(3.3)\qquad \begin{bmatrix} \bar{B}^*\bar{B} & \bar{B}^*\bar{C} \\ \bar{C}^*\bar{B} & \bar{C}^*\bar{C} \end{bmatrix} \begin{bmatrix} \bar{y} \\ \bar{z} \end{bmatrix} = \begin{bmatrix} \bar{B}^* \\ \bar{C}^* \end{bmatrix}$$

If the perturbation is rather small, it is guaranteed that $\text{rank}(\bar{B}) = \text{rank}(B)$. Now, instead of solving (3.3), we solve the "regularized" system

$$(3.4)\qquad [\bar{B}^*\bar{B},\bar{B}^*\bar{C}]\begin{bmatrix} \bar{y} \\ \bar{z} \end{bmatrix} = \bar{B}^*\bar{b}\ .$$

Denote $D = [B^*B,B^*C]$ and $\bar{D} = [\bar{B}^*\bar{B},\bar{B}^*\bar{C}]$. Then the solution to (3.1) is $x_{LS} = D^+B^*b$ and the solution to (3.4) $\hat{x}_{LS} = \bar{D}^+\bar{B}^*\bar{b}$ is considered as the computed minimal norm least squares solution to (3.1). As an application of Theorem 1, we have the error estimates as follows.

Corollary 2. Let A, $\bar{A} = A + \delta A \in L(\mathbb{C}^n,\mathbb{C}^m)$ and b, $\bar{b} = b + \delta b \in \mathbb{C}^m$. Suppose $\|\delta A\| \le \varepsilon\|A\|$ and $\|\delta b\| \le \varepsilon\|b\|$. Then there exists a constant H such that for $\kappa\varepsilon << 1$,

$$(3.5)\qquad \frac{\|\hat{x}_{LS} - x_{LS}\|}{\|x_{LS}\|} \le H\cdot\kappa\varepsilon$$

Proof. Since $x_{LS} = D^+B^*b$ and $\hat{x}_{LS} = \bar{D}^+\bar{B}^*\bar{b}$ satisfy $Dx_{LS} = B^*b$ and $\bar{D}\hat{x}_{LS} = \bar{B}^*\bar{b}$, respectively, from the inequality (2.7) in Theorem 1, if $\|\bar{D} - D\|\|D^+\| < 1$, then

$$\frac{\|\hat{x}_{LS} - x_{LS}\|}{\|x_{LS}\|} \le \|\bar{D} - D\|\|D^+\|\left[\frac{2}{1 - \|\bar{D} - D\|\|D^+\|} + 1\right].$$

Note that

$$\|\bar{D} - D\| \le \|\bar{B}^* - B^*\|\|\bar{A}\| + \|B^*\|\|\bar{A} - A\|$$
$$\le \left(\frac{1}{1 - \kappa\varepsilon} + 1\right)\|A\|^2\varepsilon = \left(\frac{1}{1 - \kappa\varepsilon} + 1\right)\frac{\|A\|^2}{\kappa}\kappa\varepsilon\ ,$$

we have for $\kappa\varepsilon$ sufficiently small

$$\frac{\|\hat{x}_{LS} - x_{LS}\|}{\|x_{LS}\|} \le \left(\frac{1}{1 - \kappa\varepsilon} + 1\right)\frac{\|A\|^2\|D^+\|}{\kappa}\kappa\varepsilon \le H\cdot\kappa\varepsilon$$

with $H = \dfrac{3\|A\|^2\|D^+\|}{\kappa}$. Q.E.D.

The key idea in our revised form of least squares problems with perturbation is that of making the perturbation keep the rank constant. We call this procedure the rank-reserved deformation and this idea may be extended to other singularity problem in numerical analysis.

In practice, we don't use the normal equations to solve the least squares problems because of the consideration of numerical stability. We usually use unitary transformations to make the matrix A into an upper triangular form. If A is full-ranked, the computed triangular matrix is still full-ranked for the small perturbations and the computed minimal norm least squares solution is close to the true solution to the unperturbed problem if the problem is well-conditioned. If A is rank-deficient, we must use the rank-reserved deformation idea to compute the solution.

Suppose for the moment that it is known that rank(A) = rank([B,C]) = rank B = r. Then after r times unitary transformation, say Householder's transformations, we have

$$(3.6)\qquad fl(U_r \cdots U_1 A) = \begin{bmatrix} \bar{R} & \bar{E} \\ 0 & \bar{F} \end{bmatrix} = \bar{A}_r ,$$

where $fl(\cdot)$ means the computed result of the expression in the parenthesis, $\bar{R}$ is an $r \times r$ nonsingular matrix of upper triangular form. Since rank(A) = r, in general, the small round-off error introduced in the calculation implies that each element of $\bar{F}$ is small. So instead of solving

$$\text{minimize}\{\|\bar{x}\| : \|\bar{A}_r\bar{x} - \bar{b}_r\| = \min_{\bar{z}\in\mathbb{C}^n} \|\bar{A}_r\bar{z} - \bar{b}_r\|\} ,$$

where $\bar{b}_r = fl(U_r \cdots U_1 b)$, we solve

$$(3.7)\qquad \text{minimize}\{\|\bar{x}\| : \|[\bar{R},\bar{E}]\bar{x} - \bar{b}_r^1\| = \min_{\bar{z}\in\mathbb{C}^n} \|[\bar{R},\bar{E}]\bar{z} - \bar{b}_r^1\|\}$$

with $\bar{b}_r^1$ the r-vector consisting of the first r elements of $\bar{b}_r$.

Since

$$U_r \cdots U_1(A,b) = \begin{bmatrix} R & E & b_r \\ 0 & 0 & b_r^2 \end{bmatrix} ,$$

the problem (3.1) is equivalent to the problem

$$(3.8)\qquad \text{minimize}\{\|x\| : \|[R,E]x - b_r^1\| = \min_{z\in\mathbb{C}^n} \|[R,E]z - b_r^1\|\} .$$

By our previous results, if $\|\bar{R} - R\|$, $\|\bar{E} - E\|$ and $\|\bar{b}_r^1 - b_r^1\|$ are sufficiently small, the solution to (3.7) is very near to that of (3.8). Hence, artificially, we have eliminated the singularity by forcing the rank unchanged.

Commonly, we don't know in advance the rank of A and furthermore, the matrix A is not of the form [B,C] with rank(A) = rank(B). Now we can combine our rank-reserved deformation principle with the pivot rule to change A into a upper triangular form. Specifically, suppose through s Householder's transformations, we have

$$fl(U_s \cdots U_1 A) = \begin{bmatrix} \bar{Q} & \bar{P} \\ 0 & \bar{S} \end{bmatrix}$$

where $\bar{Q}$ is an $s \times s$ upper triangular nonsingular matrix. If the (s + 1)-th column of A is a linear combination of the previous ones, then the first column $\bar{s}_1$ of $\bar{S}$ is very small in norm. In this case, we have $\bar{s}_1$ become 0 and continue with the next column. Otherwise, we use the next Householder's transformation to transform $\bar{s}_1$ into $(\alpha,0,\cdots,0)^T$ with $|\alpha| = \|\bar{s}_1\|$. In this way, after a permutation of columns in $fl(U_r \cdots U_1 A)$ we have the required triangular form.

Before ending this section, we say something about the condition of the problem (3.1). In numerical algebra, ill-condition of a problem is often caused by a "almost singular" matrix. It is not true, however, that the cause of ill-condition is just the singularity. In the least squares problems, for example, the condition number of the matrix A is $\kappa = \|A\|\|A^{+}\| = \sigma_1/\sigma_r$ where σ_1 and σ_r are the maximum and minimum singular values for A, respectively. It follows that if σ_r is well separated from zero then the condition number κ of A is not big even though A is rank-deficient. On the other hand, for an $n \times n$ nonsingular matrix A, if σ_n is very small, then κ may be very large to make A very ill-conditioned.

4. Conclusion

In this paper, we obtained the error bounds for general linear least squares problems under arbitrary small perturbation and thus investigated the upper semicontinuity property of the least squares solution map. To our knowledge, the results here have some advantages over other known ones. Firstly, we don't have extra restrictions to the problems and perturbations, that is, we don't require the unperturbed problem is consistent and full-ranked and we consider any small perturbation not necessary rank-keeping. Secondly, we use only the spectral norm for vectors and matrices throughout the discussion without seeking help of Frobenius norm and the results seem more concise and symmetric. Lastly, the introduction of the concept of continuity of set-valued maps gives us some means to investigate the perturbation influence in numerical analysis.

References

[1] J.P. Aubin and A. Cellina, Differential Inclusions, Springer-Verlag, New York, 1984.

[2] X. He, The Theory and Numerical Methods of Generalized Inverses and Linear Least Squares Problems, Nanjing University Press, Nanjing, 1983.

[3] T. Kato, Perturbation Theory for Linear Operators, Springer-Verlag, New York, 1980.

[4] M. Wei, "The perturbation of consistent least squares problems", Linear Algebra and its Applications 112 (1989), 231-245.

[5] J.M. Wilkinson, Algebraic Eigenvalue Problems, Clarendon Press, Oxford, 1965.

Additive decomposition of a transfer function with respect to a specified region

Bo Kågström and Paul Van Dooren

Abstract A robust and reliable method for computing the stable projection (with respect to a specified region Γ in the complex plane) of a transfer matrix given by its (generalized) state space realization $H(\lambda) = C(\lambda E - A)^{-1}B$ is presented. The sensitivity of the problem and the stability of the proposed computational algorithm are discussed and compared with a more classical approach. Numerical test results are also reported.

1. Introduction

Constructing the stable projection $H_\gamma(\lambda)$ of a given transfer matrix $H(\lambda)$ is a problem which appears in several analysis and design problems of systems and control theory ([5], [13], [4]). It could be briefly described as follows. Let Γ be a region of the complex plane and let $H_{mn}(\lambda)$ be a m×n rational transfer matrix with poles inside and outside the region Γ ($\mathbf{C} = \Gamma \cup \Gamma_c$, $\Gamma \cap \Gamma_c = \emptyset$). Then $H(\lambda)$ always has an *additive decomposition*

(1) $$H(\lambda) = H_\gamma(\lambda) + H_{\bar\gamma}(\lambda)$$

where $H_\gamma(\lambda)$ has only poles inside Γ, and $H_{\bar\gamma}(\lambda)$ has only poles outside Γ (i.e. inside Γ_c). The term $H_\gamma(\lambda)$ is then also called the *stable projection* of $H(\lambda)$ on Γ. Notice that this decomposition is not unique. Indeed, let $H_\gamma(\lambda)$ and $H_{\bar\gamma}(\lambda)$ be a solution to this problem then $H_\gamma(\lambda) + M$ and $H_{\bar\gamma}(\lambda) - M$ with M an arbitrary m×n *constant* matrix, is also a solution to (1), since adding or subtracting a constant matrix does not affect the poles of a transfer matrix. The classical construction of a solution to the above problem is the *partial fraction expansion* of $H(\lambda)$:

(2a) $$H(\lambda) = \sum_{i=1}^{k} H_{\alpha_i}(\lambda) + H_0 + H_\infty(\lambda)$$

where

(2b) $$H_\infty(\lambda) \equiv H_1\lambda + H_2\lambda^2 + \ldots + H_d\lambda^d$$

and

(2c) $$H_{\alpha_i}(\lambda) \equiv \frac{H_{\alpha_i,1}}{(\lambda - \alpha_i)} + \ldots + \frac{H_{\alpha_i,d_i}}{(\lambda - \alpha_i)^{d_i}}$$

The poles of $H(\lambda)$ are clearly the $\{\alpha_i, i = 1, \ldots, k\}$ and $\alpha_{k+1} = \infty$ (depending on the existence of the H_∞ -part). It is clear that $H(\lambda)$ and $H_\gamma(\lambda)$ *must* have coinciding polar parts $H_{\alpha_i}(\lambda)$ for $\alpha_i \in \Gamma$ and similarly $H(\lambda)$ and $H_{\bar{\gamma}}(\lambda)$ must have coinciding polar parts $H_{\alpha_i}(\lambda)$ for $\alpha_i \in \Gamma_c$. One now obtains a construction for $H_\gamma(\lambda)$ and $H_{\bar{\gamma}}(\lambda)$ by merely grouping in $H\gamma(\lambda)$ all polar parts $H_{\alpha_i}(\lambda)$ corresponding to a pole inside Γ and grouping in $H_{\bar{\gamma}}(\lambda)$ all polar parts $H_{\alpha_i}(\lambda)$ corresponding to a pole outside Γ. This immediately reflects the degree of freedom in the constant term H_0 which can arbitrarily be distributed over both $H_\gamma(\lambda)$ and $H_{\bar{\gamma}}(\lambda)$. The problem becomes uniquely defined though, when one adds the constraint

(3) $$H_\gamma(\lambda_0) = \bar{H}_0\,, \quad \lambda_0 \in \Gamma_c$$

for an arbitrary point λ_0 outside Γ, since that fixes the constant part of $H_\gamma(\lambda)$.

2. A state-space approach

Assume that $H_{mn}(\lambda)$ is given via its generalized state-space realization:

(4) $$H(\lambda) = C(\lambda E - A)^{-1}B + D$$

where $\{\lambda E - A, B, C, D\}$ is a minimal realization, i.e. *rank* E equals the McMillan degree of $H(\lambda)$ [11]. Algorithms for obtaining such realizations are e.g. given in [11]. If $H(\lambda)$ happens to be bounded at $\lambda = \infty$, then a realization can always be chosen with $E = I$, which is then a (standard) state space realization. Notice that a minimal realization such as (4) is not unique. Several such realizations can in fact be derived from (4) via the equivalence relation:

(5) $$\{\lambda E - A, B, C, D\} \overset{U,\,V}{\to} \{U^{-1}(\lambda E - A)V, U^{-1}B, CV, D\}$$

where U and V are nonsingular matrices. If we define

(6) $$S(\lambda) = \begin{bmatrix} \lambda E - A & B \\ -C & D \end{bmatrix}$$

to be the *system matrix* of the realization (4) then the equivalent realization (5) has a system matrix:

(7) $$S_{U,V}(\lambda) = \begin{bmatrix} U^{-1} & 0 \\ 0 & I_m \end{bmatrix} \begin{bmatrix} \lambda E - A & B \\ -C & D \end{bmatrix} \begin{bmatrix} V & 0 \\ 0 & I_n \end{bmatrix}$$

which is a compact notation always used in the sequel. That both $S(\lambda)$ and $S_{U,V}(\lambda)$ correspond to the same transfer function $H(\lambda)$ is easily verified from (4). By choosing U and V to be the transformation matrices reducing the regular pencil $\lambda E - A$ to its *Weierstrass canonical form*:

(8a) $$U^{-1}(\lambda E - A)V = \text{diag}\{\lambda I - J, \lambda N - I\}$$

where

(8b) $\lambda I - J$ is in Jordan canonical form (finite eigenvalues α_i)

(8c) N is nilpotent and in Jordan canonical form (infinite eigenvalue),

the corresponding realization $S_{U,V}(\lambda)$ yields information which is very close to the partial fraction expansion of $H(\lambda)$. Denoting $B_U = U^{-1}B$ and $C_V = CV$ one then has indeed

(9) $$H(\lambda) = C_V \, \text{diag}\{(\lambda I - J)^{-1}, (\lambda N - I)^{-1}\} \, B_U + D$$

Let $(\lambda - \alpha_i)I - N_i$ be the diagonal block of $\lambda I - J$ corresponding to a given eigenvalue $\lambda = \alpha_i$ (N_i is thus nilpotent) and let B_{U,α_i} be the corresponding block rows of B_U and C_{V,α_i} the corresponding block columns of C_V. Then, clearly

(10) $$H_{\alpha_i}(\lambda) = C_{V,\alpha_i}[(\lambda - \alpha_i)I - N_i]^{-1}B_{U,\alpha_i}$$

is the polar section of $H_{\alpha_i}(\lambda)$ at $\lambda = \alpha_i$ and the coefficients $H_{\alpha_{i,j}}$ of (2c) can easily be derived from appropriate columns and rows of C_{V,α_i} and B_{U,α_i}, respectively. We only pointed out this here to show the connection between the Weierstrass form of a state-space realization and the partial fraction expansion of its transfer function. At the same time we also indicate the weakness of the partial fraction expansion, since it is related to the Weierstrass canonical form: its computation may be *extremly* sensitive when $H(\lambda)$ has multiple poles or poles that are close to each other.

3. Deflating subspaces

At the same time we have suggested above that the solution to the additive decomposition is in fact by a *block decomposition* of the type:

(11) $$S_{U,V}(\lambda) = \begin{bmatrix} \lambda E_{11} - A_{11} & 0 & B_1 \\ 0 & \lambda E_{22} - A_{22} & B_2 \\ -C_1 & -C_2 & D \end{bmatrix}$$

Indeed, it is readily verified that the transfer function of $S_{U,V}(\lambda)$ is:

(12) $$H(\lambda) = C_1(\lambda E_{11} - A_{11})^{-1} B_1 + C_2(\lambda E_{22} - A_{22})^{-1} B_2 + D$$

and that it thus suffices to find a decomposition of the type (11) where $\lambda E_{11} - A_{11}$ has only eigenvalues inside Γ and $\lambda E_{22} - A_{22}$ only outside Γ. Such block decompositions boil down to the computation of two complementary deflating subspaces of the original pencil $\lambda E - A$ [3]. For explaining this we first recall the definition of a deflating subspace of a regular pencil $\lambda E - A$ [10]. Any subspace **X** satisfying

(13) $$\mathbf{Y} = E\mathbf{X} + A\mathbf{X}; \quad \dim\mathbf{Y} = \dim\mathbf{X}$$

is called a *deflating subspace* of the pencil $\lambda E - A$. The name stems from the following decomposition

(14) $$Q^*(\lambda E - A)Z = \begin{bmatrix} \lambda E_{11} - A_{11} & \lambda E_{12} - A_{12} \\ 0 & \lambda E_{22} - A_{22} \end{bmatrix}$$

which is directly obtained from (13) using any *unitary* transformations Q and Z such that the first n_1 columns of Z and Q span the subspaces **X** and **Y**, respectively [10]. The block decomposition (14) shows that the generalized eigenvalue problem $\lambda E - A$ is now "deflated" into two smaller dimensional problems $\lambda E_{ii} - A_{ii}$, i=1, 2. The spectrum of $\lambda E_{11} - A_{11}$ associated with the subspace **X** uniquely determines the space **X** if and only if it is disjoint from the spectrum of $\lambda E_{22} - A_{22}$ [10]. Indeed **X** is then the subspace spanned by the eigenvectors and principal vectors of higher grade associated with those eigenvalues. But since the spectra of $\lambda E_{11} - A_{11}$ and $\lambda E_{22} - A_{22}$ are disjoint there is also a unique deflating subspace $\mathbf{X_c}$ with the spectrum of $\lambda E_{22} - A_{22}$. In [3] it is shown that these spaces are then *complementary* (i.e. they are linear independent and add up to the whole space) as well as their ranges $\mathbf{Y} = A\mathbf{X} + E\mathbf{X}$ and $\mathbf{Y_c} = A\mathbf{X_c} + E\mathbf{X_c}$. Further, there exists a decomposition

(15) $$U^{-1}(\lambda E - A)V = \begin{bmatrix} \lambda E_{11} - A_{11} & 0 \\ 0 & \lambda E_{22} - A_{22} \end{bmatrix}$$

where V and U are no longer unitary, but are such that their first n_1 columns span **X** and **Y**, respectively, and their last n_2 columns span $\mathbf{X_c}$ and $\mathbf{Y_c}$. The complementarity of these pairs of subspaces also ensures that U and V are invertible [3], and U and V are given by:

(16) $$U = [Y \mid Y_c]; \quad V = [X \mid X_c]$$

where X, X_c, Y and Y_c are submatrices whose columns span the spaces **X**, $\mathbf{X_c}$, **Y** and $\mathbf{Y_c}$, respectively. From (15) (16) it follows that our problem is solved by the computation of the subspaces

(17) $$\mathbf{X} = \mathbf{X}_\gamma \; ; \quad \mathbf{X}_c = \mathbf{X}_{\bar{\gamma}}$$

corresponding to the complementary regions Γ and Γ_c we are interested in. These spaces are unique, as follows from our discussion and the matrices U and V derived from (16) are then invertible and yield the derived decomposition (15).

4. Algorithmic aspects

In order to construct a decomposition of the type (11) we first transform $\lambda E - A$ to generalized Schur form by applying the QZ-algorithm [7]. Then the eigenvalues of $\lambda E -$

A are reordered so that the n_1 eigenvalues inside Γ are placed in the (1,1)-block of the transformed pencil and consequently the remaining eigenvalues (outside Γ) are placed in the (2,2)-block of the transformed pencil. This reordering can also be accomplished with unitary matrices [12]. After this generalized state-space transformation (accomplished with unitary transformation matrices Q and Z) we have a system matrix of the form:

$$S_{Q,Z}(\lambda) = \begin{bmatrix} \lambda E_{11} - A_{11} & \lambda E_{12} - A_{12} & B_1 \\ 0 & \lambda E_{22} - A_{22} & B_2 \\ -C_1 & -C_2 & D \end{bmatrix} \tag{18}$$

The next step is to zero out λE_{12} - A_{12} with an equivalence transformation of the form:

$$\begin{bmatrix} I_{n_1} & -L & 0 \\ 0 & I_{n_2} & 0 \\ 0 & 0 & I_m \end{bmatrix} \begin{bmatrix} \lambda E_{11} - A_{11} & \lambda E_{12} - A_{12} & B_1 \\ 0 & \lambda E_{22} - A_{22} & B_2 \\ -C_1 & -C_2 & D \end{bmatrix} \begin{bmatrix} I_{n_1} & R & 0 \\ 0 & I_{n_2} & 0 \\ 0 & 0 & I_n \end{bmatrix} \tag{19}$$

This leads to solve for R and L in the generalized Sylvester equation

$$\begin{aligned} A_{11}R - LA_{22} &= -A_{12} \\ E_{11}R - LE_{22} &= -E_{12} \end{aligned} \tag{20}$$

which is a linear system with $2n_1n_2$ unknowns, the entries of L and R. In [6] efficient and reliable algorithms for solving this matrix equation is presented. One of the algorithms uses the generalized Schur form of both λE_{ii} - A_{ii}, $i = 1, 2$ which happen to be provided by the triangular form described in the previous step (18). Notice that B_1 and C_2 are also affected by the equivalence transformation (19). In the following we summarize the main steps of the proposed algorithm.

Algorithm ADTF (*Additive Decomposition of a Transfer Function*)

Step 1: Apply the QZ-algorithm to $\lambda E - A$.

Step 2: Reorder the eigenvalues so that the n_1 eigenvalues inside Γ are in λE_{11} - A_{11} After steps 1-2 we have the following decomposition of $\lambda E - A$:

$$Q^*(\lambda E - A)\,Z = \begin{bmatrix} \lambda E_{11} - A_{11} & \lambda E_{12} - A_{12} \\ 0 & \lambda E_{22} - A_{22} \end{bmatrix} \tag{21}$$

Step 3: Zero out λE_{12} - A_{12} by solving the generalized Sylvester equation (20).

Step 4: Complete the equivalence transformation from steps 1-2 giving a system matrix of the form (18):

(22) $$B \equiv \begin{bmatrix} B_1 \\ B_2 \end{bmatrix} := Q^* B; \quad C \equiv [C_1 \; C_2] := CZ$$

Step 5: Complete the equivalence transformation (19):

(23) $$B_1 := B_1 - LB_2; \quad C_2 := C_1 R + C_2$$

Step 6: Construct U and V in (16) from steps 1, 2 and 3:

(24) $$U \equiv [U_1 \,|\, U_2] := [Q_1 \,|\, Q_1 L + Q_2], \; V \equiv [V_1 \,|\, V_2] := [Z_1 \,|\, Z_1 R + Z_2]$$

where $Q \equiv [Q_1 \,|\, Q_2]$, $Z \equiv [Z_1 \,|\, Z_2]$, and Q_1, Z_1, U_1 and V_1 have n_1 columns, respectively, and Q_2, Z_2, U_2 and V_2 have n_2 columns, respectively.

Notice that only U_2 and V_2 are affected by the transformation matrices of the equivalence transformation (19). We have only introduced U and V in step 6 for clarity. They should be stored in Q and Z, respectively. In table 4.1 we summarize the amount of work in flops required by algorithm ADTF (one flop is the amount of work associated with one floating point add and one floating point multiply). The pencil $\lambda E - A$ is of dimension $n_1 + n_2$ by $n_1 + n_2$. By adding the flops for all steps in table 4.1 we get an estimate of the overall workcount for the ADTF-algorithm (cubic in $n_1 + n_2$).

Step	#flops	Step	#flops
1	$15(n_1 + n_2)^3$	4	$(n_1 + n_2)^2(m + n)$
2	$5(n_1 + n_2)^3$ on average	5	$n_1 n_2(m + n)$
3	$5n_1^2 n_2 + 5n_1 n_2^2 + O(n_1 n_2)$	6	$2(n_1 + n_2)n_1 n_2$

Table 4.1 The amount of work in flops for algorithm ADTF

5. The sensitivity of the problem

One way of measuring the sensitivity of the additive decomposition problem (1) is to introduce a condition number that provides information about the sensitivity of the block decomposition (11) to perturbations in data. In section 3 we showed that computing (11) is a deflating subspace problem and that the additive decomposition problem is equivalent to computing the subspaces **X** and $\mathbf{X}_c$ (17). The sensitivity of these subspaces depends on dif(A_{11}, A_{22}; E_{11}, E_{22}) [10], which is the smallest singular value of the linear operator that maps (L, R) in (20) to $(A_{11}R - LA_{22}, E_{11}R - LE_{22})$. In [3] it is shown that the dif-function also is $\sigma_{min}(Z)$, the smallest singular value of the $2n_1n_2$ by $2n_1n_2$ matrix

$$(25) \qquad Z = \begin{bmatrix} I_{n_2} \otimes A_{11} & -A_{22}^T \otimes I_{n_1} \\ I_{n_2} \otimes E_{11} & -E_{22}^T \otimes I_{n_1} \end{bmatrix}$$

where Z is one representation of the coefficient matrix of the linear system (20). The smallest singular value of Z is nonzero as long as the spectra of $\lambda E_{11} - A_{11}$ and $\lambda E_{22} - A_{22}$ are disjoint (which they are by definition in our application). A consequence of these definitions is that we can bound the solution of the generalized Sylvester equation [3] as

$$(26) \qquad \|(L, R)\|_F \leq \|(A_{12}, B_{12})\|_F / \sigma_{min}(Z)$$

So, a small $\sigma_{min}(Z)$ may cause L and R to be large, implying that the deflating subspaces **X** and $\mathbf{X_c}$ are not well separated. Clearly, the sensitivity of the block decomposition (11) for perturbations in the data is proportional to $\sigma_{min}^{-1}(Z)$. (For details of the perturbation theory for deflating subspaces see [10], [3] and for details of the perturbation theory for the generalized Sylvester equation see [6]). One interpretation of these perturbation results says that if $\sigma_{min}(Z)$ is small then small changes in the minimal realization (especially perturbations in A and/or E) can induce large changes in **X** and $\mathbf{X_c}$ (17), i.e. the additive decomposition problem is ill-conditioned. Similarly, a moderate or large value of $\sigma_{min}(Z)$ indicates a well-conditioned additive decomposition problem (1) and small changes in the minimal realization will only cause correspondingly small changes in the subspaces **X** and $\mathbf{X_c}$ (17).

In summary, $\sigma_{min}(Z)$ is the relevant condition number for the additive decomposition problem. It measures the separation of σ_1, the spectrum of $\lambda E_{11} - A_{11}$, from σ_2, the spectrum of $\lambda E_{22} - A_{22}$. It is small if only a small perturbation is needed to make an eigenvalue in σ_1 coalesce with one in σ_2 [3]. Notice that if $\lambda E_{11} - A_{11}$ and $\lambda E_{22} - A_{22}$ have well-separated spectra and $\sigma_{min}(Z)$ is small then this signals that the original regular pencil $\lambda E - A$ is close to a singular pencil ($\det(\lambda E - A) \equiv 0$ for all λ).

Notice that individual eigenvalues of $\lambda E_{ii} - A_{ii}$ and their Jordan canonical forms can be more sensitive for small perturbations in the data than the separation of the spectrum of $\lambda E_{11} - A_{11}$ from the spectrum of $\lambda E_{22} - A_{22}$. In the partial fraction expansion (2) the sensitivity of the individual eigenvalues will be reflected in the sensitivity of the computed additive decomposition. This is in contrast to our block decomposition approach, where the inherent conditioning of the additive decomposition problem (measured as $\sigma_{min}(Z)$) will be reflected in the sensitivity of the computed solution which is the best we can ask for. In other words, algorithm ADTF is numerically stable in the sense that small changes in data will only cause relative errors in the computed quantities of size proportional to the conditioning of the original problem. For steps 1, 2, 4 and 6 we can make use of Wilkinson's backward error analysis for products of unitary matrices [14]. A rounding error analysis of the generalized Schur algorithm for solving the generalized Sylvester equation is presented in [6]. The conclusion there is that the algorithm is weakly stable [1], meaning that the relative errors in the computed (L, R) is small for all well-conditioned problems. More precisely, the relative errors in the computed (L, R) is proportional to the condition number times a smooth function of the relative machine precision. These errors will finally propagate to the computations where R and L are involved (steps 4 - 6).

In [6] dif^{-1}-estimators (lower bounds on $\sigma_{min}^{-1}(Z)$ in the style of the Linpack condition estimator [2]) that are incorporated into the generalized Schur algorithm for solving the generalized Sylvester equation (20) are presented. The heuristic condition estimators need $O(n_1^2 n_2 + n_1 n_2^2)$ flops, which is the same magnitude of work as solving the generalized Sylvester equation (20) (see table 4.1). This should be compared with $O(10 n_1^3 n_2^3)$ flops for computing the exact value of $\sigma_{min}^{-1}(Z)$. Since solving the generalized Sylvester equation is one step in algorithm ADTF, the block decomposition approach also provide us with estimates of the conditioning of the original problem and the sensitivity of the computed additive decomposition.

6. Numerical experiments

Here we report some numerical results computed by our Fortran routines. To assess the reliability and robustness of algorithm ADTF with condition estimators from [6], we ran different sets of test problems, including both well-conditioned and ill-conditioned additive decomposition problems. In all examples we have chosen Γ to be the interior of the unit circle. The spectra for the examples shown here are as follows: Ex 1: {0.25, 0.5, 1.5, 2.0}; Ex 2: {-1.005, - 0.995, 0.995, 1.005}; Ex 3: 5 random real eigenvalues; Ex 4: 4 random complex conjugate pairs; Ex 5: {0., 0.995i, 1.005i, -0.995i, -1.005i}. All tests were performed using double precision *real* arithmetic (machep ≈ 1.E-17).

Ex	n_1	dif^{-1}	$\sigma_{min}(Z)\,\mathrm{dif}^{-1}$	$\kappa(U)$	$\kappa(V)$	$\|A' - UAV\|$	$\|E' - UEV\|$
1	2	1.57	0.38	1.85	1.92	3.5E-18	0.0E0
2	2	1.4E+7	0.45	1.0	1.0	1.4E-16	1.1E-16
3	2	5.99	0.32	35.7	255.2	1.0E-16	7.9E-17
4	4	47.57	0.32	399.1	2433.4	2.0E-15	3.8E-15
5	5	2.5E+5	0.36	1.0E+6	2.7E+7	1.7E-8	9.5E-9

Table 6.1 Results from algorithm ADTF with condition estimator

Discussion: In all examples the correct dimension n_1 of the stable projection of $H(\lambda)$ on Γ was computed. The condition estimator was in all cases within a factor four of the correct value. Notice the large dif^{-1} for examples 2 and 5 and the relatively large condition numbers of U and V for examples 4 and 5. What does this mean? As expected the computed residuals in table 6.1 are of the size max{κ(U), κ(V)} times machep. If dif^{-1} and/or the condition numbers of U and V are large the computed additive decomposition including the subspaces **X** and $\mathbf{X}_c$ (17) may be sensitive for small perturbations in A and/or E.

7. Conclusions and future work

We have presented a robust and numerically stable algorithm ADTF for solving the additive decomposition problem of a transfer matrix given by its generalized state space realization $H(\lambda) = C(\lambda E - A)^{-1}B + D$. We have introduced a condition number $\sigma_{min}(Z)$ that measures the sensitivity of the computed block decomposition and the subspaces $\mathbf{X}$ and $\mathbf{X_c}$ (17). Our algorithm ADTF also provides reliable lower bounds of $1 / \sigma_{min}(Z)$[6] that estimate the sensitivity of the computed additive decomposition (1). The algorithm is of course also possible to apply to the standard state space realization $H(\lambda) = C(\lambda I - A)^{-1}B + D$. The use of the Sylvester equation in this particular case was also suggested in [9]. In a future paper algorithm ADTF will be compared with a method based on two reorderings of the eigenvalues giving an optimal block-diagonalizing equivalence transformation.

Bo Kågström has been supported by STU (the Swedish Board for Technical Development) under Contract STU 86-04050P.

References

1. Bunch J. R., "The weak and strong stability of algorithms in numerical linear algebra", *Lin. Alg. Applics.*, Vol. 88/89, 1987, pp 49-66
2. Cline A., Moler C., Stewart G.W., and Wilkinson J., "An estimate for the condition number of a matrix", *SIAM J. Num. Anal.*, Vol. 16, 1979, pp 368-375
3. Demmel J. and Kågström B., "Computing stable eigendecompositions of matrix pencils", *Lin. Alg. Applics.*, Vol. 88/89, 1987, pp 139-186
4. Francis B. A., "A course in H_∞ control theory", *Lecture Notes in Control and Information Sciences*, Vol. 88, Springer Verlag, New York 1987
5. Kailath T., *Linear Systems* , New York, Prentice Hall, 1980
6. Kågström B., Westin L., "Generalized Schur Methods with Condition Estimators for Solving the Generalized Sylvester Equation", *IEEE Trans. on Autom. Contr.*, Vol. 34, No. 7, July 1989, pp 745-751
7. Moler C., Stewart G.W., "An algorithm for the generalized eigenvalue problem", *SIAM J. Num. Anal.*, Vol. 10, 1973, pp 241-256
8. Rosenbrock H., *State space and multivariable theory*, New York, Wiley, 1970
9. Safonov M., Jonckheere E., Verma M., Limebeer D., "Synthesis of positive real multivariable feedback systems", *Int. J. Control*, Vol. 45, 1987, pp 817-842
10. Stewart G.W., "Error and perturbation bounds for subspaces associated with certain eigenvalue problems", *SIAM Review*, Vol. 15, 1973, pp 727-764
11. Van Dooren P., "The generalized eigenstructure problem in linear system theory", *IEEE Trans. on Autom. Contr.*, Vol. AC-26, No. 1, 1981, pp 111-129
12. Van Dooren P., "Algorithm 590: DSUBSP and EXCHQZ, Fortran subroutines for computing deflating subspaces with specified spectrum", *ACM Trans. on Math. Software*, Vol. 4, 1982, pp 376-382
13. Vidyasagar M., *Control System Synthesis: A Factorization Approach"*, MIT Press, Cambridge, 1985
14. Wilkinson J.H., *The Algebraic Eigenvalue Problem*, Oxford University Press, 1965

Numerical Solution of Differential Algebraic Riccati Equations

PETER KUNKEL AND VOLKER MEHRMANN

Abstract
We consider Riccati matrix differential algebraic equations arising from singular or descriptor control problems. In order to apply numerical methods for differential algebraic systems one has to transform the equation. Unfortunately then these equations have a linear part, which is described by a singular pencil and thus, the usual integration methods do not apply. Under some conditions, which we discuss, these singularities can be removed by a preprocessing algorithm and the equation can then be solved by well–known methods for differential algebraic systems like DASSL of L. Petzold or LIMEX of Deuflhard, Hairer and Zugck. We discuss the numerical procedure and give a numerical example.

1. Introduction.

We study the differential algebraic Riccati equation (DARE)

$$(1.1)\quad -E^T(t)\dot{X}(t)E(t) = E^T(t)X(t)A(t) + A^T(t)X(t)E(t) + Q(t) - E^T(t)X(t)W(t)X(t)E(t)$$

with "terminal" condition

$$(1.2)\quad E^T(t_f)X(t_f)E(t_f) = M.$$

Here $X(t), E(t), A(t), Q(t), W(t), M \in \mathbf{R}^{n,n}$ and $X(t), Q(t), W(t), M$ are symmetric in the interval $[t_0, t_f] \subset \mathbf{R}$, where $\mathbf{R}^{n,n}$ denotes the real $n \times n$ matrices. Such equations arise for example from optimal regulator problems with differential algebraic equations (DAE's) or from Kalman-Bucy filtering applied to stochastical differential algebraic equations. Consider for example the optimal control problem:

$$(1.3)\quad \text{Minimize}\quad J(x,u) = \frac{1}{2}\int_{t_0}^{t_f} [x(t)^T Q(t)x(t) + u(t)^T R(t)u(t)]dt$$

subject to·

$$(1.4)\quad E(t)\dot{x}(t) = A(t)x(t) + B(t)u(t)\ ,\quad x(t_0) = x^0\ ,$$

where $E(t), A(t), Q(t) \in \mathbf{R}^{n,n}$, $B(t) \in \mathbf{R}^{n,m}$, $R(t) \in \mathbf{R}^{m,m}$, $Q(t) = Q(t)^T$, $R(t) = R(t)^T, Q(t), R(t)$ positive definite.

In is shown in [8] how under some further assumptions the optimal solution of (1.3), (1.4) can be obtained via the solution of (1.1), (1.2). In [8] also conditions are given, under which the standard maximum principle can be applied formally as if $E(t)$ were nonsingular.

For the case that the matrices E, A, Q, R, B are independent of t this is already shown in [9], [10]. In the nonautonomous case where $E(t)$ is nonsingular, this follows from standard linear control theory, e.g. [1], [7].

In this paper we discuss the numerical solution of (1.1), (1.2). The obvious approach is to vectorize this equation and, leaving out the extraneous components of the symmetric matrix, to transform it into a standard differential algebraic equation in $\mathbf{R}^{\frac{n(n+1)}{2}}$. If $E(t)$ is singular, and even in the most simple case that the DAE (1.4) has index at most 1, this approach unfortunately leads to a DAE, which is not solvable in the sense defined for example in [2], since the linear part of the equation is described by a singular pencil. It follows that none of the standard solvers for differential algebraic equations like LIMEX [3] or DASSL [11] can be applied directly. The state of the art in numerical methods for DAE's is described in a recent monograph by Brenan et al [2]. Fortunately it is possible to remove the singularities which make the system nonsolvable. We shall give conditions, when this is possible and how this can be done numerically. The basic tool to do this preprocessing is a slight generalization of a method of Rheinboldt [14] to compute a locally smooth QR–decomposition of the matrix $E(t)$. We then combine this method with standard DAE solvers to a general numerical method for (1.1). Finally we shall present some numerical examples.

2. Reduction to standard case.

For standard matrix Riccati differential equations, i.e. $E(t) = I$ in (1.1), (1.2) the theory is well established, e.g. [13]. There are several cases, where we can transform the DARE (1.1), (1.2) to a form, where the standard theory can be applied.

The first case we consider is

$$E(t) \text{ nonsingular for all } t \in [t_0, t_f] \tag{2.1}$$

In this case (1.1), (1.2) is equivalent to the standard Riccati differential equation

$$\begin{aligned} -\dot{X}(t) &= X(t)A(t)E^{-1}(t) + (E^{-1}(t))^T A(t)^T X(t) \\ &+ (E^{-1}(t))^T Q(t)E(t) - X(t)W(t)X(t), \end{aligned} \tag{2.2}$$

$$X(t_f) = (E(t_f)^{-1})^T M \; E(t_f)^{-1}. \tag{2.3}$$

Although this is a veritable approach from the theoretical point of view, it is not in general advisable for the numerical treatment, since $E(t)^{-1}$ may be very ill conditioned.

The second case we consider is

$$\begin{aligned}&E(t) \text{ continuous, } \lambda E(t) - A(t) \text{ regular, } \operatorname{rank}(E(t)) = p < n \\ (2.4)\qquad &\operatorname{ind}_\infty(E(t), A(t)) = q \leq 1 \text{ for all } t \in [t_0, t_f], \\ &p, q \text{ independent of } t,\end{aligned}$$

and if $U(t_f)\Sigma(t_f)V(t_f)^T = E(t_f)$ is the singular value decomposition of $E(t_f)$, then

$$U(t_f)^T E(t) = \begin{bmatrix} E_1(t) \\ 0 \end{bmatrix} \}\, p \quad \text{for all } t \in [t_0, t_f]. \tag{2.5}$$

Here we use the following notation:
A pencil $\lambda E - A$ is called *regular* if $\det(\lambda E - A) \not\equiv 0$. By $\operatorname{ind}_\infty(E, A)$ we denote the size of the largest block to infinite eigenvalues in the Kronecker canonical form of $\lambda E - A$, e.g. [4], [10].
Observe that this case includes the case

$$E, A \text{ constant in } t,\ \alpha E - \beta A \text{ regular },\operatorname{ind}_\infty(E, A) \leq 1.$$

In this last case, it is well–known how to use the singular value decomposition of E to partition the system into it's algebraic and it's dynamic part,e.g. [6]. For a nonautonomous $E(t)$ such an SVD has to be at least differentiable to perform the subsequent transformations. If neither of the conditions (2.1) or (2.4) holds, then the solution of these equations faces the well–known difficulties as discussed for example in [5], [12]. One may have no solution or nonunique solutions or one needs further differentiability conditions. For these reasons we discuss here only the cases, where conditions (2.1) or (2.4) and (2.5) hold. Condition (2.5) is needed to get a continous singular value decomposition of a matrix $E(t)$. We suspect that this condition can be relaxed, but this topic is still under investigation.

If (2.4),(2.5) hold then we get the continous decomposition of $E(t)$

$$U(t_f)^T E(t) V(t) = \begin{bmatrix} E_{11}(t) & 0 \\ 0 & 0 \end{bmatrix} \}\, p\ , \tag{2.7}$$

where $E_{11}(t)$ is nonsingular in $[t_0, t_f]$ by (2.4), in the following way. Let

$$E_1(t)^T = \widetilde{V}(t) \begin{bmatrix} R(t) \\ 0 \end{bmatrix} =: \left[\widetilde{V}_1(t), \widetilde{V}_2(t)\right] \begin{bmatrix} R(t) \\ 0 \end{bmatrix} \tag{2.8}$$

be the QR–decomposition of $E_1(t)^T$ in (2.5). Then the columns of $\widetilde{V}_2$ belong to the nullspace of $E_1(t)$ and the columns of $\widetilde{V}_1$ belong to its orthogonal complement. Continuity of V can be obtained via the solution of an optimization problem for the computation of orthogonal matrices $Q_1(t), Q_2(t)$ such that

$$V(t) = [\widetilde{V}_1(t)Q_1(t), \widetilde{V}_2(t)Q_2(t)], \tag{2.9}$$

as it is discussed in [14]. One obtains

$$U(t_f)^T E(t)V(t) = \begin{bmatrix} E_1(t) \\ 0 \end{bmatrix} [\widetilde{V}_1(t)Q_1(t), \widetilde{V}_2(t)Q_2(t)]$$

$$= \begin{bmatrix} E_{11}(t) & 0 \\ 0 & 0 \end{bmatrix} \}\, p =: \widetilde{E}(t), \tag{2.10}$$

since $E_1(t)\widetilde{V}_2(t) = 0$ for all $t \in [t_0, t_f]$ and hence $E_{11}(t)$ is continuous and nonsingular.
Now transform (1.1), (1.2) with $U(t_f), V(t)$. Set

$$\left\{ \begin{array}{lll} y(t) & := \begin{bmatrix} Y_{11}(t) & Y_{12}(t) \\ Y_{21}(t) & Y_{22}(t) \end{bmatrix} & := U(t_f)^T X(t) U(t_f), \\ \widetilde{A}(t) & := \begin{bmatrix} A_{11}(t) & A_{12}(t) \\ A_{21}(t) & A_{22}(t) \end{bmatrix} & := U(t_f)^T A(t) V(t), \\ \widetilde{Q}(t) & := \begin{bmatrix} Q_{11}(t) & Q_{12}(t) \\ Q_{21}(t) & Q_{22}(t) \end{bmatrix} & := V(t)^T Q(t) V(t), \\ \widetilde{W}(t) & := \begin{bmatrix} W_{11}(t) & W_{12}(t) \\ W_{21}(t) & W_{22}(t) \end{bmatrix} & := U(t_f)^T W(t) U(t_f), \\ \widetilde{M} & := \begin{bmatrix} M_{11} & M_{12} \\ M_{21} & M_{22} \end{bmatrix} & := V(t_f)^T M V(t_f). \end{array} \right. \tag{2.11}$$

Then (1.1), (1.2) decouples into the four equations:

$$-E_{11}^T \dot{Y}_{11} E_{11} = (A_{11}^T Y_{11} + A_{21}^T Y_{21}) E_{11} + E_{11}^T (Y_{11} A_{11} + Y_{12} A_{21}) + Q_{11} \tag{2.12}$$
$$-E_{11}^T ([Y_{11} W_{11} + Y_{12} W_{21}] Y_{11} + [Y_{11} W_{12} + Y_{12} W_{22}] Y_{21}) E_{11}$$

$$0 = E_{11}^T [Y_{11} A_{12} + Y_{12} A_{22}] + Q_{12} \tag{2.13}$$

$$0 = [A_{22}^T Y_{21} + A_{12}^T Y_{11}] E_{11} + Q_{21} \tag{2.14}$$

$$0 = Q_{22}. \tag{2.15}$$

$$E_{11}^T(t_f) Y_{11}(t_f) E_{11}(t_f) = M_{11}, \quad 0 = M_{12}, \quad 0 = M_{22}, \tag{2.16}$$

where we have left out the argument t in (2.12),...,(2.15) for simplification.

Equations (2.13), (2.14) are equivalent, equation (2.16) contains two consistency conditions and condition (2.15) is a solvability condition.

Theoretically we could now invert $E_{11}(t)$, solve for $X_{12}(t)$, vectorize $X_{11}(t)$, leave out the extraneous equations due to symmetry in $X_{11}(t)$ and use an initial value solver to solve for $X_{11}(t)$. For the obvious numerical reasons we would like to use a DAE solver instead, as we will will do in section 3. In [8] it is shown that whenever the matrix $E(t)$ is singular, the vectorization of problem (1.1), (1.2) leads to a singular DAE. Fortunately, as is shown in [8], if $\lambda E(t) - A(t)$ is regular and has index 1, these singularities can be removed, leaving a vectorized system which is solvable and has index 1 Thus the well established theory and methods for such DAE's apply, e.g. [2].

3. The numerical method.

The approach discussed in section 2 is now transformed into a numerical method. It is usefull to write different alorithms for the autonomous and nonautonomous case as it is done in [8]. We here only describe the general algorithm for the nonautonomous case.

3.1 ALGORITHM DAREV.

Given $A(t), E(t), M(t), Q(t), W(t) \in \mathbf{R}^{n,n}, Q = Q^T, W = W^T, M = M^T, t_0, t_f \in \mathbf{R}$ and a rank tolerance $\epsilon \in \mathbf{R}$, with $E(t), A(t)$ satisfying (2.4), (2.5), this algorithm solves the Riccati differential algebraic equation (1.1), (1.2) in the interval $[t_0, t_f]$.

Step 1 Perform a singular value decomposition of $E(t_f)$ using the LINPACK DSVDC routine, i.e. determine $U = U(t_f)$, $V(t_f) \in \mathbf{R}^{n,n}$ orthogonal, such that $U^T E(t_f) V(t_f) = \Sigma(t_f) = \mathrm{diag}\{\sigma_1(t_f), \cdots, \sigma_n(t_f)\}$, $\sigma_1(t_f) \geq \cdots \geq \sigma_n(t_f) \geq 0$.

Step 2 Determine the rank p of $\Sigma(t_f)$ as the largest index $i \in \{1, \cdots, n\}$ such that $\sigma_i(t_f) \geq \epsilon\sigma_1(t_f)$ and set $\sigma_{p+1}(t_f), \cdots, \sigma_n(t_f) = 0$.

Step 3 Perform the transformations with U, $V(t_f)$ to $A(t_f), Q(t_f), M(t_f)$, i.e. set $A(t_f) := U^T A(t_f) V(t_f)$, $Q(t_f) := V(t_f)^T Q(t_f) V(t_f)$, $M := V(t_f)^T M V(t_f)$, $E(t_f) := \begin{bmatrix} \Sigma_0(t_f) & 0 \\ 0 & 0 \end{bmatrix}$, where $\Sigma_0(t_f) = \mathrm{diag}\{\sigma_1(t_f), \cdots, \sigma_p(t_f)\}$.

Step 4 Check the consistency of system (2.12), (2.13), (2.14), (2.15), (2.16), i.e. whether $Q_{22}(t_f) = 0, M_{12} = 0, M_{22} = 0$.

If these conditions do not hold, the system is not solvable,

STOP

Step 5 Solve the algebraic equations (2.13) for $Y_{12}(t_f)$ using the QR-decomposition with column pivoting for $A_{22}(t_f)$:

$$Y_{12}(t_f) = -\Sigma_0^{-1}(M_{11}\Sigma_0^{-1}(t_f)A_{12}(t_f) + Q_{12}(t_f))A_{22}(t_f)^{-1} \ .$$

Step 6 Rewrite the relevant part of $Y(t_f) = \begin{bmatrix} Y_{11} & Y_{12} \\ Y_{21} & Y_{22} \end{bmatrix}$, given by the shaded region in 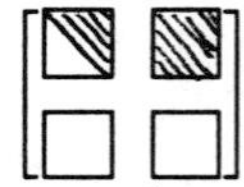, in vector form, by going through $[Y_{11}\ Y_{12}]$ rowwise . Denote this transformed system as

(3.2) $$\widetilde{E}(t)\dot{z}(t) = \widetilde{f}(t, z(t)).$$

Step 7 Solve system (3.2) for example with LIMEX, where

a) in the beginning of each time step, a new V_{ref} is determined as the Q factor in the QR factorization of $E^T(t_i)U$, and

b) for the evaluation of $V(t_i)$ continuous, use the algorithm described in [14], then transform $A(t_i), Q(t_i), W(t_i), E(t_i)$ with U and this continuous $V(t_i)$ according to the formulas

$$A(t_i) := U^T A(t_i)V(t_i), \quad Q(t_i) := V(t_i)^T Q(t_i)V(t_i),$$
$$W(t_i) = U^T W(t_i)U, \quad Y(t_i) = U^T X(t_i)U, \ E(t_i) := U^T E(t_i)V(t_i)$$

and evaluate the vectorized functions $\widetilde{E}(t_i), \widetilde{f}(t_i)$ with these transformed matrices.

Step 8 Rewrite the solution in matrix form by completing it with 0 in position Y_{22} and in a symmetric way in Y_{21}.

Step 9 Form $X(t_0) = U\ Y(t_0)U^T$.

The above algorithm has been implemented as FORTRAN subroutine on an IBM 3090VF in double precision. In the numerical tests, the common required accuracy for LIMEX was 10^{-5} and the integration was carried out from $t_f = 1$ to $t_0 = 0$.

3.3 Example.

The following differential algebraic Riccati equation with variable coefficients was obtained by adjusting the inhomogeneity $Q(t)$.

$$n = 5, \ E(t) = \begin{bmatrix} 2-t & 0 & 0 & t & 0 \\ 0 & 1 & 0 & 0 & 0 \\ 0 & 0 & 1 & 0 & t/2 \\ 0 & 0 & 0 & 0 & 0 \\ 0 & 0 & 0 & 0 & 0 \end{bmatrix},$$

$$A(t) = \text{diag}(0, t, 0, 1, 1),$$

$$Q(t) = \begin{bmatrix} 0 & 0 & q_{13} & q_{14} & q_{15} \\ 0 & q_{22} & 0 & 0 & q_{25} \\ q_{13} & 0 & 0 & q_{34} & 0 \\ q_{14} & 0 & q_{34} & q_{44} & q_{45} \\ q_{15} & q_{25} & 0 & q_{45} & 0 \end{bmatrix}$$

with

$$\begin{aligned} q_{13} &= -(2-t) \\ q_{14} &= -(2-t)e^{-t} \\ q_{15} &= -(2-t)\frac{t}{2} \\ q_{22} &= e^{2t} + (e^t-1)^2 - (2t+1)e^t \\ q_{25} &= 1-e^t \\ q_{34} &= -t \\ q_{44} &= -2te^{-t} \\ q_{45} &= -\frac{t^2}{2}, \end{aligned}$$

$$X(t) = \begin{bmatrix} 1 & 0 & t & e^{-t} & 0 \\ 0 & e^t & 0 & 0 & e^t-1 \\ t & 0 & 0 & 0 & 0 \\ e^{-t} & 0 & 0 & 0 & 0 \\ 0 & e^t-1 & 0 & 0 & 0 \end{bmatrix},$$

$$M = E(1)^T\, X(1)\, E(1).$$

After 0.315 sec CPU-time, DAREV obtained $X(0)$ with a maximum error of $1.3 \cdot 10^{-5}$. Further examples are discussed in [8].

4. Conclusion.

We have discussed the numerical solution of matrix Riccati differential algebraic equations as they arise from descriptor control problems. Under some typical conditions to the underlying pencil, we can preprocess the system, such that we can use standard packages for differential algebraic equations for the solution.

It is obvious how this approach can be carried over to the case of complex matrices $A(t), E(t), R(t), W(t), Q(t)$.

Another problem which can usually be treated analogously, is the solution of discrete Riccati difference equations, which in the analogous case come from autonomous discrete optimal descriptor control problems. In the autonomous case it takes the form

$$E^T X_k E = \tag{4.1}$$

$$Q + A^T X_{k+1} A - A^T X_{k+1} B(R + B^T X_{k+1} B)^{-1} B^T X_{k+1} A, \quad k = K-1, \cdots, 0.$$

$$E^T X_K E = M. \tag{4.2}$$

For each time step the solution for X_k is a purely algebraic problem of a different nature than in the continuous case. Other generalizations can also be considered in the control problem, investigating more general cost functionals or more complicated differential equations.

5. Acknowledgement.

We thank IBM Deutschland and in particular Dr. R. Janßen for giving us the opportunity to carry out this work.

6. References.

[1] M. Athans, P.L. Falb, *Optimal Control*, MacGraw Hill, New York (1966).

[2] K.E. Brenan, S.L. Campbell, L.R. Petzold, *Numerical Solution of Initial-Value Problems in Differential-Algebraic Equations*, Elsevier, New York, N.Y. (1989).

[3] P. Deuflhard, E. Hairer, J. Zugck, *One-step and Extrapolation Methods for Differential-Algebraic Systems*, Numer. Math. 51 (1987), 501–516.

[4] F.R. Gantmacher, Matrix Theory, Vol. II, Chelsea, New York (1959).

[5] C.W. Gear, L. Petzold, *Differential-Algebraic Systems and Matrix Pencils*, in Matrix Pencils, B. Kågström, A. Ruhe, Eds., Springer, Berlin (1983), 75–90.

[6] G.H. Golub, C.R. Van Loan, *Matrix Computation*, North Oxford Academic, Oxford (1983).

[7] H.W. Knobloch, H. Kwakernaack, *Lineare Kontrolltheorie*, Springer, Berlin (1985).

[8] P. Kunkel, V. Mehrmann, *Numerical solution of differential algebraic Riccati equations*, Technical report, IBM Science Center Heidelberg, TR 89.03.006, to appear in Lin. Alg. Appl. (1989).

[9] V. Mehrmann, *The Linear Quadratic Control Problem: Theory and Numerical Algorithms*, Habilitationsschrift, Universität Bielefeld (1987).

[10] V. Mehrmann, *Existence, Uniqueness and Stability of Solutions to Singular Linear Quadratic Optimal Control Problems*, Lin. Alg. Appl., 121 (1989), 291–331.

[11] L. Petzold, *A description of DASSL: A Differential-Algebraic System Solver*, IMACS Trans. on Scient. Comp. Vol. 1, R.S. Stepleman, Ed. (1982).

[12] L. Petzold, *Differential/Algebraic Equations are not ODE's*, SIAM J. Sci. Stat. Comput. 3 (1982), 367–384.

[13] W.T. Reid, *Riccati Differential Equations*, Academic Press, New York (1972).

[14] W. Rheinboldt, *On the Computation of Multi-Dimensional Solution Manifolds of Parametrized Equtions*, Numer. Math. 53 (1988), 165–181.

SVD updating for tracking slowly time-varying systems. A parallel implementation.

Marc Moonen, Paul Van Dooren, Joos Vandewalle

Abstract

A procedure is presented for updating the singular value decomposition of a matrix as new rows are appended successively. The updating consists in an interlacing of QR-updatings and a Jacobi-type SVD-algorithm applied to the triangular factor. At any time step an approximate decomposition is computed from a previous approximation, with a limited number of operations ($O(n^2)$). When combined with exponential weighting, the algorithm is seen to be highly applicable to tracking problems, as will be demonstrated with an example. Furthermore, this algorithm can elegantly be mapped onto a systolic array, making use of slight modifications of well known systolic implementations for the matrix-vector product, the QR-updating and the SVD.

1 Introduction

Common techniques for updating the singular value decomposition (SVD) of a matrix after appending new rows, mostly reduce to solving the corresponding modified symmetric eigenvalue problem [2, 3, 5]. A drawback is the necessary knowledge of the *exact* original eigenstructure, in order to compute the updated eigenstructure. On the one hand, this considerably increases the computational load (e.g. in an on-line environment, where at each time step an *exact* updating is thus to be performed), while on the other, round-off errors are likely to accumulate unboundedly.

In this paper, we present an SVD updating procedure that interlaces a Jacobi-type SVD algorithm [1, 7] with a QR-updating each time a new row is to be worked in. After each QR-update, the SVD algorithm reduces the R-factor to an approximately diagonal matrix, by applying a sequence of plane rotations. It may be appropriate to try and perform only a fixed number of rotations (e.g. $n-1$, see below) on the triangular factor after each QR-update. This results in a seemingly *non-iterative* algorithm that at any time step outputs an *approximate decomposition* computed from a previous approximation, with only a limited number of operations (viz. $O(n^2)$). The crux is then to show how only a few SVD-steps after each QR-update can restore an acceptable approximation at every stage.

When combined with exponential weighting, this algorithm is seen to be highly applicable to *tracking problems.* Furthermore, the updating algorithm can be mapped onto a *systolic array* with $O(n^2)$ parallelism for $O(n^2)$ operations, by appropriately combining systolic implementations for the matrix-vector product, the QR-updating and the SVD.

The SVD updating procedure is presented in section 2 and illustrated by a practical example in section 3. Finally, in section 4, a systolic implementation is presented.

2 Updating the singular value decomposition

The SVD of a matrix $A_{m \times n}$ $(m \geq n)$ is a factorization of A into a product of three matrices.

$$A_{m \times n} = U_{m \times n} \Sigma_{n \times n} V^t_{n \times n}$$

where $U^t U = I$, $V^t V = I$, and Σ is a pseudo-diagonal matrix, with the singular values on its main diagonal.

Updating the SVD after appending a new row, consists in computing the SVD of the modified matrix

$$\begin{bmatrix} \lambda.A \\ a^t \end{bmatrix} = U^\star_{(m+1) \times n} \Sigma^\star_{n \times n} V^{\star t}_{n \times n}$$

making use of the original SVD of A. The exponential weighting factor $\lambda \leq 1$ accounts for the reduction of "older" data. In on-line applications, a continuous updating after each sampling is to be performed, where in most cases only V and Σ are of interest.

We first briefly review a Jacobi-type SVD-algorithm, originally due to Kogbetliantz [7]. Next we show how an adaptive algorithm can be constructed by interlacing the SVD-diagonalization procedure with repeated QR-updates.

A Jacobi-type SVD algorithm

In order to compute the SVD of A, it is known to be computationally more efficient to first perform a QR-factorisation of A

$$A_{m \times n} = Q_{m \times n} . R_{n \times n}$$

where $Q^t Q = I$, and R is upper triangular. Obviously, V and Σ can be computed from the SVD of R as well.

$$R = U_R . \Sigma_R . V_R^t = U_R . \Sigma . V^t$$

The SVD of R can be computed by making use of a Jacobi strategy [8], that generates a sequence of triangular matrices R_k via the recurrence

$$R_k = \Theta_k^t . R_{k-1} . \Phi_k$$

with $R_0 = R$, and Θ_k and Φ_k to be defined next. As each iteration reduces the norm of the off-diagonal elements, R is gradually reduced to a diagonal matrix. Suppose that after r rotations a diagonal matrix (up to the relative precision ϵ) is obtained, then

$$\Theta_r^t . \Theta_{r-1}^t \ldots \Theta_1^t . (R) . \Phi_1 \ldots \Phi_{r-1} . \Phi_r = R_r \overset{\epsilon}{\simeq} \Sigma_R = \Sigma$$

and

$$V = V_R \overset{\epsilon}{\simeq} \Phi_1 . \ldots . \Phi_{r-1} . \Phi_r$$

Θ_k and Φ_k are plane rotations through angles θ_k and ϕ_k in the $(i, i+1)$-plane, up to a row or column permutation. The rotation angles θ_k and ϕ_k are chosen such that the $(i, i+1)$ element in R_k transforms to zero,

$$\begin{bmatrix} r_{k+1}^{i,i} & 0 \\ 0 & r_{k+1}^{i+1,i+1} \end{bmatrix} := \begin{bmatrix} \sin\theta_k & \cos\theta_k \\ \cos\theta_k & -\sin\theta_k \end{bmatrix} \begin{bmatrix} r_k^{i,i} & r_k^{i,i+1} \\ 0 & r_k^{i+1,i+1} \end{bmatrix} \begin{bmatrix} \sin\phi_k & \cos\phi_k \\ \cos\phi_k & -\sin\phi_k \end{bmatrix}$$

Each iteration thus reduces to a 2×2 SVD on the main diagonal, setting the super-diagonal ("pivot"-) element $r^{i,i+1}$ equal to zero. The off-diagonal elements can then be annihilated in a cyclic manner . An element that is being annihilated in one step, reappears in the next step, but mostly with a reduced modulus. By making use of *outer rotations* on the one hand, and an elegant *pipelining* of forward and backward sweeps on the other, an efficient implementation can be obtained [12], where furthermore the triangular structure is preserved at every stage. It then suffices to let the pivot index successively take up all possible values

$$i = 1, 2, \ldots, n-1$$

in a cyclic manner, and perform 2×2 SVD's (with outer rotations) on the main diagonal accordingly [8]. The updating scheme presented in this paper interlaces a QR-update in between such sequences of $(n-1)$ 2×2 SVD's, as will be detailed next.

SVD-updating

If the rows of A become sequentially available, as is the case in an on-line environment, the diagonalization procedure can be interlaced with QR-updates [6]. After each QR-updating, the iterative diagonalization process of the R-factor is resumed. A sequential description is as follows.

Suppose that, at a certain time step k, the $A^{(k)}$-matrix is reduced to $R^{(k)}$ (upper triangular, almost diagonal). The corresponding V-matrix is $V^{(k)}$. (The $U^{(k)}$-matrix is not stored).

$$A^{(k)} = U^{(k)}.R^{(k)}.V^{(k)^t}$$

The modified matrix then equals

$$A^{(k+1)} = \begin{bmatrix} \lambda.A^{(k)} \\ a^{(k)^t} \end{bmatrix} = \begin{bmatrix} U^{(k)} & \\ & 1 \end{bmatrix} \begin{bmatrix} \lambda.R^{(k)} \\ a^{(k)^t}.V^{(k)} \end{bmatrix} V^{(k)^t}$$

Clearly, the R-factor is to be updated with the transformed input vector $a_\star^{(k)^t} = a^{(k)'}.V^{(k)}$, resulting in

$$A^{(k+1)} = \left(\begin{bmatrix} U^{(k)} & \\ & 1 \end{bmatrix} .Q^{(k)}_{(n+1)\times n} \right) \begin{bmatrix} R_\star^{(k)} \\ 0 \end{bmatrix} V^{(k)^t}$$

The QR-updating clearly does not alter the $V^{(k)}$-matrix. Finally this updated $R_\star^{(k)}$ is further diagonalized through a series of subsequent rotations.

$$\begin{aligned} R^{(k+1)} &= \Theta_{k_r}^t.\Theta_{k_{r-1}}^t \ldots \Theta_{k_1}^t.(R_\star^{(k)}).\Phi_{k_1} \ldots \Phi_{k_{r-1}}.\Phi_{k_r} \\ V^{(k+1)} &= V^{(k)}.\Phi_{k_1}.\ldots.\Phi_{k_{r-1}}.\Phi_{k_r} \end{aligned}$$

(where k_i refers to the ith rotation in time step k.)

Thus, the QR-updatings and the diagonalization procedure are interlaced, yielding an approximate V-matrix at every stage. The *number of rotations* k_r should be chosen as a function of the off-diagonal part ("off-norm") in $R_\star^{(k)}$. Usually only very few rotations are necessary to restore a reasonable approximation at any time step. The **choice of** $(n-1)$ **rotations following each** QR**-update** (where the pivot index successively takes up all values $i = 1, \ldots, n-1$), seems particularly well suited for an efficient interlacing of QR-updates and SVD-steps. Moreover, for most practical applications, a reasonably low off-norm is observed when only $n-1$ rotations are performed after each QR-update, as will be demonstrated with an example in the next section.

3 Adaptive parameter estimation

The SVD-updating procedure can be used for adaptive (Total Least Squares) parameter estimation. An input-output sequence was generated, where the input $u^{(k)}$ $(k = 1, \ldots)$ is a white noise sequence, while the output $y^{(k)}$ at any time step k satisfies the second order difference equation

$$y^{(k)} = \alpha_1^{(k)}.y^{(k-1)} + \alpha_2^{(k)}.y^{(k-2)} + \alpha_3^{(k)}.u^{(k)} + \alpha_4^{(k)}.u^{(k-1)} + \alpha_5^{(k)}.u^{(k-2)}$$

The coefficients $\alpha_i^{(k)}$ were chosen to be sinusoidally time-varying.
Clearly, if these coefficients were constants, the exponentially weighted matrix

$$A^{(k)} = \begin{bmatrix} y^{(k)} & y^{(k-1)} & y^{(k-2)} & u^{(k)} & u^{(k-1)} & u^{(k-2)} \\ \lambda.y^{(k-1)} & \lambda.y^{(k-2)} & \lambda.y^{(k-3)} & \lambda.u^{(k-1)} & \lambda.u^{(k-2)} & \lambda.u^{(k-3)} \\ \lambda^2.y^{(k-2)} & \lambda^2.y^{(k-3)} & \lambda^2.y^{(k-4)} & \lambda^2.u^{(k-2)} & \lambda^2.u^{(k-3)} & \lambda^2.u^{(k-4)} \\ \vdots & \vdots & \vdots & \vdots & \vdots & \vdots \\ \lambda^{k-3}.y^{(3)} & \lambda^{k-3}.y^{(2)} & \lambda^{k-3}.y^{(1)} & \lambda^{k-3}.u^{(3)} & \lambda^{k-3}.u^{(2)} & \lambda^{k-3}.u^{(1)} \end{bmatrix}$$

would have its last singular value equal to zero, corresponding to a right singular vector (null space)

$$v_6 \quad \| \quad [-1 \;\; \alpha_1 \;\; \alpha_2 \;\; \alpha_3 \;\; \alpha_4 \;\; \alpha_5]^t$$

With "slowly" time-varying coefficients $\alpha_i^{(k)}$, the smallest singular value would be *approximately* equal to zero, provided the weighting factor λ is chosen sufficiently small. At each time instance k, the corresponding right singular vector v_6 is then a reasonable approximation (up to a factor of proportionality) for

$$[-1 \;\; \alpha_1^{(k)} \;\; \alpha_2^{(k)} \;\; \alpha_3^{(k)} \;\; \alpha_4^{(k)} \;\; \alpha_5^{(k)}]^t$$

The coefficients $\alpha_i^{(k)}$ can thus approximately be identified from the SVD of the above matrix, where in each time step a new row is appended. The SVD can then be computed adaptively with the SVD-updating procedure. The exponential weighting factor was set equal to $1 - 2^{-5}$, and only $n - 1 = 5$ rotations were performed after each QR-update. The true and the reconstructed coefficients are seen to be in good mutual aggreement (Figure 1).

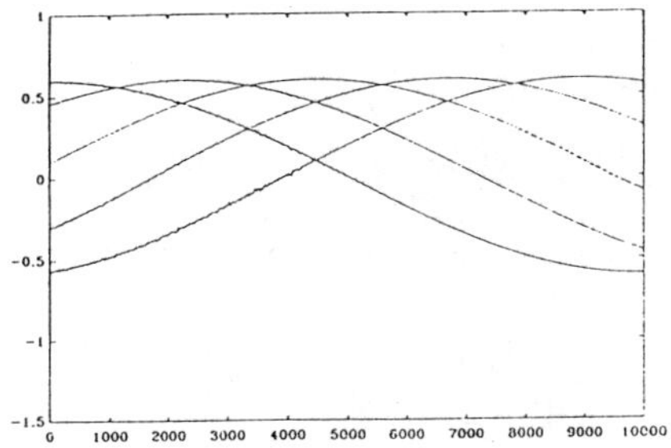

Figure 1: Reconstructed and true coefficients, as a function of time

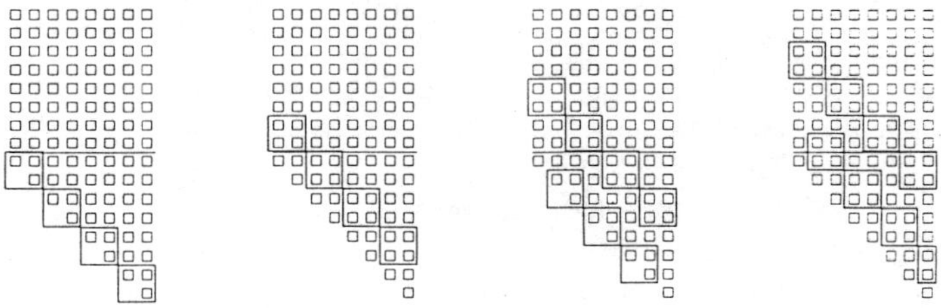

Figure 2: SVD array

4 A systolic implementation

The SVD-updating algorithm can elegantly be mapped onto a systolic array. In particular if after each QR-update, $n-1$ Jacobi rotations are performed, an efficient parallel implementation is conceivable with $O(n^2)$ parallelism for $O(n^2)$ complexity. (If the number of rotations after each QR-update is e.g. halved or doubled, the array can easily be modified accordingly.)

As we intend to combine SVD-diagonalization and QR-updating on the same array, let us first briefly review the triangular SVD-array, as described in [8] (Figure 2). Processor elements on the main diagonal perform 2×2 SVD's (combined with column and row permutations). Row transformation parameters are passed on to the right, while column transformation parameters are passed on upwards. Off-diagonal processors only apply and propagate these transformations. Column transformations are also propagated through the upper square part, containing the right singular V-matrix (V's first row in the top row, etc.).

Off-diagonal elements with odd and even row numbers are being annihilated in an alternative fashion *(odd-even ordering)*. It can easily be verified that an odd-even ordering corresponds to a cyclic by rows or columns ordering, except for a different start-up phase. The 2×2 SVD's that are performed in parallel on the main diagonal can then be thought of as corresponding to different pipelined sequences of $n-1$ rotations, where in each sequence the pivot index successively takes up the values

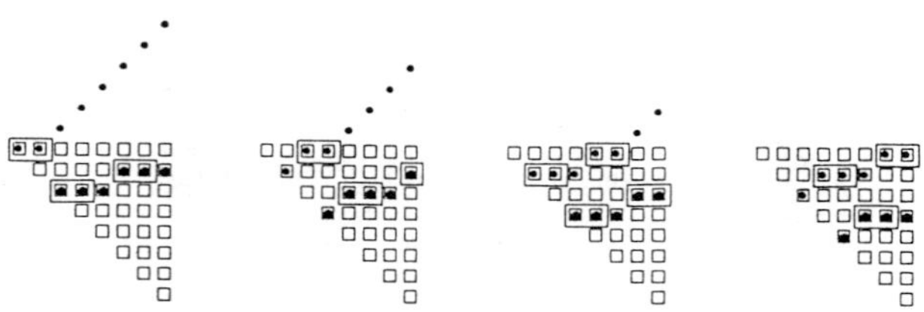

Figure 3: Modified QR-updating array

$i = 1, \ldots, n-1$. As pointed out in section 2, the QR-updatings should be inserted in between these sequences.

A QR-updating is performed by applying a sequence of orthogonal transformations *(Givens rotations)* [6]. The well known Gentleman-Kung QR-updating array [4] can be matched to the SVD-array, as shown in Figure 3. Processor elements on the main diagonal compute and propagate the rotation parameters, while off-diagonal processors apply and propagate rotation parameters and intermediate results. While all operations remain unaltered as compared to the original array, the pipelining is somewhat different, so that the data vectors run through the array in a slightly different manner. The distribution of active processors however now turns out to be the same as in the SVD-array. Framed elements in Figure 3 then constitute the upper parts of the 2×2 blocks in Figure 2.

Without going into details, we give in Figure 4 a snapshot of the overall SVD updating array (see [10] for a detailed description). New data vectors are continuously fed in, to the left-hand side of the array. The matrix-vector products $a^t.V$ are performed in the square part, in between two SVD-rotation fronts. Intermediate result are propagated upwards, while the a-components are propagated to the right. The resulting vectors $a^\star = a^t.V$ become available at the top end, and are bounced back into the array, as indicated (■ becomes •). While propagating downwards, the $a_\star$-vector crosses upgoing rotation fronts. Clearly, the corresponding transformations should be applied to the $a_\star$-vector as well (briefly : $a^t.(V.\theta) = (a^t.V).\theta = a^t_\star\theta$). Each frame in the square array then corresponds to a (column) transformation, that is applied to a 2×2 block of the V-matrix, *and* to the two available components of the $a_\star$-vector. These components are propagated one step downwards next. The downward propagation of an $a_\star$-vector is thus carried out in exactly the same manner as the propagation in the modified Gentleman-Kung array (Figure 3).

Finally, the triangular array performs the SVD-diagonalization, with interlaced QR-updatings. In each frame, column and row transformations corresponding to the SVD diagonalization are performed first (see also Figure 2), where now column transformations should be applied to the $a_\star$-components as well. In a second step, only row transformations are performed corresponding to the modified QR-updating, affecting the $a_\star$-components and the upper part of the 2×2 -blocks (see also Figure 3).

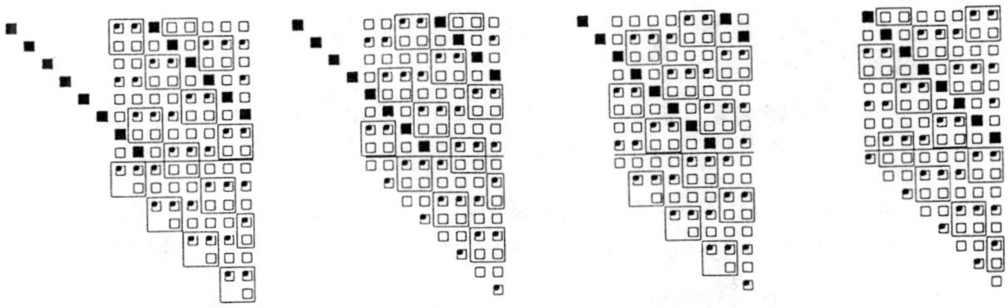

Figure 4: SVD-updating array

5 Conclusion

An SVD updating procedure was constructed as a combination of QR-updating and Kogbetliantz's SVD algorithm applied to the triangular factor. This algorithm can be mapped into a systolic array, making use of slightly modified systolic implementations for each part. The SVD-updating is then performed at a speed comparable to the speed of merely QR-updating. We refer to [9, 10] and [11] for further details, as well as for a performance analysis, square root free algorithms and extensions towards generalized decompositions.

Acknowledgement: M. Moonen is supported by the Belgian National Fund for Scientific Research (N.F.W.O.).

References

[1] BRENT R.P., LUK F.T., 1985. The solution of singular value and symmetric eigenvalue problems on multiprocessor arrays. SIAM J.Sc.Stat.Comp., Vol.6, No 1., pp 69-84.

[2] BUNCH J.R.,NIELSEN C.P. 1978. Updating the singular value decomposition Num.Math.31, pp 111-129.

[3] BUNCH J.R.,NIELSEN C.P.,SORENSEN D.C. 1978. Rank one modification of the symmetric eigenproblem. Num.Math.31, pp 31-48.

[4] GENTLEMAN W.M, KUNG H.T., 1981. *Matrix triangularization by systolic arrays.* Real-Time Signal Processing IV, Proc. SPIE, Vol. 298.

[5] GOLUB G.H. 1973. Some modified eigenvalue problems. SIAM Review, vol. 15, no. 2, pp 318-334.

[6] GILL P.E., GOLUB G.H., MURRAY W., SAUNDERS M.A., 1974. *Methods for modifying matrix factorizations.* Mathematics of Computation, Vol. 28, 126, pp 505-535.

[7] KOGBETLIANTZ E., 1955. Solution of linear equations by diagonalization of coefficient matrices. Quart. Appl. Math. 13, 123-132.

[8] LUK F.T., 1986. A triangular processor array for computing singular values. Lin. Alg. Appl., 77, pp 259-273.

[9] MOONEN M., VANDEWALLE J. 1988. (G)SVD updating for tracking slowly time-varying systems. ESAT K.U.Leuven, Internal report.

[10] MOONEN M., VAN DOOREN P. , VANDEWALLE J. *Parallel updating algorithms, part one : the ordinary singular value decomposition.* Submitted for publication.

[11] MOONEN M., VAN DOOREN P. , VANDEWALLE J. *Parallel updating algorithms, part two : generalized singular value decompositions.* Submitted for publication.

[12] STEWART G.W., 1985. A Jacobi-like algorithm for computing the Schur decomposition of a nonhermitian matrix. SIAM J. Sc. Stat. Comp., Vol. 6, No. 4, pp 853-863.

RECURSIVE LEAST SQUARES COMPUTATIONS

Robert J. Plemmons

Abstract. We consider parallel implementations of algorithms for recursive least squares computations based upon the information matrix and the covariance matrix updating methods. The target architecture is a shared-memory multiprocessor, and test results on an Alliant system with 8 vector processors demonstrate the parallel efficiencies of the algorithms. The results also show that the covariance method in a form suggested by Pan and Plemmons is easily the most efficient on the Alliant multiprocessor. However, the singular values of the factor matrix decrease in covariance updating, and we give the results of a scheme for adaptively tracking the smallest singular value. Applications include robust regression in statistics and modification of the Hessian matrix in optimization, but the primary motivation for this work is the need for fast algorithms for recursive least squares computations in signal processing.

1. INTRODUCTION

1.1. Background and Applications. The problem considered in this paper is that of designing efficient parallel algorithms for recursive least squares computations based upon updating the information matrix or the covariance matrix factorization. Applications of recursive least squares updating include modification of the Hessian matrix in certain optimization schemes, (e.g., the quasi-Newton method) [5],[13], and estimation methods in adaptive signal processing and control [1-4],[8],[11],[14]. Although many situations in which signal processing is important require special purpose systolic array type devices, our interest in this paper is adaptive signal processing for computations in non real-time applications and on general high performance parallel computers. Interesting non real-time applications of recursive least squares computations in signal processing include adaptive spectral estimation. Applications here include, for example: speech recognition [1], seismology [3], and fetal cardio-analysis [8].

1.2. Least Squares. The *linear least squares problem* can be posed as follows without reference to any specific application:

Given a real $m \times n$ matrix X with full column rank n and a real m-vector s, find the n-vector w that solves

$$\min \|s - Xw\|_2, \tag{1}$$

where $\|\cdot\|_2$ denotes the usual Euclidean norm. The solution to (1) is given by

$$w = (X^T X)^{-1} X^T s. \tag{2}$$

If R denotes the upper triangular *Cholesky factor* of the cross product matrix X^TX, *i.e.*, $R^TR = X^TX$, then w can be obtained by solving the triangular systems $R^Tv = X^Ts$, $Rw = v$, where v is an intermediate vector. However, in many applications where accuracy and stability are important [7], R is computed directly from X by a sequence of orthogonal transformations; that is

$$QX = \begin{bmatrix} R \\ 0 \end{bmatrix}, \quad Q^TQ = I. \tag{3}$$

Then setting

$$Qs = \begin{bmatrix} s_1 \\ s_2 \end{bmatrix}, \quad s_1 \text{ an } n\text{-vector}, \tag{4}$$

w is computed by solving the upper triangular system $Rw = s_1$. Such schemes are sometimes called "square root" methods in the signal processing literature. In addition, the matrix

$$P \equiv (X^TX)^{-1} \tag{5}$$

is called the *covariance matrix* for (1). It measures the expected errors in the least squares vector w. Its inverse

$$P^{-1} = X^TX, \tag{6}$$

often called the *normal equations matrix* [7], is known as the *information matrix* for (1) in the signal processing literature [1],[3],[8]. It measures the information content in the experiment leading to (1).

1.3. Recursive Least Squares. In *recursive least squares* it is required to recalculate w when observations (*i.e.*, equations) are successively added to, or deleted from, the problem (1). For example, in many applications information arrives continuously and must be incorporated into the solution w. This is called *updating.* It is sometimes important to delete old observations and have their effect excised from w. This is called *downdating.* Alternatively, an exponential forgetting factor (See, *e.g.*, [8]) may be incorporated into the updating computations.

In some important applications the model itself is not stationary, requiring that an additional equation involving w be incorporated into the model [1],[3],[8]. For simplicity, *we consider only stationary models in this paper* and will not incorporate an exponential forgetting factor into the analysis.

There are two main approaches to solving recursive least squares problems; the *information matrix method* based upon modifying $P^{-1} = X^TX$ and the *covariance matrix method* based upon modifying P [1],[3],[8]. Instead of modifying P^{-1} or P directly, it is generally preferable for stability reasons to modify their Cholesky factors [7],[14].

2. RECURSIVE LEAST SQUARES ALGORITHMS

2.1. Updating Computations. Recursive least squares (RLS) updating computations can be described as follows. Consider the least squares problem (1) and, without loss of generality, assume that the additional data vector y and scalar σ for the equation

$$y^T w = y_1 w_1 + \cdots + y_n w_n \approx \sigma$$

are appended to X and s forming

$$\widetilde{X} = \begin{bmatrix} X \\ y^T \end{bmatrix}, \quad \tilde{s} = \begin{bmatrix} s \\ \sigma \end{bmatrix}. \tag{7}$$

One now seeks to solve the modified problem

$$\min \|\tilde{s} - \widetilde{X}\tilde{w}\|_2 \tag{8}$$

for the updated least squares estimate vector $\tilde{w}$ at each recursive time step.

2.2. Information Matrix Updating. First, we review the information matrix method based upon a factorization of $P^{-1} = X^T X$ for computing X. If R is the Cholesky factor of X, as in (3), then from (7)

$$\widetilde{X}^T \widetilde{X} = X^T X + yy^T = R^T R + yy^T.$$

The Cholesky factor $\widetilde{R}$ of $\widetilde{X}^T\widetilde{X}$ can thus be calculated by applying orthogonal plane rotations to $\begin{bmatrix} R \\ y^T \end{bmatrix}$ in order to reduce the row vector y^T to zero, as described in [2],[3],[9]. Of course the same orthogonal rotations are also applied to the vector $\begin{bmatrix} s_1 \\ \sigma \end{bmatrix}$, where s_1 is the n-vector given in (4). More precisely if

$$\widetilde{Q}\begin{bmatrix} R \\ y^T \end{bmatrix} = \begin{bmatrix} \widetilde{R} \\ 0^T \end{bmatrix}, \quad \widetilde{Q}\begin{bmatrix} s_1 \\ \sigma \end{bmatrix} = \begin{bmatrix} \tilde{s}_1 \\ \tilde{\sigma} \end{bmatrix} \tag{9}$$

where $\widetilde{Q}^T\widetilde{Q} = I$, then $\widetilde{R}^T\widetilde{R} = R^T R + yy^T = \widetilde{X}^T\widetilde{X}$. Consequently, the updated least squares estimate vector $\tilde{w}$ is then obtained by solving the triangular system $\widetilde{R}\tilde{w} = \tilde{s}_1$.

The information matrix method described above can thus be implemented by standard procedures (see, e.g., [3]) for updating the Cholesky factor R of $P^{-1} = X^T X$ to the Cholesky factor $\widetilde{R}$ of $\widetilde{P}^{-1} = \widetilde{X}^T\widetilde{X}$, followed by the triangular solution for $\tilde{w}$. The standard information matrix updating algorithm, as implemented for example in LINPACK, is given next.

Algorithm 1 (*Information Updating*) Given the current n-dimensional least squares estimate vector w, the current factor R of $P^{-1} = (X^T X)$ and the observation $y^T w \approx \sigma$ being added, the algorithm computes the updated factor $\widetilde{R}$ of $\widetilde{P}^{-1}$ and the updated least squares estimate vector $\tilde{w}$. Here the n-vector s_1 is given in (4).

1. Choose orthogonal plane rotations $G_{i,n+1}$, rotating the *ith* row into the $n+1st$ row, to form

$$G_{n,n+1}\cdots G_{1,n+1}\begin{bmatrix} R & s_1 \\ y^T & \sigma \end{bmatrix} = \begin{bmatrix} \widetilde{R} & \tilde{s}_1 \\ 0^T & \tilde{\sigma} \end{bmatrix} \tag{10}$$

   by reducing y^T to 0 from the left to right preserving the upper triangular form of R in $\widetilde{R}$.
2. Solve the upper triangular system $\widetilde{R}\tilde{w} = \tilde{s}_1$.

At the next recursive time step when a new equation is added to (1), $\widetilde{R}$ replaces R, $\tilde{s}_1$ replaces s_1, $\tilde{w}$ replaces w and the process returns to step 1. The total algorithm requires up to $\frac{5}{2}n^2 + O(n)$ multiplications per time step. The $\frac{5}{2}n^2$ term includes $2n^2$ multiplications to update R and $\frac{1}{2}n^2$ multiplications for the triangular solve to compute the updated w. A parallel distributed-memory hypercube implementation of this information matrix scheme is given by Henkel, Heath and Plemmons in [9].

2.3. Covariance Matrix Updating. The process of modifying least squares computations by updating the covariance matrix P has been used in control and signal processing for some time in the context of linear sequential filtering [1],[3],[8],[11]. One begins with estimates for $P = R^{-1}R^{-T}$ and w, and updates R^{-1} to $\widetilde{R}^{-1}$ and w to $\tilde{w}$ at each recursive time step. Recently Pan and Plemmons [12] have described a parallel scheme for these computations.

Observe first that with $\widetilde{X}$ given in (7), $\widetilde{P}^{-1}$ is given by

$$\widetilde{P}^{-1} = \widetilde{X}^T\widetilde{X} = X^TX + yy^T = P^{-1} + yy^T.$$

Consequently, by the Sherman-Morrison formula (see, *e.g.*, [7]),

$$\widetilde{P} = P - \frac{1}{1+y^TPy}Pyy^TP. \tag{11}$$

The updated w is given by

$$\tilde{w} = (\widetilde{X}^T\widetilde{X})^{-1}\widetilde{X}^T\begin{bmatrix} s \\ \sigma \end{bmatrix} = \widetilde{P}\widetilde{X}^T\begin{bmatrix} s \\ \sigma \end{bmatrix}. \tag{12}$$

By substituting (7) and (11) into (12) and using the representation (2) for w, there results the following basic formula for the update:

$$\tilde{w} = w + \widetilde{P}y(\sigma - y^Tw). \tag{13}$$

The vector

$$k \equiv \widetilde{P}y \tag{14}$$

is often called the Kalman gain vector (see, *e.g.*, [1],[3],[8]). It weights the predicted residual $\sigma - y^Tw$.

The following algorithm for updating P to $\widetilde{P}$ by updating R^{-T} to $\widetilde{R}^{-T}$ and for computing the updated least squares estimate vector $\widetilde{w}$ was given in Theorem 2 and Lemma 3 of [12] (see also Morf and Kailath [11]). It was indicated in Theorem 2 of [12] that the same sequence of orthogonal transformations used to update R to $\widetilde{R}$ also updates R^{-T} to $\widetilde{R}^{-T}$. Lemma 3 in [12] shows that this sequence of orthogonal transformations is precisely given by (15) below. Here we write $P = L^T L$, where $L \equiv R^{-T}$.

Algorithm 2 (*Covariance Updating*) Given the current n-dimensional least squares estimate vector w, the current factor $L \equiv R^{-T}$ of $P = (X^T X)^{-1}$ and the observation $y^T w \approx \sigma$ being added, the algorithm computes the updated factor $\widetilde{L} \equiv \widetilde{R}^{-1}$ of $\widetilde{P}$ and the updated least squares estimate vector $\widetilde{w}$.

1. Form the matrix vector product $a = Ly$.
2. Choose orthogonal plane rotations $G_{i,n+1}$, rotating the ith component into the $n+1st$ component, to form

$$(15) \quad G_{n,n+1} \cdots G_{1,n+1} \begin{bmatrix} -a \\ 1 \end{bmatrix} = \begin{bmatrix} 0 \\ \delta \end{bmatrix}, \quad \delta = \sqrt{1 + \|a\|_2^2},$$

by reducing $-a$ to 0 from the top down, and apply

$$(16) \qquad G_{n,n+1} \cdots G_{1,n+1} \begin{bmatrix} L \\ 0^T \end{bmatrix} = \begin{bmatrix} \widetilde{L} \\ u^T \end{bmatrix},$$

preserving the lower triangular form of L in $\widetilde{L}$.

3. Form

$$(17) \qquad \widetilde{w} = w - \frac{1}{\delta} u(\sigma - y^T w).$$

Some comments about the algorithm are in order. First, the rotation parameters in (15) are computed successively using the components of the vector a. It follows from [12], Theorem 2, that $u = -\delta k$, where k is the Kalman gain vector given by (14). As the recursive least squares computation proceeds, $\widetilde{L}$ replaces L, $\widetilde{w}$ replaces w, a new equation is added, and the process returns to step 1. The total algorithm requires up to $\frac{5}{2}n^2 + O(n)$ multiplications per time step. The $\frac{5}{2}n^2$ term comes entirely from updating R^{-T} in steps 1 and 2 and no triangular solves are involved. Algorithm 2 has been successfully implemented on a distributed-memory iPSC System 2 hypercube architecture with 64 processors by Henkel and Plemmons [10].

We comment that a discussion of the stability properties and relative merits of the information and covariance matrix methods have been given by Verhaegen and Van Dooren [14]. It is pointed out there that the covariance matrix method, when properly implemented, performs as well numerically as the information matrix method in the context of Kalman filtering. In addition Ferng, Pierce and Plemmons [6] have developed adaptive schemes for tracking the condition number of the covariance matrix P.

A scheme for tracking the smallest singular value of L in Algorithm 2 is incorporated into our code and some results are reported in Figure 2.

Parallel implementations of Algorithms 1 and 2 on an Alliant shared-memory system with 8 vector processors are discussed in the next section. Some of the FORTRAN coding for this project was done by students in a graduate seminar on parallel processing taught by the author at North Carolina State University, Spring 1989.

3. NUMERICAL COMPUTATIONS

3.1. Target Architecture. This section concerns implementations of the Information Matrix and Covariance Matrix algorithms in FORTRAN on a shared-memory parallel computer. The target architecture is an Alliant shared-memory system. The algorithms have been implemented locally on an Alliant FX/40 in our Center for Research on Scientific Computation at North Carolina State University and on an Alliant FX/8 at the Argonne National Laboratory with 8 vector processors sharing 32 megabytes of memory. We will give test results only for the implementation on the larger FX/8 since the architectures are similar.

The Alliant compiler attempts to detect those sections of the FORTRAN program that have potential for vector and parallel processing, and generates machine code that will use the vectorization and concurrency features of the hardware. For example, when the loops are nested, the compiler attempts to generate code so that the innermost loop will run in vector mode, and the next-to-innermost loop in concurrent mode. The programmer can insert optimization directives into the code to control vectorization and concurrency at various levels. Essentially, the Alliant system can be thought of as a moderate-scale supercomputer sharing many features with the Cray X-MP, Cray Y-MP and Cray 2 systems.

For our implementations of Algorithms 1 and 2, the computations associated with the current square root information matrix factor R of P^{-1} or the covariance matrix factor $L \equiv R^{-T}$ of P are assigned by columns, as are the corresponding computations associated with the entries of the least squares estimate vector w^T and the update vector y^T. Our codes are written in standard FORTRAN, and so can be run on any machine with a FORTRAN 77 compiler. However the codes are structured so that they will be optimized by the Alliant compiler to best advantage. Also, use has been made of compiler directives to enhance the vectorization/concurrency. These directives are preceded by a 'C' so they appear to a standard FORTRAN 77 compiler comments, and do not interfere with the portability of the code.

3.2. Test Results. In this section we report the actual performances of the parallel algorithms on the Alliant multiprocessor. The data was generated randomly and all computations were performed in single precision. Problems ranging in size from $n = 100$ to size $n = 1500$ least squares estimates were considered.

Figure 1 reports the times in seconds for ten recursive applications of Algorithms 1 and 2 on test problems of various sizes of n. Ten complete recursive update computations for the least squares estimate vectors w are performed for each n. This reflects the recursive addition of ten equations to the least squares system, as described in Section 2.1.

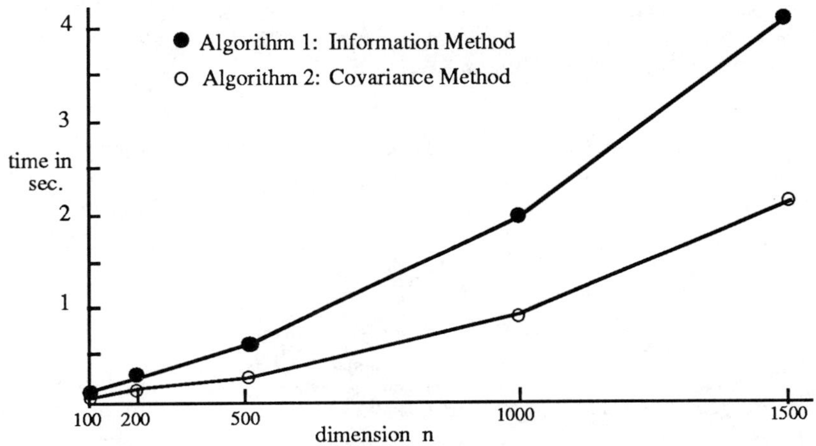

FIGURE 1. Computations on Alliant FX/8 with full vector-concurrency.

The next figure shows the results of tracking the smallest singular value of L in Algorithm 2 for fifty updates. The scheme used is adaptive and increases the cost of the algorithm by less than 20 percent.

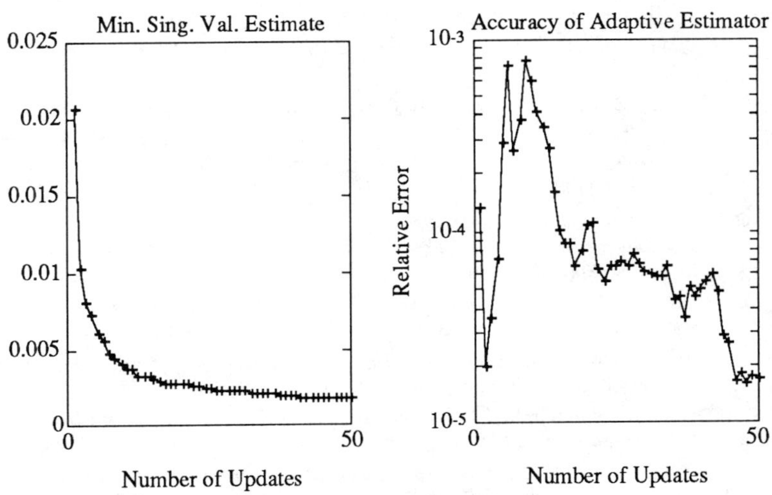

FIGURE 2. Estimation of minimum singular value in covariance method.

4. SUMMARY. For the important scientific application of recursive least squares computations, we have investigated parallel solution algorithms for shared-memory parallel computers. The algorithms are based upon the standard information matrix method and the covariance matrix updating method in a parallelizable form described in [12]. Timing results in Figure 1 indicate that the covariance method performs best on the Alliant multiprocessor. The primary reason for the superior performance of Algorithm 2 on a vector multiprocessor system is the parallelizability of the algorithm as formulated by Pan and Plemmons.

Research supported by the Air Force Office of Scientific Research under grant AFOSR-88-0285 and the National Science Foundation under grants DMS-85-21154 and DMS-89-02121.

REFERENCES

[1] S.T. Alexander, **Adaptive Signal Processing**, Springer Verlag, New York (1986).

[2] S.T. Alexander, C.-T. Pan and R.J. Plemmons, *Analysis of a recursive least squares hyperbolic rotation scheme for signal processing*, **Lin. Alg. Appl. Special Issue on Electrical Eng.** 98 (1988) 3-40.

[3] G.J. Bierman, **Factorization Methods for Discrete Sequential Estimation**, Academic Press, NY, 1977.

[4] J.M. Cioffi and T. Kailath, *Windowed fast transveral filters, adaptive algorithms with normalization*, **IEEE Trans. on Acous., Speech and Sig. Proc.** 33 (1985) 607-625.

[5] J.E. Dennis and R.B. Schnabel, **Quasi-Newton Methods for Nonlinear Problems**, Prentice-Hall, Englewood Cliffs NY, 1983.

[6] R. Ferng, D.J. Pierce and R.J. Plemmons, *ACE: An adaptive condition estimator for signal processing*, preprint (1989).

[7] G.H. Golub and C. Van Loan, **Matrix Computations**, Johns Hopkins Press, Baltimore, 1983.

[8] S. Haykin, **Adaptive Filter Theory**, Prentice-Hall, Englewood Cliffs, NJ, 1986.

[9] C.S. Henkel, M.T. Heath and R.J. Plemmons, *Cholesky downdating on a hypercube*, in **Hypercube Concurrent Computers and Applications**, Ed. by G. Fox, ACM Press (1988), 1592-1598.

[10] C.S. Henkel and R.J. Plemmons, *Recursive least squares computations on a hypercube multiprocessor using the covariance factorization*, **SIAM J. Sci. Statist. Comp.**, to appear (1990).

[11] M. Morf and T. Kailath, *Square root algorithms and least squares estimation*, **IEEE Trans. on Automatic Control** 20 (1975) 487-497.

[12] C.-T. Pan and R.J. Plemmons, *Parallel least squares modifications using inverse factorizations*, **Computational and Applied Math.**, 27 (1989) 109-127.

[13] R.J. Plemmons and S.J. Wright, *An efficient parallel scheme for minimizing a sum of Euclidean norms*, **Lin. Alg. and Appl.**, 121 (1989) 71-85.

[14] M. Verhaegen and P. Van Dooren, *Numerical aspects of different Kalman filter implementations*, **IEEE Trans. on Automatic Control** 31 (1986) 907-917.

Numerical Solution of Large Lyapunov Equations

Youcef Saad

Abstract

In this paper we propose a few methods for solving large Lyapunov equations that arise in control problems. We consider the common case where the right hand side is a small rank matrix. For the single input case, i.e., when the equation considered is of the form $AX + XA^T + bb^T = 0$, where b is a column vector, we establish the existence of approximate solutions of the form $X = VGV^T$ where V is $N \times m$ and G is $m \times m$, with m small. The first class of methods proposed is based on the use of numerical quadrature formulas, such as Gauss-Laguerre formulas, applied to the controllability Grammian. The second is based on a projection process of Galerkin type. Numerical experiments are presented to test the effectiveness of these methods for large problems.

1 Introduction

Many of the matrix equations that arise in control problems can be successfully solved by well-known numerical techniques, when the matrices involved are small. In contrast there has been very little done to provide numerical methods for solving these same problems when the associated matrices are very large. Yet, there are now several applications in control which lead to matrix equations involving very large sparse matrices. These typically arise whenever the model involves a partial differential equation in several space dimensions such as when considering large space structures [2] or when a large network model, e.g. electrical network, [3] is involved.

Recently, there has been a few efforts directed towards large scale matrix problems in control. In [5] we proposed a method for partial pole placement, which consists of placing a few of the poles of the matrix, namely only those that are unstable. The methods proposed are based on projecting the problem onto a small invariant subspace of A associated with the unstable eigenvalues.

In this paper we focus on Lyapunov's equations. The numerical solution of these equations have gained considerable importance in the last few years due to the crucial role that they play in the so-called H_∞ analysis. The basic problem addressed by this theory is to find a low dimensional ODE model that approximates the original dynamical system governed by

$$\begin{aligned} \dot{x} &= Ax + Bu \\ y &= Cx \end{aligned} \tag{1}$$

Here, A is $N \times N$, B is $N \times p$ and C is $q \times N$. In the situation we consider N may be very large while p and q are very small. The closeness of this system from its lower dimensional model is measured in terms of the singular values of the matrix XY where X and Y are solutions of the two Lyapunov equations,

$$AX + XA^T + BB^T = 0 \tag{2}$$
$$A^TY + YA + C^TC = 0 \tag{3}$$

These are termed Hankel singular values of (1). Typically, one would like to compute some of the largest Hankel singular values. In this paper we will provide the tools for computing approximations to X and Y. Moreover, these approximations are in such a form that they allow easy computation of the largest singular values of XY.

Throughout, we will assume that the eigenvalues of A have negative real parts. In this situation there is an explicit formula for the solution of (2),

$$X = \int_0^\infty e^{\tau A} BB^T e^{\tau A^T} d\tau \tag{4}$$

known as the controllability Grammian of (1).

Much of the theory in this paper will be established by using the above formula. The main observation which we will prove in the next section, is that there are accurate approximations to the system (2) *of small rank*. Extracting such approximations will be the subject of sections 3 and 4.

2 Low rank solutions to the Lyapunov equation

The purpose of this section is to identify the types of approximations that will be used in this paper. We will restrict ourselves to the case corresponding to the single input situation in (1), i.e., the equation (2) becomes

$$AX + XA^T + bb^T = 0 \tag{5}$$

where b is a single column vector, and its explicit solution is given by

$$X = \int_0^\infty e^{\tau A} bb^T e^{\tau A^T} d\tau. \tag{6}$$

One way to integrate formula (6), would be to start by replacing (6) by its approximation

$$X(s) = \int_0^s e^{\tau A} bb^T e^{\tau A^T} d\tau \tag{7}$$

where s is selected so that the error in the approximation is small enough. Note that we must integrate in the interval $[0, s]$ a function of the form $w(t)w(t)^T$ where $w(t)$ is the vector function $w(t) = e^{tA}b$. Assume that we approximate $w(t)$ in the interval $[0, s]$ as $w(t) \approx w_m(t) = q_m(At)b$, where q_m is a polynomial of degree $m - 1$. The resulting approximation for $X(s)$ is

$$X_m(s) = \int_0^s w_m(\tau)w_m(\tau)^T d\tau \tag{8}$$

This approximation can be made arbitrarily accurate by increasing the degree of q_m. Thus, by choosing a large enough s and and a large enough degree for q, we can make the approximation $X_m(s)$ to X as accurate as desired. Let $V_m = [b, Ab, \ldots, A^{m-1}b]$ and $q_m(t) = \alpha_0 + \alpha_1 t + ... + \alpha_{m-1}t^{m-1}$. Then $w_m(t)$ can be written as $w_m(t) = V_m z_m(t)$, $z_m(t) = [\alpha_0, \alpha_1 t, ..., \alpha_{m-1}t^{m-1}]^T$ and therefore

$$X_m(s) = V_m \left(\int_0^s z_m(\tau)z_m(\tau)^T d\tau \right) V_m^T \equiv V_m G_m V_m^T \tag{9}$$

where G_m is an $m \times m$ matrix whose entries can easily be shown to be equal to

$$g_{ij}^{(m)} = \frac{\alpha_{i-1}\alpha_{j-1}s^{i+j-1}}{i+j-1}. \tag{10}$$

Throughout the paper we will exploit approximations to the solution X which are of the form

$$X_m = V_m G_m V_m^T. \tag{11}$$

where $V_m = [v_1, v_2, ..., v_m]$ is a fixed set of vectors and G_m is an arbitrary $m \times m$ matrix. The set of all such matrices is clearly a subspace of the space of $N \times N$ matrices. The range of any matrix in this subspace is included in the subspace $K = span\{V_m\}$ while its kernel contains the orthogonal to K. The subspace of the matrices (11) is in fact uniquely defined by the range K of these matrices. Thus, we will denote by $\mathbf{Z}_m(K)$ the space of all matrices of the form (11), or $\mathbf{Z}_m$ if there is no ambiguity, where V_m is a basis of the subspace K. If we restrict the matrices to be symmetric, then $\mathbf{Z}_m$ is of dimension $(m(m+1))/2$. If one wants to deal with the more general situation where the right-hand-side is of a more general form bv^T rather than bb^T then G can no longer restricted to be symmetric and the dimension of $\mathbf{Z}_m$ is m^2.

What we have just shown above is that there are approximations from $\mathbf{Z}_m(K_m)$ where K_m is the Krylov subspace

$$K_m = span\{b, Ab, ..., A^{m-1}b\} \tag{12}$$

that will converge to X as m tends to infinity. Here we should mention that, rigorously speaking, this statement is trivial in the finite dimensional case since there is always an exact solution of the form (11) for $m \geq N$. However, 'convergence' is meant in the infinite dimensional context, in the same way that one speaks of the convergence of the conjugate gradient method which is known to be a finite process.

If we can find a good approximation to $w(t)$ in the interval $[0, s]$, by a low degree polynomial, we should be able to find a good approximation to X from $\mathbf{Z}_m$. However, the above formulas should not be used as a practical procedure since they are likely to be highly unstable. Section 4 presents a process that is mathematically (but not algorithmically) equivalent to the procedure outlined above with a specific choice for q_m. In the next section we exploit well-known numerical quadrature formulas to derive particular subspaces K and their corresponding approximations.

3 Use of numerical quadrature

The first procedure that we propose for approximating X is based upon evaluating the integral (6) by standard numerical quadrature formulas. We will describe two such procedures. First, assume that s has been chosen and that we can evaluate $w(t)$ at m equally spaced points

$$t_i = (i-1)\frac{s}{m-1}, i = 1, 2, ..., m \tag{13}$$

then an appropriate quadrature formula for evaluating (6) would yield,

$$X_m = \sum_{i=1}^{m} \delta_i w(t_i) w(t_i)^T \tag{14}$$

where the δ_i's are the quadrature coefficients. In other words,

$$X_m = W_m \Delta W_m^T \tag{15}$$

in which W_m is the $N \times m$ matrix

$$W_m = [w(t_1), w(t_2), ..., w(t_m)]$$

and

$$\Delta = Diag(\delta_1, ..., \delta_m).$$

Note that $w(t_1) = b$. In order to use the formula (14) we also need to compute and save $w(t_i), i = 2..., m$. One way in which this can be done is by solving the system of Ordinary Differential Equations

$$\dot{w} = Aw \tag{16}$$
$$w(0) = b \tag{17}$$

by any technique and save the vectors $w(t_i), i = 1, 2, .., m$ at the points t_i. Alternatively, one may compute $w(t_{i+1})$ from $w(t_i)$ by $w(t_{i+1}) = e^{A\Delta t_i} w(t_i)$, where $\Delta t_i = t_{i+1} - t_i$, in a number of efficient ways as has been recently suggested in [4]. The problem considered in [4] is to approximate the product of the exponential of a matrix times a vector. We will describe one such approach in some detail in Section 4.

Thus, the class of algorithms based on numerical quadrature would begin by choosing a quadrature rule and performing the following procedure.

Algorithm 1

(1) Start: Define $w(t_1) = b$;

(2) For $i = 2, ..., m$ do:

- Compute $w(t_i)$ from $w(t_{i-1})$;
- Save $w(t_i)$ in the i-th column of W_m.

It is clear that the actual matrix X_m should never be computed explicitly, since it is a dense $N \times N$ matrix in general. One only needs to store W_m and Δ_m. In fact, it is worth pointing out that one may not even need to store W_m if the original problem is to approximate Xv where v is some vector, rather than to compute an approximation to the matrix X itself. In this situation the approximation $x = X_m v$ can be accumulated as $x := x + (\delta_i w(t_i)^T v) w(t_i)$ every time that a new $w(t_i)$ is generated and one can discard the old $w(t_i)$'s.

The question that we would like to address next is which integration schemes to use and how to implement them. First, a restriction of the scheme is that the weights δ_i should be positive, since we know that the solution is semi-positive definite matrix. For example, this is not satisfied by all the Newton-Cotes formulas but it is true for the Gauss-Legendre formulas, see e.g., [6], pp. 145. We have implemented and tested two basic classes of quadrature formulas.

1. Closed Newton-Cotes formulas with 3 points (Simpson's rule), 5 points, and 7 points.

2. Gauss-Laguerre formulas with 9 points and 15 points.

Newton-Cotes formulas are fairly standard and the corresponding coefficients can be found in any elementary numerical analysis textbook. These use equally spaced points which may be an undesirable feature in our case because of the exponential behavior of the function to integrate. At the beginning of the interval $w(t)$ varies far more than for large t. Therefore it is natural to expand the intervals of integration as we proceed. One procedure we have experimented with is to double the size of the intervals $[x_i, x_{i+k}]$ on which the Newton-Cotes formula is based at every time.

For the 7-point Newton-Cotes formula we triple the size of the interval. This simple modification does provide better results than keeping a fixed interval.

The Gauss-Laguerre quadrature formula is a highly accurate scheme to evaluate integrals of functions of the form $e^{-t}g(t)$ on the infinite half line. The basic formula is,

$$\int_0^\infty e^{-\tau}g(\tau)d\tau \approx \sum_{i=1}^{m} \omega_i g(t_i) \tag{18}$$

where the t_i are the roots of the Laguerre polynomial of degree m and where the ω_i's are the quadrature weights. For general functions the formula is used as follows,

$$\int_0^\infty g(\tau)d\tau = \int_0^\infty e^{-\tau}(e^{\tau}g(\tau))d\tau \approx \sum_{i=1}^{m} \omega_i e^{t_i} g(t_i) \equiv \sum_{i=1}^{m} \delta_i g(t_i). \tag{19}$$

The roots t_i and the coefficients δ_i for our numerical tests have been taken from [1].

An aspect which we found to be critical to the performance of the Gauss-Laguerre integration is a proper scaling of the variable τ. For some problems, such as the one in the numerical experiments section, the norm of A can be very large and as a result the vector $w(t_i)$ can be tiny for most of the t_i's, thus leading to a poor evaluation of the integral (6). For example assume that A is symmetric negative definite having $\lambda_1 = -10$ as its eigenvalue closest to zero, and consider Gauss Laguerre integration with 15 points. Then for the second root $t_2 = 1.21559..$ all the eigencomponents of $w(t_2)$ will be no larger than $e^{-12.1559..} \approx 10^{-06}$. For t_3 the components do not exceed $e^{-22.6694} \approx 1.38610^{-10}$, and beyond t_3 the $w(t_i)$'s become too small. A remedy is to use a change of variable $\xi = \alpha\tau$ to ensure that the value of $w(t)$ at the last root t_m is not too small or too large. For example, if an estimate of the rightmost eigenvalue λ_1 of A is available, we might choose α so that $|e^{\alpha t_m \lambda_1}| = tol$, where tol is some prescribed tolerance.

4 A Krylov subspace technique

In this section we present a method that is based on a global approximation to $w(t)$ in $[0, +\infty)$. Explicit integration of this approximation will then yields an approximation to X. This, as will be seen, is equivalent to a projection process of Galerkin-type.

The first question that we address is how to approximate $e^{A}b$ for a given vector b. This was considered in [4] where polynomial approximation to the exponential was used. The approximation to $e^{A}b$ is taken of the form

$$e^{A}b \approx p_{m-1}(A)b \tag{20}$$

where p_{m-1} is a polynomial of degree $m-1$.

Clearly, one can use other types of approximations, e.g., rational, which are usually more accurate. The attraction of polynomials is the fact that they do not require solving linear systems. One way in which a good polynomial can be found is by attempting to minimize some norm of the error $e^{z} - p_{m-1}(z)$ on a continuum in the complex plane that encloses the spectrum of A. For example, Chebyshev approximation can be used. The disadvantage of this approach is that it requires some approximation to the spectrum of A. In [4] we considered a technique based on Arnoldi's method which does not require any eigenvalue information. This technique is now summarized.

The approximation (20) to $e^{A}b$ is an element of the Krylov subspace (12). In this approach we need to generate an orthonormal basis $V_m = [v_1, v_2, v_3, \ldots, v_m]$ of K_m via the well-known Arnoldi algorithm starting with $v_1 = b/\|b\|_2$.

Algorithm 2: Arnoldi

1. *Initialize:* Compute $v_1 := b/\|b\|_2$.
2. *Iterate:* Do $j = 1, 2, ..., m$
   1. Compute $w := Av_j$
   2. Compute a set of j coefficients h_{ij} so that $w := w - \sum_{i=1}^{j} h_{ij} v_i$ is orthogonal to all previous v_i's.
   3. Compute $h_{j+1,j} = \|w\|_2$ and $v_{j+1} = w/h_{j+1,j}$.

By construction the above algorithm produces an orthonormal basis

$$V_m = [v_1, \ldots, v_m],$$

of the Krylov subspace K_m. If we denote the $m \times m$ upper Hessenberg matrix consisting of the coefficients h_{ij} computed by the algorithm by H_m we have the relation

$$AV_m = V_m H_m + h_{m+1,m} v_{m+1} e_m^T \tag{21}$$

from which we get $H_m = V_m^T A V_m$. Therefore H_m represents the projection of the linear transformation A onto the subspace K_m, with respect to the basis V_m. As is well-known in the particular case where A is symmetric, Arnoldi's algorithm simplifies to the Lanczos process in which case H_m becomes tridiagonal symmetric.

We can write the desired approximation to $x = e^A b$ as $x_m = p_m(A)v$ or equivalently $x_m = V_m y$ where y is an m-vector. In [4], the choice $y = \beta e^{H_m} e_1$ with $\beta = \|b\|_2$ was suggested, leading to the following formula for arbitrary t,

$$e^{tA} b \approx \beta V_m e^{tH_m} e_1 \tag{22}$$

The quality of this approximation was also analyzed in [4] and the following result was shown.

Theorem 4.1 *Let A be any square matrix and let $\rho = \|A\|_2$. Then the error of the approximation (22) is such that*

$$\|e^{tA} b - \beta V_m e^{tH_m} e_1\|_2 \le 2\beta \frac{(t\rho)^m e^{t\rho}}{m!}. \tag{23}$$

Experiments reported in [4], reveal that this approximation can be very accurate even for moderate values of the degree m. The theorem shows convergence of the approximation (22) for fixed t, as m increases to infinity. However, note that the above approximation is exact when $m = N$, see [4].

Let us now substitute the expression (22) in (6). We obtain the approximation

$$X_m = V_m \left(\int_0^\infty e^{\tau H_m} (\beta e_1)(\beta e_1)^T e^{\tau H_m^T} d\tau \right) V_m^T \equiv V_m G_m V_m^T. \tag{24}$$

Assuming that H_m has eigenvalues with negative real parts, we note that G_m is the solution of the $m \times m$ Lyapunov equation

$$H_m G_m + G_m H_m^T + \beta^2 e_1 e_1^T = 0 \tag{25}$$

In other words, modulo the polynomial approximation made on the exponential of A, we have reduced the original problem into one of dimension m. This raises the question as to whether or not the process just described is mathematically equivalent to a projection-type method such as a Galerkin process.

To see that this is the case we need to define the subspace of approximants and the inner product. The subspace of approximants will be the subspace $\mathbf{Z}_m(K_m)$ as defined in Section 2. This is the same as the subspace of all matrices of size N of the form $V_m G V_m^T$ where V_m is the orthogonal basis of K_m as constructed from the Arnoldi process. Recall that from our definition V_m *is fixed* and therefore the actual variable is the $m \times m$ matrix G. For the inner product we take the usual inner product in R^{N^2}, i.e., the inner product of two matrices taken as vectors of N^2 elements. It is a simple exercise to show that this can also be defined by

$$< X, Y >= tr(XY^T) \tag{26}$$

The Galerkin condition defines an approximation $\tilde{X}$ by stipulating that $\tilde{X}$ belong to $\mathbf{Z}_m$ and that the residual $R(\tilde{X}) = A\tilde{X} + \tilde{X}A^T + bb^T$ be orthogonal to all of the subspace of approximants. This second condition gives,

$$tr[ZR^T(\tilde{X})] = 0, \quad \forall Z \in \mathbf{Z_m} \tag{27}$$

and therefore,

$$0 = tr[V_m G V_m^T R(\tilde{X})^T] = tr[G V_m^T R(\tilde{X})^T V_m] \ \ \forall \ G \tag{28}$$

Taking matrices G of the form $G = e_i e_j^T, i, j = 1, ..., m$ leads immediately to

$$V_m^T R(\tilde{X})^T V_m = 0. \tag{29}$$

Let us now substitute $\tilde{X} = V_m G_m V_m^T$ in (29). Remembering that V_m is orthogonal and that we have $b = \beta v_1$, where v_1 is the first column of V_m, we get

$$\begin{aligned} 0 &= V_m^T R(\tilde{X})^T V_m = V_m^T (A V_m G_m V_m^T + V_m G_m V_m^T A^T + bb^T) V_m \\ &= H_m G_m + G_m H_m^T + \beta^2 e_1 e_1^T \end{aligned} \tag{30}$$

which is exactly (25). We have therefore just proved the following result.

Theorem 4.2 *The Galerkin method applied to the Lyapunov equation (5) over the subspace* $\mathbf{Z}_m$ *is mathematically equivalent to approximating the solution* X *by evaluating the integral (6) in which* $w(\tau) = e^{\tau A} b$ *is replaced by its approximation (22).*

This theorem can allow one to establish error bounds for the approximation provided by a Galerkin-type process onto the subspace $\mathbf{Z}_m$.

5 Numerical Experiments

In this section we describe a few numerical experiments in order to test and compare the methods described in Sections 3 and 4. All the tests have been performed in double precision arithmetic on an Ardent Titan super workstation. Our example is derived from the discretization of a partial differential equation of the form:

$$\frac{\partial u}{\partial t} = \Delta u + F(x, y) g(t) \tag{31}$$

in a rectangular domain, with Dirichlet boundary conditions. Here $\Delta = \frac{\partial^2 u}{\partial x^2} + \frac{\partial^2 u}{\partial y^2}$ is the Laplacean operator. If we discretize the rectangle using $n_x + 2$ points in the x direction and $n_y + 2$ points in the y direction, the above equations lead to a matrix problem of the form:

m	$\|Res\|_F$	Time (sec)
5	1.10 E-04	0.18
10	5.40 E-06	0.23
15	7.92 E-07	0.35
20	1.92 E-07	0.45

Table 1: Performance of the Krylov subspace method.

$$\dot{u} = Au + bg \tag{32}$$

where A is square of dimension $N = n_x n_y$. In this experiment we took $n_x = 20$ and $n_y = 40$ leading to a matrix of size 800. The corresponding matrix has a 1-norm of $3,528.0$. We have taken b to be simply e_1 the first column of the identity. Tests with other choices for b showed similar results. First we would like to show the behavior of the residual achieved by the Krylov subspace method described in Section 4, as the degree m varies. Table 1 shows the scaled Frobenius norm of the residual, i.e., the quantity $\|AX_m + X_m A^T + bb^T\|_F$, where $\|Z\|_F = (tr[Z^T Z]/N)^{1/2}$. This is done for $m = 5, 10, 15, 20$.

We have used the Arnoldi process instead of the Lanczos algorithm on purpose, despite the fact that the matrix is symmetric. This is in order to give an idea of the increase in time in the more general nonsymmetric case. The table indicates that the accuracy of the Krylov subspace approximation to the Lyapunov equation is good for very small m and then improves slowly. The times reported in this table and in the next one are in seconds on the Titan and have been obtained using the -O3 compiling option.

Next we would like to test the methods based on numerical integration described in Section 3. Just as in the previous test we show the residual norm and the time to compute the approximate solution for various choices of accuracy and step size. In all of the methods we used the same change of variable for scaling purposes as described at the end of Section 3: we scaled the variable t by $2.5/\|A\|_1$. To approximate $e^{A\Delta t_i}u_i$ we used the formula (20) of degree 5 for the Newton Cotes formulas and 10 for the Gauss-Laguerre formulas. In general the Newton-Cotes formulas do not perform as well as those based on Gauss-Laguerre. The disadvantage of Gauss-Laguerre formulas is that if their accuracy is not sufficient one must restart all over again since the roots of the Laguerre polynomials of different degrees are different. With the Newton Cotes formulas it suffices to refine the previous intervals and the previous

Method	m	$\|Res\|_F$	Time
Laguerre(9)	9	4.21 E-06	0.57
Laguerre(15)	15	7.08 E-08	0.92
NC-3 $\Delta t = 0.3$	6	3.47 E-04	0.23
NC-3 $\Delta t = 0.1$	11	1.17 E-04	0.30
NC-3 $\Delta t = 0.05$	16	8.53 E-05	0.44
NC-5 $\Delta t = 0.40$	9	1.59 E-04	0.26
NC-5 $\Delta t = 0.20$	12	5.97 E-05	0.35
NC-5 $\Delta t = 0.20$	16	1.45 E-05	0.41
NC-7 $\Delta t = 0.5$	7	3.76 E-04	0.23
NC-7 $\Delta t = 0.25$	13	4.96 E-05	0.36
NC-7 $\Delta t = 0.20$	18	9.60 E-06	0.46

Table 2: Behavior of the integration techniques.

work can be saved and used.

6 Conclusion

We have proposed and tested two types of techniques for approximating solutions to large Lyapunov equations that arise in control problems. The attraction of the Krylov subspace method is its simplicity and its overall effectiveness. The Gauss-Laguerre formula may be particularly useful when only the product of X by a vector is desired and one cannot afford to store the vectors $w(t_i)$. The use of Newton-Cotes formulas is not recommended without incorporating a proper change of variables to improve effectiveness of the integration scheme. Although we have relied on the integral formulation of the solution which requires that the eigenvalues of A all have negative real parts, the Galerkin approach may be applied to more general situations. The theory for this is not established however. Another case which we have not addressed is the general situation of equations of the form $AX + XA^T + B = 0$, where B is may be sparse but not necessarily of small rank. A completely different approach may be required in this situation, one that might exploit any possible sparsity in the solution.

Acknowledgements.
Work herein supported was by Cooperative Agreement NCC 2-387 between the National Aeronautics and Space Administration (NASA) and the Universities Space Research Association (USRA).
The author is indebted to the referee for pointing out an important result concerning Gaussian quadrature.

References

[1] M. Abaramowitz and I. A. Stegun. *Handbook of Mathematical Functions with Formulas, Graphs, and Mathematical Tables.* National Bureau of Standards, Washington, DC, 1964.

[2] M.J. Balas. Trends in large space structure control theory: fondest dreams, wildest hopes. *IEEE Trans. Aut. Contr.*, AC-2:522–535, 1982.

[3] P. Kokotovic B.H. Krogh. Feedback control of overloaded networks. *IEEE Trans. Aut. Cont.*, AC-29:704–711, 1984.

[4] E. Gallopoulos and Y. Saad. On the parallel solution of parabolic equations. In R. De Groot, editor, *Proceedings of the International Conference on Supercomputing 1989, Heraklion, Crete, June 5-9, 1989.* ACM press, 1989.

[5] Y. Saad. Projection and deflation methods for partial pole assignment in linear state feedback. *IEEE Trans. Aut. Cont.*, 33:290–287, 1988.

[6] J. Stoer and R. Bulirsch. *Introduction to numerical analysis.* Springer Verlag, New York, Heidelberg, 1980.

Matrix Decomposition and Optimal Estimation of Rigid-Body Displacement

Renjeng Su and Wenli Xu

Abstract

We consider the problem of optimal estimation of rigid-body displacement. The input data are 3-D position vectors of feature points on the moving object measured at two time instants. Under our formulation it is shown that the estimation problems of rotation and translation may be separated. A way of decomposing square matrices, called RS decomposition, is introduced. Our main result is a closed-form solution to the problem which is obtained through RS decomposition. The closed-form solution completely characterizes the optimal estimates within our framework, and gives a necessary and sufficient condition for the existence of a unique optimal solution.

1. Introduction

Consider a rigid body moving in three dimensional space, including translation and rotation. Suppose there are n feature points on the body whose position vectors can be measured with respect to a reference coordinate system. Let $\mathbf{p}_j$ and $\mathbf{q}_j \in \Re^3$ be the position vectors at two time instants. The problem of estimating the rigid body displacement is to find a (rotation) matrix $R \in \Re^{3\times 3}$ and a (translation) vector $\mathbf{t} \in \Re^3$ such that they satisfy the equation

$$\mathbf{q}_j = R\mathbf{p}_j + \mathbf{t}, \qquad j = 1, \ldots, n. \tag{1}$$

When the measured data are precise, it is a simple linear problem. With measurement errors, it becomes a nonlinear optimization problem due to the nonlinear constraints of R being a rotation matrix, i.e. orthogonal with determinant $+1$.

This mathematical problem has arisen from a number of engineering applications, such as the spacecraft attitude estimation [6,8], and motion estimation in computer vision [1,2,4,5,7].

In this paper we study a particular least squares formulation of the optimization problem. We show that the problem can be completely solved with a matrix decomposition called the RS decomposition. The RS decomposition of a real matrix is a specialized form of the well-known decomposition of a matrix into an orthogonal matrix and a symmetric matrix. With RS decomposition the optimal displacement estimates can be completely characterized with a closed-form solution.

2. Definition of Optimal Estimates

Given two sets of position vectors $\mathbf{p}_j$ and $\mathbf{q}_j \in \Re^3, j = 1, \ldots, n$, the optimal motion estimate is defined to be a pair of a translation vector $\mathbf{t} \in \Re^3$ and a rotation matrix $R \in \Re^{3\times 3}$ which minimize the objective function

$$\sum_{j=1}^{n} \|\mathbf{q}_j - R\mathbf{p}_j - \mathbf{t}\|^2, \tag{2}$$

where the vector norm is the Euclidean norm.

Using differential calculus it is easy to see that for any rotation matrix R the optimal translation estimate is

$$\mathbf{t}^* = (1/n)\sum_{j=1}^{n}(\mathbf{q}_j - R\mathbf{p}_j) = \mathbf{q}_c - R\mathbf{p}_c \tag{3}$$

where

$$\mathbf{p}_c = (1/n)\sum_{j=1}^{n}\mathbf{p}_j, \quad \mathbf{q}_c = (1/n)\sum_{j=1}^{n}\mathbf{q}_j. \tag{4}$$

They are the centroids of the feature points at the two time instants. Therefore, the original problem can be solved by first finding a rotation matrix R which minimizes

$$\sum_{j=1}^{n}\|\mathbf{q}'_j - R\mathbf{p}'_j\|^2, \tag{5}$$

where

$$\mathbf{p}'_j = \mathbf{p}_j - \mathbf{p}_c, \quad \mathbf{q}'_j = \mathbf{q}_j - \mathbf{q}_c. \tag{6}$$

The optimal translation vector can then be obtained from Eq. (3).

In the remainder of this paper, we focus on the problem of optimal rotation. A rotation matrix has three independent parameters. In particular, it can be parameterized by a unit vector $\mathbf{k}$ as its rotation axis and a rotation angle θ (with the convention of the right-hand rule). With this parameterization the rotation matrix R can be expressed as [8]

$$\mathrm{Rot}(\mathbf{k},\theta) = \begin{bmatrix} k_x k_x \mathrm{v}\theta + \mathrm{c}\theta & k_x k_y \mathrm{v}\theta - k_z \mathrm{s}\theta & k_x k_z \mathrm{v}\theta + k_y \mathrm{s}\theta \\ k_x k_y \mathrm{v}\theta + k_z \mathrm{s}\theta & k_y k_y \mathrm{v}\theta + \mathrm{c}\theta & k_y k_z \mathrm{v}\theta - k_x \mathrm{s}\theta \\ k_x k_z \mathrm{v}\theta - k_y \mathrm{s}\theta & k_y k_z \mathrm{v}\theta + k_x \mathrm{s}\theta & k_z k_z \mathrm{v}\theta + \mathrm{c}\theta \end{bmatrix}, \tag{7}$$

where $\mathrm{c}\theta = \cos\theta$, $\mathrm{s}\theta = \sin\theta$, $\mathrm{v}\theta = 1 - \cos\theta$ and $k_x^2 + k_y^2 + k_z^2 = 1$.

3. RS Decomposition

We are interested in decomposing an $n \times n$ real matrix into a product of a rotation matrix and a symmetric matrix. Let A be an $n \times n$ real matrix. By the theorem of the singular value decomposition [3] there exist two orthogonal matrices U and V such that

$$U^T A V = D = \mathrm{diag}(d_1, \ldots, d_n), \quad d_1 \geq \cdots \geq d_n \geq 0, \tag{8}$$

where d_i are the singular values of A. Singular values are unique whereas the decomposition factors U and V are not. From (8) we have

$$A = (UV^T)(VDV^T). \tag{9}$$

The matrix VDV^T is clearly symmetric. The matrix UV^T is orthogonal, but not necessarily a rotation matrix, since we may have $\det(UV^T) = -1$ rather than $+1$. Let $r = \det(UV^T)$, then $|r| = 1$. Define

$$W = \begin{bmatrix} I_{n-1} & 0 \\ 0 & r \end{bmatrix}, \tag{10}$$

and

$$\tilde{D} = WD = \mathrm{diag}(s_1, \ldots, s_{n-1}, s_n). \tag{11}$$

Clearly, the matrix W is orthogonal and $W^2 = I$. Eq. (8) can then be rewritten

$$A = (UWV^T)(V\tilde{D}V^T) = RS, \tag{12}$$

where $R = UWV^T$ and $S = V\tilde{D}V^T$. The matrix R is a rotation matrix, since it is a product of orthogonal matrices and $\det(R) = \det(UV^T)\det(W) = r^2 = 1$. The matrix S is symmetric and similar to the diagonal matrix $\tilde{D}$. From (8), (10) and (11), we have between D and $\tilde{D}$ that $s_i = d_i, i = 1, \ldots, n-1$ and $s_n = rd_n$. Regardless of the sign of r, however, we can conclude that $s_1 \geq, \cdots, \geq s_{n-1} \geq |s_n| \geq 0$, since $|r| = 1$. Furthermore, it follows from Eq. (9) that $\det(A) = r\prod_{i=1}^{n} d_i$, and

$$r = \begin{cases} 1, & \text{if } \det(A) \geq 0; \\ -1, & \text{otherwise.} \end{cases} \tag{13}$$

In summary, we have proved the following decomposition theorem.

Theorem 1 (RS decomposition) If A is an $n \times n$ real matrix, then

(i) there exist a rotation matrix R and a symmetric matrix S such that $A = RS$;

(ii) the matrix S can always be chosen to be similar to $\mathrm{diag}(s_1, \ldots, s_n)$ such that $s_1 \geq \cdots \geq s_{n-1} \geq |s_n| \geq 0$.

An RS decomposition which satisfies condition (ii) is called a *normal* RS decomposition. Expression (12) gives a means for obtaining a normal RS decomposition. We now investigate the uniqueness problem.

Theorem 2 Let A be an $n \times n$ real matrix. If RS and $\tilde{R}\tilde{S}$ are two normal RS decompositions of A, then S and $\tilde{S}$ must be similar.

Proof: According to the definition of normal RS decomposition, there are orthogonal matrices V and $\tilde{V}$ such that

$$V^TSV = D = \mathrm{diag}(s_1, \ldots, s_n), \quad s_1 \geq \cdots \geq s_{n-1} \geq |s_n|, \tag{14}$$

$$\tilde{V}^T\tilde{S}\tilde{V} = \tilde{D} = \mathrm{diag}(\tilde{s}_1, \ldots, \tilde{s}_n), \quad \tilde{s}_1 \geq \cdots \geq \tilde{s}_{n-1} \geq |\tilde{s}_n|. \tag{15}$$

From $A = RS = \tilde{R}\tilde{S}$, we have

$$RVDV^T = \tilde{R}\tilde{V}\tilde{D}\tilde{V}^T. \tag{16}$$

Left multiplying with $(RVDV^T)^T$ and $(\tilde{R}\tilde{V}\tilde{D}\tilde{V}^T)^T$ on the left and the right sides of Eq. (16) respectively, it follows that $VD^2V^T = \tilde{V}\tilde{D}^2\tilde{V}^T$, i.e., D^2 and $\tilde{D}^2$ are similar. From Eq. (16), $\det(D) = \det(\tilde{D})$. Hence, it is clear from (14) and (15), that $D = \tilde{D}$. Therefore, S and $\tilde{S}$ must be similar. **Q.E.D.**

Corollary Let A be an $n \times n$ real matrix with singular values $d_1 \geq \cdots \geq d_n \geq 0$. If RS is a normal RS decomposition of A, then S must be similar to $\mathrm{diag}(d_1, d_2, \ldots, rd_n)$. There $r = 1$ if $\det(A) \geq 0$, and $r = -1$ otherwise.

For the 3-dimensional case the following theorem gives a necessary and sufficient condition for a unique RS decomposition.

Theorem 3 Let A be any nonzero 3×3 real matrix and $A = R^*S^*$ be a normal RS decomposition of A. Let $S^* = V^*DV^{*^T}$, where V^* is orthogonal and $D = \mathrm{diag}(s_1, s_2, s_3)$ with $s_1 \geq s_2 \geq |s_3| \geq 0$.

(1) If $s_2 + s_3 \neq 0$, then R^*S^* is the unique normal RS decomposition of A.

(2) If $s_2 + s_3 = 0$, then any pair of R and S gives a normal RS decomposition if and only if $R = R^* V^* \tilde{R}^T V^{*^T}$ and $S = V^* \tilde{R} V^{*^T} S^*$, where
(i) if $s_1 \neq s_2$, $\tilde{R}$ is a rotation about the X-axis with any angle,
(ii) if $s_1 = s_2$, $\tilde{R}$ is a rotation about any line in the X-Y plane with any angle.

Proof: Suppose $A = RS$ is another normal RS decomposition of A. By Theorem 2, S and S^* are similar. There exists an orthogonal matrix V such that

$$RVDV^T = R^* V^* D V^{*^T}. \tag{17}$$

Let

$$\tilde{R} = V^{*^T} R^T R^* V^*, \quad \tilde{V} = V^{*^T} V, \tag{18}$$

Eq. (17) is equivalent to

$$\tilde{V} D \tilde{V}^T = \tilde{R} D. \tag{19}$$

Notice that $\tilde{R}$ is a rotation matrix, and $\tilde{V}$ is orthogonal. Eq. (19) means that RS is a normal RS decomposition if and only if a rotation matrix $\tilde{R}$ defined in (18) is such that $\tilde{R}D$ is symmetric and similar to D. From Eq. (19),

$$\text{tr}(D) = \text{tr}(\tilde{R}D). \tag{20}$$

Let $\tilde{R} = \text{Rot}(\mathbf{k}, \theta)$. Recall that $D = \text{diag}(s_1, s_2, s_3)$. Eq. (20) is thus equivalent to

$$(1 - \cos\theta)[(s_2 + s_3)k_x^2 + (s_1 + s_3)k_y^2 + (s_1 + s_2)k_z^2] = 0. \tag{21}$$

Suppose $s_2 + s_3 \neq 0$. It follows that $s_2 + s_3$, $s_1 + s_3$, and $s_1 + s_2$ are all positive, since $s_1 \geq s_2 \geq |s_3| \geq 0$. In addition, being a unit vector, the components of $\mathbf{k}$ can not all vanish. Therefore, the complete set of solutions to Eq. (21) is characterized by $(1 - \cos\theta) = 0$. That is, $\theta = 2m\pi$ and $\mathbf{k}$ is an arbitrary unit vector. Equivalently, $\tilde{R} = \text{Rot}(\mathbf{k}, 2m\pi) = I$. It follows from Eq. (18) that $R = R^*$, which in turn implies $S = S^*$. So, $R^* S^*$ is the unique normal RS decomposition of A.

We now show part (2) of Theorem 3. Keep in mind the inequality $s_1 \geq s_2 \geq |s_3| \geq 0$. In case (i), we have $s_2 + s_3 = 0$, $s_1 + s_3 > 0$, and $s_1 + s_2 > 0$. A pair $(\mathbf{k}, \theta)$ is a solution to Eq. (21) if and only if $\mathbf{k} = \mathbf{x} = (1, 0, 0)^T$ with any θ, i.e., $\tilde{R} = \text{Rot}(\mathbf{x}, \theta)$. Using Eq. (7), we can get the matrix expression of $\tilde{R}$ corresponding to $\text{Rot}(\mathbf{x}, \theta)$. Recall that $D = \text{diag}(s_1, s_2, s_3)$ and $s_2 + s_3 = 0$. It is easy to check that $\tilde{R}D$ is symmetric and similar to D. So, the set of all possible rotation matrices satisfying Eq. (19) are $\tilde{R} = \text{Rot}(\mathbf{x}, \theta)$, where θ is arbitrary. In case (ii), we have $s_2 + s_3 = 0$ and $s_1 + s_3 = 0$. Since A is a nonzero matrix, $s_1 + s_2 > 0$. In terms of the same procedure as in case (i), it is easy to see that in this case the set of all possible rotation matrices satisfying Eq. (19) are $\tilde{R} = \text{Rot}(\mathbf{k}, \theta)$, where $\mathbf{k} = (k_x, k_y, 0)^T$ is a unit vector in the X-Y plane, and θ is an arbitrary angle.

From the foregoing analysis and Eq. (18), we conclude that when $s_2 + s_3 = 0$ including both cases (*i*) and (*ii*), a pair of R and S gives a normal RS decomposition of A if and only if $R = R^* V^* \tilde{R}^T V^{*^T}$. It follows that $S = V^* \tilde{R} V^{*^T} S^*$, since $RS = R^* S^*$. **Q.E.D.**

4. Optimal Rotation Estimation

We recall that the problem is to find a rotation matrix R for given n pairs $\mathbf{p}_j, \mathbf{q}_j \in \Re^3$, such that the objective function $\sum_{j=1}^n \|\mathbf{q}_j - R\mathbf{p}_j\|^2$ is minimized. To simplify the notation, let

$$P = [\mathbf{p}_1, \ldots, \mathbf{p}_n], \quad Q = [\mathbf{q}_1, \ldots . \mathbf{q}_n]. \tag{22}$$

It is easy to see that

$$\sum_{j=1}^{n} \|\mathbf{q}_j - R\mathbf{p}_j\|^2 = \|Q - RP\|^2, \tag{23}$$

where the matrix norm is the Frobenius norm. Since

$$\begin{aligned}\|Q - RP\|^2 &= \mathrm{tr}[(Q - RP)(Q - RP)^T] \\ &= \|Q\|^2 + \|P\|^2 - 2\mathrm{tr}(RPQ^T),\end{aligned} \tag{24}$$

it is clear that a rotation matrix R minimizes $\|Q - RP\|^2$ if and only if it maximizes $\mathrm{tr}(RPQ^T)$.

Notice that PQ^T is a 3×3 matrix. The following important lemma shows that the RS decomposition of PQ^T gives a closed-form solution to the optimal rotation estimation problem.

Lemma 1 Let A be any 3×3 real matrix and $A = R^* S^*$ be a normal RS decomposition of A. Then R^{*^T} maximizes $\mathrm{tr}(RA)$ and $\max[\mathrm{tr}(RA)] = \mathrm{tr}(S^*)$.

Proof: Since $R^* S^*$ is a normal RS decomposition of A, we may have $A = R^* V^* D V^{*^T}$, where V^* is an orthogonal matrix, $D = \mathrm{diag}(s_1, s_2, s_3)$ with $s_1 \geq s_2 \geq |s_3| \geq 0$. Notice that $\mathrm{tr}(RA) = \mathrm{tr}(RR^* V^* D V^{*^T}) = \mathrm{tr}(V^{*^T} RR^* V^* D)$. Let

$$\tilde{R} = V^{*^T} RR^* V^*. \tag{25}$$

We have

$$\mathrm{tr}(RA) = \mathrm{tr}(\tilde{R}D) = s_1\tilde{r}_{11} + s_2\tilde{r}_{22} + s_3\tilde{r}_{33}, \tag{26}$$

where $\tilde{r}_{ii}$ are the diagonal entries of $\tilde{R}$. Let $\tilde{R} = \mathrm{Rot}(\mathbf{k}, \theta)$, it follows from Eq. (3) that

$$\begin{aligned}\tilde{r}_{11} &= k_x^2(1 - \cos\theta) + \cos\theta, \\ \tilde{r}_{22} &= k_y^2(1 - \cos\theta) + \cos\theta, \\ \tilde{r}_{33} &= k_z^2(1 - \cos\theta) + \cos\theta,\end{aligned} \tag{27}$$

where

$$k_x^2 + k_y^2 + k_z^2 = 1. \tag{28}$$

So we have transformed $\mathrm{tr}(RA)$ from a function of R into a function of k_x, k_y, k_z and θ. Let

$$f(k_x, k_y, k_z, \theta) = \mathrm{tr}(RA) = \mathrm{tr}(\tilde{R}D) = s_1\tilde{r}_{11} + s_2\tilde{r}_{22} + s_3\tilde{r}_{33}. \tag{29}$$

Now, the problem of finding a rotation matrix to maximize $\mathrm{tr}(RA)$ becomes the maxima problem of f with search variables θ and k_x, k_y, k_z subject to (28). Using Lagrange multiplier method, let

$$F = f(k_x, k_y, k_z, \theta) + \lambda(k_x^2 + k_y^2 + k_z^2 - 1). \tag{30}$$

The following conditions on critical points are obtained

$$\begin{aligned}&[s_1k_x^2 + s_2k_y^2 + s_3k_z^2 - (s_1 + s_2 + s_3)]\sin\theta = 0, \\ &k_x^2 + k_y^2 + k_z^2 - 1 = 0, \\ &k_x[s_1(1 - \cos\theta) + \lambda] = 0, \\ &k_y[s_2(1 - \cos\theta) + \lambda] = 0, \\ &k_z[s_3(1 - \cos\theta) + \lambda] = 0.\end{aligned} \tag{31}$$

Considering the first equation in (31), we have three cases:

$$\begin{aligned}
&(a)\ s_1k_x^2 + s_2k_y^2 + s_3k_z^2 - (s_1 + s_2 + s_3) = 0,\\
&(b)\ \theta = 2m\pi,\\
&(c)\ \theta = (2m+1)\pi, \quad m \text{ is an integer.}
\end{aligned} \tag{32}$$

Combining (27), (29) and (32), the values of f in the three cases are respectively

$$\begin{aligned}
f_a &= f_b = s_1 + s_2 + s_3,\\
f_c &= 2(s_1k_x^2 + s_2k_y^2 + s_3k_z^2) - (s_1 + s_2 + s_3).
\end{aligned} \tag{33}$$

From the fact that $s_1 \geq s_2 \geq |s_3| \geq 0$, we have

$$\begin{aligned}
f_c &\leq 2s_1(k_x^2 + k_y^2 + k_z^2) - (s_1 + s_2 + s_3)\\
&= s_1 - s_2 - s_3 \leq s_1 + s_2 + s_3.
\end{aligned} \tag{34}$$

Notice that f is bounded and differentiable everywhere. So, the global maximum value of f is $s_1 + s_2 + s_3$.

Now consider the case (b) in which f takes on the maximum. It is easy to check that in this case $\theta = 2m\pi, \lambda = 0$ and an arbitrary unit vector $(k_x, k_y, k_z)^T$ satisfies all the conditions in (31). So, $\tilde{R} = \mathrm{Rot}(\mathbf{k}, 2m\pi) = I$ maximizes $\mathrm{tr}(\tilde{R}D)$. It follows by substituting $\tilde{R} = I$ into Eq. (25) that $R = R^{*^T}$ maximizes $\mathrm{tr}(RA)$. The maximum is $\mathrm{tr}(R^{*^T}A)$ which is equal to $\mathrm{tr}(S^*)$ since $A = R^*S^*$. **Q.E.D.**

Lemma 2 For any rotation matrix which maximizes $\mathrm{tr}(RA)$, its transpose must be the rotation part of a normal RS decomposition of A.

Proof: Let $A = R^*S^*$ be a normal RS decomposition of A and all the notations used in the proof of Lemma 1 carry over. Suppose a rotation matrix R maximizes $\mathrm{tr}(RA)$. It follows from (25) and (26) that $\tilde{R} = V^{*^T}RR^*V^*$ maximizes $\mathrm{tr}(\tilde{R}D)$. Let $\tilde{R} = \mathrm{Rot}(\mathbf{k}, \theta)$, then $(\mathbf{k}, \theta)$ satisfies the equations in (31). Therefore, $(\mathbf{k}, \theta)$ must satisfy one of the conditions (a), (b), and (c) in (32). Case (b) $\theta = 2m\pi$ implies $\tilde{R} = I$. It follows from (25) that $R^T = R^*$. If condition (c) is satisfied, recall that $\tilde{R}$ maximizes $\mathrm{tr}(\tilde{R}D)$ and the maximum value is $s_1 + s_2 + s_3$, then f_c in (33) should be equal to $s_1 + s_2 + s_3$. It leads to

$$s_1k_x^2 + s_2k_y^2 + s_3k_z^2 - (s_1 + s_2 + s_3) = 0,$$

which is exactly the condition (a) in (32). It can be rewritten

$$(s_2 + s_3)k_x^2 + (s_1 + s_3)k_y^2 + (s_1 + s_2)k_z^2 = 0. \tag{35}$$

The left side of (35) is positive definite unless $s_2 + s_3 = 0$. Under the condition $s_2 + s_3 = 0$, we have either $s_1 \neq s_2$ or $s_1 = s_2$. With the same arguments used for Eq. (21) in the proof of Theorem 3, if $s_1 \neq s_2$, then (35) implies that $\mathbf{k} = (1,0,0)^T$, i.e., $\tilde{R}$ is a rotation about the X-axis. If $s_1 = s_2$, then (35) implies that $\mathbf{k} = (k_x, k_y, 0)^T$, i.e., $\tilde{R}$ is a rotation about a line in the X-Y plane. It follows from $\tilde{R} = V^{*^T}RR^*V^*$ that the transpose of R must be in the form of $R^*V^*\tilde{R}^TV^{*^T}$ which is the rotation part of a normal RS decomposition of A as indicated in part (2) of Theorem 3. **Q.E.D.**

In summary, we have proved the following theorem.

Theorem 4 Suppose a pair of matrices $P, Q \in \Re^{3\times n}$ are given.

(i) A rotation matrix R^* minimizes $\|Q - RP\|^2$ if and only if R^{*^T} is the rotation part of a normal RS decomposition of PQ^T;

(ii) $\min \|Q - RP\|^2 = \|Q - R^*P\|^2 = \|Q\|^2 + \|P\|^2 - 2\mathrm{tr}(S)$, where S is the symmetric matrix in any normal RS decomposition of PQ^T.

It follows from Theorem 4 that the uniqueness of the optimal rotation estimate is completely determined by the uniqueness of the normal RS decomposition of PQ^T. A useful corollary follows immediately from Theorem 3 and Theorem 4.

Corollary The problem of finding R to minimize $\|Q-RP\|^2$ has a unique solution if $\mathrm{rank}(PQ^T) = 2$ or $\det(PQ^T) > 0$.

5. Examples

Example 1 Suppose n pairs of position data $\mathbf{p}_j, \mathbf{q}_j \in \Re^3$ are given. Let $P = [\mathbf{p}_1, \ldots, \mathbf{p}_n]$ and $Q = [\mathbf{q}_1, \ldots, \mathbf{q}_n]$. Assume that P and Q are consistent in the sense that there is a rotation matrix R^* such that $Q = R^*P$. We show that, in this case, if $\mathrm{rank}(Q) = \mathrm{rank}(P) \geq 2$, the RS decomposition method leads to the unique solution R^*.

Substituting $P = R^{*^T} Q$, we have

$$PQ^T = R^{*^T} QQ^T = R^{*^T} S, \tag{36}$$

where R^{*^T} is a rotation matrix. Clearly, $S = QQ^T$ is symmetric and non-negative definite. It follows that S must be similar to $\mathrm{diag}(s_1, s_2, s_3)$ such that $s_1 \geq s_2 \geq s_3 \geq 0$. That is, expression (36) is a normal RS decomposition of PQ^T. Furthermore, if $\mathrm{rank}(Q) \geq 2$, then at least $s_1 \geq s_2 > 0$, and hence $s_2 + s_3 > 0$. From Theorem 3 and Theorem 4, the normal RS decomposition of PQ^T is unique, and R^* is the unique optimal rotation estimate.

Example 2 Consider the same problem as in Example 1. Suppose the measured data are $\tilde{P} = [\alpha_1\mathbf{p}_1, \ldots, \alpha_n\mathbf{p}_n]$ and $\tilde{Q} = [\beta_1\mathbf{q}_1, \ldots, \beta_n\mathbf{q}_n]$, where α_i, β_i are positive scaling factors. It means that there are errors in magnitude, but not in direction, of the measured position data. We show that the optimal rotation estimate is invariant with respect to the scaling factors, although the norm of estimation error varies.

Let $A = \mathrm{diag}[\alpha_1, \ldots, \alpha_n]$, and $B = \mathrm{diag}[\beta_1, \ldots, \beta_n]$.

$$\tilde{P}\tilde{Q}^T = PABQ^T = R^{*^T} QC(QC)^T = R^{*^T} \tilde{S}. \tag{37}$$

where $C = (AB)^{1/2}$, and $\tilde{S} = QC(QC)^T$. Again, R^{*^T} is a rotation matrix, and $\tilde{S}$ is symmetric and nonnegative definite. For the same reason as in Example 1, $R^{*^T}\tilde{S}$ is a normal RS decomposition. Hence, R^* is an optimal roation estimate for given $\tilde{P}$ and $\tilde{Q}$. The norm of the estimation error can be shown as

$$\|\tilde{Q} - R^*\tilde{P}\|^2 = \sum_{j=1}^{n} (\beta_j - \alpha_j)^2 \|\mathbf{q}_j\|^2. \tag{38}$$

Example 3 Consider again the problem in Example 1. Suppose the measured position data $\tilde{P}$ and $\tilde{Q}$ are different from the actual precise position data P and Q in direction. such measurement errors generally lead to an optimal rotation estimate different from the actual rotation. Here we show a special case when the optimal estimate is not unique.

Suppose $\tilde{Q} = Q$, $\tilde{P} = FP$, where $F = \text{diag}(-1,1,1)$. That is, $\tilde{P}$ is a reflection of the true data P in the Y-Z plane. Clearly, F is orthogonal with $\det(F) = -1$ and $F^2 = I$. Substituting, we have

$$\tilde{P}\tilde{Q}^T = FPQ^T = FR^{*^T}QQ^T. \tag{39}$$

Since QQ^T is symmetric and nonnegative definite, there exists an orthogonal matrix V such that $QQ^T = VDV^T$ and $D = \text{diag}(d_3, d_2, d_1)$ with $d_1 \geq d_2 \geq d_3 \geq 0$. Eq. (39) is then rewritten as

$$\tilde{P}\tilde{Q}^T = FR^{*^T}VDV^T = (FR^{*^T}VFV^T)(VFDV^T) = \tilde{R}^{*^T}\tilde{S}, \tag{40}$$

where $\tilde{R}^{*^T} = FR^{*^T}VFV^T$ is a rotation matrix, and $\tilde{S} = VFDV^T$ is symmetric. Since $FD = \text{diag}(-d_3, d_2, d_1)$, $\tilde{S}$ is similar to $\text{diag}(d_1, d_2, -d_3)$ with $d_1 \geq d_2 \geq |-d_3| \geq 0$. Hence, $\tilde{R}^{*^T}\tilde{S}$ is a normal RS decomposition of $\tilde{P}\tilde{Q}^T$, and $\tilde{R}^*$ is an optimal rotation estimate for the given $\tilde{Q}$ and $\tilde{P}$. Notice that the optimal rotation estimate is not unnique when $d_2 + (-d_3)$ vanishes. This nonuniqueness results from such a serious measurement error as a reflection.

6. Conclusion

Due to the measurement errors, the estimation of rigid-body displacements is formulated as an optimization problem. We have shown that general displacement estimation problem can be reduced to a pure rotation estimation problem. We have applied the technique of RS decomposition to this problem, and completely characterized the rotation estimates with a closed-form solution.

This work is supported in part by NSF Grant DMC8451492, AT & T Information Systems, and Martin Marietta Denver Aerospace.

References

[1] O. D. Faugeras and M. Hebert, "The Representation, Recognition, and Locating of 3-D Objects", *The Internatinal Journal of Robotics Research*, Vol. 5, No. 3, pp. 27-52, Fall 1986.

[2] O. D. Faugeras, F. Lustman, and G. Toscani, "Motion and Structure from Motion from Point and Line Matches", *Proceedings of IEEE First International Conference on Computer Vision*, pp. 25-34, June 8-11, 1987.

[3] C. L. Lawson and R. J. Hanson, *Solving Least Squares Problems*, Prentice-Hall, Inc., Englewood Cliffs, N.J., 1974.

[4] Z. C. Lin, T. S. Huang, S. D. Blostein, H. Lee and E. A. Margerum, "Motion Estimation from 3-D Point Sets with and without Correspondences", *Proceedings of IEEE Computer Vision and Pattern Recognition*, pp. 194-201, Miami, Beach, Florida, June 22-26, 1986.

[5] A. Mitiche and P. Bouthemy, "Tracking Modeled Objects Using Binocular Images", *Computer Vision, Graphics and Image Processing*, Vol. 32, pp. 384-396, 1985.

[6] M. D. Shuster, "Approximate Algorithms for Fast Optimal Attitude Computation", *Proceedings of AIAA Guidance and Control Specialist Conference*, pp. 88-95, Palo Alto, CA, Aug. 1978.

[7] J. Y. Weng, T. S. Huang and N. Ahuja, "Error Analysis of Motion Parameter Estimation from Image Sequences", *Proceedings of IEEE First International Conference on Computer Vision*, pp. 703-707, June 8-11, 1987.

[8] J. R. Wertz, *Spacecraft Attitude Determination and Control*, D. Reidel Publication Co., 1978.

SOLUTION AND PROPERTIES OF THE RESTRICTED TOTAL LEAST SQUARES PROBLEM

Sabine Van Huffel

Abstract

The Restricted Total Least Squares (RTLS) problem is devised for solving overdetermined sets of linear equations $AX \approx B$ in which the data $[A; B]$ are perturbed by errors of the form $E^{\star} = DEC$. D and C are known, E is arbitrary but bounded. By choosing D and C appropriately, the RTLS problem formulation can handle any weighted Least Squares (LS), generalized LS, total LS and generalized total LS problem. Also, equality constraints can be imposed.
The RTLS algorithm, which solves these problems, is based on the Restricted Singular Value Decomposition (RSVD), a generalization of the ordinary SVD for triple matrix products. Using the RSVD, some properties of the RTLS problem are proven.

1 Introduction.

Every linear parameter estimation problem arising in signal processing, system identification, automatic control or in general engineering, statistics, medicine, etc.... gives rise to an overdetermined set of linear equations $AX \approx B$ which are usually solved with the ordinary Least Squares (LS) method. Very often, errors occur in both A and B. For those cases, the Total Least Squares (TLS) approach has been devised [6] and amounts to fitting a "best" subspace to $[A; B]$ when the errors in the measurements A and B are uncorrelated with zero mean and equal variance. As proven in [5], this TLS solution $\widehat{X}$ is a strongly consistent estimate of the true solution X of the corresponding unperturbed set $A_0X = B_0$ when $\lim_{m\to\infty} \frac{1}{m} A_0^T A_0$ exists and is positive definite, i.e. $\widehat{X}$ converges to X with probability one as the number of equations m is going to infinity.
However, in many linear parameter estimation problems some columns of A may be error-free. Moreover, the errors in the remaining data may be correlated and not equally sized. In order to maintain consistency of the result when solving these problems, the Generalized TLS (GTLS) problem has been devised [15,13]. This problem can now be further generalized, as shown below.

__Restricted TLS formulation__ :

$$AX \approx B \qquad A \in \mathcal{R}^{m\times n}, B \in \mathcal{R}^{m\times d} \text{ and } X \in \mathcal{R}^{n\times d} \tag{1}$$

is given where $[A;B] = [A_0;B_0] + E^\star$. A_0, B_0 are error-free and $E^\star$ is a "restricted" perturbation matrix of the form

$$E^\star = DEC \qquad D \in \mathcal{R}^{m\times p}, C \in \mathcal{R}^{q\times(n+d)} \text{ and } E \in \mathcal{R}^{p\times q} \tag{2}$$

D, C are known. E is unknown and arbitrary but bounded.
Then, a Restricted TLS (RTLS) solution of (1 -2) is any solution of the set

$$(A - \Delta\widehat{A})X = B - \Delta\widehat{B} \tag{3}$$

in which $[\Delta\widehat{A}; \Delta\widehat{B}]$ is a matrix of the form $[\Delta\widehat{A}; \Delta\widehat{B}] = D\widehat{E}C$ such that :

$$\text{Range}(B - \Delta\widehat{B}) \subseteq \text{Range}(A - \Delta\widehat{A}) \tag{4}$$

$$\text{and } \|\widehat{E}\|_F \quad \text{is minimal.} \tag{5}$$

The problem of finding $[\Delta\widehat{A}; \Delta\widehat{B}]$ satisfying (4 -5), is called the **RTLS problem**. Whenever the solution is **not unique**, RTLS singles out a **weighted minimum norm** solution, denoted by $\widehat{X}$, in the sense that $\|F_1\widehat{X}F_2\|_F$ is minimized with $F_1 \in \mathcal{R}^{n\times n}$ and $F_2 \in \mathcal{R}^{d\times d}$ appropriately chosen nonsingular weighting matrices.

The RTLS formulation is more general than the GTLS formulation given in [15,13]. Indeed, D and C may be rectangular, even rank-deficient, and error-free columns in A, if any, need not be stored in the first columns of A. Using an appropriately chosen D and C, even columns in B may be error-free and equality constraints can be imposed (see below). Finally, if the RTLS solution is not unique, then not only the solution $\widehat{X}$ with minimal $\|\widehat{X}\|_F$ can be singled out but any other $\widehat{X}$ with an appropriately "weighted" minimal norm $\|F_1\widehat{X}F_2\|_F$.

Observe that the RTLS problem is **not always solvable**. First of all, it is possible that no matrix $\widehat{E}$ exists such that $\text{rank}([A;B] - D\widehat{E}C) \leq n$, e.g. when $[A;B] = \begin{bmatrix} 0 & 1 \\ 1 & 0 \end{bmatrix}$ and $[\Delta\widehat{A}; \Delta\widehat{B}] = D\widehat{E}C = \begin{bmatrix} 0 \\ 1 \end{bmatrix} \hat{e}[0\ 1]$. Since the corrections $[\Delta\widehat{A}; \Delta\widehat{B}]$, applied to the data $[A;B]$, may only affect the perturbed entries in $[A;B]$ (specified by D and C), the rank of $[A;B]$ can not be sufficiently reduced in these problems in order to obtain a solvable set $(A - \Delta\widehat{A})X = B - \Delta\widehat{B}$.
Secondly, it is possible that there exists an $\widehat{E}$ with minimal Frobenius norm such that $\text{rank}([A;B] - D\widehat{E}C) \leq n$ but condition (4) is not satisfied. In this case, many $\mathcal{O}(\epsilon)$ perturbations $\widehat{E}_\epsilon$ of $\widehat{E}$ can make $R(B - \Delta\widetilde{B}) \subseteq R(A - \Delta\widetilde{A})$ where $[\Delta\widetilde{A}; \Delta\widetilde{B}] = D\widehat{E}_\epsilon C$, but there is no smallest value of $\|\widehat{E}_\epsilon\|_F$ for which (4) is satisfied, e.g.

$$[A;B] = \begin{bmatrix} 1 & 0 & 1 \\ 0 & 0 & 1 \\ 0 & 0 & 1 \end{bmatrix}, \quad C = I, \quad D = I \text{ and } \widehat{E}_\epsilon = \begin{bmatrix} 0 & 0 & 0 \\ 0 & \epsilon & 0 \\ 0 & \epsilon & 0 \end{bmatrix}$$

This kind of unsolvability only occurs when A is (nearly) rank-deficient or when the set of equations (1) is highly conflicting. By imposing additional constraints on the solution space, these RTLS problems can be made solvable and are called **nongeneric** RTLS problems. See [12,16] for more details.

It is easy to see that the RTLS formulation can handle any LS and generalized LS problem, as well as every TLS and generalized TLS problem. Indeed :
If D equals the $m \times m$ identity matrix, denoted by I_m, and $C = I_{n+d}$ (resp., $C = [0_{d\times n}; I_d]$), then one has the **ordinary TLS** (resp., **LS**) formulation [6,12].
If $D = I_m$ and $C = [0_{q\times n_1}; I_q]$ with $q = n + d - n_1$, an **extension** of the ordinary TLS problem is obtained which assumes the first n_1 columns of A to be error-free [14,2].
The **Generalized TLS (GTLS)** problem also assumes that the first n_1 columns A_1 of $A = [A_1; A_2]$ are error-free but moreover allows for correlations between the errors $E_2^\star$ in the noisy submatrix $[A_2; B]$ provided square, nonsingular matrices D and C_2 are known such that the errors in $D^{-1}E_2^\star C_2^{-1}$ are equilibrated, i.e. uncorrelated with equal variance. By taking $C = [0; C_2]$, C_2 and D as defined above, the RTLS problem reduces to the GTLS problem [15,13]. If $D = I_m$, then C_2 is, up to a factor of proportionality, given by the square root of the error covariance matrix which defines the correlations between the errors in each row of $[A_2; B]$.
If rank(D)=p and $C = [0_{d\times n}; I_d]$, then one obtains the **Generalized LS (GLS)** problem, formulated as follows [10] : $\min \|E\|_F$ such that $B = AX + DE$.
D represents, up to a factor of proportionality, the square root of the symmetric nonnegative definite covariance matrix which defines the correlations between the errors in each column of the right-hand side matrix B.
The RTLS formulation can also handle **any variant** of the GTLS and GLS problem allowing for error-free columns in A_2 and B as well, e.g. the RTLS problem:

$$AX \approx B,\ E^\star = [DEC_1; 0_{m\times d}] \text{ with } C = [C_1; 0_{q\times d}] \text{ and } D,\ C_1 \text{ known,} \qquad (6)$$

assumes B to be error-free. If $C_1 = [0_{1\times(i-1)}, 1, 0_{1\times(n-i)}]$, then (6) defines a GLS problem in which A and B are error-free except for the i-th column of A.
Furthermore, the RTLS formulation also allows to solve any **(under)determined** set of linear equations (i.e. $m \leq n$), defined as follows :

$$\min_{\widehat{X}} \|F_1 \widehat{X} F_2\|_F \quad \text{such that} \quad A\widehat{X} = B$$

This set always has one exact solution $\widehat{X}$. Hence, $E^\star$ must be set to zero in the RTLS formulation, e.g. by taking $D = 0$ or $C = 0$.
Finally, observe that **equality constraints** can be imposed, given by the error-free rows of $[A; B]$, e.g. if the first m_1 rows of $[A; B]$ are error-free, then $D = \begin{bmatrix} 0_{m_1 \times p} \\ D_2 \end{bmatrix}$.
By adding the assumption that also n_1 columns of $[A; B]$ are error-free, a whole class of RTLS problems $AX \approx B$ can be defined, characterized by the fact that only 1 submatrix of $[A; B]$ is perturbed and may be changed by the RTLS algorithm, e.g. $D = \begin{bmatrix} D_1 \\ 0_{m_1 \times p} \end{bmatrix}$ and $C = [C_1; 0_{q\times n_1}]$.
If $F_1 \neq I_m$ or $F_2 \neq I_d$, the corresponding **"weighted"** problems are considered, e.g. the GLS problem given in [3] ($d = 1$) :

$$\min_{e,x} \|e\|_2^2 \quad \text{such that } b = Ax + De \quad \text{with } \|F_1 x\|_2 \text{ minimal,}$$

is a special RTLS problem with $C = [0_{1\times n}; 1]$, $F_2 = 1$ and F_1, D as defined above.

2 Restricted Singular Value Decomposition.

While the ordinary TLS algorithm is strongly based on the ordinary Singular Value Decomposition (SVD) [6,12], the RTLS algorithm is strongly based on the Restricted SVD (RSVD). The RSVD, introduced by Zha [17,18] and further analyzed by De Moor and Golub [2], is described as follows :

Definition : **Restricted Singular Values (RSVs)** of (T, D, C).
Let $T \in \mathcal{R}^{m\times n}$ be perturbed by a matrix of the form $E^\star = DEC$ where $D \in \mathcal{R}^{m\times p}$, $E \in \mathcal{R}^{p\times q}$ and $C \in \mathcal{R}^{q\times n}$. Then the RSVs of the triplet (T, D, C) are defined as follows :

$$\sigma_i(T,D,C) = \min_{E\in\mathcal{R}^{p\times q}} \{\|E\|_2 \mid \operatorname{rank}(T + DEC) \le i-1\} \qquad i = 1,\dots,n$$

If, for some i, no matrix E exists such that $\operatorname{rank}(T + DEC) \le i - 1$, we simply define $\sigma_i(T, D, C) = \infty$. Also, if $m < n$, define $\sigma_i(T, D, C) = 0$ for $m + 1 \le i \le n$. Observe that the ordinary (resp., generalized) singular values are special RSVs given by $\sigma_i(T, I_m, I_n)$ (resp., $\sigma_i(T, I_m, C)$ or $\sigma_i(T, D, I_n)$) [18]. We also have $\sigma_{i-1} \ge \sigma_i\ \forall i = 2,\dots,n$. These RSVs are computed from the RSVD of (T, D, C), as shown below and proven in [18] (denote by $\operatorname{diag}(\alpha_1,\dots,\alpha_p)$, $p = \min\{m, n\}$, the $m \times n$ matrix $A = [a_{ij}]$ with $a_{ij} = 0$ whenever $i \ne j$ and $a_{ii} = \alpha_i$ for $i = 1,\dots,p$).

Theorem 1 *:* **Restricted Singular Value Decomposition** *of* (T, D, C).
If $T \in \mathcal{R}^{m\times n}$, $D \in \mathcal{R}^{m\times p}$ *and* $C \in \mathcal{R}^{q\times n}$, *then there exist orthonormal* $U \in \mathcal{R}^{p\times p}$ *and* $V \in \mathcal{R}^{q\times q}$ *and nonsingular* $P \in \mathcal{R}^{m\times m}$ *and* $Q \in \mathcal{R}^{n\times n}$ *such that :*

$$PTQ = \begin{bmatrix} \Sigma_T & 0 \\ 0 & 0 \end{bmatrix}, \quad PDU = \begin{bmatrix} \Sigma_D \\ 0 \end{bmatrix} \quad \text{and } V^TCQ = \begin{bmatrix} \Sigma_C \ ; & 0 \end{bmatrix}$$

where

$$\Sigma_T = \begin{bmatrix} I_j & 0 & 0 & 0 \\ 0 & I_k & 0 & 0 \\ 0 & 0 & I_l & 0 \\ 0 & 0 & 0 & \tilde{S}_T \end{bmatrix}, \Sigma_D = \begin{bmatrix} I_j & 0 & 0 & 0 \\ 0 & 0 & 0 & 0 \\ 0 & 0 & \tilde{S}_D & 0 \\ 0 & 0 & 0 & I_{s_2} \end{bmatrix} k+l \quad , \ \Sigma_C = \begin{bmatrix} 0 & 0 & 0 & 0 \\ 0 & I_l & 0 & 0 \\ 0 & 0 & \tilde{S}_C & 0 \\ 0 & 0 & 0 & I_{s_1} \end{bmatrix} \atop j+k$$

and $\tilde{S}_T = \operatorname{diag}(\tilde{\tau}_1,\dots,\tilde{\tau}_s)$, $\tilde{S}_D = \operatorname{diag}(\tilde{\delta}_1,\dots,\tilde{\delta}_s)$ *and* $\tilde{S}_C = \operatorname{diag}(\tilde{\gamma}_1,\dots,\tilde{\gamma}_s)$, $\tilde{\tau}_g > 0$, $\tilde{\delta}_g > 0$, $\tilde{\gamma}_g > 0$, $\forall g = 1,\dots,s$. *Let* $ljk = l + j + k$ *and define*

$$\begin{array}{llll} \tau_i = 1, & \delta_i = 1, & \gamma_i = 0 & i = 1,\dots,j \\ \ \ \ = 1, & \ \ \ = 0, & \ \ \ = 0 & i = j+1,\dots,j+k \\ \ \ \ = 1, & \ \ \ = 0, & \ \ \ = 1 & i = j+k+1,\dots,ljk \\ \ \ \ = \tilde{\tau}_g, & \ \ \ = \tilde{\delta}_g, & \ \ \ = \tilde{\gamma}_g, & i = ljk+g \quad \text{and} \quad g = 1,\dots,s \\ \ \ \ = 0, & \ \ \ = 1, & \ \ \ = 1 & i = ljk+s+1,\dots,ljk+s+\min(s_1,s_2) \end{array}$$

then the RSVs of (T, D, C) *are given by :*

$$\begin{aligned} \sigma_i(T,D,C) &= \tfrac{\tau_i}{\delta_i\gamma_i} & i &= 1,\dots,\operatorname{rank}(T) \\ &= 0 & i &= \operatorname{rank}(T)+1,\dots,n. \end{aligned}$$

Trivial RSVs, defined by $\frac{0}{0}$, *are set to zero.*

RSVs play an important role in the **rank determination** of a matrix T_0 under a restricted perturbation of the form $E^\star = DEC$ where E satisfies $\|E\|_2 \le \epsilon$ and ϵ is known. It is easy to see that the best estimate of the rank of $T_0 = T + DEC$, (T, D, C) known, is given by :

$$\text{rank}_\epsilon(T) = \max\{i \mid \sigma_i(T, D, C) > \epsilon\}$$

i.e. there exists a matrix $\widehat{E}$ satisfying $\|\widehat{E}\|_2 \le \epsilon$ such that $\text{rank}(T + D\widehat{E}C) = i$ but no $\widetilde{E}$ exists satisfying $\|\widetilde{E}\|_2 \le \epsilon$ such that $\text{rank}(T + D\widetilde{E}C) < i$. This matrix $\widehat{E}$ can be easily computed from the RSVD of (T, D, C) and allows to compute the RTLS correction matrix $[\Delta\widehat{A}; \Delta\widehat{B}] = D\widehat{E}C$ satisfying (4 -5). Indeed, comparing the definition of the RSVs with the RTLS condition (5), the following theorem can be proven [16]:

Theorem 2 *Assume that the RTLS problem $A_{m\times n}X \approx B_{m\times d}$ with given $D_{m\times p}$, $C_{q\times(n+d)}$ is solvable and $\sigma_n([A;B], D, C) > \sigma_{n+1}([A;B], D, C)$, then :*
The RSVD of the RTLS approximation $([A;B] - [\Delta\widehat{A}; \Delta\widehat{B}], D, C)$ is given by the RSVD of $([A;B], D, C)$ in which the smallest d RSVs are equal to zero.

3 Algorithm.

The basic problem is to find a matrix $[\Delta\widehat{A}; \Delta\widehat{B}] = D\widehat{E}C$ with minimal $\|\widehat{E}\|_F$ such that $\text{rank}([A;B] - [\Delta\widehat{A}; \Delta\widehat{B}]) \le n$ and (4) is satisfied. Hereto, the RSVD of the matrix triplet (T, D, C), $T = [A;B]$, is computed. This is done as follows. First of all, the regular submatrix triplet (T_r, D_r, C_r) which only contains the nontrivial finite RSVs of the general matrix triplet (T, D, C), is extracted from (T, D, C), using only orthogonal transformations. This implies that D_r and C_r are nonsingular and hence, the implicit SVD algorithm of Ewerbring and Luk [3] or Zha [17] can now be applied in order to compute the SVD of the triple matrix product $D_r^{-1}T_rC_r^{-1}$ without explicitly forming the products and without inverting D_r or C_r. This guarantees its better numerical performance. Moreover, by first performing orthogonal transformations, the RTLS algorithm only needs to compute the implicit 3-product SVD of a smaller submatrix, hereby improving its computational efficiency.
$\widehat{E}$ satisfying (4- 5) can now be computed and also a basis of the null space of $[A - \Delta\widehat{A}; B - \Delta\widehat{B}]$ of dimension at least d. From the latter, the RTLS solution $\widehat{X}$ can be deduced, as done in [15]. If the RTLS solution is not unique, the solution $\widehat{X}$ with minimal $\|F_1\widehat{X}F_2\|_F$ is singled out. See [16] for an outline of the RTLS algorithm.

4 Properties.

The following theorem, proven in [16], allows to derive the **consistency** of the solution of most RTLS problems described in section 1 :

Theorem 3 *Consider the RTLS problem, defined by (1 -2), where C is of the form* $C = \underset{\quad n_1 \qquad n+d-n_1}{\begin{bmatrix} 0 & 0 \\ 0 & C_2 \end{bmatrix}} = \underset{\quad n \qquad d}{\begin{bmatrix} C_a & C_{ab} \\ C_{ba} & C_b \end{bmatrix}} \begin{matrix} n \\ d \end{matrix}$ *and C_2, D square, known* *and nonsingular. Denote by σ' (resp., σ) the minimal restricted singular value of the matrix triplet (A, D, C_a) (resp., $([A; B], D, C)$). Let σ have multiplicity d and denote $\mathcal{D} = D^T D$ and $\mathcal{C} = C^T C = \begin{bmatrix} \mathcal{C}_a & \mathcal{C}_{ab} \\ \mathcal{C}_{ba} & \mathcal{C}_b \end{bmatrix} \begin{matrix} n \\ d \end{matrix}$ (column sizes n, d). If $\sigma' > \sigma$, the RTLS solution is given by :*

$$\widehat{X} = (A^T \mathcal{D}^{-1} A - \sigma^2 \mathcal{C}_a)^{-1} (A^T \mathcal{D}^{-1} B - \sigma^2 \mathcal{C}_{ab}) \tag{7}$$

Theorem 3 proves the correspondence between the RTLS solution and well-known expressions of consistent estimates in statistics, as well as in system identification. If $D = I_m$, (7) is a well-known expression in **linear regression** analysis. Its consistency and other statistical properties have been investigated by Gallo [4]. Gleser [5] studied the special case that $C = I_{n+d}$ and $n_1 = 0$, corresponding to the ordinary TLS problem. Using their results and assuming that $\lim_{m\to\infty} \frac{1}{m} A_0^T A_0$ (defined below) exists and is positive definite, it can be concluded that the RTLS solution is a **strongly consistent** estimate of the true parameters X of the **general errors-in-variables model**, defined as :

$$B_0 = (A_0)_{m\times n} X_{n\times d} = A_1 X_1 + (A_2)_0 X_2; \quad A_2 = (A_2)_0 + \Delta A_2 \text{ and } B = B_0 + \Delta B \tag{8}$$

A_1 is known but $(A_2)_0$ and B_0 not. The observations A_2 and B of the unknown values $(A_2)_0$ and B_0 contain measurement errors ΔA_2 and ΔB such that the rows of $[\Delta A_2; \Delta B]$ are independently and identically distributed (i.i.d.) with zero mean and known positive definite covariance matrix $\mathcal{C}_2 = C_2^T C_2$, up to a factor of proportionality.
Not only in statistics, but also in system identification, expression (7) (with $D = I_m$) is well-known, see e.g. [11]. In particular, (7) arises as a consistent estimator in transfer function modelling. These models are given by :

$$y(t) + a_1 y(t-1) + \cdots + a_{n_a} y(t-n_a) = b_1 u(t-1) + \cdots + b_{n_b} u(t-n_b) \tag{9}$$

where the $\{u(t)\}$ and $\{y(t)\}$ are the input and output sequences respectively and $\{a_j\}$ and $\{b_j\}$ are the unknown constant parameters of the system. If sufficient observations are taken, (9) gives rise to an overdetermined Toeplitz-like set of equations. If the observed in- and outputs (resp., observed outputs) are disturbed by mutually independent stationary zero-mean white noise sequences of equal variance, i.e. $C_2 = I$ and $n_1 = 0$ (resp., $= n_b$), the RTLS solution of this set, given by (7), coincides with the Koopmans-Levin estimate [9] (resp., the Compensated LS estimate [11]) and is hence strongly consistent under the same conditions as described in [1,11]. Several authors, e.g. [7,8], extended these results in order to prove consistency for multi-input multi-output systems, modelled by (9), in which the disturbances are not necessarily white provided the covariance matrix $\mathcal{C}_2 = C_2^T C_2$ of the correlated noise in the input-output data is known, up to a factor of proportionality. See [15] for a complete list of references and a detailed description of the consistency conditions.
The condition $D = I_m$ is not a real restriction for consistency. Indeed, if D is nonsingular, the same consistency conditions apply to the transformed data $[A^\star; B^\star] = D^{-1}[A; B]$. Also, the assumption that the first n_1 columns of A are

error-free is not a restriction. In fact, one can always find an appropriate permutation of the columns of A such that the consistency conditions are satisfied. Consistency of the solution of more general RTLS problems than those defined in Theorem 3, has not yet been proven and needs to be further analyzed.

Acknowledgment : The author is supported by the Belgian N.F.W.O. (National Fund of Scientific Research), as a Senior Research Assistant.

References

[1] M. Aoki and P.C. Yue, On a priori error estimates of some identification methods. IEEE Trans. Automat. Contr. AC-15 (1970), pp. 541-548.

[2] B.L.R. De Moor and G.H. Golub, The Restricted Singular Value Decomposition : Properties and Applications. Techn. Rep. NA-89-03, Dept. of Computer Science, Stanford University, Stanford (1989).

[3] L.M. Ewerbring and F.T. Luk, Canonical Correlations and Generalized SVD: Applications and New Algorithms. Proc. SPIE 977, Real Time Signal Processing XI, paper 23 (1988).

[4] P.P. Gallo, Consistency of regression estimates when some variables are subject to error. Commun. Statist.-Theor. Meth. 11 (1982), pp. 973-983.

[5] L.J. Gleser, Estimation in a multivariate "errors in variables" regression model: large sample results. Ann. Statist. 9 (1981), pp. 24-44.

[6] G.H. Golub and C.F. Van Loan, An analysis of the total least squares problem. SIAM J. Numer. Anal. 17 (1980), pp. 883-893.

[7] A. Grosjean and C. Foulard, Extensions of the Levin's method (or eigenvector method) for the identification of discrete, linear, multivariable, stochastic, time invariant, dynamic systems. Proc. 4th IFAC Symp. on Identification and System Parameter Estimation, Tbilisi, USSR (1976), pp. 2003-2010.

[8] Ü. Kotta, Structure and parameter estimation of multivariable systems using the eigenvector method. Proc. 5th IFAC Symp. on Identification and System Parameter Estimation, Darmstadt, FRG (1979), pp. 453-458.

[9] M.J. Levin, Estimation of a system pulse transfer function in the presence of noise. IEEE Trans. Automat. Contr. AC-9 (1964), pp. 229-235.

[10] C.C. Paige, The general linear model and the generalized singular value decomposition. Lin. Alg. Appl. 70 (1985), pp. 269-284.

[11] P. Stoica and T. Söderström, Bias correction in least squares identification. Int. J. Control 35 (1982), pp. 449-457.

[12] S. Van Huffel and J. Vandewalle, Analysis and solution of the nongeneric total least squares problem. SIAM J. Matrix Anal. Appl. 9 (1988), pp. 360-372.

[13] S. Van Huffel and J. Vandewalle, Reliable and efficient techniques based on total least squares for computing consistent estimators in models with errors in the variables. Proc. IMA Conf. on Mathematics in Signal Processing, Warwick, UK, 13-15 Dec. (1988), to appear.

[14] S. Van Huffel and J. Vandewalle, Comparison of total least squares and instrumental variable methods for parameter estimation of transfer function models. Int. J. Control (1989), to appear.

[15] S. Van Huffel and J. Vandewalle, Analysis and properties of the generalized total least squares problem $AX \approx B$ when some or all columns in A are subject to error. SIAM J. Matrix Anal. Appl. 10 (1989), no.3, to appear.

[16] S. Van Huffel and H. Zha, Restricted Total Least Squares : a Unified Approach for Solving (Generalized) (Total) Least Squares Problems with(out) Equality Constraints. Int. Rep. ESAT Lab., K.U.Leuven (1989), submitted to SIAM J. Matrix Anal. Appl.

[17] H. Zha, A numerical algorithm for computing the Restricted Singular value Decomposition of matrix triplets. Scientific Report 89-1, Konrad-Zuse Zentrum für Informationstechnik Berlin (1989).

[18] H. Zha, Restricted Singular Value Decomposition of matrix triplets. Scientific Report 89-2, Konrad-Zuse Zentrum für Informationstechnik Berlin (1989).

THE GEOMETRIC MEAN OF POSITIVE REAL OPERATORS

(Preliminary Report)

William N. Anderson, Jr.
and
George E. Trapp

Abstract

Various matrix operations, including the **Parallel Sum** and **Geometric Mean** have previously been studied for Hermitian matrices; this is the natural class of matrices to study in connection with resistive networks. For more general networks the class of **positive real matrices** is relevant. In this paper some of the matrix operations are extended to the more general setting.

1. Introduction

It is well known that positive real operators, that is operator-valued analytic functions which have positive real quadratic forms, are important in describing the behavior of n-port networks. When two networks are suitably connected, the impedance of the new network will be a function of the impedances of the connected networks. It thus seems of interest to consider the operations on positive real operators induced by connections of the underlying networks.

In our earlier papers [1] - [4], we laid the foundation for this study, and considered the standard network interconnnections: shorting of a single n-port network, and the series, parallel, hybrid and cascade connections of two n-port networks. These studies depended on the fact that the network connections could be described by an easily analyzed graph, and thus algebraic matrix operations could be used to perform the necessary calculations.

The **Geometric Mean** of two positive semidefinite Hermitian operators is a natural extension of the idea of network conections, but there does not seem to be a finite network connection that will actually give the geometric mean. Thus other definitions are necessary, and these have been provided for positive semidefinite Hermitian operators by Pusz and Woronowicz [9], Ando [5], and others. We have given infinite ladder networks for the geometric mean in [4].

In this paper we present, without proofs, an outline of our studies on the Geometric mean of positive real operators. We will consider a number of definitions, all of which turn out to be equivalent. Each of these definitions really defines two concepts: the geometric mean of the positive real operators, and the geometric mean of two constant matrices obtained by evaluating the positive real operators at a point in the right half plane. The geometric mean can thus be defined pointwise, as long as analyticity can be separately guaranteed.

We will consider linear operators on a finite-dimensional complex vector space, with inner product denoted by $\langle\ ,\ \rangle$. A linear operator A is called **almost right definite if**

(i) $\mathcal{Re}\ \langle Ax, x\rangle \geq 0$ for all vectors x, and
(ii) $\mathcal{Re}\ \langle Ax, x\rangle = 0$ only if $Ax = 0$.

We have shown in [3] that an equivalent definition is obtained by writing A in the form $A = A_H + A_S$ where A_H is Hermitian and A_S is skew-Hermitian. Then A is almost right definite if and only if

(iii) A_H is positive semidefinite, and
(iv) $\mathcal{Range}(A_S) \subset \mathcal{Range}(A_H)$.

An operator valued function $A(\lambda)$ is called a **positive real operator** if A is an analytic function of λ; $A(\lambda)$ is Hermitian for real $\lambda > 0$; and $A(\lambda)$ is almost right definite for $Re(\lambda) > 0$. We insist that $A(\lambda)$ is Hermitian for real λ so that we may have a useful partial order between the operators; not all authors enforce this restriction. Many important properties of positive real operators are discussed by Newcomb [8] and Weinberg and Slepian [10]. Positive real functions are considered for entirely different reasons by Donoghue [6].

In view of the range condition (iv), we will assume, for simplicity, that the matrices $A(\lambda)$ are invertible. This is not completely general, but will suffice for this preliminary report.

For positve real operators $A(\lambda)$ and $B(\lambda)$, we define, as in [3], three fundamental matrix operations . The **arithmetic mean** is $(A + B)/2$; the **parallel sum** $A:B$ is defined by $A:B = (A^{-1} + B^{-1})^{-1}$; and the **harmonic mean** is $2(A:B)$. Kubo and Ando show that these fundamental matrix operations suffice to define a wide class of matrix operations.[7]

2. The Geometric Mean

In this section we present a number of equivalent definitions of the geometric mean. These fall into five classes: the fundamental arithmetic-harmonic iteration, explicit formulas, solution of equations, ladder networks, and suprema. All of these definitions have previously been studied for the positive semidefinite Hermitian case [4]. Our work involves extending the concepts to the new setting.

Definition 0:

Given two positive real operators $A(\lambda)$ and $B(\lambda)$, we define sequences A_n and B_n by

$$A_0 = A, \quad B_0 = B, \quad \text{and iteratively}$$

$$A_{n+1} = (A_n + B_n)/2 \qquad B_{n+1} = 2(A_n : B_n).$$

The sequences A_n and B_n converge to a common limit, which is called $A\#B$, the **geometric mean** of A and B.

We note that if $B(\lambda) = I$, the identity operator, then at all stages of the iteration A_n and B_n commute, and their product is A. Thus the common limit is $A^{1/2}$, the square root of A. An equivalent computation for the square root is to observe that all the eigenvalues of the matrix A lie in the right half plane, and thus the square root may be obtained by cutting the complex plane on the negative real axis and then using the spectral theorem. This definition of the square root is used below for matrices which have no eigenvalues on the negative real axis.

Definition 1a:

$$A\#B = A\sqrt{A^{-1}B}$$

Definition 1b:

$$A\#B \;=\; (A+B)\sqrt{(A+B)^{-1}A(A+B)^{-1}B}$$

Definition 1c:

$$A\#B \;=\; A^{1/2}\sqrt{A^{-1/2}B\,A^{-1/2}}\;\;A^{1/2}$$

It is a straightforward matrix computation to show that these three definitions are equivalent. We note that definition *1b* is merely definition *1a* applied to the arithmetic and harmonic means of the initial operators.

Definition 2a:

A#B is the positive real solution to the equation

$$X = (A + X):(B + X).$$

One method to solve this equation, and thus to compute $A\#B$, is to start with $X = 0$; substitute this X into the right hand side, and iterate. This iteration will converge to the geometric mean, although more slowly than the arithmetic-harmonic iteration.

Definition 2b:

A#B is the positive real solution to the equation

$$X\,A^{-1}X = B.$$

Definition 3: $A\#(A + 4B)$ is the impedance of the ladder network

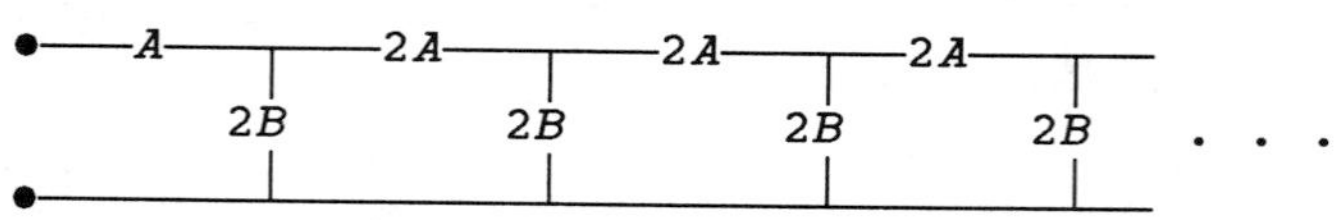

We note that this impedance can also be written as the solution to the equation $X = A + 2B:(X + A)$, and may be computed by an iteration similar to that in definition 2a. Each step in the iteration corresponds to a finite network; the infinite ladder is the limit of the finite networks. [11]

Definition 4: Let $A(\lambda)$ and $B(\lambda)$ be positive real operators. Then the geometric mean, $A\#B$, is the supremum of all positive real operators such that the block operator

$$\begin{bmatrix} A(\lambda) & X(\lambda) \\ X(\lambda) & B(\lambda) \end{bmatrix}$$

is a positive real operator.

The partial order in this definition is that induced by positive reality; that is, positive real operators $A(\lambda)$ and $B(\lambda)$ are related by $A(\lambda) \geq B(\lambda)$ if $A(\lambda) - B(\lambda)$ is positive real. This partial order is defined for the functions only; at a point off the real axis there seems to be no useful way to define a partial order between almost right definite matrices. This supremum definition is in the nature of a conjecture; we have not at present completed all the details of the proof.

References:

1. Anderson, W. N., Jr. and G. E. Trapp, "Shorted Operators II", SIAM J. Appl. Math. 28(1975), 160-171.

2. Anderson, W. N., Jr, R. J. Duffin and G. E. Trapp, "Matrix Operations induced by Network Connections", SIAM J. Control 13(1975), 446-461.

3. Anderson, W. N., Jr. and G. E. Trapp, "Analytic Operator Functions and Electrical Networks", in *Recent Applications of Generalized Inverses,* Pitman, London, 1982

4. Anderson, W. N., Jr., T. D. Morley and G. E. Trapp, "Ladder Networks, Fixed Points, and the Geometric Mean", Cir. Sys. Sig. Process 3(1983), 259-268.

5. Ando; T., *Topics on Operator Inequalities,* Lecture Notes, Hokkaido University, Sapporo, Japan 1978

6. Donoghue, W. F., Jr., *Monotone Matrix Functions and Analytic Continuation,* Springer-Verlag, New York, 1974

7. Kubo, F. and T. Ando, "Means of Positive Operators", Math. Ann. 246(1980), 205-224.

8. Newcomb, R. W., *Linear Multiport Synthesis*, McGraw-Hill, New York, 1966.

9. Pusz, W. and S. L. Wornowicz, "Functional Calculus for Sesquilinear Forms and the Purification Map", Rep. Math. Phys. 8(1975), 159-170.

10. Weinberg, L. and P. Slepian, "Positive Real Matrices", J. Math. Mech. 9(1960), 71-84.

11. Zemanian, A. H., "Infinite Electrical Networks" Proc IEEE 64(1976), 6-17.

Inequalities Among Operator Symmetric Function Means

Tsuyoshi ANDO and Fumio KUBO

Abstract

For an n-tuple of bounded positive semi-definite operators $\boldsymbol{A} = (A_1, \cdots, A_n)$, in terms of usual addition and parallel addition, two series of operator symmetric function means have been defined; $S_{k,n}(\boldsymbol{A})$ and $s_{k,n}(\boldsymbol{A})$, $k = 1, 2, \cdots, n$. These operator symmetric function means admit network realization. We discuss order relations between $S_{k,n}(\boldsymbol{A})$ and $S_{k+1,n}(\boldsymbol{A})$, or between $s_{k,n}(\boldsymbol{A})$ and $s_{k+1,n}(\boldsymbol{A})$, and for $n = 4$ we establish the inequalities

$$S_{2,4}(\boldsymbol{A}) \geq S_{3,4}(\boldsymbol{A}) \qquad \text{and} \qquad s_{2,4}(\boldsymbol{A}) \geq s_{3,4}(\boldsymbol{A}).$$

1. Operator symmetric function means.

For a pair of invertible, positive (semi-definite) operators A, B its *parallel sum* $A : B$ is defined as $(A^{-1} + B^{-1})^{-1}$. If A or B is not invertible, $A : B$ is defined as the strong limit of the net $(A + \varepsilon I) : (B + \varepsilon I)$ as $\varepsilon \downarrow 0$, where I is the identity operator. For the matrix case, the notion of parallel sum was originally introduced by Anderson and Duffin [1], in studying the impedance matrix of the network, obtained by connecting two resistive multiport linear networks in parallel. Remark that the impedance matrix of the network, obtained by connecting two multiport networks in series, is given by the usual sum of the impedance matrices of both component networks. We refer to the paper [1] for details.

In the sequel, for a positive operator A and a vector x we shall use $A[x]$ for the quadratic form $\langle Ax|x\rangle$. Order relation $A \geq B$ between two positive operators A and B is defined, as usual, by requiring $A[x] \geq B[x]$ for all vector x. When A denotes the impedance of a linear network, $A[x]$ is the power or energy when a current x flows into the network.

Anderson and Trapp [3] gave a variational description for $(A : B)[x]$;

$$(A : B)[x] = \inf \{A[u] + B[v]; u + v = x\}. \tag{1}$$

This formula admits a physical interpretation. According to the Kirchhoff's law, a current x flowing into the network, obtained by connected two networks in parallel, is divided to two branch currents u, v ; $x = u + v$. Then formula (1) means that actual division of the current x occurs so as to make the energy of the resulting network minimum.

Since parallel addition is associative, for an n-tuple of positive operators $(A_1, \cdots, A_n)$, we can write, without any ambiguity,

$$\prod_{i=1}^{n}: A_i = A_1 : A_2 : \cdots : A_n.$$

Operator symmetric function means for an n-tuple of positive operators will be defined in terms of usual addition and parallel addition. Let us begin with the scalar case. For an n-tuple of positive numbers $\boldsymbol{\alpha} = (\alpha_1, \cdots, \alpha_n)$, Marcus and Lopes [7] defined symmetric function means (or Marcus-Lopes means) $E_{k,n}(\boldsymbol{\alpha})$ by

$$E_{k,n}(\boldsymbol{\alpha}) = e_{k,n}(\boldsymbol{\alpha})/e_{k-1,n}(\boldsymbol{\alpha}) \qquad \text{for } 1 \leq k \leq n, \tag{2}$$

where $e_{k,n}(\boldsymbol{\alpha})$ is the normalized kth elementary symmetric function of $\boldsymbol{\alpha}$, that is, $e_{0,n}(\boldsymbol{\alpha}) = 1$ and

$$e_{k,n}(\boldsymbol{\alpha}) = \Big\{ \sum_{1 \leq i_1 < \cdots < i_k \leq n} \prod_{j=1}^{k} \alpha_{i_j} \Big\} \Big/ \binom{n}{k} \qquad \text{for } 1 \leq k \leq n.$$

Using an equivalent version of definition (2), Anderson, Morley and Trapp [2] (see also [8]) introduced two kinds of *operator symmetric function means* for an n-tuple of positive operators $\boldsymbol{A} = (A_1, \cdots, A_n)$;

$$S_{1,n}(\boldsymbol{A}) \equiv \Big\{ \sum_{i=1}^{n} A_i \Big\} / n \qquad \text{(arithmetic mean)}, \tag{3}$$

$$s_{n,n}(\boldsymbol{A}) \equiv n \Big\{ \prod_{i=1}^{n}: A_i \Big\} \qquad \text{(harmonic mean)}, \tag{4}$$

and recurrently

$$S_{k+1,n}(\boldsymbol{A}) \equiv \sum_{i=1}^{n} \{ A_i/(n-k) : S_{k,n-1}(\boldsymbol{A}_{(i)})/k \} \qquad \text{for } 1 \leq k \leq n-1 \tag{5}$$

and

$$s_{k,n}(\boldsymbol{A}) \equiv \prod_{i=1}^{n}: \{ kA_i + (n-k) s_{k,n-1}(\boldsymbol{A}_{(i)}) \} \qquad \text{for } 1 \leq k \leq n-1, \tag{6}$$

where $\boldsymbol{A}_{(i)}$ denotes the $(n-1)$-tuple $(A_1, \cdots, A_{i-1}, A_{i+1}, \cdots, A_n)$.

Since

$$A/k = \overbrace{A : \cdots : A}^{k \text{ times}} \qquad \text{and} \qquad kA = \overbrace{A + \cdots + A}^{k \text{ times}}, \tag{7}$$

starting with $S_{1,n}(\boldsymbol{A})$ and $s_{n,n}(\boldsymbol{A})$, in view of (5) and (6), we can give network interpretations to $S_{k,n}(\boldsymbol{A})$ and $s_{k,n}(\boldsymbol{A})$. The resistive network with impedance operator $S_{k,n}(\boldsymbol{A})$ or $s_{k,n}(\boldsymbol{A})$ can be constructed from those with impedance $A_1, \cdots, A_n$ by using series and parallel connections only. See Figure (I) for $S_{2,4}(\boldsymbol{A})$ and Figure (II) for $S_{3,4}(\boldsymbol{A})$.

If all A_i are scalars, that is, $\boldsymbol{A} = \boldsymbol{\alpha}$, then both $S_{k,n}(\boldsymbol{\alpha})$ and $s_{k,n}(\boldsymbol{\alpha})$ coincide with the Marcus-Lopes mean $E_{k,n}(\boldsymbol{\alpha})$.

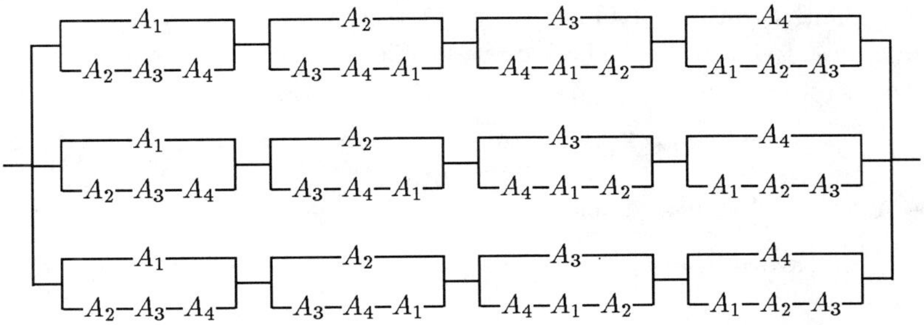

FIGURE (I) $S_{2,4}(\boldsymbol{A})$

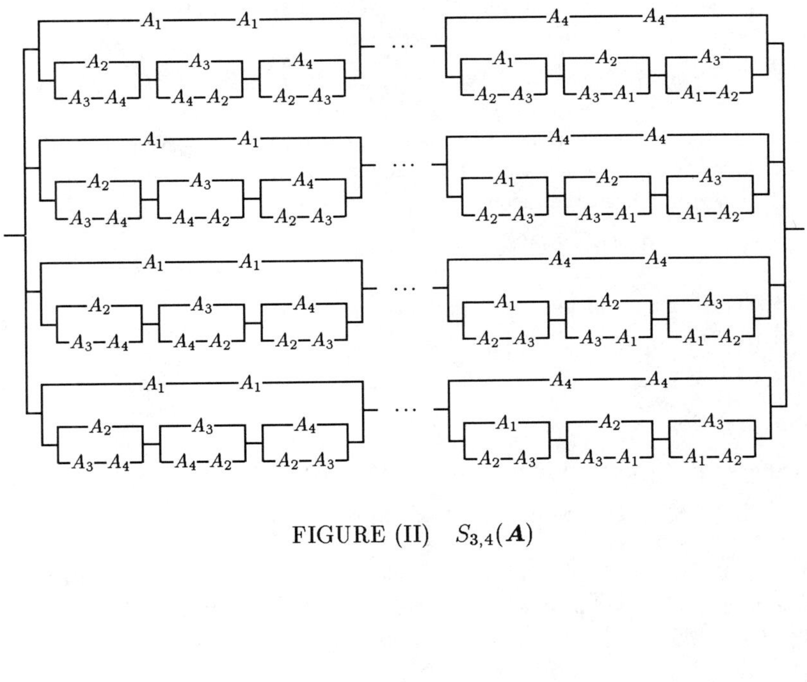

FIGURE (II) $S_{3,4}(\boldsymbol{A})$

By definition both $S_{k,n}(\boldsymbol{A})$ and $s_{k,n}(\boldsymbol{A})$ are invariant under permutations of indices of $(A_1, \cdots, A_n)$. The following relations hold for any n-tuple of invertible positive operators $\boldsymbol{A} = (A_1, \cdots, A_n)$;

$$S_{k,n}(\boldsymbol{A}) = s_{n-k+1,n}(\boldsymbol{A}^{-1})^{-1} \qquad \text{for } 1 \le k \le n \tag{8}$$

where $\boldsymbol{A}^{-1} = (A_1^{-1}, \cdots, A_n^{-1})$, and

$$S_{n,n}(\boldsymbol{A}) = s_{n,n}(\boldsymbol{A}) \qquad \text{and} \qquad s_{1,1}(\boldsymbol{A}) = S_{1,1}(\boldsymbol{A}). \tag{9}$$

Besides the easily proved inequalities

$$S_{1,n}(\boldsymbol{A}) \ge S_{k,n}(\boldsymbol{A}) \ge S_{n,n}(\boldsymbol{A}) \tag{10}$$

and

$$s_{1,n}(\boldsymbol{A}) \ge s_{k,n}(\boldsymbol{A}) \ge s_{n,n}(\boldsymbol{A}) \tag{11}$$

not much is known about order relations among $S_{i,n}(\boldsymbol{A})$ and $s_{j,n}(\boldsymbol{A})$ $(i,j = 2,3,\cdots,n-1)$.

In [2], [4] and [5] order relation between $S_{k,n}(\boldsymbol{A})$ and $s_{k,n}(\boldsymbol{A})$ for same k was studied. We proved in [5] that

$$S_{2,n}(\boldsymbol{A}) \ge s_{2,n}(\boldsymbol{A}) \qquad \text{and} \qquad S_{n-1,n}(\boldsymbol{A}) \ge s_{n-1,n}(\boldsymbol{A}). \tag{12}$$

In the present paper we discuss order relation between $S_{k,n}(\boldsymbol{A})$ and $S_{k+1,n}(\boldsymbol{A})$, and one between $s_{k,n}(\boldsymbol{A})$ and $s_{k+1,n}(\boldsymbol{A})$. Remark that if $\boldsymbol{A} = (A_1, \cdots, A_n)$ is a commuting n-tuple, from the inequality

$$E_{k,n}(\boldsymbol{\alpha}) \ge E_{k+1,n}(\boldsymbol{\alpha}) \qquad \text{for } k = 2, \cdots, n-2$$

(see [7]), by using a simultaneous diagonalization we can deduce that

$$S_{k,n}(\boldsymbol{A}) = s_{k,n}(\boldsymbol{A}) \ge S_{k+1,n}(\boldsymbol{A}) = s_{k+1,n}(\boldsymbol{A}).$$

2. Inequalities.

Theorem. *For any quartet of positive operators* $\boldsymbol{A} = (A_1, A_2, A_3, A_4)$

$$S_{2,4}(\boldsymbol{A}) \ge S_{3,4}(\boldsymbol{A}) \qquad \textit{and} \qquad s_{2,4}(\boldsymbol{A}) \ge s_{3,4}(\boldsymbol{A}). \tag{13}$$

By (8) the second inequality of (13) follows from the first for $\boldsymbol{A}^{-1}$. A proof of the first inequality will be given in the next section. According to (12) and (13) for $\boldsymbol{A} = (A_1, A_2, A_3, A_4)$ there are the following order relations among $S_{i,4}(\boldsymbol{A})$ and $s_{j,4}(\boldsymbol{A})$ $(i,j = 1,2,3,4)$

$$\begin{array}{ccccccc} S_{1,4}(\boldsymbol{A}) & \ge & S_{2,4}(\boldsymbol{A}) & \ge & S_{3,4}(\boldsymbol{A}) & \ge & S_{4,4}(\boldsymbol{A}) \\ \parallel & & \text{VI} & & \text{VI} & & \parallel \\ s_{1,4}(\boldsymbol{A}) & \ge & s_{2,4}(\boldsymbol{A}) & \ge & s_{3,4}(\boldsymbol{A}) & \ge & s_{4,4}(\boldsymbol{A}). \end{array}$$

Incidentally we should mention that there is no order relation between $S_{3,4}(\boldsymbol{A})$ and $s_{2,4}(\boldsymbol{A})$ in general. That $S_{3,4}(\boldsymbol{A}) \not\ge s_{2,4}(\boldsymbol{A})$ in general is seen even for the

Marcus-Lopes means. To see that $s_{2,4}(\boldsymbol{A}) \not\geq S_{3,4}(\boldsymbol{A})$ in general, let us consider a quartet of 2×2 positive matrices;

$$A_1 = \begin{bmatrix} 1 & 0 \\ 0 & 0 \end{bmatrix}, \ A_2 = \begin{bmatrix} 0 & 0 \\ 0 & 1 \end{bmatrix}, \ A_3 = \begin{bmatrix} 1 & 1 \\ 1 & 1 \end{bmatrix}, \quad \text{and} \quad A_4 = \begin{bmatrix} 1 & -1 \\ -1 & 1 \end{bmatrix}.$$

Since $A_i : A_j = 0$ for $i \neq j$, it is easy to see that $s_{2,4}(\boldsymbol{A}) = 0$. On the other hand, since

$$A_i + A_j \geq \varepsilon I \qquad \text{with } \varepsilon = \frac{3-\sqrt{5}}{2},$$

we have

$$\begin{aligned} S_{3,4}(\boldsymbol{A}) &= \frac{1}{4}\sum_{i=1}^{4}\Big\{2A_i : \Big(\sum_{j\neq i}(A_j : (\sum_{k\neq i,j} A_k))\Big)\Big\} \\ &\geq \frac{1}{4}\sum_{i=1}^{4}\Big\{2A_i : \Big(\frac{\varepsilon}{\varepsilon+2}\sum_{j\neq i} A_j\Big)\Big\} \\ &\geq \frac{1}{4}\sum_{i=1}^{4}\Big\{2A_i : \frac{\varepsilon^2}{\varepsilon+2} I\Big\} \geq \frac{\varepsilon^3}{\varepsilon^2+4\varepsilon+8} I. \end{aligned}$$

3. Proof.

We may assume that all A_i are invertible. By definition of order, for the proof we have to show that

$$S_{2,4}(\boldsymbol{A})[x] \geq S_{3,4}(\boldsymbol{A})[x] \qquad \text{for all } x. \tag{14}$$

Since $S_{3,4}(\boldsymbol{A}) = s_{2,4}(\boldsymbol{A}^{-1})^{-1}$ by (8), and since for any invertible positive operator T

$$T^{-1}[x] = \sup_y \frac{|\langle x|y\rangle|^2}{T[y]},$$

inequality (14) is equivalent to the following inequality

$$S_{2,4}(\boldsymbol{A})[x] \cdot s_{2,4}(\boldsymbol{A}^{-1})[y] \geq |\langle x|y\rangle|^2 \qquad \text{for all } x, y. \tag{15}$$

Further, in view of the arithmetic-geometric means inequality, (15) is reduced to the following

$$S_{2,4}(\boldsymbol{A})[x] + s_{2,4}(\boldsymbol{A}^{-1})[y] \geq 2|\langle x|y\rangle| \qquad \text{for all } x, y. \tag{16}$$

For computational convenience, replacing x by $\sqrt{3}\,x$ and y by $\frac{1}{2}y$, instead of (16), we will prove the following

$$3S_{2,4}(\boldsymbol{A})[x] + \frac{1}{4}s_{2,4}(\boldsymbol{A}^{-1})[y] \geq \sqrt{3}\,|\langle x|y\rangle| \qquad \text{for all } x, y. \tag{17}$$

It follows from definitions (5) and (6) that

$$3S_{2,4}(\boldsymbol{A}) = \sum_{i=1}^{4}\Big\{A_i : (\sum_{j\neq i} A_j)\Big\} \tag{18}$$

and

$$\frac{1}{4}s_{2,4}(\boldsymbol{A}^{-1}) = \prod_{i=1}^{4}: \left\{ \frac{1}{2}A_i^{-1} + \prod_{j\neq i}:(A_j^{-1} + \prod_{k\neq i,j}: A_k^{-1}) \right\}. \tag{19}$$

By the variational formula (1), for each i

$$\left\{A_i : (\sum_{j\neq i} A_j)\right\}[x] = \inf_{x_i} \left\{A_i[x - x_i] + \sum_{j\neq i} A_j[x_i]\right\}.$$

Then we can conclude from (18), after a suitable change of order of summation, that $3S_{2,4}(\boldsymbol{A})[x]$ coincides with the infimum of the expression

$$\Phi(\boldsymbol{A}; \{x_i\}) \equiv \sum_{i=1}^{4}\{A_i[x - x_i] + \sum_{j\neq i} A_i[x_j]\} \tag{20}$$

when each x_i runs independently over all vectors. In a similar way, using the variational formula (1) at each step where parallel addition appears, we can see from (19) that $\frac{1}{4}s_{2,4}(\boldsymbol{A}^{-1})[y]$ coincides with the infimum of the expression

$$\begin{aligned} &\Psi(\boldsymbol{A}; \{y_j\}, \{y_{j,k}\}, \{y_{j,k,l}\}) \\ &\equiv \sum_{i=1}^{4}\left\{A_i^{-1}[\frac{1}{\sqrt{2}}y_i] + \sum_{j\neq i}\{A_i^{-1}[y_{j,i}] + \sum_{k\neq i,j} A_i^{-1}[y_{j,k,i}]\}\right\} \end{aligned} \tag{21}$$

when y_j, $y_{j,k}$ and $y_{j,k,l}$ $(j, k, l = 1, 2, 3, 4)$ run over all vectors satisfying

$$\sum_{j=1}^{4} y_j = y, \quad \sum_{k\neq j} y_{j,k} = y_j, \quad \sum_{l\neq j,k} y_{j,k,l} = y_{j,k}. \tag{22}$$

Therefore we have to show that for any choice of $\boldsymbol{A} = (A_1, A_2, A_3, A_4)$, $\{x_i\}$, $\{y_j\}$, $\{y_{j,k}\}$ and $\{y_{j,k,l}\}$

$$\Phi(\boldsymbol{A}; \{x_i\}) + \Psi(\boldsymbol{A}; \{y_j\}, \{y_{j,k}\}, \{y_{j,k,l}\}) \geq \sqrt{3}\,|\langle x|y\rangle|. \tag{23}$$

To this end, fix $\{x_i\}$, $\{y_j\}$, $\{y_{j,k}\}$ and $\{y_{j,k,l}\}$, and let us prove

$$\inf_{\boldsymbol{A}} \left\{\Phi(\boldsymbol{A}; \{x_i\}) + \Psi(\boldsymbol{A}; \{y_j\}, \{y_{j,k}\}, \{y_{j,k,l}\})\right\} \geq \sqrt{3}\,|\langle x|y\rangle|. \tag{24}$$

Here we use the Flanders' formula (see [6]): for any m-tuple of vectors $(u_1, \cdots, u_m)$ and any n-tuple of vectors $(v_1, \cdots, v_n)$

$$\inf_{T\geq 0}\left\{\sum_{i=1}^{m} T[u_i] + \sum_{j=1}^{n} T^{-1}[v_j]\right\} = 2||R||_1 \tag{25}$$

where R is the $m \times n$ Gramian

$$R = \left[\langle u_i|v_j\rangle\right] \tag{26}$$

and $||R||_1$ is the *trace norm* of R;

$$||R||_1 = \mathrm{tr}\big((TT^*)^{1/2}\big). \tag{27}$$

Then it follows from (18) and (19) that

$$\inf_{\boldsymbol{A}} \left\{ \Phi(\boldsymbol{A};\{x_i\}) + \Psi(\boldsymbol{A};\{y_j\},\{y_{j,k}\},\{y_{j,k,l}\}) \right\} = 2\sum_{i=1}^{4} ||R_i||_1 \tag{28}$$

where the Gramian R_1 is constructed from the following vectors;

$$u_1 = x - x_1, \qquad u_j = x_j \quad \text{for } j = 2,3,4 \tag{29}$$

and

$$\begin{aligned} v_1 = \frac{1}{\sqrt{2}} y_1, \quad v_2 &= y_{2,1}, \quad v_3 = y_{3,2,1}, \quad v_4 = y_{4,2,1}, \\ v_5 &= y_{3,1}, \quad v_6 = y_{4,3,1}, \quad v_7 = y_{2,3,1}, \\ v_8 &= y_{4,1}, \quad v_9 = y_{2,4,1}, \quad v_{10} = y_{3,4,1}, \end{aligned} \tag{30}$$

while the Gramians R_i $(i = 2,3,4)$ are obtained by permuting indices in (29) and (30) cyclically. Thus it remains to prove that

$$2\sum_{i=1}^{4} ||R_i||_1 \geq \sqrt{3}\,|\langle x|y\rangle|. \tag{31}$$

Consider the 4×10 matrix S defined by

$$S = \begin{bmatrix} 3\sqrt{2} & 2 & 1 & 1 & 2 & 1 & 1 & 2 & 1 & 1 \\ \sqrt{2} & 2 & 1 & 1 & 0 & -1 & 1 & 0 & 1 & -1 \\ \sqrt{2} & 0 & 1 & -1 & 2 & 1 & 1 & 0 & -1 & 1 \\ \sqrt{2} & 0 & -1 & 1 & 0 & 1 & -1 & 2 & 1 & 1 \end{bmatrix}.$$

It is easy to show that

$$SS^* = \begin{bmatrix} 36 & 12 & 12 & 12 \\ 12 & 12 & 0 & 0 \\ 12 & 0 & 12 & 0 \\ 12 & 0 & 0 & 12 \end{bmatrix}$$

and that the eigenvalues of SS^* consist of 48, 12, 12 and 0. Therefore the *operator* (*spectral*) *norm* $||S||_\infty$ of S is equal to $4\sqrt{3}$. Now it follows from a general duality formula that

$$\sum_{i=1}^{4} ||R_i||_1 \geq \left| \sum_{i=1}^{4} \operatorname{tr}(R_i S^*) \right| / ||S||_\infty. \tag{32}$$

Though a little tedious, we can check that

$$\sum_{i=1}^{4} \operatorname{tr}(R_i S^*) = 6\langle x|y\rangle. \tag{33}$$

Finally it follows from (32) and (33), with $||S||_\infty = 4\sqrt{3}$, that

$$2\sum_{i=1}^{4} ||R_i||_1 \geq \sqrt{3}\,|\langle x|y\rangle|,$$

proving (31). This completes the proof of the theorem.

The research of the first author was supported in part by Grant-in-Aid for Scientific Research no. 01540091.

References

[1] W. N. Anderson, Jr. and R. Duffin, Series and parallel addition of matrices, J. Math. Anal. Appl., 26 (1969), 576–594.

[2] W. N. Anderson, Jr., T. D. Morley and G. E. Trapp, Symmetric function means of positive operators, Linear Algebra Appl., 60 (1984), 129–143.

[3] W. N. Anderson, Jr. and G. E. Trapp, Shorted operators II, SIAM J. Appl. Math., 28 (1975), 60–71.

[4] T. Ando, An inequality between symmetric function means of positive operators, Acta Sci. Math., 45 (1983), 19–22.

[5] T. Ando and F. Kubo, Some matrix inequalities in multiport network connections, to appear.

[6] H. Flanders, An extremal problem on the space of positive definite matrices, Linear Multilinear Algebra, 3 (1975), 33–39.

[7] M. Marcus and L. Lopes, Inequalities for symmetric functions and Hermitian matrices, Canad. J. Math., 9 (1957), 305–312.

[8] G. E. Trapp, Hermitian semidefinite matrix means and related matrix inequalities — An introduction, Linear Multilinear Algebra, 16 (1984), 113–123.

OPERATOR MEANS AND SYMMETRIC NORMED IDEALS

J. ARAZY

ABSTRACT

Let E be a symmetric sequence space and let C_E be the associated symmetric normed ideal of all compact operators T on Hilbert space H for which the sequence $(s_n(T))$ of singular numbers belongs to E, with the norm

$$\| T \|_{C_E} = \| (s_n(T))\|_E.$$

Let $M:B(H)^+ \times B(H)^+ \to B(H)^+$ be an operator mean of two variables. We show that $M:C_E^+ \to C_E^+$ and that

$$\| M(A, B) \|_{C_E} \le M(\| A \|_{C_E}, \| B \|_{C_E})$$

for all $A, B \in C_E^+$. This result extends to a certain class of operator means of 3 or more variables, and to the context of Banach modules over von-Neumann algebras.

1. *The case of operator means of 2 variables.*

Let H be a Hilbert space, $B(H)$ the space of bounded linear operators on H with the usual norm $\| \cdot \|$ and let $B(H)^+$ denote the cone of positive operators T on H (namely, $(T x,x) \ge 0$ for all $x \in H$). $S \le T$ means that $T - S \in B(H)^+$ and $A_n \downarrow A$ in $B(H)^+$ means that $A_1 \ge A_2 \ge \cdots \ge A_n \ge \cdots$ and A_n converges strongly to A.

1.1. *Definition.*

A map $M:B(H)^+ \times B(H)^+ \to B(H)^+$ is an *operator mean* (of 2 variables) if it satisfies the following axioms

(a) *monotonicity* $0 \le A_1 \le B_1$ and $0 \le A_2 \le B_2$ imply $M(A_1, A_2) \le M(B_1, B_2)$;

(b) *weak continuity on decreasing sequences:* if $A_{1,j} \downarrow A_1$ and $A_{2,j} \downarrow A_2$ in $B(H)^+$ then $M(A_{1,j}, A_{2,j}) \downarrow M(A_1, A_2)$;

(c) *transformer inequality:* for every $A_1, A_2 \in B(H)^+$ and $T \in B(H)$, $T^*M(A_1, A_2)T \leq M(T^*A_1T, T^*A_2T)$

(d) *normalization:* $M(I, I) = I$.

The theory of operator means is presented in [6], see also [4], and [7]. Notice that the transformer inequality (c) becomes an equality if T is invertible. This shows that M is invariant under congruence, and also shows that M is positively homogeneous. It is also easy to show that if M satisfies (a), (b), and (c) then for any non-negative numbers a_1, a_2, $M(a_1 I, a_2 I)$ is a scalar operator which we denote by $M(a_1, a_2)I$.

We denote by C_∞ the ideal of compact operators on H. A *symmetric normed ideal of* $B(H)$ is a Banach space $(X, \|\cdot\|_X)$ of compact operators on H which is a bimodule over $B(H)$ and satisfies $\| ATB \|_X \leq \| A \| \| T \|_X \| B \|$ for all $A, B \in B(H)$, $T \in X$; and $\| T \|_X = \| T \|$ for every rank-one partial isometry T, see [13]. A Banach space E of sequences is a *symmetric sequence space* if its norm is invariant under permutations and changes of complex signs of the coordinates, and if E contains the standard unit vectors $e_n = (0, ..., 0, 1, 0, ...)$ and $\| e_n \| = 1$. The symmetric normed ideal C_E associated with E is the Banach space of all compact operators T on H whose sequences $s(T) = (s_n(T))$ of singular numbers belongs to E, with the norm $\| T \|_{C_E} = \| s(T) \|_E$. It is known (see [13]) that every symmetric normed ideal has this form for a unique E. In what follows we shall assume that E is either *minimal* (i.e. $\text{span}\{e_n\}_{n=1}^{\infty}$ is dense in E), of *maximal* (i.e. a dual of minimal symmetric space). See [13] and [15].

We denote $X^+ = X \cap B(H)^+$.

1.2 *Theorem: Let* M *be an operator mean and let* X *be a symmetric normed ideal. Then* $A_0, A_1 \in X^+$ *implies* $M(A_0, A_1) \in X$, *and* $\| M(A_0, A_1) \|_X \leq M(\| A_0 \|_X, \| A_1 \|_X)$.

Theorem 1.1 was announced in [9]. In the special case where M is the harmonic mean it was proved in [16] for $X = C_E$ where $E = \ell_p (1 \leq p < \infty)$ or $E = c_0$, and in [17] for E a special Lorentz space. Our proof is rather elementary.

We need more information on the structure of operator means. For every $A_0, A_1 \in B(H)^+$ there is unique positive operator $A_0{:}A_1$ on H defined via

$$((A_0:A_1)x,x) = \inf_{x=y+z} (A_0y,y) + (A_1z,z)$$

$A_0:A_1$ is the *parallel sum* of A_0 and A_1. See [3], [1], [16], and [17].

The *weighted harmonic means* $H_t(A_0, A_1)$ of A_0, A_1 are defined for $0 < t < 1$ by

$$H_t(A_0, A_1) = ((1 - t)^{-1} A_0): (t^{-1} A_1)$$

and for $t = 0, 1$ by

$$H_0(A_0, A_1) = A_0 , \qquad H_1(A_0, A_1) = A_1.$$

It is known that H_t is an operator mean for any $0 \le t \le 1$. Moreover, it is known that the operator means are in one-to-one correspondence with the probability measures ν on $[0, 1]$, via $M_\nu(A_0, A_1) = \int_0^1 H_t(A_0, A_1)\, d\nu(t)$ (where the integral converges weakly). The measure is ν the *representing measure* of M_ν. Since the weighted harmonic means satisfy $H_t(A, A) = A$ for all $A \in B(H)^+$, one gets a strengthening of the normalization axiom (d) in Definition 1.1 for every operator mean M:

(d') $$M(A, A) = A, \qquad A \in B(H)^+$$

Proof of Theorem 1.2 Let $A_0, A_1 \in X^+$, then $M(A_0, A_1) \le A_0 + A_1$ and so $M(A_0, A_1) \in X^+$. For any $x \in H$

$$((A_0:A_1)x,x) = \inf_{x=y+z} (A_0y,y) = (A_1z,z)$$
$$\le \inf_{\substack{\alpha+\beta=1 \\ \alpha,\beta\ge 0}} \alpha^2(A_0x,x) + \beta^2(A_1x,x)$$

by choosing $y = \alpha x$, $z = \beta x$. It follows that for every $\alpha \in [0, 1]$, $A_0:A_1 \le \alpha^2 A_0 + (1 - \alpha)^2 A_1$ and so $\| A_0:A_1 \|_X \le \alpha^2\| A_0 \|_X + (1 - \alpha)^2\| A_1 \|_X$. Taking infimum over all $0 \le \alpha \le 1$, we get $\| A_0:A_1 \|_X \le \| A_0 \|_X:\| A_1 \|_X$. It follows that for the weighted harmonic means we have $\| H_t(A_0, A_1) \|_X \le H_t(\| A_0 \|_X, \| A_1 \|_X)$. Let ν be the representing measure of M. Then

$$\| M(A_0, A_1) \|_X = \| \int_0^1 H_t(A_0, A_1)\, d\nu(t) \|_X \le \int_0^1 \| H_t(A_0, A_1) \|_X\, d\nu(t)$$

$$\le \int_0^1 H_t(\| A_0 \|_X, \| A_1 \|_X)\, dv(t) = M(\| A_0 \|_X, \| A_1 \|_X).$$

□

2. *Extension to operator means of 3 or more variables.*

We define the notion of operator mean of $n \ge 3$ variables in complete analogy with Definition 1.1. Unlike the case $n = 2$, no general description of such means is known. Motivated by series-parallel networks, Anderson, Morley, and Trapp study in [2] symmetric means of $n \ge 3$ operator. These means are studied also in [5].

We now generalize the construction of [2], and show that the corresponding operator means satisfy the inequality of Theorem 1.2.

Let $G = (V, E)$ be a directed graph, where B is the set of vertices an $E \subseteq V \times V$ is the set of directed edges. For $v \in V$ let $I(v)$ and $T(v)$ be the sets of edges $e \in E$ for which v is the initial, respectively terminal vertex. We assume that there is a unique initial vertex v_0, $(T(v_0) = \emptyset)$ and a unique terminal vertex v_1, $(I(v_1) = \emptyset)$, and that any edge $e \in E$ is a part of a path from v_0 to v_1. We assume that both V and E are finite.

An *admissible current* is a map $c{:}E \to H$ obeying *Kirkoff current* law

$$\sum_{e \in I(v)} c(e) = \sum_{e \in T(v)} c(e); \qquad v \in V\backslash\{v_0, v_1\}.$$

Notice that $s(c) := \sum_{e \in I(v_0)} c(e) = \sum_{e \in T(v_1)} c(e)$ if c is admissible. For any $x \in H$ let K_x be the set of all admissible currents $c{:}E \in H$ for which $s(c) = x$.

2.1 ***Theorem*** *[8]. Let* $A{:}\ E \to B(H)^+$. *Then there exists a unique positive operator* $\tilde{M}_G(A)$ *on* H, *satisfying*

$$(\tilde{M}_G(A)\, x,x) = \inf_{c \in K_x} \sum_{e \in E} (A(e)c(e),c(e)).$$

The map $A \mapsto \tilde{M}_G(A)$ *is a connection, i.e. satisfies the axioms analogous to (a), (b), and (c) of Definition 1.1. Thus there exists a positive number* γ *so that* $M(A) := \gamma\tilde{M}_G(A)$ *is an operator mean of the* $n = |E|$ *variables* $(A(e))_{e \in E}$.

Clearly, if $A(e) = I$ for every $e \in E$, then $\tilde{M}_G(A) = \gamma^{-1}I$.

If $E = \bigcup_{i=1}^{m} E_j$ is the partition of E and $A(e) = A_j$ for $e \in E_j$ $(j = 1, 2, ..., m)$, then $N(A_1, ..., A_m) = M_G(A)$ defines an operator mean of m variables. The symmetric means of [2] and [5] are obtained in this way for some parallel-series networks.

2.2 *Theorem: Let* $G = (V, E)$ *be as above and let* M_G *be the corresponding operator mean. Let* X *be any symmetric normed ideal of* $B(H)$. *If* $A:E \to X^+$ *then* $M_G(A) \in X^+$ *and* $\| M_G(A) \|_X \le M_G(\| A \|_X)$ *where* $\| A \|_X: E \to [0, \infty)$ *is defined by* $\| A \|_X(e) = \| A(e) \|_X$.

Proof. Let $x \in H$. If $\alpha: E \to \mathbb{R}$ is a one-dimensional admissible current with $s(\alpha) = 1$, then $c(e) = \alpha(e)x$ defines an admissible current $c:E \to H$ with $s(c) = x$. Thus

$$(\tilde{M}_G(A)\, x,x) \le \sum_{e\in E} (A(e)c(e),\ c(e)) = \sum_{e\in E} \alpha(e)^2(A(e)x,x)$$

$$= ((\sum_{e\in E} \alpha(e)^2 A(e))\ x,x).$$

Thus,

$$\tilde{M}_G(A) \le \sum_{e\in E} \alpha(e)^2 A(e).$$

It follows that $\tilde{M}_G(A) \in X^+$ and $\| \tilde{M}_G(A) \|_X \le \sum_{e\in E} \alpha(e)^2 \| A(e) \|_X$. Taking infimum over all one-dimensional admissible current $\alpha: E \to \mathbb{R}$ with $s(x) = 1$, we get $\| \tilde{M}_G(A) \|_X \le \tilde{M}_G(\| A \|_X)$. Multiplying by γ, we get the desired inequality. □

3. *Extension to Banach modules over von-Neumann algebras.*

Recall [14] that a von-Neumann algebra W is a weakly closed, self adjoint subalgebra W of $B(H)$ for some Hilbert space H. We assume that W is semifinite. We fix a normal, semi-finite, faithful (n.s.f) trace τ on W. The generalized singular numbers $\mu_t(A)$ of a (possibly unbounded) τ-measurable operator A affiliated with W are studied in [11] and [12]. For $t > 0$ one has $\mu_t(A) = \inf_P \| A(I - P) \|$ where the infimum is taken over the set of projections P in W with $\tau(P) < t$. Let X be a rearrangement invariant Banach function space on $[0, \infty)$ (where if τ is finite we replace $[0, \infty)$ by $[0, \tau(I)]$, (see [15]). Notice that according to the definition in [15], X is either minimal or maximal. The associated *Banach module* $X(W)$ over W is the space of all τ-measurable (possibly unbounded) operators A affiliated with W for which the function $\mu(A)(t) := \mu_t(A)$ $(t > 0)$ belongs to X, with the norm $\| A \|_{X(W)} := \| \mu(A) \|_X$.
Clearly, $X(W)$ is a bimodule over W, and for any $A, C \in W$ and $B \in X(W)$ $\| ABC \|_{X(W)} \le \| A \| \, \| B \|_{X(W)} \| C \|$. For instance, if $X = L^p$ then $X(W) = L^p(W, \tau)$.

3.1 *Proposition: Let* M *be an operator mean of* n *variables. Then* M *maps* $(W^+)^n := W^+ \times W^+ \times \cdots \times W^+$ into W^+.

Proof: Let $A_1, \ldots, A_n \in W^+$, and denote $\vec{A} = (A_1, \ldots, A_n)$. Let P be a projection in the commutant W' or W and let $S = 2P - I$. The $S = S^* = S^{-1}$. Thus the transformer inequality becomes an equality:

$$SM(\vec{A})S = M(SA_1S, \ldots, SA_nS) = M(S^2A_1, \ldots, S^2A_n)) = M(\vec{A}),$$

since S commutes with the A_j and $S^2 = I$. It follows that $M(\vec{A})$ commutes with S and hence also with P. Since W' is generated by its projections it follows that $M(\vec{A}) \in (W')' = W^* = W$.

□

3.2 Lemma [8]. *Let* $G = (V, E)$ *be a directed graph as in section 2, and let* M_G *be the corresponding operator mean. For every* $e \in E$ *let* A(e) *be a (possibly unbounded) positive operator on* H, *which is affiliated with* W *and* τ-*measurable. Then there is a unique (possibly unbounded) operator* $M_G(A)$ *which is affiliated with* W *and* τ-*measurable, and so that on a dense subspace* $\tilde{H}$

$$(M_G(A)\, x,x) = \gamma.\inf \sum_{e \in E} (A(e)c(e), \quad c(e))$$

where the infimum is taken over all admissible currents $c: E \to \tilde{H}$ *with* $c(e) = x$.

3.3 *Theorem. Let* X(W) *be a Banach module over* W, *and let* M_G *be the operator mean associated with the directed graph* $G = (V, E)$ *as in section 2. If* $A: E \to X(W)^+$ *then* $M_G(A) \in X(W)^+$ *and* $\| M_G(A) \|_{X(W)} \leq M_G(\| A \|_{X(W)})$.

Proof: As in the proof of Theorem 2.2, $\tilde{M}_G(A) \leq \sum_{e \in E} \alpha(e)^2 A(e)$ for every admissible current $\alpha: E \to \mathbb{R}$ with $s(\alpha) = 1$. This shows that $\tilde{M}_G(A) \in X(W)^+$ and $\| \tilde{M}_G(A) \|_{X(W)} \leq \sum_{\alpha \in E} \alpha(e)^2 \| A(e) \|_{X(W)}$. From this the desired inequality follows as in the proof of Theorem 2.2.

□

3.4 *Remark.* We conjecture that the inequality of Theorem 3.3 actually holds for general operator mean. It clearly holds for means obtained from the means of the form M_G by the following operations (see [8] for details):

(i) superposition, and in particular convex combinations;

(ii) weak, monotone decreasing limits;

(iii) $N(A_1, \ldots, A_n) = M(a_1 A_1, \ldots, \alpha_n A_n)$ (where $M(\alpha_1, \ldots, \alpha_n) = 1$);

(iv) integration of Means M_α, $\alpha \in \Omega$ against a probability measure ν on Ω: $M = \int_\Omega M_\alpha \, d\nu(\alpha)$;

(v) $N(A_1, \ldots, A_n) = M(B_1, \ldots, B_m)$, where each of the A_i is some B_j.

We do not know whether this class of means exhausts all the means in $n \geq 3$ variables.

3.5 *Remark.* Theorem 3.3 holds for the rearrangement-invariant function space X. The proof is essentially the same. Certainly, Theorems 1.2 and 2.2 hold for symmetric sequence spaces instead of symmetric normed ideals.

References

[1] W. N. Anderson, Jr. and G. E. Trapp, Shorted operators II, SIAM J. Appl. Math., Vol. 28 (1975), 60-71.

[2] W. N. Anderson, Jr., T. D. Morley and G. E. Trapp, Symmetric Function Means of Positive Operators, Linear Alg. Appl. 60 (1984), 129-143.

[3] W. N. Anderson, Jr. and R. J. Duffin, Series and Parallel Addition of Matrices, J. Math. Anal. Appl. 26 (1969), 576-594.

[4] T. Ando, Topics in operator inequalities, preprint (1982), Division of Applied Mathematics, Research Institute of Applied Electricity, Hokkaido University, Sapporo, Japan.

[5] T. Ando, An inequality between symmetric functions means of positive operators, Acta. Sc. Math., 45 (1983), 19-22

[6] T. Ando and F. Kubo, Means of Positive Linear Operators, Math. Ann. 246 (1980), 205-224.

[7] T. Ando and K. Nishio, Characterization of Operations Derived from Network Connections, J. Math. Anal. Appl., 53 (1976), 539-549.

[8] J. Arazy, Operator means and networks, in preparation.

[9] J. Arazy, T. Cleason, S. Janson and J. Peetre, Means and their iterations, Proceedings of the Nineteenth Congress of Mathematicians, Rejkajavik 1984, 191-212.

[10] W. F. Donoghue, Monotone matrix functions and analytic continuation, Grundlehren 207, Springer Verlag, 1974.

[11] T. Fack, Sur la notion de valeur caractéristique, J. Operator Theory, 7 (1982), 307-333.

[12] T. Fack and H. Kosaki, Generalized s-numbers of τ-measurable operators, Pacific J. Math. 123 (1986), 269-300.

[13] I. C. Gohberg and M. G. Krein, Introduction to the theory of linear nonselfadjoint operators, Amer. Math. Soc. Translations, Vol. 18, Providence, R. I., 1969.

[14] R. V. Kadison and J. R. Ringrose, Fundamentals of the Theory of Operator Algebras, vol. II, Academic Press (1986).

[15] J. Lindenstrauss and L. Tzafriri, Classical Banach spaces II. Function spaces. Springer Verlag, Ergebnisse 97 (1979).

[16] T. D. Morley, Parallel Summation, Maxwell's Principle and the Infimum of Projections, J. Math. Anal. Appl. 70 (1979), 33-41.

[17] E. L. Pekarev and Ju. L. Smulijan, Parallel additon and parallel subtraction of operators, Math. USSR Izvestia, vol. 10 (1976), No. 2, 351-370.

Operator Means, Norm Convergence and Matrix Functions

W. L. Green and T. D. Morley

Abstract

Operator means are nonlinear matrix functions that arise in the study of the interconnection of n-ports or Hilbert ports. In this paper, we study some of the relations among operator means, shorted operators, Dini's theorem, and norm convergence. Matrix valued functions are used to present examples of many diverse convergence behaviors.

§1: Introduction

The purpose of this paper is to illustrate the role of algebras of matrices in the development of the theory of operator means, and in that of the related theory of the shorted operator. Throughout the paper we shall write $M_n(A)$ for the algebra of all n by n matrices with entries from an algebra A, and we shall be interested chiefly in two choices of A:

1) A = C(**X**), the algebra (with pointwise operations) of all continuous complex-valued functions on a compact Hausdorff space **X**;

and

2) A= $\mathcal{B}(\mathcal{H})$, the algebra of all bounded linear operators on a complex Hilbert space $\mathcal{H}$.

The first of these choices, with **X** equal to the one point compactification of the positive integers, played an implicit but pivotal role in [3] , where a convergent sequence of two by two matrices was used to build an operator whose short behaves badly with respect to norm convergence. As we shall show below, $M_2(C(\mathbf{X}))$ has continued to provide a testing ground and a ready source of theorems, examples, and counterexamples, many connected again with norm convergence properties. The second choice, $\mathcal{B}(\mathcal{H})$, is an important tool in showing that the parallel sum (or equivalently the harmonic mean) and the shorted operator determine one another. (See the next section below.) Indeed, one can write the parallel sum of two operators on $\mathcal{B}(\mathcal{H})$ as the short of an element of $M_2(\mathcal{B}(\mathcal{H}))$ (considered as $\mathcal{B}(\mathcal{H}\oplus\mathcal{H})$) to the subspace $\mathcal{H}\oplus\{0\}$. We note in passing that although we shall concentrate here on the harmonic and geometric means, much of our discussion is applicable in principle to any operator mean in the sense of Ando [4].

In the next section we sketch the definition and basic properties of the shorted operator, the harmonic and geometric means, and the parallel sum. In the final section we give some examples which illustrate explicitly the use of matrix algebras to study these operations. Throughout this paper all algebras and all Hilbert spaces will be vector spaces over the field **C** of complex numbers, and all algebras will have identities, usually denoted by I. We shall always regard $M_n(C(\mathbf{X}))$ as a complete normed algebra (in fact as a C*-algebra [10]) , equipped with the norm given by

$$\|f\| = \sup\{\ \|f(x)\| : x \in X\},$$

where the norm of a matrix f(x) is its norm when acting by matrix multiplication on the usual n-dimensional Hilbert space of column vectors. Note that convergence with respect to this norm is just uniform convergence in $M_n(C(\mathbf{X}))$, regarded in the obvious way as a space of continuous functions on **X** to $M_n(\mathbf{C})$. We shall have many occasions to use the following easy extension of Dini's Theorem. An element T of $M_n(C(\mathbf{X}))$ is **positive** if T(x) is a positive semidefinite matrix for each x in **X**.

Theorem (Dini): Let $\{A_k\}$ be a monotone increasing bounded net of positive elements of $M_n(C(\mathbf{X}))$. For each x in X, let A(x) denote the limit of $A_k(x)$, computed in the algebra $M_n(\mathbf{C})$ of all n by n matrices over the complex numbers. Then the net A_k is norm convergent in $M_n(C(\mathbf{X}))$ if and only if $A : X \to M_n(\mathbf{C})$ is a continuous function.

§2: An introduction to operator means and the shorted operator.

Operator means are closely connected with *shorted operators*, or *generalized Schur complements*. Let A be a positive operator on a complex Hilbert space $\mathcal{H}$. Given a closed subspace $\mathcal{S}$, we may write $\mathcal{H}$ as the orthogonal direct sum $\mathcal{H} = \mathcal{S} \oplus \mathcal{S}^{\perp}$. With respect to this decomposition we may write A as $\begin{bmatrix} A_{11} & A_{12} \\ A_{21} & A_{22} \end{bmatrix}$, with $A_{11} : \mathcal{S} \to \mathcal{S}$, $A_{12} : \mathcal{S} \to \mathcal{S}^{\perp}$, etc. If the operator $A_{22} : \mathcal{S}^{\perp} \to \mathcal{S}^{\perp}$ is invertible, then the **shorted operator** of A to $\mathcal{S}$ is defined to be the **classical Schur complement**:

$$\mathcal{S}(A) = A_{11} - A_{12} A_{22}^{-1} A_{21}.$$

Even if A_{22} is not invertible, $A_{22} + \varepsilon I$ is invertible for all $\varepsilon > 0$. This allows to define the shorted operator in general as the monotone strong operator limit

$$\mathcal{S}(A) = \lim_{\varepsilon \downarrow 0} A_{11} - A_{12} (A_{22} + \varepsilon I)^{-1} A_{21} .$$

It can be shown that this defines a positive linear operator $\mathcal{S}(A)$, termed the **shorted operator of A to $\mathcal{S}$**, see Anderson and Trapp [3]. The name shorted operator comes from the fact that it gives the joint impedance of a resistive n-port (or Hilbert port) some of whose ports have been short circuited. Although $\mathcal{S}(A)$ as define above is an operator on $\mathcal{S}$, we shall often regard it, in the obvious way, as an operator on $\mathcal{H}$. Note that if A is a positive element of $M_2(C(\mathbf{X}))$, and if A_{22} is invertible in C(X), then we may define the short of A to its upper left hand corner in precisely the same way, even in the absence of a Hilbert space: $\mathcal{S}(A) = A_{11} - A_{12} A_{22}^{-1} A_{21}$. For any positive A in $M_2(C(\mathbf{X}))$, we may then define the short of A as before

$$\mathcal{S}(A) = \lim_{\varepsilon \downarrow 0} A_{11} - A_{12}(A_{22} + \varepsilon I)^{-1} A_{21}$$

provided that the pointwise limit of the net $\{A_{11} - A_{12} (A_{22} + \varepsilon I)^{-1} A_{21}\}_{\varepsilon > 0}$ lies in $C(\mathbf{X})$. By Dini's Theorem, this last is the case if and only if this net is norm convergent in C (X).

If A and B are invertible positive operators on $\mathcal{H}$, then we define their *parallel sum* A:B and *geometric mean* A#B as follows:

$$A{:}B = A(A + B)^{-1} B$$

$$A\#B = A^{1/2}(A^{-1/2} B A^{-1/2})^{1/2} A^{1/2} .$$

In the (possibly) singular case, we can take strong operator limits as before:

$$A : B = \lim_{\varepsilon \downarrow 0} (A + \varepsilon I) : (B + \varepsilon I)$$

$$A \# B = \lim_{\varepsilon \downarrow 0} (A + \varepsilon I) \# (B + \varepsilon I).$$

We define the **harmonic mean** of A and B to be 2(A:B).

Remark: Another way of getting the geometric mean (see [5] or [3]) is to iterate the harmonic and arithmetic means, i.e., define $A_0 = A$, $B_0 = B$, and $A_{n+1} = (A_n + B_n)/2$, $B_{n+1} = 2 (A_n{:}B_n)$. In this case we have $A\#B = \lim A_n = \lim B_n$. Moreover the A_n's converge monotonically downward to A#B, and the B_n's converge monotonically upward to A#B. We shall return to this pair of iterations in the last section.

It can be shown that these specifications define positive operators A:B and A#B. See Anderson and Trapp, [4], Ando [5]. See Trapp[12] for a recent survey of this and related matters. For related papers on operator means, we refer the reader to [1,2,6,7,11]. The name parallel sum comes from the fact that A:B is the joint impedance of two resistive n-ports (or Hilbert ports) connected together in parallel. The name geometric mean comes from the fact that if A and B commute (e.g. if the Hilbert space is one dimensional) then $A\#B = \sqrt{AB}$. Note however that the geometric mean cannot be derived from the interconnection of finitely many circuit elements, as the geometric mean of rational functions need not be rational, while the joint impedance of a finite LRC circuit (of 1-ports) is rational, see [13]. Note too that the parallel sum is not an operator mean in the sense of [5], since $A:A = A/2 \neq A$. The harmonic mean 2(A:B) is however an operator mean.

These operations are all intimately related, and in fact the parallel sum and the shorted operator determine one another (see [1] and [4]): if B has closed range $\mathcal{S}$, then $\mathcal{S}(A)$ is the monotone norm limit (as n tends to $+\infty$) of the sequence $\{A : nB\}$, i.e.,

$$\mathcal{S}(A) = \lim_{n\to\infty} (A : nB),$$

On the other hand if A and B are positive, then the parallel sum, A : B, is the short to the subspace $\mathcal{H} \oplus \{0\}$ of the following operator defined on $\mathcal{H} \oplus \mathcal{H}$:

$$\begin{bmatrix} A & A \\ A & A+B \end{bmatrix}.$$

Moreover, by an integral representation theorem of Ando [5], every operator mean (including the geometric mean) can be recovered from weighted parallel sums and an appropriate measure.

§3: C* representations, operator means, and Dini's Theorem

To illustrate properties of operator means, particularly properties which involve norm convergence, we shall want to allow algebras of matrix valued functions to act on various Hilbert spaces of functions. As a brief paradigm of what we have in mind, we quote here an example of W. N. Anderson from [1]. Let $\{A(i)\}$ and $\{B(i)\}$ be the sequences defined by

$$A(i) = \begin{bmatrix} 1 & i^{-1} \\ i^{-1} & i^{-2} \end{bmatrix}, \quad B(i) = \begin{bmatrix} 1 & 0 \\ 0 & 0 \end{bmatrix}$$

and let A and B denote the block diagonal matrices formed from the A(i)'s and the B(i)'s in the usual way. Thus A and B operate (by matrix multiplication) on the ℓ^2 - sum of countably many copies of $\mathbf{C}^2$. The short $\mathcal{S}(A)$ of A to the range $\mathcal{S}$ of B is zero, and the short $\mathcal{S}(A+\varepsilon I)$ of $A+\varepsilon I$ converges to zero in the strong operator topology. However $\|\mathcal{S}(A+\varepsilon I)\| \geq 1$ for all $\varepsilon > 0$, so that $\mathcal{S}(A+\varepsilon I)$ fails to converge in norm, even though $A + \varepsilon I$ converges in norm to A. Since $\lim_i A(i) = \lim_i B(i) = B(1)$, we may regard A and B as elements of $M_2(C(\mathbf{N}_\infty))$, where $\mathbf{N}_\infty$ denotes the one point compactification of the positive integers. A moment's reflection on Dini's Theorem (see above) shows that this failure of norm convergence can be traced directly to a discontinuity at $+\infty$ in the pointwise limit of the net $\{\mathcal{S}(A + \varepsilon I)\}_{\varepsilon>0}$, regarded as a net of functions in $M_2(C(\mathbf{N}_\infty))$. It is not difficult to show that for the same reasons, the geometric means $(A + \varepsilon I)\#B$ and the harmonic means $2((A + \varepsilon I):B)$ also fail to converge in norm as $\varepsilon > 0$ tends down to zero, even though $(A + \varepsilon I)\#B$ and $2\{(A + \varepsilon I):B\}$ converge respectively to A#B and 2(A:B) in the strong operator topology.

Observe that when we consider matrices with functional elements, it is important to distinguish between two related constructs:

1) The function which sends t into the short of operator A(t) to the range of B(t). It may happen that this function is continuous, in which case it is actually in $M_n(C(\mathbf{X}))$.

and

2) The shorted operator of A to the range of B, computed by considering the functions $A(\cdot)$ and $B(\cdot)$ in $M_n(C(\mathbf{X}))$ as operators acting on some Hilbert space. It may happen that the resulting shorted operator can be expressed as matrix with functional elements acting on the same Hilbert space. If these functional elements are continuous on **X**, then we may regard the result as being in $M_n(C(\mathbf{X}))$.

These two constructs need not be the same. It turns out that they agree, by Dini's Theorem, if both produce elements of $M_n(C(\mathbf{X}))$. Similar remarks apply with parallel sums, or with geometric or harmonic means in place of shorts.

A somewhat simpler version of Anderson's example can be obtained by replacing $\mathbf{N}_\infty$ by the closed unit interval [0,1] and the reciprocal of i by t; we then take

$$A(t) = \begin{bmatrix} 1 & t \\ t & t^2 \end{bmatrix}, \quad B(t) = \begin{bmatrix} 1 & 0 \\ 0 & 0 \end{bmatrix}.$$

We may then let A and B act by matrix multiplication on a two-fold copy of $L^2[0,1]$. Since the function which carries t in [0,1] to $\mathcal{S}(A(t) + \varepsilon I)$ is not continuous at zero, Dini's Theorem implies once again that the shorts $\mathcal{S}(A + \varepsilon I)$ to the range of B fail to converge in norm in $M_2(C([0,1]))$.

To see what is going on in general we recall the following result from [8].

Theorem 1: Let A and B be positive operators on a Hilbert space, then the following are equivalent (See [10] for terminology) :

i) $(A + \varepsilon I) : (B + \varepsilon I)$ converges **in norm** to $A{:}B$.
ii) $\pi(A{:}B) = \pi(A){:}\pi(B)$ for all representations π of the C*-algebra $\mathcal{B}(\mathcal{H})$.
iii) $\pi(A{:}B) = \pi(A){:}\pi(B)$ for the universal representation π of $\mathcal{B}(\mathcal{H})$.

Two operators A and B satisfying the equivalent conditions of the above theorem are called **uniformly and universally parallelizable** [8]. If the range of A + B is closed, or if the range of A is contained in the range of A + B, then A and B are uniformly and universally parallelizable. (See [8] and [11].) If the ranges of A and B are closed, then all these conditions are equivalent [8].

By combining the version of Dini's theorem from section one and the above theorem one can obtain:

Theorem 2: Let A and B be operators in $M_n(C(\mathbf{X}))$, where **X** is compact. Then A and B are uniformly and universally parallelizable if and only if the pointwise limit of $(A(t) + \varepsilon I) : (B(t) + \varepsilon I)$ is continuous.

Let us now consider some examples in view of theorem 2. Let

$$A_1(t) = \begin{bmatrix} 1 & t \\ t & t^2 \end{bmatrix} \quad \text{and} \quad B_1(t) = \begin{bmatrix} 1 & 0 \\ 0 & 0 \end{bmatrix},$$

then an elementary computation shows that pointwise we have $A_1(t) : B_1(t) = \mathbf{0}$, for $t \neq 0$, and $A_1(0){:}B_1(0) = \frac{1}{2}\begin{bmatrix} 1 & 0 \\ 0 & 0 \end{bmatrix}$. Thus A(t) and B(t) are not uniformly and universally parallelizable. This example is nothing more than the continuous analogue of Anderson's example with parallel sums in place of shorts.

Two other useful examples are:

$$A_2(t) = \frac{t}{1 + t^2} \begin{bmatrix} 1 & t \\ t & t^2 \end{bmatrix} \quad \text{and} \quad B_2(t) = \begin{bmatrix} 1 & 0 \\ 0 & 0 \end{bmatrix}$$

$$A_3(t) = \begin{bmatrix} 2 & t \\ t & t^2 \end{bmatrix} \quad \text{and } B_3(t) = \begin{bmatrix} 1 & 0 \\ 0 & 0 \end{bmatrix},$$

Computing, we see that $A_2(t) : B_2(t) = 0$ for all t, and thus A_2 and B_2 are uniformly and universally parallelizable. Note however that if we set E_k equal to the range projection of A_k, and F_k equal to the range projection of B_k, then $E_2(t) = t^{-1} A(t)$, $F_k = B_k$. Computing we get that the pointwise limit of $(E_2(t) + \varepsilon I):(F_2(t) + \varepsilon I)$ is zero when $t \neq 0$, and is nonzero when $t = 0$. On the other hand, using theorem 2 it is not difficult to see that $A_3(t)$ and $B_3(t)$ are not uniformly and universally parallelizable, but E_3 and F_3 are. The operators A_2 and B_2 can also be used to show (see [9]) that uniform and universal parallelizability does not imply the containment of range(A) or range(B) in range(A + B).

We conclude with a new result on the geometric mean. This result sharpens a result from [8].

Theorem 3: Let A and B be positive operators on a Hilbert space. As before we define $A_0 = A$, $B_0 = B$, and iterate: $A_{n+1} = (A_n + B_n)/2$, $B_{n+1} = 2(A_n{:}B_n)$. Then $\text{range}(A_i^{1/2}) = \text{range}(A_j^{1/2})$, and $\text{range}(B_i^{1/2}) = \text{range}(B_j^{1/2})$, for all i and j ≥ 1. If, in addition, the range of A + B is closed, then range($A_n + B_n$) is closed for all $n \geq 1$. Also in this case the ranges of the A_n are closed and are equal to the ranges of ($A_n + B_n$). In particular, A_n and B_n are uniformly and universally parallelizable and thus A#B is the norm limit of $(A + \varepsilon I) \# (B + \varepsilon I)$.

Outline of Proof: Using Crimmins's Theorem [4], and Douglas's Theorem (see for example [7]) and various properties of the range of sums and parallel sums, see [4], we compute:

$$\begin{aligned}
\text{range}(A_2^{1/2}) &= \text{range}((A_1 + B_1)^{1/2}) \\
&= \text{range}(A_1^{1/2}) + \text{range}(B_1^{1/2}) \\
&= \text{range}(A_0^{1/2}) + \text{range}(B_0^{1/2}) + \text{range}(B_1^{1/2}) \\
&= \text{range}(A_0^{1/2}) + \text{range}(B_0^{1/2}) + (\text{range}(A_0^{1/2}) \cap \text{range}(B_0^{1/2})) \\
&= \text{range}(A_0^{1/2}) + \text{range}(B_0^{1/2}) \\
&= \text{range}((A_0 + B_0)^{1/2}) \\
&= \text{range}(A_1^{1/2}).
\end{aligned}$$

If range(A_1) is closed, $\text{range}(A_1^{1/2}) = \text{range}(A_1) = \text{range}(A_2^{1/2})$. Using the fact that the range of a positive operator C is closed if and only if $\text{range}(C^{1/2})$ is closed, it follows that the range of A_2 is closed. The proof for the B's is similar.

For example let A be multiplication by x on $L_2[0,1]$ and let B be multiplication by 1 (i.e., the identity). Then A_0 does not have closed range, but each A_i for $i > 0$ is invertible. On the other hand, B_0 is invertible, but B_i has nonclosed range for $i > 0$. The geometric mean A#B is multiplication by $\sqrt{x}$, which also has nonclosed range.

Remark: It can be easily shown that $\text{range}(B_n^{1/2})$ stabilizes at $\text{range}[(A\#B)^{1/2}]$, and the range of $A_n^{1/2}$ stabilizes at $\text{range}[(A+B)^{1/2}]$.

Suppose now that A and B are positive elements of $M_n(C(X))$, and that $M_n(C(X))$ acts faithfully on a Hilbert space, i.e., that there exists an injective identity preserving representation π of $M_n(C(X))$ into the bounded linear operators on the Hilbert space. Then the spectra of A + B and

$\pi(A) + \pi(B)$ coincide, and $\pi(A) + \pi(B)$ has closed range if and only if $sp(A + B) \setminus \{0\}$ is compact. It follows that if $sp(A + B) \setminus \{0\}$ is compact, then the iterated means of Theorem 3 remain in $M_n(C(X))$ and converge in norm. Clearly this is the case if, for example, each of the sets

$$\{ t \in X: \mathrm{rank}(A(t) + B(t)) = k\}$$

is compact, where $k = 1,2,\ldots,n$.

References

1. W. N. Anderson, *Shorted operators*, SIAM J. Appl. Math. **20** (1970), pp. 520-525.

2. W. N. Anderson and R. J. Duffin, *Series and parallel addition of matrices*, J. Math Anal. Appl. **26** (1969), pp. 576-594.

3. W. N. Anderson, T. D. Morley, *A characterization of parallel subtraction,* Proc. Natl. Acad. Sci. USA **76** (1979), pp. 3599-3601.

4. W. N. Anderson and G. E. Trapp, *Shorted operators II*, SIAM J. Appl. Math. **28** (1975), pp. 60-71.

5. T. Ando, **Topics in operator inequalities**, Lecture Notes, Hokkiado University, Sapporo, Japan, 1978.

6. R. S. Bucy, *A priori bounds for the Riccati equation,* in **Proceedings of the Sixth Berkley Symposium on Mathematical Statistics and Probability, Vol III: Probability Theory**, pp. 645-656, U. of Cal. Press, Berkley, CA, 1972.

7. P. A. Fillmore and J. P. Williams, *On operator ranges*, Adv. in Math. **7** (1971), pp. 254-281.

8. W. L. Green and T. D. Morley, *Operator means, fixed points, and the norm convergence of monotone approximants*, Math. Scand. **60** (1987), pp. 202-218.

9. W. L. Green and T. D. Morley, *Parallel sums and norm convergence,* to appear, Circuits, Systems and Signal Processing, special issue on infinite networks.

10. R. V. Kadison and J. R. Ringrose, **Fundamentals of Operator Algebras, vol. I, Elementary Theory**, 1983, Academic Press, New York.

11. M. R. Khadivi, **Operator Theory and Infinite Networks**, thesis, Georgia Institute of Technology, 1988.

12. G. E. Trapp, *Hermitian semidefinite matrix means and related matrix inequalities - an introduction*, Linear and Multilinear Alg. **16** (1984), pp. 113-123.

13. L. Weinberg, **Network Anlysis and Synthesis**, 1962, Mc Graw Hill, New York.

Operator Means Determined by an Operator Equation

Fumio KUBO

Abstract.
Operator equations of ladder type will be discussed from the viewpoint of fixed point theorem. The solution of the equation yields a binary operation among operator means, and the equation can be thought as equation of operator means. The network connection associated with the solution is realizable by using the initial pair of network connection. Some new operator monotone functions can be obtained as the representing functions of the operator mean solution. The proofs of those results will be published elsewhere.

1. Introduction.

Let $\mathcal{H}$ be a Hilbert space. In the algebra $B(\mathcal{H})$ of all bounded linear operators on $\mathcal{H}$, the cone of all positive (semidefinite) operators will be denoted as $P(\mathcal{H})$, namely, an operator $T \in B(\mathcal{H})$ is positive if and only if

$$< Tf \mid f > \geq 0 \qquad (f \in \mathcal{H}).$$

It is well known that the impedance of a passive linear network is represented by positive operators and that series and parallel connection are fundamental connection of those networks. It will be easily agreed that the usual addition represents the impedance of the series connection of networks.

Motivated by these fundamental network connection, mathematicians of Duffin's school [1] introduced an operation among $P(\mathcal{H})$ called the parallel sum. This operation has been studied by many mathematicians and several characterizations are known. In this paper, the following definition will be adopted: $A : B = \lim_{\epsilon\downarrow 0}(A_\epsilon^{-1} + B_\epsilon^{-1})^{-1} \quad (A, B \in P(\mathcal{H}))$, where $X_\epsilon = X + \epsilon I \quad (X \in P(\mathcal{H}))$.

By suitable normalizations, the usual sum and the parallel sum may be considered as the arithmetic mean and the harmonic mean of positive operators respectively: $A \nabla B = \frac{A+B}{2}$, $A \,!\, B = 2(A : B)$. By an analogy of numerical means, the geometric mean for positive operators is expected. W. Pusz and S.L. Woronowicz [11] introduced the notion of operator geometric mean. The following characterization was given by T. Ando [2]: $A \# B = \lim_{\epsilon\downarrow 0} A_\epsilon^{1/2}(A_\epsilon^{-1/2}BA_\epsilon^{-1/2})^{1/2}A_\epsilon^{1/2}$. These three operator means are called classical in the following lines. The numerical analogy brought another operator mean, the arithmetic-geometric operator mean of Gauss type, discussed by J.I. Fujii [5].

An systematic treatment of these operator means was discussed by F. Kubo and T. Ando [9], where, among others, an axiom of operator means was given.

DEFINITION. An *operator connection* is a binary operation σ among positive operators satisfying the following conditions:

(1) $A \le B,\ C \le D \ \Rightarrow\ A\sigma C \le B\sigma D$,

(2) $T^*(A\sigma B)T \le (T^*AT)\sigma(T^*BT) \qquad (T \in B(\mathcal{H}))$,

and

(3) $A_n \downarrow A,\ B_n \downarrow B \Rightarrow A_n\sigma B_n \downarrow A\sigma B$,

The inequality (2) turns out to be an equality if the operator T has a bounded inverse, which implies the numerical homogeneity of operator connections.

DEFINITION An *operator mean* is an operator connection σ satisfying the normalizing condition:

(4) $A\sigma A = A$.

It was one of the main purposes of [9] to point out the parallelism between the theory of operator means and the classical Löwner theory on operator (matrix) monotone functions. The relation between them was formulated as follows:

$$\sigma : \text{operator connection} \longleftrightarrow f : \text{operator monotone function},$$

where the function f is determined by the equation $f(t)I = I\sigma tI \qquad (t \ge 0)$. Conversely, if $f : (0,\infty) \to [0,\infty)$ is an operator monotone function, then an operator connection σ is given by $A\sigma B = \lim_{\epsilon\downarrow 0} A_\epsilon^{1/2} f(A_\epsilon^{-1/2}B_\epsilon A_\epsilon^{-1/2})A_\epsilon^{1/2}$. The correspondence is bijective and preserves their affine order structures. Especially, an operator mean corresponds to a normalized operator monotone function, where an operator monotone function is normalized if and only if $f(1) = 1$.

The operator monotone function associated with an operator connection σ will be referred to as the representing function of σ.

The integral representation of the operator monotone function due to K. Löwner yields an integral representation of operator connections.

$$A\sigma B = \alpha A + \beta B + \int_{(0,+\infty)} \frac{1+t}{t}[(tA) : B]d\mu(t),$$

where μ is a positive Radon measure on $[0, +\infty]$ and $\alpha = \mu(\{0\})$, $\beta = \mu(\{\infty\})$. The integral representation of operator connections or operator means may be considered as a ideal realization of networks using the parallel addition and an infinite addition with suitable weights. Another realization was given in [7], which will be discussed in the last section of this paper.

The order relation among operator connections and operator means is defined in a natural way: Let σ_1, σ_2 be a pair of operator connections. The order relation $\sigma_1 \geq \sigma_2$ holds if and only if $A\sigma_1 B \geq A\sigma_2 B \quad (\forall A, B \in P(\mathcal{H}))$.

The topology among the operator connections is determined by their convergence structure, cf., [9]. For a generalized sequence (σ_n of operator connections and an operator connection σ, the convergence $\lim_{n\to\infty} \sigma_n = \sigma$ is defined by $\forall$ invertible $A, B \in P(\mathcal{H})$; w–$\lim_{n\to\infty} A\sigma_n B = A\sigma B$.

If σ is a non zero operator connection with representing function f, then the following equations determine a group of operator means associated with σ:

$$\begin{array}{ll} \text{(transpose)} & A\sigma' B = B\sigma A \\ \text{(adjoint)} & A\sigma^* B = \lim_{\epsilon\downarrow 0}(A_\epsilon^{-1}\sigma B_\epsilon^{-1})^{-1} \qquad (A, B \in P(\mathcal{H})). \\ \text{(dual)} & A\sigma^\perp B = A(\sigma')^* B \end{array}$$

And their representing functions are given by

$$\begin{aligned} f'(t) &= tf(1/t) \\ f^*(t) &= 1/f(1/t) \qquad (t \geq 0), \\ f^\perp(t) &= t/f(t) \end{aligned}$$

respectively with the convention that $0 \times \infty = 0$.

An operator connection or operator mean is called symmetric if and only if

$$A\sigma B = B\sigma A \qquad (A, B \in P(\mathcal{H})),$$

or $\sigma = \sigma'$. The representing function of a symmetric operator mean or connection satisfies the equation

$$f(t) = f'(t) = tf(1/t) \qquad (t \geq 0).$$

It is obvious that $\sigma = \sigma' \Longleftrightarrow \sigma^* = \sigma^\perp$.

In [3], T. Ando discussed the asymptotic behavior of two ladder syntheses of the geometric operator mean. He defined a pair of monotone maps on the convex cone $P(\mathcal{H})$ of positive operators by

$$\Phi_\alpha(X) = (A \nabla X) \,!\, (B \nabla X) \qquad (X \in P(\mathcal{H})),$$

and

$$\Phi_\omega(X) = (A \,!\, X) \nabla (B \,!\, X) \qquad (X \in P(\mathcal{H})).$$

These are the representations of the following connections of passive resistive electrical networks:

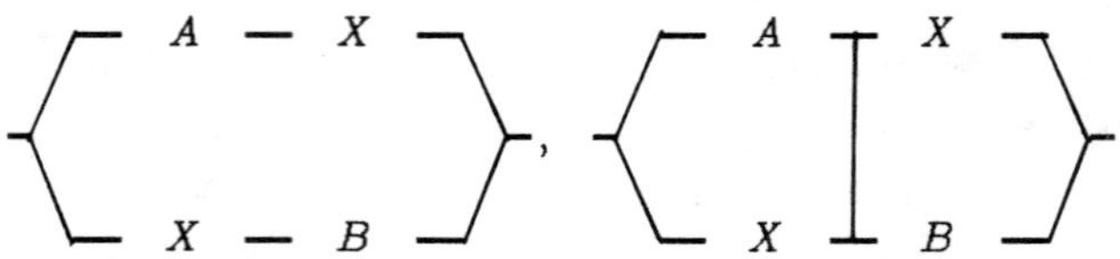

(Fig. 1).

In each cases, consider a pair of those sequences $(X_n), (Y_n)$ that start from the usual arithmetic mean and harmonic mean of A and B respectively;

$$X_0 = A \nabla B, \qquad Y_0 = A \,!\, B,$$

and defined by their iterates;

$$X_{n+1} = \Phi_\alpha(X_n), \qquad Y_{n+1} = \Phi_\omega(Y_n) \qquad (n = 0, 1, 2, \ldots),$$

respectively. These represent the ladder connection of networks. What are the networks iterated sufficiently many times ? The answer was summarized in the following theorems.

THEOREM A. *The geometric mean $A \# B$ is the unique fixed point of the map Φ_α and $A \# B = \lim_{n\to\infty} \Phi_\alpha^n(A \nabla B) = \lim_{n\to\infty} \Phi_\alpha^n(A \,!\, B)$.*

THEOREM Ω. *The geometric mean $A \# B$ is the unique fixed point of the map Φ_ω among those operators X that satisfy $X \geq c(A \,!\, B)$ for some $c > 0$, and $A \# B = \lim_{n\to\infty} \Phi_\omega^n(A \nabla B) = \lim_{n\to\infty} \Phi_\omega^n(A \,!\, B)$.*

These theorems suggest the possibility to determine a new operator mean from a pair of symmetric operator means as a solution of operator equation of ladder type: Let σ_1, σ_2 be a pair of operator means and

$$\Phi_{\sigma_1,\sigma_2}(x) = \Phi(X) = (A\sigma_1 X)\sigma_2(B\sigma_1 X) = X \quad (X \in P(\mathcal{H})).$$

The solution of the preceding equations, where classical means only are used, remains a classical mean. There may be the possibility to obtain more operator means in this way. This is the motivation of this paper.

The plan of this paper is as follows. In the next section, it will be shown that any equation of ladder type has a unique solution and that the solution is represented as an operator mean of the known operators in the equation. Thus the equation of ladder type can be considered not only as operator equation but also mean equation. In the network theoretical point of view, the ladder iteration of connecting networks yields not only another ideal network but also another ideal network connection. This process getting a binary operation $\otimes$ of operator means is different from the known ones.

In several examples among classical means, the solution of the ladder equation will be determined concretely. One of these examples will show that the new binary operation among operator means is different from the operation obtained by F. Kubo and T. Ando [9].

In the third section, the properties of the binary operation $\otimes$ among operator means obtained in the second section will be discussed.

The last section contains some results on non-negative operator monotone functions on the positive half-axis.

2. Means obtained by ladder synthesis.

In this section, it is discussed that given a pair σ_1, σ_2 of symmetric operator means and A, B of positive operators, the ladder equation

$$(A\sigma_1 X)\sigma_2(B\sigma_1 X) = X \quad (X \in P(\mathcal{H})),$$

will be discussed. Consider the (nonlinear) map $\Phi = \Phi_{\sigma_1,\sigma_2}$ and the iterates Φ^n on $P(\mathcal{H})$ defined by

$$\Phi(X) = \Phi_{\sigma_1,\sigma_2}(X) = (A\sigma_1 X)\sigma_2(B\sigma_1 X) \qquad (X \in P(\mathcal{H})),$$

$$\Phi^0(X) = X \qquad (X \in P(\mathcal{H})),$$

and

$$\Phi^{n+1} = \Phi(\Phi^n(X)) \qquad (X \in P(\mathcal{H})).$$

The first goal of this section is the following theorem.

THEOREM 1. *There exists a unique fixed point X of the map Φ among those operators X that satisfy $X \geq c(A \,!\, B)$ for some $c > 0$, and*

$$X = \lim_{n\to\infty} \Phi^n(A \nabla B) = \lim_{n\to\infty} \Phi^n(A \,!\, B).$$

Now the equation $\Phi(X) = X$ has a unique operatorial solution. The next step is to ask whether it determines a mean or not. Strictly speaking, the question is to ask whether the binary operation

$$(A, B) \mapsto \{\text{the unique solution of } (A\ \sigma_1\ X)\ \sigma_2\ (B\ \sigma_1 X) = X \quad X \in P(\mathcal{H})\}$$

is a symmetric operator mean or not. The next theorem shows that it has an affirmative answer.

THEOREM 2. *For any pair σ_1, σ_2 of symmetric operator means, there exists a unique symmetric operator mean ρ such that*

$$[A\ \sigma_1\ (A\ \rho\ B)]\ \sigma_2\ [B\ \sigma_1(A\ \rho\ B)] = A\ \rho\ B \qquad (A, B \in P(\mathcal{H})).$$

The prescribed ladder type construction of symmetric operator mean will be called $\otimes$-composition and denoted by $\sigma_1 \otimes \sigma_2$. Then the fixed point theorems of T. Ando will be written in the following simple forms: $\nabla \otimes ! = \#$, $\quad \otimes \nabla = \#$.

There are more examples among classical operator means.

PROPOSITION. (1) $\nabla \otimes \nabla = \nabla$, (2) $\# \otimes \nabla = \nabla^2$, (3) $\# \otimes \# = \#$, (4) $\# \otimes ! = \nabla_2$, *and* (5) $! \otimes ! = !$, *where ∇^2 and ∇_2 are the power means defined by $A \nabla^2\ B = (A\ \nabla\ B)\ \nabla\ (A\ \#\ B)$, and $A \nabla_2\ B = (A\ !\ B)\ !\ (A\ \#\ B)$, respectively.*

REMARK. The means ∇^2 and ∇_2 deserve their name because $a \nabla^2\ b = (a^{1/2}\ \nabla\ b^{1/2})^2$, and $a \nabla_2\ b = (a^{1/2}\ !\ b^{1/2})^2$, where a, b are positive real numbers.

REMARK. The second example shows that the $\otimes$-composition is different from the composition of Gaussian arithmetico-geometric type that was discussed generally in [9]. The examples (2) and (4) show that $\otimes$-composition is not symmetric. In fact a numerical manipulation shows that ∇^2 is not the same as $\nabla \otimes \#$.

There are two more combinations of the classical operator means beyond the preceding PROPOSITION, namely $\nabla \otimes \#$ and $! \otimes \#$. To represent these operator means by combinations of classical means is a routine but is a little complicated. The operator monotone functions associated with the operator means $\nabla \otimes \#$ and $! \otimes \#$ are given by; $\frac{(t+1)+\sqrt{t^2+14t+1}}{6}$, and $\frac{6t}{(t+1)+\sqrt{t^2+14t+1}}$, respectively.

COROLLARY. *If $\sigma_1 \leq \sigma_2$, and $\tau_1 \leq \tau_2$, then*

$$\sigma_1 \otimes \tau_1 \leq \sigma_2 \otimes \tau_2.$$

3. Adjoint mean and other properties.

Concerning to the adjoint operation on operator means, it is also easy to see that $(\sigma_1 \otimes \sigma_2)^* = (\sigma_1)^* \otimes (\sigma_2)^*$. The examples of the preceding sections suggest the following theorem.

THEOREM 3. *If σ is a symmetric operator mean, then the unique solution τ, among those symmetric operator means that satisfy the equation*

$$\sigma \otimes \tau = \#$$

is the adjoint σ^* of σ.

The semicontinuity of the $\otimes$-operation holds:

PROPOSITION. *If monotone decreasing sequences of symmetric operator means (σ_n), and (τ_n) have the limits σ, and τ respectively in the system of symmetric operator means, then the sequence $(\sigma_n \otimes \tau_n)$ converges to $\sigma \otimes \tau$.*

REMARK. Despite the monotonity and the semicontinuity of $\otimes$-operation, the $\otimes$-operation is not induced by simply taking a mean of the pair of means. For example, the normalization identity $\sigma \otimes \sigma = \sigma$ does not hold in general. If f, f^* denote the operator monotone functions associated with σ, σ^* respectively, then the identity $\sigma \otimes \sigma = \sigma$ is equivalent to say that $f\left(f^*(f(t)) \cdot f(f^*(t))\right) = f^*(f(t)) \quad (t \geq 0)$. But the symmetric operator mean σ defined by $A\sigma B = (A \nabla B)\nabla(A \,!\, B)$ and the associated operator monotone function $f(t) = (t^2 + 6t + 1)/4(t+1)$ does not satisfy the identity.

The representing operator monotone functions of $\# \otimes \nabla = \nabla^2$ and $\nabla \otimes \#$ are given by $\left(\frac{t^{1/2}+1}{2}\right)^2$, and $\frac{(t+1)+\sqrt{t^2+14t+1}}{6}$, respectively, which shows that the $\otimes$-operation is not symmetric.

4. Applications to functional equation.

It will be easily understood that the theory of equation among operator means is applicable to the theory of operator monotone functions.

In this section, the theorem of ladder equation of operator means is applied to the study of operator monotone functions.

In the words of operator monotone functions, the second theorem is stated as follows.

THEOREM 2'. *If f, g is a pair of symmetric normalized operator monotone functions, then there exists a unique solution of the equation*

$$g(f^*(h(t)) \cdot f(h^*(t))) = f^*(h(t)) \qquad (t \geq 0),$$

among symmetric normalized operator monotone functions h.

This theorem determines the unique solution of several functional equations.

COROLLARY. *If f is a symmetric operator monotone function, then there exists uniquely up to positive constant a symmetric operator monotone function h such that*

$$f(h^*(t)) = f(1)^2 f^*(h(t)) \qquad (t \geq 0).$$

COROLLARY. *If g is a symmetric operator monotone function, then so is h defined by*

$$h(t) = g(t^{1/2})^2 \qquad (t \geq 0).$$

REMARK. This COROLLARY has an equivalent form in terms of operator means: If h is the representing operator monotone function of a symmetric operator mean τ, then the representing operator monotone function of the operator mean $\tau \otimes \#$ is given by $h(t^{1/2})^2$.

This COROLLARY is a special case of the well-known result on operator monotone functions, cf., [6].

THEOREM 4. *If $\{f_j \ : \ j = 1, 2, ..., n\}$ and $\{g_j \ : \ j = 1, 2, ..., n\}$ is a pair of finite families of operator monotone functions such that*

$$\prod_{j=1}^{n} g_j(t) = t \qquad (t \geq 0),$$

then the function h defined by

$$h(x) = \prod_{j=1}^{n} f_j(g_j(t)) \qquad (t \geq 0)$$

is also operator monotone.

There is another class of symmetric operator means, called logarithmic type. This class was studied in numerical case by K.B. Stolarsky [12]. The extension of these means to operator means is less known, cf., [8],[10]. In this section, only the logarithmic mean will be discussed. In the heat engineering, the numerical logarithmic mean is well known. It is an average temperature of heat radiator with some width.

The logarithmic mean for a pair of positive real numbers x, y is defined by

$$\lambda(x,y) = \begin{cases} \frac{x-y}{\log x - \log y} & \text{if } x \neq y, \\ x & \text{if } x = y. \end{cases}$$

Remark also the following identity: $\lim_{y \to x} \lambda(x,y) = x$.

Recall the quasilinearization for operator means. The bijective correspondence $\mu \longleftrightarrow \sigma$ defined by

$$A\sigma B = \lim_{\epsilon \downarrow 0} \left(\int_0^1 [(1-u)A + uB + \epsilon I]^{-1} \, d\mu(u) \right)^{-1} \qquad (A, B \in P(\mathcal{H}))$$

gives a one-to-one correspondence between the class of positive normalized Radon measures on the interval $[0,1]$ and the class of operator means, cf.,[7].

And the short-pass from numerical logarithmic mean to operatorial one will be found in a paper of B.C. Carlson [4]: $\frac{1}{\lambda(x,y)} = \int_0^1 [(1-u)x + uy]^{-1} du \quad (x, y > 0)$. Thus in the abbreviated form, the logarithmic operator mean $A\lambda B$ is defined by

$$A\lambda B = \left(\int_0^1 [(1-u)A + uB]^{-1} du \right)^{-1} \qquad (A, B \in P(\mathcal{H})),$$

for positive semidefinite operators $A, B \in P(\mathcal{H})$. The representing operator monotone function is given by

$$\lambda(t) = \begin{cases} \frac{t-1}{\log t} & \text{if } t \neq 1, \\ 1 & \text{if } t = 1. \end{cases}$$

From the definition of logarithmic operator mean, the following properties are immediate.

Just in the same way as the proof of the first corollary of THEOREM 2', one has the following theorem by replacing $g(t)$ with the operator monotone function associated with the logarithmic operator mean.

COROLLARY. *Let $\phi : [0, \infty) \longrightarrow [0, \infty)$ is the function defined by $\phi(u) = \frac{\exp u}{u}$. If f is a symmetric operator monotone function, then there exists uniquely up to positive constant a symmetric operator monotone function h such that*

$$\phi(f(h^*(t))) = \phi\left(\frac{1}{f^*(h(t))} \right) \qquad (t \geq 0).$$

References

[1] Anderson, Jr.,W.N. and R.J. Duffin, Series and parallel addition of matrices, J. Math. Anal. Appl., 26 (1969), 576-594.

[2] Ando, T., *Topics on Operator Inequalities*, Lecture Note in Hokkaido University, 1978.

[3] Ando, T., Limit of iterates of cascade addition of matrices, Numer. Funct. Anal. Optimiz., 2 (1980), 579-589.

[4] Carlson, B.C., Some inequalities for hypergeometric functions, Proc. Amer. Math. Soc., 17 (1966), 32-39.

[6] Fujii, J.I., F. Kubo and K. Kubo, Parametrization between operator means, Math. Japon., 33 (1988), 201-208.

[7] Kubo, F., Conditional expectations and operations derived from network connections, J. Math. Anal. Appl., 80 (1981), 477-489.

[8] Kubo, F., On logarithmic operator means, Proceedings of the Tenth Symposium on Applied Functional Analysis (Information Theory and Related Topics), ed. by H. Umegaki, (1987).

[9] Kubo, F. and T. Ando, Means of positive linear operators, Math. Ann. 246(1980), 205-224.

[10] Nakamura, Y. Stieltjes functions and classes of operator monotone functions, preprint.

[11] Pusz, W. and S.L. Woronowicz, Functional calculus for sesquilinear forms and the purification map, Rep. Math. Phys., 8 (1975), 159-170 (54# 8316).

[12] Stolarsky, K.B. The power and generalized logarithmic means, Amer. Math. Monthly, 87 (1980), 545-548.

ADDED IN PROOF. The author would like to express his hearty thanks to the referee to inform him the following paper that discusses another ladder type of matrix operation.

[13] Anderson, Jr., W.N., Morley, T.D. and Trapp, G.E., Ladder networks, fixed points and the geometric mean, Circuits, Systems and Signal Processing, 2 (1983), 259-268.

Progress in Systems and Control Theory

Series Editor

Christopher I. Byrnes
Department of Systems Science and Mathematics
Washington University
Campus P.O. 1040
One Brookings Drive
St. Louis, MO 63130-4899
U.S.A.

Progress in Systems and Control Theory is designed for the publication of workshops and conference proceedings, sponsored by various research centers in all areas of systems and control theory, and lecture notes arising from ongoing research in theory and applications of control.

We encourage preparation of manuscripts in such forms as LATEX or AMS TEX for delivery in camera ready copy which leads to rapid publication, or in electronic form for interfacing with laser printers.

Proposals should be sent directly to the editor or to: Birkhäuser Boston, 675 Massachusetts Avenue, Suite 601, Cambridge, MA 02139, U.S.A.

PSCT1 Computation and Control: Proceedings of the Bozeman Conference, Bozeman, Montana, 1988
K. Bowers and J. Lund

PSCT2 Perspectives in Control Theory: Proceedings of the Sielpia Conference, Sielpia, Poland, 1988
B. Jakubczyk, K. Malanowski, and W. Respondek

PSCT3 Realization and Modelling in System Theory: Proceedings of the International Symposium MTNS-89, Volume I
M.A. Kaashoek, J.H. van Schuppen, and A.C.M. Ran

PSCT4 Robust Control of Linear Systems and Nonlinear Control: Proceedings of the International Symposium MTNS-89, Volume II
M.A. Kaashoek, J.H. van Schuppen, and A.C.M. Ran

PSCT5 Signal Processing, Scattering and Operator Theory, and Numerical Methods: Proceedings of the International Symposium MTNS-89, Volume III
M.A. Kaashoek, J.H. van Schuppen, and A.C.M. Ran